Further Pure Mathematics

also by L. Bostock and S. Chandler:

APPLIED MATHEMATICS I
APPLIED MATHEMATICS II
PURE MATHEMATICS I
PURE MATHEMATICS II
MATHEMATICS – THE CORE COURSE FOR A-LEVEL
MATHEMATICS – MECHANICS AND PROBABILITY
FURTHER MECHANICS AND PROBABILITY

Further Pure Mathematics

L. Bostock, B.Sc.
formerly Senior Lecturer, Southgate Technical College

S. Chandler, B.Sc.
formerly of the Godolphin and Latymer School

C. Rourke, Ph.D.
Reader in Mathematics, University of Warwick

Stanley Thornes (Publishers) Ltd.

First published 1982 by
Stanley Thornes (Publishers) Ltd
Old Station Drive
Leckhampton
CHELTENHAM GL53 0DN

Reprinted 1984
Reprinted 1985 with minor corrections
Reprinted 1987 with corrections
Reprinted 1989
Reprinted 1990

British Library Cataloguing in Publication Data

Bostock, L.
Further pure mathematics.
1. Mathematics–1961–
I. Title II. Chandler, S. III. Rourke, C.
510 QA39.2

ISBN 0-85950-103-5

Typeset by Tech-Set, Gateshead, Tyne & Wear.
Printed and bound in Great Britain at The Bath Press, Avon.

PREFACE

Further Pure Mathematics continues the work covered in *Mathematics – The Core Course for A-level* and is intended to complete a full two-year course in Pure Mathematics at sixth-form level.

This book contains much material already published in our book *Pure Mathematics II*, but large sections have been revised and new topics introduced to cover new syllabus developments at A-level. In particular there is now a more formal treatment of functions and new sections on numerical solutions of differential equations, transformations in the complex plane and vector spaces. Colin Rourke has contributed the chapters on group theory and matrix algebra.

Many worked examples are incorporated in the text to illustrate each main development of a topic and a set of straightforward problems follows each section. A selection of more challenging questions is given in the miscellaneous exercise at the end of each chapter and, where appropriate, we have included multiple choice exercises.

We are grateful to the following Examination Boards for permission to reproduce questions from past examination papers:

University of London (U of L)
Joint Matriculation Board (JMB)
University of Cambridge Local Examinations Syndicate (C)
Oxford Delegacy of Local Examinations (O)
The Associated Examining Board (AEB)

1982

L. Bostock
S. Chandler
C. Rourke

CONTENTS

NOTES ON USE OF THE BOOK

Notation

$=$	is equal to	$\in$	is a member of
$\equiv$	is identical to	$:$	such that
$\simeq$	is approximately equal to*	$\mathbb{N}$	the natural numbers
$>$	is greater than	$\mathbb{Z}$	the integers
$\geqslant$	is greater than or equal to	$\mathbb{Q}$	the rational numbers
$<$	is less than	$\mathbb{R}$	the real numbers
$\leqslant$	is less than or equal to	$\mathbb{R}^+$	the positive real numbers excluding zero
∞	infinitely large		
$\longmapsto$	maps to	$\mathbb{C}$	the complex numbers
$\Rightarrow$	implies	$[a,b]$	the interval $\{x : a \leqslant x \leqslant b\}$
$\Leftarrow$	is implied by	$(a,b]$	the interval $\{x : a < x \leqslant b\}$
$\Longleftrightarrow$	implies and is implied by	(a,b)	the interval $\{x : a < x < b\}$
$\cong$	is isomorphic to		

$|x|$ the modulus of x, $\begin{cases} x = x & \text{for } x \geqslant 0 \\ x = -x & \text{for } x < 0 \end{cases}$

$\binom{n}{r}$ the binomial coefficient $\dfrac{n!}{r!(n-r)!}$

A stroke through a symbol negates it, i.e. $\neq$ means 'is not equal to'.

Abbreviations

$\parallel$	parallel
+ ve	positive
− ve	negative
w.r.t.	with respect to

*Practical problems rarely have exact answers. Where numerical answers are given they are correct to two or three decimal places depending on their context, e.g. π is 3.142 correct to 3 d.p. and although we write $\pi = 3.142$ it is understood that this is not an exact value. We reserve the symbol $\simeq$ for those cases where the approximation being made is part of the method used.

Instructions for Answering Multiple Choice Exercises

These exercises are at the end of most chapters. The questions are set in groups, each group representing one of the variations that may arise in examination papers. The answering techniques are different for each type of question and are classified as follows:

TYPE I

These questions consist of a problem followed by several alternative answers, only *one* of which is correct.

Write down the letter corresponding to the correct answer.

TYPE II

In this type of question some information is given and is followed by a num-number of possible responses. *One or more* of the suggested responses follow(s) directly and necessarily from the information given

Write down the letter(s) corresponding to the correct response(s).

e.g. PQR is a triangle

(a) $\widehat{P} + \widehat{Q} + \widehat{R} = 180°$

(b) PQ + QR is less than PR

(c) if $\widehat{P}$ is obtuse, $\widehat{Q}$ and $\widehat{R}$ must both be acute

(d) $\widehat{P} = 90°$, $\widehat{Q} = 45°$, $\widehat{R} = 45°$.

The correct responses are (a) and (c).

(b) is definitely incorrect and (d) may or may not be true of triangle PQR, i.e. it does not follow directly and necessarily from the information given. Responses of this kind should not be regarded as correct.

TYPE III

Each problem contains two independent statements (a) and (b).

1) If (a) always implies (b) but (b) does not always imply (a) write A.
2) If (b) always implies (a) but (a) does not always imply (b) write B.
3) If (a) always implies (b) *and* (b) always implies (a) write C.
4) If (a) denies (b) and (b) denies (a), i.e. if (a) and (b) are mutually incompatible write D.
5) If none of the first four relationships apply write E.

TYPE IV

A problem is introduced and followed by a number of pieces of information. You are not required to solve the problem but to decide whether:

1) the given information is *all* needed to solve the problem. In this case write A;

2) the total amount of information is insufficient to solve the problem. If so write I;
3) the problem can be solved without using one or more of the given pieces of information. In this case write down the letter(s) corresponding to the items not needed.

TYPE V

A single statement is made. Write T if it is true and F if it is false.

CHAPTER 1

TRANSFORMATIONS, MATRICES AND DETERMINANTS

COLUMN VECTORS

The position vector $\overrightarrow{OP}$ of a point $P(x, y)$ can be denoted either by the sum of its Cartesian components,

i.e. $$x\mathbf{i} + y\mathbf{j}$$

or by the column vector $$\begin{pmatrix} x \\ y \end{pmatrix}$$

In this chapter we are going to use the column vector form of notation. Hence if $\overrightarrow{OA}$ is the position vector of $A(1, 3)$ we write

$$\overrightarrow{OA} = \begin{pmatrix} 1 \\ 3 \end{pmatrix}$$

and if $\overrightarrow{OB}$ is the position vector of $B(-2, 3)$ we write

$$\overrightarrow{OB} = \begin{pmatrix} -2 \\ 3 \end{pmatrix}$$

Using vector addition gives

$$\overrightarrow{OA} + \overrightarrow{OB} = \begin{pmatrix} 1 \\ 3 \end{pmatrix} + \begin{pmatrix} -2 \\ 3 \end{pmatrix} = \begin{pmatrix} -1 \\ 6 \end{pmatrix}$$

Similarly $$\overrightarrow{OB} - \overrightarrow{OA} = \begin{pmatrix} -2 \\ 3 \end{pmatrix} - \begin{pmatrix} 1 \\ 3 \end{pmatrix} = \begin{pmatrix} -3 \\ 0 \end{pmatrix}$$

EXERCISE 1a

Given A(2, 5), B(− 1, − 3), C(0, 4), D(− 2, 3) find the column vector which represents:

1) $\overrightarrow{OA}+\overrightarrow{OC}$ 2) $\overrightarrow{OA}+\overrightarrow{OD}$ 3) $\overrightarrow{OC}-\overrightarrow{OB}$ 4) $\overrightarrow{OA}+\overrightarrow{OB}+\overrightarrow{OC}$

5) the position vector of the midpoint of AB

6) the position vector of the midpoint of OC

7) the position vector of the centroid of $\triangle$ABC.

TRANSFORMATIONS IN TWO DIMENSIONS

Suppose that a rubber balloon has a picture printed on it. If the balloon is stretched, the picture is distorted. If the balloon is blown up the picture is enlarged. If the balloon is turned over, the picture changes position. Mathematically, such changes of position and distortions of shape are called transformations. In this section we are going to investigate some transformations of the xy plane.

In particular we are going to look at linear transformations under which straight lines stay straight and the origin does not change position.
Suppose that the xy plane is rotated through an angle $\pi/6$ about the origin so that Ox and Oy are rotated to the positions shown in diagram (ii).

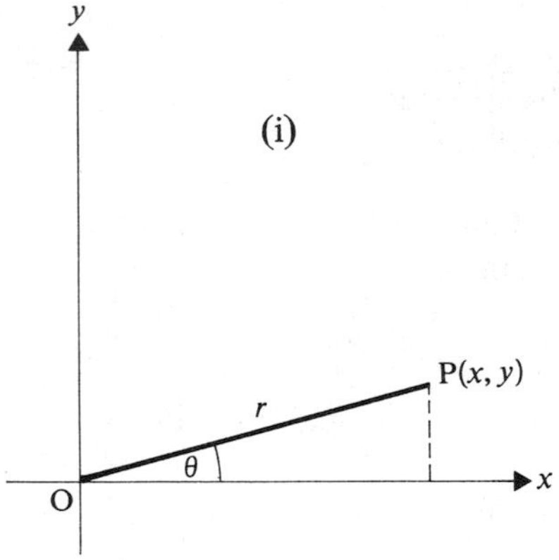

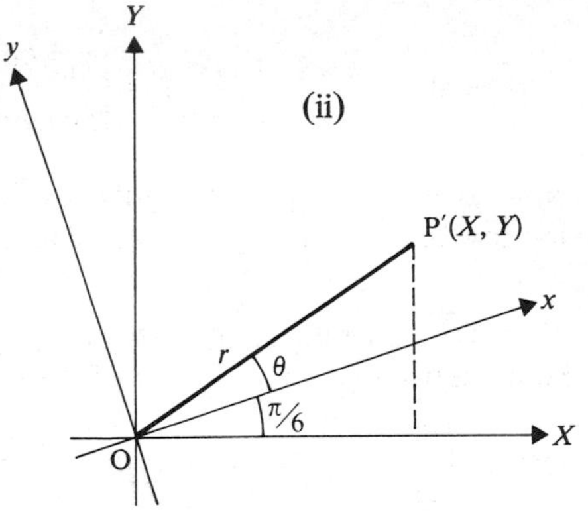

If P is the point (x, y) in diagram (i) then $\overrightarrow{OP}$ becomes $\overrightarrow{OP'}$ in diagram (ii).
P′ is called the image of P and we say that the transformation maps P to P′.
Referring P′ to coordinate axes OX and OY we see from the diagrams that

$$\left.\begin{aligned} x &= r\cos\theta \\ y &= r\sin\theta \end{aligned}\right\} \quad\text{and}\quad \left\{\begin{aligned} X &= r\cos(\theta+\pi/6) = \tfrac{\sqrt{3}}{2}r\cos\theta-\tfrac{1}{2}r\sin\theta \\ Y &= r\sin(\theta+\pi/6) = \tfrac{1}{2}r\cos\theta+\tfrac{\sqrt{3}}{2}r\sin\theta \end{aligned}\right.$$

Combining these two sets of equations gives

$$\frac{\sqrt{3}}{2}x - \tfrac{1}{2}y = X \qquad [1]$$

$$\tfrac{1}{2}x + \frac{\sqrt{3}}{2}y = Y \qquad [2]$$

This pair of equations enables us to transform the coordinates of any point $P(x, y)$ to the coordinates of the point $P'(X, Y)$ where $\overrightarrow{OP'}$ is the vector obtained by rotating $\overrightarrow{OP}$ anticlockwise through an angle of $\pi/6$ radians about O.

i.e. $\overrightarrow{OP} \longmapsto \overrightarrow{OP'}$ where '$\longmapsto$' means 'maps to'.
This pair of scalar equations can be written as a single vector equation,

i.e.
$$\begin{pmatrix} \frac{\sqrt{3}}{2}x + (-\frac{1}{2})y \\ \frac{1}{2}x + \frac{\sqrt{3}}{2}y \end{pmatrix} = \begin{pmatrix} X \\ Y \end{pmatrix}$$

The four constants on the left hand side form a square array of numbers, known as a *matrix*. Using matrix notation the equation above is expressed as

$$\begin{pmatrix} \frac{\sqrt{3}}{2} & -\frac{1}{2} \\ \frac{1}{2} & \frac{\sqrt{3}}{2} \end{pmatrix}\begin{pmatrix} x \\ y \end{pmatrix} = \begin{pmatrix} X \\ Y \end{pmatrix}$$

In general,
$$\begin{pmatrix} X \\ Y \end{pmatrix} = \begin{pmatrix} ax+by \\ cx+dy \end{pmatrix} = \begin{pmatrix} a & b \\ c & d \end{pmatrix}\begin{pmatrix} x \\ y \end{pmatrix}$$

Now $ax + by$ is a scalar quantity which is a combination of the entries in the top row of the matrix, $\begin{pmatrix} a & b \end{pmatrix}$, and the entries in the column vector $\begin{pmatrix} x \\ y \end{pmatrix}$.

So we define the *scalar product* of the row vector $\begin{pmatrix} a & b \end{pmatrix}$ and the column vector $\begin{pmatrix} x \\ y \end{pmatrix}$, in that order, as the scalar quantity obtained by

adding {the product of the 1st entry in the row vector and the 1st entry in the column vector} **to** {the product of the 2nd entry in the row vector and the 2nd entry in the column vector}

e.g.
$$\begin{pmatrix} p & q \end{pmatrix}\begin{pmatrix} r \\ s \end{pmatrix} = pr + qs$$

Note that we have not redefined the scalar product of vectors since

$$(p\mathbf{i} + q\mathbf{j}) \cdot (r\mathbf{i} + s\mathbf{j}) = pr + qs$$

Note also that $\begin{pmatrix} p \\ q \end{pmatrix} \begin{pmatrix} r & s \end{pmatrix}$ is not defined at this stage and cannot be considered as a scalar product.

It follows that the product of the matrix $\begin{pmatrix} a & b \\ c & d \end{pmatrix}$ and the column vector $\begin{pmatrix} x \\ y \end{pmatrix}$, taken in that order, is the column vector whose top entry is the scalar $\begin{pmatrix} a & b \end{pmatrix} \begin{pmatrix} x \\ y \end{pmatrix}$ and whose bottom entry is the scalar $\begin{pmatrix} c & d \end{pmatrix} \begin{pmatrix} x \\ y \end{pmatrix}$

i.e.
$$\begin{pmatrix} a & b \\ c & d \end{pmatrix} \begin{pmatrix} x \\ y \end{pmatrix} = \begin{pmatrix} ax+by \\ cx+dy \end{pmatrix}$$

e.g.
$$\begin{pmatrix} 3 & 2 \\ 0 & 1 \end{pmatrix} \begin{pmatrix} 2 \\ 4 \end{pmatrix} = \begin{pmatrix} (3)(2)+(2)(4) \\ (0)(2)+(1)(4) \end{pmatrix} = \begin{pmatrix} 14 \\ 4 \end{pmatrix}$$

Note that $\begin{pmatrix} x \\ y \end{pmatrix} \begin{pmatrix} a & b \\ c & d \end{pmatrix}$ is not defined and has no meaning.

Returning to the transformation that we began this section with,

i.e.
$$\begin{pmatrix} \frac{\sqrt{3}}{2} & -\frac{1}{2} \\ \frac{1}{2} & \frac{\sqrt{3}}{2} \end{pmatrix} \begin{pmatrix} x \\ y \end{pmatrix} = \begin{pmatrix} X \\ Y \end{pmatrix}$$

we may regard the matrix as an operator which rotates $\begin{pmatrix} x \\ y \end{pmatrix}$ anticlockwise through $\frac{\pi}{6}$ radians to $\begin{pmatrix} X \\ Y \end{pmatrix}$.

Thus, if we want to rotate the square ABCD through $\frac{\pi}{6}$ about O, we can obtain the position vectors of its vertices after the rotation, by operating on $\overrightarrow{OA}, \overrightarrow{OB}, \overrightarrow{OC}, \overrightarrow{OD}$ with the matrix $\begin{pmatrix} \frac{\sqrt{3}}{2} & -\frac{1}{2} \\ \frac{1}{2} & \frac{\sqrt{3}}{2} \end{pmatrix}$ as follows.

If A, B, C, D are the points $(2, 0), (4, 0), (4, 2), (2, 2)$, then

$$\overrightarrow{OA} \longmapsto \begin{pmatrix} \frac{\sqrt{3}}{2} & -\frac{1}{2} \\ \frac{1}{2} & \frac{\sqrt{3}}{2} \end{pmatrix}\begin{pmatrix} 2 \\ 0 \end{pmatrix} = \begin{pmatrix} \sqrt{3} \\ 1 \end{pmatrix} = \overrightarrow{OA}'$$

$$\overrightarrow{OB} \longmapsto \begin{pmatrix} \frac{\sqrt{3}}{2} & -\frac{1}{2} \\ \frac{1}{2} & \frac{\sqrt{3}}{2} \end{pmatrix}\begin{pmatrix} 4 \\ 0 \end{pmatrix} = \begin{pmatrix} 2\sqrt{3} \\ 2 \end{pmatrix} = \overrightarrow{OB}'$$

$$\overrightarrow{OC} \longmapsto \begin{pmatrix} \frac{\sqrt{3}}{2} & -\frac{1}{2} \\ \frac{1}{2} & \frac{\sqrt{3}}{2} \end{pmatrix}\begin{pmatrix} 4 \\ 2 \end{pmatrix} = \begin{pmatrix} 2\sqrt{3}-1 \\ 2+\sqrt{3} \end{pmatrix} = \overrightarrow{OC}'$$

$$\overrightarrow{OD} \longmapsto \begin{pmatrix} \frac{\sqrt{3}}{2} & -\frac{1}{2} \\ \frac{1}{2} & \frac{\sqrt{3}}{2} \end{pmatrix}\begin{pmatrix} 2 \\ 2 \end{pmatrix} = \begin{pmatrix} \sqrt{3}-1 \\ 1+\sqrt{3} \end{pmatrix} = \overrightarrow{OD}'$$

The result of this transformation is shown in the diagram below.

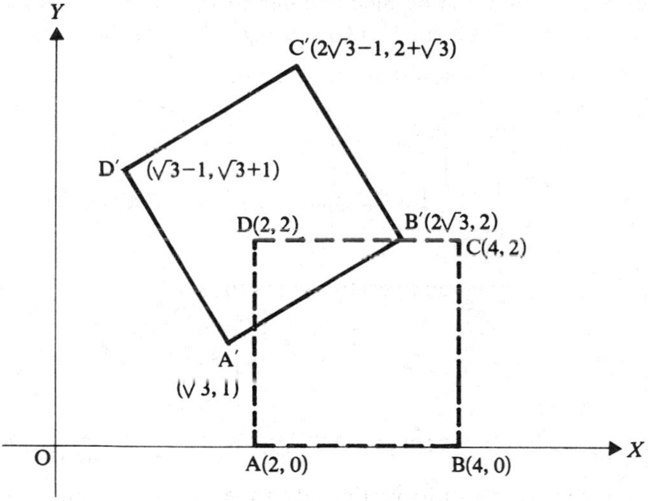

EXERCISE 1b

1) Evaluate

(a) $\begin{pmatrix} 1 & 3 \end{pmatrix}\begin{pmatrix} -1 \\ 4 \end{pmatrix}$ (b) $\begin{pmatrix} 2 & 7 \end{pmatrix}\begin{pmatrix} 3 \\ -4 \end{pmatrix}$ (c) $\begin{pmatrix} -1 & 0 \end{pmatrix}\begin{pmatrix} 3 \\ -1 \end{pmatrix}$.

2) Evaluate

(a) $\begin{pmatrix} 2 & 1 \\ 3 & -1 \end{pmatrix}\begin{pmatrix} 2 \\ 4 \end{pmatrix}$ (b) $\begin{pmatrix} -1 & 3 \\ 5 & 2 \end{pmatrix}\begin{pmatrix} -1 \\ -2 \end{pmatrix}$ (c) $\begin{pmatrix} -1 & 0 \\ -1 & 2 \end{pmatrix}\begin{pmatrix} 4 \\ 5 \end{pmatrix}$.

3) The points A(1, 0), B(2, 0), C(2, 3) form a triangle ABC. Use the matrix operator $\begin{pmatrix} 1 & 0 \\ 0 & -1 \end{pmatrix}$ to transform the position vectors $\overrightarrow{OA}, \overrightarrow{OB}, \overrightarrow{OC}$ to the position vectors $\overrightarrow{OA'}, \overrightarrow{OB'}, \overrightarrow{OC'}$. On the same diagram, draw triangle ABC and triangle A′B′C′ and hence describe the result of the transformation

$$\Delta ABC \longmapsto \Delta A'B'C'.$$

4) Repeat Question (3) with the matrix operator $\begin{pmatrix} -1 & 0 \\ 0 & 1 \end{pmatrix}$.

FURTHER TRANSFORMATIONS OF THE *xy* PLANE

In the previous section, the matrix which performed a rotation of $\pi/6$ radians about O was derived from the linear equations for the transformation.

Any transformation of $\begin{pmatrix} x \\ y \end{pmatrix}$ to $\begin{pmatrix} X \\ Y \end{pmatrix}$ that can be expressed by the linear equations $\begin{cases} ax + by = X \\ cx + dy = Y \end{cases}$ is called a *linear transformation* and is expressed by the matrix equation

$$\begin{pmatrix} a & b \\ c & d \end{pmatrix}\begin{pmatrix} x \\ y \end{pmatrix} = \begin{pmatrix} X \\ Y \end{pmatrix}$$

$\begin{pmatrix} a & b \\ c & d \end{pmatrix}$ is called the matrix of the transformation and we denote it by **M**.

Note that, under the operator **M**, $\begin{pmatrix} a & b \\ c & d \end{pmatrix}\begin{pmatrix} 0 \\ 0 \end{pmatrix} = \begin{pmatrix} 0 \\ 0 \end{pmatrix}$,

i.e. the origin does not change position.

So for any transformation matrix of the form $\begin{pmatrix} a & b \\ c & d \end{pmatrix}$ the origin is invariant.

Examples of such transformations are

1) *Rotation through any angle about the origin*

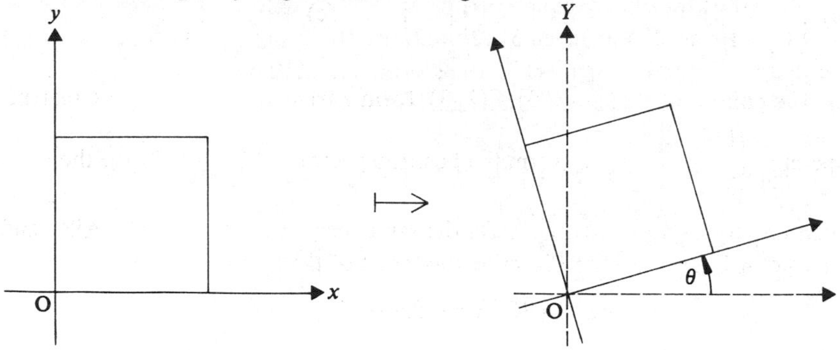

2) *Reflection in any line through O*

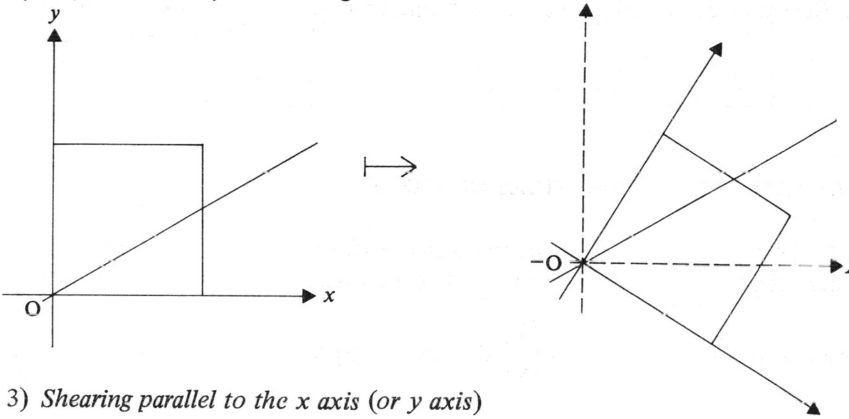

3) *Shearing parallel to the x axis (or y axis)*

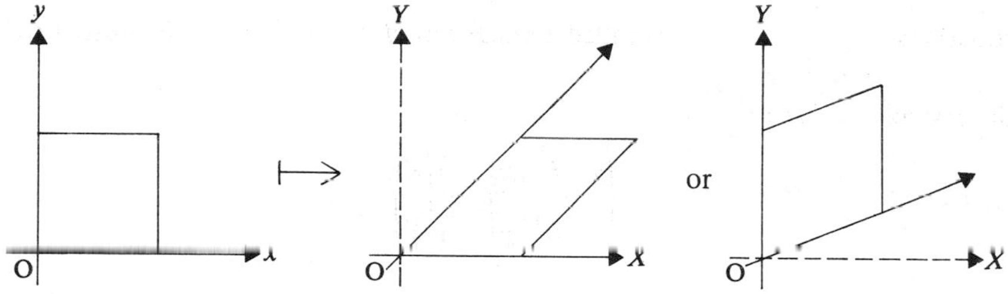

4) *Enlargement (or reduction)*

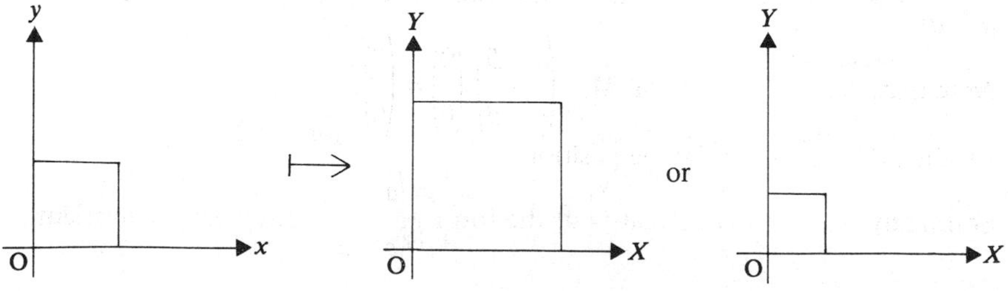

The matrix operator which produces each of these linear transformations can be found by the method used in the previous section, i.e. by first finding the appropriate pair of linear equations.
However, the algebra involved can be tedious, so we now develop a more direct method for obtaining a particular transformation matrix.
Suppose that under a given linear transformation

$$\mathbf{i} \text{ maps to } \mathbf{p} \text{ where } \mathbf{p} = \begin{pmatrix} a \\ c \end{pmatrix} \quad \text{and} \quad \mathbf{j} \text{ maps to } \mathbf{q} \text{ where } \mathbf{q} = \begin{pmatrix} b \\ d \end{pmatrix}$$

i.e.

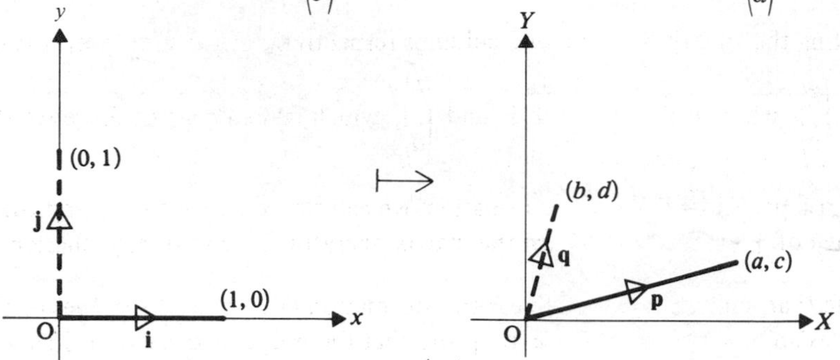

then $\quad \lambda\mathbf{i}$ maps to $\lambda\mathbf{p}$ where $\quad \lambda\mathbf{p} = \begin{pmatrix} \lambda a \\ \lambda c \end{pmatrix}$

and $\quad \mu\mathbf{j}$ maps to $\mu\mathbf{q}$ where $\quad \mu\mathbf{q} = \begin{pmatrix} \mu b \\ \mu d \end{pmatrix}$

i.e.

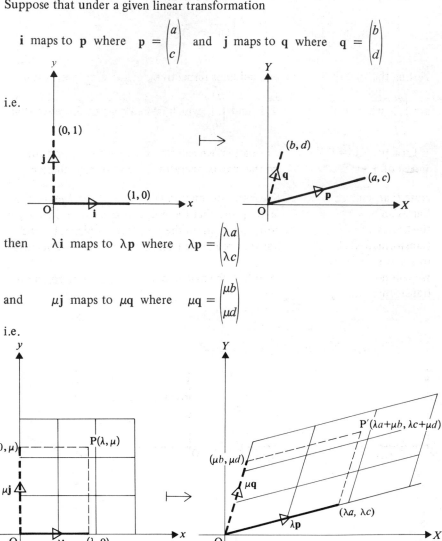

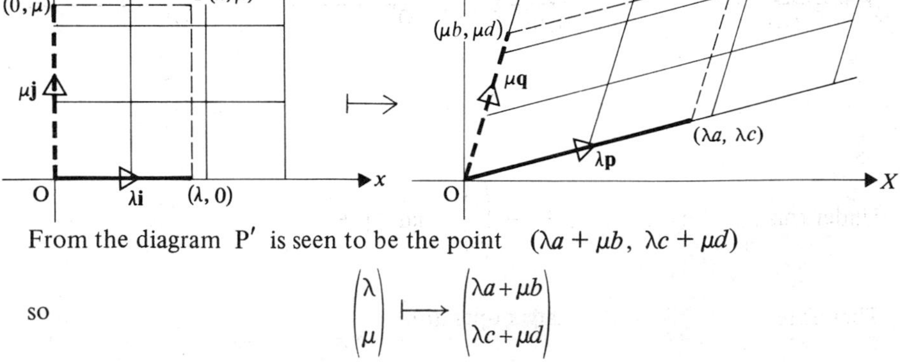

From the diagram P′ is seen to be the point $(\lambda a + \mu b,\ \lambda c + \mu d)$

so
$$\begin{pmatrix} \lambda \\ \mu \end{pmatrix} \longmapsto \begin{pmatrix} \lambda a + \mu b \\ \lambda c + \mu d \end{pmatrix}$$

Now
$$\begin{pmatrix}\lambda a+\mu b\\ \lambda c+\mu d\end{pmatrix} = \begin{pmatrix}a & b\\ c & d\end{pmatrix}\begin{pmatrix}\lambda\\ \mu\end{pmatrix}$$

Hence the mapping of $\overrightarrow{\text{OP}}$ to $\overrightarrow{\text{OP}'}$ can be expressed by the equation

$$\begin{pmatrix}a & b\\ c & d\end{pmatrix}\overrightarrow{\text{OP}} = \overrightarrow{\text{OP}'}$$

Noting that the first and second columns respectively of the matrix operator are $\begin{pmatrix}a\\ c\end{pmatrix}$, which is the image of $\mathbf{i}$, and $\begin{pmatrix}b\\ d\end{pmatrix}$, which is the image of $\mathbf{j}$, we see that

if for a particular linear transformation, we can find the image of $\mathbf{i}$ and the image of $\mathbf{j}$ we can *write down* the matrix operator for that transformation.

Note that under the general linear transformation described, the images of the Cartesian base vectors $\mathbf{i}, \mathbf{j}$ are $\mathbf{p}, \mathbf{q}$ and that the image of the vector $\lambda\mathbf{i} + \mu\mathbf{j}$ in the Cartesian plane is of the form $\lambda\mathbf{p} + \mu\mathbf{q}$ in the transformed plane, i.e. the transformation changes the frame of reference from the Cartesian base vectors to a new set of base vectors.
We will now use this approach to find the matrix operators for some common transformations.

1) *Reflection in the x-axis*

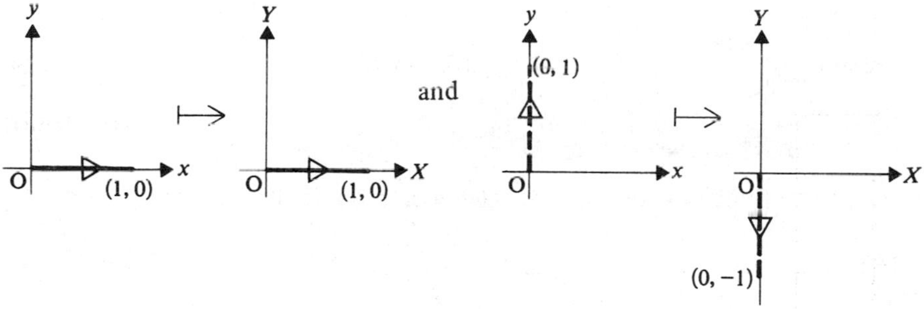

Under this transformation $\mathbf{i} \longmapsto \begin{pmatrix}1\\ 0\end{pmatrix}$ and $\mathbf{j} \longmapsto \begin{pmatrix}0\\ -1\end{pmatrix}$.

Therefore $\begin{pmatrix}1 & 0\\ 0 & -1\end{pmatrix}$ is the matrix operator.

2) *Reflection in the line* $y = x$

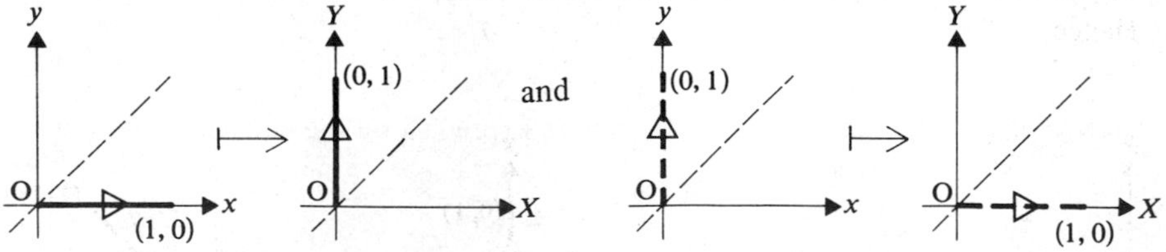

Under this transformation $\mathbf{i} \longmapsto \begin{pmatrix} 0 \\ 1 \end{pmatrix}$ and $\mathbf{j} \longmapsto \begin{pmatrix} 1 \\ 0 \end{pmatrix}$.

Therefore $\begin{pmatrix} 0 & 1 \\ 1 & 0 \end{pmatrix}$ is the matrix operator.

3) *Rotation through an angle* θ *about* O

Note that a positive angle means an anticlockwise rotation.

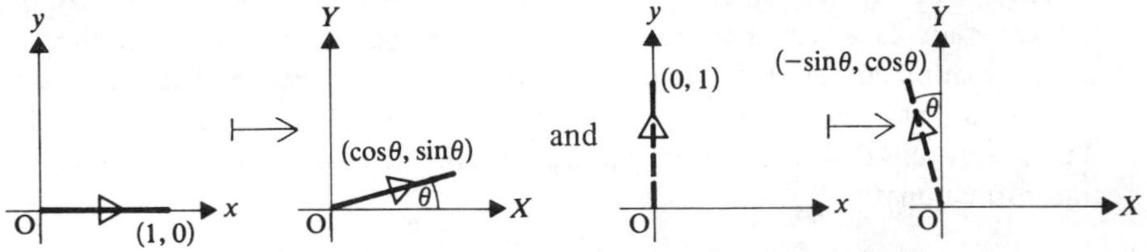

Under this transformation $\mathbf{i} \longmapsto \begin{pmatrix} \cos\theta \\ \sin\theta \end{pmatrix}$ and $\mathbf{j} \longmapsto \begin{pmatrix} -\sin\theta \\ \cos\theta \end{pmatrix}$.

Therefore $\begin{pmatrix} \cos\theta & -\sin\theta \\ \sin\theta & \cos\theta \end{pmatrix}$ is the matrix operator.

4) *Shear of 45° in the direction* Ox

The effect of this shear on a grid of squares is shown in the diagram below:

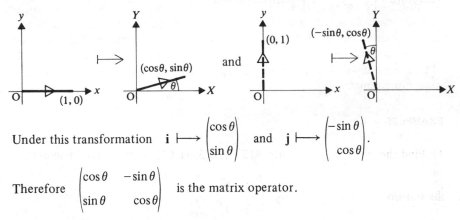

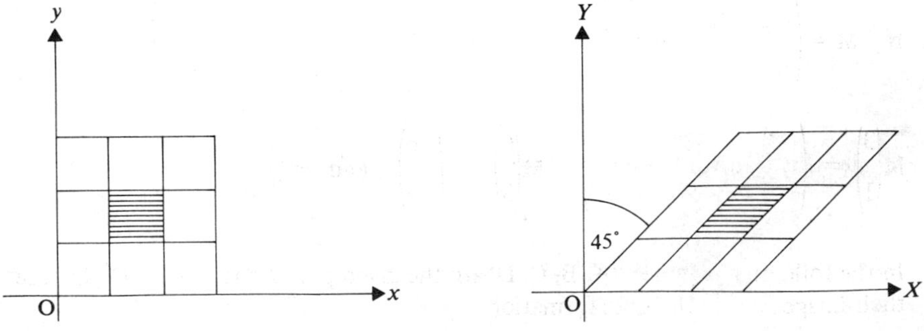

i.e. lines parallel to Oy are tilted at $45°$ and stretched to maintain height. A shear can be thought of as a stretch in one direction.
Hence

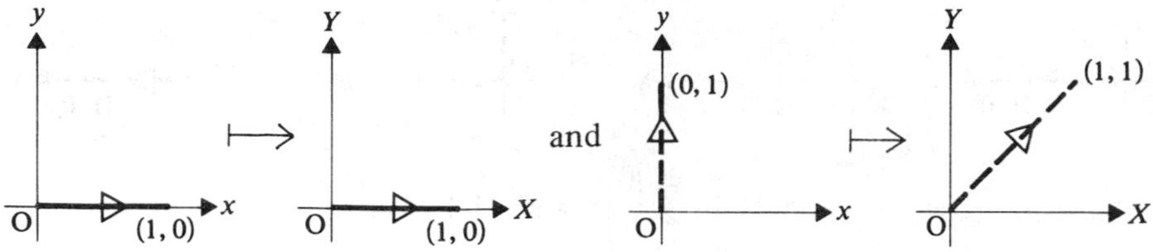

i.e. $\qquad \mathbf{i} \longmapsto \begin{pmatrix} 1 \\ 0 \end{pmatrix}$ and $\mathbf{j} \longmapsto \begin{pmatrix} 1 \\ 1 \end{pmatrix}$

Therefore $\begin{pmatrix} 1 & 1 \\ 0 & 1 \end{pmatrix}$ is the matrix operator.

EXAMPLES 1c

1) Find the images of the points A(3, 1), B(3, 3), C(6, 3), and D(6, 1) under the transformation produced by $\begin{pmatrix} 1 & 0 \\ -2 & 1 \end{pmatrix}$.

Illustrate the effect of the transformation and describe it.

If $\mathbf{M} = \begin{pmatrix} 1 & 0 \\ -2 & 1 \end{pmatrix}$ then

$$\mathbf{M}\begin{pmatrix} 3 \\ 1 \end{pmatrix} = \begin{pmatrix} 3 \\ -5 \end{pmatrix}, \quad \mathbf{M}\begin{pmatrix} 3 \\ 3 \end{pmatrix} = \begin{pmatrix} 3 \\ -3 \end{pmatrix}, \quad \mathbf{M}\begin{pmatrix} 6 \\ 3 \end{pmatrix} = \begin{pmatrix} 6 \\ -9 \end{pmatrix} \quad \text{and} \quad \mathbf{M}\begin{pmatrix} 6 \\ 1 \end{pmatrix} = \begin{pmatrix} 6 \\ -11 \end{pmatrix}$$

In the following diagram, A, B, C, D are the given points and A′, B′, C′, D′ are their images under the transformation.

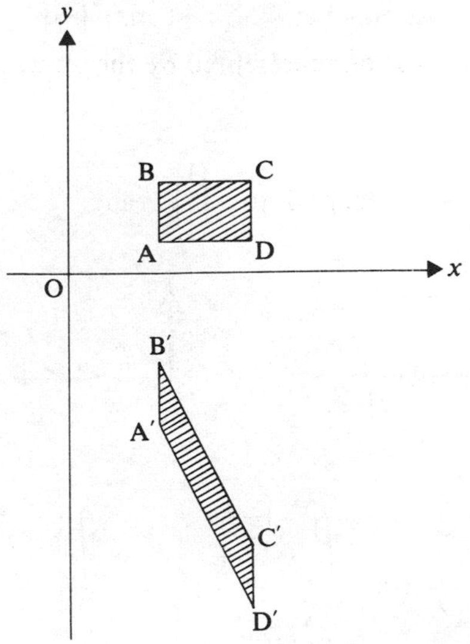

By inspection we see that the transformation is a shear in the direction of the negative y axis.

Alternatively, the effect of a transformation can be described from the matrix operator as follows:

$$\begin{pmatrix} 1 & 0 \\ -2 & 1 \end{pmatrix} \text{ maps } \mathbf{i} \text{ to } \begin{pmatrix} 1 \\ -2 \end{pmatrix} \text{ and } \mathbf{j} \text{ to } \begin{pmatrix} 0 \\ 1 \end{pmatrix}$$

i.e.

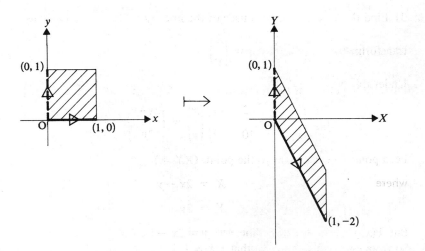

from which we see that the transformation is a shear.

2) Describe the transformation represented by the matrix $\begin{pmatrix} 1 & 1 \\ 1 & 1 \end{pmatrix}$.

$$\begin{pmatrix} 1 & 1 \\ 1 & 1 \end{pmatrix} \text{ maps } \mathbf{i} \text{ to } \begin{pmatrix} 1 \\ 1 \end{pmatrix} \text{ and } \mathbf{j} \text{ to } \begin{pmatrix} 1 \\ 1 \end{pmatrix}$$

i.e.

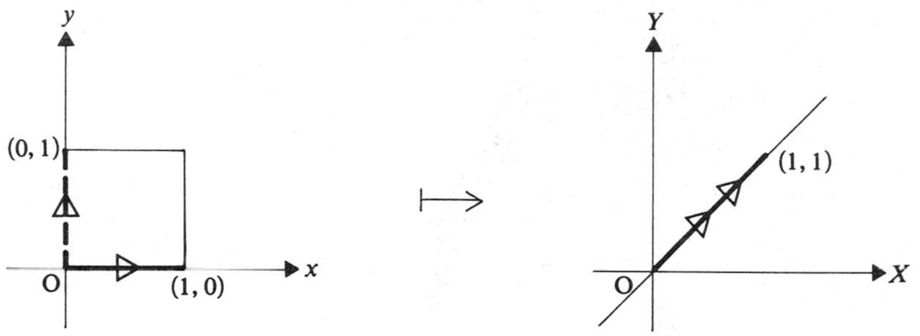

Further, for any point (x, y)

$$\begin{pmatrix} 1 & 1 \\ 1 & 1 \end{pmatrix}\begin{pmatrix} x \\ y \end{pmatrix} = \begin{pmatrix} x+y \\ x+y \end{pmatrix} = \begin{pmatrix} X \\ Y \end{pmatrix}$$

i.e. $X = Y$ so this transformation maps all points in the xy plane to points on the line $Y = X$.
This transformation is called *singular* because it maps the plane to a line.
Any linear transformation which maps the plane to a line or to the origin is called singular.

3) Find the equation of the image of the line $y = 2x - 1$ under the transformation $\begin{pmatrix} 2 & -1 \\ 0 & 2 \end{pmatrix}\begin{pmatrix} x \\ y \end{pmatrix} = \begin{pmatrix} X \\ Y \end{pmatrix}$.

Under this transformation,

$$\begin{pmatrix} x \\ y \end{pmatrix} \longmapsto \begin{pmatrix} 2 & 1 \\ 0 & 2 \end{pmatrix}\begin{pmatrix} x \\ y \end{pmatrix} = \begin{pmatrix} 2x \quad y \\ 2y \end{pmatrix} = \begin{pmatrix} X \\ Y \end{pmatrix}$$

i.e. a point $P(x, y)$ maps to the point $Q(X, Y)$

where $\quad X = 2x - y \quad$ [1]

$\quad Y = 2y \quad$ [2]

But $P(x, y)$ is on the given line $\iff y = 2x - 1$ (or $2x - y = 1$).
So from equation [1] we see that $X = 1$.
Therefore the image of $y = 2x - 1$ is the line $X = 1$.

In more general problems of this type, the equation of the locus of Q(X, Y) is found by eliminating x and y from the equation of the locus of P, using equations [1] and [2].

Alternatively we find the equation of the image of $y = 2x - 1$ by

(a) finding the points where $y = 2x - 1$ cuts the x and y axes: these are $(0, -1)$ and $(\frac{1}{2}, 0)$

(b) finding the images of these points:

these are $\begin{pmatrix} 2 & -1 \\ 0 & 2 \end{pmatrix}\begin{pmatrix} 0 \\ -1 \end{pmatrix}$ and $\begin{pmatrix} 2 & -1 \\ 0 & 2 \end{pmatrix}\begin{pmatrix} \frac{1}{2} \\ 0 \end{pmatrix}$

i.e. $(1, -2)$ and $(1, 0)$

(c) finding the equation of the line through the transformed points, i.e. $X = 1$

i.e.

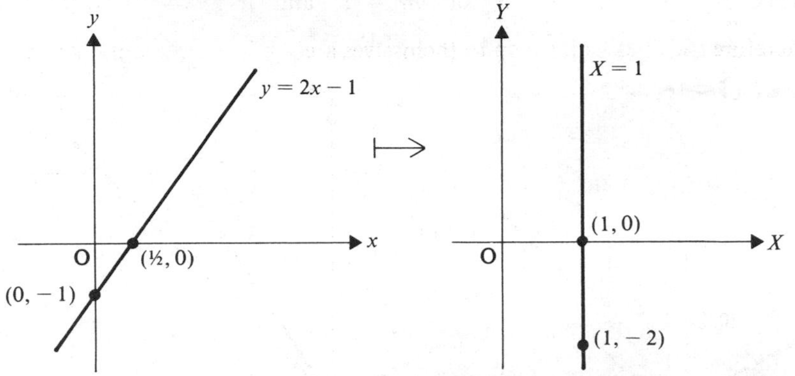

4) Find the equations of the lines that are mapped on to themselves under the transformation

$$\begin{pmatrix} 2 & 1 \\ 3 & 0 \end{pmatrix}\begin{pmatrix} x \\ y \end{pmatrix} = \begin{pmatrix} X \\ Y \end{pmatrix}$$

$\begin{pmatrix} x \\ y \end{pmatrix}$ is mapped to $\begin{pmatrix} X \\ Y \end{pmatrix}$ where

$$\begin{pmatrix} X \\ Y \end{pmatrix} = \begin{pmatrix} 2 & 1 \\ 3 & 0 \end{pmatrix}\begin{pmatrix} x \\ y \end{pmatrix} = \begin{pmatrix} 2x + y \\ 3x \end{pmatrix} \qquad [1]$$

If the line $y = mx + c$ in the xy plane maps to itself, then its equation in the transformed plane is $Y = mX + c$.

From [1] $\quad X = 2x + y$ and $Y = 3x$.

Hence $\quad Y = mX + c \Rightarrow 3x = m(2x + y) + c$

$$\Rightarrow \quad y = \frac{3 - 2m}{m}x - \frac{c}{m}$$

So, if $y = mx + c$ maps to $Y = mX + c$ then $y = \dfrac{3-2m}{m}x - \dfrac{c}{m}$

Comparing coefficients we see that

$$m = \frac{(3-2m)}{m} \quad \text{and} \quad c = -\frac{c}{m}$$

So $\quad m^2 = -2m+3 \quad \Rightarrow \quad m = -3 \text{ or } 1$

and $\quad c = -\dfrac{c}{m} \quad \Rightarrow \quad c(m+1) = 0 \quad \Rightarrow \quad c = 0 \quad \text{or} \quad m = -1$

$$\text{but } m \neq -1; \text{ therefore } c = 0$$

Hence $\quad m = -3 \quad \text{or} \quad m = 1 \quad \text{and} \quad c = 0$

Therefore the lines which map to themselves are $y = -3x$ and $y = x$

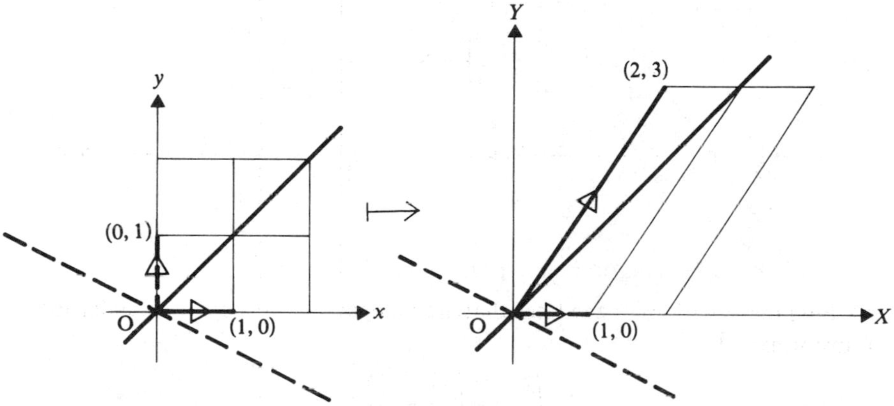

Note that although $y = x$ maps to $Y = X$, a particular point on $y = x$

does *not*, in general, map to the same point on $Y = X$, e.g. $\begin{pmatrix}1\\1\end{pmatrix} \longmapsto \begin{pmatrix}3\\3\end{pmatrix}$

EXERCISE 1e

1) Find the matrices for the following transformations:
(a) reflection in the y axis,
(b) a rotation about the origin of $45°$,
(c) a stretch by a factor of 2 parallel to Oy,
(d) reflection in the line $y = 2x$,
(e) a shear of $30°$ in the direction of Oy,
(f) a stretch by a factor of 3 parallel to Ox,
(g) an enlargement by a factor of 3,
(h) reflection in the x axis combined with a stretch by a factor of 2 parallel to the x axis.

2) Describe the transformations represented by the following matrices.

(a) $\begin{pmatrix} -1 & 0 \\ 0 & 1 \end{pmatrix}$ (b) $\begin{pmatrix} -1 & 0 \\ 0 & -1 \end{pmatrix}$ (c) $\begin{pmatrix} -1 & 1 \\ -1 & 1 \end{pmatrix}$

(d) $\begin{pmatrix} 1 & 0 \\ 0 & 1 \end{pmatrix}$ (e) $\begin{pmatrix} 2 & 1 \\ 0 & 2 \end{pmatrix}$ (f) $\begin{pmatrix} 1 & 1 \\ 0 & 0 \end{pmatrix}$

(g) $\begin{pmatrix} 0 & 0 \\ 0 & 0 \end{pmatrix}$ (h) $\begin{pmatrix} \sin\theta & -\cos\theta \\ \cos\theta & \sin\theta \end{pmatrix}$ (i) $\begin{pmatrix} 1 & 0 \\ -2 & 1 \end{pmatrix}$.

3) A triangle ABC has its vertices at the points A(1, 1), B(2, 1), C(2, 4). Draw diagrams to represent the image of triangle ABC under the transformations

(a) $\begin{pmatrix} 2 & 0 \\ 0 & 2 \end{pmatrix}$ (b) $\begin{pmatrix} -3 & 0 \\ 0 & -3 \end{pmatrix}$ (c) $\begin{pmatrix} -1 & 0 \\ 0 & 3 \end{pmatrix}$

(d) $\begin{pmatrix} 1 & 0 \\ 1 & 0 \end{pmatrix}$ (e) $\begin{pmatrix} 1 & 2 \\ 3 & 4 \end{pmatrix}$ (f) $\begin{pmatrix} 0 & 2 \\ 0 & -3 \end{pmatrix}$.

4) Find the reflections of **i** and **j** in the line $y = x\tan\theta$.

Hence show that any matrix of the form $\begin{pmatrix} \cos 2\theta & \sin 2\theta \\ \sin 2\theta & -\cos 2\theta \end{pmatrix}$ represents a reflection in the line $y = mx$ where $m = \tan\theta$.

Write down the matrices representing a reflection in
(a) $3y = 4x$ (b) $y = -2x$.

5) Find the images of **i** and **j** under a rotation about the origin of θ radians.

Hence show that any matrix of the form $\begin{pmatrix} \cos\theta & -\sin\theta \\ \sin\theta & \cos\theta \end{pmatrix}$ represents a rotation about the origin.

Write down the matrices representing a rotation about the origin of

(a) $\dfrac{\pi}{3}$ (b) $-\dfrac{\pi}{4}$.

6) The image of the point (x, y) is (X, Y) under the transformation

$$\begin{pmatrix} 0 & -1 \\ -1 & 0 \end{pmatrix}\begin{pmatrix} x \\ y \end{pmatrix} = \begin{pmatrix} X \\ Y \end{pmatrix}$$

Find the equation of the images under this transformation of the lines
(a) $2y = x$ (b) $x = 2$ (c) $x + y + 2 = 0$
Illustrate your results on a diagram.

7) The point $\begin{pmatrix} x \\ y \end{pmatrix}$ is mapped to the point $\begin{pmatrix} X \\ Y \end{pmatrix}$ under the transformation $\mathbf{M}\begin{pmatrix} x \\ y \end{pmatrix} = \begin{pmatrix} X \\ Y \end{pmatrix}$. Find the equations of the lines which are mapped on to themselves when $\mathbf{M}$ is

(a) $\begin{pmatrix} 0 & 1 \\ 5 & -4 \end{pmatrix}$ (b) $\begin{pmatrix} -2 & 2 \\ -2 & -3 \end{pmatrix}$ (c) $\begin{pmatrix} 0 & 3 \\ 1 & -2 \end{pmatrix}$

Illustrate your results on diagrams.

MATRICES

In the previous section a form of notation was introduced to deal with the particular problem of linear transformations in the xy plane. This form of notation has far wider applications and may be used for a variety of problems involving linear relationships between any number of variables. In this section the notation is defined in more general terms, and the laws for operations between matrices are introduced; i.e. we develop a matrix algebra.

Definition of a Matrix

A matrix is a rectangular array of numbers, e.g. $\begin{pmatrix} 1 & 3 & 7 \\ 2 & 0 & -1 \end{pmatrix}$.

A particular matrix is denoted by a bold capital letter,

e.g. $\mathbf{A}$ ($\underline{A}$ when handwritten).

The size of a matrix is described by the number of its rows and the number of its columns,

e.g. $\begin{pmatrix} 1 & 3 & 7 \\ 2 & 0 & -1 \end{pmatrix}$ has 2 rows and 3 columns and is described as a 2 by 3 (or 2×3) matrix.

In general if a matrix, $\mathbf{A}$, has m rows and n columns then $\mathbf{A}$ is called an $m \times n$ matrix.

A matrix with just one column, e.g. $\begin{pmatrix} 2 \\ -1 \\ 3 \end{pmatrix}$, is called a *column vector* and a matrix with just one row, e.g. $\begin{pmatrix} -1 & 0 & 3 & -4 \end{pmatrix}$, is called a *row vector*.
Column vectors and row vectors are denoted by lower case bold letters, e.g.

$$\mathbf{a} = \begin{pmatrix} 1 & 2 & 3 \end{pmatrix}, \qquad \mathbf{b} = \begin{pmatrix} 2 \\ 5 \\ 1 \end{pmatrix}.$$

A matrix with an equal number of rows and columns is called a *square matrix*.
The numbers in a matrix are called the *entries* of that matrix. A particular entry can be identified by using the following notation.
A general 2×3 matrix is written as

$$\mathbf{A} = \begin{pmatrix} a_{11} & a_{12} & a_{13} \\ a_{21} & a_{22} & a_{23} \end{pmatrix}$$

where the suffixes attached to a particular entry refer first to its row and second to its column.

In general, the entry in the ith row and jth column of a matrix is denoted by a_{ij}.

Matrices of the Same Size

Two matrices, **A** and **B**, are the same size if the number of rows and columns in **A** is the same as the number of rows and columns in **B**.

Hence $\begin{pmatrix} 1 & 4 \\ 2 & 1 \\ 3 & -1 \end{pmatrix}$ and $\begin{pmatrix} 1 & 3 \\ 0 & 1 \\ -1 & 5 \end{pmatrix}$ are the same size

but $\begin{pmatrix} 2 & 1 \\ 3 & 4 \\ 1 & -1 \end{pmatrix}$ and $\begin{pmatrix} 1 & 0 & 1 \\ -2 & 5 & 0 \end{pmatrix}$ are not.

Equal Matrices

Two matrices **A** and **B** are equal if each entry of **A** is equal to the corresponding entry of **B**;

i.e. $\mathbf{A} = \mathbf{B} \iff a_{ij} = b_{ij}$ for all i and j.

So $\begin{pmatrix} 1 & 3 \\ 2 & 5 \end{pmatrix} = \begin{pmatrix} 1 & 3 \\ 2 & 5 \end{pmatrix}$ but $\begin{pmatrix} 1 & 3 \\ 2 & 5 \end{pmatrix} \neq \begin{pmatrix} 1 & 2 \\ 3 & 5 \end{pmatrix}$ although they are the same size.

Addition of Matrices

If two matrices are the same size they can be added by summing the corresponding entries.

e.g. if $\mathbf{A} = \begin{pmatrix} 1 & 3 & 0 & 2 \\ -1 & 4 & 0 & -3 \end{pmatrix}$ and $\mathbf{B} = \begin{pmatrix} 0 & 1 & 4 & 3 \\ 7 & 4 & 9 & 10 \end{pmatrix}$

then

$$\mathbf{A}+\mathbf{B} = \begin{pmatrix} (1+0) & (3+1) & (0+4) & (2+3) \\ (-1+7) & (4+4) & (0+9) & (-3+10) \end{pmatrix} = \begin{pmatrix} 1 & 4 & 4 & 5 \\ 6 & 8 & 9 & 7 \end{pmatrix}.$$

Note that two matrices of equal size are said to be conformable for addition.

Note that if **A** and **B** are not the same size, then $\mathbf{A}+\mathbf{B}$ has no meaning.

Note also that if $\mathbf{A}+\mathbf{B}$ exists then $\mathbf{A}+\mathbf{B} = \mathbf{B}+\mathbf{A}$; i.e., addition of matrices is commutative.

Multiplication by a Scalar

The matrix $2\mathbf{A}$ is such that each of its entries is twice the corresponding entry of **A**,

e.g. if $\mathbf{A} = \begin{pmatrix} 1 & 3 \\ -2 & 4 \\ 0 & -1 \end{pmatrix}$ then $2\mathbf{A} = \begin{pmatrix} 2 & 6 \\ -4 & 8 \\ 0 & -2 \end{pmatrix}$.

In general, if a_{ij} is a typical entry of **A** then $\lambda\mathbf{A}$ is the matrix whose corresponding entry is λa_{ij}.

EXERCISE 1d

Each of the questions below refers to the following matrices:

$\mathbf{A} = \begin{pmatrix} 1 & 6 & 0 \\ 2 & 1 & 5 \end{pmatrix}$ $\mathbf{b} = \begin{pmatrix} 0 & 1 & 3 \end{pmatrix}$ $\mathbf{C} = \begin{pmatrix} -3 & 3 \\ 2 & 5 \\ 7 & 4 \end{pmatrix}$

$\mathbf{D} = \begin{pmatrix} 1 & 0 \\ -4 & 7 \\ 0 & -10 \end{pmatrix}$ $\mathbf{E} = \begin{pmatrix} -3 & 0 & 10 \\ 12 & -9 & -1 \end{pmatrix}$ $\mathbf{F} = \begin{pmatrix} 1 & 3 \\ 2 & 5 \end{pmatrix}$

$\mathbf{G} = \begin{pmatrix} -5 & 9 \\ 0 & -3 \end{pmatrix}$ $\mathbf{h} = \begin{pmatrix} -7 & 0 & -3 \end{pmatrix}$.

1) Describe the size of each of the matrices given above.

2) Write down the pairs of matrices which are conformable for addition.

3) Find, where it exists, $\mathbf{A}+\mathbf{E}$, $\mathbf{b}+\mathbf{h}$, $\mathbf{F}+\mathbf{G}$, $\mathbf{A}+\mathbf{D}$, $\mathbf{E}+\mathbf{F}$, $\mathbf{D}+\mathbf{C}$.

4) Find: (a) $4\mathbf{A}$ (b) $-2\mathbf{E}$ (c) $5\mathbf{G}$ (d) $\lambda\mathbf{b}$.

5) If $a_{ij}, b_{ij}, c_{ij}, \ldots$ are typical elements of $\mathbf{A}, \mathbf{b}, \mathbf{C}, \ldots$ write down the values of

$$a_{22}, b_{13}, c_{31}, d_{12}, e_{23}, f_{21}, g_{11}, h_{12}.$$

6) Find: (a) $2\mathbf{A}+3\mathbf{E}$ (b) $3\mathbf{C}-2\mathbf{D}$ (c) $\mathbf{F}-\mathbf{G}$.

Multiplication of Matrices

When the matrix notation was introduced to deal with linear transformations, the product of the row vector $\begin{pmatrix} a & b \end{pmatrix}$ and the column vector $\begin{pmatrix} x \\ y \end{pmatrix}$ was defined by

$$\begin{pmatrix} a & b \end{pmatrix}\begin{pmatrix} x \\ y \end{pmatrix} = ax+by$$

This can be extended to cover the product of any row vector and column vector *provided that the row vector and the column vector contain the same number of entries*

e.g.
$$\begin{pmatrix} 3 & 4 & -1 \end{pmatrix}\begin{pmatrix} 2 \\ -3 \\ 4 \end{pmatrix} = (3)(2)+(4)(-3)+(-1)(4) = -10$$

In general

$$\begin{pmatrix} a_{11} & a_{12} & a_{13} \end{pmatrix}\begin{pmatrix} b_{11} \\ b_{21} \\ b_{31} \end{pmatrix} = a_{11}b_{11}+a_{12}b_{21}+a_{13}b_{31}$$

The product $\begin{pmatrix} a & b \\ c & d \end{pmatrix}\begin{pmatrix} x \\ y \end{pmatrix}$ is defined as the column vector

whose top entry is $\begin{pmatrix} a & b \end{pmatrix}\begin{pmatrix} x \\ y \end{pmatrix}$ and whose bottom entry is $\begin{pmatrix} c & d \end{pmatrix}\begin{pmatrix} x \\ y \end{pmatrix}$.

Using suffix notation we have

$$\begin{pmatrix} a_{11} & a_{12} \\ a_{21} & a_{22} \end{pmatrix}\begin{pmatrix} b_{11} \\ b_{21} \end{pmatrix} = \begin{pmatrix} a_{11}b_{11} + a_{12}b_{21} \\ a_{21}b_{11} + a_{22}b_{21} \end{pmatrix}$$

This definition may be extended to cover the product of any matrix and column vector *provided that the number of entries in each row of the matrix is the same as the number of entries in the column vector.*

e.g.

$$\begin{pmatrix} 2 & -1 & 0 \\ 1 & 3 & 4 \end{pmatrix}\begin{pmatrix} 3 \\ -5 \\ 9 \end{pmatrix} = \begin{pmatrix} (2)(3) + (-1)(-5) + (0)(9) \\ (1)(3) + (3)(-5) \quad + (4)(9) \end{pmatrix} = \begin{pmatrix} 11 \\ 24 \end{pmatrix}$$

and

$$\begin{pmatrix} 1 & 3 & -2 & 4 \\ -1 & 0 & 2 & -1 \\ 3 & 1 & 0 & 2 \end{pmatrix}\begin{pmatrix} 2 \\ -1 \\ 0 \\ 2 \end{pmatrix} = \begin{pmatrix} 7 \\ -4 \\ 9 \end{pmatrix}$$

Note that the product $\begin{pmatrix} a_1 \\ a_2 \\ a_3 \end{pmatrix}\begin{pmatrix} b_{11} & b_{12} & b_{13} \\ b_{21} & b_{22} & b_{23} \\ b_{31} & b_{32} & b_{33} \end{pmatrix}$ is *not* defined and has no meaning.

EXERCISE 1e

Find the following products:

1) $\begin{pmatrix} 1 & 0 & -3 \end{pmatrix}\begin{pmatrix} 2 \\ -4 \\ 1 \end{pmatrix}$

2) $\begin{pmatrix} -3 & 0 & 4 & 1 \end{pmatrix}\begin{pmatrix} 5 \\ 3 \\ 0 \\ 4 \end{pmatrix}$

3) $\begin{pmatrix} 5 & 7 & -10 \end{pmatrix}\begin{pmatrix} 4 \\ -1 \\ 0 \end{pmatrix}$

4) $\begin{pmatrix} p & q & r \end{pmatrix}\begin{pmatrix} 2p \\ -1 \\ 3r \end{pmatrix}$

5) $\begin{pmatrix} 3 & 2 \\ -1 & -2 \\ 4 & 0 \end{pmatrix}\begin{pmatrix} -3 \\ 0 \end{pmatrix}$

6) $\begin{pmatrix} -5 & 4 & -1 \\ 1 & 0 & -1 \end{pmatrix}\begin{pmatrix} -2 \\ 4 \\ -1 \end{pmatrix}$

7) $\begin{pmatrix} -1 & 3 & -1 \\ 4 & 0 & -2 \\ 1 & 3 & -2 \end{pmatrix}\begin{pmatrix} -3 \\ 5 \\ 3 \end{pmatrix}$

8) $\begin{pmatrix} 2 & -1 \\ -1 & 3 \\ -2 & 0 \\ 4 & -5 \end{pmatrix}\begin{pmatrix} -3 \\ 4 \end{pmatrix}$

9) $\begin{pmatrix} -5 & 0 & -1 \\ 4 & 0 & -2 \\ 0 & -1 & 1 \\ 1 & -2 & 1 \end{pmatrix}\begin{pmatrix} 1 \\ 2 \\ 3 \end{pmatrix}$

10) $\begin{pmatrix} \cos\theta & \sin\theta \\ -\sin\theta & \cos\theta \end{pmatrix}\begin{pmatrix} \cos\theta \\ \sin\theta \end{pmatrix}$

11) $\begin{pmatrix} x & 3x \\ 2y & y \end{pmatrix}\begin{pmatrix} 3x \\ 2y \end{pmatrix}$

12) $\begin{pmatrix} \cos\theta & \sin\theta & 1 \\ \sin\theta & \cos\theta & 2 \end{pmatrix}\begin{pmatrix} \cos\theta \\ \sin\theta \\ -1 \end{pmatrix}$

13) $\begin{pmatrix} t & t^2 \\ 2t & -t^2 \\ -t & 2t^2 \end{pmatrix}\begin{pmatrix} -t \\ 1 \end{pmatrix}$

14) $\begin{pmatrix} i & j \\ 1 & 1 \end{pmatrix}\begin{pmatrix} i \\ j \end{pmatrix}$.

The Product of Two Matrices

Consider the line joining the points $A(2,1)$ and $B(-1,2)$ which is transformed by the matrix $\begin{pmatrix} 3 & 4 \\ 5 & 6 \end{pmatrix}$ into the line joining A' and B' where A' and B' are the images of A and B.

The position vectors of A' and B' are given by

$$\begin{pmatrix} 3 & 4 \\ 5 & 6 \end{pmatrix}\begin{pmatrix} 2 \\ 1 \end{pmatrix} = \begin{pmatrix} 10 \\ 16 \end{pmatrix} \qquad [1]$$

and

$$\begin{pmatrix} 3 & 4 \\ 5 & 6 \end{pmatrix}\begin{pmatrix} -1 \\ 2 \end{pmatrix} = \begin{pmatrix} 5 \\ 7 \end{pmatrix} \qquad [2]$$

By adjoining the column vectors in [1] and [2], these two equations can be expressed as the single matrix equation

$$\begin{pmatrix} 3 & 4 \\ 5 & 6 \end{pmatrix}\begin{pmatrix} 2 & -1 \\ 1 & 2 \end{pmatrix} = \begin{pmatrix} 10 & 5 \\ 16 & 7 \end{pmatrix} \qquad [3]$$

As equation [3] is equivalent to equations [1] and [2], this equivalence defines the product

$$\begin{pmatrix} 3 & 4 \\ 5 & 6 \end{pmatrix}\begin{pmatrix} 2 & -1 \\ 1 & 2 \end{pmatrix}$$

as a 2×2 matrix each of whose entries is the product of the appropriate row from the first matrix and the appropriate column from the second matrix.

So if $\mathbf{A} = \begin{pmatrix} a_{11} & a_{12} \\ a_{21} & a_{22} \end{pmatrix}$ and $\mathbf{B} = \begin{pmatrix} b_{11} & b_{12} \\ b_{21} & b_{22} \end{pmatrix}$

then $\mathbf{AB}$ is $\begin{pmatrix} a_{11} & a_{12} \\ a_{21} & a_{22} \end{pmatrix}\begin{pmatrix} b_{11} & b_{12} \\ b_{21} & b_{22} \end{pmatrix}$

$$= \begin{pmatrix} \text{(1st row of }\mathbf{A})\cdot\text{(1st column of }\mathbf{B}) & \text{(1st row }\mathbf{A})\cdot\text{(2nd column }\mathbf{B}) \\ \text{(2nd row of }\mathbf{A})\cdot\text{(2nd column of }\mathbf{B}) & \text{(2nd row }\mathbf{A})\cdot\text{(2nd column }\mathbf{B}) \end{pmatrix}$$

e.g. if $\mathbf{A} = \begin{pmatrix} -1 & 3 \\ -4 & 2 \end{pmatrix}$ and $\mathbf{B} = \begin{pmatrix} 0 & -1 \\ 3 & -2 \end{pmatrix}$

then

$$\mathbf{AB} = \begin{pmatrix} -1 & 3 \\ -4 & 2 \end{pmatrix}\begin{pmatrix} 0 & -1 \\ 3 & -2 \end{pmatrix}$$

$$= \begin{pmatrix} (-1)(0)+(3)(3) & (-1)(-1)+(3)(-2) \\ (-4)(0)+(2)(3) & (-4)(-1)+(2)(-2) \end{pmatrix}$$

$$= \begin{pmatrix} 9 & -5 \\ 6 & 0 \end{pmatrix}$$

Similarly

$$\mathbf{BA} = \begin{pmatrix} 0 & -1 \\ 3 & -2 \end{pmatrix}\begin{pmatrix} -1 & 3 \\ -4 & 2 \end{pmatrix} = \begin{pmatrix} 4 & -2 \\ 5 & 5 \end{pmatrix}$$

and we note that $\mathbf{AB} \neq \mathbf{BA}$

i.e. matrix multiplication is not commutative.

This definition may now be extended to cover the product, **AB**, of any two matrices, provided that the number of columns in **A** is the same as the number of rows in **B**.

Hence if $\mathbf{A} = \begin{pmatrix} a_{11} & a_{12} \\ a_{21} & a_{22} \\ a_{31} & a_{32} \end{pmatrix}$ and $\mathbf{B} = \begin{pmatrix} b_{11} & b_{12} & b_{13} & b_{14} \\ b_{21} & b_{22} & b_{23} & b_{24} \end{pmatrix}$

then as **A** has two columns and **B** has two rows we define the product **AB** as

$$\begin{pmatrix} a_{11} & a_{12} \\ \boxed{a_{21} \quad a_{22}} \\ a_{31} & a_{32} \end{pmatrix} \begin{pmatrix} b_{11} & b_{12} & \boxed{\begin{matrix} b_{13} \\ b_{23} \end{matrix}} & b_{14} \\ b_{21} & b_{22} & & b_{24} \end{pmatrix} = \begin{pmatrix} c_{11} & c_{12} & c_{13} & c_{14} \\ c_{21} & c_{22} & \boxed{c_{23}} & c_{24} \\ c_{31} & c_{32} & c_{33} & c_{34} \end{pmatrix}$$

where c_{23} is the product of $\begin{cases} \text{the 2nd row vector of } \mathbf{A} \\ \text{and the 3rd column vector of } \mathbf{B} \end{cases}$

and we call this operation 'row–column' multiplication.

Note that under this definition, **BA** has no meaning because **B** has four columns whereas **A** has only three rows.

EXAMPLE 1f

Find **AB** where $\mathbf{A} = \begin{pmatrix} 3 & -1 & 2 \\ 0 & 4 & -1 \end{pmatrix}$ and $\mathbf{B} = \begin{pmatrix} 2 & 0 \\ 4 & 3 \\ 1 & -1 \end{pmatrix}$.

Now **AB** exists because the number of columns in **A** and rows in **B** are the same.

Hence $\mathbf{AB} = \begin{pmatrix} 3 & -1 & 2 \\ 0 & 4 & -1 \end{pmatrix} \begin{pmatrix} 2 & 0 \\ 4 & 3 \\ 1 & -1 \end{pmatrix}$

$$= \begin{pmatrix} (3)(2) + (-1)(4) + (2)(1) & (3)(0) + (-1)(3) + (2)(-1) \\ (0)(2) + (4)(4) + (-1)(1) & (0)(0) + (4)(3) + (-1)(-1) \end{pmatrix}$$

$$= \begin{pmatrix} 4 & -5 \\ 15 & 13 \end{pmatrix}$$

Now **BA** also exists as **B** has two columns and **A** has two rows.

Hence $\quad \mathbf{BA} = \begin{pmatrix} 2 & 0 \\ 4 & 3 \\ 1 & -1 \end{pmatrix}\begin{pmatrix} 3 & -1 & 2 \\ 0 & 4 & -1 \end{pmatrix} = \begin{pmatrix} 6 & -2 & 4 \\ 12 & 8 & 5 \\ 3 & -5 & 3 \end{pmatrix}.$

Note again that $\mathbf{AB} \neq \mathbf{BA}$, in fact they are not even of the same size.

Because the order of the product matters, it is ambiguous to refer to 'the product of **A** and **B**' without specifying the order. For the order **AB** we say that **A** *premultiplies* **B** or that **B** *postmultiplies* **A**.

Note also that the product **AB** results in a matrix with the same number of rows as **A** and the same number of columns as **B**. Hence a 2×4 matrix premultiplying a 4×3 matrix results in a 2×3 matrix.
In general if **A** is of size $m \times n$, and **B** is of size $n \times p$ then **AB** is of size $m \times p$.

To summarize:

AB exists if the number of columns in **A** equals the number of rows in **B**, i.e. if **A** is an $m \times n$ matrix and **B** is an $n \times t$ matrix, **A** and **B** are said to be *compatible* for the product **AB**, and **AB** is an $m \times t$ matrix.
If **AB** exists, **BA** does not necessarily exist.
If **AB** exists and **BA** exists, then in general $\mathbf{AB} \neq \mathbf{BA}$.

If **AB** exists and is equal to **C**, then c_{ij} is the product of the ith row of **A** and jth column of **B**.

In those cases where $\mathbf{AB} = \mathbf{BA}$, **A** and **B** are said to commute.

EXERCISE 1f

Express as a single matrix the product **AB**, and, where it exists, **BA**. State also any pair of matrices that commute.

1) $\mathbf{A} = \begin{pmatrix} 2 & -1 \\ 4 & 2 \end{pmatrix} \qquad \mathbf{B} = \begin{pmatrix} -1 & 0 \\ 2 & -3 \end{pmatrix}$

2) $\mathbf{A} = \begin{pmatrix} 0 & -5 \\ 6 & -1 \end{pmatrix} \qquad \mathbf{B} = \begin{pmatrix} 3 & -1 \\ -4 & 0 \end{pmatrix}$

3) $\mathbf{A} = \begin{pmatrix} -3 & 2 \\ 4 & 1 \end{pmatrix} \qquad \mathbf{B} = \begin{pmatrix} 6 & -2 & 7 \\ 0 & 1 & -2 \end{pmatrix}$

4) $\mathbf{A} = \begin{pmatrix} 3 & -7 & 2 \\ -4 & 0 & -1 \\ 1 & -2 & 4 \end{pmatrix} \qquad \mathbf{B} = \begin{pmatrix} 2 & 1 \\ -1 & 3 \\ 0 & -2 \end{pmatrix}$

5) $\mathbf{A} = \begin{pmatrix} 3 & -4 \\ 1 & 6 \\ 8 & 10 \end{pmatrix}$ $\mathbf{B} = \begin{pmatrix} 2 & -1 \\ 4 & -2 \end{pmatrix}$

6) $\mathbf{A} = \begin{pmatrix} -1 & 0 & 1 \\ -1 & 1 & 2 \\ 2 & 0 & 1 \end{pmatrix}$ $\mathbf{B} = \begin{pmatrix} 3 & -2 & 1 \\ 5 & 0 & 3 \\ 0 & -1 & 1 \end{pmatrix}$

7) $\mathbf{A} = \begin{pmatrix} 1 & 3 \\ 0 & -1 \end{pmatrix}$ $\mathbf{B} = \begin{pmatrix} 1 & 0 & -1 & 2 \\ 3 & -4 & 2 & 0 \end{pmatrix}$

8) $\mathbf{A} = \begin{pmatrix} 3 & 2 \\ 4 & -1 \\ 0 & 2 \end{pmatrix}$ $\mathbf{B} = \begin{pmatrix} 0 & 1 & 3 \\ -1 & 0 & 2 \end{pmatrix}$

9) $\mathbf{A} = \begin{pmatrix} 2 & -1 & 4 \\ 0 & -2 & 5 \\ 2 & 1 & 3 \end{pmatrix}$ $\mathbf{B} = \begin{pmatrix} 2 & 0 & 1 \\ 1 & 3 & -1 \\ 0 & 1 & 0 \end{pmatrix}$

10) $\mathbf{A} = \begin{pmatrix} 1 & 0 & 0 \\ 0 & 1 & 0 \\ 0 & 0 & 1 \end{pmatrix}$ $\mathbf{B} = \begin{pmatrix} 1 & 2 & 3 \\ 4 & 5 & 6 \\ 7 & 8 & 9 \end{pmatrix}$

11) $\mathbf{A} = \begin{pmatrix} 5 & 2 \\ 10 & 4 \end{pmatrix}$ $\mathbf{B} = \begin{pmatrix} 2 & 4 \\ -5 & -10 \end{pmatrix}$

12) $\mathbf{A} = \begin{pmatrix} 2 \\ 7 \end{pmatrix}$ $\mathbf{B} = \begin{pmatrix} 3 & 2 \end{pmatrix}$

COMPOUND TRANSFORMATIONS

Consider the transformation, A, followed by the transformation, B, where A is a stretch by a factor of 2 in the direction of the x axis, and B is a rotation of $\pi/2$ radians about the origin.

A

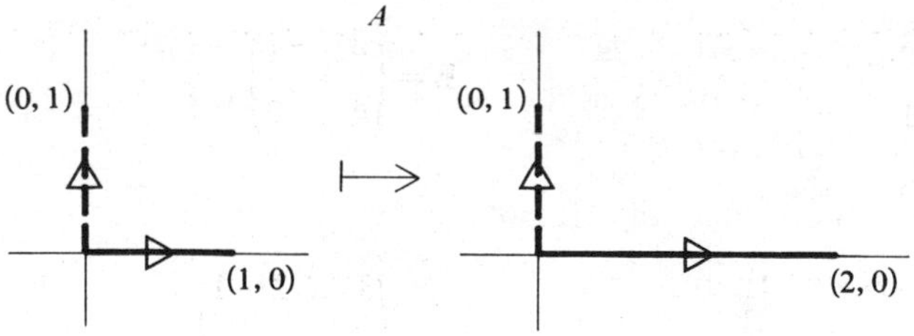

For *A* we see from the diagram that

i maps to $\begin{pmatrix}2\\0\end{pmatrix}$ and **j** maps to $\begin{pmatrix}0\\1\end{pmatrix}$, i.e. the matrix operator **A** is $\begin{pmatrix}2 & 0\\0 & 1\end{pmatrix}$

B

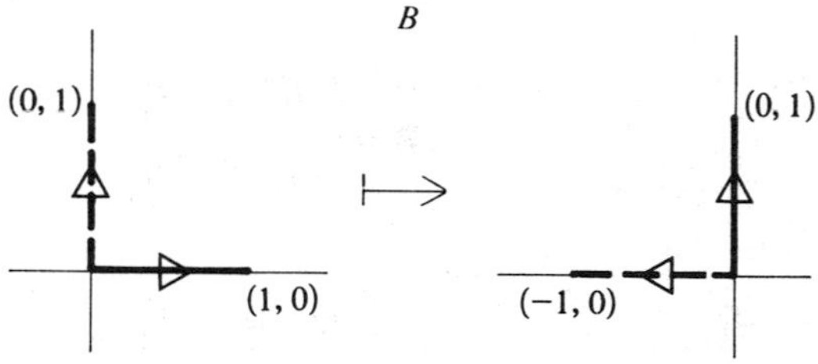

For *B*, the diagram above shows that **i** maps to $\begin{pmatrix}0\\1\end{pmatrix}$ and **j** to $\begin{pmatrix}-1\\0\end{pmatrix}$,

i.e. $\mathbf{B} = \begin{pmatrix}0 & -1\\1 & 0\end{pmatrix}$ is the matrix operator.

Now transformation *A* followed by transformation *B* results in

$$\mathbf{i} \longmapsto \begin{pmatrix}2\\0\end{pmatrix} \longmapsto \begin{pmatrix}0 & -1\\1 & 0\end{pmatrix}\begin{pmatrix}2\\0\end{pmatrix} = \begin{pmatrix}0\\2\end{pmatrix} \qquad [1]$$

and

$$\mathbf{j} \longmapsto \begin{pmatrix}0\\1\end{pmatrix} \longmapsto \begin{pmatrix}0 & -1\\1 & 0\end{pmatrix}\begin{pmatrix}0\\1\end{pmatrix} = \begin{pmatrix}-1\\0\end{pmatrix} \qquad [2]$$

Hence the combined effect of these two transformations is expressed by the operator

$$\mathbf{C} = \begin{pmatrix}0 & -1\\2 & 0\end{pmatrix} \qquad [3]$$

But

$$\mathbf{BA} = \begin{pmatrix}0 & -1\\1 & 0\end{pmatrix}\begin{pmatrix}2 & 0\\0 & 1\end{pmatrix} = \begin{pmatrix}0 & -1\\2 & 0\end{pmatrix} = \mathbf{C}$$

Hence when transformation B is carried out after the transformation A the compound transformation has as its matrix operator the product **BA** where **B** is the matrix operator for B and **A** is the matrix operator for A.

This result is confirmed by the diagram below.

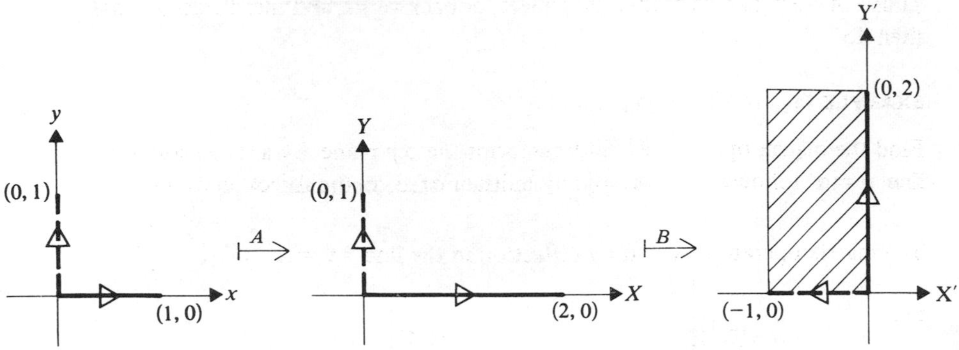

Here we see that, when **B** is carried out after **A**, $\mathbf{i} \longmapsto \begin{pmatrix} 0 \\ 2 \end{pmatrix}$ and $\mathbf{j} \longmapsto \begin{pmatrix} -1 \\ 0 \end{pmatrix}$

$$\Longrightarrow \quad \text{the matrix operator} \quad \mathbf{C} = \begin{pmatrix} 0 & -1 \\ 2 & 0 \end{pmatrix}$$

If we reverse the order of the transformations, i.e. A is carried out after B, to give the transformation D, we get a different result as the following diagrams show.

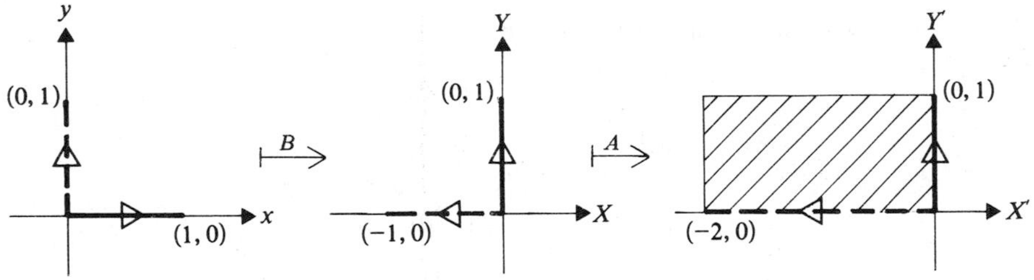

Here we see that the matrix operator is **D** where

$$\mathbf{D} = \begin{pmatrix} 0 & -2 \\ 1 & 0 \end{pmatrix} \quad \text{and that} \quad \mathbf{AB} = \begin{pmatrix} 2 & 0 \\ 0 & 1 \end{pmatrix}\begin{pmatrix} 0 & -1 \\ 1 & 0 \end{pmatrix} = \begin{pmatrix} 0 & -2 \\ 1 & 0 \end{pmatrix} = \mathbf{D}.$$

So the order in which the transformations are performed is important and must be stated.

Note that the operation of combining transformations in this way is called the *composition* of transformations.

If a transformation A follows a transformation B, this composition is written as

$$A \circ B$$

where '$\circ$' means 'following'.

Thus $A \circ B$ is read from right to left, or backwards, and means 'do B first then do A'.

EXAMPLE 1g

Find the matrix operator which transforms the xy plane by a reflection in the line $y = x$ followed by a stretch by a factor of 3 in the direction of Oy.

The matrix operator, **A**, for a reflection in the line $y = x$ is $\begin{pmatrix} 0 & 1 \\ 1 & 0 \end{pmatrix}$

i.e.

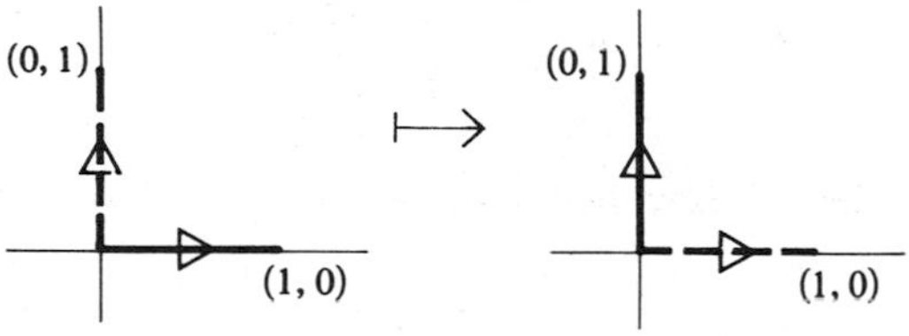

The matrix operator, **B**, for a stretch by a factor of 3 in the direction Oy is $\begin{pmatrix} 1 & 0 \\ 0 & 3 \end{pmatrix}$

i.e.

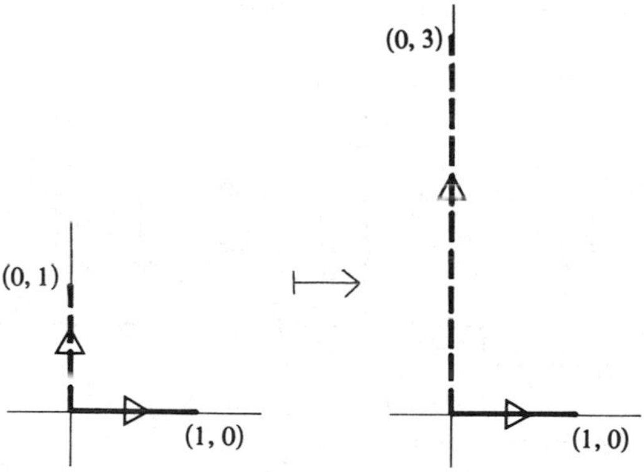

Hence the matrix operator for the given composition of transformations is given by

$$\mathbf{BA} = \begin{pmatrix} 1 & 0 \\ 0 & 3 \end{pmatrix} \begin{pmatrix} 0 & 1 \\ 1 & 0 \end{pmatrix} = \begin{pmatrix} 0 & 1 \\ 3 & 0 \end{pmatrix}$$

Note that the order in which the transformations are performed is given by reading the product **BA** from right to left.

EXERCISE 1g

The following examples refer to the transformations

A, a reflection in Ox,
B, an enlargement by a factor of 2,
C, a rotation of $\pi/2$ about O,
D, a stretch by a factor of 3 in the direction Oy,
E, a reflection in the line $y = x$.

Find the matrix operators for

1) A followed by C.

2) A following C.

3) $B \circ D$.

4) C carried out after E.

5) $A \circ B$.

6) B followed by C.

7) A followed by B followed by C.

The Meaning of $\mathbf{A}^n$ where n is a Positive Integer

For any matrix **A**, **AA** can exist only if the number of columns of **A** and the number of rows of **A** are the same, i.e. **AA** exists only if **A** is a square matrix. Hence if

$$\mathbf{A} = \begin{pmatrix} 1 & 2 \\ 3 & 4 \end{pmatrix}$$

then

$$\mathbf{A}^2 = \mathbf{AA} = \begin{pmatrix} 1 & 2 \\ 3 & 4 \end{pmatrix}\begin{pmatrix} 1 & 2 \\ 3 & 4 \end{pmatrix} = \begin{pmatrix} 7 & 10 \\ 15 & 22 \end{pmatrix}$$

Now

$$\mathbf{AA}^2 = \begin{pmatrix} 1 & 2 \\ 3 & 4 \end{pmatrix}\begin{pmatrix} 7 & 10 \\ 15 & 22 \end{pmatrix} = \begin{pmatrix} 37 & 54 \\ 81 & 118 \end{pmatrix}$$

and

$$\mathbf{A}^2\mathbf{A} = \begin{pmatrix} 7 & 10 \\ 15 & 22 \end{pmatrix}\begin{pmatrix} 1 & 2 \\ 3 & 4 \end{pmatrix} = \begin{pmatrix} 37 & 54 \\ 81 & 118 \end{pmatrix}$$

i.e. $\mathbf{AAA} = \mathbf{AA}^2 = \mathbf{A}^2\mathbf{A}$.

This property can be shown to be true for any square matrix.
Hence we define $\mathbf{A}^3$ as $\mathbf{AAA}$
and $\mathbf{A}^n$ as $\mathbf{AAA}\ldots\mathbf{A}$.

The Associative Law

We showed above that $\mathbf{A} \times (\mathbf{A} \times \mathbf{A}) = (\mathbf{A} \times \mathbf{A}) \times \mathbf{A}$.
In fact for *any* matrices **A**, **B** and **C** that are compatable for multiplication we find that $\mathbf{A} \times (\mathbf{B} \times \mathbf{C}) = (\mathbf{A} \times \mathbf{B}) \times \mathbf{C}$.
This is known as the associative law and it means that for a product of three matrices, *provided that their order is not changed,* we can choose to begin by multiplying either the first pair or the second pair. It also means that we can write the product of three matrices as **ABC** without ambiguity.

For example, if $\mathbf{A} = \begin{pmatrix} 2 & 1 \\ 3 & 2 \end{pmatrix}$, $\mathbf{B} = \begin{pmatrix} 1 & 3 & -1 \\ 0 & 1 & 0 \end{pmatrix}$, $\mathbf{C} = \begin{pmatrix} -1 & 1 & 0 \\ -1 & 2 & 1 \\ 0 & 0 & 1 \end{pmatrix}$

$$\mathbf{A}(\mathbf{BC}) = \begin{pmatrix} 2 & 1 \\ 3 & 2 \end{pmatrix} \left[\begin{pmatrix} 1 & 3 & -1 \\ 0 & 1 & 0 \end{pmatrix} \begin{pmatrix} -1 & 1 & 0 \\ -1 & 2 & 1 \\ 0 & 0 & 1 \end{pmatrix} \right]$$

$$= \begin{pmatrix} 2 & 1 \\ 3 & 2 \end{pmatrix} \begin{pmatrix} -4 & 7 & 2 \\ -1 & 2 & 1 \end{pmatrix} = \begin{pmatrix} -9 & 16 & 5 \\ -14 & 25 & 8 \end{pmatrix}$$

$$(\mathbf{AB})\mathbf{C} = \left[\begin{pmatrix} 2 & 1 \\ 3 & 2 \end{pmatrix} \begin{pmatrix} 1 & 3 & -1 \\ 0 & 1 & 0 \end{pmatrix} \right] \begin{pmatrix} -1 & 1 & 0 \\ -1 & 2 & 1 \\ 0 & 0 & 1 \end{pmatrix}$$

$$= \begin{pmatrix} 2 & 7 & -2 \\ 3 & 11 & -3 \end{pmatrix} \begin{pmatrix} -1 & 1 & 0 \\ -1 & 2 & 1 \\ 0 & 0 & 1 \end{pmatrix} = \begin{pmatrix} -9 & 16 & 5 \\ -14 & 25 & 8 \end{pmatrix}$$

i.e. $$\mathbf{A}(\mathbf{BC}) = (\mathbf{AB})\mathbf{C}$$

THE UNIT MATRIX

Consider the matrix $\begin{pmatrix} 1 & 0 \\ 0 & 1 \end{pmatrix}$.

Taking the product of this matrix and a general 2×2 matrix we find that

$$\begin{pmatrix} 1 & 0 \\ 0 & 1 \end{pmatrix} \begin{pmatrix} a & b \\ c & d \end{pmatrix} = \begin{pmatrix} a & b \\ c & d \end{pmatrix}$$

and $$\begin{pmatrix} a & b \\ c & d \end{pmatrix}\begin{pmatrix} 1 & 0 \\ 0 & 1 \end{pmatrix} = \begin{pmatrix} a & b \\ c & d \end{pmatrix}$$

i.e. both premultiplication and postmultiplication by

$\begin{pmatrix} 1 & 0 \\ 0 & 1 \end{pmatrix}$ leaves $\begin{pmatrix} a & b \\ c & d \end{pmatrix}$ unchanged.

Hence $\begin{pmatrix} 1 & 0 \\ 0 & 1 \end{pmatrix}$ has the same effect in matrix multiplication that unity has in the multiplication of real numbers and, for this reason, $\begin{pmatrix} 1 & 0 \\ 0 & 1 \end{pmatrix}$ is called the *unit matrix* of size 2×2 and is denoted by **I**.

Similarly it can be shown that any 3×3 matrix is unchanged by either premultiplication or postmultiplication by $\begin{pmatrix} 1 & 0 & 0 \\ 0 & 1 & 0 \\ 0 & 0 & 1 \end{pmatrix}$, and this matrix is therefore called the unit 3×3 matrix and it, also, is denoted by **I**.

In fact, a unit matrix of any size is denoted by **I**, the size either being obvious from the context, or the size being stated, i.e. $\mathbf{I}_n$ is the $n \times n$ unit matrix.

Now consider **I** as a transformation matrix.

Under $\mathbf{I} = \begin{pmatrix} 1 & 0 \\ 0 & 1 \end{pmatrix}$, $\mathbf{i} \longmapsto \mathbf{i}$ and $\mathbf{j} \longmapsto \mathbf{j}$,

i.e. the xy plane and its image under **I** are identical, i.e. **I** represents the identity transformation, so **I** is also called an *identity matrix.*

It is interesting to note that if $\mathbf{C} = \begin{pmatrix} 1 & 2 & 3 \\ 4 & 5 & 6 \end{pmatrix}$

then $$\mathbf{IC} = \begin{pmatrix} 1 & 0 \\ 0 & 1 \end{pmatrix}\begin{pmatrix} 1 & 2 & 3 \\ 4 & 5 & 6 \end{pmatrix} = \begin{pmatrix} 1 & 2 & 3 \\ 4 & 5 & 6 \end{pmatrix}$$

That is, premultiplying any 2 row matrix by **I** leaves that matrix unchanged. Also, postmultiplying any 2 column matrix by **I** leaves that matrix unchanged.

Similarly for an $n \times p$ matrix, $\mathbf{I}_n$ is a premultiplying identity and $\mathbf{I}_p$ is a postmultiplying identity.

The Null Matrix

Any matrix, all of whose entries are zero, is called a *null* or *zero matrix*, and is denoted by $\mathbf{0}$.
$\mathbf{0}$ may be any size, i.e. $\mathbf{0}$ is not unique.

If $\mathbf{0} = \begin{pmatrix} 0 & 0 \\ 0 & 0 \end{pmatrix}$ then $\begin{pmatrix} 0 & 0 \\ 0 & 0 \end{pmatrix}\begin{pmatrix} 1 & 2 \\ 3 & 4 \end{pmatrix} = \begin{pmatrix} 0 & 0 \\ 0 & 0 \end{pmatrix} = \begin{pmatrix} 1 & 2 \\ 3 & 4 \end{pmatrix}\begin{pmatrix} 0 & 0 \\ 0 & 0 \end{pmatrix}$.

If $\mathbf{0} = \begin{pmatrix} 0 \\ 0 \end{pmatrix}$ then $\begin{pmatrix} 1 & 2 \\ 3 & 4 \end{pmatrix}\begin{pmatrix} 0 \\ 0 \end{pmatrix} = \begin{pmatrix} 0 \\ 0 \end{pmatrix}$.

In fact, if $\mathbf{0}$ is conformable for premultiplying $\mathbf{A}$, $\mathbf{0A} = \mathbf{0}$,
and if $\mathbf{0}$ is conformable for postmultiplying $\mathbf{A}$, $\mathbf{A0} = \mathbf{0}$.

For real numbers a and b, $ab = 0 \Rightarrow a = 0$ or $b = 0$.
However, from Question 11 of Exercise 1f, it is seen that the product of two non-zero matrices can be a zero matrix

i.e. $\mathbf{AB} = \mathbf{0} \nRightarrow \mathbf{A} = \mathbf{0}$ or $\mathbf{B} = \mathbf{0}$

Further for real numbers, a, b and c, $ab = ac \Rightarrow$ either $a = 0$ or $b = c$ but this is not true for matrix products as we see from the example below.

For the matrices $\mathbf{A}$, $\mathbf{B}$ and $\mathbf{C}$ as shown,

$$\mathbf{AB} = \begin{pmatrix} 1 & 1 \\ 1 & 1 \end{pmatrix}\begin{pmatrix} 1 & 0 \\ 0 & 1 \end{pmatrix} = \begin{pmatrix} 1 & 1 \\ 1 & 1 \end{pmatrix}$$

$$\mathbf{AC} = \begin{pmatrix} 1 & 1 \\ 1 & 1 \end{pmatrix}\begin{pmatrix} 0 & 1 \\ 1 & 0 \end{pmatrix} = \begin{pmatrix} 1 & 1 \\ 1 & 1 \end{pmatrix}$$

i.e. $\mathbf{AB} = \mathbf{AC}$ but $\mathbf{A} \neq \mathbf{0}$ and $\mathbf{B} \neq \mathbf{C}$,

i.e. $\mathbf{AB} = \mathbf{AC} \nRightarrow \mathbf{A} = \mathbf{0}$ or $\mathbf{B} = \mathbf{C}$

EXAMPLES 1h

1) If $\mathbf{M} = \frac{1}{2}\begin{pmatrix} -1 & \sqrt{3} \\ -\sqrt{3} & -1 \end{pmatrix}$ find $\mathbf{M}^3$ and describe the transformation $\mathbf{M}\begin{pmatrix} x \\ y \end{pmatrix} = \begin{pmatrix} X \\ Y \end{pmatrix}$.

$$\mathbf{M}^2 = \tfrac{1}{4}\begin{pmatrix} -1 & \sqrt{3} \\ -\sqrt{3} & -1 \end{pmatrix}\begin{pmatrix} -1 & \sqrt{3} \\ -\sqrt{3} & -1 \end{pmatrix} = \tfrac{1}{4}\begin{pmatrix} -2 & -2\sqrt{3} \\ 2\sqrt{3} & -2 \end{pmatrix}$$

$$= \tfrac{1}{2}\begin{pmatrix} -1 & -\sqrt{3} \\ \sqrt{3} & -1 \end{pmatrix}$$

$$\mathbf{M}^3 = \mathbf{M}\mathbf{M}^2 = \frac{1}{4}\begin{pmatrix} -1 & \sqrt{3} \\ -\sqrt{3} & -1 \end{pmatrix}\begin{pmatrix} -1 & -\sqrt{3} \\ \sqrt{3} & -1 \end{pmatrix} = \frac{1}{4}\begin{pmatrix} 4 & 0 \\ 0 & 4 \end{pmatrix} = \begin{pmatrix} 1 & 0 \\ 0 & 1 \end{pmatrix}$$

Hence $\mathbf{M}^3 = \mathbf{I}$ i.e. $\mathbf{M}^3$ is the identity transformation.
Hence the transformation represented by $\mathbf{M}$, when repeated three times, maps the xy plane on to itself.
One transformation that would give such a result is a rotation through one third of a revolution about O i.e. through $\pm 2\pi/3$ radians, as shown below.

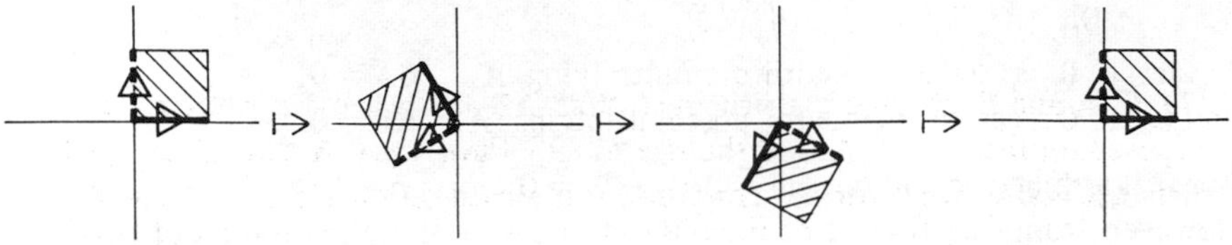

The transformation matrix for an anticlockwise rotation through an angle θ about O is

$$\begin{pmatrix} \cos\theta & -\sin\theta \\ \sin\theta & \cos\theta \end{pmatrix}$$

Now $$\mathbf{M} = \begin{pmatrix} -\frac{1}{2} & \frac{\sqrt{3}}{2} \\ -\frac{\sqrt{3}}{2} & -\frac{1}{2} \end{pmatrix} = \begin{pmatrix} \cos(-2\pi/3) & -\sin(-2\pi/3) \\ \sin(-2\pi/3) & \cos(-2\pi/3) \end{pmatrix}$$

So we see that $\mathbf{M}$ produces a clockwise rotation of $2\pi/3$ radians about O.

EXERCISE 1h

1) If $\mathbf{A} = \begin{pmatrix} 1 & 0 \\ 2 & -1 \end{pmatrix}$ find $\mathbf{A}^2$ and $\mathbf{A}^3$.

2) If $\mathbf{B} = \begin{pmatrix} 1 & 0 & -1 \\ 0 & 1 & 1 \\ -1 & 0 & 0 \end{pmatrix}$ find $\mathbf{B}^2$ and $\mathbf{B}^3$.

3) Find a 2×2 matrix, $\mathbf{B}$, such that $\mathbf{B}^2 = \mathbf{I}$.
(*Hint*. Think in terms of a double transformation that maps **i** to **i** and **j** to **j**.)

4) Find a 2×2 matrix, $\mathbf{C}$, such that $\mathbf{C}^4=\mathbf{I}$ and $\mathbf{C}^2\neq\mathbf{I}$.

5) If $\mathbf{A}=\begin{pmatrix} i & 0\\ 0 & i\end{pmatrix}$ where $i=\sqrt{(-1)}$, find $\mathbf{A}^2$ and $\mathbf{A}^4$.

6) If $\mathbf{D}=\begin{pmatrix}\cos\theta & 0 & 0\\ 0 & \sin\theta & 0\\ 0 & 0 & -\cos\theta\end{pmatrix}$ find $\mathbf{D}^2$.

7) If $\mathbf{A}$ and $\mathbf{B}$ are two matrices such that $\mathbf{AB}$ and $\mathbf{BA}$ exist, what condition is imposed on the size of $\mathbf{A}$ and the size of $\mathbf{B}$? Further, if $\mathbf{A}$ and $\mathbf{B}$ are such that $\mathbf{AB}=\mathbf{BA}$, what further condition is imposed on the sizes of $\mathbf{A}$ and $\mathbf{B}$?

8) If $\mathbf{M}=\begin{pmatrix}1 & 0\\ 1 & 0\end{pmatrix}$, describe the transformation $\mathbf{M}\begin{pmatrix}x\\ y\end{pmatrix}=\begin{pmatrix}X\\ Y\end{pmatrix}$ and the transformation $\mathbf{M}^2\begin{pmatrix}x\\ y\end{pmatrix}=\begin{pmatrix}X'\\ Y'\end{pmatrix}$.

9) If $\mathbf{A}=\begin{pmatrix}1 & 1\\ 2 & 2\end{pmatrix}$ and $\mathbf{B}=\begin{pmatrix}1 & -3\\ -1 & 3\end{pmatrix}$, describe, with the help of a diagram, the transformations

(a) $\mathbf{AB}\begin{pmatrix}x\\ y\end{pmatrix}=\begin{pmatrix}X\\ Y\end{pmatrix}$ (b) $\mathbf{BA}\begin{pmatrix}x\\ y\end{pmatrix}=\begin{pmatrix}X\\ Y\end{pmatrix}$.

Discuss the statement $\mathbf{AB}=\mathbf{0}\Rightarrow\mathbf{BA}=\mathbf{0}$.

10) If $\mathbf{A}=\begin{pmatrix}1 & 0\\ 2 & -1\end{pmatrix}$ and $\mathbf{B}=\begin{pmatrix}-2 & 1\\ 3 & 0\end{pmatrix}$ find

(a) $(\mathbf{A}+\mathbf{B})^2$ (b) $\mathbf{A}^2+\mathbf{AB}+\mathbf{BA}+\mathbf{B}^2$.

Hence show that $(\mathbf{A}+\mathbf{B})^2=\mathbf{A}^2+\mathbf{AB}+\mathbf{BA}+\mathbf{B}^2$.

11) The matrix $\mathbf{B}$ is said to be a real square root of the matrix $\mathbf{A}$ if $\mathbf{A}=\mathbf{B}^2$ and all the entries of $\mathbf{B}$ are real. Find two real and different square roots of $\mathbf{A}$ where $\mathbf{A}=\begin{pmatrix}9 & 0\\ 0 & 9\end{pmatrix}$.

DETERMINANTS

Consider the effect upon the area of a figure produced by the transformation matrix $\mathbf{M} = \begin{pmatrix} a & b \\ c & d \end{pmatrix}$.

$$\mathbf{M} \text{ maps } \mathbf{i} \text{ to } \begin{pmatrix} a \\ c \end{pmatrix} \text{ and } \mathbf{j} \text{ to } \begin{pmatrix} b \\ d \end{pmatrix}.$$

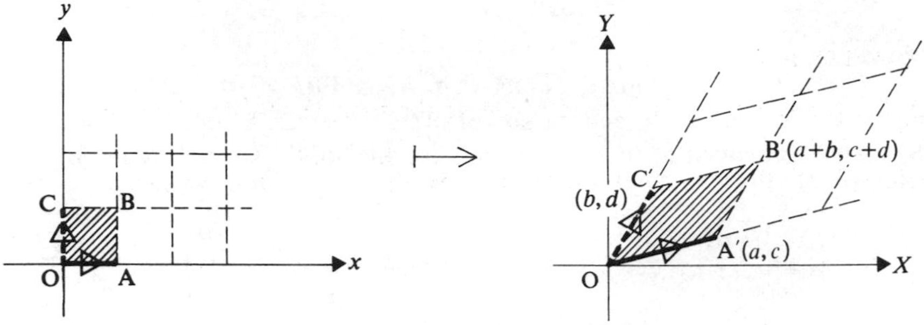

From the diagram, the area of OABC is 1 square unit
and the area of OA′B′C′ is $(ad-bc)$ square units.

Hence $\mathbf{M}$ maps an area of 1 square unit in the xy plane to an area of $(ad-bc)$ square units in the transformed plane,

i.e. $\begin{pmatrix} a & b \\ c & d \end{pmatrix}$ changes area by a factor $(ad-bc)$

This factor, $(ad-bc)$, is a number obviously related to the matrix $\begin{pmatrix} a & b \\ c & d \end{pmatrix}$ and it is called the *determinant* of the matrix $\mathbf{M}$.

The determinant of the matrix $\mathbf{M}$ is denoted by $|\mathbf{M}|$, or by $\det \mathbf{M}$, and we write

$$|\mathbf{M}| = \begin{vmatrix} a & b \\ c & d \end{vmatrix} = ad - bc$$

In general, if $\mathbf{A} = \begin{pmatrix} a_{11} & a_{12} \\ a_{21} & a_{22} \end{pmatrix}$

then $$|\mathbf{A}| = \begin{vmatrix} a_{11} & a_{12} \\ a_{21} & a_{22} \end{vmatrix} = a_{11}a_{22} - a_{12}a_{21}$$

e.g.
$$\begin{vmatrix} 4 & 2 \\ -1 & 3 \end{vmatrix} = (4)(3)-(2)(-1) = 14$$

and
$$\begin{vmatrix} -7 & -4 \\ 3 & 6 \end{vmatrix} = -42-(-12) = -30$$

Note that $|\mathbf{A}|$ is defined for a 2×2 matrix *only*, at this stage, and that the value of $|\mathbf{A}|$ is the product of the entries of $\mathbf{A}$ on the leading diagonal (top left to lower right) *minus* the product of the entries on the other diagonal.

EXAMPLES 1i

1) The triangle OAB is mapped by $\mathbf{M}$ to the triangle OA′B′ where A is the point $(0, 2)$, B is the point $(3, 0)$ and $\mathbf{M} = \begin{pmatrix} 3 & 2 \\ 5 & 8 \end{pmatrix}$. Find the area of triangle OA′B′.

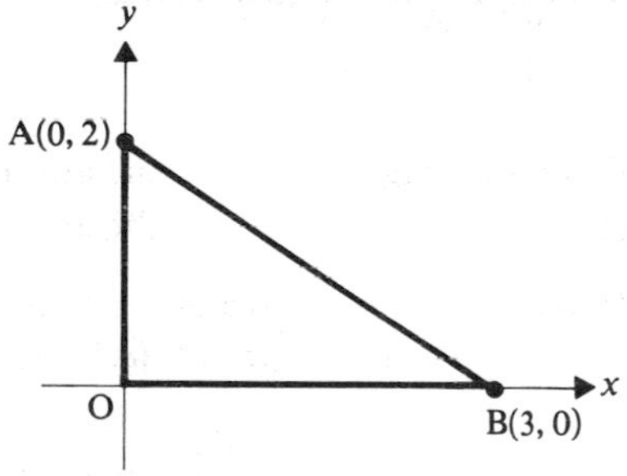

From the diagram, we see that the area of triangle OAB is 3 square units.

Also, $|\mathbf{M}| = \begin{vmatrix} 3 & 2 \\ 5 & 8 \end{vmatrix} = 14$

So $\mathbf{M}$ changes area by a factor of 14.

Therefore the area of triangle OA′B′ is $14\times3 = 42$ square units.

2) The square whose vertices are the points A(0, 3), B(1, 1), C(3, 2), D(2, 4) is mapped to the plane figure A′B′C′D′ by $\mathbf{M} = \begin{pmatrix} 1 & 2 \\ 1 & -2 \end{pmatrix}$.

Find the area of A′B′C′D′.

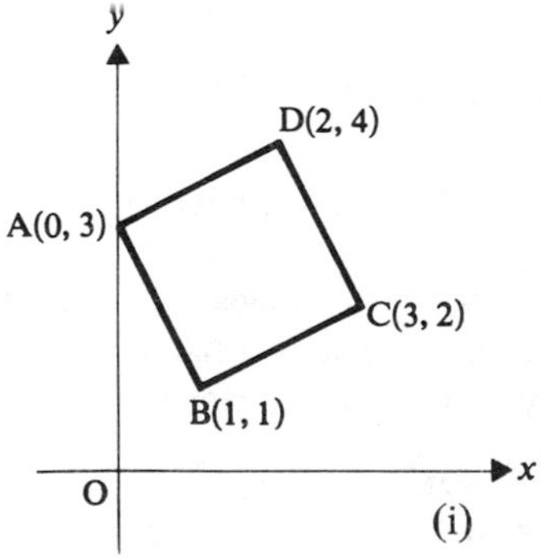

(i)

From the diagram, the area of ABCD is seen to be 5 square units.

Now $\mathbf{M}$ changes area by a factor $|\mathbf{M}|$,

and $|\mathbf{M}| = \begin{vmatrix} 1 & 2 \\ 1 & -2 \end{vmatrix} = -4$

To interpret the meaning of the negative sign we look at the image figure A′B′C′D′. Under **M**, the images of A, B, C and D are respectively

$$(6,-6), (3,-1), (7,-1) \text{ and } (10,-6).$$

Comparing ABCD in diagram (i), with its image A′B′C′D′ in diagram (ii), we see that not only is the shape of ABCD distorted and rotated under **M**, it is also *turned over*; i.e. in (ii) we are looking at an area which is the image of the *reverse side* of the area in (i).
So the negative value of $|\mathbf{M}|$ indicates that a reflection is involved.
So **M** changes area by a factor 4.
Hence the area of A′B′C′D′ is $5 \times 4 = 20$ square units.

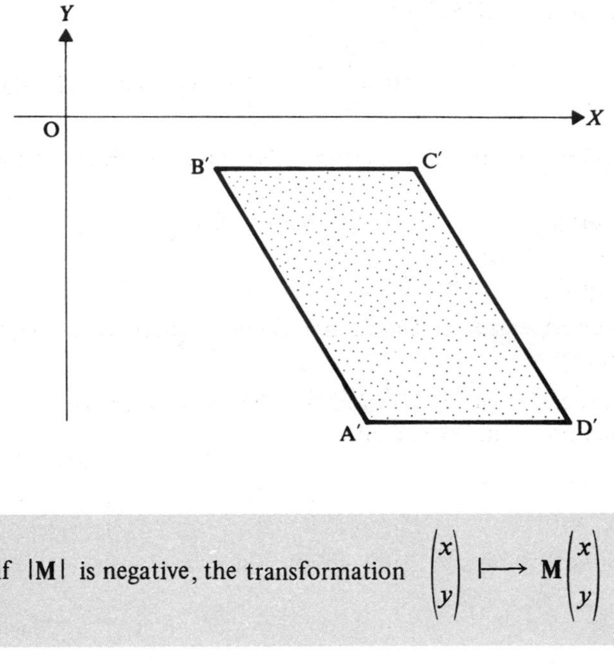

(ii)

In general, if $|\mathbf{M}|$ is negative, the transformation $\begin{pmatrix} x \\ y \end{pmatrix} \longmapsto \mathbf{M}\begin{pmatrix} x \\ y \end{pmatrix}$ turns an area over.

We saw earlier on page 13 that a matrix, **M**, is called singular if the transformation $\mathbf{M}\begin{pmatrix} x \\ y \end{pmatrix} = \begin{pmatrix} X \\ Y \end{pmatrix}$ maps the xy plane on to a line or on to the origin. Under such a transformation, area is destroyed, i.e. any area in the xy plane maps to zero area in the transformed plane. So a singular matrix **M** changes area by a factor of zero. Conversely if $|\mathbf{M}| = 0$ then area is destroyed so **M** must map the basis vectors **i** and **j** to collinear vectors or to the origin.

So $$\mathbf{M} \text{ is singular} \iff |\mathbf{M}| = 0$$

EXERCISE 1i

Evaluate

1) $\begin{vmatrix} 3 & 2 \\ 1 & 5 \end{vmatrix}$. 2) $\begin{vmatrix} -1 & 3 \\ 2 & -1 \end{vmatrix}$. 3) $\begin{vmatrix} 2 & -1 \\ 4 & -2 \end{vmatrix}$. 4) $\begin{vmatrix} 2 & -1 \\ -4 & -1 \end{vmatrix}$.

5) If each of the determinants in Questions 1 to 4 is the determinant of a transformation matrix, describe the effect that the transformation has on the square OABC where $\overrightarrow{OA} = \mathbf{i}$ and $\overrightarrow{OC} = \mathbf{j}$. Illustrate your answers with a diagram.

6) If $\mathbf{A} = \begin{pmatrix} 1 & 3 \\ -2 & 1 \end{pmatrix}$ and $\mathbf{B} = \begin{pmatrix} 2 & 4 \\ -1 & 5 \end{pmatrix}$ find

(a) $\mathbf{AB}$ (b) $|\mathbf{A}|$ (c) $|\mathbf{B}|$ (d) $|\mathbf{AB}|$.
Is $|\mathbf{AB}|$ equal to $|\mathbf{A}||\mathbf{B}|$?

7) If $\mathbf{A} = \begin{pmatrix} 3 & 0 \\ -1 & 4 \end{pmatrix}$ and $\mathbf{B} = \begin{pmatrix} -2 & 1 \\ 0 & -3 \end{pmatrix}$ find $|\mathbf{AB}|$ and $|\mathbf{BA}|$.

8) If **A** and **B** are any 2×2 matrices, determine which of the following is/are true.
(a) $|\mathbf{AB}| = |\mathbf{A}|\,|\mathbf{B}|$ (b) $|\mathbf{AB}| = |\mathbf{BA}|$
(c) $|\mathbf{AB}| = 0 \Rightarrow |\mathbf{A}| = 0$ or $|\mathbf{B}| = 0$
(d) $|\mathbf{AB}| = |\mathbf{BA}| \Rightarrow \mathbf{AB} = \mathbf{BA}$.

9) For those statements in Question 8 which you decide are true, write down a general proof.

10) For those statements in Question 8 which you decide are false, find examples that prove them to be false.

11) Expand and simplify

(a) $\begin{vmatrix} \cos\theta & \sin\theta \\ -\sin\theta & \cos\theta \end{vmatrix}$ (b) $\begin{vmatrix} x & y \\ x^2 & y^2 \end{vmatrix}$ (c) $\begin{vmatrix} a & b \\ b & a \end{vmatrix}$

(d) $\begin{vmatrix} a & a+b \\ b & a \end{vmatrix}$ (e) $\begin{vmatrix} \ln 2 & \ln 4 \\ \ln 5 & \ln 6 \end{vmatrix}$ (f) $\begin{vmatrix} \cos 2\theta & \sin\theta \\ -\sin 2\theta & \cos\theta \end{vmatrix}$.

12) Solve for x and y the equations
$$a_1x + b_1y = c_1$$
$$a_2x + b_2y = c_2.$$

Hence show that $x = \dfrac{\begin{vmatrix} c_1 & b_1 \\ c_2 & b_2 \end{vmatrix}}{\begin{vmatrix} a_1 & b_1 \\ a_2 & b_2 \end{vmatrix}}$ and $y = \dfrac{\begin{vmatrix} a_1 & c_1 \\ a_2 & c_2 \end{vmatrix}}{\begin{vmatrix} a_1 & b_1 \\ a_2 & b_2 \end{vmatrix}}$.

13) Use the general result obtained in Question 12 to solve the following equations simultaneously for x and y

(a) $3x-2y=1$, $x+3y=2$ (b) $3x+4y=3$, $2x-7y=-1$ (c) $7x-3y=-2$, $4x+2y=3$.

THE DETERMINANT OF A 3 x 3 MATRIX

Any square matrix has an associated value which is called its determinant and is denoted by $|\mathbf{A}|$, e.g. if $\mathbf{A}$ is a 3×3 matrix where

$$\mathbf{A} = \begin{pmatrix} 1 & 2 & 3 \\ 4 & 5 & 6 \\ 7 & 8 & 9 \end{pmatrix}, \quad \text{then} \quad |\mathbf{A}| = \begin{vmatrix} 1 & 2 & 3 \\ 4 & 5 & 6 \\ 7 & 8 & 9 \end{vmatrix}$$

The value of $|\mathbf{A}|$ is found, basically, by extracting 2×2 determinants from the 3×3 determinant.

If the row and column through a particular entry are crossed out, four entries are left which form a 2×2 determinant,

e.g. if, in $|\mathbf{A}|$, we cross out the row and column through the entry 4,

i.e. $\begin{vmatrix} 1 & 2 & 3 \\ \textcircled{4} & 5 & 6 \\ 7 & 8 & 9 \end{vmatrix}$, we are left with the determinant $\begin{vmatrix} 2 & 3 \\ 8 & 9 \end{vmatrix}$.

This determinant is known as the *minor* of the entry 4.

In general, for a 3×3 determinant, Δ, where

$$\Delta = \begin{vmatrix} a_{11} & a_{12} & a_{13} \\ a_{21} & a_{22} & \textcircled{a_{23}} \\ a_{31} & a_{32} & a_{33} \end{vmatrix}, \quad \text{the minor of } a_{23} \text{ is } \begin{vmatrix} a_{11} & a_{12} \\ a_{31} & a_{32} \end{vmatrix}.$$

These minors have associated signs, + or −, depending on the position in the determinant of the entry of which they are a minor.

These associated signs are shown in this diagram.

$$\begin{vmatrix} + & - & + \\ - & + & - \\ + & - & + \end{vmatrix}$$

So the associated sign for the minor of a_{23} is $-$, that for the minor of a_{31} is $+$, and so on.

The minor of a particular entry, together with its associated sign, is called the cofactor of that entry.

e.g. if $\Delta = \begin{vmatrix} 1 & 2 & 3 \\ 4 & 5 & 6 \\ 7 & 8 & 9 \end{vmatrix}$, the cofactor of 4 is $-\begin{vmatrix} 2 & 3 \\ 8 & 9 \end{vmatrix}$

and the cofactor of 3 is $+\begin{vmatrix} 4 & 5 \\ 7 & 8 \end{vmatrix}$.

In general, if $\Delta = \begin{vmatrix} a_{11} & a_{12} & a_{13} \\ a_{21} & a_{22} & a_{23} \\ a_{31} & a_{32} & a_{33} \end{vmatrix}$

the cofactor of a_{21}, say, is denoted by A_{21} where $A_{21} = -\begin{vmatrix} a_{12} & a_{13} \\ a_{32} & a_{33} \end{vmatrix}$

We can now define the value of Δ as the sum of the products of the entries of the first row with their respective cofactors.

e.g., if $\Delta = \begin{vmatrix} 1 & 2 & 3 \\ 4 & 5 & 6 \\ 7 & 8 & 9 \end{vmatrix}$

the cofactor of 1 is $+\begin{vmatrix} 5 & 6 \\ 8 & 9 \end{vmatrix}$, the cofactor of 2 is $-\begin{vmatrix} 4 & 6 \\ 7 & 9 \end{vmatrix}$

and the cofactor of 3 is $+\begin{vmatrix} 4 & 5 \\ 7 & 8 \end{vmatrix}$.

Hence
$$\Delta = (1)\begin{vmatrix} 5 & 6 \\ 8 & 9 \end{vmatrix} - (2)\begin{vmatrix} 4 & 6 \\ 7 & 9 \end{vmatrix} + (3)\begin{vmatrix} 4 & 5 \\ 7 & 8 \end{vmatrix}$$
$$= (1)(-3) - (2)(-6) + (3)(-3)$$
$$= -3 + 12 - 9 = 0$$

In general,

$$\Delta = \begin{vmatrix} a_1 & b_1 & c_1 \\ a_2 & b_2 & c_2 \\ a_3 & b_3 & c_3 \end{vmatrix}$$

$$= a_1 \begin{vmatrix} b_2 & c_2 \\ b_3 & c_3 \end{vmatrix} - b_1 \begin{vmatrix} a_2 & c_2 \\ a_3 & c_3 \end{vmatrix} + c_1 \begin{vmatrix} a_2 & b_2 \\ a_3 & b_3 \end{vmatrix}$$

$$= a_1b_2c_3 - a_1c_2b_3 - b_1a_2c_3 + b_1c_2a_3 + c_1a_2b_3 - c_1b_2a_3 \qquad [1]$$

Clearly the determinant notation $\begin{vmatrix} a_1 & b_1 & c_1 \\ a_2 & b_2 & c_2 \\ a_3 & b_3 & c_3 \end{vmatrix}$ expresses the expansion [1]

much more neatly. So a 3×3 determinant may be considered as a shorthand form for expressions like [1]. If the elements of a determinant are real, the determinant itself has a real value and so is a member of the set of real numbers.

At the beginning of this section, a 3×3 determinant was introduced as the associated value of a 3×3 matrix and an apparantly arbitrary definition was given for its value.

However expansions of this form occur frequently in many branches of mathematics. An example arises in coordinate geometry when finding the area of the triangle whose vertices are the points $(x_1, y_1), (x_2, y_2), (x_3, y_3)$.

Using the trapeziums SACT, TCBU and SABU, we find the area of triangle ABC to be

$$\tfrac{1}{2}(x_2y_3 - y_2x_3 - x_1y_3 + x_3y_1 + x_1y_2 - x_2y_1).$$

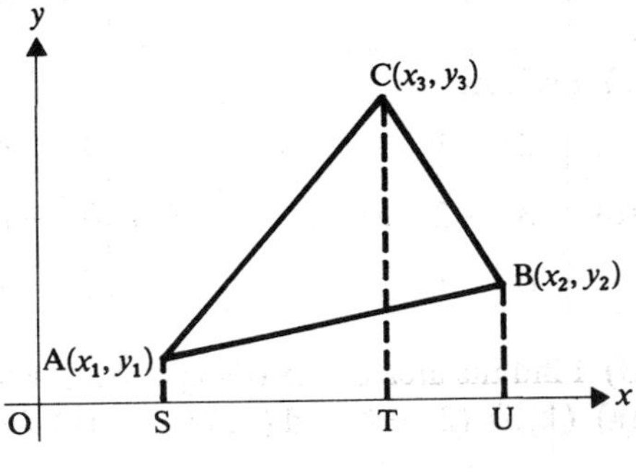

Inserting a factor of one into each term we have

$$\text{Area of } \triangle\text{ABC} = \tfrac{1}{2}(1x_2y_3 - 1y_2x_3 - 1x_1y_3 + 1x_3y_1 + 1x_1y_2 - 1x_2y_1).$$

Comparing with the expansion of the general determinant we see that the expression obtained for the area of ΔABC can be written

$$\frac{1}{2}\begin{vmatrix} 1 & 1 & 1 \\ x_1 & x_2 & x_3 \\ y_1 & y_2 & y_3 \end{vmatrix}$$

i.e. the area of the triangle whose vertices are the points $(x_1, y_1), (x_2, y_2), (x_3, y_3)$ is

$$\frac{1}{2}\begin{vmatrix} 1 & 1 & 1 \\ x_1 & x_2 & x_3 \\ y_1 & y_2 & y_3 \end{vmatrix}$$

If the area of the triangle is zero, then the three points are collinear.

Hence the condition for three points, $(x_1, y_1), (x_2, y_2), (x_3, y_3)$ to be collinear is

$$\begin{vmatrix} 1 & 1 & 1 \\ x_1 & x_2 & x_3 \\ y_1 & y_2 & y_3 \end{vmatrix} = 0$$

EXERCISE 1j

1) Find the cofactors of each term of the determinants

(a) $\begin{vmatrix} 3 & 2 & -1 \\ 0 & 4 & -6 \\ 2 & -1 & 3 \end{vmatrix}$ (b) $\begin{vmatrix} -1 & 0 & 2 \\ -4 & 1 & -3 \\ 7 & 0 & -2 \end{vmatrix}$.

2) Evaluate

(a) $\begin{vmatrix} -2 & 1 & 4 \\ 3 & -2 & 5 \\ 0 & 1 & 3 \end{vmatrix}$ (b) $\begin{vmatrix} 0 & 1 & 3 \\ 0 & -1 & 4 \\ 2 & 6 & -2 \end{vmatrix}$ (c) $\begin{vmatrix} -2 & 0 & 1 \\ 3 & -4 & 5 \\ -7 & -3 & 2 \end{vmatrix}$.

3) Find the area of the triangle ABC where A, B and C are the points
(a) $(1, 3), (2, -4)$ and $(5, 7)$ (b) $(-6, 2), (-1, -1)$ and $(-3, 5)$.

4) Determine which of the following sets of points are collinear.

(a) $(0, \frac{1}{3}), (-1, 0), (5, 2)$ (b) $(0, 1), (1, 0), (1, -1)$
(c) $(0, -6), (1, -3), (3, 3)$.

5) Expand and simplify the determinants

(a) $\begin{vmatrix} 1 & 1 & 1 \\ \cos\theta & \cos^2\theta & 1 \\ \sin\theta & \sin^2\theta & 1 \end{vmatrix}$ (b) $\begin{vmatrix} a & b & c \\ a^2 & b^2 & c^2 \\ a^3 & b^3 & c^3 \end{vmatrix}$

(c) $\begin{vmatrix} \cos\theta & 1 & 1 \\ \sin\theta & 1 & 1 \\ 1 & \sin\theta & \cos\theta \end{vmatrix}$ (d) $\begin{vmatrix} 1 & n & n^2 \\ n & n^2 & n^3 \\ 1/n & 1/n^2 & 1/n^3 \end{vmatrix}$.

6) Verify, by expansion, that $\begin{vmatrix} a_1 & a_2 & a_3 \\ b_1 & b_2 & b_3 \\ c_1 & c_2 & c_3 \end{vmatrix}$

$$= a_1A_1 + a_2A_2 + a_3A_3 = b_1B_1 + b_2B_2 + b_3B_3 = c_1C_1 + c_2C_2 + c_3C_3.$$

SIMPLIFICATION OF DETERMINANTS

If the entries of a determinant are large numbers or complicated algebraic expressions, the evaluation of the determinant can involve some tedious work. There are, however, various properties of determinants which can be used to reduce the complexity of the entries without altering the value of the determinant. These properties are set out below and, in all cases, Δ refers to the general 3×3 determinant $\begin{vmatrix} a_1 & a_2 & a_3 \\ b_1 & b_2 & b_3 \\ c_1 & c_2 & c_3 \end{vmatrix}$.

(1) The value of a determinant is unaltered when the rows and columns are completely interchanged.

This property can be proved as follows.
Interchanging the rows and columns of Δ results in the determinant

$$\begin{vmatrix} a_1 & b_1 & c_1 \\ a_2 & b_2 & c_2 \\ a_3 & b_3 & c_3 \end{vmatrix} = \Delta'$$

(i.e. Δ' is the determinant whose rows are the columns of Δ.)

$$\begin{aligned}\Delta' &= a_1(b_2c_3 - b_3c_2) - b_1(a_2c_3 - a_3c_2) + c_1(a_2b_3 - a_3b_2)\\ &= a_1(b_2c_3 - b_3c_2) - a_2(b_1c_3 - b_3c_1) + a_3(b_2c_1 - b_1c_2)\\ &= \Delta\end{aligned}$$

This property does not help to reduce the complexity of the entries but it does mean that

any property proved for rows is also valid for columns.

Question 6 in Exercise 1h showed that

a determinant can be expanded by using any row and its cofactors.
As $\Delta' = \Delta$ it therefore follows that a determinant can also be expanded using any column and its cofactors,

e.g. if $\Delta = \begin{vmatrix} 2 & -1 & 7\\ 0 & 8 & -2\\ 0 & -4 & 2\end{vmatrix}$, we see that column 1 contains two zeros, so it is sensible to use this column and its cofactors to evaluate Δ,

i.e.

$$\begin{aligned}\Delta &= 2\begin{vmatrix} 8 & -2\\ -4 & 2\end{vmatrix} + 0 + 0\\ &= 16\end{aligned}$$

The next property, together with the remaining properties, is stated without proof and any of them can be demonstrated to be true by adapting the approach used in (1).

(2) If any row (or column) is added to or subtracted from any other row (or column), the value of Δ is not changed.

e.g. $$\begin{vmatrix} a_1 & a_2 & a_3\\ b_1 & b_2 & b_3\\ c_1 & c_2 & c_3\end{vmatrix} = \begin{vmatrix} a_1 & a_2 & a_3\\ b_1+a_1 & b_2+a_2 & b_3+a_3\\ c_1 & c_2 & c_3\end{vmatrix}$$

(3) The value of Δ is unaltered if a multiple of any row (or column) is added to any other row (or column).

e.g. $$\begin{vmatrix} a_1 & a_2 & a_3\\ b_1 & b_2 & b_3\\ c_1 & c_2 & c_3\end{vmatrix} = \begin{vmatrix} a_1 & a_2 & a_3\\ \lambda a_1+b_1 & \lambda a_2+b_2 & \lambda a_3+b_3\\ c_1 & c_2 & c_3\end{vmatrix}$$

This is a most useful property for simplifying determinants before expansion, as the following examples illustrate.

(a)

$$\Delta = \begin{vmatrix} 4 & 6 & 2 \\ 3 & 7 & 1 \\ 3 & 5 & 2 \end{vmatrix}$$

Subtracting row 3 from row 1 gives

$$\Delta = \begin{vmatrix} 1 & 1 & 0 \\ 3 & 7 & 1 \\ 3 & 5 & 2 \end{vmatrix}$$

Subtracting column 1 from column 2 gives

$$\Delta = \begin{vmatrix} 1 & 0 & 0 \\ 3 & 4 & 1 \\ 3 & 2 & 2 \end{vmatrix}$$

which may now be evaluated easily as follows:

$$\Delta = 1\begin{vmatrix} 4 & 1 \\ 2 & 2 \end{vmatrix} + 0 + 0 = 6$$

(b)

$$\Delta = \begin{vmatrix} 10 & 42 & -7 \\ 2 & 10 & 1 \\ -3 & -9 & 4 \end{vmatrix}$$

Subtracting 3 times column 1 from column 2 gives

$$\Delta = \begin{vmatrix} 10 & 12 & -7 \\ 2 & 4 & 1 \\ -3 & 0 & 4 \end{vmatrix}$$

Subtracting 3 times row 2 from row 1 gives

$$\Delta = \begin{vmatrix} 4 & 0 & -10 \\ 2 & 4 & 1 \\ -3 & 0 & 4 \end{vmatrix} = 4(16-30) = -56$$

(4) If two rows (or columns) are interchanged the determinant changes sign.

e.g.
$$\begin{vmatrix} a_1 & a_2 & a_3 \\ b_1 & b_2 & b_3 \\ c_1 & c_2 & c_3 \end{vmatrix} = -\begin{vmatrix} b_1 & b_2 & b_3 \\ a_1 & a_2 & a_3 \\ c_1 & c_2 & c_3 \end{vmatrix}$$

(5) A determinant, Δ, may be expressed as the sum or difference of two determinants Δ_1 and Δ_2 where two columns (or rows) of Δ, Δ_1 and Δ_2 are identical and the entries in the remaining column of Δ are the sum (or difference) of the corresponding entries in Δ_1 and Δ_2.

e.g.
$$\begin{vmatrix} a_1 & b_1 & c_1+d_1 \\ a_2 & b_2 & c_2+d_2 \\ a_3 & b_3 & c_3+d_3 \end{vmatrix} = \begin{vmatrix} a_1 & b_1 & c_1 \\ a_2 & b_2 & c_2 \\ a_3 & b_3 & c_3 \end{vmatrix} + \begin{vmatrix} a_1 & b_1 & d_1 \\ a_2 & b_2 & d_2 \\ a_3 & b_3 & d_3 \end{vmatrix}$$

Note that the aim in simplifying a determinant is to obtain as many zeros as possible in one row or column. However, it is easy to get carried away; the evaluation of a determinant is usually a small part of any problem and simplification is meant to minimize the risk of arithmetic mistakes. Mistakes are just as likely to occur when combining multiples of rows or columns as they are when expanding the determinant. So in general only straightforward combinations are worth using.
Note also that when solving problems, the simplification of a determinant can be written down directly without explaining the combination of rows or columns used.

FACTORIZATION OF DETERMINANTS

There are some further properties of determinants that are useful for factorization, particularly when the elements are algebraic expressions.

(6) If one row (or column) of Δ is multiplied by λ, the resulting determinant is equal to $\lambda\Delta$.

e.g.
$$\begin{vmatrix} a_1 & a_2 & a_3 \\ \lambda b_1 & \lambda b_2 & \lambda b_3 \\ c_1 & c_2 & c_3 \end{vmatrix} = \lambda\begin{vmatrix} a_1 & a_2 & a_3 \\ b_1 & b_2 & b_3 \\ c_1 & c_2 & c_3 \end{vmatrix}$$

It follows that if all three rows are multiplied by λ, the determinant is multiplied by a factor λ^3,

i.e.
$$\begin{vmatrix} \lambda a_1 & \lambda a_2 & \lambda a_3 \\ \lambda b_1 & \lambda b_2 & \lambda b_3 \\ \lambda c_1 & \lambda c_2 & \lambda c_3 \end{vmatrix} = \lambda^3 \begin{vmatrix} a_1 & a_2 & a_3 \\ b_1 & b_2 & b_3 \\ c_1 & c_2 & c_3 \end{vmatrix}$$

It also follows that a common factor of the entries of one row (or column) is a factor of the determinant.

For example,
$$\Delta = \begin{vmatrix} x^3 & x^2y & xy^2 \\ x^2 & xy^2 & y^3 \\ x & y^3 & 1 \end{vmatrix}$$

has common factors x in the first column and y in the second column.

Therefore
$$\Delta = xy \begin{vmatrix} x^2 & x^2 & xy^2 \\ x & xy & y^3 \\ 1 & y^2 & 1 \end{vmatrix}$$

There is also a common factor x in the top row

so
$$\Delta = x^2y \begin{vmatrix} x & x & y^2 \\ x & xy & y^3 \\ 1 & y^2 & 1 \end{vmatrix}$$

(7) If all the entries in one row (or column) of a determinant are zero, the determinant is zero.

Also, if two rows (or columns) of a determinant are identical, it follows from property (2) that a row (or column) of zeros can be obtained. Hence

if two rows (or columns) of a determinant are identical then $\Delta = 0$.

This property, in conjunction with the factor theorem, is very useful for factorizing determinants.

e.g. if
$$f(x) \equiv \begin{vmatrix} x & a & b \\ x^2 & a^2 & b^2 \\ x^3 & a^3 & b^3 \end{vmatrix},$$

then
$$f(a) = \begin{vmatrix} a & a & b \\ a^2 & a^2 & b^2 \\ a^3 & a^3 & b^3 \end{vmatrix} = 0$$
as two columns are identical.

So $x-a$ is a factor of $f(x)$.
Similarly $f(b) = 0 \Rightarrow (x-b)$ is a factor of $f(x)$.

The following examples illustrate how these properties can be used to evaluate or factorize a determinant. It should be noted that the first step in the simplification of a determinant should be the removal of any common factors.

EXAMPLES 1k

1) Evaluate $\begin{vmatrix} -7 & 14 & 7 \\ 2 & -8 & 6 \\ 9 & -3 & 12 \end{vmatrix}$.

Removing common factors from the three rows gives

$$\begin{vmatrix} -7 & 14 & 7 \\ 2 & -8 & 6 \\ 9 & -3 & 12 \end{vmatrix} = (7)(2)(3)\begin{vmatrix} -1 & 2 & 1 \\ 1 & -4 & 3 \\ 3 & -1 & 4 \end{vmatrix}$$

Adding row 2 to row 1 gives

$$42\begin{vmatrix} 0 & -2 & 4 \\ 1 & -4 & 3 \\ 3 & -1 & 4 \end{vmatrix} = 42\left\{2\begin{vmatrix} 1 & 3 \\ 3 & 4 \end{vmatrix} + 4\begin{vmatrix} 1 & -4 \\ 3 & -1 \end{vmatrix}\right\}$$

$$= 1428$$

2) Factorize $\begin{vmatrix} x & 1 & 2 \\ x^2 & 1 & 4 \\ x^3 & 1 & 8 \end{vmatrix}$.

Removing common factors from the first and last columns gives

$$\begin{vmatrix} x & 1 & 2 \\ x^2 & 1 & 4 \\ x^3 & 1 & 8 \end{vmatrix} = 2x\begin{vmatrix} 1 & 1 & 1 \\ x & 1 & 2 \\ x^2 & 1 & 4 \end{vmatrix}$$

If $f(x) \equiv \begin{vmatrix} 1 & 1 & 1 \\ x & 1 & 2 \\ x^2 & 1 & 4 \end{vmatrix}$ then

$$f(1) = \begin{vmatrix} 1 & 1 & 1 \\ 1 & 1 & 2 \\ 1 & 1 & 4 \end{vmatrix} = 0 \quad \text{(two columns identical).}$$

So $(x - 1)$ is a factor.

$$\text{Also} \quad f(2) = \begin{vmatrix} 1 & 1 & 1 \\ 2 & 1 & 2 \\ 4 & 1 & 4 \end{vmatrix} = 0. \quad \text{So } (x-2) \text{ is a factor.}$$

So far we have found three linear factors, viz., $2x, (x-1), (x-2)$.
By inspection (i.e. without expanding) it is clear that the given determinant is a polynomial of degree 3. So the only other possible factor is a constant, K say.

$$\text{Then} \qquad \begin{vmatrix} x & 1 & 2 \\ x^2 & 1 & 4 \\ x^3 & 1 & 8 \end{vmatrix} \equiv 2Kx(x-1)(x-2)$$

K can be evaluated by comparing the coefficients of a particular power of x. For instance, in the expansion of the determinant the coefficient of x^3 is 2 while the corresponding coefficient in the factorized form is $2K$,

$$\text{i.e.} \quad 2 = 2K \quad \Rightarrow \quad K = 1 \quad \text{and} \quad \begin{vmatrix} x & 1 & 2 \\ x^2 & 1 & 4 \\ x^3 & 1 & 8 \end{vmatrix} \equiv 2x(x-1)(x-2).$$

Alternatively we may proceed as follows:

$$f(x) \equiv \begin{vmatrix} x & 1 & 2 \\ x^2 & 1 & 4 \\ x^3 & 1 & 8 \end{vmatrix} \equiv 2x \begin{vmatrix} 1 & 1 & 1 \\ x & 1 & 2 \\ x^2 & 1 & 4 \end{vmatrix}$$

$$f(1) = 2\begin{vmatrix} 1 & 1 & 1 \\ 1 & 1 & 2 \\ 1 & 1 & 4 \end{vmatrix} \Rightarrow (x-1) \quad \text{is a factor.}$$

Subtracting column 2 from column 1 gives

$$f(x) \equiv 2x \begin{vmatrix} 0 & 1 & 1 \\ x-1 & 1 & 2 \\ x^2-1 & 1 & 4 \end{vmatrix}$$

(As $x-1$ is a known factor we have looked for a combination of rows or columns to give $x-1$ as a common factor of a row or column.)

Removing $(x-1)$ from column 1 gives

$$f(x) \equiv 2x(x-1)\begin{vmatrix} 0 & 1 & 1 \\ 1 & 1 & 2 \\ x+1 & 1 & 4 \end{vmatrix} \equiv 2x(x-1)\begin{vmatrix} 0 & 1 & 0 \\ 1 & 1 & 1 \\ x+1 & 1 & 3 \end{vmatrix}$$

The remaining determinant is now easily expanded to give

$$f(x) \equiv 2x(x-1)\left\{(-1)\begin{vmatrix} 1 & 1 \\ x+1 & 3 \end{vmatrix}\right\} = 2x(x-1)(x-2)$$

This method has the advantage that all the factors (including K) are found directly. It has the disadvantage that it is not always easy to see the combination or rows (or columns) that will produce a common factor.

Note that it is also possible to expand Δ without any preliminary work and to factorize the result by using the factor theorem.

EXERCISE 1k

Evaluate the following determinants.

1) $\begin{vmatrix} 2 & -7 & 12 \\ 9 & -3 & 21 \\ 2 & 4 & 6 \end{vmatrix}$

2) $\begin{vmatrix} 1 & 8 & -10 \\ 2 & 4 & 15 \\ 1 & 12 & 5 \end{vmatrix}$

3) $\begin{vmatrix} 150 & 200 & -100 \\ 80 & -90 & 50 \\ 70 & 10 & -20 \end{vmatrix}$

4) $\begin{vmatrix} -5 & 15 & 7 \\ 6 & 9 & 2 \\ -3 & 8 & -5 \end{vmatrix}$

Factorize the following determinants.

5) $\begin{vmatrix} x & x^2 & 1 \\ x^2 & x & 1 \\ x^3 & x^3 & 1 \end{vmatrix}$

6) $\begin{vmatrix} x-1 & 1 & x+1 \\ -1 & 1 & 1 \\ x+1 & 1 & x-1 \end{vmatrix}$

7) $\begin{vmatrix} \sin\theta & \cos\theta & 1 \\ \sin^2\theta & \cos^2\theta & 1 \\ \sin^3\theta & \cos^3\theta & 1 \end{vmatrix}$

8) $\begin{vmatrix} 1 & a & a+1 \\ a+1 & 1 & a \\ a & a+1 & 1 \end{vmatrix}$

9) $\begin{vmatrix} 1 & 1 & 1 \\ x^2+4 & x^2+9 & x^2+16 \\ 2 & 3 & 4 \end{vmatrix}$

10) Solve the equation $\begin{vmatrix} 1 & 1 & 1 \\ x & x+1 & x-1 \\ x-1 & 2x & x+1 \end{vmatrix} = 0.$

SUMMARY

If $\mathbf{M} = \begin{pmatrix} a & b \\ c & d \end{pmatrix}$ and $\mathbf{M}$ maps (x, y) to (X, Y)

then $\mathbf{M}$ maps $\mathbf{i}$ to $\begin{pmatrix} a \\ c \end{pmatrix}$ and $\mathbf{j}$ to $\begin{pmatrix} b \\ d \end{pmatrix}$.

$\mathbf{M}$ alters area by a factor $|\mathbf{M}| = ad - bc$.

$\mathbf{M} = \begin{pmatrix} \cos\theta & -\sin\theta \\ \sin\theta & \cos\theta \end{pmatrix}$ represents a rotation about the origin through an angle θ.

$\mathbf{M} = \begin{pmatrix} \cos 2\theta & \sin 2\theta \\ \sin 2\theta & -\cos 2\theta \end{pmatrix}$ represents a reflection in the line $y = x \tan\theta$.

If $|\mathbf{M}| = 0$, $\mathbf{M}$ is singular and maps two dimensional space to a line or a point.

$\mathbf{A} = \mathbf{B} \iff a_{ij} = b_{ij}$ for all i and j.

$\mathbf{A} = \mathbf{0} \iff a_{ij} = 0$ for all i and j.

$\mathbf{A}$ and $\mathbf{B}$ are the same size if both $\mathbf{A}$ and $\mathbf{B}$ are $n \times m$ matrices.

$\mathbf{A} + \mathbf{B}$ exists if $\mathbf{A}$ and $\mathbf{B}$ are the same size, when $\mathbf{A} + \mathbf{B}$ is given by adding the corresponding entries of $\mathbf{A}$ and $\mathbf{B}$.

$\mathbf{A} + \mathbf{B} = \mathbf{B} + \mathbf{A}$.

$$\lambda\mathbf{A} = \begin{pmatrix} \lambda a_{11} & \cdots & \\ & \cdots & \lambda a_{ij} \end{pmatrix}$$

$\mathbf{AB}$ is defined if $\mathbf{A}$ is $m \times n$ and $\mathbf{B}$ is $n \times p$ and then $\mathbf{AB}$ is $m \times p$.

If $\mathbf{AB} = \mathbf{C}$, $c_{ij} = \big(i\text{th row of } \mathbf{A}\big)\begin{pmatrix} j\text{th} \\ \text{column} \\ \text{of } \mathbf{B} \end{pmatrix}$.

In general $\mathbf{AB} \neq \mathbf{BA}$

$$\mathbf{A} \times (\mathbf{B} \times \mathbf{C}) = (\mathbf{A} \times \mathbf{B}) \times \mathbf{C}$$

$\mathbf{A}^n$ exists if $\mathbf{A}$ is square, when $\mathbf{A}^n = \mathbf{A} \times \mathbf{A} \times \ldots \mathbf{A}$.

The unit, or identity, matrix $\mathbf{I}$ is square with unit entries in the leading diagonal and zeros elsewhere.

$$\Delta = \begin{vmatrix} a_1 & a_2 \\ b_1 & b_2 \end{vmatrix} = a_1 b_2 - a_2 b_1$$

$$\Delta = \begin{vmatrix} a_1 & a_2 & a_3 \\ b_1 & b_2 & b_3 \\ c_1 & c_2 & c_3 \end{vmatrix} = a_1 A_1 + a_2 A_2 + a_3 A_3$$

where A_1, A_2, A_3 are the cofactors of a_1, a_2, a_3

and $A_1 = +\begin{vmatrix} b_2 & b_3 \\ c_2 & c_3 \end{vmatrix}$, $A_2 = -\begin{vmatrix} b_1 & b_3 \\ c_1 & c_3 \end{vmatrix}$, $A_3 = +\begin{vmatrix} b_1 & b_2 \\ c_1 & c_2 \end{vmatrix}$

$$\Delta' = \begin{vmatrix} a_1 & b_1 & c_1 \\ a_2 & b_2 & c_2 \\ a_3 & b_3 & c_3 \end{vmatrix}$$

If a multiple of any row (column) is added to any other row (column), the value of Δ is unaltered.

A common factor of any row (column) is a factor of the determinant.

MULTIPLE CHOICE EXERCISE 1

(Instructions for answering these questions are given on page xii.)

TYPE 1

1) $\begin{pmatrix} 2 & 4 & 6 \end{pmatrix} \begin{pmatrix} 3 \\ 2 \\ -1 \end{pmatrix} =$

(a) 20 (b) -8 (c) 8 (d) $\begin{pmatrix} 6 \\ 8 \\ -6 \end{pmatrix}$ (e) none of these.

2) Under the transformation $\begin{pmatrix} 1 & 2 \\ 1 & 2 \end{pmatrix}\begin{pmatrix} x \\ y \end{pmatrix} = \begin{pmatrix} X \\ Y \end{pmatrix}$, the area of a unit square is mapped to an area

(a) twice the size,
(b) the same size but which has been reflected in a line through O,
(c) which is destroyed, (d) none of these.

3) $\begin{pmatrix} 3 & 7 \\ -1 & 4 \end{pmatrix}\begin{pmatrix} 2 & 7 & -3 \\ 1 & 0 & 1 \end{pmatrix} =$

(a) $\begin{pmatrix} 13 & 21 & -2 \\ 2 & -7 & 7 \end{pmatrix}$ (b) $\begin{pmatrix} 15 & 2 \\ 21 & -7 \\ -2 & 7 \end{pmatrix}$

(c) $\begin{pmatrix} 1 & 0 \\ 0 & 1 \end{pmatrix}$ (d) $\begin{pmatrix} -1 & 21 & -16 \\ -6 & -7 & -1 \end{pmatrix}$.

4) $\begin{pmatrix} 3 & 1 & 2 \end{pmatrix} + \begin{pmatrix} 2 & 1 & 3 \end{pmatrix} =$

(a) $\begin{pmatrix} 3 & 1 & 2 \\ 2 & 1 & 3 \end{pmatrix}$ (b) $2\begin{pmatrix} 3 & 1 & 2 \end{pmatrix}$ (c) 12 (d) $\begin{pmatrix} 5 & 2 & 5 \end{pmatrix}$

(e) has no meaning.

5) $\begin{vmatrix} 1 & 0 & 1 \\ 0 & 1 & 0 \\ 1 & 0 & 1 \end{vmatrix} =$

(a) 0 (b) 1 (c) 2 (d) -1 (e) 5.

6) If $\mathbf{A} = \begin{pmatrix} 1 & 0 \\ -1 & 1 \end{pmatrix}$, $\mathbf{b} = \begin{pmatrix} 1 \\ 2 \end{pmatrix}$, then $\mathbf{bA} =$

(a) $\begin{pmatrix} 1 \\ -1 \end{pmatrix}$ (b) $\begin{pmatrix} 0 \\ 0 \end{pmatrix}$ (c) 2 (d) $\begin{pmatrix} 1 & 1 \end{pmatrix}$ (e) has no meaning.

7) The transformation represented by $\mathbf{M} = \begin{pmatrix} 1 & -1 \\ 1 & 1 \end{pmatrix}$ is:

(a) a rotation of $\pi/4$ about O, (b) a reflection in the y axis,
(c) a rotation of $\pi/4$ together with an enlargement by a factor $\sqrt{2}$,
(d) a shear parallel to Ox, (e) none of these.

8) If $\mathbf{A} = \begin{pmatrix} 1 & 2 \\ 3 & 4 \end{pmatrix}$, then $\mathbf{A}^2 =$

(a) $\begin{pmatrix} 1 & 4 \\ 9 & 16 \end{pmatrix}$ (b) $\begin{pmatrix} 5 & 5 \\ 25 & 25 \end{pmatrix}$ (c) $\begin{pmatrix} 7 & 10 \\ 15 & 22 \end{pmatrix}$ (d) $\begin{pmatrix} 7 & 10 \\ 3 & 5 \end{pmatrix}$

(e) $\mathbf{A}^2$ has no meaning.

9) The cofactor of the element 6 in $\begin{vmatrix} 1 & 2 & 3 \\ 4 & 5 & 6 \\ 7 & 8 & 9 \end{vmatrix}$ is:

(a) $\begin{vmatrix} 1 & 2 \\ 7 & 8 \end{vmatrix}$ (b) $\begin{vmatrix} 7 & 8 \\ -1 & -2 \end{vmatrix}$ (c) $\begin{vmatrix} -1 & -2 \\ 7 & 8 \end{vmatrix}$ (d) $\begin{vmatrix} -1 & -2 \\ -7 & -8 \end{vmatrix}$.

10) $\begin{vmatrix} 2 & -1 & 4 \\ -2 & 1 & 4 \\ 2 & 1 & 0 \end{vmatrix} =$

(a) $\begin{vmatrix} -3 & -1 & 4 \\ -3 & 1 & 3 \\ 3 & 1 & 0 \end{vmatrix}$ (b) $\begin{vmatrix} 0 & -1 & 4 \\ 0 & 1 & 4 \\ 4 & 1 & 0 \end{vmatrix}$ (c) $\begin{vmatrix} 1 & -1 & 2 \\ -1 & 1 & 2 \\ 1 & 1 & 0 \end{vmatrix}$

(d) $\begin{vmatrix} 0 & 0 & 0 \\ -2 & 1 & 4 \\ 2 & 1 & 0 \end{vmatrix}$.

TYPE II

11) The area of the triangle ABC, where A, B, C are the points (1,2), (3,1), (− 2, 1) is given by:

(a) $\frac{1}{2}\begin{vmatrix} 1 & 1 & 1 \\ 1 & 3 & -2 \\ 2 & 1 & 1 \end{vmatrix}$ (b) $\frac{1}{2}\begin{vmatrix} 1 & 1 & 2 \\ 1 & 3 & 1 \\ 1 & -2 & 1 \end{vmatrix}$ (c) $\frac{1}{2}\begin{vmatrix} 1 & 1 & 1 \\ 2 & 1 & 1 \\ 1 & 3 & -2 \end{vmatrix}$.

12) $\mathbf{M} = \begin{pmatrix} \cos\theta & -\sin\theta \\ \sin\theta & \cos\theta \end{pmatrix}$.

(a) $\mathbf{M}$ is singular.
(b) $\mathbf{M}$ represents a reflection in $y = mx$, where $m = \tan\theta$.
(c) $\mathbf{M}$ represents a rotation of θ about O.

13) $\mathbf{I} = \begin{pmatrix} 1 & 0 \\ 0 & 1 \end{pmatrix}$ and $\mathbf{A} = \begin{pmatrix} 1 & 1 & 1 \\ 2 & 2 & 1 \end{pmatrix}$.

(a) $\mathbf{IA} = \mathbf{A}$.
(b) $\mathbf{AI} = \mathbf{0}$.
(c) $\mathbf{A}^2 = \begin{pmatrix} 1 & 1 & 1 \\ 4 & 4 & 1 \end{pmatrix}$.

14) $\mathbf{A} = \begin{pmatrix} a_1 & a_2 & a_3 \\ b_1 & b_2 & b_3 \\ c_1 & c_2 & c_3 \end{pmatrix}$.

(a) $|\mathbf{A}| = \begin{vmatrix} a_1 & b_1 & c_1 \\ a_2 & b_2 & c_2 \\ a_3 & b_3 & c_3 \end{vmatrix}$. (b) $\mathbf{A}^2$ exists. (c) $|\mathbf{A}| = \begin{vmatrix} c_1 & c_2 & c_3 \\ a_1 & a_2 & a_3 \\ b_1 & b_2 & b_3 \end{vmatrix}$.

15) $\mathbf{A} = \begin{pmatrix} a_{11} & a_{12} \\ a_{21} & a_{22} \end{pmatrix}$, $\mathbf{B} = \begin{pmatrix} b_{11} & b_{12} \\ b_{21} & b_{22} \end{pmatrix}$.

(a) $\mathbf{AB} = \mathbf{BA}$. (b) $|\mathbf{AB}| = |\mathbf{BA}|$. (c) $\mathbf{A}^2\mathbf{B}^2 = (\mathbf{AB})^2$.

TYPE III

16) (a) $\mathbf{AB} = \mathbf{BA}$.
(b) $\mathbf{A} = \mathbf{I}$ or $\mathbf{B} = \mathbf{I}$.

17) (a) $\mathbf{A}$ and $\mathbf{B}$ are the same size.
(b) $\mathbf{A} + \mathbf{B} = \mathbf{B} + \mathbf{A}$.

18) (a) $\mathbf{A} = \mathbf{0}$ or $\mathbf{B} = \mathbf{0}$.
(b) $\mathbf{AB} = \mathbf{0}$.

19) (a) $\mathbf{A}$ is an $n \times p$ matrix, $\mathbf{B}$ is an $m \times n$ matrix.
(b) $\mathbf{AB}$ exists.

20) (a) $\Delta = \begin{vmatrix} 1 & 2 \\ 3 & 4 \end{vmatrix}$.

(b) $\frac{1}{\Delta} = \begin{vmatrix} 1 & \frac{1}{2} \\ \frac{1}{3} & \frac{1}{4} \end{vmatrix}$.

21) (a) $\mathbf{IA} = \mathbf{A}$.
(b) $\mathbf{A}$ is square.

22) (a) The matrix $\mathbf{M}$ maps $\mathbf{i}$ to $\begin{pmatrix} 1 \\ 2 \end{pmatrix}$ and $\mathbf{j}$ to $\begin{pmatrix} 2 \\ 1 \end{pmatrix}$.

(b) The matrix $\mathbf{M}$ maps $\begin{pmatrix} 2 \\ 2 \end{pmatrix}$ to $\begin{pmatrix} 6 \\ 6 \end{pmatrix}$.

23) (a) $\mathbf{A} = \begin{pmatrix} \cos\theta & \cos\theta \\ \sin\theta & -\sin\theta \end{pmatrix}$.

(b) $|\mathbf{A}| = 0$.

TYPE IV

24) Find the equations of the lines which map to themselves under the transformation $\mathbf{M}\begin{pmatrix} x \\ y \end{pmatrix} = \begin{pmatrix} X \\ Y \end{pmatrix}$.

(a) $\mathbf{M}\begin{pmatrix} 0 \\ 0 \end{pmatrix} = \begin{pmatrix} 0 \\ 0 \end{pmatrix}$. (b) $\mathbf{M}$ maps $\mathbf{i}$ to $\begin{pmatrix} 1 \\ 1 \end{pmatrix}$.

(c) $\mathbf{M}$ maps $\mathbf{j}$ to $\begin{pmatrix} 0 \\ -1 \end{pmatrix}$.

25) Find the factor by which the transformation $\mathbf{M}\begin{pmatrix} x \\ y \end{pmatrix} = \begin{pmatrix} X \\ Y \end{pmatrix}$ changes area.

(a) $\mathbf{M}$ represents a rotation of $\pi/3$ about O.

(b) $\mathbf{M}$ maps $\mathbf{i}$ to $\frac{1}{2}\begin{pmatrix} 1 \\ \sqrt{3} \end{pmatrix}$.

(c) The origin is invariant under $\mathbf{M}$.

26) Find a square root of the matrix **A**.
(a) **A** is a 2×2 matrix. (b) $\begin{pmatrix}a_{11} & a_{12}\end{pmatrix} = \begin{pmatrix}1 & 2\end{pmatrix}$.

(c) $\begin{pmatrix}a_{11}\\ a_{21}\end{pmatrix} = \begin{pmatrix}1\\ 3\end{pmatrix}$.

TYPE V

27) $\lambda|\mathbf{A}| = |\lambda\mathbf{A}|$.

28) $\begin{vmatrix}a_1 & a_2\\ b_1 & b_2\end{vmatrix} = \begin{vmatrix}b_1 & b_2\\ a_1 & a_2\end{vmatrix}$.

29) $\begin{vmatrix}a_1 & a_2\\ b_1 & b_2\end{vmatrix} = \begin{vmatrix}a_1 & b_1\\ a_2 & b_2\end{vmatrix}$.

30) If $\mathbf{AB} = \mathbf{BA}$ then **A** and **B** must both be square.

31) If **AB** and **BA** both exist then **A** and **B** must both be square.

32) If the plane defined by the base vectors **i** and **j** is transformed to the plane defined by the base vectors **p** and **q** then the vector $\lambda\mathbf{i}+\mu\mathbf{j}$ is transformed to the vector $\lambda\mathbf{p}+\mu\mathbf{q}$.

33) If the xy plane is rotated about the point $(1,1)$, the image of any point (x, y) may be obtained from an equation of the form $\begin{pmatrix}a & b\\ c & d\end{pmatrix}\begin{pmatrix}x\\ y\end{pmatrix} = \begin{pmatrix}X\\ Y\end{pmatrix}$.

MISCELLANEOUS EXERCISE 1

1) Evaluate the determinant

$$\begin{vmatrix}1 & 2 & 5\\ 2 & -5 & 3\\ 4 & -1 & 7\end{vmatrix}.$$

(U of L)

2) Given the matrices

$$\mathbf{A} = \begin{pmatrix}1 & 2\\ 0 & 1\end{pmatrix}, \quad \mathbf{I} = \begin{pmatrix}1 & 0\\ 0 & 1\end{pmatrix}.$$

obtain constants p and q such that

$$\mathbf{A}^2 = p\mathbf{A} + q\mathbf{I}.$$

(U of L)

3) Show that the determinant

$$\begin{vmatrix} 1 & 1 & 1 \\ x & y & z \\ yz & zx & xy \end{vmatrix} = (x-y)(y-z)(z-x).$$ (U of L)p

4) If two dimensional space is transformed by

$$\begin{pmatrix} x \\ y \end{pmatrix} \longmapsto \begin{pmatrix} 4 & -1 \\ 6 & -3 \end{pmatrix} \begin{pmatrix} x \\ y \end{pmatrix}$$

find the equations of the straight lines which are mapped on to themselves.
(U of L)p

5) Show that the transformation

$$\begin{pmatrix} x_2 \\ y_2 \end{pmatrix} = \begin{pmatrix} \cos\theta & -\sin\theta \\ \sin\theta & \cos\theta \end{pmatrix} \begin{pmatrix} x_1 \\ y_1 \end{pmatrix}$$

represents a rotation about the origin.
Show also that the transformation

$$\begin{pmatrix} x_2 \\ y_2 \end{pmatrix} = \begin{pmatrix} -\frac{3}{5} & \frac{4}{5} \\ \frac{4}{5} & \frac{3}{5} \end{pmatrix} \begin{pmatrix} x_1 \\ y_1 \end{pmatrix}$$

represents a reflection in a fixed line through the origin. (U of L)p

6) Show, with the help of a diagram, that the matrix **P** of the linear transformation which rotates the plane in the counter-clockwise sense through an angle θ about the origin is

$$\begin{pmatrix} \cos\theta & -\sin\theta \\ \sin\theta & \cos\theta \end{pmatrix}$$

Find the matrix **Q** of the linear transformation which reflects the points of the plane in the line $x = y$. Find the values of θ for which $\mathbf{PQ} = \mathbf{QP}$.
(U of L)p

7) Find the 2×2 matrices corresponding to:
(a) the reflection in the line through the origin making an angle of $60°$ with the positive x axis,
(b) the rotation about the origin through an angle of $90°$,
(c) the reflection in the line through the origin making an angle of $120°$ with the positive x axis.
(All angles are measured anticlockwise.)
Describe geometrically the resultant of the three transformations, taken in the given order. (O)

8) Show that, if $\mathbf{A} = \begin{pmatrix} 1 & -1 \\ 2 & -1 \end{pmatrix}$ and $\mathbf{B} = \begin{pmatrix} 1 & 1 \\ 4 & -1 \end{pmatrix}$ then

$$(\mathbf{A}+\mathbf{B})^2 = \mathbf{A}^2 + \mathbf{B}^2.$$ (U of L)p

9) Show that $(a-b)$ is a factor of the determinant

$$\begin{vmatrix} 1+a^2 & a & 1 \\ 1+b^2 & b & 1 \\ 1+c^2 & c & 1 \end{vmatrix}$$

and express the determinant in a completely factorized form. (JMB)

10) Simplify the determinant

$$\begin{vmatrix} 1 & z & z+1 \\ z+1 & 1 & z \\ z & z+1 & 1 \end{vmatrix}$$

Hence prove that the value of the determinant is a real number if z^3 is real. (JMB)

11) Show that $\sin 4\theta - \sin\theta$ is a factor of the determinant

$$D = \begin{vmatrix} 1 & 1 & 1 \\ \frac{1}{2} & \sin\theta & \sin 4\theta \\ \frac{1}{4} & \sin^2\theta & \sin^2 4\theta \end{vmatrix}$$

and express D as the product of three factors, each of which depends on θ. Find all the values of θ for which $D = 0$. (JMB)

12) Prove that the value of a 3×3 determinant is unaltered if λ times the first column is added to the last column.

13) If **A** and **B** are any 3×3 matrices, prove that $|\mathbf{AB}| = |\mathbf{BA}|$.

14) The point (x, y) is transformed to the point (X, Y) under the transformation

$$\begin{pmatrix} X \\ Y \end{pmatrix} = \begin{pmatrix} 2 \\ 1 \end{pmatrix} + \begin{pmatrix} -1 & 0 \\ 0 & 1 \end{pmatrix}\begin{pmatrix} x \\ y \end{pmatrix}$$

Find the images of A(1, 2) and O under this transformation and draw a diagram to illustrate the effect of the transformation on AO. Is the origin invariant under this transformation?

15) Show that the transformation of the plane given by the matrix

$\mathbf{S} = \begin{pmatrix} 1 & 0 \\ 0 & -1 \end{pmatrix}$ is a reflection in the x axis.

Show also that the transformation of the plane given by the matrix

$\mathbf{R}_\alpha = \begin{pmatrix} \cos\alpha & -\sin\alpha \\ \sin\alpha & \cos\alpha \end{pmatrix}$ is a rotation about the origin through an angle α.

Form the product $\mathbf{R}_\alpha\mathbf{S}\mathbf{R}_{-\alpha}$ and show that the transformation of the plane given by this matrix is a reflection in the line $y = x\tan\alpha$. (U of L)

16) Find the two numerical values of λ such that

$$\begin{pmatrix} 4 & 3 \\ 1 & 2 \end{pmatrix}\begin{pmatrix} u \\ 1 \end{pmatrix} = \lambda\begin{pmatrix} u \\ 1 \end{pmatrix}$$

Hence, or otherwise, find the equations of the two lines through the origin which are invariant under the transformation of the plane defined by

$$\begin{pmatrix} x' \\ y' \end{pmatrix} = \begin{pmatrix} 4 & 3 \\ 1 & 2 \end{pmatrix}\begin{pmatrix} x \\ y \end{pmatrix}.$$

(C)

17) $\mathbf{M}$ is the matrix $\begin{pmatrix} a & b \\ c & d \end{pmatrix}$, and $\mathbf{i}, \mathbf{j}$ are the vectors $\begin{pmatrix} 1 \\ 0 \end{pmatrix}, \begin{pmatrix} 0 \\ 1 \end{pmatrix}$ respectively. Write down the vectors $\mathbf{u}$ and $\mathbf{v}$, where $\mathbf{u} = \mathbf{Mi}$ and $\mathbf{v} = \mathbf{Mj}$, and show that:

(a) if $\mathbf{u}$ and $\mathbf{v}$ are perpendicular then $ab + cd = 0$,

(b) if $\mathbf{u}$ and $\mathbf{v}$ have equal magnitudes then $a^2 + c^2 = b^2 + d^2$.

$\mathbf{r}_1 = x_1\mathbf{i} + y_1\mathbf{j}$ and $\mathbf{r}_2 = x_2\mathbf{i} + y_2\mathbf{j}$ are any two vectors, and when multiplied by $\mathbf{M}$ they are transformed to $\mathbf{s}_1$ and $\mathbf{s}_2$ respectively. Given that a, b, c, d satisfy $ab + cd = 0$ and $a^2 + c^2 = b^2 + d^2$, show that the angle between $\mathbf{s}_1$ and $\mathbf{s}_2$ is the same as that between $\mathbf{r}_1$ and $\mathbf{r}_2$. (C)

18) It is known that three non-null 2×2 matrices, $\mathbf{P}, \mathbf{Q}, \mathbf{R}$ satisfy the equation $\mathbf{PQ} = \mathbf{RQ}$. State whether the deduction that $\mathbf{P} = \mathbf{R}$ is true or false. If you think the deduction is true, prove it, if you think it is false, give an example of three non-null matrices $\mathbf{P}, \mathbf{Q}$ and $\mathbf{R}$ which disproves it. (C)p

19) Show that the determinant

$$\begin{vmatrix} 1 & 1 & 1 \\ \cos^2 a & \cos^4 a & \sec^2 a \\ \sin^2 a & \sin^4 a & \tan^2 a \end{vmatrix}$$

is equal to $2\sin^4 a\cos^2 a$. (JMB)

20) Evaluate

$$\begin{vmatrix} x & x-y & x+y \\ x-y & x+y & x \\ x+y & x & x-y \end{vmatrix}$$

Hence, or otherwise, show that

$$\begin{vmatrix} 4 & 11 & -3 \\ 11 & -3 & 4 \\ -3 & 4 & 11 \end{vmatrix} = \begin{vmatrix} 4 & -3 & 11 \\ -3 & 11 & 4 \\ 11 & 4 & -3 \end{vmatrix}.$$ (JMB)

21) Express $\begin{vmatrix} 1 & 1 & n \\ n+1 & n-1 & 2 \\ n(n-1) & n(n+1) & 0 \end{vmatrix}$ as the product of factors linear in n.

Hence show that for all integer values of n, the determinant is divisible by 24. (JMB)

22) Let $\mathbf{A}$ be the matrix $\begin{pmatrix} a & b \\ c & d \end{pmatrix}$, where no one of a, b, c, d is zero.

It is required to find a non-zero 2×2 matrix $\mathbf{X}$ such that $\mathbf{AX} + \mathbf{XA} = \mathbf{0}$, where $\mathbf{0}$ is the zero 2×2 matrix. Prove that either

(a) $a + d = 0$, in which case the general solution for $\mathbf{X}$ depends on two parameters, or

(b) $ad - bc = 0$, in which case the general solution for $\mathbf{X}$ depends on one parameter. (O)

23) $\mathbf{A}$ and $\mathbf{X}$ are the matrices $\begin{pmatrix} a & b \\ c & d \end{pmatrix}$ and $\begin{pmatrix} x & y \\ u & v \end{pmatrix}$ respectively, where b is not equal to zero. Prove that if $\mathbf{AX} = \mathbf{XA}$ then $u = \dfrac{cy}{b}$ and $v = x + \dfrac{(d-a)y}{b}$.

Hence prove that if $\mathbf{AX} = \mathbf{XA}$ then there are numbers p and q such that $\mathbf{X} = p\mathbf{A} + q\mathbf{I}$, where $\mathbf{I}$ is the unit matrix $\begin{pmatrix} 1 & 0 \\ 0 & 1 \end{pmatrix}$, and find p and q in terms of a, b, x, y. (O)

24) When are two matrices conformable for multiplication?

Express $\begin{pmatrix} 1 & 2 & -3 \end{pmatrix}\begin{pmatrix} 2 & 1 & 4 \\ 1 & 0 & 3 \\ 4 & 3 & 5 \end{pmatrix}\begin{pmatrix} 1 \\ 2 \\ -3 \end{pmatrix}$ as a single matrix. (U of L)p

25) If **A** is the matrix $\begin{pmatrix} \cos\dfrac{\pi}{n} & -\sin\dfrac{\pi}{n} \\ \sin\dfrac{\pi}{n} & \cos\dfrac{\pi}{n} \end{pmatrix}$, where n is a positive integer,

prove that $\mathbf{A}^{2n} = \mathbf{I}$, where **I** is the identity (or unit) matrix of order 2.
(U of L)p

26) Find the most general form for the matrix **P** if $\mathbf{PQ} = \mathbf{QP}$, where **Q** is the matrix $\begin{pmatrix} 2 & 1 \\ 4 & 5 \end{pmatrix}$.

If the non-zero column vectors **x** and **y** are such that $\mathbf{Qx} = \mathbf{x}$ and $\mathbf{Qy} = 6\mathbf{y}$, obtain the particular matrix **P** such that $\mathbf{PQ} = \mathbf{QP}$, $\mathbf{Px} = -\mathbf{x}$ and $\mathbf{Py} = 4\mathbf{y}$. (U of L)p

27) The point (x, y) is transformed to the point (x', y') by means of the transformation

$$\begin{pmatrix} x' \\ y' \end{pmatrix} = \begin{pmatrix} 3 & 0 \\ 0 & 4 \end{pmatrix}\begin{pmatrix} x \\ y \end{pmatrix} + \begin{pmatrix} 1 \\ 1 \end{pmatrix}$$

Find the image of the line $y = 2x$ under this transformation. (U of L)

CHAPTER 2

FURTHER TRANSFORMATIONS

All the work in this chapter is related to three dimensional space and it is assumed that the reader is familiar with the equations of lines and planes in three dimensions.

We will start this chapter with an operation on vectors which has useful applications in three dimensional space.

VECTOR PRODUCT

The vector product of two vectors **a** and **b** which are inclined at an angle θ is written as $\mathbf{a} \times \mathbf{b}$ (or sometimes $\mathbf{a} \wedge \mathbf{b}$) and is defined as a vector of magnitude $ab \sin \theta$ in a direction perpendicular to the plane containing **a** and **b** in the sense of a right-handed screw turned from **a** to **b**.

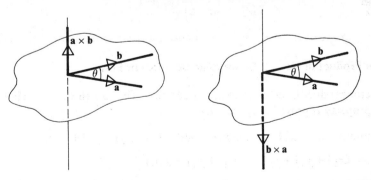

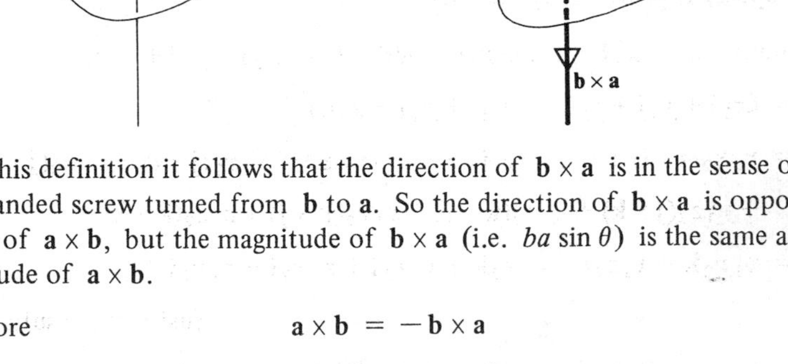

From this definition it follows that the direction of $\mathbf{b} \times \mathbf{a}$ is in the sense of a right-handed screw turned from **b** to **a**. So the direction of $\mathbf{b} \times \mathbf{a}$ is opposite to that of $\mathbf{a} \times \mathbf{b}$, but the magnitude of $\mathbf{b} \times \mathbf{a}$ (i.e. $ba \sin \theta$) is the same as the magnitude of $\mathbf{a} \times \mathbf{b}$.

Therefore $$\mathbf{a} \times \mathbf{b} = -\mathbf{b} \times \mathbf{a}$$

Thus vector product is *not* commutative.

Vector Product of Parallel Vectors

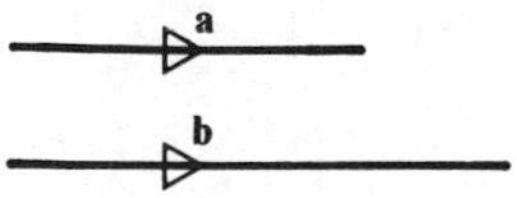

If **a** and **b** are parallel vectors

$$|\mathbf{a} \times \mathbf{b}| = ab \sin\theta$$

But $\sin\theta = 0$

Therefore $\mathbf{a} \times \mathbf{b} = \mathbf{0}$

Vector Product of Perpendicular Vectors

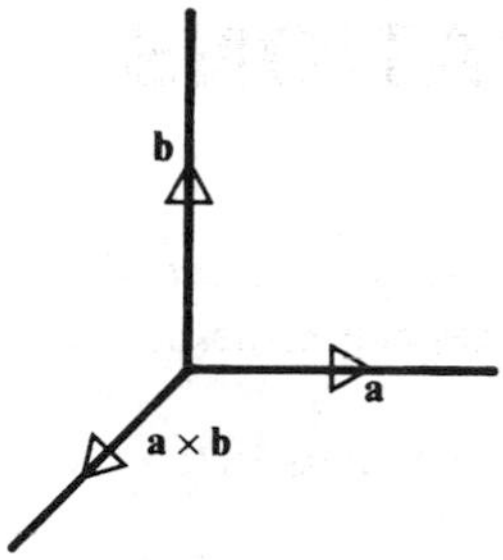

If **a** and **b** are perpendicular vectors, $\sin\theta = 1$ and $|\mathbf{a} \times \mathbf{b}| = ab$.
In this case **a**, **b** and **a** × **b** form a right-handed set of three mutually perpendicular vectors as shown in the diagram.

This result is particularly important in the case of the unit vectors **i**, **j** and **k**.

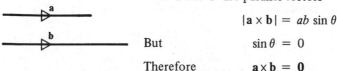

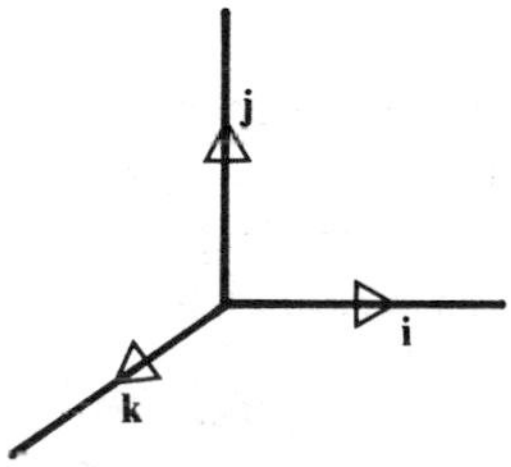

Thus $\mathbf{i} \times \mathbf{j} = \mathbf{k}$ and $\mathbf{j} \times \mathbf{i} = -\mathbf{k}$

$\mathbf{j} \times \mathbf{k} = \mathbf{i}$ and $\mathbf{k} \times \mathbf{j} = -\mathbf{i}$

$\mathbf{k} \times \mathbf{i} = \mathbf{j}$ and $\mathbf{i} \times \mathbf{k} = -\mathbf{j}$

Also $\mathbf{i} \times \mathbf{i} = \mathbf{j} \times \mathbf{j} = \mathbf{k} \times \mathbf{k} = 0$

Vector Product of Vectors in Cartesian Component Form

Vector product is distributive across addition: i.e. $\mathbf{a} \times (\mathbf{b} + \mathbf{c}) = \mathbf{a} \times \mathbf{b} + \mathbf{a} \times \mathbf{c}$. (This property is proved on page 70.)

Therefore if $\mathbf{a} = x_1\mathbf{i} + y_1\mathbf{j} + z_1\mathbf{k}$ and $\mathbf{b} = x_2\mathbf{i} + y_2\mathbf{j} + z_2\mathbf{k}$

$$\begin{aligned}
\mathbf{a} \times \mathbf{b} &= (x_1\mathbf{i} + y_1\mathbf{j} + z_1\mathbf{k}) \times (x_2\mathbf{i} + y_2\mathbf{j} + z_2\mathbf{k}) \\
&= x_1x_2(\mathbf{i} \times \mathbf{i}) + x_1y_2(\mathbf{i} \times \mathbf{j}) + x_1z_2(\mathbf{i} \times \mathbf{k}) + y_1x_2(\mathbf{j} \times \mathbf{i}) + y_1y_2(\mathbf{j} \times \mathbf{j}) \\
&\quad + y_1z_2(\mathbf{j} \times \mathbf{k}) + z_1x_2(\mathbf{k} \times \mathbf{i}) + z_1y_2(\mathbf{k} \times \mathbf{j}) + z_1z_2(\mathbf{k} \times \mathbf{k}) \\
&= x_1y_2\mathbf{k} - x_1z_2\mathbf{j} - y_1x_2\mathbf{k} + y_1z_2\mathbf{i} + z_1x_2\mathbf{j} - z_1y_2\mathbf{i}
\end{aligned}$$

(using the results above)

$$= (y_1z_2 - z_1y_2)\mathbf{i} - (x_1z_2 - z_1x_2)\mathbf{j} + (x_1y_2 - y_1x_2)\mathbf{k}$$

This expression is the expansion of the determinant $\begin{vmatrix} \mathbf{i} & \mathbf{j} & \mathbf{k} \\ x_1 & y_1 & z_1 \\ x_2 & y_2 & z_2 \end{vmatrix}$

Therefore $(x_1\mathbf{i}+y_1\mathbf{j}+z_1\mathbf{k})\times(x_2\mathbf{i}+y_2\mathbf{j}+z_2\mathbf{k}) = \begin{vmatrix} \mathbf{i} & \mathbf{j} & \mathbf{k} \\ x_1 & y_1 & z_1 \\ x_2 & y_2 & z_2 \end{vmatrix}$

For example,
$$(2\mathbf{i}+\mathbf{j}-2\mathbf{k})\times(\mathbf{j}+3\mathbf{k}) = \begin{vmatrix} \mathbf{i} & \mathbf{j} & \mathbf{k} \\ 2 & 1 & -2 \\ 0 & 1 & 3 \end{vmatrix}$$
$$= 5\mathbf{i}-6\mathbf{j}+2\mathbf{k}$$

Some calculations involve a mixture of vector and scalar product, e.g. $\mathbf{a}\times\mathbf{b}\cdot\mathbf{c}$. Brackets are unnecessary in expressions of this type as the cross product *must* be calculated first. If $\mathbf{b}\cdot\mathbf{c}$ were worked first this would lead to the vector product of $\mathbf{a}$ and a scalar quantity, which is meaningless.

To summarize:

$\mathbf{a}\times\mathbf{b} = ab\sin\theta\,\hat{\mathbf{n}}$ where $\hat{\mathbf{n}}$ is a unit vector perpendicular to $\mathbf{a}$ and $\mathbf{b}$ in the direction of a right-handed screw turned from $\mathbf{a}$ to $\mathbf{b}$.

$\mathbf{a}\times\mathbf{b} = -\mathbf{b}\times\mathbf{a}$ so the order of the vectors in the product is important.

If $\mathbf{a}$ and $\mathbf{b}$ are parallel $\mathbf{a}\times\mathbf{b} = \mathbf{0}$.

If $\mathbf{a}$ and $\mathbf{b}$ are perpendicular $\mathbf{a}$, $\mathbf{b}$ and $\mathbf{a}\times\mathbf{b}$ form a right-handed set of mutually perpendicular vectors.

EXAMPLES 2a

1) Simplify (a) $\mathbf{a}\times(\mathbf{a}-\mathbf{b})$, (b) $\mathbf{a}\times\mathbf{b}\cdot\mathbf{a}$.

(a)
$$\mathbf{a}\times(\mathbf{a}-\mathbf{b}) = \mathbf{a}\times\mathbf{a}-\mathbf{a}\times\mathbf{b}$$
$$= \mathbf{0}-\mathbf{a}\times\mathbf{b}$$
$$= \mathbf{b}\times\mathbf{a}$$

(b) By definition $\mathbf{a}\times\mathbf{b}$ is perpendicular to $\mathbf{a}$ and the scalar product of perpendicular vectors is zero.
Therefore $\mathbf{a}\times\mathbf{b}\cdot\mathbf{a} = 0$.

2) $\mathbf{a}, \mathbf{b}$ and $\mathbf{c}$ are three vectors such that $\mathbf{a} \times \mathbf{b} = \mathbf{c} \times \mathbf{a}$, $\mathbf{a} \neq \mathbf{0}$.
Find a linear relationship between $\mathbf{a}, \mathbf{b}$ and $\mathbf{c}$.

$$\mathbf{a} \times \mathbf{b} = \mathbf{c} \times \mathbf{a} \Rightarrow \mathbf{a} \times \mathbf{b} = -\mathbf{a} \times \mathbf{c}$$

$$\Rightarrow \quad \mathbf{a} \times \mathbf{b} + \mathbf{a} \times \mathbf{c} = \mathbf{0}$$

$$\Rightarrow \quad \mathbf{a} \times (\mathbf{b} + \mathbf{c}) = \mathbf{0} \qquad \text{(distributive law)}$$

so either $\mathbf{a}$ and $\mathbf{b} + \mathbf{c}$ are parallel vectors,
or $\mathbf{b} + \mathbf{c} = \mathbf{0}$,
i.e either $\mathbf{a} = k(\mathbf{b} + \mathbf{c})$ where k is a scalar quantity
or $\mathbf{b} = -\mathbf{c}$

EXERCISE 2a

1) Simplify the following:
(a) $(\mathbf{a} + \mathbf{b}) \times \mathbf{b}$ (b) $(\mathbf{a} + \mathbf{b}) \times (\mathbf{a} + \mathbf{b})$
(c) $(\mathbf{a} - \mathbf{b}) \times (\mathbf{a} + \mathbf{b})$ (d) $\mathbf{a} \times (\mathbf{b} + \mathbf{c}) . \mathbf{b}$
(e) $\mathbf{a} . (\mathbf{b} + \mathbf{c}) \times \mathbf{a}$ (f) $\mathbf{a} \times \mathbf{b} . \mathbf{a} + \mathbf{b} . \mathbf{a} \times \mathbf{b}$.

2) If $\mathbf{a} = \mathbf{i} + \mathbf{j} - \mathbf{k}$ and $\mathbf{b} = 2\mathbf{i} - \mathbf{j} + \mathbf{k}$ find
(a) $\mathbf{a} \times \mathbf{b}$ (b) $\mathbf{a} \times (\mathbf{a} + \mathbf{b})$
and verify that $\mathbf{a} . \mathbf{a} \times \mathbf{b} = 0$.

3) If $\mathbf{a} = \mathbf{i} + 2\mathbf{j} - \mathbf{k}$ and $\mathbf{b} = \mathbf{j} + \mathbf{k}$ find the unit vector perpendicular to both $\mathbf{a}$ and $\mathbf{b}$. Calculate also the sine of the angle between $\mathbf{a}$ and $\mathbf{b}$.

4) If $\mathbf{a} = \mathbf{i} + \mathbf{j} - \mathbf{k}$, $\mathbf{b} = \mathbf{i} - \mathbf{j}$ and $\mathbf{c} = 2\mathbf{i} + \mathbf{k}$, find $(\mathbf{a} \times \mathbf{b}) \times \mathbf{c}$ and $\mathbf{a} \times (\mathbf{b} \times \mathbf{c})$.
Verify also that $\mathbf{a} . \mathbf{b} \times \mathbf{c} = \mathbf{a} \times \mathbf{b} . \mathbf{c}$

5) If $\mathbf{a} = \mathbf{i} + \mathbf{j}$ and $\mathbf{b} = 2\mathbf{i} + \mathbf{k}$ find the sine of the angle between $\mathbf{a}$ and $\mathbf{b}$, and the unit vector perpendicular to both $\mathbf{a}$ and $\mathbf{b}$.

6) A, B and C are the points $(0, 1, 2)$, $(3, 2, 1)$ and $(1, -1, 0)$ respectively. Find the unit vector which is perpendicular to the plane ABC.

7) Three vectors $\mathbf{a}, \mathbf{b}$ and $\mathbf{c}$ are such that $\mathbf{a} \times \mathbf{b} = \mathbf{a} \times \mathbf{c}$ $(\mathbf{a} \neq \mathbf{0})$.
Show that $\mathbf{b} - \mathbf{c} = k\mathbf{a}$ where k is a scalar.

8) Three vectors $\mathbf{a}, \mathbf{b}$ and $\mathbf{c}$ are such that $\mathbf{a} \times 3\mathbf{b} = 2\mathbf{a} \times \mathbf{c}$. Find a linear relationship between $\mathbf{a}, \mathbf{b}$ and $\mathbf{c}$.

9) Show that $\mathbf{u} = (\mathbf{i} + \mathbf{j})$ is a solution of the equation

$$\mathbf{u} \times (\mathbf{i} + 4\mathbf{j}) = 3\mathbf{k}$$

Show also that the general solution to this equation is

$$\mathbf{u} = -3\mathbf{j} + t(\mathbf{i} + 4\mathbf{j}).$$

10) Find a general solution to the equation

$$\mathbf{u} \times (\mathbf{i} - 3\mathbf{k}) = 2\mathbf{j}.$$

APPLICATIONS OF THE VECTOR PRODUCT

1. Area of a Parallelogram

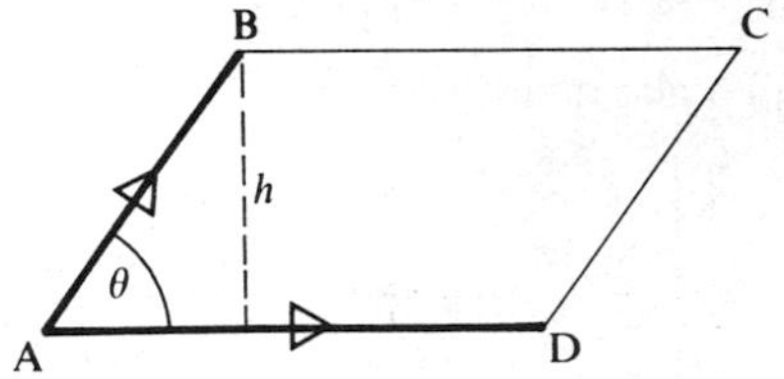

The area of a parallelogram

$$= \text{base} \times \text{height}$$
$$= (AD)(h)$$
$$= (AD)(AB \sin\theta)$$
$$= |\overrightarrow{AD} \times \overrightarrow{AB}|$$

Therefore the area of a parallelogram is the magnitude of the vector product of two adjacent sides.

2. Area of a Triangle

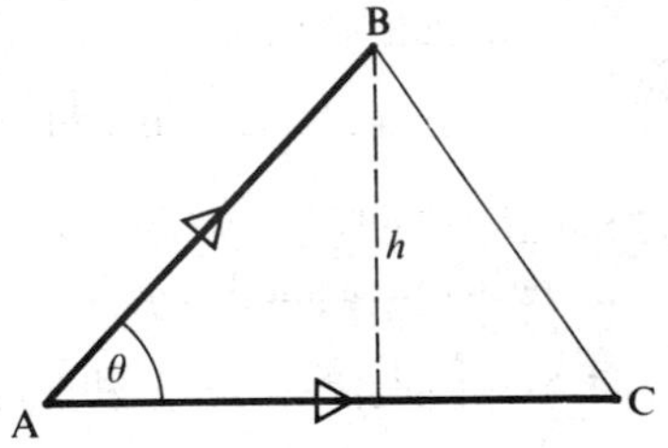

The area of $\triangle ABC$

$$= \tfrac{1}{2}\text{base} \times \text{height}$$
$$= \tfrac{1}{2}(AC)(h)$$
$$= \tfrac{1}{2}(AC)(AB \sin\theta)$$
$$= \tfrac{1}{2}|\overrightarrow{AC} \times \overrightarrow{AB}|$$

Therefore the area of a triangle is half the magnitude of the vector product of two sides.

EXAMPLES 2b

1) A triangle ABC has its vertices at the points A(1, 2, 1), B(1, 0, 3), C(− 1, 2, − 1).
Find the area of $\triangle ABC$.

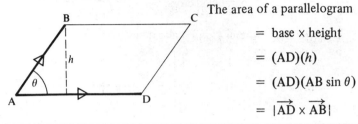

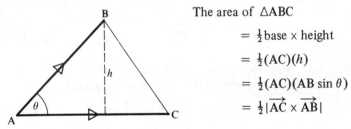

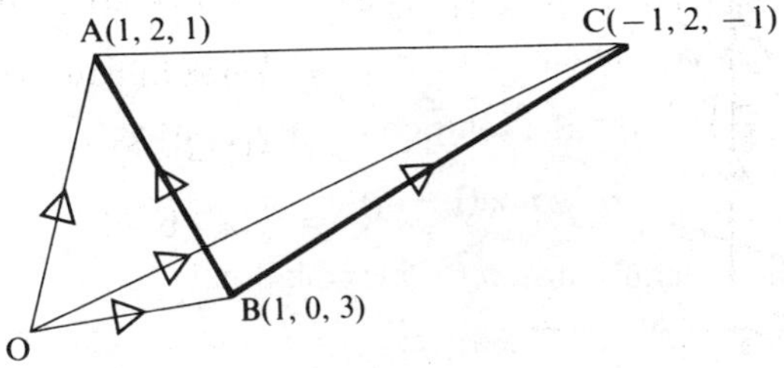

If O is the origin then $\overrightarrow{OA} = \mathbf{i} + 2\mathbf{j} + \mathbf{k}$

$$\overrightarrow{OB} = \mathbf{i} + 3\mathbf{k}$$

$$\overrightarrow{OC} = -\mathbf{i} + 2\mathbf{j} - \mathbf{k}$$

Therefore

$$\overrightarrow{BA} = \overrightarrow{OA} - \overrightarrow{OB} = 2\mathbf{j} - 2\mathbf{k}$$

$$\overrightarrow{BC} = \overrightarrow{OC} - \overrightarrow{OB} = -2\mathbf{i} + 2\mathbf{j} - 4\mathbf{k}$$

$$\text{Area of } \Delta ABC = \tfrac{1}{2}|\overrightarrow{BA} \times \overrightarrow{BC}|$$

$$\overrightarrow{BA} \times \overrightarrow{BC} = \begin{vmatrix} \mathbf{i} & \mathbf{j} & \mathbf{k} \\ 0 & 2 & -2 \\ -2 & 2 & -4 \end{vmatrix} = -4\mathbf{i} + 4\mathbf{j} + 4\mathbf{k}$$

Therefore area of $\Delta ABC = |\tfrac{1}{2}|-4\mathbf{i} + 4\mathbf{j} + 4\mathbf{k}| = 2\sqrt{3}$.

3. Volume of a Parallelepiped

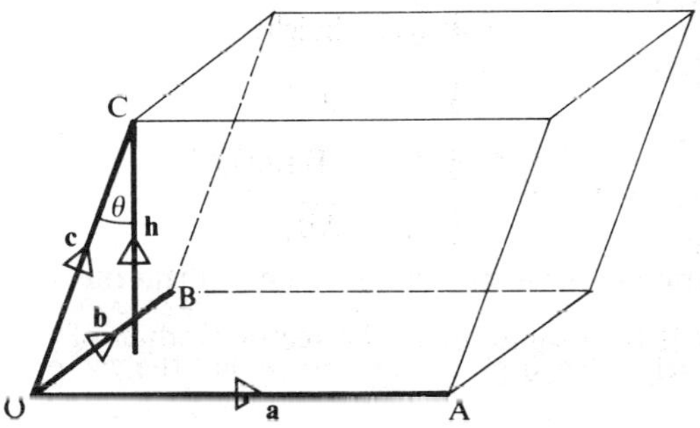

The volume of a parallelepiped

$= \text{area of base} \times \text{height}$

$= |\mathbf{a} \times \mathbf{b}|h$

$= |\mathbf{a} \times \mathbf{b}||\mathbf{c}| \cos\theta$

θ is the angle between **c** and **h** and as **h** is perpendicular to **a** and **b**, θ is also the angle between $\mathbf{a} \times \mathbf{b}$ and **c**.

$\therefore$ $|\mathbf{a} \times \mathbf{b}||\mathbf{c}| \cos\theta$ is the scalar product of the vectors $\mathbf{a} \times \mathbf{b}$ and **c**.

Therefore the volume of the parallelepiped is $|\mathbf{a} \times \mathbf{b} . \mathbf{c}|$

4. Volume of a Tetrahedron

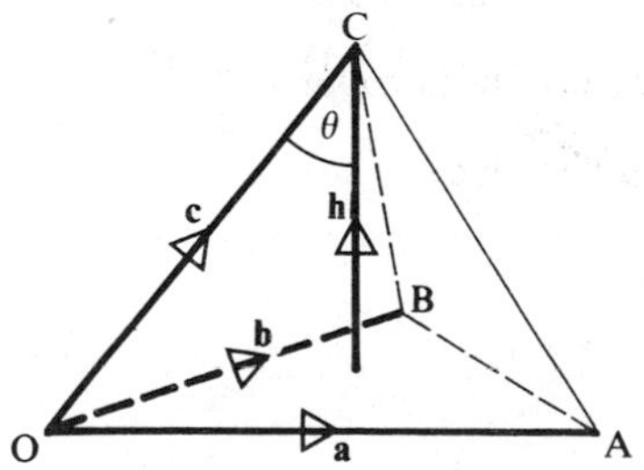

The volume of a tetrahedron

$= \tfrac{1}{3} \text{area of base} \times \text{height}$

$= \tfrac{1}{3}(\tfrac{1}{2}|\mathbf{a} \times \mathbf{b}|)|\mathbf{c}| \cos\theta$

$= \tfrac{1}{6}|\mathbf{a} \times \mathbf{b} . \mathbf{c}|$

EXAMPLES 2b (continued)

2) Find the volume of the tetrahedron OABC where O is the origin and A, B, C are the points (2, 1, 1), (0, −1, 1), (−1, 3, 0).

The volume of OABC is $\frac{1}{6}|\overrightarrow{OA} \times \overrightarrow{OB} . \overrightarrow{OC}|$

$$= \tfrac{1}{6}|(2\mathbf{i}+\mathbf{j}+\mathbf{k}) \times (-\mathbf{j}+\mathbf{k}).(-\mathbf{i}+3\mathbf{j})|$$

$$= \tfrac{1}{6}|(2\mathbf{i}-2\mathbf{j}-2\mathbf{k}).(-\mathbf{i}+3\mathbf{j})|$$

$$= 1\tfrac{1}{3}$$

Triple Scalar Product and the Proof of the Distributive Law for Vector Products

Expressions of the form $\mathbf{a} \times \mathbf{b}.\mathbf{c}$ are called triple scalar products.
The volume of the parallelepiped above could have been obtained by considering the side defined by $\mathbf{b}$ and $\mathbf{c}$ as the base, in which case the result would have been in the form $\mathbf{b} \times \mathbf{c}.\mathbf{a}$,

i.e. $$\mathbf{a} \times \mathbf{b}.\mathbf{c} = \mathbf{b} \times \mathbf{c}.\mathbf{a}$$

The order in which a scalar product is performed does not matter

i.e. $$\mathbf{b} \times \mathbf{c}.\mathbf{a} = \mathbf{a}.\mathbf{b} \times \mathbf{c}$$

Therefore $$\mathbf{a} \times \mathbf{b}.\mathbf{c} = \mathbf{a}.\mathbf{b} \times \mathbf{c}$$

Thus in a triple scalar product, the 'cross' and 'dot' may be interchanged without altering the value of the expression. This property will be referred to as the *triple scalar product property*.
(But if the order of the vectors is altered, the expression may not remain the same: e.g. $\mathbf{a} \times \mathbf{b}.\mathbf{c} = -\mathbf{b} \times \mathbf{a}.\mathbf{c}$.)
We will now use this property of a triple scalar product to prove that the vector product is distributive.

Consider $\mathbf{d}.(\mathbf{a} \times \mathbf{b} + \mathbf{a} \times \mathbf{c})$ where $\mathbf{d}$ is any vector

$$= \mathbf{d}.\mathbf{a} \times \mathbf{b} + \mathbf{d}.\mathbf{a} \times \mathbf{c} \quad \text{(scalar product is distributive)}$$

$$= \mathbf{d} \times \mathbf{a}.\mathbf{b} + \mathbf{d} \times \mathbf{a}.\mathbf{c} \quad \text{(triple scalar product property)}$$

$$= \mathbf{d} \times \mathbf{a}.(\mathbf{b} + \mathbf{c}) \quad \text{(scalar product is distributive)}$$

$$= \mathbf{d}.\mathbf{a} \times (\mathbf{b} + \mathbf{c}) \quad \text{(triple scalar product property),}$$

i.e. $$\mathbf{d}.(\mathbf{a} \times \mathbf{b} + \mathbf{a} \times \mathbf{c}) = \mathbf{d}.\mathbf{a} \times (\mathbf{b} + \mathbf{c})$$

Since $\mathbf{d}$ is any vector it follows that

$$\mathbf{a} \times \mathbf{b} + \mathbf{a} \times \mathbf{c} = \mathbf{a} \times (\mathbf{b} + \mathbf{c}),$$

i.e. the vector product is distributive.

EXERCISE 2b

1) The vectors $(2\mathbf{i}+3\mathbf{j}-\mathbf{k})$ and $(\mathbf{i}+2\mathbf{j}+\mathbf{k})$ represent two sides of a triangle. Find the area of the triangle.

2) The triangle ABC has its vertices at the points A(0, 0, 1), B(1, 0, 1), C(2, 1, 3). Find the area of the triangle ABC.

3) The vertices of a triangle are at the points with position vectors **a**, **b** and **c**. Prove that the area of the triangle is $\frac{1}{2}|\mathbf{a}\times\mathbf{b}+\mathbf{b}\times\mathbf{c}+\mathbf{c}\times\mathbf{a}|$.

4) A parallelogram OABC has one vertex O at the origin and the vertices A and B at the points (0, 1, 3), (0, 2, 5). Find the area of OABC.

5) The parallelogram ABCD has three of its vertices A, B and C at the points (1, 2, −1), (1, 3, 2) and (−1, 3, −1). Find the area of the parallelogram.

6) The vectors $\overrightarrow{OA}, \overrightarrow{OB}, \overrightarrow{OC}$ are three edges of a parallelepiped where O is the origin and A, B and C are the points (2, 1, 0), (−1, −1, 1), (0, 2, −1). Find the volume of the parallelepiped.

7) Find the volume of the tetrahedron OABC where O is the origin and A, B and C are the points (2, 0, 1), (3, 1, 2) and (−1, 3, 0).

8) ABCD is a tetrahedron and A, B, C and D are the points (0, 1, 0), (0, 0, 4), (1, 1, 1) and (−1, 3, 2). Find the volume of the tetrahedron.

9) The four vertices of a tetrahedron are at the points with position vectors **a**, **b**, **c**, **d**. Find the volume of the tetrahedron.

TRANSFORMATIONS OF THREE DIMENSIONAL SPACE

This section extends to transformations of three dimensional space, the work already covered on transformations of the two dimensional xy plane. As before, we are going to consider only those transformations which leave the origin unchanged and which can be expressed by linear equations, i.e. transformations which map the point $P(x, y, z)$ to the point $P'(X, Y, Z)$ where the relationship between the coordinates of P and P′ can be expressed in the form

$$a_1x+b_1y+c_1z = X$$
$$a_2x+b_2y+c_2z = Y$$
$$a_3x+b_3y+c_3z = Z$$

These three equations can be expressed as the single matrix equation

$$\begin{pmatrix} a_1 & b_1 & c_1 \\ a_2 & b_2 & c_2 \\ a_3 & b_3 & c_3 \end{pmatrix}\begin{pmatrix} x \\ y \\ z \end{pmatrix} = \begin{pmatrix} X \\ Y \\ Z \end{pmatrix} \qquad [1]$$

Thus any linear transformation of the xyz space may be expressed in the form

$$\mathbf{M}\begin{pmatrix} x \\ y \\ z \end{pmatrix} = \begin{pmatrix} X \\ Y \\ Z \end{pmatrix}$$

where $\mathbf{M}$ is a 3×3 matrix.

So $\mathbf{M}$ may be regarded as an operator which maps the point P whose position vector is $\begin{pmatrix} x \\ y \\ z \end{pmatrix}$ to the point P′ whose position vector is $\begin{pmatrix} X \\ Y \\ Z \end{pmatrix}$.

Two properties of linear transformations follow from equation [1].

(1) If $\mathbf{M}$ maps $\mathbf{r} = \begin{pmatrix} x \\ y \\ z \end{pmatrix}$ to $\mathbf{r}' = \begin{pmatrix} X \\ Y \\ Z \end{pmatrix}$ then any scalar multiple of $\mathbf{r}$ maps to the same scalar multiple of $\mathbf{r}'$.

i.e. if $\begin{pmatrix} x \\ y \\ z \end{pmatrix} \longmapsto \begin{pmatrix} X \\ Y \\ Z \end{pmatrix}$ then $\lambda\begin{pmatrix} x \\ y \\ z \end{pmatrix} \longmapsto \lambda\begin{pmatrix} X \\ Y \\ Z \end{pmatrix}$

(2) If $\mathbf{M}$ maps $\begin{pmatrix} x_1 \\ y_1 \\ z_1 \end{pmatrix}$ to $\begin{pmatrix} X_1 \\ Y_1 \\ Z_1 \end{pmatrix}$ and $\begin{pmatrix} x_2 \\ y_2 \\ z_2 \end{pmatrix}$ to $\begin{pmatrix} X_2 \\ Y_2 \\ Z_2 \end{pmatrix}$

then $\begin{pmatrix} x_1 \\ y_1 \\ z_1 \end{pmatrix} + \begin{pmatrix} x_2 \\ y_2 \\ z_2 \end{pmatrix} \longmapsto \begin{pmatrix} X_1 \\ Y_1 \\ Z_1 \end{pmatrix} + \begin{pmatrix} X_2 \\ Y_2 \\ Z_2 \end{pmatrix}$

Both of these properties can be verified using the definitions of matrix multiplication in Chapter 1.

Note that a linear transformation may be visualized by considering a foam rubber cuboid which is fixed at one corner (the origin). If the rubber block is then uniformly stretched or reduced in any direction, rotated about any line (which must obviously pass through its fixed corner), reflected in any plane (which again must obviously pass through its fixed corner) or is subjected to any combination of these distortions, the block is undergoing a linear transformation in which one point (its fixed corner) does not change position. If however the block is bent or *non*-uniformly stretched so that, for example, some planes contained in the block are transformed to curved surfaces, such transformations are *not* linear.

Examples of some linear transformations are

(a) rotation about any line through the origin,

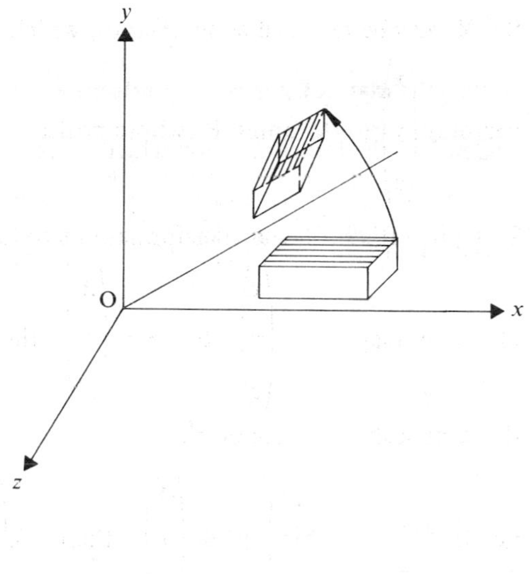

(b) reflection in any plane through the origin,

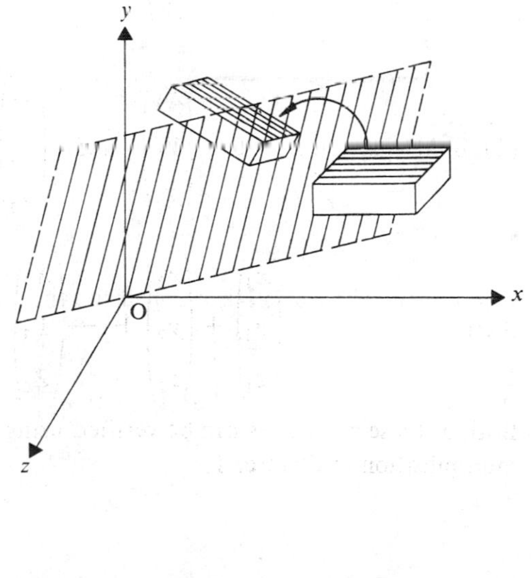

(c) enlargement or reduction (the origin is the centre of the enlargement).

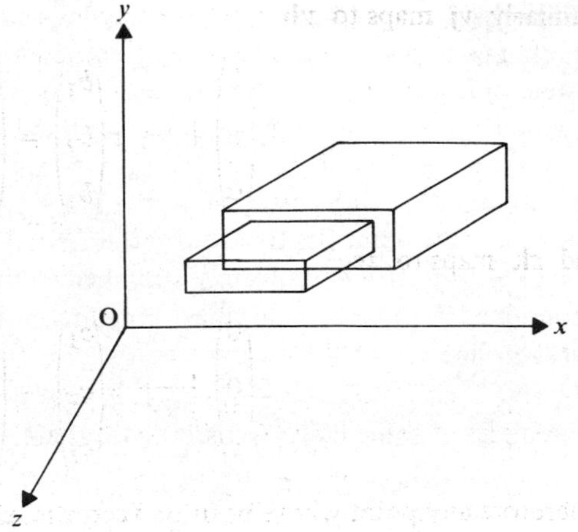

Finding the Matrix Operator to Perform a Given Transformation

Consider the transformation which maps **i** to **a**, **j** to **b**, **k** to **c**

where $\mathbf{a} = \begin{pmatrix} a_1 \\ a_2 \\ a_3 \end{pmatrix}$, $\mathbf{b} = \begin{pmatrix} b_1 \\ b_2 \\ b_3 \end{pmatrix}$ and $\mathbf{c} = \begin{pmatrix} c_1 \\ c_2 \\ c_3 \end{pmatrix}$ respectively.

i.e.

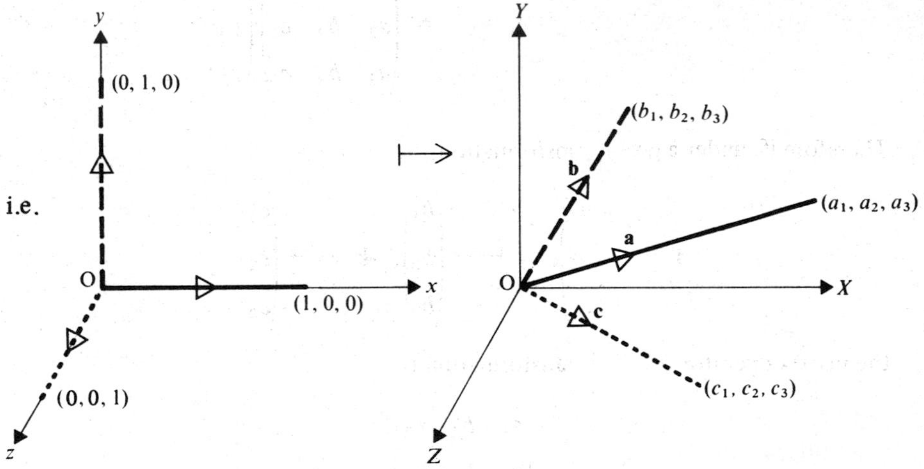

As **i** is mapped to **a**, $x\mathbf{i}$ is mapped to $x\mathbf{a}$,

i.e. $$\begin{pmatrix} x \\ 0 \\ 0 \end{pmatrix} \longmapsto x \begin{pmatrix} a_1 \\ a_2 \\ a_3 \end{pmatrix} = \begin{pmatrix} xa_1 \\ xa_2 \\ xa_3 \end{pmatrix}$$

Similarly $y\mathbf{j}$ maps to $y\mathbf{b}$,

i.e. $$\begin{pmatrix}0\\y\\0\end{pmatrix} \longmapsto y\begin{pmatrix}b_1\\b_2\\b_3\end{pmatrix} = \begin{pmatrix}yb_1\\yb_2\\yb_3\end{pmatrix}$$

and $z\mathbf{k}$ maps to $z\mathbf{c}$,

i.e. $$\begin{pmatrix}0\\0\\z\end{pmatrix} \longmapsto z\begin{pmatrix}c_1\\c_2\\c_3\end{pmatrix} = \begin{pmatrix}zc_1\\zc_2\\zc_3\end{pmatrix}$$

Therefore any point whose position vector is $x\mathbf{i} + y\mathbf{j} + z\mathbf{k}$ is mapped to the point whose position vector is $x\mathbf{a} + y\mathbf{b} + z\mathbf{c}$.

Thus $$\begin{pmatrix}x\\y\\z\end{pmatrix} \longmapsto \begin{pmatrix}xa_1\\xa_2\\xa_3\end{pmatrix} + \begin{pmatrix}yb_1\\yb_2\\yb_3\end{pmatrix} + \begin{pmatrix}zc_1\\zc_2\\zc_3\end{pmatrix} = \begin{pmatrix}xa_1+yb_1+zc_1\\xa_2+yb_2+zc_2\\xa_3+yb_3+zc_3\end{pmatrix}$$

$$= \begin{pmatrix}a_1 & b_1 & c_1\\a_2 & b_2 & c_2\\a_3 & b_3 & c_3\end{pmatrix}\begin{pmatrix}x\\y\\z\end{pmatrix}$$

Therefore if, under a given transformation,

$$\mathbf{i} \longmapsto \begin{pmatrix}a_1\\a_2\\a_3\end{pmatrix}, \quad \mathbf{j} \longmapsto \begin{pmatrix}b_1\\b_2\\b_3\end{pmatrix}, \quad \mathbf{k} \longmapsto \begin{pmatrix}c_1\\c_2\\c_3\end{pmatrix}$$

the matrix operator for that transformation is

$$\begin{pmatrix}a_1 & b_1 & c_1\\a_2 & b_2 & c_2\\a_3 & b_3 & c_3\end{pmatrix}$$

and the images of **i**, **j** and **k** form the columns of the matrix.

Note that the Cartesian frame of reference, whose base vectors are **i**, **j** and **k** is transformed to the frame of reference whose base vectors are **a**, **b** and **c**

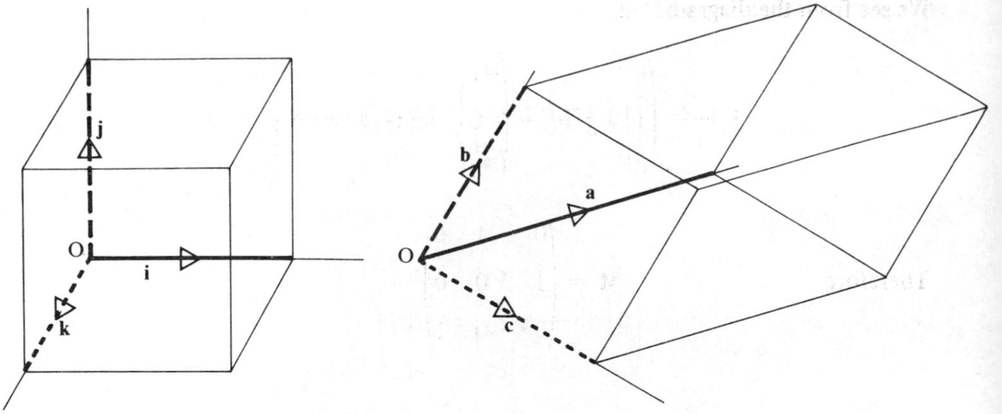

and that a three dimensional grid of unit volume cubes in xyz space is transformed to a three dimensional grid of parallelepipeds in the image space.

EXAMPLES 2c

1) Find the matrix **M** such that the point $P(x, y, z)$ is rotated through $90°$ about Oz to the point $P'(X, Y, Z)$ by the transformation

$$\begin{pmatrix} X \\ Y \\ Z \end{pmatrix} = \mathbf{M}\begin{pmatrix} x \\ y \\ z \end{pmatrix}$$

Note that a positive rotation about Oz is in the sense $\mathbf{i} \longmapsto \mathbf{j}$, a positive rotation about Ox is in the sense $\mathbf{j} \longmapsto \mathbf{k}$ and a positive rotation about Oy is in the sense $\mathbf{k} \longmapsto \mathbf{i}$.

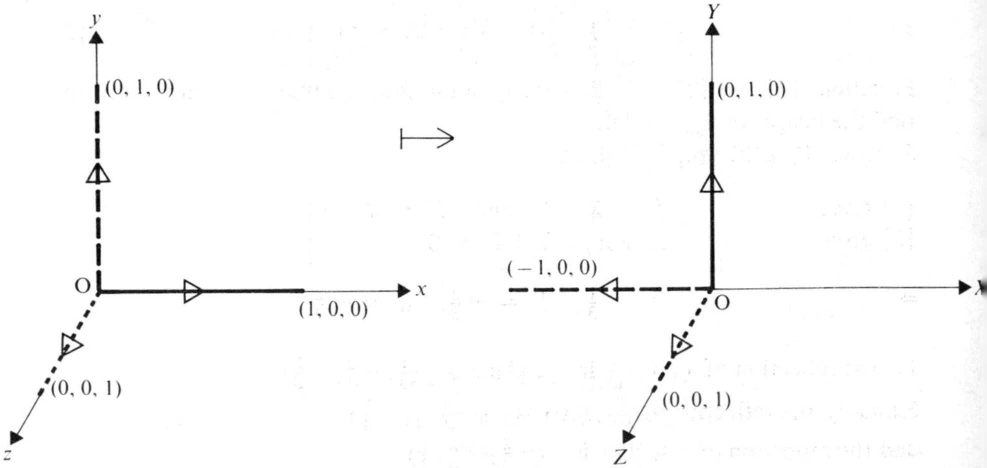

We see from the diagram that

$$\mathbf{i} \longmapsto \begin{pmatrix}0\\1\\0\end{pmatrix}, \quad \mathbf{j} \longmapsto \begin{pmatrix}-1\\0\\0\end{pmatrix} \quad \text{and} \quad \mathbf{k} \longmapsto \begin{pmatrix}0\\0\\1\end{pmatrix}$$

Therefore
$$\mathbf{M} = \begin{pmatrix}0 & -1 & 0\\1 & 0 & 0\\0 & 0 & 1\end{pmatrix}$$

2) Find the matrix $\mathbf{M}$ which transforms the point $P(x, y, z)$ to the point $P'(X, Y, Z)$ by the equation $\mathbf{M}\,\overrightarrow{OP} = \overrightarrow{OP'}$ where P' is the reflection of P in the plane $x+y+z=0$.

If P' is the reflection of P in the plane $x+y+z=0$ then PP' is normal to the plane and the midpoint of PP' is on the plane.
The direction ratios of the normal to the plane are

$1:1:1$ from the equation of the plane

and $X-x:Y-y:Z-z$ from the coordinates of P and P',

so
$$X-x = Y-y = Z-z \qquad [1]$$

The midpoint of PP' is $\{\tfrac{1}{2}(x+X), \tfrac{1}{2}(y+Y), \tfrac{1}{2}(z+Z)\}$ and it is on the plane,

so
$$\tfrac{1}{2}(x+X)+\tfrac{1}{2}(y+Y)+\tfrac{1}{2}(z+Z) = 0 \qquad [2]$$

Equations [1] and [2] are valid for any point $P(x, y, z)$ so they can be used to find the images of $\mathbf{i}, \mathbf{j}$ and $\mathbf{k}$.
So when P is the point $(1, 0, 0)$

[1] gives $\quad Y = X-1$ and $Z = X-1$
[2] gives $\quad (1+X)+Y+Z = 0$

$$\Rightarrow \quad X = \tfrac{1}{3}, \quad Y = -\tfrac{2}{3}, \quad Z = -\tfrac{2}{3}$$

i.e. the reflection of $(1,0,0)$ in the plane is $(\tfrac{1}{3}, -\tfrac{2}{3}, -\tfrac{2}{3})$.
Similarly the reflection of $(0,1,0)$ is $(-\tfrac{2}{3}, \tfrac{1}{3}, -\tfrac{2}{3})$
and the reflection of $(0,0,1)$ is $(-\tfrac{2}{3}, -\tfrac{2}{3}, \tfrac{1}{3})$.

Therefore

$$\begin{pmatrix}1\\0\\0\end{pmatrix} \longmapsto \begin{pmatrix}\frac{1}{3}\\-\frac{2}{3}\\-\frac{2}{3}\end{pmatrix}, \quad \begin{pmatrix}0\\1\\0\end{pmatrix} \longmapsto \begin{pmatrix}-\frac{2}{3}\\\frac{1}{3}\\-\frac{2}{3}\end{pmatrix}, \quad \begin{pmatrix}0\\0\\1\end{pmatrix} \longmapsto \begin{pmatrix}-\frac{2}{3}\\-\frac{2}{3}\\\frac{1}{3}\end{pmatrix}$$

so
$$\mathbf{M} = \begin{pmatrix}\frac{1}{3} & -\frac{2}{3} & -\frac{2}{3}\\-\frac{2}{3} & \frac{1}{3} & -\frac{2}{3}\\-\frac{2}{3} & -\frac{2}{3} & \frac{1}{3}\end{pmatrix} = \frac{1}{3}\begin{pmatrix}1 & -2 & -2\\-2 & 1 & -2\\-2 & -2 & 1\end{pmatrix}$$

3) A mapping of three dimensional space is defined by the matrix

$$\mathbf{M} = \begin{pmatrix}0 & 0 & -1\\0 & 1 & 0\\1 & 0 & 0\end{pmatrix}$$

Interpret geometrically the mappings defined by $\mathbf{M}$ and by $\mathbf{M}^2$.

Under $\mathbf{M}$
$$\mathbf{i} \longmapsto \begin{pmatrix}0\\0\\1\end{pmatrix}, \quad \mathbf{j} \longmapsto \begin{pmatrix}0\\1\\0\end{pmatrix}, \quad \mathbf{k} \longmapsto \begin{pmatrix}-1\\0\\0\end{pmatrix},$$

i.e.

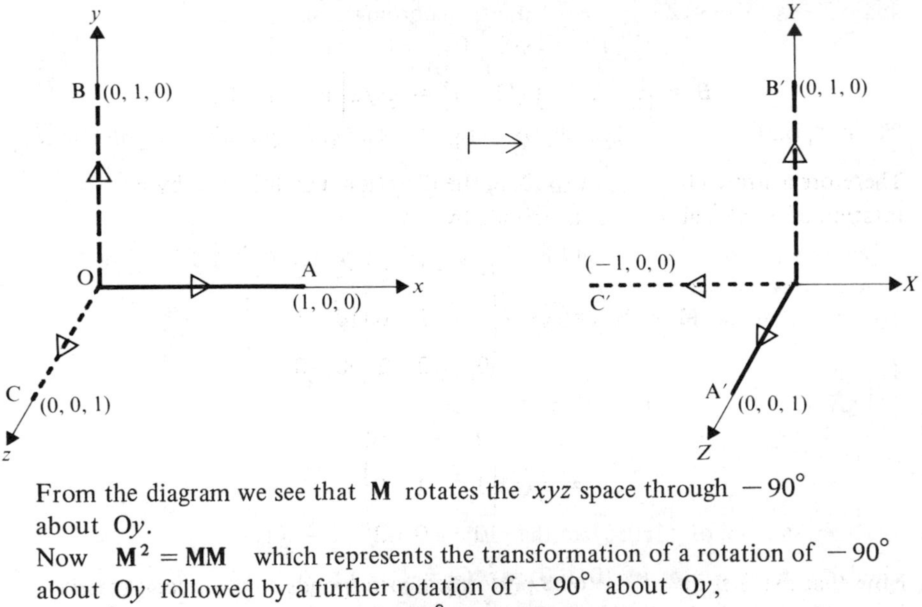

From the diagram we see that $\mathbf{M}$ rotates the xyz space through -90° about Oy.
Now $\mathbf{M}^2 = \mathbf{MM}$ which represents the transformation of a rotation of -90° about Oy followed by a further rotation of -90° about Oy,
i.e. $\mathbf{M}^2$ represents a rotation of 180° about Oy.

4) A transformation of three dimensional space is defined as: a stretch by a factor of 2 in the direction Ox, followed by a rotation of 45° about Oz. Find the matrix which defines this mapping.

The matrix defining a transformation of a stretch by a factor of 2 in the direction Ox is

$$\mathbf{A} = \begin{pmatrix} 2 & 0 & 0 \\ 0 & 1 & 0 \\ 0 & 0 & 1 \end{pmatrix}$$

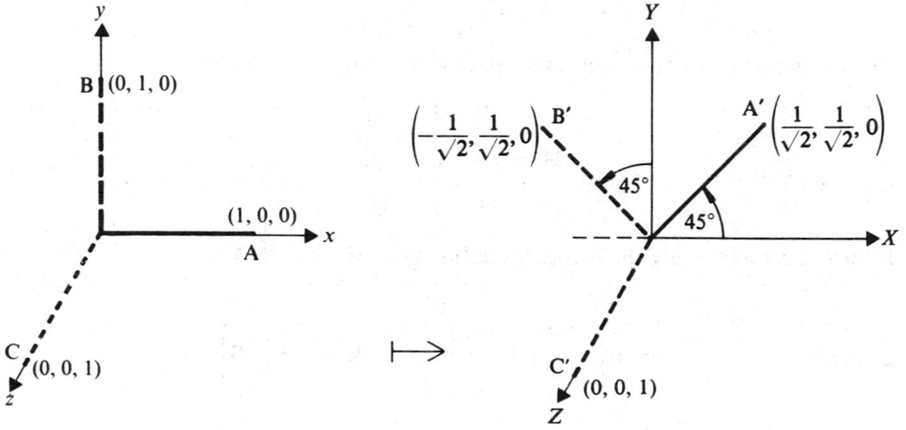

and the matrix defining a rotation of -45° about Oz is

$$\mathbf{B} = \begin{pmatrix} \frac{1}{2}\sqrt{2} & -\frac{1}{2}\sqrt{2} & 0 \\ \frac{1}{2}\sqrt{2} & \frac{1}{2}\sqrt{2} & 0 \\ 0 & 0 & 1 \end{pmatrix} = \tfrac{1}{2}\sqrt{2}\begin{pmatrix} 1 & -1 & 0 \\ 1 & 1 & 0 \\ 0 & 0 & 2 \end{pmatrix}$$

Therefore a stretch by a factor of 2 in the direction Ox followed by a rotation of -45° about Oz is defined by

$$\mathbf{M} = \mathbf{BA} = \tfrac{1}{2}\sqrt{2}\begin{pmatrix} 1 & -1 & 0 \\ 1 & 1 & 0 \\ 0 & 0 & 2 \end{pmatrix}\begin{pmatrix} 2 & 0 & 0 \\ 0 & 1 & 0 \\ 0 & 0 & 1 \end{pmatrix}$$

$$= \tfrac{1}{2}\sqrt{2}\begin{pmatrix} 2 & -1 & 0 \\ 1 & 1 & 0 \\ 0 & 0 & 2 \end{pmatrix}$$

Note that the first operation to be performed corresponds to the second matrix, i.e. we 'read' from right to left.

EXERCISE 2c

1) Find the matrices which perform the following transformations of three dimensional space:

(a) a rotation of $\pi/6$ about Oy, (b) a rotation of $\pi/4$ about Ox,

(c) a reflection in the xy plane, (d) a reflection in the xz plane,

(e) a rotation of π about the line $x = y$, $z = 0$,

(f) a reflection in the plane $x - y - z = 0$,

(g) an enlargement by a factor of 2,

(h) a contraction by a factor of $\frac{1}{3}$,

(i) a stretch by a factor of 2 in the directions Ox and Oz.

2) Describe the effect on three dimensional space of the transformation defined by $\mathbf{M}$ where $\mathbf{M}$ is

(a) $\begin{pmatrix} 0 & 0 & 1 \\ 0 & 1 & 0 \\ 1 & 0 & 0 \end{pmatrix}$ (b) $\begin{pmatrix} -2 & 0 & 0 \\ 0 & -2 & 0 \\ 0 & 0 & 1 \end{pmatrix}$ (c) $\begin{pmatrix} 1 & 0 & 1 \\ 0 & 1 & 0 \\ 0 & 0 & 0 \end{pmatrix}$.

3) Using the transformations defined in Question 1, find the matrices which define the composite transformations

(a) (b) followed by (c) (b) (c) followed by (b)

(c) (g) followed by (a) (d) (e) followed by (i).

4) Describe the effects of the transformations defined by $\mathbf{AB}$, $\mathbf{AC}$ and $\mathbf{BA}$ where

$$\mathbf{A} = \begin{pmatrix} -1 & 0 & 0 \\ 0 & -1 & 0 \\ 0 & 0 & -1 \end{pmatrix}, \quad \mathbf{B} = \begin{pmatrix} 1 & 0 & 0 \\ 0 & 3 & 0 \\ 0 & 0 & 1 \end{pmatrix}, \quad \mathbf{C} = \begin{pmatrix} 2 & 0 & 0 \\ 0 & 2 & 0 \\ 0 & 0 & 2 \end{pmatrix}.$$

The Effect on Volume

The transformation defined by $\mathbf{M} = \begin{pmatrix} a_1 & b_1 & c_1 \\ a_2 & b_2 & c_2 \\ a_3 & b_3 & c_3 \end{pmatrix}$ maps a cube of unit volume in xyz space to a parallelepiped in the image space.

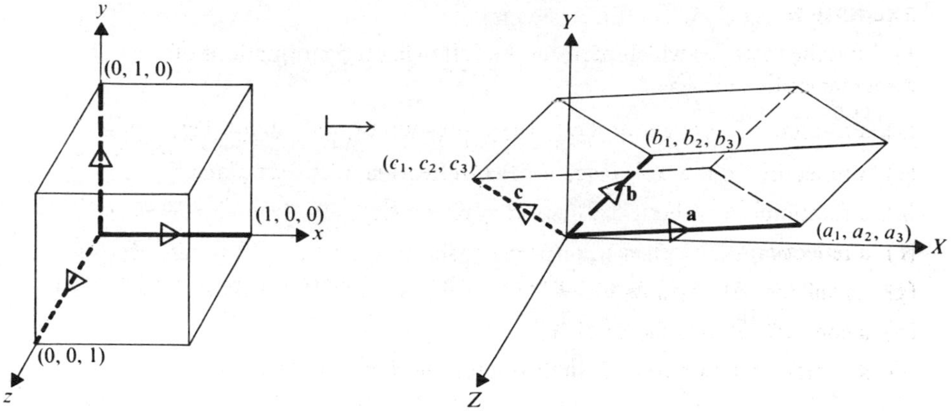

The volume of the parallelepiped is

$$\mathbf{a}\cdot\mathbf{b}\times\mathbf{c} = (a_1\mathbf{i}+a_2\mathbf{j}+a_3\mathbf{k})\cdot\begin{vmatrix} \mathbf{i} & \mathbf{j} & \mathbf{k} \\ b_1 & b_2 & b_3 \\ c_1 & c_2 & c_3 \end{vmatrix}$$

$$= a_1\begin{vmatrix} b_2 & b_3 \\ c_2 & c_3 \end{vmatrix} - a_2\begin{vmatrix} b_1 & b_3 \\ c_1 & c_3 \end{vmatrix} + a_3\begin{vmatrix} b_1 & b_2 \\ c_1 & c_2 \end{vmatrix}$$

$$= \begin{vmatrix} a_1 & a_2 & a_3 \\ b_1 & b_2 & b_3 \\ c_1 & c_2 & c_3 \end{vmatrix} = \begin{vmatrix} a_1 & b_1 & c_1 \\ a_2 & b_2 & c_2 \\ a_3 & b_3 & c_3 \end{vmatrix} = |\mathbf{M}|,$$

i.e. a volume of 1 cubic unit is mapped to a volume of $|\mathbf{M}|$ cubic units.

Hence $\mathbf{M}$ alters volume by a factor $|\mathbf{M}|$.

For example, if $\mathbf{A} = \begin{pmatrix} 1 & 0 & 2 \\ -1 & 1 & 4 \\ 0 & 0 & 4 \end{pmatrix}$, $|\mathbf{A}| = 4$

so that under the transformation defined by $\mathbf{A}$, a volume of V cubic units is mapped to a volume of $4V$ cubic units.

If $|\mathbf{M}| = 0$ then volume is destroyed and $\mathbf{M}$ is called singular,

i.e. the volume of the parallelepiped is zero, from which it follows that **a**, **b** and **c** are coplanar.

Thus if **a, b** and **c** are any three vectors

$$|\mathbf{M}| = 0 \iff \mathbf{a}, \mathbf{b} \text{ and } \mathbf{c} \text{ are coplanar}$$

and conversely

$$|\mathbf{M}| \neq 0 \iff \mathbf{a}, \mathbf{b} \text{ and } \mathbf{c} \text{ are not coplanar and so provide a set of basis vectors for three dimensional space.}$$

Now, as **M** maps three dimensional *xyz* space to the space whose base vectors are **a, b** and **c**, when **M** is singular the entire *xyz* space is mapped to the set of points lying in the plane of **a, b** and **c.**

If, further, **a, b** and **c** are parallel, **M** maps three dimensional space to a line, as in this case $\mathbf{a} = \lambda\mathbf{b} = \mu\mathbf{c}$, so any point (x, y, z) has an image point whose position vector, **r**, is $\mathbf{r} = s\mathbf{a}$.
In the exceptional case when $\mathbf{a} = \mathbf{b} = \mathbf{c} = \mathbf{0}$, **M** maps all points to the origin.

So $|\mathbf{M}| = 0 \iff$ **M** collapses three dimensional space to a plane or a line or a point.

Note that, as the origin is invariant under any transformation of three dimensional space defined by a 3×3 matrix, when three dimensional space is mapped to a plane or a line, the plane or line concerned contains the origin.

EXAMPLES 2d

1) Describe the transformation defined by $\mathbf{M} = \begin{pmatrix} 1 & -1 & 2 \\ 2 & -2 & 4 \\ 1 & -1 & 2 \end{pmatrix}$.

If $\mathbf{i} \longmapsto \mathbf{a}$, $\mathbf{j} \longmapsto \mathbf{b}$ and $\mathbf{k} \longmapsto \mathbf{c}$ then, from **M**, we see that

$$\mathbf{a} = \begin{pmatrix} 1 \\ 2 \\ 1 \end{pmatrix}, \quad \mathbf{b} = \begin{pmatrix} -1 \\ -2 \\ -1 \end{pmatrix} = -\mathbf{a}, \quad \mathbf{c} = \begin{pmatrix} 2 \\ 4 \\ 2 \end{pmatrix} = 2\mathbf{a}$$

So **a, b** and **c** are parallel, showing that **M** maps *xyz* space to a line.
Now any point (x, y, z) is mapped to the point (X, Y, Z) where

$$\mathbf{M}\begin{pmatrix} x \\ y \\ z \end{pmatrix} = \begin{pmatrix} X \\ Y \\ Z \end{pmatrix} \Rightarrow \begin{cases} x - y + 2z = X \\ 2x - 2y + 4z = Y \\ x - y + 2z = Z \end{cases}$$

$$\Rightarrow 2X = Y = 2Z$$

and these are the equations of a line.
So all points in *xyz* space map to points on this line.

2) Describe the transformation which maps (x, y, z) to (X, Y, Z) by the relation

$$\begin{pmatrix} 1 & 2 & -1 \\ 3 & 4 & -5 \\ 5 & 8 & -7 \end{pmatrix}\begin{pmatrix} x \\ y \\ z \end{pmatrix} = \begin{pmatrix} X \\ Y \\ Z \end{pmatrix}$$

From the matrix, $\mathbf{M}$, of the transformation, we see that $\mathbf{i} \longmapsto \mathbf{a}, \mathbf{j} \longmapsto \mathbf{b}$, $\mathbf{k} \longmapsto \mathbf{c}$, where

$$\mathbf{a} = \begin{pmatrix} 1 \\ 3 \\ 5 \end{pmatrix}, \quad \mathbf{b} = \begin{pmatrix} 2 \\ 4 \\ 8 \end{pmatrix}, \quad \mathbf{c} = \begin{pmatrix} -1 \\ -5 \\ -7 \end{pmatrix}$$

As $\mathbf{a}, \mathbf{b}$ and $\mathbf{c}$ do not have equal direction ratios, they are not parallel, so $\mathbf{M}$ does not map xyz space to a line.

However
$$|\mathbf{M}| = \begin{vmatrix} 1 & 2 & -1 \\ 3 & 4 & -5 \\ 5 & 8 & -7 \end{vmatrix} = \begin{vmatrix} 1 & 2 & -1 \\ 3 & 4 & -5 \\ 1 & 2 & -1 \end{vmatrix} = 0$$

So $\mathbf{M}$ maps three dimensional space to a plane.
To find the equation of this plane we can use the facts that
(a) the plane contains the origin,
(b) $\mathbf{a}$ and $\mathbf{b}$ (or $\mathbf{a}$ and $\mathbf{c}$, or $\mathbf{b}$ and $\mathbf{c}$) are the position vectors of two points in the plane.

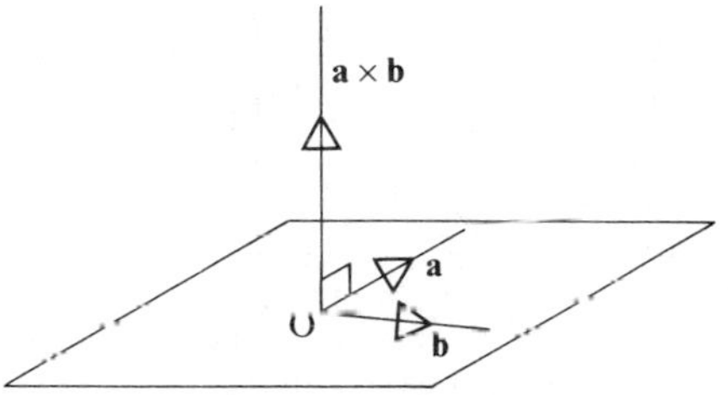

Now $\mathbf{a} \times \mathbf{b}$ is a normal to the plane and

$$\mathbf{a} \times \mathbf{b} = \begin{vmatrix} \mathbf{i} & \mathbf{j} & \mathbf{k} \\ 1 & 3 & 5 \\ 2 & 4 & 8 \end{vmatrix} = 4\mathbf{i} + 2\mathbf{j} - 2\mathbf{k}.$$

So the vector equation of the plane is $\quad \mathbf{r}.(2\mathbf{i} + \mathbf{j} - \mathbf{k}) = 0$
and the Cartesian equation of the plane is $\quad 2x + y - z = 0$.

EXERCISE 2d

1) Find the effect on volume of the transformations defined by the following matrices:

(a) $\begin{pmatrix} 2 & -1 & 1 \\ 0 & 1 & 4 \\ 0 & -2 & 0 \end{pmatrix}$ (b) $\begin{pmatrix} -1 & 3 & -1 \\ 0 & 1 & 2 \\ -1 & 0 & 1 \end{pmatrix}$ (c) $\begin{pmatrix} 0 & 0 & 1 \\ 0 & 1 & 0 \\ 1 & 0 & 0 \end{pmatrix}$ (d) $\begin{pmatrix} 2 & -1 & 4 \\ 3 & -2 & 5 \\ 4 & 1 & 2 \end{pmatrix}$.

2) Show, in each of the following cases, that the transformation defined by the given matrix maps three dimensional space to a plane and find the Cartesian equation of the plane:

(a) $\begin{pmatrix} 2 & 4 & 6 \\ 1 & -2 & 1 \\ 3 & 2 & 7 \end{pmatrix}$ (b) $\begin{pmatrix} 2 & 0 & 4 \\ 0 & 1 & -1 \\ 1 & 2 & 0 \end{pmatrix}$ (c) $\begin{pmatrix} 3 & -1 & 5 \\ -1 & 2 & 4 \\ 0 & 0 & 0 \end{pmatrix}$ (d) $\begin{pmatrix} 6 & 3 & 4 \\ 2 & 12 & 5 \\ 4 & 9 & 5 \end{pmatrix}$.

3) Show that **M** maps three dimensional space to a line and find the Cartesian equations of the line, where **M** is

(a) $\begin{pmatrix} 2 & 1 & 3 \\ 4 & 2 & 6 \\ -2 & -1 & -3 \end{pmatrix}$ (b) $\begin{pmatrix} 5 & -5 & 10 \\ -2 & 2 & -4 \\ 1 & -1 & 2 \end{pmatrix}$ (c) $\begin{pmatrix} 0 & 0 & 0 \\ 1 & 3 & -2 \\ 0 & 0 & 0 \end{pmatrix}$

(d) $\begin{pmatrix} 1 & 2 & -1 \\ 3 & 6 & -3 \\ -1 & -2 & 1 \end{pmatrix}$.

4) Describe the effect on three dimensional space of the transformation represented by **M**, where **M** is

(a) $\begin{pmatrix} 3 & -1 & 5 \\ 2 & 3 & -3 \\ 5 & 2 & 2 \end{pmatrix}$ (b) $\begin{pmatrix} 6 & 5 & 1 \\ 2 & -3 & -2 \\ -4 & 8 & 5 \end{pmatrix}$ (c) $\begin{pmatrix} 0 & 1 & 1 \\ 1 & 0 & 1 \\ 1 & 1 & 0 \end{pmatrix}$

(d) $\begin{pmatrix} 1 & -2 & 1 \\ -3 & 6 & -3 \\ 2 & -4 & 2 \end{pmatrix}$ (e) $\begin{pmatrix} 0 & 0 & 0 \\ 0 & 0 & 0 \\ 0 & 0 & 0 \end{pmatrix}$ (f) $\begin{pmatrix} 0 & 0 & 1 \\ 0 & 0 & 0 \\ 0 & 0 & 0 \end{pmatrix}$.

5) Find the values of λ for which the vectors $\mathbf{a} = \begin{pmatrix} 1 \\ 2 \\ \lambda \end{pmatrix}$, $\mathbf{b} = \begin{pmatrix} \lambda \\ 2 \\ -1 \end{pmatrix}$,

$\mathbf{c} = \begin{pmatrix} 0 \\ 1 \\ \lambda \end{pmatrix}$ are coplanar.

6) Determine which of the following sets of vectors provide a basis set for three dimensions:

(a) $\begin{pmatrix}1\\-1\\2\end{pmatrix}, \begin{pmatrix}3\\5\\-2\end{pmatrix}, \begin{pmatrix}1\\0\\1\end{pmatrix}$ (b) $\begin{pmatrix}-1\\4\\6\end{pmatrix}, \begin{pmatrix}2\\-1\\3\end{pmatrix}, \begin{pmatrix}3\\-5\\-3\end{pmatrix}$ (c) $\begin{pmatrix}4\\1\\2\end{pmatrix}, \begin{pmatrix}-1\\4\\0\end{pmatrix}, \begin{pmatrix}2\\1\\1\end{pmatrix}$.

INVERSE TRANSFORMATIONS IN TWO DIMENSIONS

Consider the transformation which rotates the xy plane through 90° about the origin.

The matrix for this transformation is $\mathbf{A} = \begin{pmatrix}0 & -1\\1 & 0\end{pmatrix}$.

Suppose that $\mathbf{A}$ maps a point (x, y) to the point $(2,3)$,

i.e. $$\begin{pmatrix}0 & -1\\1 & 0\end{pmatrix}\begin{pmatrix}x\\y\end{pmatrix} = \begin{pmatrix}2\\3\end{pmatrix} \qquad [1]$$

If we want to find the point (x, y) from which its image, $(2, 3)$, originates, we must reverse the transformation. In this case we must rotate the plane through $-90°$ about O to map $(2,3)$ back to (x, y). The matrix for this transformation is $\begin{pmatrix}0 & 1\\-1 & 0\end{pmatrix}$

so $$\begin{pmatrix}x\\y\end{pmatrix} = \begin{pmatrix}0 & 1\\-1 & 0\end{pmatrix}\begin{pmatrix}2\\3\end{pmatrix} \qquad [2]$$

This transformation is called the *inverse* of the original transformation.
Note that the matrices that define the original transformation and its inverse are such that

$$\begin{pmatrix}0 & 1\\-1 & 0\end{pmatrix}\begin{pmatrix}0 & -1\\1 & 0\end{pmatrix} = \begin{pmatrix}1 & 0\\0 & 1\end{pmatrix} = \mathbf{I}$$

In general, if a transformation defined by a matrix $\mathbf{A}$ maps $\begin{pmatrix}x\\y\end{pmatrix}$ to $\begin{pmatrix}X\\Y\end{pmatrix}$, the transformation which maps $\begin{pmatrix}X\\Y\end{pmatrix}$ back to $\begin{pmatrix}x\\y\end{pmatrix}$ is called the inverse transformation and the matrix which defines it is called the inverse of $\mathbf{A}$ and is denoted by $\mathbf{A}^{-1}$.

It follows from this that the composite transformation defined by $\mathbf{A}^{-1}\mathbf{A}$ first maps $\begin{pmatrix} x \\ y \end{pmatrix}$ to $\begin{pmatrix} X \\ Y \end{pmatrix}$ and then maps $\begin{pmatrix} X \\ Y \end{pmatrix}$ to $\begin{pmatrix} x \\ y \end{pmatrix}$

i.e. $\mathbf{A}^{-1}\mathbf{A}$ maps $\begin{pmatrix} x \\ y \end{pmatrix}$ to $\begin{pmatrix} x \\ y \end{pmatrix}$ $\Rightarrow$ $\mathbf{A}^{-1}\mathbf{A} = \mathbf{I}$

Similarly $\mathbf{A}\mathbf{A}^{-1}$ maps $\begin{pmatrix} X \\ Y \end{pmatrix}$ to $\begin{pmatrix} X \\ Y \end{pmatrix}$ $\Rightarrow$ $\mathbf{A}\mathbf{A}^{-1} = \mathbf{I}$

Hence $$\mathbf{A}\mathbf{A}^{-1} = \mathbf{A}^{-1}\mathbf{A} = \mathbf{I}$$

Now $\mathbf{A}^{-1}\mathbf{A} = \mathbf{I} \Rightarrow |\mathbf{A}^{-1}\mathbf{A}| = 1$ so if $\mathbf{A}$ changes a given area by a factor λ, then $\mathbf{A}^{-1}$ must restore the transformed area to its original size,
i.e. $\mathbf{A}^{-1}$ changes area by a factor $1/\lambda$.
So if $|\mathbf{A}| = \lambda$ then $|\mathbf{A}^{-1}| = 1/\lambda$.
But if $\lambda = 0$, $|\mathbf{A}^{-1}|$ does not exist so clearly we cannot reverse the transformation defined by $\mathbf{A}$.
i.e. if $|\mathbf{A}| = 0$, $\mathbf{A}$ has no inverse.

THE INVERSE OF A 2×2 MATRIX

If $\mathbf{M} = \begin{pmatrix} a & b \\ c & d \end{pmatrix}$ and $\mathbf{M}\begin{pmatrix} x \\ y \end{pmatrix} = \begin{pmatrix} X \\ Y \end{pmatrix}$,

i.e. $$\begin{pmatrix} a & b \\ c & d \end{pmatrix}\begin{pmatrix} x \\ y \end{pmatrix} = \begin{pmatrix} X \\ Y \end{pmatrix}$$

then $$\left.\begin{array}{l} ax + by = X \\ cx + dy = Y \end{array}\right\} \Rightarrow \begin{cases} x = (Xd - Yb)/(ad - bc) & [1] \\ y = (-Xc + Ya)/(ad - bc). & [2] \end{cases}$$

Noting that $ad - bc = |\mathbf{M}|$, we can write equations [1] and [2] in the form

$$x = \frac{1}{|\mathbf{M}|}\begin{pmatrix} d & -b \end{pmatrix}\begin{pmatrix} X \\ Y \end{pmatrix}, \quad y = \frac{1}{|\mathbf{M}|}\begin{pmatrix} -c & a \end{pmatrix}\begin{pmatrix} X \\ Y \end{pmatrix}$$

which may be adjoined to give

$$\begin{pmatrix} x \\ y \end{pmatrix} = \frac{1}{|\mathbf{M}|}\begin{pmatrix} d & -b \\ -c & a \end{pmatrix}\begin{pmatrix} X \\ Y \end{pmatrix}$$

i.e. $$\text{if} \quad \mathbf{M} = \begin{pmatrix} a & b \\ c & d \end{pmatrix} \quad \text{then} \quad \mathbf{M}^{-1} = \frac{1}{|\mathbf{M}|}\begin{pmatrix} d & -b \\ -c & a \end{pmatrix}$$

and we see that if $|\mathbf{M}| = 0$, $\mathbf{M}$ has no inverse
and the transformation defined by $\mathbf{M}$ has no inverse.

Note that $\mathbf{M}^{-1}$ is obtained from $\mathbf{M}$ by transposing the entries in the major diagonal, changing the sign of the entries in the minor diagonal and then dividing the resulting matrix by $|\mathbf{M}|$. The steps in this operation are best expressed in a flow diagram:

$\mathbf{M} = \begin{pmatrix} a & b \\ c & d \end{pmatrix}$

↓

Transpose entries in leading diagonal to give $\begin{pmatrix} d & b \\ c & a \end{pmatrix}$

↓

Change signs of entries in minor diagonal to give $\begin{pmatrix} d & -b \\ -c & a \end{pmatrix}$

↓

Divide by $|\mathbf{M}|$

↓

$\mathbf{M}^{-1} = \dfrac{1}{|\mathbf{M}|}\begin{pmatrix} d & -b \\ -c & a \end{pmatrix}$

For example, if

$$\mathbf{A} = \begin{pmatrix} 2 & 1 \\ 3 & 2 \end{pmatrix}, \quad |\mathbf{A}| = 1$$

so $$\mathbf{A}^{-1} = \begin{pmatrix} 2 & -1 \\ -3 & 2 \end{pmatrix}$$

and if $$\mathbf{B} = \begin{pmatrix} 2 & 5 \\ -1 & 4 \end{pmatrix}, \quad \mathbf{B}^{-1} = \tfrac{1}{13}\begin{pmatrix} 4 & -5 \\ 1 & 2 \end{pmatrix} = \begin{pmatrix} \frac{4}{13} & -\frac{5}{13} \\ \frac{1}{13} & \frac{2}{13} \end{pmatrix}$$

But if $\mathbf{C} = \begin{pmatrix} 2 & 1 \\ 4 & 2 \end{pmatrix}$, $|\mathbf{C}| = 0$, so $\mathbf{C}^{-1}$ does not exist.

EXAMPLES 2e

1) Find the point whose image is $(3, -1)$ under the composite transformation of a clockwise rotation of 90° about O followed by a stretch in the direction of Ox by a factor of 2.

Now $\mathbf{A} = \begin{pmatrix} 2 & 0 \\ 0 & 1 \end{pmatrix}$ stretches the xy plane by a factor 2 in the direction Ox,

and $\mathbf{B} = \begin{pmatrix} 0 & 1 \\ -1 & 0 \end{pmatrix}$ rotates the xy plane clockwise through an angle of 90° about O.

So $\mathbf{AB}$ maps (x, y) to $(3, -1)$.

Hence $(\mathbf{AB})^{-1}$ maps $(3, -1)$ to (x, y).

Now $\mathbf{AB} = \begin{pmatrix} 0 & 2 \\ -1 & 0 \end{pmatrix}$ so $(\mathbf{AB})^{-1} = \begin{pmatrix} 0 & -1 \\ \frac{1}{2} & 0 \end{pmatrix}$

Therefore $$\begin{pmatrix} x \\ y \end{pmatrix} = \begin{pmatrix} 0 & -1 \\ \frac{1}{2} & 0 \end{pmatrix}\begin{pmatrix} 3 \\ -1 \end{pmatrix} = \begin{pmatrix} 1 \\ \frac{3}{2} \end{pmatrix}$$

Alternatively we can find (x, y) by considering only inverse transformations. To map $(3, -1)$ to (x, y) we must reverse the given transformation, i.e. first shrink the xy plane by a factor of 2 in the direction Ox, and this is performed by $\begin{pmatrix} \frac{1}{2} & 0 \\ 0 & 1 \end{pmatrix}$

and then rotate the xy plane anticlockwise through an angle of 90° about O, and this is performed by $\begin{pmatrix} 0 & -1 \\ 1 & 0 \end{pmatrix}$.

So $\begin{pmatrix} 0 & -1 \\ 1 & 0 \end{pmatrix}\begin{pmatrix} \frac{1}{2} & 0 \\ 0 & 1 \end{pmatrix} = \begin{pmatrix} 0 & -1 \\ \frac{1}{2} & 0 \end{pmatrix}$ maps $\begin{pmatrix} 3 \\ -1 \end{pmatrix}$ to $\begin{pmatrix} x \\ y \end{pmatrix}$.

Comparing these two methods it is interesting to note that $\begin{pmatrix} 0 & -1 \\ 1 & 0 \end{pmatrix} = \mathbf{B}^{-1}$ and $\begin{pmatrix} 0 & -1 \\ \frac{1}{2} & 0 \end{pmatrix} = \mathbf{A}^{-1}$.

So $$(\mathbf{AB})^{-1} = \mathbf{B}^{-1}\mathbf{A}^{-1}$$

This is a result that is generally true as we shall show in the next section.

2) Find the image of the line $y = 2x + 1$ under the transformation of the xy plane defined by the relation

$$\begin{pmatrix} -2 & 1 \\ 3 & -1 \end{pmatrix}\begin{pmatrix} x \\ y \end{pmatrix} = \begin{pmatrix} X \\ Y \end{pmatrix}$$

If $\mathbf{A} = \begin{pmatrix} -2 & 1 \\ 3 & -1 \end{pmatrix}$ then $\mathbf{A}^{-1} = (-1)\begin{pmatrix} -1 & -1 \\ -3 & -2 \end{pmatrix} = \begin{pmatrix} 1 & 1 \\ 3 & 2 \end{pmatrix}$

so $\begin{pmatrix} x \\ y \end{pmatrix} = \begin{pmatrix} 1 & 1 \\ 3 & 2 \end{pmatrix}\begin{pmatrix} X \\ Y \end{pmatrix} \Rightarrow \begin{cases} x = X + Y \\ y = 3X + 2Y \end{cases}$

Hence the line $y = 2x + 1$ is mapped by $\mathbf{A}$ to the line

$$(3X + 2Y) = 2(X + Y) + 1, \quad \text{i.e. to} \quad X = 1$$

EXERCISE 2e

1) Find, where it exists, the inverse of

(a) $\begin{pmatrix} 2 & 0 \\ 0 & 1 \end{pmatrix}$ (b) $\begin{pmatrix} 3 & -1 \\ 4 & 2 \end{pmatrix}$ (c) $\begin{pmatrix} 3 & -1 \\ -6 & 2 \end{pmatrix}$ (d) $\begin{pmatrix} 0 & 1 \\ 0 & 2 \end{pmatrix}$

(e) $\begin{pmatrix} 5 & 1 \\ 2 & 1 \end{pmatrix}$ (f) $\begin{pmatrix} -2 & 4 \\ -1 & -3 \end{pmatrix}$ (g) $\begin{pmatrix} \sin\theta & \cos\theta \\ -\cos\theta & \sin\theta \end{pmatrix}$ (h) $\begin{pmatrix} p & q \\ r & s \end{pmatrix}$.

2) *Write down* the matrix which represents the inverse transformation of
(a) a rotation of θ about the origin,
(b) an enlargement by a factor of 2,
(c) a stretch by a factor of 3 parallel to Ox,
(d) a shear of $45°$ parallel to Oy.

3) Find the equations of the images of the lines
(a) $y = 3x + 2$, (b) $x - 2y + 4 = 0$

under the transformation which maps $\begin{pmatrix} x \\ y \end{pmatrix}$ to $\begin{pmatrix} 2 & -1 \\ 5 & 3 \end{pmatrix}\begin{pmatrix} x \\ y \end{pmatrix}$.

4) Show that the transformation which maps the point (x, y) to the point (X, Y) by the relation

$$\begin{pmatrix} 2 & 1 \\ 4 & 2 \end{pmatrix}\begin{pmatrix} x \\ y \end{pmatrix} = \begin{pmatrix} X \\ Y \end{pmatrix}$$

maps all points in the xy plane to points on a line in the image plane. Show also that all the points on the line $2x + y - 2 = 0$ map to the point $X = 2$, $Y = 4$.

5) A transformation of the xy plane is defined by the matrix $\begin{pmatrix} 2 & -4 \\ -1 & 2 \end{pmatrix}$.

Show that this transformation maps all points in the xy plane to a line and find the equation of this line. Show further that all the points lying on a particular line in the xy plane map to the point $(6, -3)$ and find the equation of that line.

INVERSE TRANSFORMATIONS IN THREE DIMENSIONS

If $\mathbf{M}$ is a 3×3 matrix that maps $\mathbf{r}$ to $\mathbf{r}'$

i.e. $$\mathbf{Mr} = \mathbf{r}'$$

then the inverse transformation, if it exists, maps $\mathbf{r}'$ to $\mathbf{r}$.
The inverse transformation is defined by the inverse matrix $\mathbf{M}^{-1}$,

i.e. $$\mathbf{M}^{-1}\mathbf{r}' = \mathbf{r}$$

Using similar arguments to those used for transformations of a plane, it follows that

$$\mathbf{M}^{-1} \text{ exists only if } |\mathbf{M}| \neq 0$$

and $$\mathbf{M}^{-1}\mathbf{M} = \mathbf{M}\mathbf{M}^{-1} = \mathbf{I}$$

Calculation of $\mathbf{M}^{-1}$

Suppose that $$\mathbf{M} = \begin{pmatrix} a_1 & a_2 & a_3 \\ b_1 & b_2 & b_3 \\ c_1 & c_2 & c_3 \end{pmatrix}$$

and that A_1 is the cofactor of a_1, A_2 is the cofactor of a_2, etc., then the inverse matrix $\mathbf{M}^{-1}$ is given by proceeding as follows.

$$\mathbf{M} = \begin{pmatrix} a_1 & a_2 & a_3 \\ b_1 & b_2 & b_3 \\ c_1 & c_2 & c_3 \end{pmatrix}$$

↓

Transpose rows and columns of $\mathbf{M}$ $\Rightarrow$ $\begin{pmatrix} a_1 & b_1 & c_1 \\ a_2 & b_2 & c_2 \\ a_3 & b_3 & c_3 \end{pmatrix}$ which is called the transpose matrix $\mathbf{M}^T$

↓

Replace elements by their cofactors $\Rightarrow$ $\begin{pmatrix} A_1 & B_1 & C_1 \\ A_2 & B_2 & C_2 \\ A_3 & B_3 & C_3 \end{pmatrix}$ which is called the adjoint or adjugate matrix, adj $\mathbf{M}$

↓

Divide by $|\mathbf{M}|$ $\Rightarrow$ $\mathbf{M}^{-1}$

i.e.

$$\mathbf{M}^{-1} = \frac{1}{|\mathbf{M}|}\,\text{adj}\,\mathbf{M}$$

That $\mathbf{M}^{-1}$ is $\dfrac{1}{|\mathbf{M}|}\,\text{adj}\,\mathbf{M}$ is shown in the appendix at the end of this chapter.

Note that, if $|\mathbf{M}| = 0$, $\mathbf{M}^{-1}$ does not exist.

Note also that a matrix of any size can be transposed, e.g.

$$\begin{pmatrix} 2 & -1 & 4 \\ 1 & 3 & 0 \end{pmatrix}^T = \begin{pmatrix} 2 & 1 \\ -1 & 3 \\ 4 & 0 \end{pmatrix}$$

but only a square matrix can have an inverse.

Thus to find the inverse of $\mathbf{A} = \begin{pmatrix} 1 & -2 & 0 \\ 3 & 1 & 5 \\ -1 & 2 & 3 \end{pmatrix}$ we proceed as follows.

$$\mathbf{A}^T = \begin{pmatrix} 1 & 3 & -1 \\ -2 & 1 & 2 \\ 0 & 5 & 3 \end{pmatrix}, \quad \text{adj}\,\mathbf{A} = \begin{pmatrix} -7 & 6 & -10 \\ -14 & 3 & -5 \\ 7 & 0 & 7 \end{pmatrix}, \quad |\mathbf{A}| = 21$$

Hence

$$\mathbf{A}^{-1} = \frac{1}{21}\begin{pmatrix} -7 & 6 & -10 \\ -14 & 3 & -5 \\ 7 & 0 & 7 \end{pmatrix}$$

Checking, we have

$$\mathbf{A}^{-1}\mathbf{A} = \frac{1}{21}\begin{pmatrix} -7 & 6 & -10 \\ -14 & 3 & -5 \\ 7 & 0 & 7 \end{pmatrix}\begin{pmatrix} 1 & -2 & 0 \\ 3 & 1 & 5 \\ -1 & 2 & 3 \end{pmatrix} = \frac{1}{21}\begin{pmatrix} 21 & 0 & 0 \\ 0 & 21 & 0 \\ 0 & 0 & 21 \end{pmatrix} = \mathbf{I}$$

Note that this check on the calculation of $\mathbf{A}^{-1}$ is well worthwhile, because of the high probability of arithmetic mistakes.

Note also that later in this chapter, we give another method for calculating an inverse matrix, which is often shorter.

PROPERTIES OF INVERSE AND TRANSPOSE MATRICES

1) $$\mathbf{A}\mathbf{A}^{-1} = \mathbf{A}^{-1}\mathbf{A} = \mathbf{I}$$

By considering transformations we demonstrated on page 86 that this property is true.

2) If $\mathbf{AB} = \mathbf{I}$ then $\mathbf{B} = \mathbf{A}^{-1}$ and $\mathbf{A} = \mathbf{B}^{-1}$.

Proof: Premultiplying $\mathbf{AB} = \mathbf{I}$ by $\mathbf{A}^{-1}$ gives

$$(\mathbf{A}^{-1}\mathbf{A})\mathbf{B} = \mathbf{A}^{-1} \quad \Rightarrow \quad \mathbf{B} = \mathbf{A}^{-1}$$

Similarly postmultiplying $\mathbf{AB} = \mathbf{I}$ by $\mathbf{B}^{-1}$ gives

$$\mathbf{A}(\mathbf{BB}^{-1}) = \mathbf{B}^{-1} \quad \Rightarrow \quad \mathbf{A} = \mathbf{B}^{-1}$$

Note also that $\mathbf{AB} = \mathbf{I} \Rightarrow |\mathbf{A}||\mathbf{B}| = 1 \Rightarrow |\mathbf{A}| \neq 0$ and $|\mathbf{B}| \neq 0$.

3) $$(\mathbf{A}^{-1})^{-1} = \mathbf{A}$$

Proof: As $\mathbf{A}^{-1}\mathbf{A} = \mathbf{AA}^{-1}$, $\mathbf{A}$ is the inverse of $\mathbf{A}^{-1}$,

i.e. $$(\mathbf{A}^{-1})^{-1} = \mathbf{A}$$

4) $$(\mathbf{AB})^{-1} = \mathbf{B}^{-1}\mathbf{A}^{-1}$$

Proof: Multiplying $\mathbf{B}^{-1}\mathbf{A}^{-1}$ on the left by $\mathbf{AB}$ gives

$$(\mathbf{AB})(\mathbf{B}^{-1}\mathbf{A}^{-1}) = \mathbf{A}(\mathbf{BB}^{-1})\mathbf{A}^{-1} = \mathbf{AIA}^{-1} = \mathbf{AA}^{-1} = \mathbf{I},$$

i.e. the inverse of $\mathbf{AB}$ is $\mathbf{B}^{-1}\mathbf{A}^{-1}$,
hence $(\mathbf{AB})^{-1} = \mathbf{B}^{-1}\mathbf{A}^{-1}$.

5) $$(\mathbf{AB})^T = \mathbf{B}^T\mathbf{A}^T$$

We will demonstrate that this property is true for a pair of 3×3 matrices. A general proof of this property appears in Chapter 15.

If $\mathbf{A}$ has *rows* $\mathbf{a}_1, \mathbf{a}_2, \mathbf{a}_3$,

i.e. $$\mathbf{A} = \begin{pmatrix} \mathbf{a}_1 \\ \mathbf{a}_2 \\ \mathbf{a}_3 \end{pmatrix}, \quad \text{then} \quad \mathbf{A}^T = \begin{pmatrix} \mathbf{a}_1 & \mathbf{a}_2 & \mathbf{a}_3 \end{pmatrix}$$

and if $\mathbf{B}$ has *columns* $\mathbf{b}_1, \mathbf{b}_2, \mathbf{b}_3$,

i.e. $$\mathbf{B} = \begin{pmatrix} \mathbf{b}_1 & \mathbf{b}_2 & \mathbf{b}_3 \end{pmatrix}, \quad \text{then} \quad \mathbf{B}^T = \begin{pmatrix} \mathbf{b}_1 \\ \mathbf{b}_2 \\ \mathbf{b}_3 \end{pmatrix}$$

Now $$\mathbf{AB} = \begin{pmatrix} \mathbf{a}_1 \\ \mathbf{a}_2 \\ \mathbf{a}_3 \end{pmatrix} \begin{pmatrix} \mathbf{b}_1 & \mathbf{b}_2 & \mathbf{b}_3 \end{pmatrix} = \begin{pmatrix} \mathbf{a}_1 \cdot \mathbf{b}_1 & \mathbf{a}_1 \cdot \mathbf{b}_2 & \mathbf{a}_1 \cdot \mathbf{b}_3 \\ \mathbf{a}_2 \cdot \mathbf{b}_1 & \mathbf{a}_2 \cdot \mathbf{b}_2 & \mathbf{a}_2 \cdot \mathbf{b}_3 \\ \mathbf{a}_3 \cdot \mathbf{b}_1 & \mathbf{a}_3 \cdot \mathbf{b}_2 & \mathbf{a}_3 \cdot \mathbf{b}_3 \end{pmatrix}$$

and

$$\mathbf{B}^T\mathbf{A}^T = \begin{pmatrix} \mathbf{b}_1 \\ \mathbf{b}_2 \\ \mathbf{b}_3 \end{pmatrix} \begin{pmatrix} \mathbf{a}_1 & \mathbf{a}_2 & \mathbf{a}_3 \end{pmatrix} = \begin{pmatrix} \mathbf{b}_1 \cdot \mathbf{a}_1 & \mathbf{b}_1 \cdot \mathbf{a}_2 & \mathbf{b}_1 \cdot \mathbf{a}_3 \\ \mathbf{b}_2 \cdot \mathbf{a}_1 & \mathbf{b}_2 \cdot \mathbf{a}_2 & \mathbf{b}_2 \cdot \mathbf{a}_3 \\ \mathbf{b}_3 \cdot \mathbf{a}_1 & \mathbf{b}_3 \cdot \mathbf{a}_2 & \mathbf{b}_3 \cdot \mathbf{a}_3 \end{pmatrix} = (\mathbf{AB})^T$$

i.e. $(\mathbf{AB})^T = \mathbf{B}^T\mathbf{A}^T$.

EXERCISE 2f

For each of the matrices given in Questions 1–5, find (a) the transpose matrix, (b) the adjoint matrix, (c) where it exists, the inverse matrix.

1) $\mathbf{A} = \begin{pmatrix} 1 & 2 & 0 \\ -1 & 1 & 0 \\ 2 & 5 & 1 \end{pmatrix}$ 2) $\mathbf{B} = \begin{pmatrix} 2 & 3 & 1 \\ 5 & 3 & 4 \\ -1 & 2 & 5 \end{pmatrix}$

3) $\mathbf{C} = \begin{pmatrix} 1 & -1 & 3 \\ 2 & 0 & 4 \\ 6 & -2 & 22 \end{pmatrix}$ 4) $\mathbf{D} = \begin{pmatrix} -1 & 0 & 4 \\ 0 & 5 & -2 \\ 1 & 4 & -1 \end{pmatrix}$

5) $\mathbf{E} = \begin{pmatrix} 2 & -1 & 4 \\ 3 & 2 & -1 \\ 5 & -2 & 9 \end{pmatrix}$

6) For the matrices **A** and **B** given above, confirm that $(\mathbf{AB})^T = \mathbf{B}^T\mathbf{A}^T$.

7) For the matrices **B** and **D** given above, confirm that $(\mathbf{BD})^{-1} = \mathbf{D}^{-1}\mathbf{B}^{-1}$.

8) Using the matrix **A** given above, show that $(\mathbf{A}^{-1})^{-1} = \mathbf{A}$.

9) If $\mathbf{A} = \begin{pmatrix} 2 & 1 & 4 \\ 3 & -1 & 0 \end{pmatrix}$ and $\mathbf{B} = \begin{pmatrix} 4 & 1 \\ 0 & 3 \\ 1 & 0 \end{pmatrix}$

write down the transpose of **A** and of **B**.
Calculate the products **AB** and $\mathbf{B}^T\mathbf{A}^T$ and use your results to confirm that

$$(\mathbf{AB})^T = \mathbf{B}^T\mathbf{A}^T.$$

10) Prove, for three nonsingular matrices, **A**, **B**, **C** that

$$(\mathbf{ABC})^{-1} = \mathbf{C}^{-1}\mathbf{B}^{-1}\mathbf{A}^{-1}.$$

11) Prove that $(\mathbf{A}^T)^{-1} = (\mathbf{A}^{-1})^T$.

12) If $\mathbf{A} = \begin{pmatrix} 1 & 2 \\ 3 & 4 \end{pmatrix}$, find $\mathbf{A}^{-1}$. Find the values of the real constants a and b such that $\mathbf{A} + a\mathbf{A}^{-1} = b\mathbf{I}$. Hence show that $\mathbf{A}$ satisfies the quadratic equation $\mathbf{A}^2 - b\mathbf{A} + a\mathbf{I} = \mathbf{0}$.

APPLICATIONS OF INVERSE MATRICES

We will now look at some problems where the calculation of an inverse matrix gives one method of solution.

EXAMPLES 2g

1) Find the coordinates of the point in the xyz space from which the point (2, 1, 5) in the XYZ space originates under the transformation given by

$$\begin{pmatrix} 2 & -1 & 4 \\ 1 & 0 & 0 \\ 1 & -2 & 0 \end{pmatrix}\begin{pmatrix} x \\ y \\ z \end{pmatrix} = \begin{pmatrix} X \\ Y \\ Z \end{pmatrix}$$

If (x_1, y_1, z_1) is the point from which (2, 1, 5) originates, then

$$\begin{pmatrix} 2 & -1 & 4 \\ 1 & 0 & 0 \\ 1 & -2 & 0 \end{pmatrix}\begin{pmatrix} x_1 \\ y_1 \\ z_1 \end{pmatrix} = \begin{pmatrix} 2 \\ 1 \\ 5 \end{pmatrix} \qquad [1]$$

If $\mathbf{A} = \begin{pmatrix} 2 & -1 & 4 \\ 1 & 0 & 0 \\ 1 & -2 & 0 \end{pmatrix}$, then $\mathbf{A}^{-1} = -\dfrac{1}{8}\begin{pmatrix} 0 & -8 & 0 \\ 0 & -4 & 4 \\ -2 & 3 & 1 \end{pmatrix}$

so $$\begin{pmatrix} x_1 \\ y_1 \\ z_1 \end{pmatrix} = \frac{1}{8}\begin{pmatrix} 0 & 8 & 0 \\ 0 & 4 & -4 \\ 2 & -3 & -1 \end{pmatrix}\begin{pmatrix} 2 \\ 1 \\ 5 \end{pmatrix} = \frac{1}{8}\begin{pmatrix} 8 \\ -16 \\ -4 \end{pmatrix}$$

$\Rightarrow$ $$x_1 = 1, \quad y_1 = -2, \quad z_1 = -\tfrac{1}{2}$$

Note that [1] is a matrix representation of the set of simultaneous equations

$$\begin{aligned} 2x_1 - y_1 + 4z_1 &= 2 \\ x_1 \qquad\qquad &= 1 \\ x_1 - 2y_1 \qquad &= 5 \end{aligned}$$

and that this set of equations can be solved by observation.
In this case this second method is quicker and more direct but it is not always so.

2) Find the image of the plane $x + 2y - 7z = 2$ under the transformation defined by $\begin{pmatrix} -1 & 2 & 1 \\ -3 & 1 & 4 \\ 0 & 1 & 2 \end{pmatrix}$.

If **M** maps $\begin{pmatrix} x \\ y \\ z \end{pmatrix}$ to $\begin{pmatrix} X \\ Y \\ Z \end{pmatrix}$ then $\mathbf{M}^{-1}$ maps $\begin{pmatrix} X \\ Y \\ Z \end{pmatrix}$ to $\begin{pmatrix} x \\ y \\ z \end{pmatrix}$

Now $$\mathbf{M}^{-1} = \frac{1}{11}\begin{pmatrix} -2 & -3 & 7 \\ 6 & -2 & 1 \\ -3 & 1 & 5 \end{pmatrix}$$

So $$\begin{pmatrix} x \\ y \\ z \end{pmatrix} = \frac{1}{11}\begin{pmatrix} -2 & -3 & 7 \\ 6 & -2 & 1 \\ -3 & 1 & 5 \end{pmatrix}\begin{pmatrix} X \\ Y \\ Z \end{pmatrix} = \frac{1}{11}\begin{pmatrix} -2X-3Y+7Z \\ 6X-2Y+Z \\ -3X+Y+5Z \end{pmatrix}$$

i.e. $$\begin{aligned} x &= \tfrac{1}{11}(-2X - 3Y + 7Z) \\ y &= \tfrac{1}{11}(6X - 2Y + Z) \\ z &= \tfrac{1}{11}(-3X + Y + 5Z) \end{aligned}$$

Hence the plane $x + 2y - 7z = 2$ maps to the plane

$$\tfrac{1}{11}(-2X - 3Y + 7Z) + \tfrac{2}{11}(6X - 2Y + Z) - \tfrac{7}{11}(-3X + Y + 5Z) = 2$$

$$\Rightarrow \qquad 31X - 14Y - 26Z = 22$$

Note that a normal to the given plane is $\begin{pmatrix} 1 \\ 2 \\ -7 \end{pmatrix}$ and, under **M**,

$$\begin{pmatrix}1\\2\\-7\end{pmatrix} \longmapsto \begin{pmatrix}-1 & 2 & 1\\-3 & 1 & 4\\0 & 1 & 2\end{pmatrix}\begin{pmatrix}1\\2\\-7\end{pmatrix} = \begin{pmatrix}-4\\-29\\-12\end{pmatrix}$$

But a normal to the image plane is $\begin{pmatrix}31\\-14\\-26\end{pmatrix}$,

i.e. in general, the normal to a plane is *not* mapped to the normal of the image plane.

However, any point in a plane *is* mapped to a point in the image plane and an alternative method for finding the image plane uses this fact.

For example, $A(2, 0, 0)$, $B(0, 1, 0)$, $C(0, 0, -\frac{2}{7})$ are points in the given plane. The images of these points under $\mathbf{M}$ are $A'(-2, -6, 0)$, $B'(2, 1, 1)$, $C'(-\frac{2}{7}, -\frac{8}{7}, -\frac{4}{7})$ respectively.

So the image of the given plane $\mathbf{M}$ is the plane containing A', B' and C', and we can find its equation from the general form $\mathbf{r.n} = \mathbf{a.n}$ noting that $\overrightarrow{A'B'} \times \overrightarrow{B'C'}$ is normal to the plane.

So the equation of the plane $A'B'C'$ is

$$\mathbf{r}.(\overrightarrow{A'B'} \times \overrightarrow{B'C'}) = \overrightarrow{OA'}.(\overrightarrow{A'B'} \times \overrightarrow{B'C'})$$

$$\Rightarrow \qquad \mathbf{r}.\begin{vmatrix}\mathbf{i} & \mathbf{j} & \mathbf{k}\\-4 & -7 & -1\\16 & 15 & 11\end{vmatrix} = (-2\mathbf{i}-6\mathbf{j}).\begin{vmatrix}\mathbf{i} & \mathbf{j} & \mathbf{k}\\-4 & -7 & -1\\16 & 15 & 11\end{vmatrix}$$

$$\Rightarrow \qquad \mathbf{r}.(62\mathbf{i}-28\mathbf{j}-52\mathbf{k}) = 44$$

$$\Rightarrow \qquad 31x - 14y - 26z = 22$$

There is little to choose between these two methods with regard to the calculation involved. However, if the image of more than one plane is required under a given transformation, the first method is preferable because once the inverse matrix is found, the relationship it provides between (x, y, z) and its image (X, Y, Z) can be used to transform as many planes as required.

3) Find the image of the line L with equations $\dfrac{x-2}{1} = \dfrac{y+1}{2} = \dfrac{z-1}{3}$

under the transformation defined by $\mathbf{A} = \begin{pmatrix} 2 & -1 & 1 \\ 0 & 1 & 0 \\ 1 & 0 & 2 \end{pmatrix}$

Any two points on L are transformed by **A** to two points on L′, the image of L.
Taking $(2, -1, 1)$ and $(3, 1, 4)$ as two points on L, the images of these two points are

$$\begin{pmatrix} 2 & -1 & 1 \\ 0 & 1 & 0 \\ 1 & 0 & 2 \end{pmatrix}\begin{pmatrix} 2 \\ -1 \\ 1 \end{pmatrix} = \begin{pmatrix} 6 \\ -1 \\ 4 \end{pmatrix} \quad \text{and} \quad \begin{pmatrix} 2 & -1 & 1 \\ 0 & 1 & 0 \\ 1 & 0 & 2 \end{pmatrix}\begin{pmatrix} 3 \\ 3 \\ 10 \end{pmatrix} = \begin{pmatrix} 9 \\ 1 \\ 11 \end{pmatrix}$$

Hence, using $\dfrac{x-x_1}{x_2-x_1} = \dfrac{y-y_1}{y_2-y_1} = \dfrac{z-z_1}{z_2-z_1}$, the equations of L′ are

$$\frac{x-6}{3} = \frac{y+1}{2} = \frac{z-4}{7}$$

4) Find the equations of the planes that map to themselves under the

transformation defined by $\mathbf{T} = \begin{pmatrix} -1 & 2 & 0 \\ 0 & -3 & 1 \\ 0 & 1 & 0 \end{pmatrix}$.

Under **T**, (x, y, z) is mapped to (X, Y, Z) where

$$\begin{pmatrix} 1 & 2 & 0 \\ 0 & -3 & 1 \\ 0 & 1 & 0 \end{pmatrix}\begin{pmatrix} x \\ y \\ z \end{pmatrix} = \begin{pmatrix} X \\ Y \\ Z \end{pmatrix}$$

Now $$\mathbf{T}^{-1} = \begin{pmatrix} -1 & 0 & 2 \\ 0 & 0 & 1 \\ 0 & 1 & 3 \end{pmatrix}$$

Hence
$$\begin{pmatrix} x \\ y \\ z \end{pmatrix} = \begin{pmatrix} -1 & 0 & 2 \\ 0 & 0 & 1 \\ 0 & 1 & 3 \end{pmatrix} \begin{pmatrix} X \\ Y \\ Z \end{pmatrix}$$

$$\Rightarrow \quad \begin{cases} x = -X + 2Z \\ y = Z \\ z = Y + 3Z \end{cases}$$

Hence any plane $ax + by + cz = D$

maps to $\quad a(-X + 2Z) + b(Z) + c(Y + 3Z) = D$

$$\Rightarrow \quad -aX + cY + (2a + b + 3c)Z = D$$

If these equations represent the same plane then comparing

$$x + \frac{b}{a}y + \frac{c}{a}z = \frac{D}{a}$$

with
$$X - \frac{c}{a}Y - \left(\frac{2a + b + 3c}{a}\right)Z = -\frac{D}{a}$$

we see that

$$\frac{b}{a} = -\frac{c}{a} \quad \Rightarrow \quad b = -c \qquad [1]$$

$$\frac{c}{a} = -\left(\frac{2a + b + 3c}{a}\right) \quad \Rightarrow \quad c = -2a - b - 3c \qquad [2]$$

$$\frac{D}{a} = -\frac{D}{a} \quad \Rightarrow \quad D = 0 \qquad [3]$$

From [1] and [2] we have $a:b:c = 3:2:-2$.
Hence there is only one plane that maps to itself, and its equation is

$$3x + 2y - 2z = 0$$

Note that, although the plane $3x + 2y - 2z = 0$ maps to the same plane $3X + 2Y - 2Z = 0$ in the image space, a particular point on the plane $3x + 2y - 2z = 0$ will *not*, in general, map to the same point on $3X + 2Y - 2Z = 0$,

e.g. $\begin{pmatrix} 2 \\ 1 \\ 4 \end{pmatrix}$ is on $3x + 2y - 2z = 0$

and $$\begin{pmatrix}2\\1\\4\end{pmatrix} \longmapsto \begin{pmatrix}-1 & 2 & 0\\0 & -3 & 1\\0 & 1 & 0\end{pmatrix}\begin{pmatrix}2\\1\\4\end{pmatrix} = \begin{pmatrix}0\\1\\1\end{pmatrix}.$$

Note also that it may appear, from the examples done so far, that planes and lines that map to themselves must necessarily pass through the origin. However this is not so. Consider, for example, a rotation about the line $x=y=z$. *Any* plane that is perpendicular to $x=y=z$ is unchanged. If further, the rotation is $180°$, any line that is perpendicular to, and intersects, $x=y=z$ is unchanged.

EXERCISE 2g

1) Find the coordinates of the point from which the point $(2, 1, 1)$ originates under the transformation defined by $\mathbf{M} = \begin{pmatrix}2 & -1 & 0\\3 & 1 & 2\\-1 & 1 & 0\end{pmatrix}$.

2) Find the inverse of the matrix $\begin{pmatrix}1 & -2 & 1\\3 & 1 & 2\\-1 & 4 & 1\end{pmatrix}$.

Hence solve the equations

$$\begin{aligned}x - 2y + z &= 1\\3x + y + 2z &= 4\\-x + 4y + z &= 2.\end{aligned}$$

3) If $\mathbf{A} = \begin{pmatrix}1 & 2 & 3\\-2 & 1 & 2\\4 & -1 & 0\end{pmatrix}$, find $\mathbf{A}^{-1}$.

Hence solve the equation $\mathbf{A}\begin{pmatrix}x\\y\\z\end{pmatrix} = \begin{pmatrix}2\\1\\3\end{pmatrix}$.

4) If $\mathbf{M}=\begin{pmatrix}2 & 1 & 4\\ 3 & 5 & 1\\ 1 & 2 & 0\end{pmatrix}$, find $\mathbf{M}^{-1}$.

Hence find the images, under the transformation $\mathbf{M}\begin{pmatrix}x\\ y\\ z\end{pmatrix}=\begin{pmatrix}X\\ Y\\ Z\end{pmatrix}$, of the planes

(a) $x+2y-z=4$, (b) $3x+5y+z=2$, (c) $-2x+y-3z=1$.

5) Find the images of the lines

(a) $\mathbf{r}=\mathbf{i}-\mathbf{j}+\lambda(\mathbf{i}+\mathbf{j}+\mathbf{k})$, (b) $\mathbf{r}=2\mathbf{i}-\mathbf{k}+\mu(3\mathbf{i}-\mathbf{j}+\mathbf{k})$,

under the transformation defined in Question 4, giving their equations in vector form.

6) Find the equations of the planes that map to themselves under the transformation defined by $\begin{pmatrix}1 & 2 & 0\\ 0 & 1 & -1\\ 0 & 2 & 1\end{pmatrix}$.

7) Find the equations of the lines that map to themselves under the transformation which maps $\begin{pmatrix}x\\ y\\ z\end{pmatrix}$ to $\begin{pmatrix}X\\ Y\\ Z\end{pmatrix}$ by the relation

$$\begin{pmatrix}2 & 0 & 0\\ 0 & 1 & 0\\ 0 & 0 & 2\end{pmatrix}\begin{pmatrix}x\\ y\\ z\end{pmatrix}=\begin{pmatrix}X\\ Y\\ Z\end{pmatrix}.$$

8) If $\mathbf{A}=\begin{pmatrix}0 & 1 & 0\\ 0 & 0 & -1\\ 1 & 0 & 0\end{pmatrix}$ interpret the mapping of xyz space represented by $\mathbf{A}^{-1}$ and find $\mathbf{A}^{-1}$. Find the values of the constant λ for which $\mathbf{A}^2+\lambda\mathbf{A}^{-1}=\mathbf{0}$. Find also the equations of the lines that map to themselves under the transformation defined by $\mathbf{A}^2$.

SYSTEMS OF LINEAR EQUATIONS

Two Variables

Consider the pair of equations

$$a_1x + b_1y + c_1 = 0 \qquad [1]$$

$$a_2x + b_2y + c_2 = 0 \qquad [2]$$

The solution set of this pair of equations is the complete set of values (x, y) that satisfy both equations. The nature of the solution set may be determined by interpreting equations [1] and [2] geometrically as the equations of two straight lines, L_1 and L_2, in the xy plane.
These lines may be such that

(a) L_1 and L_2 intersect, in which case there is only one point whose coordinates satisfy both equations, and
the equations have a unique solution.

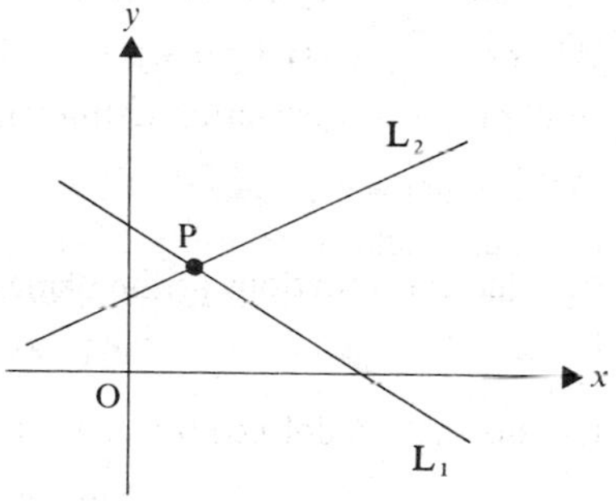

(b) L_1 and L_2 are parallel.
In this case there are no common points so
the equations have no solution.
Such a pair of equations are said to be *inconsistent*.

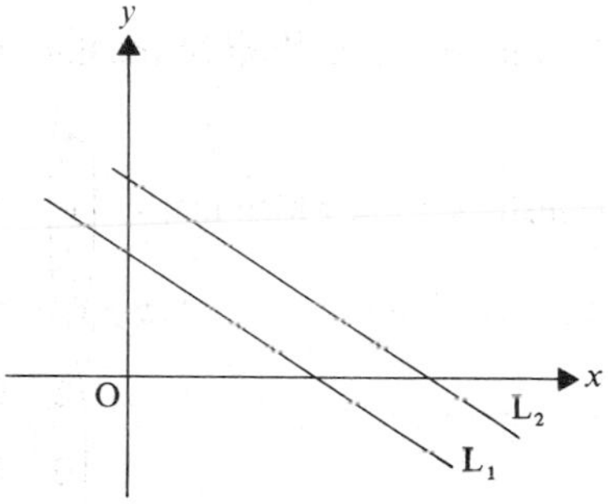

(c) L_1 and L_2 are the same line.
This time, all the points on the line are common to L_1 and L_2 so
the equations have an infinite set of solutions. Such equations are said to be *linearly dependent*.

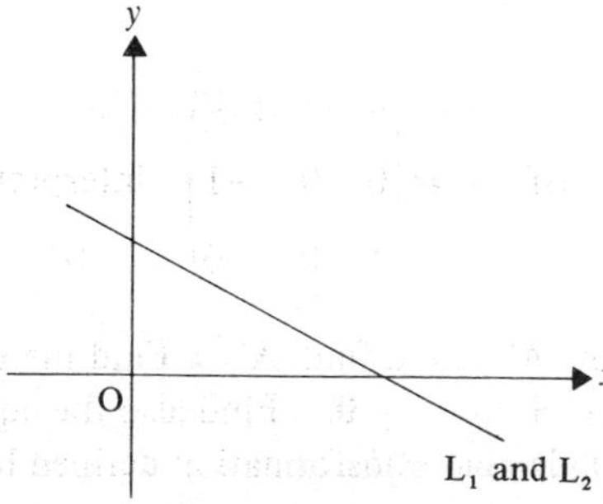

All these properties can easily be observed from the equations. For example

(a)
$$2x - 3y = 1$$
$$x + 4y = 2$$

The lines represented by these two equations are not parallel, so they intersect in one point,

i.e. this pair of equations has a unique solution, which is $x = \frac{10}{11}$, $y = \frac{3}{11}$.

(b)
$$2x - 3y = 1$$
$$4x - 6y = 1$$

The lines represented by these equations are parallel and distinct. So there is no solution to this pair of equations, i.e. they are inconsistent.

(c)
$$2x - 3y = 1$$
$$4x - 6y = 2$$

These equations represent the same line. Hence the equations are linearly dependent and their solution set contains the coordinates of all the points on the line. We may express the solution in parametric form as follows.
If $x = \lambda$, then, from either equation, $y = \frac{1}{3}(2\lambda - 1)$.
Hence the solution is
$$x = \lambda, \quad y = \tfrac{1}{3}(2\lambda - 1) \quad \text{where } \lambda \text{ is any real number.}$$
The solution of a pair of linear simultaneous equations may also be thought of in terms of linear transformations of the plane.

The equations $\begin{matrix} a_1x + b_1y = d_1 \\ a_2x + b_2y = d_2 \end{matrix}$ may be written as the matrix equation

$$\begin{pmatrix} a_1 & b_1 \\ a_2 & b_2 \end{pmatrix}\begin{pmatrix} x \\ y \end{pmatrix} = \begin{pmatrix} d_1 \\ d_2 \end{pmatrix}$$

The solution can then be thought of as

the set of vectors $\begin{pmatrix} x \\ y \end{pmatrix}$ which map to $\begin{pmatrix} d_1 \\ d_2 \end{pmatrix}$ under the transformation.

As before this set may be empty or have one member or an infinite number of members.

The Condition for Three Lines to be Concurrent

Consider the equations

$$a_1x + b_1y + c_1 = 0 \qquad [1]$$
$$a_2x + b_2y + c_2 = 0 \qquad [2]$$
$$a_3x + b_3y + c_3 = 0 \qquad [3]$$

representing the lines L_1, L_2 and L_3 in the xy plane.

Assuming that none of the lines are parallel, solving equations [2] and [3] gives

$$x = \frac{b_2c_3 - b_3c_2}{a_2b_3 - a_3b_2}, \qquad y = -\frac{a_2c_3 - a_3c_2}{a_2b_3 - a_3b_2}$$

If the lines are concurrent, the equations have a common solution, i.e. the values of x and y satisfying equations [2] and [3] also satisfy equation [1],

i.e. $$a_1(b_2c_3 - b_3c_2) - b_1(a_2c_3 - a_3c_2) + c_1(a_2b_3 - a_3b_2) = 0 \qquad [4]$$

But the LHS of [4] is the expansion of the determinant $\begin{vmatrix} a_1 & b_1 & c_1 \\ a_2 & b_2 & c_2 \\ a_3 & b_3 & c_3 \end{vmatrix}$

Hence the lines whose equations are

$$\left.\begin{aligned} a_1x + b_1y + c_1 &= 0 \\ a_2x + b_2y + c_2 &= 0 \\ a_3x + b_3y + c_3 &= 0 \end{aligned}\right\} \text{ are concurrent if } \begin{vmatrix} a_1 & b_1 & c_1 \\ a_2 & b_2 & c_2 \\ a_3 & b_3 & c_3 \end{vmatrix} = 0$$

Three Variables

Now consider the set of equations

$$\begin{aligned} a_1x + b_1y + c_1z &= d_1 \\ a_2x + b_2y + c_2z &= d_2 \\ a_3x + b_3y + c_3z &= d_3 \end{aligned}$$

which may be written as $\mathbf{A}\begin{pmatrix} x \\ y \\ z \end{pmatrix} = \begin{pmatrix} d_1 \\ d_2 \\ d_3 \end{pmatrix}$ where $\mathbf{A} = \begin{pmatrix} a_1 & b_1 & c_1 \\ a_2 & b_2 & c_2 \\ a_3 & b_3 & c_3 \end{pmatrix}$.

Provided that $\mathbf{A}^{-1}$ exists, the solution of this set of equations is

$$\begin{pmatrix} x \\ y \\ z \end{pmatrix} = \mathbf{A}^{-1}\begin{pmatrix} d_1 \\ d_2 \\ d_3 \end{pmatrix}$$

The use of this method for solving a particular set of equations means that $\mathbf{A}^{-1}$ has to be found and so it involves a tedious amount of arithmetic. We now look at another method which considerably reduces the numerical work involved.

Systematic Elimination

Consider the following set of equations and their corresponding matrix representation.

$$\begin{cases} x+y+\ z = 7 & [1] \\ x-y+2z = 9 & [2] \\ 2x+y-\ z = 1 & [3] \end{cases} \quad \text{or} \quad \begin{pmatrix} 1 & 1 & 1 \\ 1 & -1 & 2 \\ 2 & 1 & -1 \end{pmatrix}\begin{pmatrix} x \\ y \\ z \end{pmatrix} = \begin{pmatrix} 7 \\ 9 \\ 1 \end{pmatrix}$$

To solve these equations by elimination we can proceed as follows. Eliminating z from equations [1] and [2], i.e. [1] + [3] and [2] + 2[3] gives

$$\begin{cases} 3x+2y = 8 & [4] \\ 5x+\ y = 11 & [5] \\ 2x+\ y-z = 1 & [6] \end{cases} \quad \text{or} \quad \begin{pmatrix} 3 & 2 & 0 \\ 5 & 1 & 0 \\ 2 & 1 & -1 \end{pmatrix}\begin{pmatrix} x \\ y \\ z \end{pmatrix} = \begin{pmatrix} 8 \\ 11 \\ 1 \end{pmatrix}$$

Eliminating y from equation [4], i.e. [4] -2[5] gives

$$\begin{cases} -7x = -14 \\ 5x+y = 11 \\ 2x+y-z = 1 \end{cases} \quad \text{or} \quad \begin{pmatrix} -7 & 0 & 0 \\ 5 & 1 & 0 \\ 2 & 1 & -1 \end{pmatrix}\begin{pmatrix} x \\ y \\ z \end{pmatrix} = \begin{pmatrix} -14 \\ 11 \\ 1 \end{pmatrix}$$

from which we have $x = 2$ from the first equation and by direct substitution into the second and then third equations we have $y = 1$ and $z = 4$.

Examining the matrix representation of the equations at each stage of the elimination process shows that the matrices are produced by performing the same operations on their rows as is performed on the equations themselves. So we can work directly on the matrix representation, noting that the row operations must be performed on *both* the matrix on the L.H.S. *and* the vector on the R.H.S. *but not* on the vector $\begin{pmatrix} x \\ y \\ z \end{pmatrix}$.

Thus, for the purpose of elimination, we can represent the equations

$$\left.\begin{aligned} x+y+\ z &= 7 \\ x-y+2z &= 9 \\ 2x+y-\ z &= 1 \end{aligned}\right\} \text{ by the } \textit{augmented} \text{ matrix } \left(\begin{array}{ccc:c} 1 & 1 & 1 & 7 \\ 1 & -1 & 2 & 9 \\ 2 & 1 & -1 & 1 \end{array}\right)$$

Denoting the rows by $\mathbf{r}_1, \mathbf{r}_2, \mathbf{r}_3$ and this time choosing to eliminate x from equations [2] and [3], we have

$$\mathbf{r}_3 - 2\mathbf{r}_1 \quad \Rightarrow \quad \begin{pmatrix} 1 & 1 & 1 & \vdots & 7 \\ 1 & -1 & 2 & \vdots & 9 \\ 0 & -1 & -3 & \vdots & -13 \end{pmatrix}$$

$$\mathbf{r}_2 - \mathbf{r}_1 \quad \Rightarrow \quad \begin{pmatrix} 1 & 1 & 1 & \vdots & 7 \\ 0 & -2 & 1 & \vdots & 2 \\ 0 & -1 & -3 & \vdots & -13 \end{pmatrix}$$

$$2\mathbf{r}_3 - \mathbf{r}_2 \quad \Rightarrow \quad \begin{pmatrix} 1 & 1 & 1 & \vdots & 7 \\ 0 & -2 & 1 & \vdots & 2 \\ 0 & 0 & -7 & \vdots & -28 \end{pmatrix} \quad \text{or} \quad \begin{bmatrix} x + y + z = 7 \\ -2y + z = 2 \\ -7z = -28 \end{bmatrix}$$

$$\Rightarrow \quad z = 4, \quad y = 1 \quad \text{and} \quad x = 2$$

Note that the first solution resulted in a matrix with zeros above the leading diagonal and that in the second solution the matrix has zeros below the leading diagonal.

In both cases, the final form of the matrix is called the reduced or echelon form (echelon is the Greek word for ladder).

Note that to produce the echelon form, we aim for combinations of rows (*not* columns) giving zeros either above or below the leading diagonal.

This systematic method of elimination (or reduction as it is sometimes called) can be extended to the solution of larger numbers of equations containing more variables. It is basically this method that is used in computer programmes for the solution of such sets of equations.

However, we are not computers and so should not feel bound to produce an echelon form matrix. We can instead use the augmented matrix and reduce it to the point where we can easily solve the equations.

For example, to solve the equations

$$\begin{aligned} x - 2y + z &= 6 \\ x + 5y + z &= -1 \\ 2x - y + 4z &= 15 \end{aligned}$$

we can reduce the augmented matrix $\begin{pmatrix} 1 & -2 & 1 & \vdots & 6 \\ 1 & 5 & 1 & \vdots & -1 \\ 2 & -1 & 4 & \vdots & 15 \end{pmatrix}$ as follows.

$$\mathbf{r}_1 - \mathbf{r}_2 \quad \Rightarrow \quad \begin{pmatrix} 0 & -7 & 0 & \vdots & 7 \\ 1 & 5 & 1 & \vdots & -1 \\ 2 & -1 & 4 & \vdots & 15 \end{pmatrix}$$

$$2\mathbf{r}_2 - \mathbf{r}_3 \quad \Rightarrow \quad \begin{pmatrix} 0 & -7 & 0 & \vdots & 7 \\ 0 & 11 & -2 & \vdots & -17 \\ 2 & -1 & 4 & \vdots & 15 \end{pmatrix} \Rightarrow y = -1, \quad z = 3, \quad x = 1$$

Remember also that, unless a solution by reduction is asked for, a simple algebraic solution is often the best method.
For example, we can solve the same set of equations as follows without any matrix notation.

$$x - 2y + z = 6 \qquad [1]$$
$$x + 5y + z = -1 \qquad [2]$$
$$2x - y + 4z = 15 \qquad [3]$$

[2] − [1] $\Rightarrow$ $7y = -7$ $\Rightarrow$ $y = -1$

[1] and [3] $\Rightarrow$ $3y + 2z = 3$ $\Rightarrow$ $z = 3$

then, from any equation, $x = 1$.

EXERCISE 2h

Solve for x, y, z the following sets of equations, using systematic reduction to produce an echelon matrix. Check your answers by an algebraic solution.

1) $2x + y - 3z = 4$
$x - 2y + z = 1$
$2x + y - z = 0$

2) $3x - y - z = 2$
$x + y + z = 4$
$4x - y + z = 7$

3) $4x - y + 5z = 8$
$5x + 7y - 3z = 42$
$3x + 4y + z = 27$

4) $2x - 5y + 2z = 14$
$9x + 3y - 4z = 13$
$7x + 3y - 2z = 3$

Solve for x, y, z the following sets of equations by reducing the augmented matrix.

5) $x - y + 2z = -1$
$4x + y + z = 13$
$5x - y + 8z = 5$

6) $x - y - 4z = 1$
$2x + 5y - z = 2$
$3x + 2y - 3z = -1$

7) $4x + 2y - z = 24$
$2x + 3y + 2z = 17$
$6x - 5y + 7z = -21$

8) $4x - 7y + 6z = -18$
$5x + y - 4z = -9$
$3x - 2y + 3z = 12$

THE SOLUTION SETS FOR LINEAR EQUATIONS IN THREE VARIABLES

The sets of equations in the last section all had unique solutions. However, there are other possibilities.
Consider again the set of equations

$$a_1x + b_1y + c_1z = d_1$$

$$a_2x + b_2y + c_2z = d_2$$

$$a_3x + b_3y + c_3z = d_3$$

In the xyz coordinate system these equations represent planes Π_1, Π_2 and Π_3. The coordinates of any point common to the three planes is a solution of the equations.

These equations may be such that

(a) Π_1, Π_2 and Π_3 intersect in one point, in which case the set of equations representing them have a unique solution.

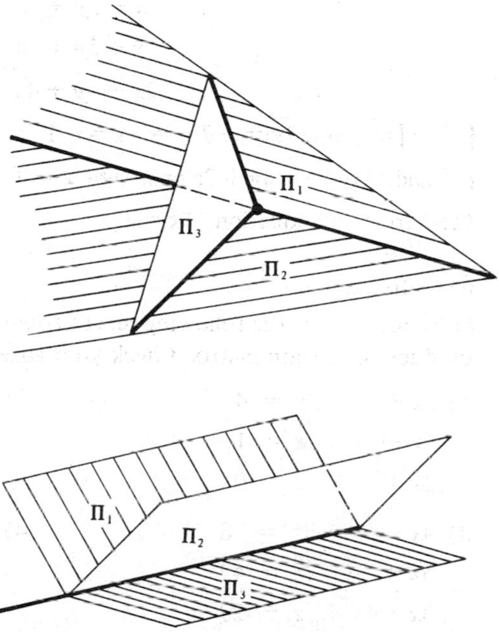

(b) Π_1, Π_2 and Π_3 intersect in a line, (in this case the equation of Π_3, say, is a linear combination of the equations of Π_1 and Π_2),

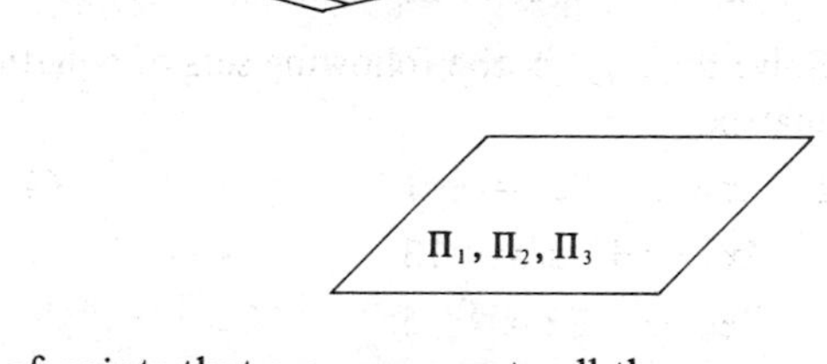

or

Π_1, Π_2, Π_3 are identical.

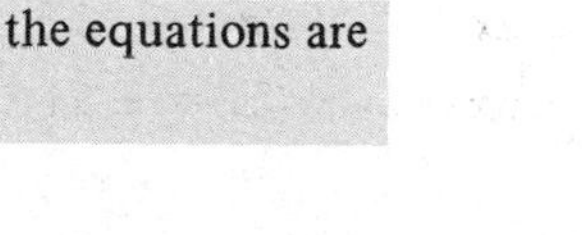

In either case there is an infinite set of points that are common to all three planes.
So the equations have an infinite set of solutions and the equations are linearly dependent.

(c) Π_1, Π_2 and Π_3 are all parallel and distinct,

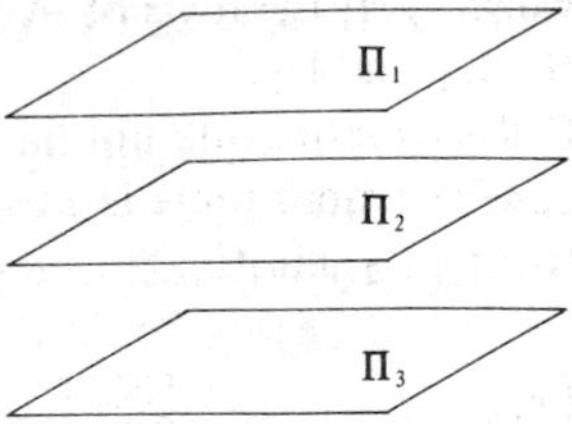

or any two are parallel but distinct,

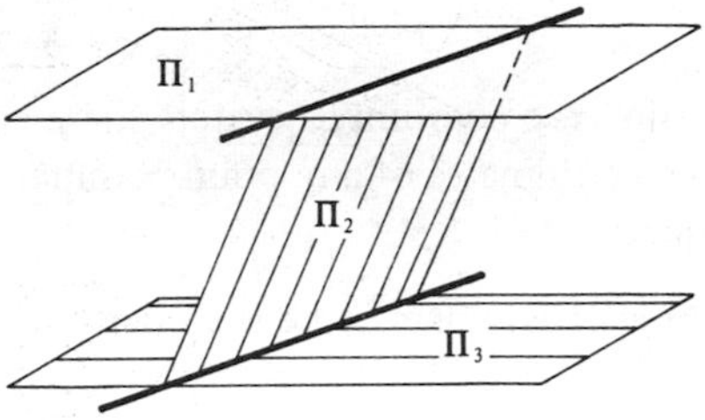

or one plane is parallel to the *line* of intersection of the other two.

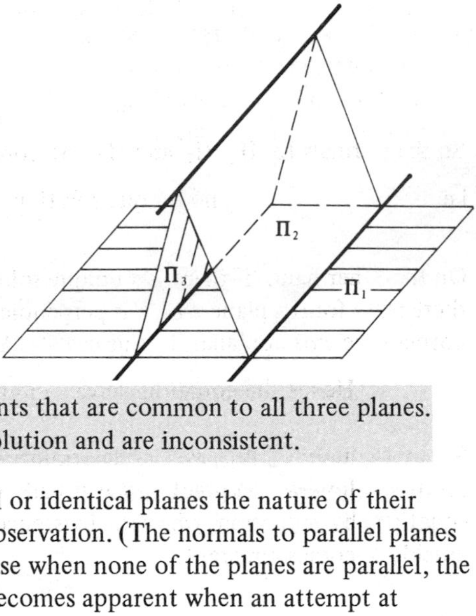

In any of these cases there are no points that are common to all three planes. Hence the set of equations have no solution and are inconsistent.

When the equations represent parallel or identical planes the nature of their solution sets can be determined by observation. (The normals to parallel planes have equal direction ratios.) In the case when none of the planes are parallel, the nature of their solution set quickly becomes apparent when an attempt at solution by reduction is made.

A Unique Solution

In the matrix representation of the set of equations

$$\mathbf{A}\begin{pmatrix} x \\ y \\ z \end{pmatrix} = \begin{pmatrix} d_1 \\ d_2 \\ d_3 \end{pmatrix} \quad \text{where} \quad \mathbf{A} = \begin{pmatrix} a_1 & b_1 & c_1 \\ a_2 & b_2 & c_2 \\ a_3 & b_3 & c_3 \end{pmatrix}$$

we see that the rows of $\mathbf{A}$ represent vectors which are *normal* to the planes Π_1, Π_2 and Π_3.
If the equations do not have a unique solution, then in all five of their possible configurations there is a fourth plane that can be drawn which is perpendicular to Π_1, Π_2 and Π_3.

i.e.

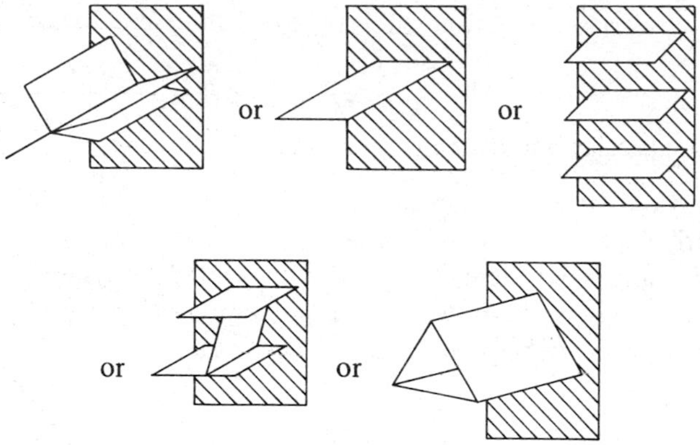

So the normals to Π_1, Π_2 and Π_3 are *coplanar*, i.e. $|\mathbf{A}| = 0$,

i.e. no unique solution $\Rightarrow$ $|\mathbf{A}| = 0$

On the other hand, if there is a unique solution, the planes meet in a point and there is *no* fourth plane which is perpendicular to Π_1, Π_2 and Π_3 and so their normals are *not* coplanar. In this case $|\mathbf{A}| \neq 0$.

Hence the equations have a unique solution $\Longleftrightarrow$ $|\mathbf{A}| \neq 0$.

So, by evaluating $|\mathbf{A}|$, we can determine whether or not there is a unique solution. However, in practice it is usually quicker to actually solve the equations by reduction, when the existence, or otherwise, of a unique solution quickly becomes apparent.

An Infinite Set of Solutions

In this case, Π_1, Π_2 and Π_3 have a line in common. So the equation of Π_1 (or Π_2, or Π_3) may be expressed as a linear combination of the equations of the other two planes,
i.e. for some value of k,

$$a_1x + b_1y + c_1z - d_1 = (a_2x + b_2y + c_2z - d_2) + k(a_3x + b_3y + c_3z - d_3)$$

and the equations are *linearly dependent*.

Considering the augmented matrix

$$\begin{pmatrix} a_1 & b_1 & c_1 & \vdots & d_1 \\ a_2 & b_2 & c_2 & \vdots & d_2 \\ a_3 & b_3 & c_3 & \vdots & d_3 \end{pmatrix}$$

we see that, in this case, reduction will lead to *one complete row of zeros.*

Inconsistent Equations

If the equations are inconsistent, the complete rows of the augmented matrix are not linearly dependent but, as $|\mathbf{A}| = 0$, the rows of $\mathbf{A}$ are linearly dependent. So, in this case, reduction of the augmented matrix eventually results in a row of zeros *only* in $\mathbf{A}$. However, it is not always necessary to reduce the augmented matrix to this stage, because evidence of inconsistency often appears earlier.

The following examples show how observation of the given equations for parallel planes, followed if necessary by reduction of the augmented matrix, quickly leads to the determination of the nature of their solution set.

EXAMPLES 2i

1) Solve the equations

$$\begin{aligned} x - 2y + 4z &= 1 \\ 2x - 4y + 8z &= 2 \\ 3x - 6y + 12z &= 3. \end{aligned}$$

By observation we see that the planes Π_1, Π_2 and Π_3 represented by these equations are identical.
(Dividing the second equation by 2 and the third equation by 3 gives the first equation in each case.)
Hence the solution set is the coordinates of all points contained in the plane, i.e. the values of x, y and z such that $x - 2y + 4z = 1$. This solution may be expressed in parametric form, e.g. $x = \lambda$, $y = \mu \Rightarrow z = \frac{1}{4}(1 - \lambda + 2\mu)$.

2) Solve the equations

$$\begin{aligned} x - y + z &= 4 \\ 2x + y - 2z &= 1 \\ 5x - 2y + z &= 13. \end{aligned}$$

None of the planes represented by these equations are parallel.

Reducing the augmented matrix

$$\begin{pmatrix} 1 & -1 & 1 & \vdots & 4 \\ 2 & 1 & -2 & \vdots & 1 \\ 5 & -2 & 1 & \vdots & 13 \end{pmatrix}$$

gives $\quad \mathbf{r}_1 - \mathbf{r}_3 \Rightarrow \begin{pmatrix} -4 & 1 & 0 & \vdots & -9 \\ 2 & 1 & -2 & \vdots & 1 \\ 5 & -2 & 1 & \vdots & 13 \end{pmatrix}$

$$\mathbf{r}_2 + 2\mathbf{r}_3 \Rightarrow \begin{pmatrix} -4 & 1 & 0 & \vdots & -9 \\ 12 & -3 & 0 & \vdots & 27 \\ 5 & -2 & 1 & \vdots & 13 \end{pmatrix}$$

Now we see that $\mathbf{r}_2 = -3\mathbf{r}_1$, so the next step would give a complete row of zeros.

Hence the equations are linearly dependent and the planes they represent have a line in common. So the solution of the equations is the set of coordinates of all the points on this line. The equations of this line are any pair of the three given equations.

So the solution set of the given equations is the set of values of x, y, z such that

$$x - y + z = 4$$
$$2x + y - 2z = 1$$

which may be expressed in parametric form as

$$x = \lambda, \quad y = 4\lambda - 9, \quad z = 3\lambda - 5$$

Alternatively, attempting an algebraic solution gives

$$\left.\begin{aligned} x - y + z &= 4 \\ 2x + y - 2z &= 1 \end{aligned}\right\} \Rightarrow 3x - z = 5 \Rightarrow z = 3x - 5$$

$$\left.\begin{aligned} 2x + y - 2z &= 1 \\ 5x - 2y + z &= 13 \end{aligned}\right\} \Rightarrow 9x - 3z = 15 \Rightarrow z = 3x - 5$$

So these equations are linearly dependent. The solution is given by

$$z = 3x - 5$$

and by substitution into the first equation

$$z = \tfrac{1}{4}(3y + 7)$$

Hence $\quad 3x - 5 = \dfrac{3y + 7}{4} = z$

and these are the equations of the line common to the three given planes.

3) Solve the equations

$$\begin{aligned} x+y-\ z &= 1 \\ x-y+2z &= 5 \\ 2x+y-\ z &= 3 \end{aligned}$$

None of the planes represented by these equations are parallel.

The augmented matrix is $\begin{pmatrix} 1 & 1 & -1 & \vdots & 1 \\ 1 & -1 & 2 & \vdots & 5 \\ 2 & 1 & -1 & \vdots & 3 \end{pmatrix}$

and $\mathbf{r}_3 - \mathbf{r}_1$ gives $\begin{pmatrix} 1 & 1 & -1 & \vdots & 1 \\ 1 & -1 & 2 & \vdots & 5 \\ 1 & 0 & 0 & \vdots & 2 \end{pmatrix}$

From the third row, $x = 2$, but this does not necessarily mean that there is a unique solution. One more step is necessary to establish the nature of the solution set.

$$\mathbf{r}_1 + \mathbf{r}_2 \quad \Rightarrow \quad \begin{pmatrix} 2 & 0 & 1 & \vdots & 6 \\ 1 & -1 & 2 & \vdots & 5 \\ 1 & 0 & 0 & \vdots & 2 \end{pmatrix} \quad \Rightarrow \quad x = 2, \quad z = 2, \quad y = 1$$

i.e. there is a unique solution.

4) Find the solution set, if there is one, of the equations

$$\begin{aligned} 2x+\ y-\ z &= 1 \\ x-\ y+\ z &= 3 \\ x+5y-5z &= 2 \end{aligned}$$

Again, none of these equations represent parallel planes.

Now

$$\begin{pmatrix} 2 & 1 & -1 & \vdots & 1 \\ 1 & -1 & 1 & \vdots & 3 \\ 1 & 5 & -5 & \vdots & 2 \end{pmatrix} \Rightarrow \begin{pmatrix} 3 & 0 & 0 & \vdots & 4 \\ 1 & -1 & 1 & \vdots & 3 \\ 1 & 5 & -5 & \vdots & 2 \end{pmatrix} \Rightarrow x = \frac{4}{3} \quad \text{(top row)}$$

then

$$(\mathbf{r}_3 + 5\mathbf{r}_2) \Rightarrow \begin{pmatrix} 3 & 0 & 0 & \vdots & 4 \\ 1 & -1 & 1 & \vdots & 3 \\ 6 & 0 & 0 & \vdots & 17 \end{pmatrix} \Rightarrow x = \frac{4}{3} \text{ and } x = \frac{17}{6} \quad \text{(3rd row).}$$

The values obtained for x from the final matrix are inconsistent. Hence the equations are inconsistent and the planes they represent form a triangular prism.

5) If it is possible, solve the equations

$$\begin{aligned} x+y+z &= 1\\ 2x-y+z &= 4\\ x+y+z &= 2 \end{aligned}$$

Two of these equations (the first and last) represent parallel, but distinct, planes. Hence the equations are inconsistent and there is no solution.

EXERCISE 2i

In Questions 1–8 do not solve the equations, but determine whether there is a unique solution, no solution, or an infinite set of solutions, in which case state whether this set is dependent on one or two parameters.

1) $$\begin{aligned} x-3y+z &= 4\\ 2x-y+z &= 2\\ x+2y &= -2 \end{aligned}$$

2) $$\begin{aligned} x+y-2z &= 1\\ 2x-y+z &= 4\\ 3x+y-z &= 2 \end{aligned}$$

3) $$\begin{aligned} 2x+y-z &= 1\\ 6x+3y-3z &= 4\\ 4x+2y-2z &= 2 \end{aligned}$$

4) $$\begin{aligned} 2x+y-z &= 1\\ 6x+3y-3z &= 3\\ x+y-z &= 2 \end{aligned}$$

5) $$\begin{aligned} x-y+z &= 0\\ x+y-2z &= 0\\ 3x+y-3z &= 0 \end{aligned}$$

6) $$\begin{aligned} x+y+z &= 1\\ 2x+2y+2z &= 2\\ 3x+3y+3z &= 3 \end{aligned}$$

7) $$\begin{aligned} x-y+z &= 1\\ 2x+y-z &= 2\\ 3x-2y+z &= 3 \end{aligned}$$

8) $$\begin{aligned} x+y-2z &= 1\\ 3x+y-4z &= 2\\ 4x+2y-6z &= 5 \end{aligned}$$

9) Solve, where possible, the sets of equations in Questions 1–8.

10) By considering the solution to the set of equations

$$\begin{aligned} x+y-z &= 2\\ 2x-y+z &= 3\\ x+4y-4z &= 3 \end{aligned}$$

or otherwise, find the set of points which map to the point $(2,3,3)$ under the transformation defined by $\mathbf{M} = \begin{pmatrix} 1 & 1 & -1\\ 2 & -1 & 1\\ 1 & 4 & -4 \end{pmatrix}$.

11) Find the set of points which are mapped to the point $(1,-1,-1)$ by the transformation $\begin{pmatrix} 1 & -2 & 4\\ 3 & 4 & 6\\ 1 & 3 & 1 \end{pmatrix}\begin{pmatrix} x\\ y\\ z \end{pmatrix} = \begin{pmatrix} X\\ Y\\ Z \end{pmatrix}$.

12) Show that the matrix $\begin{pmatrix} 1 & 0 & 1 \\ -1 & 2 & 1 \\ 1 & 0 & 2 \end{pmatrix}$, which maps $\begin{pmatrix} x \\ y \\ z \end{pmatrix}$ to $\begin{pmatrix} X \\ Y \\ Z \end{pmatrix}$

defines a transformation in which any point in the image space arises from a unique point in xyz space.

13) Find the set of points, which under the transformation defined by

$\mathbf{M} = \begin{pmatrix} 1 & -1 & 2 \\ -1 & 1 & -2 \\ 1 & -1 & 2 \end{pmatrix}$ map to the point $(1, -1, 1)$.

14) Show that the matrix $\mathbf{M} = \begin{pmatrix} 1 & 2 & -1 \\ 2 & 4 & -2 \\ 3 & 6 & -3 \end{pmatrix}$ maps all points in

xyz space to a line in XYZ space. Find the equations of this line and show that any point on this line arises from a set of coplanar points in xyz space.

15) Find the values of a for which the lines

$$\begin{aligned} 2x - y + a &= 0 \\ ax + y + 1 &= 0 \\ x - y - a &= 0 \end{aligned}$$

are concurrent.

16) What is the condition that the equations

$$\begin{aligned} a_1x + b_1y + c_1z &= 0 \\ a_2x + b_2y + c_2z &= 0 \\ a_3x + b_3y + c_3z &= 0 \end{aligned}$$

should have solutions other than $x = y = z = 0$?

(**Note** that such equations are called homogeneous.)

17) By writing the homogeneous equations

$$\left.\begin{aligned} a_1x + b_1y + c_1z &= 0 \\ a_2x + b_2y + c_2z &= 0 \\ a_3x + b_3y + c_3z &= 0 \end{aligned}\right\} \qquad [1]$$

as

$$\left.\begin{aligned} a_1(x/z) + b_1(y/z) + c_1 &= 0 \\ a_2(x/z) + b_2(y/z) + c_2 &= 0 \\ a_3(x/z) + b_3(y/z) + c_3 &= 0 \end{aligned}\right\} \qquad [2]$$

write down the condition for the set of equations [2] to have a common solution for $\dfrac{x}{z}$ and $\dfrac{y}{z}$.

What is the geometrical significance when this condition is applied to the set of equations [1]?

18) Considering the equations

$$\left.\begin{aligned} a_1x + b_1y + c_1 &= 0 \\ a_2x + b_2y + c_2 &= 0 \\ a_3x + b_3y + c_3 &= 0 \end{aligned}\right\} \qquad [1]$$

as the set of equations

$$\left.\begin{aligned} a_1x + b_1y + c_1z &= 0 \\ a_2x + b_2y + c_2z &= 0 \\ a_3x + b_3y + c_3z &= 0 \end{aligned}\right\} \quad \text{with} \quad z = 1$$

find the condition that the first set should have a unique solution and interpret this solution geometrically.

19) Find the value of a for which the following equations are consistent

$$\begin{aligned} x - 3y + 5z &= 2 \\ x + 4y - z &= 1 \\ 7y - 6z &= a. \end{aligned}$$

20) Find a 3×3 matrix which maps all points of xyz space to points on the plane $x + y + z = 0$.

21) Write down the condition on a_1, b_1, a_2 and b_2 for the equations

$$\left.\begin{aligned} a_1x + b_1y + c_1 &= 0 \\ a_2x + b_2y + c_2 &= 0 \end{aligned}\right\} \quad \text{to have a unique solution}$$

and show that

$$y = \frac{\begin{vmatrix} a_1 & -c_1 \\ a_2 & -c_2 \end{vmatrix}}{\begin{vmatrix} a_1 & b_1 \\ a_2 & b_2 \end{vmatrix}}$$

Hence show that the quadratic equations

$$a_1y^2 + b_1y + c_1 = 0$$
$$a_2y^2 + b_2y + c_2 = 0$$

have a common root if

$$(a_2c_1 - a_1c_2)^2 = (b_1c_2 - b_2c_1)(a_1b_2 - a_2b_1).$$

The Evaluation of an Inverse Matrix by Reduction

The systematic reduction of a set of equations leads to an alternative method for the calculation of an inverse matrix.

Consider the equation $\begin{pmatrix} 1 & 1 & 1 \\ 1 & -1 & 2 \\ 2 & 1 & -1 \end{pmatrix}\begin{pmatrix} x \\ y \\ z \end{pmatrix} = \begin{pmatrix} 7 \\ 9 \\ 1 \end{pmatrix}$

which we can write as $\begin{pmatrix} 1 & 1 & 1 \\ 1 & -1 & 2 \\ 2 & 1 & -1 \end{pmatrix}\begin{pmatrix} x \\ y \\ z \end{pmatrix} = \begin{pmatrix} 1 & 0 & 0 \\ 0 & 1 & 0 \\ 0 & 0 & 1 \end{pmatrix}\begin{pmatrix} 7 \\ 9 \\ 1 \end{pmatrix}$.

We find that any row operation performed using

either **I** *or* $\begin{pmatrix} 7 \\ 9 \\ 1 \end{pmatrix}$ but *not* both, produces the same result on the R.H.S.,

e.g. $\quad (\mathbf{r}_2 - \mathbf{r}_1)$ on $\mathbf{I} \quad \Rightarrow \quad \begin{pmatrix} 1 & 0 & 0 \\ -1 & 1 & 0 \\ 0 & 0 & 1 \end{pmatrix}\begin{pmatrix} 7 \\ 9 \\ 1 \end{pmatrix} = \begin{pmatrix} 7 \\ 2 \\ 1 \end{pmatrix}$

and $\quad (\mathbf{r}_2 - \mathbf{r}_1)$ on $\begin{pmatrix} 7 \\ 9 \\ 1 \end{pmatrix} \quad \Rightarrow \quad \begin{pmatrix} 1 & 0 & 0 \\ 0 & 1 & 0 \\ 0 & 0 & 1 \end{pmatrix}\begin{pmatrix} 7 \\ 2 \\ 1 \end{pmatrix} = \begin{pmatrix} 7 \\ 2 \\ 1 \end{pmatrix}$

So if we choose to operate on **I** on the R.H.S. we work with the matrix

$$\left(\begin{array}{ccc:ccc} 1 & 1 & 1 & 1 & 0 & 0 \\ 1 & -1 & 2 & 0 & 1 & 0 \\ 2 & 1 & -1 & 0 & 0 & 1 \end{array}\right)$$

The elimination process can then be continued until the L.H.S. is reduced to the form $\mathbf{I}\begin{pmatrix}x\\y\\z\end{pmatrix}$,

$$(\mathbf{r}_3-2\mathbf{r}_1) \Rightarrow \left(\begin{array}{ccc:ccc}1 & 1 & 1 & 1 & 0 & 0\\ 1 & -1 & 2 & 0 & 1 & 0\\ 0 & -1 & -3 & -2 & 0 & 1\end{array}\right)$$

$$(\mathbf{r}_2-\mathbf{r}_1) \Rightarrow \left(\begin{array}{ccc:ccc}1 & 1 & 1 & 1 & 0 & 0\\ 0 & -2 & 1 & -1 & 1 & 0\\ 0 & -1 & -3 & -2 & 0 & 1\end{array}\right)$$

$$(2\mathbf{r}_3-\mathbf{r}_2) \Rightarrow \left(\begin{array}{ccc:ccc}1 & 1 & 1 & 1 & 0 & 0\\ 0 & -2 & 1 & -1 & 1 & 0\\ 0 & 0 & -7 & -3 & -1 & 2\end{array}\right)$$

$$(7\mathbf{r}_1+\mathbf{r}_3) \text{ and } (7\mathbf{r}_2+\mathbf{r}_3) \Rightarrow \left(\begin{array}{ccc:ccc}7 & 7 & 0 & 4 & -1 & 2\\ 0 & -14 & 0 & -10 & 6 & 2\\ 0 & 0 & -7 & -3 & -1 & 2\end{array}\right)$$

$$2\mathbf{r}_1+\mathbf{r}_2 \Rightarrow \left(\begin{array}{ccc:ccc}14 & 0 & 0 & -2 & 4 & 6\\ 0 & -14 & 0 & -10 & 6 & 2\\ 0 & 0 & -7 & -3 & -1 & 2\end{array}\right)$$

$$\tfrac{1}{14}\mathbf{r}_1,\ -\tfrac{1}{14}\mathbf{r}_2,\ -\tfrac{1}{7}\mathbf{r}_3 \Rightarrow \left(\begin{array}{ccc:ccc}1 & 0 & 0 & -\frac{1}{7} & \frac{2}{7} & \frac{3}{7}\\ 0 & 1 & 0 & \frac{5}{7} & -\frac{3}{7} & -\frac{1}{7}\\ 0 & 0 & 1 & \frac{3}{7} & \frac{1}{7} & -\frac{2}{7}\end{array}\right)$$

So we have now reduced the original set of equations to

$$\begin{pmatrix}1 & 0 & 0\\ 0 & 1 & 0\\ 0 & 0 & 1\end{pmatrix}\begin{pmatrix}x\\y\\z\end{pmatrix} = \begin{pmatrix}-\frac{1}{7} & \frac{2}{7} & \frac{3}{7}\\ \frac{5}{7} & -\frac{3}{7} & -\frac{1}{7}\\ \frac{3}{7} & \frac{1}{7} & -\frac{2}{7}\end{pmatrix}\begin{pmatrix}7\\9\\1\end{pmatrix}$$

i.e. $\dfrac{1}{7}\begin{pmatrix}-1 & 2 & 3\\ 5 & -3 & -1\\ 3 & 1 & -2\end{pmatrix}$ is the inverse of $\begin{pmatrix}1 & 1 & 1\\ 1 & -1 & 2\\ 2 & 1 & -1\end{pmatrix}$

So to find the inverse of $\mathbf{A} = \begin{pmatrix} a_1 & b_1 & c_1 \\ a_2 & b_2 & c_2 \\ a_3 & b_3 & c_3 \end{pmatrix}$

we apply row operations to the matrix

$$\left(\begin{array}{ccc:ccc} a_1 & b_1 & c_1 & 1 & 0 & 0 \\ a_2 & b_2 & c_2 & 0 & 1 & 0 \\ a_3 & b_3 & c_3 & 0 & 0 & 1 \end{array}\right)$$

until the left hand section of this matrix is reduced to $\mathbf{I}$.
The right hand section of this matrix is then $\mathbf{A}^{-1}$.
If $\mathbf{A}$ has no inverse this fact quickly becomes apparent because the attempted reduction of $\mathbf{A}$ produces a row of zeros.
This method has many advantages over the 'cofactor' method introduced earlier in this chapter, because it greatly simplifies the numerical work involved, and so is less likely to give rise to mistakes.
(But mistakes are still possible, so checking the product of $\mathbf{A}$ and the calculated $\mathbf{A}^{-1}$ is still recommended!)
It is interesting to note that computers use this reduction method for calculating inverse matrices.

EXERCISE 2j

Find the inverse, where it exists, of each of the following matrices:

1) $\begin{pmatrix} 1 & 2 & 4 \\ -1 & 2 & 1 \\ 1 & 5 & 3 \end{pmatrix}$ 2) $\begin{pmatrix} 2 & -1 & 1 \\ 1 & -2 & 3 \\ 5 & 1 & 4 \end{pmatrix}$ 3) $\begin{pmatrix} 5 & 7 & 9 \\ -5 & 4 & 6 \\ 0 & 10 & 15 \end{pmatrix}$

4) $\begin{pmatrix} 2 & -1 & 4 \\ 2 & -1 & 5 \\ 1 & 2 & 3 \end{pmatrix}$ 5) $\begin{pmatrix} -1 & 6 & 9 \\ 11 & -12 & -1 \\ 5 & -3 & 4 \end{pmatrix}$ 6) $\begin{pmatrix} 2 & 6 & 3 \\ 5 & -1 & 4 \\ 9 & 11 & 11 \end{pmatrix}$.

VECTOR SPACES

We have now looked at several mathematical concepts, for example complex numbers, series, geometric vectors in two and three dimensions, different types of functions, matrices. Many of these appear to be unrelated but we are now going to introduce a more general framework which brings together, under one roof, many of the diverse systems mentioned above. This more general framework is the concept of vector spaces but, before defining precisely what is meant by a vector space, we will look again at some of the properties of two dimensional geometric vectors.

Consider the set of *all* geometric vectors in two dimensions.

If $\mathbf{a} = \begin{pmatrix} a_1 \\ a_2 \end{pmatrix}$ and $\mathbf{b} = \begin{pmatrix} b_1 \\ b_2 \end{pmatrix}$ are any two members of this set then

we can *add* them, i.e. $\mathbf{a} + \mathbf{b} = \begin{pmatrix} a_1 + b_1 \\ a_2 + b_2 \end{pmatrix}$

This addition has the following properties.

(a) We always get another vector in the set, and we say that the *set is closed under addition*.

(b) The order in which we add the vectors does not matter, i.e. $\mathbf{a}+\mathbf{b} = \mathbf{b} + \mathbf{a}$ and we say that the *addition is commutative.*

(c) When adding three vectors it does not matter which pair we add first, i.e. $(\mathbf{a} + \mathbf{b}) + \mathbf{c} = \mathbf{a} + (\mathbf{b} + \mathbf{c})$ and we say that the *addition is associative.*

(d) The zero vector, $\mathbf{0} = \begin{pmatrix} 0 \\ 0 \end{pmatrix}$ is a member of the set and when added to any other vector it does not alter it, i.e. *there is a vector* $\mathbf{0}$ *in the set such that* $\mathbf{0} + \mathbf{a} = \mathbf{a} + \mathbf{0} = \mathbf{a}$.

(e) For any vector $\mathbf{a}$ there is always another vector, $-\mathbf{a}$, in the set which we can add to $\mathbf{a}$ to give the zero vector, i.e. for every $\mathbf{a}$ in the set there is a vector $-\mathbf{a}$ such that $\mathbf{a} + (-\mathbf{a}) = \mathbf{0}$ and we call $-\mathbf{a}$ *the inverse of* $\mathbf{a}$.

We can also *multiply vectors by a scalar*,

i.e. if λ is any real number then $\lambda\mathbf{a} = \begin{pmatrix} \lambda a_1 \\ \lambda a_2 \end{pmatrix}$.

This multiplication has the following properties.

(a) We always get another vector in the set (so the set is closed under scalar multiplication).

(b) Multiplication by 1 leaves any vector unchanged.

(c) When multiplying by two real numbers, it does not matter in which order the multiplication is done, e.g. $(2)(3)\begin{pmatrix} a_1 \\ a_2 \end{pmatrix} = 2\begin{pmatrix} 3a_1 \\ 3a_2 \end{pmatrix} = \begin{pmatrix} 6a_1 \\ 6a_2 \end{pmatrix} = 6\begin{pmatrix} a_1 \\ a_2 \end{pmatrix}$.

(d) When multiplying the sum of two vectors by a real number we can
either add the vectors first and then multiply
or multiply each vector first and then add
i.e. $\lambda(\mathbf{a} + \mathbf{b}) = \lambda\mathbf{a} + \lambda\mathbf{b}$

Similarly when multiplying a vector by the sum of two scalars we can
either add the numbers first and then multiply
or multiply by the vector first and then add
i.e. $(\lambda + \mu)\mathbf{a} = \lambda\mathbf{a} + \mu\mathbf{a}$

This list of properties looks very formal because we have spelt out what is usually taken as obvious. Most of these properties seem obvious because they stem from the properties of addition and multiplication of real numbers and these are so familiar that they are taken for granted. However, it is dangerous to assume that these properties are satisfied by all sets of elements. Consider, for example, the set of real numbers from 0 to 1 inclusive. We can add the elements of this set but we find that under addition the set is not closed, e.g. $0.7 + 0.7 = 1.4$ which is not in the set. The fifth addition property is not satisfied either as, for example, there is no member of the set to which we can add 0.7 to give zero. We can also multiply the elements of the set by any real number but the set is not closed under this multiplication, e.g. $2(0.7) = 1.4$ which is not in the set. This list is formal because it forms an important part in the definition of a vector space.

The list contains five properties associated with addition and four properties associated with multiplication by a scalar and we will refer to these as the five addition properties and the four multiplication properties respectively. We now define a vector space as follows.

A vector space is a set of elements such that:
(a) we can add the elements and the five addition properties are satisfied,
(b) we can multiply the elements by a scalar and the four multiplication properties are satisfied.

Thus we can now say that the set of all two dimensional geometric vectors is a vector space.

We will now show that the set of all polynomials of degree less than three is also a vector space.
Any polynomial of degree less than three is of the form $a_1x^2 + a_2x + a_3$ where a_1, a_2 and a_3 are real numbers.

Such polynomials can be added,

i.e. $(a_1x^2 + a_2x + a_3) + (b_1x^2 + b_2x + b_3) = (a_1 + b_1)x^2 + (a_2 + b_2)x + (a_3 + b_3)$

and

(a) we always get another polynomial of degree two or less,
(b) the order in which they are added does not matter,
(c) when three are added it does not matter which pair are added first,
(d) the set contains a zero element, i.e. the polynomial $0x^2 + 0x + 0$,
(e) each polynomial has an inverse, i.e. $-a_1x^2 - a_2x - a_3$ is the inverse of $a_1x^2 + a_2x + a_3$.

Such polynomials can be multiplied by a scalar,

i.e. $$\lambda(a_1x^2 + a_2x + a_3) = (\lambda a_1)x^2 + (\lambda a_2)x + (\lambda a_3)$$

and

(a) we always get another polynomial of degree less than three,
(b) multiplication by 1 does not change a polynomial,
(c) multiplication by two scalars can be done in any order, e.g.

$$(4)(5)(2x^2 - 3x + 4) = 4(10x^2 - 15x + 20) = 20(2x^2 - 3x + 4)$$

(d) for all expressions of the type
$(2 + 3)(a_1x^2 + a_2x + a_3)$ or $(2)[(a_1x^2 + a_2x + a_3) + (b_1x^2 + b_2x + b_3)]$
it does not matter whether the addition or the multiplication is done first.

So the set of all polynomials of degree less than three is a vector space.
Note that expressions of the form $a_1x^2 + a_2x + a_3$ are usually called quadratic functions, but it is often assumed that $a_1 \neq 0$ so that the polynomial is of degree *equal* to two.
For this reason the functions above were defined more carefully.

The examples above show that in order to prove that a given set is a vector space, we have to show that the elements can be added and multiplied by a scalar and that all nine properties are satisfied. But to show that a set is not a vector space, we need only to demonstrate that *one* property does not hold.

EXERCISE 2k

Decide which of the following sets are vector spaces.

1) The set of all complex numbers.

2) The set of all linear functions of x (i.e. polynomials of degree one or zero).

3) The set of all three dimensional vectors.

4) The set of coordinates of all points on the line $y = 2x$. (Addition of coordinates is defined by $(x_1, y_1) + (x_2, y_2) = (x_1 + x_2, y_1 + y_2)$ and multiplication by a scalar as $\lambda(x_1, y_1) = (\lambda x_1, \lambda y_1)$.)

5) The set of all infinite arithmetic progressions.

6) The set of coordinates of all points on the line $y = x + 1$.

7) The set of all real numbers between -1 and 1 inclusive.

8) The set of all infinite geometric progressions.

9) The set of coordinates of all points on the plane $x + y + z = 0$.

NOTATION FOR VECTOR SPACES

The further development of the concept of a vector space needs new notation and terminology in which to express new ideas. In the remainder of this section we will introduce some of these new ideas by illustration, using sets that have already been shown to be vector spaces.
Consider again the following examples of vector spaces,

the set of all complex numbers,
the set of all two dimensional geometric vectors,
the set of all linear polynomials.

A general element from each of these sets can be written as

$$a + bi, \quad \begin{pmatrix} a \\ b \end{pmatrix} \quad \text{and} \quad ax + b \quad \text{respectively.}$$

Now note that, in each of these examples, we need exactly two real numbers, a and b, to specify a general member of the set.
So all these sets are examples of a vector space of the same kind and they are interchangeable as far as vector space concepts are concerned. (The mathematical term is 'they are all *isomorphic*'.)
We call such a vector space ***a real vector space of dimension two*** and denote such a vector space by V_2.
It is called 'real' because the elements are defined by real numbers and we say it has 'dimension two' because two real numbers, that can independently take any value, are needed to describe the elements of the set.
An alternative name is ***a vector 2-space over the field of real numbers.***

Next consider the following examples of a vector space,

the set of all polynomials of degree less than three,
the set of all three dimensional geometric vectors.

A general element from each of these sets can be written as

$$ax^2 + bx + c \quad \text{and} \quad \begin{pmatrix} a \\ b \\ c \end{pmatrix} \quad \text{respectively.}$$

This time we need three independent real numbers, a, b and c to specify any one member of the set.
These are examples of a real vector space of dimension three, or a real vector space of triples. Such a vector space is denoted by V_3.
An example of a vector space needing four real numbers to specify any one member of the set is the set of all polynomial functions of x of degree less than 4, i.e. the set $\{ax^3 + bx^2 + cx + d\}$.
Such a vector space is denoted by V_4.

Vectors

The elements of a vector space are called *vectors.*
The constants needed to define an element, or vector, are called the *components of the vector.*
The number of components is the size of the vector, and a vector with n components is called an n-vector.
A general n-vector can be written as

$$\begin{pmatrix} x_1 \\ x_2 \\ \vdots \\ x_n \end{pmatrix} \quad \text{or} \quad \left| x_1, x_2, \ldots, x_n \right|$$

or, when we do not need to refer to the individual components, it can be written simply as x.
A *zero vector* is one in which all the components are zero and it is denoted by **0**.
A *unit vector* is such that all but one of its components are zero, the remaining one being unity, e.g. $|0, 0, 1, 0|$ i.e. a unit 4-vector.

Vector Addition

Vectors of the same size only can be added.

If $\qquad \mathbf{x} = |x_1, x_2, x_3| \quad \text{and} \quad \mathbf{y} = |y_1, y_2, y_3|$

then $\qquad \mathbf{x} + \mathbf{y} = |x_1 + y_1, x_2 + y_2, x_3 + y_3|.$

This definition can be extended to vectors of any size.

Multiplication by a Scalar

If $\mathbf{x} = |x_1, x_2, x_3|$ and λ is any real number then

$$\lambda\mathbf{x} = |\lambda x_1, \lambda x_2, \lambda x_3|$$

Again this definition can be extended to vectors of any size.

Note that

$$\begin{aligned} (-1)\mathbf{x} &= |-x_1, -x_2, -x_3| = -\mathbf{x} \\ (0)\mathbf{x} &= \mathbf{0} \\ \lambda\mathbf{0} &= \mathbf{0} \end{aligned}$$

These are the familiar operations that we have already used with two and three dimensional geometric vectors, and they obviously satisfy the properties listed on pages 118–20.

We can now define a vector space in a more abstract way as follows.

A real vector n-space is the set of all n-vectors with real components.

It is interesting to note that the real vector space of 1-vectors is just the set of real numbers. Also the real vector 2-space is defined using exactly the same terminology as is used for two dimensional geometric vectors. So what we have shown to be true for these geometric vectors is also true for any set that is a vector 2-space. The same is true for three dimensional geometric vectors and a real vector 3-space. However, for vector spaces of dimension greater than three there is no such direct analogy.

EXAMPLES 2I

1) In three dimensional space, the equation of a plane is $x+y+z=0$ and the equations of a line are $x=y, z=0$.
The vector space V is the set {the coordinates of all points on the plane}. The vector space V' is the set {the coordinates of all points on the line}.
Determine the dimension of V and the dimension of V'.

If the point (x,y,z) is on the plane $x+y+z=0$ then $z=-x-y$.
So the coordinates of any point on the plane can be written as $(x,y,-x-y)$.

Thus any vector, $\mathbf{a}$, in V is of the form given by $\mathbf{a}=\begin{pmatrix} a_1 \\ a_2 \\ -a_1-a_2 \end{pmatrix}$

and we see that we need just two independent real numbers to describe a vector in V. Hence V is a two dimensional vector space.

If (x,y,z) is a point on the line $x=y, z=0$, any point on that line has coordinates of the form $(a,a,0)$ where a is any real number,

i.e. if $\mathbf{a}$ is a vector in V', then $\mathbf{a}=\begin{pmatrix} a \\ a \\ 0 \end{pmatrix}$

from which we see that V' is a one dimensional vector space.

Note that both V and V' are contained in the vector space V_3, the set {the coordinates of all points in three dimensions}.
V and V' are called *subspaces* of V_3.

Linear Dependence

A *linear combination* of $\mathbf{a}$ and $\mathbf{b}$ is of the form $\lambda\mathbf{a}+\mu\mathbf{b}$ where λ and μ are real numbers.
We can have linear combinations of as many vectors as we choose.
For example if, in a vector space V_2,

$$\mathbf{a}=\begin{pmatrix} 2 \\ 1 \end{pmatrix}, \quad \mathbf{b}=\begin{pmatrix} -1 \\ 5 \end{pmatrix} \quad \text{and} \quad \mathbf{c}=\begin{pmatrix} 0 \\ -2 \end{pmatrix}$$

then the linear combination $2\mathbf{a} + \mathbf{b} - \mathbf{c}$ is

$$2\begin{pmatrix}2\\1\end{pmatrix} + \begin{pmatrix}-1\\5\end{pmatrix} - \begin{pmatrix}0\\-2\end{pmatrix} = \begin{pmatrix}3\\10\end{pmatrix}$$

If $\mathbf{d} = \begin{pmatrix}3\\10\end{pmatrix}$, i.e. $2\mathbf{a} + \mathbf{b} - \mathbf{c} = \mathbf{d}$ or $2\mathbf{a} + \mathbf{b} - \mathbf{c} - \mathbf{d} = \mathbf{0}$

then $\mathbf{a}, \mathbf{b}, \mathbf{c}, \mathbf{d}$ is a set of linearly dependent vectors, because we can express any one vector as a linear combination of the others.
In general if $\mathbf{x}_1, \mathbf{x}_2, \ldots, \mathbf{x}_n$ is a set of vectors such that

$$\alpha\mathbf{x}_1 + \beta\mathbf{x}_2 + \ldots + \lambda\mathbf{x}_n = \mathbf{0}$$

where $\alpha, \beta, \ldots, \lambda$ are real numbers and *not all zero,* then the set of vectors is linearly dependent.

Linear Independence

Consider the vectors $\mathbf{i} = \begin{pmatrix}1\\0\\0\end{pmatrix}$, $\mathbf{j} = \begin{pmatrix}0\\1\\0\end{pmatrix}$ and $\mathbf{k} = \begin{pmatrix}0\\0\\1\end{pmatrix}$.

If we try to find a linear combination of these vectors which gives the zero vector,

i.e.

$$\alpha\begin{pmatrix}1\\0\\0\end{pmatrix} + \beta\begin{pmatrix}0\\1\\0\end{pmatrix} + \gamma\begin{pmatrix}0\\0\\1\end{pmatrix} = \begin{pmatrix}0\\0\\0\end{pmatrix}$$

we find this is possible only if $\alpha = \beta = \gamma = 0$.
So we cannot express $\mathbf{i}$ (or $\mathbf{j}$ or $\mathbf{k}$) as a linear combination of the other two vectors and we say that the set of vectors $\mathbf{i}, \mathbf{j}$ and $\mathbf{k}$ are *linearly independent.*
In general if, for a given set of vectors,

$$\alpha\mathbf{x}_1 + \beta\mathbf{x}_2 + \ldots + \lambda\mathbf{x}_n = \mathbf{0}$$

only when all the real numbers $\alpha, \beta, \ldots \lambda$ are zero (i.e. it is impossible to express any vector in the set as a linear combination of the others) then the vectors are said to be *linearly independent.*

There is an obvious geometric interpretation of linearly independent vectors in a vector 2-space.

For example, consider the geometric vectors $\mathbf{a} = \begin{pmatrix}a_1\\a_2\end{pmatrix}$ and $\mathbf{b} = \begin{pmatrix}b_1\\b_2\end{pmatrix}$.

If **a** and **b** are linearly dependent then

$\mathbf{a} = \lambda\mathbf{b} \quad \Rightarrow \quad \mathbf{a}$ and $\mathbf{b}$ are parallel. i.e.

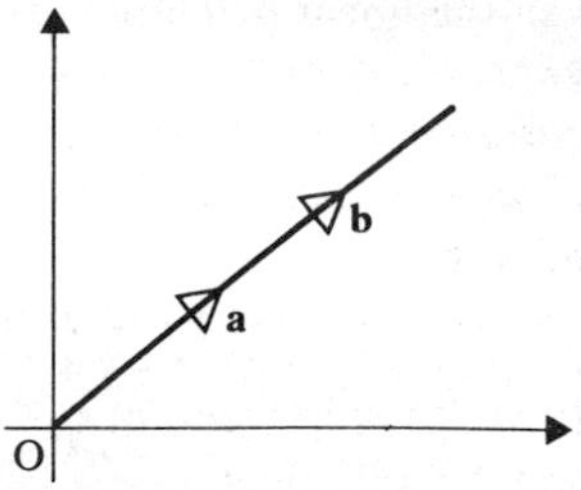

If **a** and **b** are linearly independent then, for all real values of λ

$\mathbf{a} \neq \lambda\mathbf{b} \quad \Rightarrow \quad \mathbf{a}$ and $\mathbf{b}$ are *not* parallel. i.e.

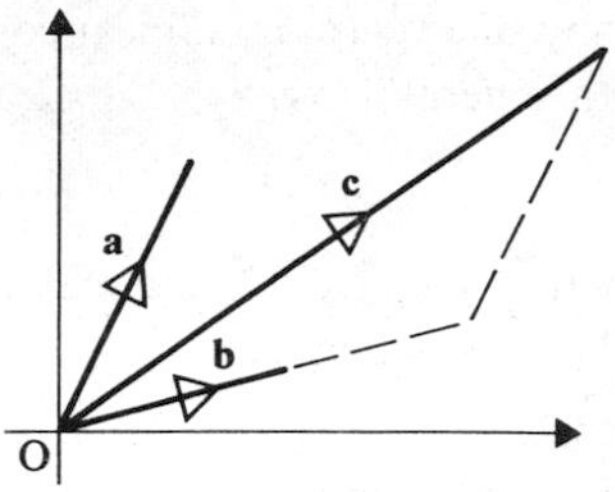

Also, we have seen in *Mathematics – The Core Course for A-level*, that all other vectors in the plane can be expressed in terms of **a** and **b**,
i.e. for any vector **c**, there are real numbers α and β (not both zero) such that

$$\mathbf{c} = \alpha\mathbf{a} + \beta\mathbf{b}$$

This means that *three (or more) 2-vectors will always be linearly dependent.*

There is a similar geometric interpretation of a vector 3-space:

Consider the geometric vectors $\mathbf{a} = \begin{pmatrix} a_1 \\ a_2 \\ a_3 \end{pmatrix}$, $\mathbf{b} = \begin{pmatrix} b_1 \\ b_2 \\ b_3 \end{pmatrix}$ and $\mathbf{c} = \begin{pmatrix} c_1 \\ c_2 \\ c_3 \end{pmatrix}$.

If **a**, **b** and **c** are linearly dependent then there are real numbers α, β and γ (not all zero) such that

$$\alpha\mathbf{a} + \beta\mathbf{b} + \gamma\mathbf{c} = \mathbf{0}$$

i.e. at least one of the vectors in the set can be expressed as a linear combination of the other two.
If, say, $\mathbf{b} = p\mathbf{a} + q\mathbf{c}$, then **b** must be in the same plane as **a** and **c**,
i.e. **a**, **b** and **c** are coplanar.

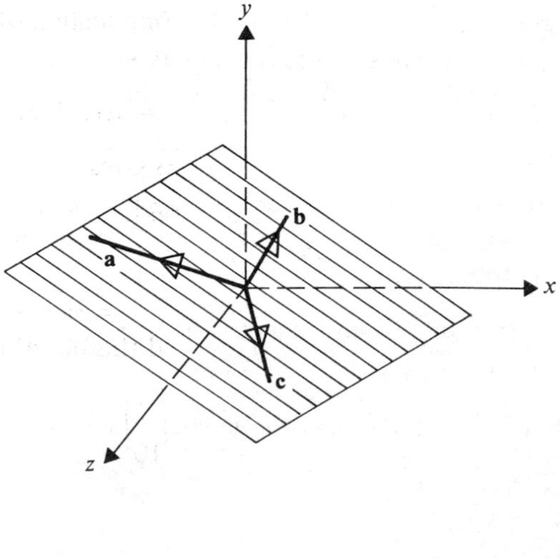

Conversely if **a, b** and **c** are linearly independent, then it is impossible to express any one of **a, b** or **c** in terms of the other two. So **a, b** and **c** are not coplanar and hence form a set of base vectors for three dimensional space. And, as we have seen in *The Core Course*, all other vectors can be expressed in terms of **a, b** and **c**
i.e. *four (or more) 3-vectors must be linearly dependent.*

If we think in terms of the transformation matrix, **M**, that maps **i** to **a**, **j** to **b** and **k** to **c**, then

$$\mathbf{M} = \begin{pmatrix} a_1 & b_1 & c_1 \\ a_2 & b_2 & c_2 \\ a_3 & b_3 & c_3 \end{pmatrix}$$

If **a, b** and **c** are coplanar then **M** collapses three dimensional space to a plane, i.e. **M** is singular.
Conversely, if **a, b** and **c** are not coplanar then **M** is not singular.

So $$\begin{pmatrix} a_1 \\ a_2 \\ a_3 \end{pmatrix}, \begin{pmatrix} b_1 \\ b_2 \\ b_3 \end{pmatrix}, \begin{pmatrix} c_1 \\ c_2 \\ c_3 \end{pmatrix} \text{ are linearly independent } \iff \begin{vmatrix} a_1 & b_1 & c_1 \\ a_2 & b_2 & c_2 \\ a_3 & b_3 & c_3 \end{vmatrix} \neq 0$$

EXAMPLES 2I (continued)

2) V_2 is the vector space of complex numbers and **a, b** and **c** are three vectors in V_2.

If $\mathbf{a} = 2 + 3\text{i}$, $\mathbf{b} = 5 - \text{i}$ and $\mathbf{c} = 1 + 10\text{i}$ show that **a** and **b** are linearly independent and express **c** as a linear combination of **a** and **b**.

It is obvious from inspection that $2 + 3\text{i}$ and $5 - \text{i}$ are linearly independent as it is clear that there is no value of λ for which $2 + 3\text{i} = \lambda(5 - \text{i})$.
To express $1 + 10\text{i}$ as a linear combination of $2 + 3\text{i}$ and $5 - \text{i}$ we need to find values of α and β such that

$$1 + 10\text{i} = \alpha(2 + 3\text{i}) + \beta(5 - \text{i})$$

Equating real and imaginary parts gives

$$\left.\begin{aligned} 1 &= 2\alpha + 5\beta \\ 10 &= 3\alpha - \beta \end{aligned}\right\} \Rightarrow \alpha = 3, \quad \beta = -1$$

So $$\mathbf{c} = 3\mathbf{a} - \mathbf{b}$$

Note that we could have answered this question in vector form by writing

a as $\begin{pmatrix} 2 \\ 3 \end{pmatrix}$, **b** as $\begin{pmatrix} 5 \\ -1 \end{pmatrix}$ and **c** as $\begin{pmatrix} 1 \\ 10 \end{pmatrix}$.

Basis Vectors

We have seen that in V_2, the vector space of two dimensions, when given *a pair of linearly independent vectors* **a** and **b**, all other vectors in V_2 can be expressed as a linear combination of **a** and **b**. We say that *the set* $\{\mathbf{a}, \mathbf{b}\}$ *is a basis for* V_2.

Similarly in V_3, given *a set of three linearly independent vectors* **a**, **b** *and* **c**, all other vectors in V_3 can be expressed as a linear combination of **a**, **b** and **c**, so *the set* $\{\mathbf{a}, \mathbf{b}, \mathbf{c}\}$ *is called a basis for* V_3.
Note that a basis for a vector space is not unique. All that is required for a basis for V_2 is a pair of linearly independent vectors and there is obviously an infinite number of such pairs, any one of which can be used as a basis. The standard basis for V_2 is the pair of unit vectors $\begin{pmatrix}1\\0\end{pmatrix}$ and $\begin{pmatrix}0\\1\end{pmatrix}$.

Similarly in V_3 there is an infinite number of sets of three linearly independent vectors and any one such set is a basis for V_3. The *standard basis* for V_3 is the set of unit vectors $\begin{pmatrix}1\\0\\0\end{pmatrix}, \begin{pmatrix}0\\1\\0\end{pmatrix}, \begin{pmatrix}0\\0\\1\end{pmatrix}$.

Note also that in V_3 we do need *three* linearly independent vectors for a basis. Two linearly independent vectors, **a** and **b**, cannot be a basis for V_3 as linear combinations of **a** and **b** give vectors only in the plane of **a** and **b** and so cannot give all the vectors in V_3.

To generalize, a basis of a vector space, V, is a set of linearly independent vectors in V such that linear combinations of this set give all other vectors in the vector space.

The *dimension* of a vector space can now be defined as the *number of elements* in a basis of that vector space.

EXAMPLES 2I (continued)

3) The real vector space V is the set of all polynomial functions of x of degree less than three. Write down a basis set for V and state the dimension of V.

Any element of V is of the form $ax^2 + bx + c$; $a, b, c \in \mathbb{R}$
and so can be expressed as a linear combination of x^2, x and 1.
Thus $\{x^2, x, 1\}$ is a basis for V.
As the basis has three elements in it, the dimension of V is three.

4) V is the real vector space of triples and

$$\mathbf{a} = \begin{pmatrix} 2 \\ 0 \\ 0 \end{pmatrix}, \quad \mathbf{b} = \begin{pmatrix} 1 \\ -1 \\ 0 \end{pmatrix}, \quad \mathbf{c} = \begin{pmatrix} 0 \\ 1 \\ 1 \end{pmatrix}, \quad \mathbf{d} = \begin{pmatrix} 1 \\ 2 \\ 0 \end{pmatrix}$$

are vectors in V.
Show that **c** cannot be expressed as a linear combination of **a, b** and **d**. State with reasons, whether $\{\mathbf{a}, \mathbf{b}, \mathbf{c}\}$ is a basis for V.

The third components of **a, b** and **d** are all zero, and there is clearly no linear combination of three zeros which gives the third component of **c**, which is 1. So **c** cannot be expressed as a linear combination of **a, b** and **d**. This becomes even clearer if we try to express **c** as a linear combination of **a, b** and **d**, i.e. if we try to find values of α, β and γ for which

$$\alpha \begin{pmatrix} 2 \\ 0 \\ 0 \end{pmatrix} + \beta \begin{pmatrix} 1 \\ -1 \\ 0 \end{pmatrix} + \gamma \begin{pmatrix} 1 \\ 2 \\ 0 \end{pmatrix} = \begin{pmatrix} 0 \\ 1 \\ 1 \end{pmatrix} \quad \Rightarrow \quad \begin{aligned} 2\alpha + \beta + \gamma &= 0 \\ -\beta + 2\gamma &= 1 \\ 0 &= 1 \end{aligned}$$

The last equation is impossible so there are no values of α, β and γ for which $\mathbf{c} = \alpha\mathbf{a} + \beta\mathbf{b} + \gamma\mathbf{d}$.
a, b and **c** form a basis for V if they are linearly independent.
The determinant whose columns are **a, b** and **c** is

$$\begin{vmatrix} 2 & 1 & 0 \\ 0 & -1 & 1 \\ 0 & 0 & 1 \end{vmatrix} = -2 \qquad \text{(i.e. not zero)}$$

So $\{\mathbf{a}, \mathbf{b}, \mathbf{c}\}$ is a basis for V.

5) The vector space V is the set of coordinates of all the points on the plane $x + 2y - z = 0$. Find a basis for V.

If (x, y, z) is any point on the plane $x + 2y - z = 0$, then $z = x + 2y$.

So all elements in V are of the form $(a, b, a + 2b)$ or $\begin{pmatrix} a \\ b \\ a+2b \end{pmatrix}$.

As V is a two dimensional vector space (only two independent real numbers are needed to describe the vectors in V) we need just two vectors for a basis.
If $a = 1$ and $b = 0$ we get the vector $(1, 0, 1)$.
If $a = 0$ and $b = 1$ we get the vector $(0, 1, 2)$

and any vector $\begin{pmatrix} a \\ b \\ a+2b \end{pmatrix}$ in V is the linear combination $a\begin{pmatrix} 1 \\ 0 \\ 1 \end{pmatrix} + b\begin{pmatrix} 0 \\ 1 \\ 2 \end{pmatrix}$.

Hence $\{(1, 0, 1), (0, 1, 2)\}$ is a basis for V.

6) V_3 is the real vector space of triples $\mathbf{x} = (x_1, x_2, x_3)$ and

f is a function such that f maps $\begin{cases} x_1 \text{ to } x_1 + x_2 \\ x_2 \text{ to } x_2 + x_3 \\ x_3 \text{ to } x_1 + 3x_2 + 2x_3 \end{cases}$

Determine the dimension of the image of V_3 under the function f.

The given function is $f\colon \begin{pmatrix} x_1 \\ x_2 \\ x_3 \end{pmatrix} \longmapsto \begin{pmatrix} x_1+x_2 \\ x_2+x_3 \\ x_1+3x_2+x_3 \end{pmatrix}$

If V' is the image of V_3 under f, then V' is the set of all vectors of the form $\mathbf{x}' = \begin{pmatrix} x_1+x_2 \\ x_2+x_3 \\ x_1+3x_2+x_3 \end{pmatrix}$

To determine the dimension of V' we need to know if the three components of $\mathbf{x}'$ are independent or if they are related in some way. If there is a relationship between the components of $\mathbf{x}'$ it must be linear in form (there is clearly no other relationship between the components).
By inspection we can see that the third component is the sum of the first and twice the second. If such a relationship cannot be seen by inspection, it can be found, if it exists, as follows.
If there is a linear relationship then constants α, β and γ (not all zero) can be found such that

$$\alpha(x_1 + x_2) + \beta(x_2 + x_3) + \gamma(x_1 + 3x_2 + 2x_3) = 0 \qquad [1]$$

or, rearranging

$$(\alpha + \gamma)x_1 + (\alpha + \beta + 3\gamma)x_2 + (\beta + 2\gamma)x_3 = 0$$

As this relationship holds for all values of x_1, x_2 and x_3, α, β and γ satisfy the three equations

$$\left.\begin{aligned} \alpha + \gamma &= 0 \\ \alpha + \beta + 3\gamma &= 0 \\ \beta + 2\gamma &= 0 \end{aligned}\right\} \quad \Rightarrow \quad \alpha = \lambda, \quad \beta = 2\lambda, \quad \gamma = -\lambda$$

for all values of λ. So $\alpha = 1, \beta = 2$ and $\gamma = -1$ is a solution.
Substituting these values in equation [1] gives the third component of $\mathbf{x}'$ as the sum of the first and twice the second.

Hence $\mathbf{x}'$ is of the form $\begin{pmatrix} a \\ b \\ a+2b \end{pmatrix}$ and V' is a vector space of dimension two.

EXERCISE 2I

In Questions 1–8 determine whether the given sets of vectors are linearly independent. If they are not, find a linear relationship between them.

1) $\mathbf{a} = (1, 4)$, $\mathbf{b} = (-2, 8)$ 2) $\mathbf{a} = (1, 0, 3)$, $\mathbf{b} = (2, 0, 6)$

3) $\mathbf{a} = (1, -2)$, $\mathbf{b} = (-2, 4)$ 4) $\mathbf{a} = (0, 1, 4)$, $\mathbf{b} = (2, 0, 1)$

5) $\mathbf{a} = (1, 2)$, $\mathbf{b} = (2, 1)$, $\mathbf{c} = (0, 1)$

6) $\mathbf{a} = (3, -1, 2)$, $\mathbf{b} = (7, -1, 1)$, $\mathbf{c} = (4, 0, -1)$

7) $\mathbf{a} = (1, 1, 0)$, $\mathbf{b} = (1, 0, 1)$, $\mathbf{c} = (0, 1, 1)$

8) $\mathbf{a} = (1, 1, 0)$, $\mathbf{b} = (2, 0, 1)$, $\mathbf{c} = (3, 2, 1)$, $\mathbf{d} = (1, 1, 1)$

9) V is the vector space of all linear functions of x (i.e. polynomials of degree one or zero). If $2x - 1$ and $-x + 2$ are two vectors in V, show that they are linearly independent and express $5x + 3$ as a linear combination of $2x - 1$ and $-x + 2$.
Write down a basis for V other than $\{2x - 1, -x + 2\}$.

10) V is the vector space of all polynomial functions of x of degree less than four. State the dimension of V and write down a basis for V.

11) V_2 is a real vector space of dimension two. The vectors $\mathbf{a}, \mathbf{b}, \mathbf{c}$ and $\mathbf{d}$ in V_2 are given by

$$\mathbf{a} = \begin{pmatrix} 0 \\ 1 \end{pmatrix}, \quad \mathbf{b} = \begin{pmatrix} 1 \\ 0 \end{pmatrix}, \quad \mathbf{c} = \begin{pmatrix} 1 \\ 1 \end{pmatrix}, \quad \mathbf{d} = \begin{pmatrix} 2 \\ 2 \end{pmatrix}.$$

Which of the following sets is a basis for V_2?
(i) $\{\mathbf{a}, \mathbf{b}\}$, (ii) $\{\mathbf{b}, \mathbf{c}\}$, (iii) $\{\mathbf{c}, \mathbf{d}\}$, (iv) $\{\mathbf{c}\}$, (v) $\{\mathbf{a}, \mathbf{d}\}$, (vi) $\{\mathbf{a}, \mathbf{b}, \mathbf{c}\}$.

12) V_3 is a real vector space of dimension three. The vectors $\mathbf{a}, \mathbf{b}, \mathbf{c}$ and $\mathbf{d}$ in V_3 are given by

$$\mathbf{a} = \begin{pmatrix} 1 \\ 1 \\ 0 \end{pmatrix}, \quad \mathbf{b} = \begin{pmatrix} 1 \\ 0 \\ 1 \end{pmatrix}, \quad \mathbf{c} = \begin{pmatrix} 0 \\ 1 \\ 1 \end{pmatrix}, \quad \mathbf{d} = \begin{pmatrix} 2 \\ 2 \\ 1 \end{pmatrix}.$$

Which of the following sets is a basis for V_3?
(i) $\{\mathbf{a}, \mathbf{b}, \mathbf{c}\}$, (ii) $\{\mathbf{a}, \mathbf{c}\}$, (iii) $\{\mathbf{b}, \mathbf{c}, \mathbf{d}\}$ (iv) $\{\mathbf{a}, \mathbf{b}, \mathbf{d}\}$, (v) $\{\mathbf{a}, \mathbf{b}, \mathbf{c}, \mathbf{d}\}$.

13) V_2 is the real vector space of the coordinates of all points in the xy plane. A subspace of V_2 is the vector space, V, of the coordinates of all points on the

line $y = x$. State the dimension of V and give a basis for V. Explain why the set of coordinates of all points on the line $y = x + 1$ is not a vector space.

14) A is the set of coordinates of all points on the plane $x + y - z = 0$ and B is the set of coordinates of all points on the plane $x + y + z = 2$. One of these sets is a vector space and the other is not. State which set is not a vector space. For the set which is a vector space, state its dimension and give a basis for the vector space.

15) V_3 is a real vector space of dimension three and $\mathbf{x} = (x_1, x_2, x_3)$ is a vector in V_3.
f_1 and f_2 are two functions defined by

$$f_1\colon (x_1, x_2, x_3) \longmapsto (2x_1, 2x_2, 2x_3)$$

$$f_2\colon (x_1, x_2, x_3) \longmapsto (x_1, x_2, x_1 + x_2)$$

V' is the image of V_3 under f_1, and V'' is the image of V_3 under f_2.

(a) State the dimension of V' and of V''.
(b) Write down a basis for V' and for V''.
(c) Define the functions $f_1 f_2$ and $f_2 f_1$.
(d) State the dimension of the image of V_3 under the function $f_2 f_1$.

16) V_3 is a real vector space of triples $\mathbf{x} = (x_1, x_2, x_3)$ and f is a function defined by $f\colon (x_1, x_2, x_3) \longmapsto (x_1^2, x_2^2, x_3^2)$.
Explain why the image of V_3 under f is not a vector space.

17) V_3 is a real vector space of triples $\mathbf{x} = (x_1, x_2, x_3)$ and f is a function defined by

(a) $f\colon (x_1, x_2, x_3) \longmapsto (x_1, x_1, x_1)$
(b) $f\colon (x_1, x_2, x_3) \longmapsto (-x_1, -x_2, -x_3)$
(c) $f\colon (x_1, x_2, x_3) \longmapsto (x_1 + 1, x_2 + 1, x_3 + 1)$
(d) $f\colon (x_1, x_2, x_3) \longmapsto (x_1, x_2^2, x_3)$.

Write down the dimension of the image of V_3 under f.
(*Hint*. In one case the image is not a vector space.)

18) V_4 is a vector space of quadruples $\mathbf{x} = (x_1, x_2, x_3, x_4)$.
Write down a basis for V_4.
The matrix $\mathbf{M}$ maps V_4 to V_3, a vector space of dimension three, where

$$\mathbf{M} = \begin{pmatrix} 1 & 0 & 0 & 1 \\ 0 & 1 & 0 & 1 \\ 0 & 0 & 1 & 1 \\ 1 & 0 & 0 & 1 \end{pmatrix}$$

Write down the image of $\mathbf{x} = (x_1, x_2, x_3, x_4)$ under $\mathbf{M}$.

Subspaces

We end this section with a brief introduction to two further concepts associated with vector spaces, those of subspaces and spanning sets.
We have already seen that the set of coordinates of all points on the plane $x+y+z=0$ is a vector space. This is a subset of the vector space, V_3, the set of coordinates of all points in three dimensional space. Such a subset is called a subspace. In general any subset, S, of a vector space V, where S is itself a vector space is called a subspace of V. However, to show that a subset of V is a subspace we do not have to show that all the properties of a vector space are satisfied, we can use the following definition.

If V is a vector space and S is a (non-empty) subset of V, then S is a subspace of V if
(a) S is closed under addition,
(i.e. if $\mathbf{x} \in S$ and $\mathbf{y} \in S$ then $\mathbf{x}+\mathbf{y} \in S$),
(b) S is closed under multiplication by a scalar,
(i.e. if $\mathbf{x} \in S$ and $\alpha \in \mathbb{R}$ then $\alpha\mathbf{x} \in S$).

That this definition is sufficient to ensure that S is a vector space follows from (b), because $0\mathbf{x} \in S$, so S includes the zero vector and $(-1)\mathbf{x} \in S$, so every vector in S has an inverse which is also in S. The remaining properties of a vector space are satisfied by the vectors in S because they are satisfied by all the vectors in V.
Thus to show that a subset of a vector space is a subspace we have only to show that the vectors in the subset satisfy the two properties given above.

Spanning Sets

If $\{\mathbf{x}_1, \mathbf{x}_2, \ldots \mathbf{x}_r\}$ is a set of vectors in a vector space then *all* possible linear combinations of the set is called their span.
For example, consider the set of coordinates $\{(1,1),(2,2)\}$.
Their span is the set $\{(\alpha+2\beta, \alpha+2\beta),\ \alpha \text{ and } \beta \in \mathbb{R}\}$

$=$ the set of coordinates of points on the line $y=x$.

A spanning set of a vector space V is a set of vectors whose span is the whole of V.

Note that a spanning set is not the same as a basis, it can contain more vectors than a basis,

for example, $\left\{\begin{pmatrix}1\\0\end{pmatrix}, \begin{pmatrix}0\\1\end{pmatrix}, \begin{pmatrix}1\\1\end{pmatrix}\right\}$ is a spanning set for V_2, a vector space of dimension two.

But any pair of three vectors in this set is a basis for V_2. So we can think of a basis as the most efficient spanning set we can find, i.e. a spanning set with the least number of vectors possible.

SUMMARY

Any transformation of three dimensional space which is linear can be expressed in the form $\mathbf{Mr} = \mathbf{r}'$ where $\mathbf{M} = \begin{pmatrix} a_1 & b_1 & c_1 \\ a_2 & b_2 & c_2 \\ a_3 & b_3 & c_3 \end{pmatrix}$ and $\mathbf{r}'$ is the image of $\mathbf{r}$ under $\mathbf{M}$.

Under $\mathbf{M}$ $\quad \mathbf{i} \longmapsto \begin{pmatrix} a_1 \\ a_2 \\ a_3 \end{pmatrix}, \quad \mathbf{j} \longmapsto \begin{pmatrix} b_1 \\ b_2 \\ b_3 \end{pmatrix}, \quad \mathbf{k} \longmapsto \begin{pmatrix} c_1 \\ c_2 \\ c_3 \end{pmatrix}$

$\mathbf{M}$ multiplies volume by a factor $|\mathbf{M}|$.

If $|\mathbf{M}| = 0$, volume is destroyed, and $\mathbf{M}$ maps xyz space to a line or a plane through O.

Three vectors $\mathbf{a} = \begin{pmatrix} a_1 \\ a_2 \\ a_3 \end{pmatrix}$, $\mathbf{b} = \begin{pmatrix} b_1 \\ b_2 \\ b_3 \end{pmatrix}$, $\mathbf{c} = \begin{pmatrix} c_1 \\ c_2 \\ c_3 \end{pmatrix}$ are coplanar if

$$\begin{vmatrix} a_1 & b_1 & c_1 \\ a_2 & b_2 & c_2 \\ a_3 & b_3 & c_3 \end{vmatrix} = 0$$

If $\mathbf{A}^T$ is the transpose of any matrix $\mathbf{A}$, then $\mathbf{A}^T$ is the matrix whose rows are the columns of $\mathbf{A}$.

For any two matrices $\mathbf{A}$ and $\mathbf{B}$ that are compatible for multiplication $(\mathbf{AB})^T = \mathbf{B}^T\mathbf{A}^T$.

For any *square* matrix $\mathbf{A}$, the adjoint matrix, adj $\mathbf{A}$, is the matrix whose elements are the cofactors of $\mathbf{A}^T$.

For any square matrix $\mathbf{A}$, the inverse matrix, $\mathbf{A}^{-1}$, exists if and only if $|\mathbf{A}| \neq 0$ and in this case $\mathbf{A}^{-1} = \dfrac{1}{|\mathbf{A}|} \operatorname{adj} \mathbf{A}$.

$\mathbf{A}^{-1}\mathbf{A} = \mathbf{I} = \mathbf{AA}^{-1}$ and $(\mathbf{A}^T)^{-1} = (\mathbf{A}^{-1})^T$.

If $\mathbf{A}$ and $\mathbf{B}$ both have inverses and are of the same size then

$$(\mathbf{AB})^{-1} = \mathbf{B}^{-1}\mathbf{A}^{-1}$$

If a transformation, defined by $\mathbf{M}$, maps $\mathbf{r}$ to $\mathbf{r}'$
the inverse transformation, defined by $\mathbf{M}^{-1}$, maps $\mathbf{r}'$ back to $\mathbf{r}$.

A set of linear equations has either

a unique solution in which case $|\mathbf{A}| \neq 0$

or an infinite set of solutions in which case $|\mathbf{A}| = 0$ and the equations are linearly dependent

or no solution, in which case $|\mathbf{A}| = 0$ and the equations are inconsistent.

Three lines

$$\left.\begin{aligned} y - m_1x - c_1 &= 0 \\ y - m_2x - c_2 &= 0 \\ y - m_3x - c_3 &= 0 \end{aligned}\right\} \text{ are concurrent if } \begin{vmatrix} 1 & -m_1 & -c_1 \\ 1 & -m_2 & -c_2 \\ 1 & -m_3 & -c_3 \end{vmatrix} = 0$$

MULTIPLE CHOICE EXERCISE 2

(*The instructions for answering these questions are on p. xii.*)

TYPE I

1) If $\mathbf{M} = \begin{pmatrix} 1 & 0 & 0 \\ 1 & 1 & 0 \\ 0 & 1 & 0 \end{pmatrix}$, $\mathbf{M}^T$ is

(a) $\begin{pmatrix} 1 & 1 & 0 \\ 1 & 1 & 0 \\ 0 & 1 & 0 \end{pmatrix}$ (b) $\begin{pmatrix} 1 & 1 & 0 \\ 0 & 1 & 1 \\ 0 & 0 & 0 \end{pmatrix}$ (c) $\begin{pmatrix} 0 & 0 & 1 \\ 0 & 0 & -1 \\ 0 & 0 & 1 \end{pmatrix}$ (d) $\begin{pmatrix} 0 & 1 & 0 \\ 1 & 1 & 0 \\ 1 & 0 & 0 \end{pmatrix}$.

2) The inverse of the matrix $\begin{pmatrix} 1 & -1 \\ 2 & 1 \end{pmatrix}$ is

(a) $\begin{pmatrix} 1 & 1 \\ -2 & 1 \end{pmatrix}$ (b) $\dfrac{1}{3}\begin{pmatrix} 1 & 2 \\ -1 & 1 \end{pmatrix}$ (c) $\begin{pmatrix} 1 & 0 \\ 0 & 1 \end{pmatrix}$

(d) $\dfrac{1}{3}\begin{pmatrix} 1 & 1 \\ -2 & 1 \end{pmatrix}$ (e) $1\bigg/\begin{pmatrix} 1 & -1 \\ 2 & 1 \end{pmatrix}$.

3) If $\mathbf{M} = \begin{pmatrix} 1 & 0 & 0 \\ 1 & 1 & 1 \\ 0 & 1 & 2 \end{pmatrix}$, a reduced echelon form may be

(a) $\begin{pmatrix} 1 & 0 & 0 \\ 0 & 0 & -1 \\ 0 & 1 & 2 \end{pmatrix}$ (b) $\begin{pmatrix} 1 & 0 & 0 \\ 1 & \frac{1}{2} & 0 \\ 0 & 1 & 2 \end{pmatrix}$ (c) $\begin{pmatrix} 0 & 0 & 0 \\ 1 & 1 & 1 \\ 0 & 1 & 2 \end{pmatrix}$ (d) $\begin{pmatrix} 1 & 0 & 0 \\ 1 & 1 & 0 \\ 0 & 1 & 2 \end{pmatrix}$

4) The set of equations $\begin{cases} x - 2y + 4 = 0 \\ 2x - 4y - 3 = 0 \end{cases}$

(a) are inconsistent, (b) have solutions depending on one parameter,
(c) have a unique solution, (d) represent a pair of intersecting lines.

5) If $\mathbf{M} = \begin{pmatrix} 1 & 0 & 0 \\ 1 & 1 & 0 \\ 0 & 0 & 1 \end{pmatrix}$, adj $\mathbf{M}$ is

(a) $\begin{pmatrix} 1 & -1 & 0 \\ 0 & 1 & 0 \\ 0 & 0 & 1 \end{pmatrix}$ (b) $\begin{pmatrix} 1 & 1 & 0 \\ 0 & 1 & 0 \\ 0 & 0 & 1 \end{pmatrix}$ (c) $\begin{pmatrix} 1 & 0 & 0 \\ 0 & 1 & 0 \\ 0 & 0 & 1 \end{pmatrix}$ (d) $\begin{pmatrix} 1 & 0 & 0 \\ -1 & 1 & 0 \\ 0 & 0 & 1 \end{pmatrix}$.

6) The transformation of xyz space defined by $\begin{pmatrix} 2 & 0 & 0 \\ 1 & 2 & 4 \\ 1 & -1 & 3 \end{pmatrix}$.

(a) is singular, (b) preserves volume,
(c) maps 3-D space to a 2-D plane, (d) multiplies volume by a factor 20.

TYPE II

7) A transformation of xyz space maps

$$x \text{ to } 2x + y - z$$
$$y \text{ to } x - y + z$$
$$z \text{ to } x + y.$$

(a) The transformation matrix is $\begin{pmatrix} 2 & 1 & -1 \\ 1 & -1 & 1 \\ 1 & 1 & 0 \end{pmatrix}$.

(b) Volume is unchanged by the transformation.

(c) The point $(2, 0, 0)$ is mapped to the point $(-3, 0, 0)$.

8) If $\mathbf{M} = \begin{pmatrix} 1 & 0 & 0 \\ 0 & 0 & 1 \\ 0 & 1 & 0 \end{pmatrix}$.

(a) $\mathbf{M}$ has an inverse, (b) $\mathbf{M}$ represents a rotation about Ox of $90°$,
(c) $\mathbf{M}$ represents a reflection in the plane $y = z$.

9) $\mathbf{M} = \begin{pmatrix} 2 & -1 & 0 \\ 2 & 1 & 4 \\ 3 & 0 & 2 \end{pmatrix}$.

(a) The equation $\mathbf{M}\begin{pmatrix} x \\ y \\ z \end{pmatrix} = \begin{pmatrix} 0 \\ 0 \\ 0 \end{pmatrix}$ has a unique solution,

(b) $\mathbf{M}$ is singular, (c) $\mathbf{M}$ maps $\begin{pmatrix} 0 \\ 1 \\ 0 \end{pmatrix}$ to $\begin{pmatrix} -1 \\ 1 \\ 0 \end{pmatrix}$.

10) $\mathbf{A} = \begin{pmatrix} 1 & 2 \\ 3 & 4 \end{pmatrix}$.

(a) $\mathbf{A}^T = \begin{pmatrix} 1 & 3 \\ 2 & 4 \end{pmatrix}$ (b) adj $\mathbf{A} = \begin{pmatrix} 4 & -3 \\ -2 & 1 \end{pmatrix}$ (c) $(\mathbf{A}^T)^{-1} = \begin{pmatrix} -2 & \frac{3}{2} \\ 1 & -\frac{1}{2} \end{pmatrix}$.

11) $2x + y - z = 4$
$x + y + z = 1$
$3x - 2y - z = 2$.

(a) The planes represented by these equations are parallel.

(b) The solution is the same as the solution of $\begin{pmatrix} 11 & 0 & 0 \\ 4 & -1 & 0 \\ 3 & -2 & -1 \end{pmatrix}\begin{pmatrix} x \\ y \\ z \end{pmatrix} = \begin{pmatrix} 11 \\ 3 \\ 2 \end{pmatrix}$.

(c) The solution is unique.

TYPE III

12) $a_1x + b_1y + c_1z = d_1, \quad a_2x + b_2y + c_2z = d_2, \quad a_3x + b_3y + c_3z = d_3.$

(a) $\begin{vmatrix} a_1 & b_1 & c_1 \\ a_2 & b_2 & c_2 \\ a_3 & b_3 & c_3 \end{vmatrix} = 0.$

(b) The planes represented by these equations are parallel.

13) $\mathbf{A}$ is a non-singular square matrix.

(a) $|\mathbf{A}^{-1}| = \Delta$ (b) $|\mathbf{A}| = \dfrac{1}{\Delta}$.

14) $\mathbf{A}$ is a non-singular square matrix.
(a) $\mathbf{A}$ is of size 3×3.
(b) Under the transformation defined by $\mathbf{A}$, any point in the image space originates from a unique point in xyz space.

15) $\mathbf{a} = \begin{pmatrix} a_1 \\ a_2 \\ a_3 \end{pmatrix}, \quad \mathbf{b} = \begin{pmatrix} b_1 \\ b_2 \\ b_3 \end{pmatrix}, \quad \mathbf{c} = \begin{pmatrix} c_1 \\ c_2 \\ c_3 \end{pmatrix}.$

(a) $\mathbf{a} = \lambda\mathbf{b} + \mu\mathbf{c}$ where λ and μ are scalars. (b) $\begin{vmatrix} a_1 & a_2 & a_3 \\ b_1 & b_2 & b_3 \\ c_1 & c_2 & c_3 \end{vmatrix} = 0.$

16) $2x - 5y + 6 = 0$ and $x + y - 2 = 0$ represents lines L_1 and L_2.
(a) The line $ax + by + c = 0$ is concurrent with L_1 and L_2,

(b) $\begin{vmatrix} 2 & -5 & 6 \\ 1 & 1 & -2 \\ a & b & c \end{vmatrix} = 0$

TYPE V

17) If $\mathbf{A}$ is singular, then $\mathbf{A}$ has no inverse.

18) If three planes form a triangular prism, the equations representing them have no solution.

19) Under a particular transformation, a plane Π is mapped to its image plane Π'. If the normal to Π is mapped to the vector $\mathbf{n}$, then $\mathbf{n}$ is normal to Π'.

20) The vectors $(2, 1), (3, -1), (0, 4)$ are linearly independent.

21) Three linearly independent 3-vectors form a basis for a vector space of dimension three.

22) The set $\{x : x \in \mathbb{N}\}$ is a vector space.

23) The set of matrices spanned by

$$\begin{pmatrix} 1 & 0 \\ 0 & 1 \end{pmatrix} \quad \text{and} \quad \begin{pmatrix} 0 & 1 \\ 1 & 0 \end{pmatrix}$$

is a vector space.

MISCELLANEOUS EXERCISE 2

1) The planes

$$\begin{aligned} 2x + y + z &= 4 \\ x + 2y + z &= 2 \\ x + y + 2z &= 6 \end{aligned}$$

meet only in the point $(1, -1, 3)$. The x, y, z coordinate system is transformed by the linear transformation

$$\begin{pmatrix} x \\ y \\ z \end{pmatrix} = \frac{1}{3}\begin{pmatrix} 1 & 2 & 2 \\ 2 & -2 & 1 \\ 2 & 1 & -2 \end{pmatrix}\begin{pmatrix} X \\ Y \\ Z \end{pmatrix}.$$

In the X, Y, Z system, obtain the equations of the planes and the coordinates of the point(s) in which they meet. (U of L)

2) The equations

$$\begin{aligned} 2\lambda x - 3y + \lambda - 3 &= 0 \\ 3x - 2y + 1 &= 0 \\ 4x - \lambda y + 2 &= 0 \end{aligned}$$

represent three straight lines in the xy plane. Find the values of λ for which the lines are concurrent. For each of these values of λ, find the coordinates of the point at which the lines are concurrent. (U of L)p

3) If $\mathbf{M} = \begin{pmatrix} 0 & 1 & 0 \\ -1 & 0 & 0 \\ 0 & 0 & 1 \end{pmatrix}$, find $\mathbf{M}^2$.

Interpret geometrically the transformations of xyz space defined by $\mathbf{M}$ and by $\mathbf{M}^2$. (U of L)

4) (a) Solve for x, y and z the equations

$$\begin{aligned} 2x + 6y + z &= 0 \\ -x + 2y - z &= 10 \\ 4x + 3y + z &= 1. \end{aligned}$$

(b) If $\mathbf{A}$ and $\mathbf{B}$ are non-singular matrices, show that
(i) $(\mathbf{AB})^{-1} = \mathbf{B}^{-1}\mathbf{A}^{-1}$ (ii) $(\mathbf{AB})^T = \mathbf{B}^T\mathbf{A}^T$. (U of L)

5) The point $P(x, y, z)$ is transformed to the point $Q(X, Y, Z)$ by the relation

$$\begin{pmatrix} X \\ Y \\ Z \end{pmatrix} = \mathbf{M}\begin{pmatrix} x \\ y \\ z \end{pmatrix}.$$

(a) If $\mathbf{M} = \begin{pmatrix} 6 & 8 & 4 \\ 9 & 12 & 6 \\ 4 & -1 & 3 \end{pmatrix}$, show that for all P the corresponding point Q lies on a plane and gives an equation of this plane.

(b) If $\mathbf{M} = \begin{pmatrix} 1 & 2 & -1 \\ 3 & 6 & -3 \\ 5 & 10 & -5 \end{pmatrix}$, show that for all P the corresponding point Q lies on a line and give equations for this line.

(c) If $\mathbf{M} = \begin{pmatrix} 0 & -1 & 0 \\ 1 & 0 & 0 \\ 0 & 0 & 1 \end{pmatrix}$, show that for all P the corresponding point Q is in the position P would reach if it were rotated through 90° about Oz and state the inverse matrix in this case. (U of L)

6) Find numbers a, b and c so that the product $\mathbf{BA}$ of the matrices

$$\mathbf{B} = \begin{pmatrix} 1 & 0 & 0 \\ a & 1 & 0 \\ b & c & 1 \end{pmatrix}, \quad \mathbf{A} = \begin{pmatrix} 1 & 2 & 3 \\ -2 & 1 & 4 \\ 2 & 1 & 1 \end{pmatrix}$$

should have only zeros below the leading diagonal.

Hence, or otherwise, solve the equation $\mathbf{Ax} = \boldsymbol{\rho}$ where $\mathbf{x}$ is the column vector $\begin{pmatrix} x_1 \\ y_1 \\ z_1 \end{pmatrix}$ and $\boldsymbol{\rho}$ is the column vector $\begin{pmatrix} 1 \\ -1 \\ 1 \end{pmatrix}$. (U of L)

7) Find the inverse $\mathbf{A}^{-1}$ of the matrix $\mathbf{A} = \begin{pmatrix} 1 & 0 & 0 \\ -1 & 1 & 0 \\ 3 & 2 & 1 \end{pmatrix}$.

Find also $\mathbf{B}^{-1}$ and $(\mathbf{AB})^{-1}$ where $\mathbf{B} = \begin{pmatrix} 1 & 4 & -2 \\ 0 & 1 & 3 \\ 0 & 0 & 1 \end{pmatrix}$.

Given that $\mathbf{AB}\begin{pmatrix} x_1 \\ x_2 \\ x_3 \end{pmatrix} = \begin{pmatrix} 1 \\ -2 \\ 1 \end{pmatrix}$ find $\begin{pmatrix} x_1 \\ x_2 \\ x_3 \end{pmatrix}$. (U of L)

8) Given that $\mathbf{A} = \begin{pmatrix} 0 & 2 & 3 \\ 2 & 0 & 0 \\ 1 & -1 & 0 \end{pmatrix}$ evaluate $\mathbf{A}^{-1}$ and $\mathbf{A}^2$, and show that $\mathbf{A}^2 + 6\mathbf{A}^{-1} = 7\mathbf{I}$, where $\mathbf{I}$ is the unit matrix of order 3.
Deduce that $\mathbf{A}^3 - 7\mathbf{A} + 6\mathbf{I} = 0$ and use this equation to find $\mathbf{A}^3$. (U of L)

9) Calculate the inverse $\mathbf{A}^{-1}$ of the matrix $\mathbf{A} = \begin{pmatrix} 2 & 2 & 1 \\ 2 & 4 & 1 \\ 3 & 2 & 0 \end{pmatrix}$.

Find the values of λ for which the determinant of the matrix $(\mathbf{A} - \lambda\mathbf{I})$ equals 0, where $\mathbf{I}$ is the unit matrix of order 3.
Show that $\mathbf{A}^2 - 6\mathbf{A} - \mathbf{I}$ is a multiple of $\mathbf{A}^{-1}$. (U of L)

10) If $\mathbf{B}$ is the row vector $\begin{pmatrix} 1 & 1 & 1 \end{pmatrix}$ and $\mathbf{C}$ is the matrix

$$\begin{pmatrix} 2 & 1 & 0 \\ 0 & 3 & 4 \end{pmatrix}$$

find the matrix $\mathbf{A}$ such that $\mathbf{BA}$ is the row vector $\begin{pmatrix} \frac{3}{2} & -\frac{1}{4} \end{pmatrix}$ and $\mathbf{CA}$ is a unit matrix. (U of L)

11) If $\mathbf{M}=\begin{pmatrix}1 & -1 & k\\ 4 & 7 & 3\\ -1 & 12 & -2\end{pmatrix}$, evaluate, in terms of k, the determinant of the matrix $\mathbf{M}$.

If $\mathbf{x}=\begin{pmatrix}x\\ y\\ z\end{pmatrix}$, solve the equations

(a) $\mathbf{Mx}=\begin{pmatrix}1\\ 11\\ 21\end{pmatrix}$ when $k=2$, (b) $\mathbf{Mx}=\begin{pmatrix}0\\ 0\\ 0\end{pmatrix}$ when $k=1$.

Interpret the result (b) geometrically. (U of L)

12) For the matrix equation $\begin{pmatrix}1 & 1 & 1\\ 1 & 2 & 3\\ 1 & 3 & k\end{pmatrix}\begin{pmatrix}x\\ y\\ z\end{pmatrix}=\begin{pmatrix}3\\ 6\\ 4+k\end{pmatrix}$ find the value of k for which the equation does not have a unique solution. For this value of k, solve the equation and interpret the solution geometrically. (C)p

13) $\mathbf{M}$ denotes the matrix $\begin{pmatrix}1 & 2\\ 3 & k\end{pmatrix}$ and $\mathbf{v}$ denotes the vector $\begin{pmatrix}x\\ y\end{pmatrix}$.

Find the solution set of vectors to which $\mathbf{v}$ must belong in each of the following cases:

(a) $k=5$ and $\mathbf{Mv}=\begin{pmatrix}3\\ 1\end{pmatrix}$ (b) $k=5$ and $\mathbf{Mv}=\begin{pmatrix}0\\ 0\end{pmatrix}$

(c) $k=6$ and $\mathbf{Mv}=\begin{pmatrix}1\\ 3\end{pmatrix}$ (d) $k=6$ and $\mathbf{Mv}=\begin{pmatrix}1\\ 1\end{pmatrix}$.

Comment briefly on the connection between these solution sets and the existence or otherwise of the inverse matrix $\mathbf{M}^{-1}$. (C)

14) Find the matrix $\mathbf{X}$ given that $\mathbf{AXA}^{-1} = \mathbf{B}$ where

$$\mathbf{A} = \begin{pmatrix} 2 & 1 \\ 3 & 2 \end{pmatrix}, \quad \mathbf{B} = \begin{pmatrix} 1 & 0 \\ 0 & 2 \end{pmatrix}.$$

(C)p

15) Find the inverse of the matrix

$$\begin{pmatrix} 2 & -1 & 3 \\ 5 & 4 & -3 \\ 3 & -2 & -1 \end{pmatrix}.$$

Hence, or otherwise, solve the equations

$$\begin{aligned} 2x - y + 3z &= -25 \\ 5x + 4y - 3z &= -1 \\ 3x - 2y - z &= -17. \end{aligned}$$

(C)

16) Use any method to find the inverse of the matrix

$$\begin{pmatrix} 3 & -1 & 5 \\ 4 & 3 & 3 \\ 5 & -4 & -2 \end{pmatrix}$$

Solve the equations

$$\begin{aligned} 3x - y + 5z &= 4 \\ 4x + 3y + 3z &= 3 \\ 5x - 4y - 2z &= 3. \end{aligned}$$

(C)

17) A transformation in three dimensional space takes the point (x, y, z) to (x_1, y_1, z_1) where

$$\begin{pmatrix} x_1 \\ y_1 \\ z_1 \end{pmatrix} = \begin{pmatrix} 0 & 0 & 1 \\ 1 & 0 & 0 \\ 0 & 1 & 0 \end{pmatrix} \begin{pmatrix} x \\ y \\ z \end{pmatrix}$$

Prove that the transformation leaves unaltered
(a) the distance between two points;
(b) the points of the line $x = y = z$.
Assuming that the transformation is a rotation about a line, find the angle of rotation.

(O)

18) If the system of equations

$$2x + y + z = \lambda x$$
$$x + 2y + z = \lambda y$$
$$x + y + 2z = \lambda z$$

has solutions in which x, y, z are not all zero, find the possible values of λ. For each such value of λ, find the general solution of the system. (O)

19) Show that any real 2×2 matrix $\mathbf{A} = \begin{pmatrix} a & b \\ c & d \end{pmatrix}$ satisfies a certain quadratic equation $\mathbf{A}^2 + p\mathbf{A} + q\mathbf{I} = \mathbf{O}$, where $\mathbf{I}$ and $\mathbf{O}$ are the unit and zero 2×2 matrices, and p, q are certain numbers depending on a, b, c and d. If $q \neq 0$, find, in the form $\alpha\mathbf{A} + \beta\mathbf{I}$, the inverse of $\mathbf{A}$ (that is, a matrix $\mathbf{B}$ such that $\mathbf{AB} = \mathbf{I}$).

20) The lines u, v, w are the bisectors of the angles between Oy and Oz, Oz and Ox, Ox and Oy respectively. P is the typical point (x, y, z). Prove that if the reflection P_1 of P in the line u is (x_1, y_1, z_1) then

$$\begin{pmatrix} x_1 \\ y_1 \\ z_1 \end{pmatrix} = \begin{pmatrix} -1 & 0 & 0 \\ 0 & 0 & 1 \\ 0 & 1 & 0 \end{pmatrix} \begin{pmatrix} x \\ y \\ z \end{pmatrix}$$

The reflection of P_1 in y is P_2 and the reflection of P_2 in w is P_3. Find the coordinates of P_3 and show that P_3 coincides with P only if P lies on the y axis. (O)

21) Find the values of k for which the matrix

$$\begin{pmatrix} k+2 & 2k+3 & 0 \\ -4 & k-5 & 2-2k \\ 3 & 4 & k-1 \end{pmatrix}$$

has no inverse. (JMB)

22) By reducing the equation

$$\begin{pmatrix} 1 & 2 & -3 \\ 2 & 6 & -11 \\ 1 & -2 & 7 \end{pmatrix} \begin{pmatrix} x \\ y \\ z \end{pmatrix} = \begin{pmatrix} a \\ b \\ c \end{pmatrix}$$

to echelon form, or otherwise, prove that the equation is soluble only if $c + 2b - 5a = 0$. (JMB)p

23) The vectors **a, b, c, d** are given by

$$\mathbf{a} = \begin{pmatrix} 1 \\ 2 \\ -1 \end{pmatrix}, \quad \mathbf{b} = \begin{pmatrix} 0 \\ 3 \\ 4 \end{pmatrix}, \quad \mathbf{c} = \begin{pmatrix} 1 \\ 2 \\ 0 \end{pmatrix}, \quad \mathbf{d} = \begin{pmatrix} 3 \\ -3 \\ -14 \end{pmatrix}.$$

(a) Show that **a, b** and **c** form a basis for the space of three dimensional vectors.
(b) Express **d** as a linear combination of **a, b** and **c**. (JMB)

24) Let V be the vector space of triples of real numbers over the field of real numbers. Show that the vectors

$$\mathbf{a} = \begin{pmatrix} 1 \\ 0 \\ 0 \end{pmatrix}, \quad \mathbf{b} = \begin{pmatrix} 1 \\ 1 \\ 0 \end{pmatrix}, \quad \mathbf{c} = \begin{pmatrix} 1 \\ 1 \\ 1 \end{pmatrix}$$

are linearly independent.

Given any vector $\mathbf{x} = \begin{pmatrix} x_1 \\ x_2 \\ x_3 \end{pmatrix} \in V$, find real numbers α, β, γ such that

$$\alpha\mathbf{a} + \beta\mathbf{b} + \gamma\mathbf{c} = \mathbf{x}$$

giving α, β, γ in terms of x_1, x_2, x_3.

Given that $\mathbf{d} = \begin{pmatrix} 0 \\ 0 \\ 1 \end{pmatrix}$, explain why $\{\mathbf{b}, \mathbf{c}, \mathbf{d}\}$ is not a basis for V.

Show that $\begin{pmatrix} 2 \\ 2 \\ 1 \end{pmatrix}$ can be expressed as a linear combination of **b, c** and **d**, but that $\begin{pmatrix} 2 \\ 3 \\ 1 \end{pmatrix}$ cannot be so expressed. (U of L) (specimen)

25) Let V be the subspace of $\mathbb{R}^3$ (three dimensional space) spanned by

$$\mathbf{v}_1 = \begin{pmatrix} 1 \\ -1 \\ 1 \end{pmatrix}, \quad \mathbf{v}_2 = \begin{pmatrix} 2 \\ 1 \\ -1 \end{pmatrix}, \quad \mathbf{v}_3 = \begin{pmatrix} -1 \\ -5 \\ -1 \end{pmatrix}.$$

Determine the dimension of V. (JMB)p

26) Show that the matrix $\mathbf{X}$, where $\mathbf{X} = \begin{pmatrix} -2 & 3 \\ 5 & 4 \end{pmatrix}$, belongs to the real vector space spanned by the matrices $\mathbf{A}, \mathbf{B}, \mathbf{C}$ where

$$\mathbf{A} = \begin{pmatrix} 1 & 0 \\ 2 & 1 \end{pmatrix}, \quad \mathbf{B} = \begin{pmatrix} 1 & -1 \\ 0 & 0 \end{pmatrix}, \quad \mathbf{C} = \begin{pmatrix} -1 & 0 \\ 1 & 2 \end{pmatrix}.$$

Find the dimension of this space. (C)p

27) Show that the subspace S of $\mathbb{R}^4$ spanned by the vectors $(1, 0, 1, 0)$ and $(1, 2, 5, 6)$ is the same as the subspace T spanned by the vectors $(0, 1, 2, 3)$ and $(2, -1, 0, -3)$. (JMB)

28) The matrix $\mathbf{A}$ is given by

$$\mathbf{A} = \begin{pmatrix} 2 & 0 & -1 & 1 \\ 0 & 1 & 3 & 1 \\ 4 & -1 & -5 & 1 \end{pmatrix}.$$

Find a basis of the subspace V of $\mathbb{R}^4$ where

$$V = \{\mathbf{X} \colon \mathbf{X} \in \mathbb{R}^4, \ \mathbf{AX} = 0\}$$

(**Note** that $\mathbb{R}^4$ is the real vector space of dimension four.) (C)

29) Given that $\mathbf{a}, \mathbf{b}, \mathbf{c}$ are linearly independent vectors, determine whether the following vectors are linearly independent:

(i) $\mathbf{a}, \mathbf{0}$ (ii) $\mathbf{a} + \mathbf{b},\ \mathbf{b} + \mathbf{c},\ \mathbf{c} + \mathbf{a}$

(iii) $\mathbf{a} + 2\mathbf{b} + \mathbf{c},\ \mathbf{a} - \mathbf{b} - \mathbf{c},\ 5\mathbf{a} + \mathbf{b} - \mathbf{c}$ (JMB)

30) Show that if $\mathbf{u}_1, \mathbf{u}_2, \mathbf{u}_3$ are linearly independent vectors, then $\mathbf{u}_1 + \mathbf{u}_2$, $\mathbf{u}_2 + \mathbf{u}_3$, $\mathbf{u}_3 + \mathbf{u}_1$ are also linearly independent. (JMB)p

31) Let S be the vector space of ordered triples of real numbers defined by

$$S = \{(x, y, z);\quad x - 2y - 2z = 0\}.$$

Find a basis for S and state the dimension of S. (JMB)

32) The set P_2 consists of all polynomials in x, of degree less than or equal to 2, and having real coefficients, i.e.

$$P_2 = \{ax^2 + bx + c;\ a, b, c, \in \mathbb{R}\}$$

Show that, with the usual operations of addition and of multiplication by a real number, P_2 is a vector space over $\mathbb{R}$ of dimension three.

For each of the following subsets of P_2, determine whether or not it is a subspace, giving brief reasons for your answers. Give a basis for each subset which you consider to be a subspace.

(i) $\{f(x) \in P_2\colon\ f(0) = 0\}$ (ii) $\{f(x) \in P_2\colon\ f(0) = 1\}$

(iii) $\{f(x) \in P_2\colon\ f(1) = 0\}$ (iv) $\{f(x) \in P_2\colon\ f(-x) = f(x)$ for all $x \in \mathbb{R}\}$.

(C)

33) The set Q consists of the functions f defined for all real x by

$$f\colon x \longmapsto e^x(a \cos 2x + b \sin 2x),$$

where a and b are real. Thus Q is a vector space over the real numbers, with the functions f_1 and f_2 as a basis,

where

$$f_1(x) = e^x \cos 2x \quad \text{and} \quad f_2(x) = e^x \sin 2x.$$

With this basis for Q the element f may be represented by the vector $\begin{pmatrix} a \\ b \end{pmatrix}$.

Find a 2×2 matrix which premultiplies this vector to give the vector representing the derived function f' of f.
Use this matrix to find in vector form
(i) the second derived function f'' of f,
(ii) the element g of Q such that

$$g'(x) = e^x(a \cos 2x + b \sin 2x).$$

(JMB)

34) Let V_4 denote the real vector space of quadruples $\mathbf{x} = (x_1, x_2, x_3, x_4)$ of real numbers and let f be the mapping of V_4 into itself defined as follows:

$$f(x) = (x_2 - x_4,\ x_1 + 3x_2 + x_3,\ x_1 + x_2 + x_3 + 2x_4,\ x_1 + 2x_2 + x_3 + x_4)$$

Show that the vectors $\{\mathbf{y}_1, \mathbf{y}_2, \mathbf{y}_3, \mathbf{y}_4\}$ for which $f(\mathbf{y}_1, \mathbf{y}_2, \mathbf{y}_3, \mathbf{y}_4) = (\mathbf{0}, \mathbf{0}, \mathbf{0}, \mathbf{0})$ form a two dimensional vector space V_2 and find a basis for V_2. (U of L)p

APPENDIX: PROOF THAT $\dfrac{1}{|\mathbf{M}|}\,\text{adj}\,\mathbf{M} = \mathbf{M}^{-1}$

If

$$\mathbf{M} = \begin{pmatrix} a_1 & b_1 & c_1 \\ a_2 & b_2 & c_2 \\ a_3 & b_3 & c_3 \end{pmatrix}$$

then
$$\operatorname{adj}\mathbf{M} = \begin{pmatrix} A_1 & A_2 & A_3 \\ B_1 & B_2 & B_3 \\ C_1 & C_2 & C_3 \end{pmatrix}$$

where $A_1, \ldots,$ are the cofactors of $a_1, \ldots$.

$$\text{Now } (\operatorname{adj}\mathbf{M})(\mathbf{M}) = \begin{pmatrix} A_1 & A_2 & A_3 \\ B_1 & B_2 & B_3 \\ C_1 & C_2 & C_3 \end{pmatrix}\begin{pmatrix} a_1 & b_1 & c_1 \\ a_2 & b_2 & c_2 \\ a_3 & b_3 & c_3 \end{pmatrix}.$$

But the product of any column of $\mathbf{M}$ and its cofactors
(e.g. $a_1A_1 + a_2A_2 + a_3A_3$) is equal to $|\mathbf{M}|$
whereas the product of any column of $\mathbf{M}$ with the cofactors of another

column, e.g. $b_1A_1 + b_2A_2 + b_3A_3 = \begin{vmatrix} b_1 & b_2 & b_3 \\ b_1 & b_2 & b_3 \\ c_1 & c_2 & c_3 \end{vmatrix}$, is zero.

Hence
$$(\operatorname{adj}\mathbf{M})(\mathbf{M}) = \begin{pmatrix} |\mathbf{M}| & 0 & 0 \\ 0 & |\mathbf{M}| & 0 \\ 0 & 0 & |\mathbf{M}| \end{pmatrix} = |\mathbf{M}|\begin{pmatrix} 1 & 0 & 0 \\ 0 & 1 & 0 \\ 0 & 0 & 1 \end{pmatrix}$$

i.e.
$$(\operatorname{adj}\mathbf{M})(\mathbf{M}) = |\mathbf{M}|\mathbf{I} \quad \text{so} \quad \frac{\operatorname{adj}\mathbf{M}}{|\mathbf{M}|}\mathbf{M} = \mathbf{I}$$

$$\Rightarrow \frac{\operatorname{adj}\mathbf{M}}{|\mathbf{M}|} = \mathbf{M}^{-1}$$

So if $\mathbf{M} = \begin{pmatrix} a_1 & b_1 & c_1 \\ a_2 & b_2 & c_2 \\ a_3 & b_3 & c_3 \end{pmatrix}$ then $\mathbf{M}^{-1} = \dfrac{1}{|\mathbf{M}|}\begin{pmatrix} A_1 & A_2 & A_3 \\ B_1 & B_2 & B_3 \\ C_1 & C_2 & C_3 \end{pmatrix}.$

Note that the cofactors in adj $\mathbf{M}$ are not in the same positions as their corresponding elements in $\mathbf{M}$, but they are in the position reached by transposing the rows and columns of $\mathbf{M}$.
Note also that if $|\mathbf{M}| = 0$, $\mathbf{M}^{-1}$ does not exist.

CHAPTER 3

MATHEMATICAL PROOF

FORMAL MATHEMATICS

Anyone who has studied both elementary Euclidean geometry and an experimental science should be aware of the very different ways in which the propositions of these two disciplines are established. In an experimental science, the propositions or 'laws' are accepted because they are confirmed by observation. For example, in Mechanics there is a law that states that an object falling freely has a constant acceleration. This law is accepted because observation of falling bodies verifies its truth.

In Euclidean geometry, the propositions or 'theorems' are accepted because they are deduced by means of a logical proof from previously established, or accepted, truths. For example, the theorem that states that 'the sum of the interior angles of a triangle is two right angles' is accepted because it is deduced logically from previously accepted properties of parallel lines and angles.

It was the ancient Greeks who first used this method to give geometry a formal structure. Later, many other branches of mathematics were subjected to the same formal treatment which is known as the axiomatic method. This consists of accepting *without proof* certain propositions, known as *axioms* (e.g. in Euclidean geometry, the statement that there is a unique straight line through two given points, is an axiom). Then all the other statements (called theorems) of the system are derived from the axioms by the principles of logic.

Any game, chess for example, provides a good analogy for this formal structure. If you play the game you accept the rules without question, so the rules are the 'axioms'. The strategies involved in playing the game are then built on these foundations and so form the 'theorems' of the game.

Later in this chapter we look at some of the ways in which Mathematical theorems can be proved, but first we introduce some of the logic concepts that can be used in a proof.

STATEMENTS

A statement, or proposition, is a sentence which is either true or false, but not both.

For example 'It is now 10 o'clock' is a statement
but 'What is the time?' is not a statement.

Statements, or propositions, can be denoted by small letters $p, q, r, \ldots$.

Examples are:

p: For all real values of x, $\int x \, dx = \frac{1}{2}x^2 + k$. (true)

q: For any two real numbers a and b, $a - b = b - a$. (false)

r: For all real values of x, $\frac{d}{dx} \sin x = \cos x$. (true)

s: Any quadratic equation in one variable has two real roots. (false)

Because a statement is either true or false, it must contain enough information for us to be able to decide whether it is correct or not. This information must be either in the sentence itself or in the context in which the sentence appears. This is particularly important in the case of mathematical sentences involving variables.

Consider for example the function $\ln x$. We know that this function is defined for real positive values of x but not for negative values of x.

Now consider the sentence: $\int \frac{1}{x} \, dx = \ln x + k$.

If this sentence appears in a context where x has positive values only, it is true. But if it appears in a context where x can take any value it is not true, the correct statement in this case is $\int \frac{1}{x} \, dx = \ln |x| + k$. Taken out of context, $\int \frac{1}{x} \, dx = \ln x + k$ cannot be said to be either right or wrong.

Similarly the sentence 'If $x^2 = 4$ then $x = 2$' is true if $x \in \mathbb{R}^+$, but untrue if $x \in \mathbb{R}$.

EXERCISE 3a

State whether the following sentences are

(a) true for all real values of x,
(b) true for at least one real value of x, but not all x,
(c) true for no real value of x.

1) $x + 1 = 3$

2) $(x + 1)^2 = x^2 + 2x + 1$

3) $x^2 + x + 1 = 0$

4) $(x - 1)(x - 2) \geqslant 0$

5) $(x-1)^2 \geqslant 0$

6) $\begin{pmatrix} x & 1 \\ 0 & 4 \end{pmatrix}$ has an inverse

7) $\cos x > 0$

8) $\begin{pmatrix} x & 1 \\ 0 & 4 \end{pmatrix}$ has no inverse

Which of the following statements are true? (x is real.)

9) There is at least one value of x for which $x^2+1=0$.

10) There are no values of x for which $x^2+x-1=0$.

11) For all values of x, $\sin^2 x + \cos^2 x = 1$.

12) For at least one value of x, $\sin^2 x - \cos^2 x = 1$.

NEGATION

Consider the statement p: 'I have a headache'.
The statement 'I do not have a headache' is called the *negation* of p and is denoted by $\sim p$.
i.e. $\sim p$ (read as 'not p') is the negation of the statement p.
Some examples of statements and their negations are:

a: It is snowing.	$\sim a$: It is not snowing.
b: All Englishmen have brown hair.	$\sim b$: It is not true to say that all Englishmen have brown hair.
c: 71 is a prime number.	$\sim c$: 71 is not a prime number.
d: For all real values of x, $x^2+1>0$.	$\sim d$: It is not true that for all real values of x, $x^2+1>0$.

Looking at the statements and their negations above we see that either the given statement, or its negation, is true. In fact, for any statement p,

if p is true, $\sim p$ is false

if p is false, $\sim p$ is true.

This property is useful when writing down the negation of a given statement. Looking again at the examples given above, we see that some statements are negated by inserting the word 'not' and others by prefacing the given statement with 'It is not true that . . . '. Any statement can be negated by this preface but great care must be taken with any other form of wording. For example, consider again the statement

b: All Englishmen have brown hair. (false)

Now consider the following attempts at the statement $\sim b$.

b_1: All Englishmen do not have brown hair.

b_2: No Englishman has brown hair.

b_3: At least one Englishman does not have brown hair.

b_1 is ambiguous because its meaning depends on the emphasis put on certain words. If the emphasis is on '*All*', b_1 can mean that all Englishmen have a hair colour that is not brown. But if the emphasis is on '*not*', it could mean that some Englishmen have, and some have not, brown hair.
b_2 has a clear meaning and is false (it is equivalent to b_1 with the emphasis on 'all').
b_3 has a clear meaning and is true (it is equivalent to b_1 with the emphasis on 'not').
On closer examination of $\sim b$: 'It is not true to say that all Englishmen have brown hair', we see that $\sim b$ is equivalent to b_3,
i.e. the negation of 'All Englishmen have brown hair' can be written as 'At least one Englishman does not have brown hair'.

Note that there are other correct negations of b, e.g. 'Not all Englishmen have brown hair'.

Now consider a similar mathematical statement, e.g. 'For all real values of x, $f(x) \equiv (x-1)(x-2) > 0$ (which is not true).

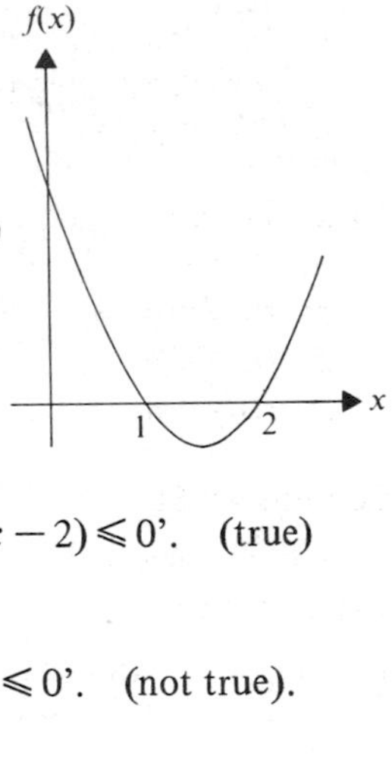

The negation of this statement can be written as

'For at least one real value of x, $f(x) \equiv (x-1)(x-2) \leqslant 0$'. (true)

but *not* as

'For all real values of x, $f(x) \equiv (x-1)(x-2) \leqslant 0$'. (not true).

EXERCISE 3b

In each of the following questions, state which of the statements b, c, d, e is/are the negation of statement a.

1) a: θ is any positive angle.
   b: θ is any negative angle.
   c: There is at least one value of θ which is not positive.

2) a: 2 is a positive integer.
   b: 2 is not a positive integer.
   c: 2 is a negative integer.

3) a: Marmalade is made from oranges.
b: Marmalade is not made from oranges.
c: There is at least one variety of marmalade that is not made from oranges.

4) a: At least one make of car does not have four wheels.
b: All cars have four wheels.
c: All cars do not have four wheels.
d: At least one make of car does have four wheels.

5) a: All quadratic equations in one variable have two real roots.
b: There is no quadratic equation in one variable that has two real roots.
c: At least one quadratic equation in one variable does not have real roots.
d: At least one quadratic equation in one variable has two real roots.

6) a: There is no real value of x for which x^2 is negative.
b: $x^2 \geqslant 0$ for all real values of x.
c: There is no real value of x for which x^2 is positive.
d: x^2 is negative for at least one real value of x.

7) a: $f(x) > x$ for all real values of $x > 1$.
b: $f(x) \ngtr x$ for all real values of $x > 1$.
c: $f(x) \leqslant x$ for at least one real value of $x \leqslant 1$.
d: $f(x) \leqslant x$ for at least one real value of $x > 1$.

8) a: For all integral values of n, $f(n) > n$.
b: $f(n) \leqslant n$ for all integral values of n.
c: $f(n) > n$ for no integral values of n.
d: $f(n) < n$ for at least one integral value of n.
e: $f(n) \leqslant n$ for at least one integral value of n.

Conditional Statements

Working in a context in which x can take any real value, consider the statements

$$a: \quad x = 3$$

$$b: \quad x^2 = 9$$

We do not know whether either of these statements is true or false but we can say that

$$\textit{if} \quad x = 3 \quad \textit{then} \quad x^2 = 9$$

This is known as an implication, or conditional statement, which we write symbolically as

$$x = 3 \quad \Rightarrow \quad x^2 = 9$$

or

$$a \quad \Rightarrow \quad b$$

There are several other ways of writing the linguistic equivalent of $a \Rightarrow b$,

e.g. $x = 3$ implies that $x^2 = 9$

$x = 3$ therefore $x^2 = 9$

$x = 3$ is a sufficient condition for $x^2 = 9$

$x = 3$ only if $x^2 = 9$.

In the conditional statement $a \Rightarrow b$, a is called the hypothesis and b is called the conclusion.
The implication $a \Rightarrow b$ will clearly be untrue if a false conclusion is drawn from a true hypothesis.

For example

$$\left.\begin{aligned} x+1 &= 0 &\Rightarrow\quad x &= 3\\ \sin A + \sin B &= 1 &\Rightarrow\quad \sin(A+B) &= 1\\ \ln(A+B) &= 2 &\Rightarrow\quad \ln A + \ln B &= 2 \end{aligned}\right\}\ \text{are all false,}$$

because our mathematical knowledge tells us that the conclusions do not follow from the hypotheses.
When this is not the case and an implication involves a variable, we must know whether the variable can take *all* real values, or only *some* real values, before we can determine the truth of the implication.
For example, $x^2 = 4 \Rightarrow x = 2$ is not true within the context given at the beginning of this section, but would be true in a context where x can take only positive values.

Note that a correct implication does not depend on the truth of the component statements. For example, consider the following solution to the problem 'Solve the equation $\sin\theta + \sin 2\theta = 1$ for values of θ in the range $0 \leqslant \theta \leqslant 90°$'.

$\sin\theta + \sin 2\theta = 1 \Rightarrow \sin 3\theta = 1$ (which is a false implication).

$\sin 3\theta = 1 \Rightarrow 3\theta = 90°$ (which is a true implication even though $\sin 3\theta = 1$ is false.)

EXERCISE 3c

All variables in this exercise can take any real value. Determine which of the following implications are true and which are false.

1) $(x+1)(x-2) = 0 \Rightarrow x = -1$ or 2

2) $x^2 = 16 \Rightarrow x = 4$

3) $f(x) \equiv (x+2)^2 \Rightarrow f(x) \geqslant 0$

4) $\sin\theta = 0 \Rightarrow \theta = 0$

5) $ax^2 + bx + c = 0$ has real roots $\Rightarrow b^2 - 4ac \geqslant 0$

6) $\dfrac{dy}{dx} = 2x \Rightarrow y = x^2$

7) $(x+1)(x-2) = 1 \Rightarrow x + 1 = 1$ or $x - 2 = 1$

8) $\cos\alpha\cos\alpha = 1 \Rightarrow \cos 2\alpha = 1$

9) $\sin\alpha\cos\alpha = 1 \Rightarrow \sin 2\alpha = \frac{1}{2}$

10) $\cos\theta = 0 \Rightarrow \sin\theta = \pm 1$

The Converse of a Conditional Statement

We have seen that, when a statement involves a variable, it is important to know what values that variable can take. For the rest of this Chapter it is to be assumed that any variable is free to take all real values *unless otherwise stated.*

If in the implication $x = 3 \Rightarrow x + 1 = 4$, we reverse the arrow,

we get $x = 3 \Leftarrow x + 1 = 4$

which might be read as 'If $x + 1 = 4$ then $x = 3$'

or '$x = 3$ is implied by $x + 1 = 4$'

or '$x = 3$ if $x + 1 = 4$'

or '$x = 3$ because $x + 1 = 4$'

or '$x = 3$ is a necessary condition for $x + 1 = 4$'.

$x = 3 \Leftarrow x + 1 = 4$ is called the *converse* of $x = 3 \Rightarrow x + 1 = 4$.

In general $p \Leftarrow q$ (or $q \Rightarrow p$) is the *converse* of $p \Rightarrow q$.

In the example given above both the implication and its converse are true. But the statement $x = 3 \Rightarrow x^2 = 9$ is true whereas its converse, $x = 3 \Leftarrow x^2 = 9$, is false because x can also have the value -3. Similarly, if a is the statement '$\sin x = 0$' and b is the statement '$\cos x = 1$' then $a \Leftarrow b$ is true but $a \Rightarrow b$ is false because $\sin x = 0 \Rightarrow \cos x = \pm 1$.

Thus a statement and its converse are not necessarily both true, nor both false.

Hence it is unsound to argue from a statement to its converse, e.g. from

'If this boy is a pupil at school X he wears a green blazer'

to

'If this boy wears a green blazer he is a pupil at school X'.

(This is a common error of reasoning.)

The Negation of a Conditional Statement

The negation of $p \Rightarrow q$ is $\sim(p \Rightarrow q)$.

Note that the *whole statement* is negated, and not p or q separately.
So the negation of 'If it is raining it is wet'
could be 'It is not true to say that if it is raining it is wet'

but not
- 'If it is not raining it is not wet'
- 'If it is raining it is not wet'
- 'If it is not raining it is wet'.

Similarly the negation of 'If I am eating then I am hungry' can be written as 'It is not true to say that if I am eating then I am hungry'.
From these two examples we see that the linguistic equivalent of $\sim(p \Rightarrow q)$ is clumsy, so we will consider them again.
If p is 'It is raining' and q is 'It is wet' then $p \Rightarrow q$ is 'If it is raining then it is wet'.
Experience tells us that $p \Rightarrow q$ is true so $\sim(p \Rightarrow q)$ is false.
It is also false to say 'It is raining and it is not wet', which can be written symbolically as 'p and $\sim q$'.
So we can replace $\sim(p \Rightarrow q)$ by the equivalent statement (p and $\sim q$).
In the same way, if p is 'I am eating' and q is 'I am hungry'
then $p \Rightarrow q$ is 'If I am eating then I am hungry'.
This time $p \Rightarrow q$ is false, so $\sim(p \Rightarrow q)$ is true.
The statement 'I am eating and I am not hungry' is also true, i.e.
(p and $\sim q$) is true.
The two examples indicate that

> a statement $\sim(p \Rightarrow q)$ can be replaced by the equivalent statement (p and $\sim q$).

Now consider '$x^2 = 9 \Rightarrow x = 3$'.

We know this is false in a context in which x can take all real values.
So $\sim$'$x^2 = 9 \Rightarrow x = 3$' is true in the same context because it is the negation of a false statement.
But we also know that this is true because there is *one* value of x (-3) for which $x^2 = 9$ and $x \neq 3$.
So, instead of the clumsy statement 'It is not true to say that if x can take any value, then $x^2 = 9$ implies that $x = 3$', we can use 'There is a value of x such that $x^2 = 9$ and $x \neq 3$'.

Now consider the true statement

$$x + 1 = 4 \Rightarrow x = 3$$

The negation of this implication, i.e. the statement

'It is not true to say that if x can take any value then $x + 1 = 4$ implies that $x = 3$',

is false.

Now there is no value of x for which $x+1=4$ and $x \neq 3$, so the statement

'There is a value of x such that $x+1=4$ and $x \neq 3$'

is also false.

These two examples indicate that

when a variable can take any value, a statement $\sim(p \Rightarrow q)$ involving just that one variable, can be replaced by the equivalent statement:

'There is at least one value of the variable such that p and $\sim q$'

as either both statements are true or both are false.

For example if $p \Rightarrow q$ is

'If all the cars in this showroom are made in Britain then they are all good cars'

then, because a variable number of cars is involved, $\sim(p \Rightarrow q)$ is

'It is not true that if all the cars in this showroom are made in Britain they are all good cars'

which can be replaced by the equivalent

'There is at least one car in this showroom that is made in Britain and which is not a good car'.

Biconditional Statements

Consider the two statements a: $x = 3$, and b: $x+1=4$.

Again, we do not know the truth of either a or b, but we can say:

If $x = 3$ then $x+1 = 4$ *and* $x = 3$ because $x+1 = 4$,

i.e. $a \Rightarrow b$ *and* $a \Leftarrow b$.

This is known as a *biconditional statement* and may be written symbolically as

$$x = 3 \Longleftrightarrow x+1 = 4$$

i.e.

$$a \Longleftrightarrow b$$

Some other linguistic equivalents for $a \Longleftrightarrow b$ are

'$x = 3$ implies and is implied by $x+1 = 4$'

'$x = 3$ if and only if $x+1 = 4$'

'$x = 3$ is a necessary and sufficient condition for $x+1 = 4$'.

In general $p \Longleftrightarrow q$ means p *if and only if* q.

$p \Longleftrightarrow q$ is true only if the arrow in the *true* statement $p \Rightarrow q$ can be *truthfully* reversed, i.e. if $p \Rightarrow q$ and $p \Leftarrow q$ are both true.

For example, for unrestricted values of θ,

$\sin\theta = 0 \Rightarrow \tan\theta = 0$ is true

$\sin\theta = 0 \Leftarrow \tan\theta = 0$ is true

hence $\sin\theta = 0 \iff \tan\theta = 0$ is true

but $\cos\theta = 1 \Rightarrow \sin\theta = 0$ is true

$\cos\theta = 1 \Leftarrow \sin\theta = 0$ is false

so $\cos\theta = 1 \iff \sin\theta = 0$ is false

Contrapositive Statements

Consider the statements $\begin{cases} a: & x = 3 \\ b: & x^2 = 9 \end{cases}$

and their negations, i.e. $\begin{cases} \sim a: & x \neq 3 \\ \sim b: & x^2 \neq 9. \end{cases}$

The true conditional relationship between a and b is

$$a \Rightarrow b \quad (\text{i.e. if} \quad x = 3 \quad \text{then} \quad x^2 = 9)$$

whereas the true conditional relationship between $\sim a$ and $\sim b$ is

$$\sim b \Rightarrow \sim a \quad (\text{i.e. if} \quad x^2 \neq 9 \quad \text{then} \quad x \neq 3).$$

(**Note** that $b \Rightarrow a$ is false, and so is $\sim a \Rightarrow \sim b$.)

Now consider the statements

c: It is raining

d: It is cloudy

and the following conditional statements

$c \Rightarrow d$: If it is raining then it is cloudy

$d \Rightarrow c$: If it is cloudy it is raining

$\sim d \Rightarrow \sim c$: If it is not cloudy it is not raining

$\sim c \Rightarrow \sim d$: If it is not raining it is not cloudy.

From our observations of the weather we see that

$c \Rightarrow d$ and $\sim d \Rightarrow \sim c$ are both true

but $d \Rightarrow c$ and $\sim c \Rightarrow \sim d$ are both false.

These two examples (the reader can no doubt think of many others) indicate that

> if $p \Rightarrow q$ is true, $\sim q \Rightarrow \sim p$ is also true,
>
> if $p \Rightarrow q$ is false, $\sim q \Rightarrow \sim p$ is also false.

The two statements $p \Rightarrow q$ and $\sim q \Rightarrow \sim p$ are each called the *contrapositive* of the other and we have seen that

a statement and its contrapositive are either both true or both false so that $p \Rightarrow q$ and $\sim q \Rightarrow \sim p$ are equivalent statements.

Thus it is sound reasoning to argue from a statement $p \Rightarrow q$ to the contrapositive statement $\sim q \Rightarrow \sim p$,
e.g. from

'If it is snowing it is cold'

to

'If it is not cold it is not snowing'.

Similarly it is valid to argue from

'If $f(x)$ has a stationary value when $x = a$ then $f'(a) = 0$'

to

'If $f'(a) \neq 0$ then $f(x)$ does not have a stationary value when $x = a$'.

Note. The soundness of an argument does not depend on the truth of the statements involved,
e.g. to argue from 'If I am doing A-level maths then I am mad'
to 'If I am not mad then I am not doing A-level maths'
is sound, but the implications are (we hope) false.
Note that if we compare the statements $p \Rightarrow q$ and $\sim p \Rightarrow \sim q$ (called the *inverse* of $p \Rightarrow q$) we find that they are not necessarily both true nor both false,
e.g. if $p \Rightarrow q$ is

'If $x = 3$ then $x^2 = 9$' (true)

then the inverse, $\sim p \Rightarrow \sim q$, is

'If $x \neq 3$ then $x^2 \neq 9$' (false because $x = -3 \Rightarrow x^2 = 9$).

So it is not reasonable to argue from an implication to its inverse,
e.g. from 'If it is raining there are clouds in the sky'
to 'If it is not raining there are no clouds in the sky'.

EXERCISE 3d

In Questions 1–16 insert the correct conditional symbol between the given statements p and q (i.e. $p \Rightarrow q$, $p \Leftarrow q$, or $p \Leftrightarrow q$). In any question involving variables it is to be assumed that the variables can take any value.

1) p: $x + 2 = 4$; q: $x = 2$

2) p: $x^2 = 4$; q: $x = 2$

3) p: $\theta = \frac{\pi}{4}$; q: $\sin\theta = \frac{\sqrt{2}}{2}$.

4) p: $P(x, y)$ is a point on the circle, centre O, radius 2; q: $x^2 + y^2 = 4$.

5) p: $\tan\theta = \pm 1$; q: $\sin\theta = -\frac{\sqrt{2}}{2}$.

6) p: $\frac{x-1}{(x+2)^2} = 0$; q: $x = 1$.

7) p: x^2 is even; q: x is even.

8) p: x is rational; q: x^2 is rational.

9) α and β are the roots of the equation $x^2 + ax + b = 0$.
p: a and b are real; q: $\alpha\beta$ is real.

10) In triangle ABC p: $\sin C = \sin 2A$; q: In triangle ABC, $a = b$.

11) p: α, β are the roots of the equation $z^2 = -1$; q: $\alpha + \beta$ is real.

12) p: $\mathbf{AB} = \mathbf{BA}$; q: $\mathbf{A} = \mathbf{B}$ ($\mathbf{A}$ and $\mathbf{B}$ are square matrices).

13) p: $x > y$; q: $-y > -x$.

14) p: $x^2 + y^2 < 4$; q: (x, y) lies inside the circle $x^2 + y^2 = 4$.

15) p: $f(x) \equiv f(-x)$; q: $f(x) \equiv e^{x^2} \cos x$.

16) Write down a linguistic statement which is (a) the converse, (b) the contrapositive, (c) the inverse, (d) the negation, of the following statement: 'If this car is made in Britain then it is a good car'.

17) Repeat Question 16 for the statement:
'If the equation $ax^2 + bx + c = 0$ has two real roots then $b^2 - 4ac \geqslant 0$'.

18) p is the statement 'Henry is doing A-level mathematics'.
q is the statement 'Henry is mad'.
Give linguistic equivalents for the following statements:
(a) $p \Rightarrow q$ (b) $p \Leftarrow q$ (c) $\sim p \Rightarrow q$ (d) $p \Leftarrow \sim q$
(e) $\sim (p \Rightarrow q)$ (f) $p \Leftrightarrow q$ (g) $\sim p \Rightarrow \sim q$ (h) $\sim p \Leftarrow \sim q$.

19) If p is the statement '$x = 1$' and q is the statement '$x^2 = 1$' write down the following statements in symbolic form:

(a) If $x = 1$ then $x^2 = 1$. (b) If $x^2 = 1$ then $x \neq 1$.

(c) $x^2 \neq 1$ because $x \neq 1$.

(d) It is not true to say that if $x^2 = 1$ then $x = 1$.

(e) $x \neq 1$ because $x^2 = 1$.

(f) $x = 1$ is a necessary and sufficient condition for $x^2 = 1$.

(g) If $x^2 \neq 1$ then $x \neq 1$.

Which of your statements are true?

20) The conditional statement $p \Rightarrow q$ is 'If $x^2 = -1$ then x is not real'. Write down sentences which are
(a) the converse, (b) the negation, (c) the inverse,
(d) the contrapositive of $p \Rightarrow q$.

21) Determine which of the following statements is equivalent to the statement 'If $f(x) \equiv x^2 + 5$ then $f(x) \nless 5$'.
(a) There is at least one value of x for which $f(x) \equiv x^2 + 5$ and $f(x) \geqslant 5$.
(b) If $f(x) \not\equiv x^2 + 5$ then $f(x) < 5$.
(c) If $f(x) < 5$ for at least one value of x then $f(x) \not\equiv x^2 + 5$.

MATHEMATICAL PROOF

In this section we look at some of the ways in which a theorem can be proved. A mathematical theorem is a result that is accepted because it is derived by correct implications and sound arguments from axioms.
(Sound argument is difficult to define, except as an argument that convinces most mathematicians!)
Most mathematical statements which require proof can be expressed in the form of a conditional statement, i.e. $p \Rightarrow q$.
For example:
If, in $\triangle ABC$, $AB = BC$, then $\angle A = \angle B$.
If $x^2 + 2x - 3 = 0$ then $x = -3$ or 1.
If θ is any angle then $\cos^2\theta + \sin^2\theta = 1$.
If n is any positive integer then $\frac{d}{dx}x^n = nx^{n-1}$.

Direct Proof by Deduction

This is probably the form of proof that the reader is most familiar with, and it is illustrated in the following example:
To prove that 'If $x^2 + 2x - 3 = 0$ then $x = -3$ or 1' is true, we proceed as follows.

$$x^2 + 2x - 3 = 0$$

$$\Rightarrow \quad (x+3)(x-1) = 0$$

$$\Rightarrow \quad x + 3 = 0 \quad \text{or} \quad x - 1 = 0$$

$$\Rightarrow \quad x = -3 \quad \text{or} \quad x = 1$$

Hence $x^2 + 2x - 3 = 0$ *does* imply $x = -3$ or 1.

The symbolic equivalent of this argument is:
To prove by deduction that $p \Rightarrow q$ is true, start with p then deduce $p \Rightarrow r \Rightarrow s \Rightarrow q$ so $p \Rightarrow q$ is true.
Note that this is a proof of the truth of the *implication* $p \Rightarrow q$.
Whether p is true, or q is true, is another question.

Proof by Induction

This method of proof was introduced in the *Core Course* but was used only in connection with the sum of a series. However, many other results are found by extrapolation from a few particular cases.
For example, consider the statement

'If n is any positive integer then $\dfrac{d}{dx}x^n = nx^{n-1}$'.

We have not proved that this is true. From first principles we *have* proved that

$$\frac{d}{dx}(x^2) = 2x, \qquad \frac{d}{dx}(x^3) = 3x^2, \qquad \frac{d}{dx}(x^4) = 4x^3$$

These results *suggest* that $\dfrac{d}{dx}x^n = nx^{n-1}$, but they certainly *do not prove* that $\dfrac{d}{dx}x^n = nx^{n-1}$.

That such a result is correct can often be proved by induction.
As a reminder of the steps involved in a proof by induction we summarize the argument below.
Let p_n be a statement involving n, where n is any positive integer.
1 Prove directly that the implication $p_k \Rightarrow p_{k+1}$ is correct.
2 Prove directly that p_1 is correct.
3 Combine steps 1 and 2 to show that $p_2, p_3, \ldots, p_n$ are correct. This third step is usually expressed in the form 'therefore, by induction, the result is correct for all positive integral values of n'.
This method of proof is illustrated in the following examples, which include some series as a reminder of the work covered in *The Core Course.*

EXAMPLES 3e

1) Prove by induction that $9^n - 1$ is a multiple of 8 for all positive integral values of n.

Let p_n be the statement
'For all positive integral values of n, $9^n - 1$ is a multiple of 8'.
If the formula is valid when $n = k$
i.e. if $9^k - 1 = 8a$, say,

then
$$\begin{aligned} 9^{k+1} - 1 &= (9^k)(9) - 1 \\ &= (9^k)(9) - 9 + 8 \\ &= 9(9^k - 1) + 8 \\ &= 9(8a) + 8 \\ &= 8(9a + 1) \quad \text{which is a multiple of } 8 \end{aligned}$$

i.e. if $9^k - 1$ is a multiple of 8 then so is $9^{k+1} - 1$.
But if $n = 1$, $9^1 - 1 = 8$, so p_1 is true.
So p_2 is true, p_3 is true, . . . and so on for all positive integral values of n.
Therefore by induction, p_n is correct for $n \in \mathbb{N}$

2) Prove by induction that $\sum_{r=1}^{n} r^2 = \frac{n}{6}(n+1)(2n+1)$ for all positive integral values of n.

Let p_n be the statement

$$\sum_{r=1}^{n} r^2 = \frac{n}{6}(n+1)(2n+1) \quad \text{for all positive integral values of } n\text{'}.$$

If the formula is valid when $n = k$

i.e. if $$\sum_{r=1}^{k} r^2 = \frac{k}{6}(k+1)(2k+1)$$

then adding $(k+1)^2$ to both sides gives

$$\sum_{r=1}^{k+1} r^2 = \frac{k}{6}(k+1)(2k+1) + (k+1)^2$$

$$= \frac{1}{6}(k+1)[k(2k+1) + 6(k+1)]$$

$$= \frac{1}{6}(k+1)(2k^2 + 7k + 6)$$

$$= \frac{1}{6}(k+1)(k+2)(2k+3)$$

which is the formula when $n = k + 1$.
Therefore if p_n is valid when $n = k$, p_n is also valid when $n = k + 1$.
When $n = 1$,

L.H.S. of p_n is $1^2 = 1$

R.H.S. of p_n is $\frac{1}{6}(1+1)(2+1) = 1$

So p_n is valid when $n = 1$, i.e. p_1 is true.
Therefore p_2 is true, p_3 is true, . . . and so on. Therefore by induction p_n is true.

3) Use proof by induction to show that $\frac{\mathrm{d}}{\mathrm{d}x}x^n = nx^{n-1}$.

Let p_n be the statement 'If n is any positive integer then $\frac{\mathrm{d}}{\mathrm{d}x}(x^n) = nx^{n-1}$'

If the formula is valid when $n = k$

i.e. if $$\frac{d}{dx}(x^k) = kx^{k-1}$$

then $$\frac{d}{dx}(x^{k+1}) = \frac{d}{dx}(x^k)(x)$$

$$= (kx^{k-1})(x) + (x^k)(1) \quad \text{(using the product rule)}$$

$$= (k+1)x^k$$

which is the formula when $n = k+1$.
So if the formula is valid when $n = k$ it is valid when $n = k+1$.
Now if $n = 1$, from first principles

$$\frac{d}{dx}(x^1) = \lim_{\delta x \to 0}\left[\frac{(x+\delta x)-x}{\delta x}\right]$$

$$= 1$$

$$= 1(x^0)$$

Hence p_1 is true.
Therefore by induction p_n is true.

4) Prove by induction that, for $n \in \mathbb{N}$ $n > 1$

$$|z_1 + z_2 + \ldots + z_n| \leqslant |z_1| + |z_2| + \ldots + |z_n|$$

Let p_n be the statement

$$|z_1 + \ldots + z_n| \leqslant |z_1| + \ldots + |z_n|$$

for all integral values of $n > 1$.
If the inequality is valid when $n = k$

i.e. if $$|z_1 + \ldots + z_k| \leqslant |z_1| + \ldots + |z_k| \qquad [1]$$

then from the diagram,

$$\overrightarrow{OB} = z_1 + \ldots + z_k + z_{k+1}$$

$$\overrightarrow{OA} = z_1 + \ldots + z_k$$

$$\overrightarrow{AB} = z_{k+1}$$

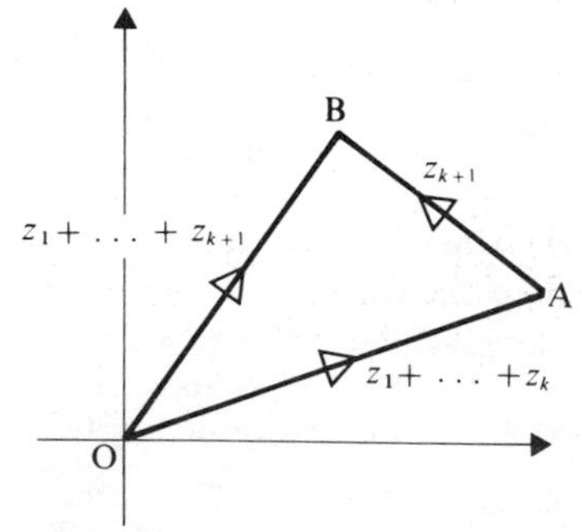

The points OAB form either a triangle or a line.

So $$|\overrightarrow{OB}| \leqslant |\overrightarrow{OA}| + |\overrightarrow{AB}|$$

i.e. $$|z_1 + \ldots + z_k + z_{k+1}| \leqslant |z_1 + z_2 + \ldots + z_k| + |z_{k+1}|$$

But if [1] is true then

$$|z_1 + \dots + z_{k+1}| \leqslant |z_1| + |z_2| + \dots + |z_k| + |z_{k+1}|$$

which is the inequality when $n = k + 1$.
So if the inequality is valid when $n = k$ then it is valid when $n = k + 1$.

When $n = 2$,

L.H.S. of p_n is $|z_1 + z_2|$

R.H.S. of p_n is $|z_1| + |z_2|$

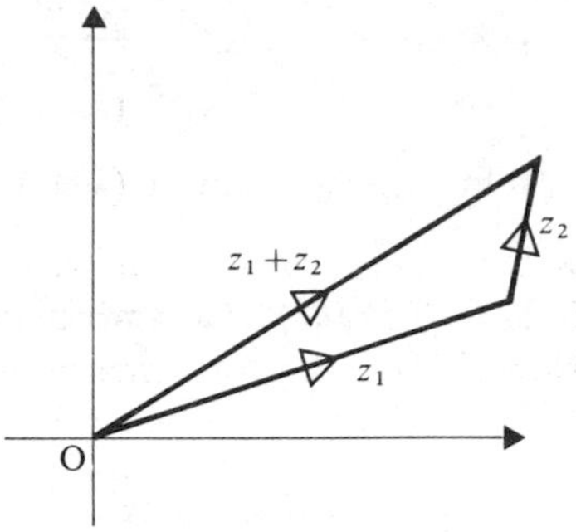

From the diagram, again using the property of the lengths of the sides of a triangle, it follows that

$$|z_1 + z_2| \leqslant |z_1| + |z_2|$$

So p_n is valid when $n = 2$.
Therefore, by induction, p_n is true for $n \in \mathbb{N}$, $n > 1$.

5) Prove by induction that $(1 + x)^n \equiv 1 + {}^nC_1x + {}^nC_2x^2 + \dots + {}^nC_nx^n$ for all positive integral values of n.

Let p_n be the statement

'If n is any positive integer then

$$(1 + x)^n \equiv 1 + {}^nC_1x + {}^nC_2x^2 + \dots + {}^nC_nx^n\text{'}$$

If the formula is valid when $n = k$

i.e. if $\quad (1 + x)^k \equiv 1 + {}^kC_1x + {}^kC_2x^2 + \dots + {}^kC_kx^k$

then, multiplying both sides by $(1 + x)$,

$(1 + x)^{k+1}$

$\equiv (1 + {}^kC_1x + {}^kC_2x^2 + \dots + {}^kC_kx^k)(1 + x)$

$\equiv 1 + ({}^kC_1 + 1)x + ({}^kC_2 + {}^kC_1)x^2 + \dots + ({}^kC_r + {}^kC_{r-1})x^r + \dots + {}^kC_kx^{k+1}$

$\equiv 1 + {}^{k+1}C_1x + {}^{k+1}C_2x^2 + \dots + {}^{k+1}C_rx^r + \dots + {}^{k+1}C_{k+1}x^{k+1}$

(since ${}^kC_r + {}^kC_{r-1} = {}^{k+1}C_r$ and ${}^kC_k = {}^{k+1}C_{k+1} = 1$)

which is the formula when $n = k + 1$.
So if the formula is valid when $n = k$ then it is valid when $n = k + 1$.
Now if $n = 1$,

$$(1 + x)^1 \equiv 1 + x \equiv 1 + {}^1C_1x^1$$

So p_1 is true.
Therefore by induction p_n is true.

EXERCISE 3e

In Questions 1–6 rephrase the statements in the form $p \Rightarrow q$ and give a direct proof to show that they are true. Also state, without proof, if the converse statement (i.e. $p \Leftarrow q$) is true.

1) When $x = \sqrt{2}$, $\dfrac{2+x}{2-x} = 3 + 2\sqrt{2}$.

2) In a triangle ABC in which AB = BC, $\angle A = \angle C$.

3) If $y = f(x)$ has a maximum point at $x = x_1$, then $\dfrac{dy}{dx} = 0$ when $x = x_1$.

4) For a parabolic mirror:
rays of light parallel to its axis are reflected through its focus.

In Questions 5-10, use proof by induction to verify the statements.

5) $\displaystyle\sum_{r=1}^{n} r = \frac{n}{2}(n+1)$.

6) $n^3 - n$ is a multiple of 6 for all positive integral values of n.

7) $\displaystyle\sum_{r=1}^{n} \frac{1}{r(r+1)} = \frac{n}{n+1}$.

8) For all integral values of n, $2^{n+2} + 3^{2n+1}$ is exactly divisible by 7.

9) $\dfrac{d^n}{d\theta^n}(\sin a\theta) = a^n \sin\left(a\theta + \dfrac{n\pi}{2}\right)$

10) $2^n > 2n$ for all integral values of n greater than 2.

11) In the sequence $u_1, u_2, u_3, \ldots, u_n$, $u_1 = 1$ and $u_{r+1} = \dfrac{2u_r - 1}{3}$.
Write down the values of u_2, u_3 and u_4 and prove by induction that $u_n = 3(\frac{2}{3})^n - 1$.

12) Prove by induction that $\displaystyle\int x^n \, dx = \frac{x^{n+1}}{n+1} + k$ for all positive integral values of n. (*Hint.* Use integration by parts.)

13) Prove (a) by direct deduction, (b) by induction, that

$\sum_{r=1}^{n} ap^r = \dfrac{a(1-p^n)}{1-p}$ for all positive integral values of n.

14) Prove by induction that for all positive integral values of n

$\dfrac{d}{dx}(x^{-n}) = -nx^{-n-1}$.

15) If $\mathbf{A} = \begin{pmatrix} 2 & a \\ 0 & 1 \end{pmatrix}$, prove by induction that for every positive integer n

$$\mathbf{A}^n = \begin{pmatrix} 2^n & (2^n-1)a \\ 0 & 1 \end{pmatrix}$$

Determine whether this relation holds for $n = -1$. (JMB)

Indirect Proof

Sometimes, when it is difficult to prove a statement in the form $p \Rightarrow q$, a direct proof of the contrapositive statement $\sim q \Rightarrow \sim p$ can be used. This provides a sound proof of $p \Rightarrow q$ because we know that
($p \Rightarrow q$ true) $\Longleftrightarrow$ ($\sim q \Rightarrow \sim p$ true).
Statements whose truth is very nearly obvious can often be proved by this form of indirect proof as illustrated in the following examples.

EXAMPLES 3f

1) Prove that if n is a natural number such that n^2 is even then n is even.

The contrapositive of the statement
'n is a natural number such that n^2 is even $\Rightarrow$ n is even'

is 'n is an odd integer $\Rightarrow$ n^2 is an odd integer'

Now n is an odd integer $\Rightarrow$ $n = 2k+1$ where k is an integer

$$\Rightarrow n^2 = (2k+1)^2$$

$$\Rightarrow n^2 = 4k^2+4k+1$$

$$\Rightarrow n^2 = 2(2k^2+2k)+1$$

$\Rightarrow$ n^2 is an odd integer.

Hence if n is an odd integer then n^2 is an odd integer.
Hence if n is a natural number such that n^2 is an even integer then n is an even integer.

2) In ΔABC, prove that if $\angle A = \angle C$ then $AB = BC$.

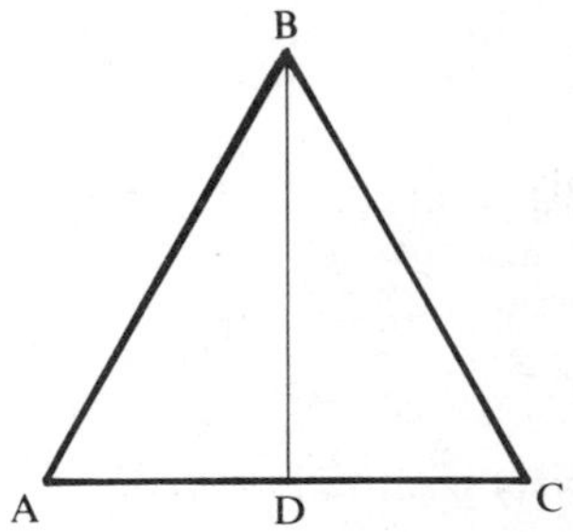

The contrapositive statement is
if $AB \neq BC$ then $\angle A \neq \angle C$.
Now

$$AB \neq BC \Rightarrow \frac{BD}{AB} \neq \frac{BD}{BC}$$
$$\Rightarrow \sin A \neq \sin C$$
$$\Rightarrow \angle A \neq \angle C.$$

Hence if $AB \neq BC$ then $\angle A \neq \angle C$.
Hence if $\angle A = \angle C$ then $AB = BC$.

Proof by Contradiction (*Reductio Ad Absurdum*)

A contradiction is a statement that is always false, regardless of the truth of its component statements.
For example, '$x = 2$ and $x \neq 2$' is false regardless of whether x is 2 or not, as x cannot be both 2 and not 2. So any statement of the form 'q and $\sim q$' is false.
To verify a statement p by contradiction we start with $\sim p$ and deduce a statement of the form 'q and $\sim q$'. As this is false we can argue that $\sim p$ is false, hence that p is true. This is illustrated in the following examples.

To prove that there is an infinite number of prime numbers
We start with the negated statement

the number of primes is finite

$\Rightarrow$ there exists an integer p, such that p is the largest prime.

Now $p! + 1 > p$, and $(p! + 1)$ is not divisible by p or by any number less than p (see footnote)

so either $(p! + 1)$ is not divisible by an integer other than 1 or $(p! + 1)$, in which case $(p! + 1)$ is prime,

or $(p! + 1)$ is divisible by a number between p and $(p! + 1)$

in either case there is a prime number larger than p.

We have now arrived at the implication

(the number of primes is finite) $\Rightarrow$ (p is the largest prime and there is a prime larger than p)

which contains a contradiction.

If 2 is a factor of a number n, 2 is not a factor of $n + 1$. Similarly if 3 is a factor of n, 3 is not a factor of $n + 1$. Now $2, 3, \ldots, p$ are all factors of $p!$ so none of these are factors of $p! + 1$.

Hence 'the number of primes is finite' is false
so 'the number of primes is infinite' is true.

To prove that $\sqrt{2}$ is irrational

The negation of '$\sqrt{2}$ is irrational' is '$\sqrt{2}$ is rational'.

Now $\sqrt{2}$ is rational $\Rightarrow$ $\sqrt{2} = \dfrac{p}{q}$

where p and q are integers with no common factor

$$\Rightarrow \quad 2 = \frac{p^2}{q^2}$$
$$\Rightarrow \quad 2q^2 = p^2$$
$$\Rightarrow \quad p^2 \text{ is even}$$
$$\Rightarrow \quad p \text{ is even}$$
$$\Rightarrow \quad p = 2k \quad \text{where } k \text{ is an integer.}$$
$$\Rightarrow \quad 2q^2 = 4k^2$$
$$\Rightarrow \quad q^2 = 2k^2$$
$$\Rightarrow \quad q^2 \text{ is even}$$
$$\Rightarrow \quad q \text{ is even}$$
$$\Rightarrow \quad q = 2m$$

Hence we have

$$(\sqrt{2} \text{ is rational}) \Rightarrow \left(\begin{array}{l}\sqrt{2} = p/q \quad \text{where } p \text{ and } q \text{ are integers with} \\ \text{no common factors and} \quad p = 2k, \quad q = 2m\end{array}\right)$$

which contains a contradiction.

As the conclusion of this implication is false, the hypothesis that $\sqrt{2}$ is rational is also false.

i.e. $\sqrt{2}$ is irrational.

Both the examples of proof by contradiction given above were proofs of an unconditional statement. We will now look at how the same basic logic can be applied to the proof of a conditional statement.

For example, if $f(x) \equiv x^2 + bx + c$, we will prove by contradiction that

$$b^2 - 4c < 0 \Rightarrow f(x) > 0 \quad \text{for all real values of } x \qquad [1]$$

Starting with the negation of this statement, i.e.
it is not true that $b^2 - 4c < 0 \Rightarrow f(x) > 0$ for all real values of x,
we replace this by the equivalent

$$b^2 - 4c < 0 \quad \text{and} \quad f(x) \leqslant 0 \quad \text{for at least one real value of } x \qquad [2]$$

We now take each component statement and make deductions from it.

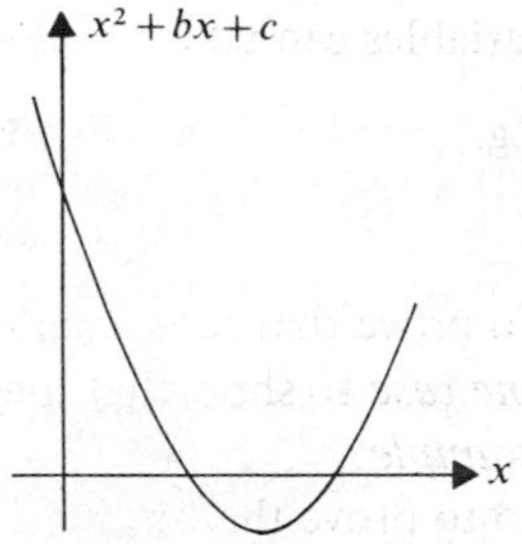

Now '$b^2-4ac<0 \Rightarrow f(x)=0$ has complex roots'

and '$f(x) \leqslant 0$ for at least one real value of x
$\Rightarrow f(x)=0$ for at least one real value of x
$\Rightarrow f(x)$ has real roots'.

Hence

$$\begin{bmatrix} b^2-4c<0 \text{ and } f(x) \leqslant 0 \\ \text{for at least one value of } x \end{bmatrix} \Rightarrow \begin{bmatrix} f(x)=0 \text{ has complex roots} \\ \text{and } f(x)=0 \text{ has real roots} \end{bmatrix}$$

which contains a contradiction.
So [2] is false, therefore [1] is true.
This argument can be written symbolically as follows.

To prove $p \Rightarrow q$ by contradiction, start with the negation, $\sim(p \Rightarrow q)$, and replace by the equivalent, p and $\sim q$.
Deduce independently that

$$p \Rightarrow r \quad \text{and} \quad \sim q \Rightarrow \sim r$$

Thus $\quad (p \text{ and } \sim q) \Rightarrow (r \text{ and } \sim r)$

As r and $\sim r$ is a contradiction, p and $\sim q$ is false, so $\sim(p \Rightarrow q)$ is false.
Hence $p \Rightarrow q$ is true.

The Use of a Counter Example

We have already seen that if a statement $p \Rightarrow q$ is true, then its converse $p \Leftarrow q$ is not necessarily true.

In the case of mathematical theorems, it is important to know whether the converse is true or false.
For example, in $\triangle ABC$,

'if $AB = BC$ then $\angle A = \angle C$' is true

and the converse

'if $\angle A = \angle C$ then $AB = BC$' is also true.

However 'If $u_1 + u_2 + \ldots$ converges then $\lim_{n\to\infty} u_n = 0$' is true

but its converse,

'If $\lim_{n\to\infty} u_n = 0$ then $u_1 + u_2 + \ldots$ converges' is false.

To prove that a general implication is false is usually very much easier than proving that it is true. A general statement is always in a context where any variables can take all values,

e.g. 'All Englishmen have brown hair'.

'For all values of x and y, $x^2 = y^2 \Rightarrow x = y$'.

To prove that such a statement is false, all that we have to do is to produce *just one case* to show that the statement is indeed false. This is called a *counter example.*

So to prove that the statement 'All Englishmen have brown hair' is false, all we have to do is to produce just one Englishman with (say) blond hair.

To prove that 'For all values of x and y, $x^2 = y^2 \Rightarrow x = y$' is false consider the counter example $x = 2$ and $y = -2$. With these values of x and y, $x^2 = y^2$ and $x \neq y$.

(**Note** that a counter example proves that the negation of the given statement is true, i.e. for at least one value of the variable, $\sim p$ is true, so showing p to be false.)

EXAMPLES 3f (continued)

3) Find a counter example to show that the following statement is false.

'If, for any series, $\lim_{n \to \infty} u_n = 0$ then $u_1 + u_2 + \ldots$ converges'.

Consider the series

$$\tfrac{1}{2} + \tfrac{1}{3} + \tfrac{1}{4} + \tfrac{1}{5} + \tfrac{1}{6} + \ldots$$

$$= \tfrac{1}{2} + (\tfrac{1}{3} + \tfrac{1}{4}) + (\tfrac{1}{5} + \tfrac{1}{6} + \tfrac{1}{7} + \tfrac{1}{8}) + (\tfrac{1}{9} + \tfrac{1}{10} + \tfrac{1}{11} + \tfrac{1}{12} + \tfrac{1}{13} + \tfrac{1}{14} + \tfrac{1}{15} + \tfrac{1}{16}) + \ldots$$

This sum is greater than the following sum:

$$\tfrac{1}{2} + (\tfrac{1}{4} + \tfrac{1}{4}) + (\tfrac{1}{8} + \tfrac{1}{8} + \tfrac{1}{8} + \tfrac{1}{8}) + (\tfrac{1}{16} + \tfrac{1}{16} + \tfrac{1}{16} + \tfrac{1}{16} + \tfrac{1}{16} + \tfrac{1}{16} + \tfrac{1}{16} + \tfrac{1}{16}) + \ldots$$

$$= \tfrac{1}{2} + \tfrac{1}{2} + \tfrac{1}{2} + \tfrac{1}{2} + \ldots$$

which clearly diverges.

Hence $\tfrac{1}{2} + \tfrac{1}{3} + \tfrac{1}{4} + \tfrac{1}{5} + \ldots$ also diverges *and* $\lim_{n \to \infty} \dfrac{1}{n} = 0$

so proving that the given statement is false.

EXERCISE 3f

In Questions 1–5 prove the given statements by a direct proof of the contrapositive statement.

1) If n is a natural number such that n^2 is odd, then n is odd.

2)

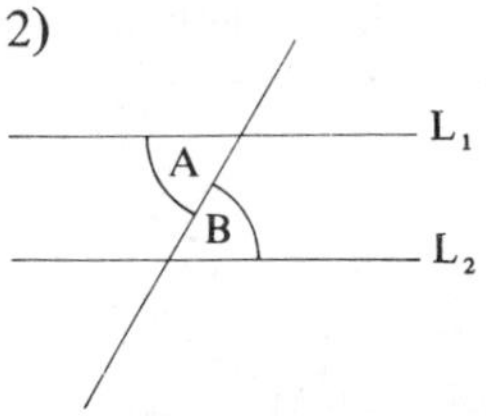

If $\angle A = \angle B$, L_1 and L_2 do not intersect.

3) If n is a perfect number, then n is not a prime number.
(A perfect number is equal to the sum of its prime factors, e.g. $6 = 3 + 2 + 1$.)

4) If two lines are perpendicular, the product of their gradients is -1.

5) If $(x-2)(1-x) > 0$ then $1 < x < 2$.

In Questions 6–8 use proof by contradiction to verify the given statements.

6) $\sqrt{3}$ is irrational.

7) If $x^2 - 3x + 2 < 0$ then $1 < x < 2$.

8) For all $x > 0$, $x + \dfrac{1}{x} \geqslant 2$.

In Questions 9–15, find a counter example to show that the given statements are false.

9) For any two real numbers a and b, $a \div b = b \div a$.

10) For any two vectors **a** and **b**, $\mathbf{a} \times \mathbf{b} = \mathbf{b} \times \mathbf{a}$.

11) For all real values of a and b, $a - b > 0 \Rightarrow a^2 - b^2 > 0$.

12) If, in the equation $ax^2 + bx + c = 0$, a, b and c are real then the roots of the equation are real.

13) For any two vectors **a** and **b**, $\mathbf{a} \times \mathbf{b} = \mathbf{0} \Rightarrow \mathbf{a} = \mathbf{0}$ or $\mathbf{b} = \mathbf{0}$.

14) If $f''(x) = 0$ when $x = a$ then $f(x)$ has a point of inflexion when $x = a$.

15) For any three matrices **A**, **B** and **C** that are compatible for the products **AB** and **AC**, $\mathbf{AB} = \mathbf{AC} \Rightarrow \mathbf{A} = \mathbf{0}$ or $\mathbf{B} - \mathbf{C} = \mathbf{0}$.

16) Give counter examples to show that the following statements are false.
(i) The sum of two unequal, positive, irrational numbers is irrational.
(ii) The product of two unequal, positive, irrational numbers is irrational. (JMB)

CHAPTER 4

FUNCTIONS

This chapter begins with a more formal treatment of work covered in *The Core Course*.

FUNCTIONS

Consider the set of all two dimensional geometric vectors, which we will call $\mathbb{R}_2$.

If any vector, $\begin{pmatrix} x \\ y \end{pmatrix}$, in $\mathbb{R}_2$ is premultiplied by the row vector $\begin{pmatrix} 2 & 0 \end{pmatrix}$ the result is a real number,

i.e. $$\begin{pmatrix} 2 & 0 \end{pmatrix}\begin{pmatrix} x \\ y \end{pmatrix} = 2x \qquad [1]$$

Now $2x$ is an element of the set of real numbers, $\mathbb{R}$.
So equation [1] gives a relationship between the elements of $\mathbb{R}_2$ and the elements of $\mathbb{R}$ and we say that the relationship maps $\mathbb{R}_2$ to $\mathbb{R}$.
Also this relationship is such that

every element, $\begin{pmatrix} x \\ y \end{pmatrix}$, in $\mathbb{R}_2$ corresponds to a unique element in $\mathbb{R}$.

A relationship such as this is called a function.

We can now express the mapping examined above as
the function that maps $\mathbb{R}_2$ to $\mathbb{R}$ such that

any element, $\begin{pmatrix} x \\ y \end{pmatrix}$, in $\mathbb{R}_2$ maps to the element $\begin{pmatrix} 2 & 0 \end{pmatrix}\begin{pmatrix} x \\ y \end{pmatrix}$ in $\mathbb{R}$.

This is written symbolically as

$$f\colon \mathbb{R}_2 \to \mathbb{R}$$

$$\begin{pmatrix} x \\ y \end{pmatrix} \longmapsto \begin{pmatrix} 2 & 0 \end{pmatrix}\begin{pmatrix} x \\ y \end{pmatrix} = 2x$$

$\mathbb{R}_2$ is called the *domain* of the function.
$\mathbb{R}$ is called the *codomain* of the function.

$2x$ is called the *image* of $\begin{pmatrix} x \\ y \end{pmatrix}$.

The set of all possible values of $2x$ is called the *image-set*.
Note that, if x is any real number then $2x$ is also any real number, so in this case the image-set is the same as the codomain. This is not always the case as we will see later.

In general a function can be defined as follows.

A function maps a set A (the domain) to a set B (the codomain) in such a way that for each element in A there is a unique image in B.

It follows from this definition that every element in A *has* an image in B, i.e. a function is defined for *every* element in the domain.

Now consider the relationship between real numbers and their squares.
The square of a real number is itself a real number, so this relationship maps real numbers to real numbers.
Also the square of any one real number has a unique value and all real numbers have squares.
Hence this relationship is a function which we write formally as

$$f\colon \mathbb{R} \to \mathbb{R}$$

$$x \longmapsto x^2$$

The domain of this function is $\mathbb{R}$, the codomain is $\mathbb{R}$ but the image-set is the set of all possible values of x^2 and this is the set of positive real numbers and zero, i.e. $\mathbb{R}^+ + \{0\}$.

Next consider the relationship between real numbers and their reciprocals.
The reciprocal of any real number is a real number except for the reciprocal of zero, i.e. $1/0$, which is meaningless. So if zero is excluded from the domain, we have a function which can be written as

$$f\colon \mathbb{R} - \{0\} \to \mathbb{R}$$

$$x \longmapsto \frac{1}{x}$$

As an example of a relationship that is not a function, consider the relationship between positive real numbers and their square roots.

The square root of a positive real number is not a single real number, e.g. $+2$ and -2 are both square roots of 4.
So this relationship is not a function.
Most functions considered at this level are relationships between real numbers so the codomain is not usually given. In this case a function is defined less formally.

For example $$f: x \longmapsto x^2, \quad x \in \mathbb{R}$$

or $$f(x) = x^2, \quad x \in \mathbb{R}$$

The domain is not always given explicitly either, in which case the function is written as

$$f(x) = x^2 \quad \text{or even just} \quad x^2$$

Note that the word *range*, when used in connection with a function, usually means the *image-set* of a function. However, it is also used by some authors to mean codomain.

Functions can be combined in a variety of ways.
For example, if g is defined by $g: x \longmapsto x^2$
and if h is defined by $h: x \longmapsto x-1$
then g and h can be added, subtracted, multiplied and divided,

i.e.
$$g(x)+h(x) = x^2+x-1$$
$$g(x)-h(x) = x^2-x+1$$
$$g(x)\times h(x) = x^2(x-1)$$
$$g(x)/h(x) = x^2/(x-1)$$

We can also take the function g of the function h, which is denoted by gh and is called the *composition* of the functions g and h.

i.e. $$gh(x) = (x-1)^2$$

Note that $gh(x)$ means 'first take the function h of x and then take the function g of the result'.
Thus $hg(x) = x^2-1$.
A function may also be a combination of two, or more, relationships defined for adjacent domains. For example

$$f: \begin{cases} x \longmapsto -x, & x < 0 \\ x \longmapsto x^2, & x \geqslant 0 \end{cases} \quad \Rightarrow$$

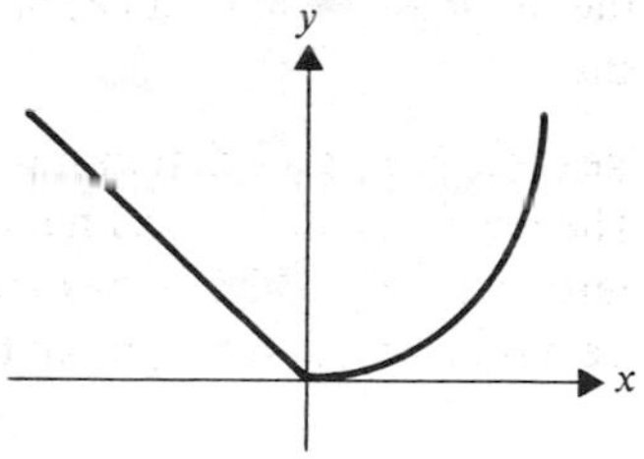

Inverse Functions

Consider $f: x \longmapsto f(x)$, $x \in \mathbb{R}$.

If this mapping can be reversed, i.e. $f(x) \longmapsto x$

and the resulting relationship is a function, it is called the inverse of the original function, and is denoted by f^{-1}.

For example, if $f: x \longmapsto x+1, \quad x \in \mathbb{R}$
the reverse relationship is $x+1 \longmapsto x$ or $x \longmapsto x-1$
which is a function.

So $f: x \longmapsto x+1, \quad x \in \mathbb{R}$ has an inverse, $f^{-1}: x \longmapsto x-1, \quad x \in \mathbb{R}$.

But if $f: x \longmapsto x^2, \quad x \in \mathbb{R}$,
the reverse relationship is $x^2 \longmapsto x$ or $x \longmapsto \pm\sqrt{x}$,
and this is not a function.
So $f: x \longmapsto x^2, \quad x \in \mathbb{R}$ does not have an inverse.
However, if we choose a different domain for f, so that f becomes a one-one mapping,

i.e. $$f: x \longmapsto x^2, \quad x \in \mathbb{R} \text{ and } x \geqslant 0$$

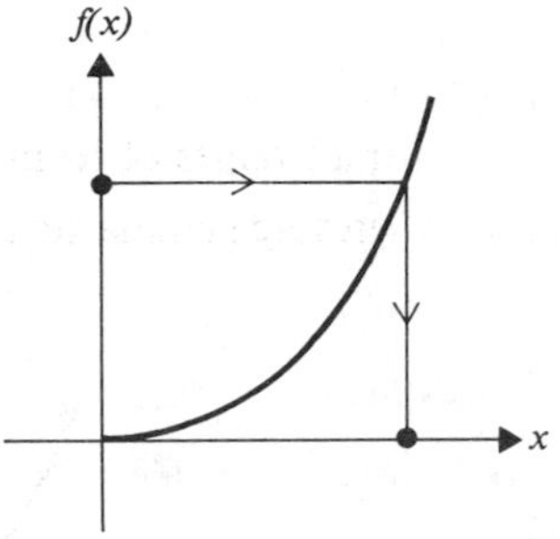

the reverse relationship is $x \longmapsto \sqrt{x}$ which is a function.

So $f(x) = x^2, \quad x \in \mathbb{R}, \quad x \geqslant 0$
has an inverse, $f^{-1}(x) = \sqrt{x}, \quad x \in \mathbb{R}, \quad x \geqslant 0$.

In general, if $y = f(x)$ so that $x \longmapsto y$ and the domain of $f(x)$ is such that $f(x) = y$ has just one value of x corresponding to one value of y, then the reverse relationship, $y \longmapsto x$, is the inverse function, where

$$x = f^{-1}(y).$$

The Graph of a Function and its Inverse

$$\left.\begin{aligned} f(x) &\longmapsto x \\ x &\longmapsto f^{-1}(x) \end{aligned}\right\} \text{ express the same relationship}$$

So $$\left.\begin{aligned} y &= f^{-1}(x) \\ x &= f(y) \end{aligned}\right\} \text{ are the same curve}$$

i.e. the curve $y = f^{-1}(x)$ is the same as the curve $y = f(x)$ with x and y interchanged.

So the curve $y = f^{-1}(x)$ is obtained by reflecting the curve $y = f(x)$ in the line $y = x$.

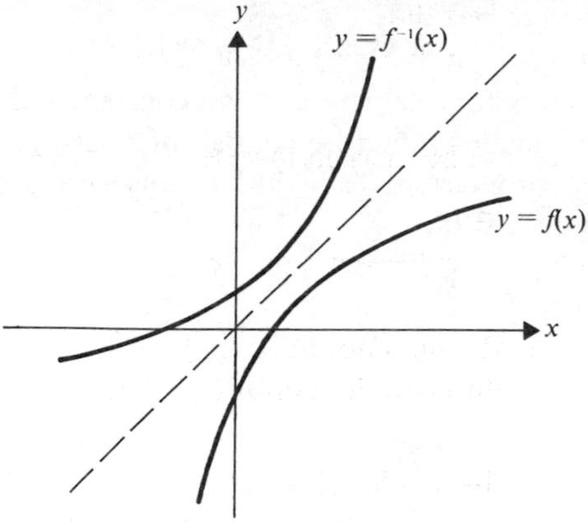

Even Functions

An even function is such that

$$f(-x) = f(x) \quad \text{for all values of } x \text{ in its domain.}$$

The graph of such a function is therefore symmetric about the vertical axis,

e.g. $f: x \longmapsto x^2, \quad x \in \mathbb{R}$

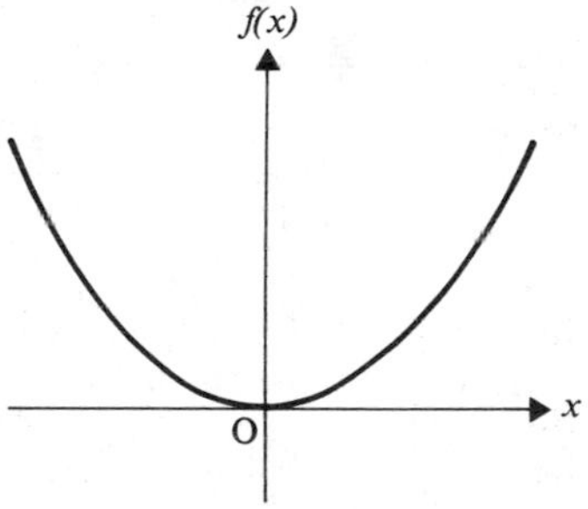

Odd Functions

An odd function is such that

$$f(-x) = -f(x) \quad \text{for all values of } x \text{ in its domain.}$$

The graph of such a function looks the same when rotated through half a revolution about O. (This is called rotational symmetry.)

e.g. $f: x \longmapsto x^3, \quad x \in \mathbb{R}$

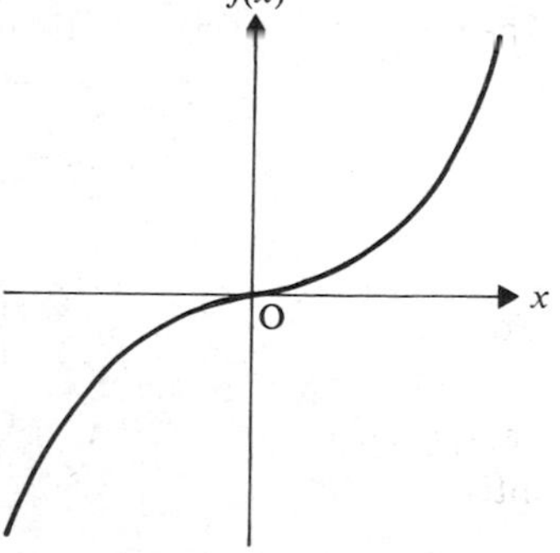

Periodic Functions

A periodic function is such that

$$f(x) = f(x + ka)$$

for all values of x in its domain, where a is a constant and k is any integer. The graph of such a function therefore consists of a basic pattern of width a which repeats at regular intervals. The width of this basic pattern is called the period of the function.

e.g. $f: x \longmapsto \sin x, \quad x \in \mathbb{R}$
is periodic with period 2π as $\sin x = \sin(x + 2n\pi)$ for $n \in \mathbb{Z}$.

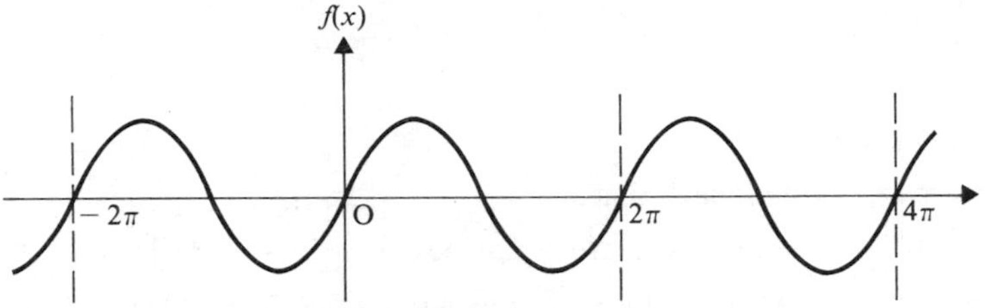

A periodic function is often specified by defining the function within one period and, separately, stating its period,

e.g. $\qquad f(x) = x(1-x)$ for $0 \leqslant x < 1$

and $\qquad f(x) = f(x+1)$ for all real values of x.

So if we draw the curve $y = x(1-x)$ in the interval $0 \leqslant x < 1$, this pattern then repeats at unit intervals.

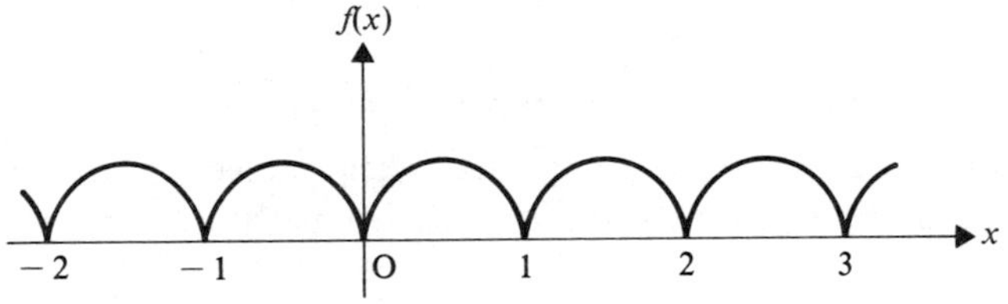

Continuity

A function f is said to be continuous at a point a in its domain if and only if

$$\lim_{x \to a} f(x) = f(a).$$

Interpreting this graphically, we note that, as a is in the domain of the function, $f(a)$ must be defined so there must be a point on the graph where $x = a$.

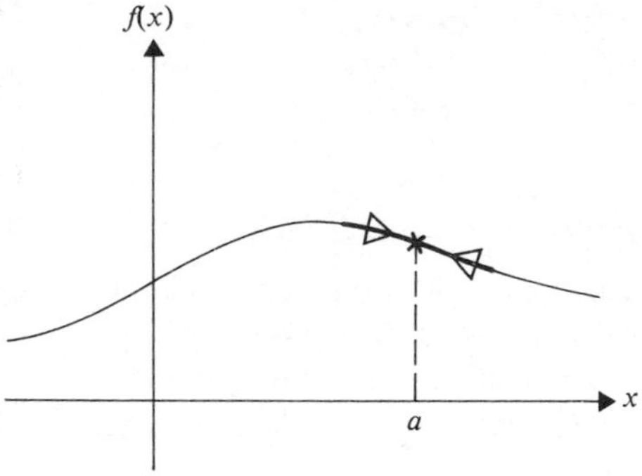

Now $\lim\limits_{x \to a} f(x) = f(a)$ means that the graph must come towards that point from each side of it (i.e. $f(x) \to f(a)$ as $x \to a$ from the right and from the left). So the curve must 'pass through' the point where $x = a$.

A continuous function is one which is continuous at all points in its domain. This means that the graph of a continuous function passes through all points in its domain, although it may have sudden changes of direction, and it may consist of two, or more, parts.

For example,

$$f: x \longmapsto x^2, \quad x \in \mathbb{R}$$

is obviously continuous,

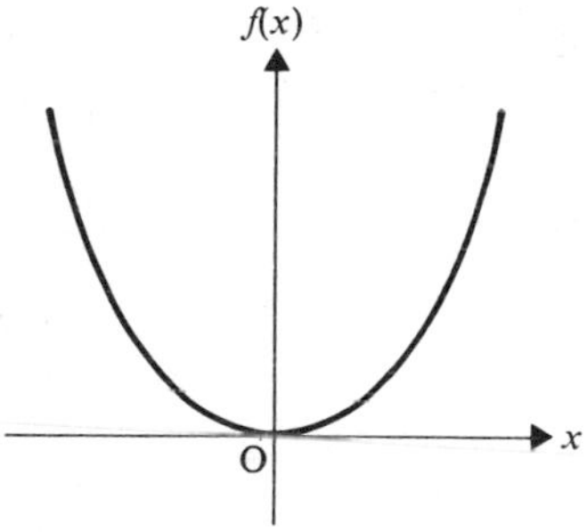

$$f: x \longmapsto |x|, \quad x \in \mathbb{R}$$

is obviously continuous at all points except perhaps at $x = 0$.
But $f(x) \to 0$ as $x \to 0$ from +ve and from −ve values.
So the function is continuous at *all* points in its domain.

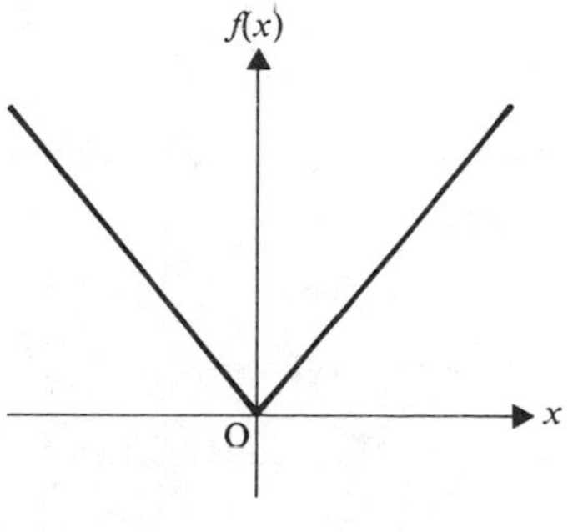

$$f: x \longmapsto \frac{1}{x}, \quad x \in \mathbb{R}, \quad x \neq 0$$

is continuous.

(This curve is in two parts but neither part has an 'end'. There is no point on the curve at $x = 0$, but $x = 0$ is not in the domain of the function.)

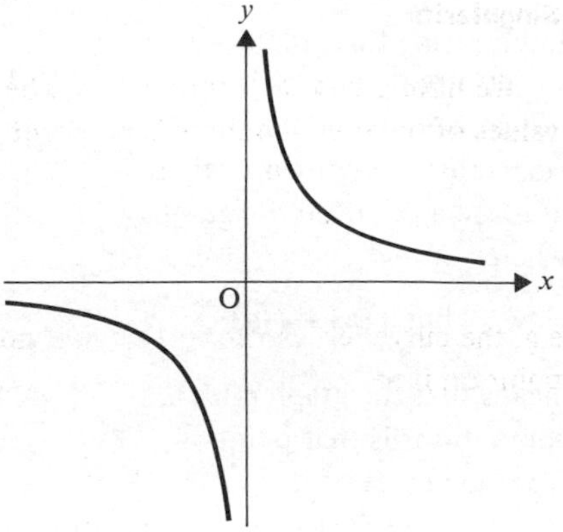

Discontinuity

If the condition for continuity is not satisfied at a particular value of x, the function is said to have a discontinuity at that value of x.
For example, consider the function

$$f: x \longmapsto [x], \quad x \in \mathbb{R}$$

where $[x]$ is used to denote 'the greatest integer $\leqslant x$',

e.g. if $x = 2.4$ $[x] = 2$

if $x = -0.8$ $[x] = -1$

if $x = 4$ $[x] = 4$

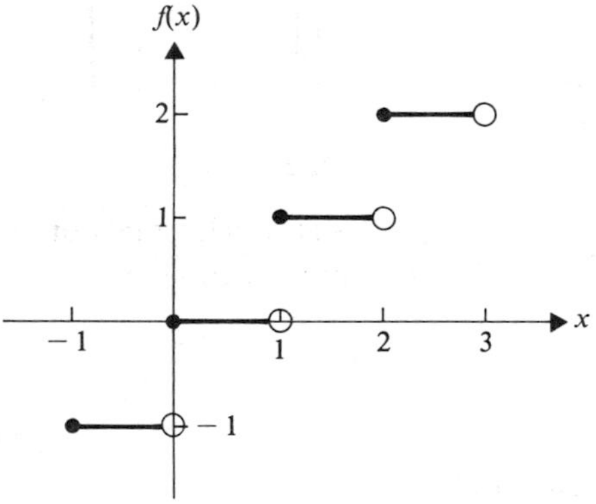

This function has a discountinuity at each integral value of x because, for example where $x = 2$,

$f(x) \to 2$ as $x \to 2$ from values greater than 2

$f(x) \to 1$ as $x \to 2$ from values less than 2

So $\lim_{x \to 2} f(x)$ does not exist.

Singularity

We have seen that a function can be continuous and yet be such that there are values of x at which there is no point on the graph of the function,

e.g. the curve representing $1/x^2$ has no point on it at $x = 0$,

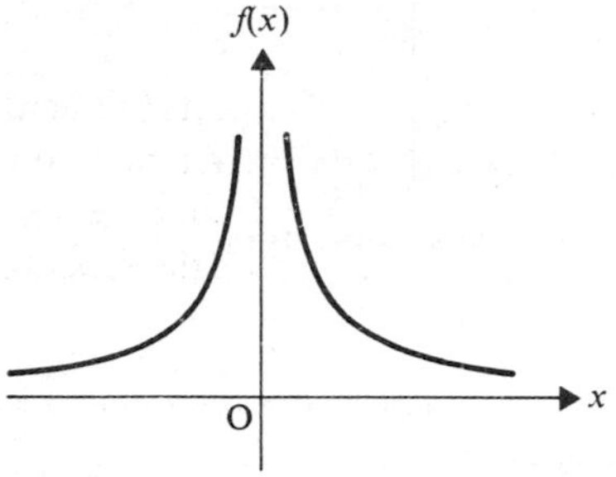

and the curve representing $\tan x$ has no point on it at any value of x that is an odd multiple of $\pi/2$.

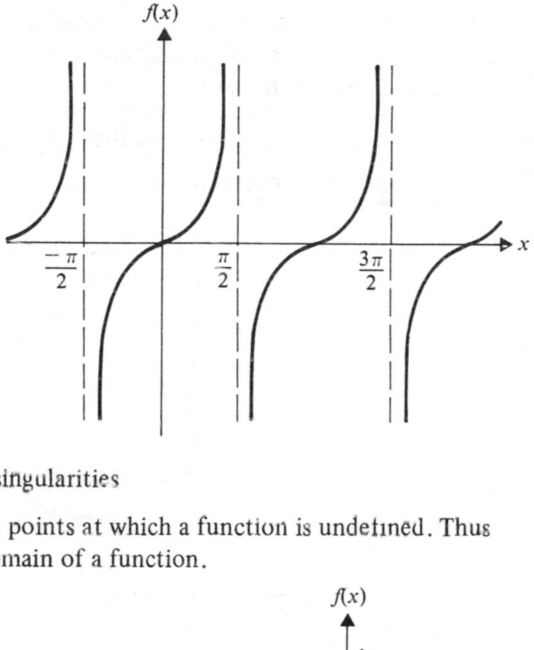

Points such as these are called singularities

Note that singularities occur at points at which a function is undefined. Thus singularities are never in the domain of a function.

For example,
for the function defined by

$$f: x \longmapsto \frac{1}{x}, \quad x \in \mathbb{R} \quad x \neq 0$$

there is a singularity at $x = 0$,

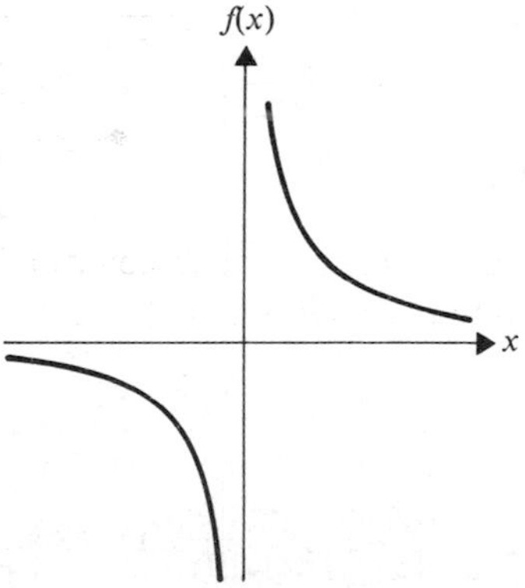

but for the function defined by

$$f: \begin{cases} x \longmapsto \dfrac{1}{x}, & x \in \mathbb{R},\ x \neq 0 \\ x \longmapsto 0, & x = 0 \end{cases}$$

there is a discontinuity at $x = 0$.

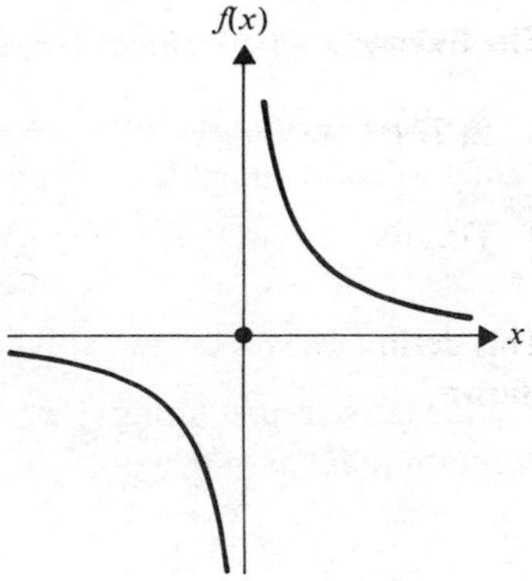

Differentiability of a Function at a Point

The derivative of $f(x)$ at the point where $x = a$ is given by

$$f'(a) = \lim_{h \to 0} \left[\frac{f(a+h) - f(a)}{h}\right]$$

So, for $f'(a)$ to exist, $f(a+h) - f(a)$ must approach zero as h approaches zero from values greater than, and less than, zero. This means that $f(x)$ must be continuous at $x = a$.

Thus a function is not differentiable at a discontinuity.

But continuity, although a necessary condition, is *not* sufficient on its own to guarantee differentiability at a point.

Consider, for example, $f: x \longmapsto |x|,\ x \in \mathbb{R}$

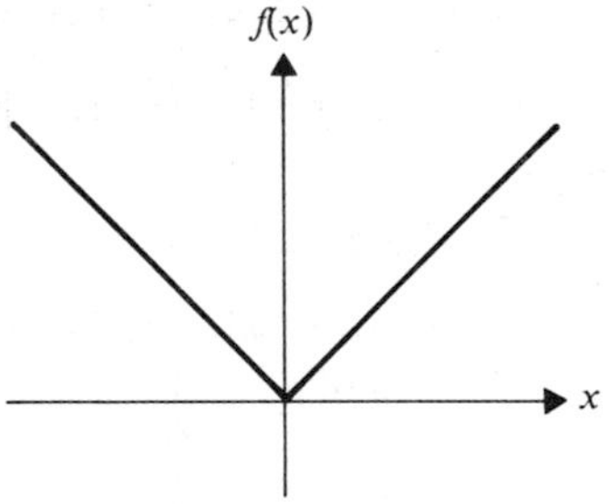

$f(x) = |x|$ is continuous for all values of x, including $x = 0$.

But $$f'(0) = \lim_{h \to 0} \left[\frac{|0+h| - |0|}{h}\right] = \lim_{h \to 0} \left[\frac{|h|}{h}\right]$$

and $\dfrac{|h|}{h} \to 1$ as $h \to 0$ from positive values

and $\dfrac{|h|}{h} \to -1$ as $h \to 0$ from negative values.

So $f'(0)$ does not exist.
Thus $|x|$ is not differentiable at $x = 0$ although it is continuous at $x = 0$.

The Existence of a Definite Integral

In *The Core Course* we defined the definite integral

$\int_a^b f(x)\,dx$ as $F(b)-F(a)$ where $F'(x)=f(x)$.

This definition was arrived at by considering the area under $y=f(x)$ as shown,

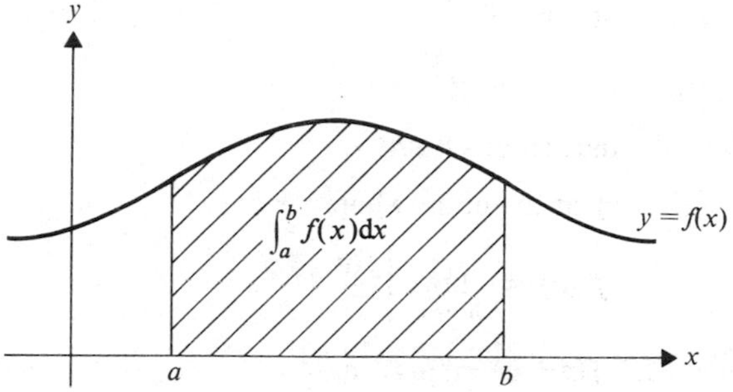

and the areas found have been only those for which $f(x)$ is continuous in the interval $a \leqslant x \leqslant b$. Now consider $\int_a^b f(x)\,dx$ when $f(x)$ has discontinuities in the range $a \leqslant x \leqslant b$.
Provided that $a \leqslant x \leqslant b$ is in the domain of $f(x)$, $f(x)$ is defined for these values of x, the part of the curve $y=f(x)$ under consideration has no points missing. So the area representing $\int_a^b f(x)\,dx$ is fully defined,

i.e. $\int_a^b f(x)\,dx$ exists.

For example, consider

$$\int_1^3 [x]\,dx$$

$[x]$ has discontinuities at $x=1, 2$ and 3.
The area represented by the integral can be found by dividing it up into sections separated by the value of x at each discontinuity.

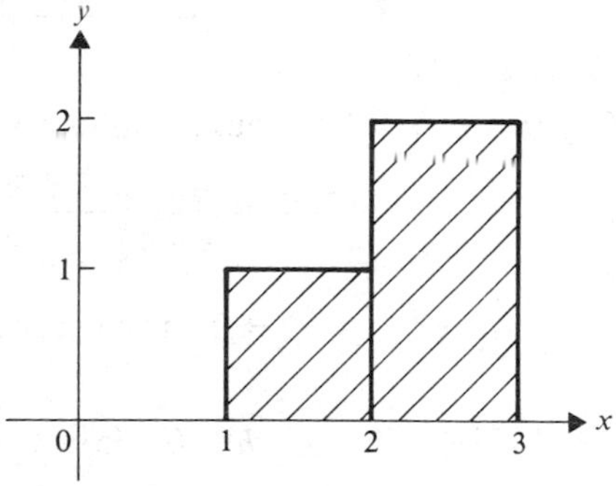

i.e. $$\int_1^3 [x]\,dx = \int_1^2 [x]\,dx + \int_2^3 [x]\,dx = \int_1^2 1\,dx + \int_2^3 2\,dx = 3$$

Now consider $\int_0^2 \frac{1}{x}\,dx$.

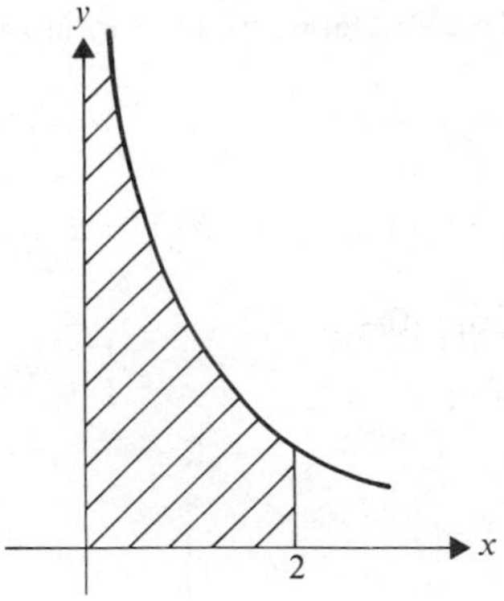

$\frac{1}{x}$ has a singularity at $x = 0$,
i.e. it is not defined for this value of x, so the 'top edge' of the area is not complete and we cannot, at this stage, find $\int_0^2 \frac{1}{x}\,dx$.

(A discussion of this type of definite integral is in Chapter 9.)

In general, provided $f(x)$ is defined for $a \leqslant x \leqslant b$,

$$\int_a^b f(x)\,dx \text{ exists,}$$

even if $f(x)$ has discontinuities in the interval $a \leqslant x \leqslant b$.

In this case the integral can usually be found by sectional integration in a manner similar to that used in the example above.
Sectional integration can also be used to find the definite integral of a function that suddenly changes direction,

e.g. $\int_{-1}^{2} |x|\,dx$ is found using

$$\int_{-1}^{0} |x|\,dx + \int_0^2 |x|\,dx$$

$$= \int_{-1}^{0} (-x)dx + \int_0^2 x\,dx$$

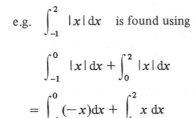

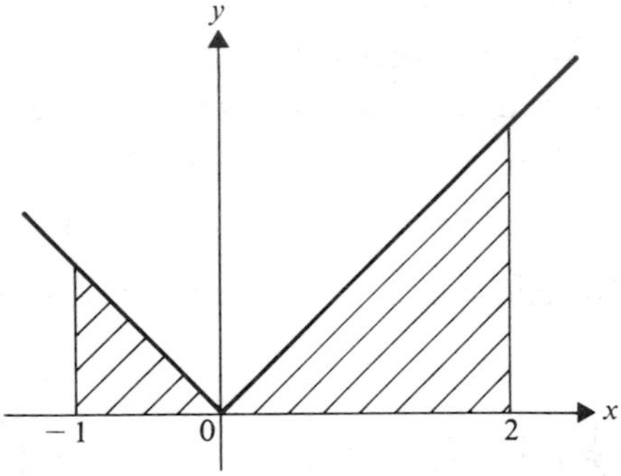

EXAMPLES 4a

1) Given that $\quad f(x) \equiv 9 - x^2 \quad$ for $\quad 0 < x \leqslant 2$

$\qquad f(x) \equiv 3x - 1 \quad$ for $\quad 2 < x \leqslant 4$

and $\quad f(x+4) = f(x) \quad$ for all values of x

sketch $f(x)$ for the range $-5 < x \leqslant 13$. Evaluate

(a) $f(23)$ (b) $\int_{-2}^{8} f(x)\,dx$.

Because $f(x+4) = f(x)$, we know that the function is periodic with a period of 4.

For $0<x\leqslant 2$

$f(x)\equiv 9-x^2 \quad \Rightarrow$

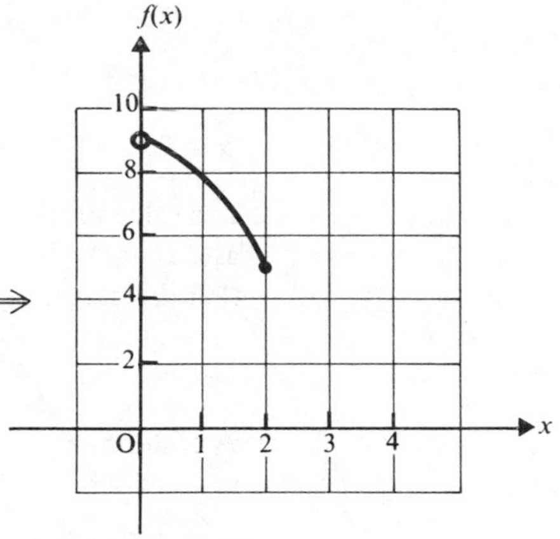

For $2<x\leqslant 4$

$f(x)\equiv 3x-1 \quad \Rightarrow$

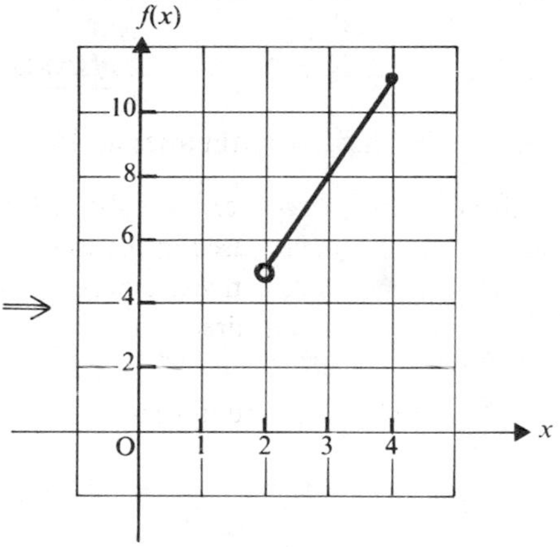

Combining these sections and repeating them at intervals of 4 units gives:

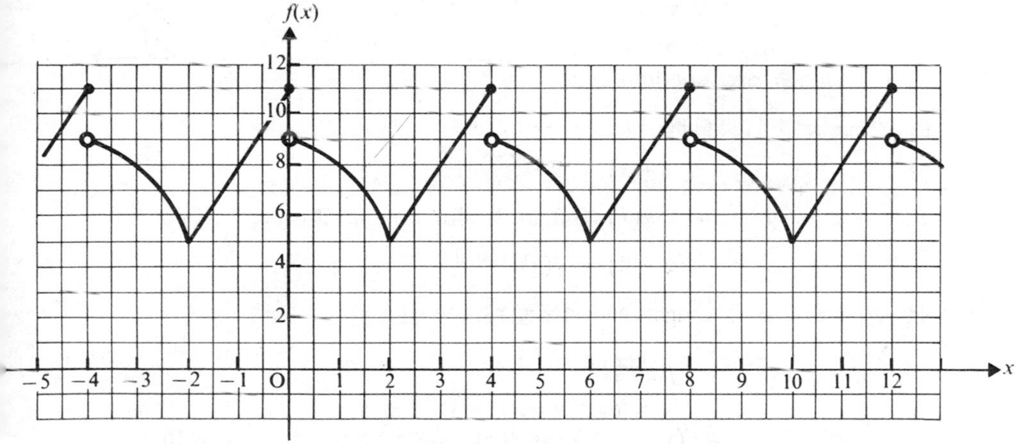

(a) As any value of $f(x)$ occurs at regular intervals of 4 units we can say

$$f(x) = f(x+4n) \quad \text{where } n \text{ is an integer}$$

So $\qquad f(23) = f(3+4\times 5) = f(3) = 8$

(b)

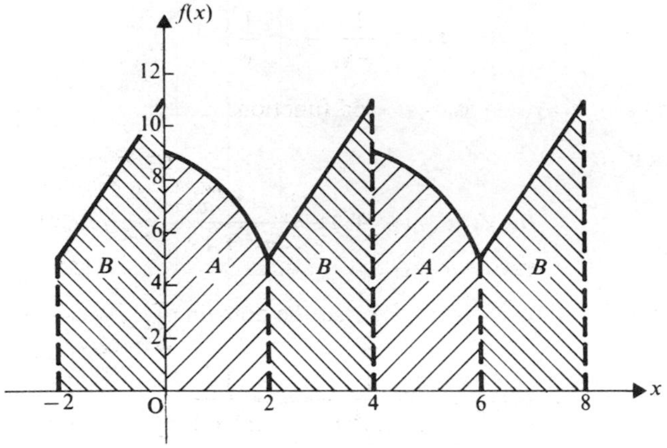

$$\int_{-2}^{8} f(x)\,dx = 2A + 3B$$

where $\qquad A = \int_0^2 (9-x^2)\,dx = \left[9x - \frac{x^3}{3}\right]_0^2 = \frac{46}{3}$

and $\qquad B = \int_2^4 (3x-1)\,dx = \left[\frac{3x^2}{2} - x\right]_2^4 = 16$

Hence $\qquad \int_{-2}^{8} f(x)\,dx = 2\left(\frac{46}{3}\right) + 3(16) = 78\tfrac{2}{3}$

2) If g and h are functions defined by

$g: x \longmapsto x^2+1,\ x \in \mathbb{R}; \quad h: x \longmapsto \frac{1}{x},\ x \in \mathbb{R},\ x \neq 0$

prove that g is an even function and that h is an odd function.
Also state the domain of $hg(x)$ and determine whether it is even, odd or neither.

To prove that $g(x)$ is even, consider $g(x)$ and $g(-x)$

$$g(x) = x^2+1$$

$$g(-x) = (-x)^2+1 = x^2+1$$

Hence $\quad g(x) = g(-x) \quad$ so g is an even function.

For the function h,

$$h(x) = \frac{1}{x}$$

$$h(-x) = \frac{1}{-x} = -\frac{1}{x}$$

Hence $h(-x) = -h(x)$ so h is an odd function.

For the function hg,

$$hg(x) = h(x^2 + 1) = \frac{1}{x^2 + 1}$$

$\frac{1}{x^2+1}$ is defined for all real values of x, so the domain of $hg(x)$ is $\mathbb{R}$.

Also $$hg(-x) = \frac{1}{(-x)^2 + 1} = \frac{1}{x^2 + 1}$$

Hence $hg(x) = hg(-x)$ so hg is an even function.

A Note on Terminology

The meaning of the word function has changed subtly over the last twenty years or so, so that it is now much more precisely defined. However, there are still many books in print which refer to (and many people who think of) a function of real variables as being *any* relationship between real variables.
So, for example, if $y = \pm\sqrt{x}$, $\pm\sqrt{x}$ would be called a function of x. If attention needed to be drawn to the fact that for any one value of x there are two values of $\pm\sqrt{x}$, it would be called a two-valued function.
A function of x was also understood to be a relationship such that x could take all real values, the function being called undefined for those values of x for which the relationship was meaningless. Now, however, a relationship is called a function only if it has a defined value for all values of x in its domain. So that, for example, $1/x$ is now only called a function if $x = 0$ is excluded from its domain.
This means that other words associated with functions, such as continuity and discontinuity, also have a more precise meaning. So if $1/x$ is thought of as a function for all real values of x then, as $1/x$ is undefined when $x = 0$, its curve has a break at the origin which would be called a discontinuity and $1/x$ would not be described as a continuous function. However, as $x = 0$ is now excluded from the domain of $1/x$, the function f such that $f(x) = 1/x$ is now called continuous and what happens at the origin is called a singularity, *not* a discontinuity.

EXERCISE 4a

Questions 1-7 refer to the following functions of x.

(a) $\sin x$ (b) $\cos x$ (c) $\arctan x$ (d) e^{-x}

(e) $|x|-1$ (f) $(x-1)^2$ (g) $\ln|x|$ (h) $x-[x]$

(i) $\dfrac{1}{x^2+1}$ (j) x^5 (k) $\left.\begin{array}{l} x^2 \text{ for } 0 \leqslant x < 1 \\ 2-x \text{ for } 1 \leqslant x < 2 \end{array}\right\} f(x)=f(x+2)$

1) State the domain of each function.

2) State which functions are even.

3) State which functions are odd.

4) State which functions are periodic and give the period.

5) State which functions are continuous.

6) State which functions have domains which include values of x at which the function is not differentiable and give these values of x.

7) State which functions have singularities.

8) State which functions are not differentiable at some value(s) of x in their domain and give the value(s) of x at which $f'(x)$ does not exist.

9) Show that $|x|-1$ is an even function and that $\sin x$ is an odd function.

10) For $f: x \longmapsto |x|-1$ and $g: x \longmapsto \sin x$, determine whether the following functions are even, odd, periodic or none of these.

(a) $f(x)g(x)$ (b) $\dfrac{f(x)}{g(x)}$ (c) $fg(x)$ (d) $gf(x)$.

11) If $f(x) \equiv 2x+1$ for $0 < x \leqslant 2$
and $f(x+2) = f(x)$ for all values of x, calculate

(a) $f(-11)$ (b) $f(17)$ (c) $\int_{-4}^{4} f(x)\,dx$.

12) Give full definitions of the periodic functions represented by the following graphs.

(a)

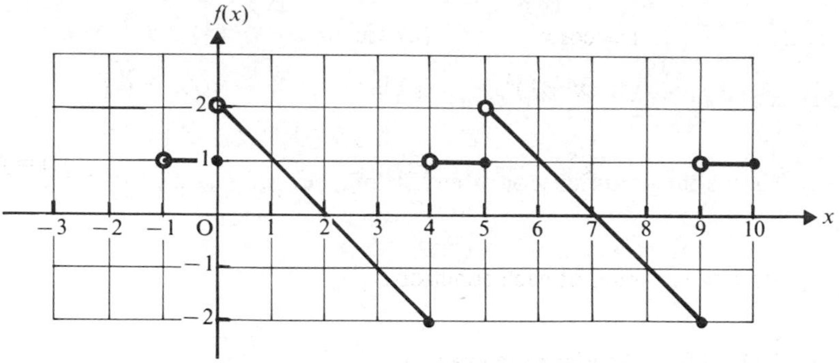

(b)

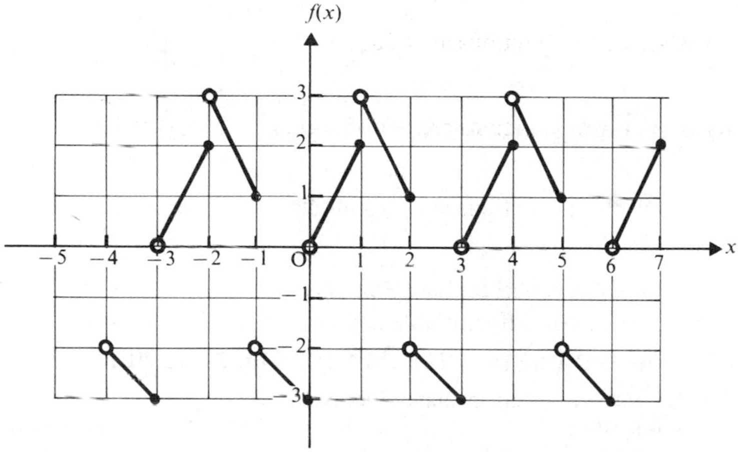

13) Which of the following definite integrals can be evaluated? (Do not carry out the integration.)

(a) $\int_0^{\pi/2} \tan x \, dx$ (b) $\int_1^3 \frac{1}{x-1} dx$ (c) $\int_{-1}^1 \frac{1}{x} dx$

(d) $\int_{\pi/6}^{\pi/3} \operatorname{cosec} x \, dx$ (e) $\int_0^4 |x-2| \, dx$.

Evaluate the following definite integrals.

14) $\int_0^2 |x-1| \, dx$

15) $\int_1^3 \{x - [x]\} \, dx$

16) $\int_{-2}^2 |x^2 - 1| \, dx$

Give the values of x at which the following functions have singularities.

17) $\dfrac{x}{(x-1)(x-2)}$ 18) $\dfrac{(x-3)(x-4)}{x}$ 19) $\dfrac{1}{x}+\dfrac{1}{x-1}$

20) $\cos(1/x)$ 21) $x\sin(1/x)$

22) Discuss the continuity or otherwise of $f(x)$ and $f'(x)$ where $f(x)$ is

(a) $\dfrac{1}{|x|}$ (b) $|\cos x|$ (c) $e^{|x|}$.

23) Prove that:

(a) if $f(x)$ is an even function,

$$\int_{-a}^{a} f(x)\,dx = 2\int_{0}^{a} f(x)\,dx$$

(b) if $f(x)$ is an odd function,

$$\int_{-a}^{a} f(x)\,dx = 0.$$

A NEW APPROACH TO THE LOGARITHMIC FUNCTION

The logarithmic function has, up to now, been regarded as the inverse of the exponential function. The logarithmic laws were derived from the laws of indices, the derivative of the log function was derived from the derivative of the exponential function; even the property $\int \frac{1}{x}\,dx = \ln|kx|$ was obtained indirectly from the exponential function. In fact the log function has not yet been defined as an independent function.
We are now going to take a new look at this function, assuming none of the properties previously obtained from the exponential function, but *defining* the function $\ln x$ in the following way,

$$\ln x = \int_{1}^{x} \frac{1}{t}\,dt, \quad x \in \mathbb{R}, \quad x > 0$$

i.e. $\ln x$ is represented by the area bounded by the curve $f(t) \equiv \dfrac{1}{t}$, the t axis and the lines $t = 1$ and $t = x$ where $x > 0$.

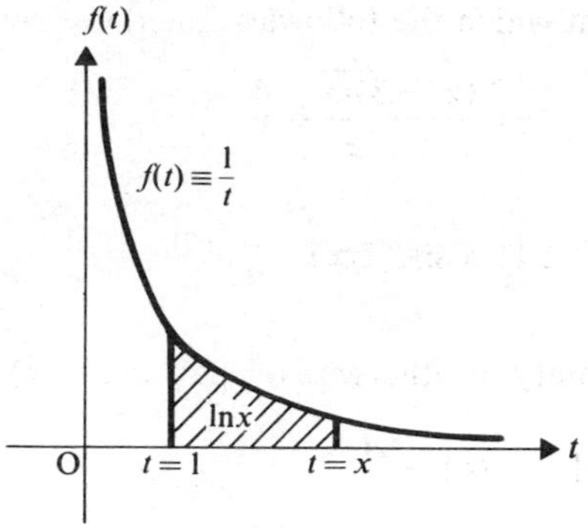

From this definition, it is now possible to derive properties of the log function.

To Prove that $\ln a + \ln b = \ln ab$

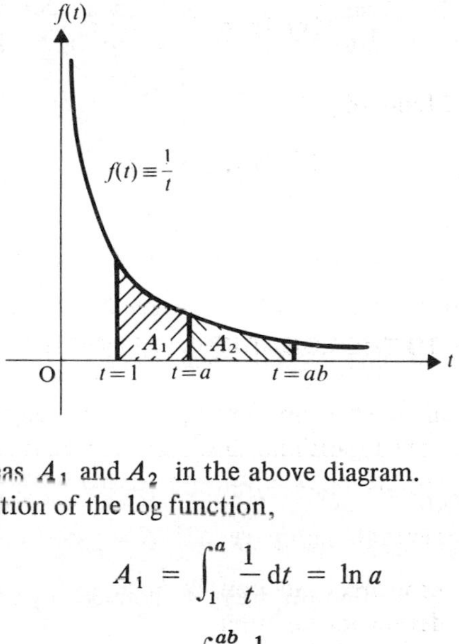

Consider the two areas A_1 and A_2 in the above diagram.
Using the new definition of the log function,

$$A_1 = \int_1^a \frac{1}{t}\,dt = \ln a \qquad [1]$$

and

$$A_1 + A_2 = \int_1^{ab} \frac{1}{t}\,dt = \ln ab \qquad [2]$$

But

$$A_2 = \int_a^{ab} \frac{1}{t}\,dt$$

which, in this form, is not a log function (the lower limit is a instead of 1).
However if we make the substitution

$$t \equiv au$$

so that

$$\ldots dt \equiv \ldots a\,du$$

and

t	a	ab
u	1	b

then $$A_2 = \int_a^{ab} \frac{1}{t}\,\mathrm{d}t \equiv \int_1^b \frac{1}{au} a\,\mathrm{d}u \equiv \int_1^b \frac{1}{u}\,\mathrm{d}u$$

Now, by definition,

$$A_2 = \ln b \qquad [3]$$

So, combining [1], [2] and [3],

$$A_1 + A_2 = \ln a + \ln b = \ln ab$$

It is important to appreciate that, when a function is defined, no properties must be assumed unless probed from that definition. The reader is given the opportunity, in Exercise 4b, to prove some more of the familiar laws of logarithms from the definition $\ln x = \int_1^x \frac{1}{t}\,\mathrm{d}t$.

EXAMPLES 4b

1) Use the trapezium rule with 7 ordinates to find an approximate value for $\int_1^4 \frac{1}{t}\,\mathrm{d}t$. Hence estimate $\ln 4$ to 3 decimal places.

t	1	1.5	2	2.5	3	3.5	4
$\frac{1}{t}$	1	0.6667	0.5000	0.4000	0.3333	0.2857	0.2500

Using the trapezium rule to find the area, A, bounded by the curve $y = \frac{1}{t}$, the t axis and the lines $t = 1$, $t = 4$ we have:

$$A \simeq \frac{0.5}{2}\left\{1 + 2(0.6667 + 0.5000 + 0.4000 + 0.3333 + 0.2857) + 0.2500\right\}$$

$$= 1.4054$$

Now, by definition,

$$\ln x = \int_1^x \frac{1}{t}\,\mathrm{d}t$$

so $$\ln 4 = \int_1^4 \frac{1}{t}\,dt \simeq 1.4054.$$

i.e. $$\ln 4 \simeq 1.405 \qquad \text{(to 3 d.p.)}.$$

2) By considering the areas bounded by the x axis, the lines $t = 1$, $t = x$ and

(a) the line $y = \dfrac{1}{x}$, (b) the curve $y = \dfrac{1}{t}$,

(c) the line joining the points $(1, 1)$ and $\left(x, \dfrac{1}{x}\right)$ on the curve $y = \dfrac{1}{t}$,

prove that, for $x \geqslant 1$, $\left(\dfrac{x}{x-1}\right) \ln x \to 1$ as $x \to 1$.

(a)

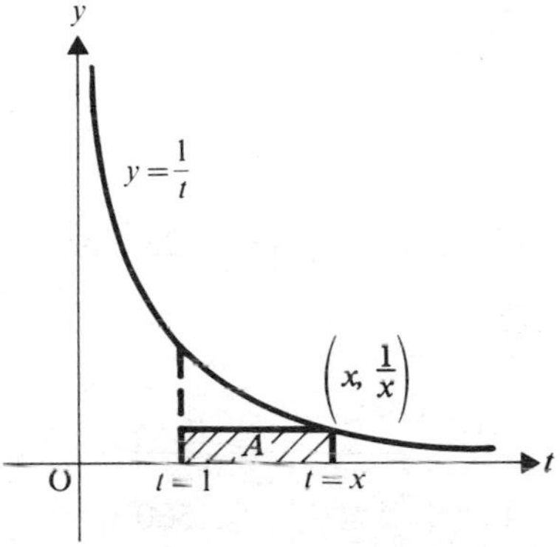

Area A is the area of a rectangle of height $\dfrac{1}{x}$ and base $(x-1)$.

So $$\text{Area } A = \frac{1}{x}(x-1)$$

(b)

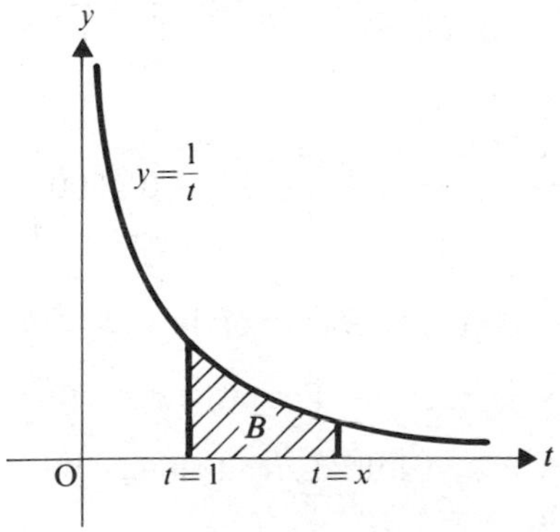

$$\text{Area } B = \int_1^x \frac{1}{t}\,dt$$
$$= \ln x \quad \text{by definition}$$

So $\text{Area } B = \ln x$

(c)

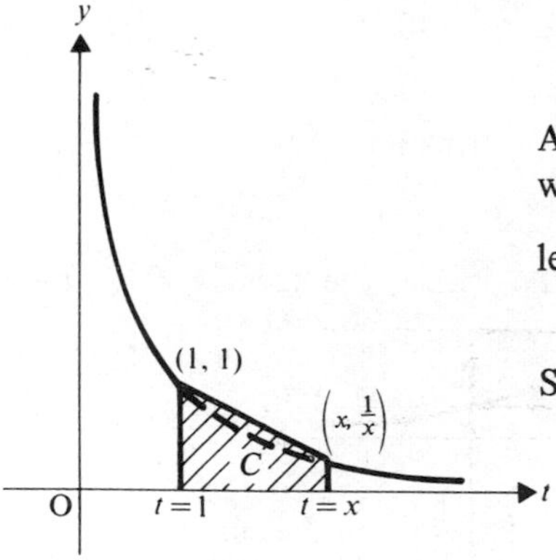

Area C is the area of a trapezium of width $(x-1)$ and with parallel sides of lengths 1 and $\frac{1}{x}$.

So Area $C = \left(\frac{x-1}{2}\right)\left(1+\frac{1}{x}\right)$

$$= \frac{x^2-1}{2x}$$

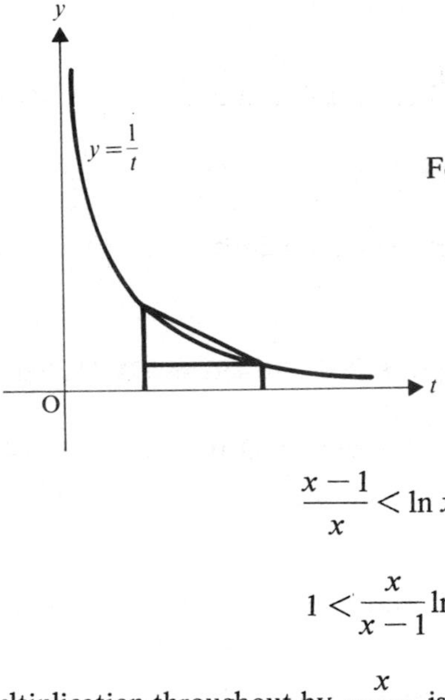

For all $x > 1$,

Area $A <$ Area $B <$ Area C

So
$$\frac{x-1}{x} < \ln x < \frac{x^2-1}{2x}$$

or
$$1 < \frac{x}{x-1}\ln x < \frac{x+1}{2}$$

(Multiplication throughout by $\frac{x}{x-1}$ is valid because $x > 1$.)

Now as $x \to 1$, $\frac{x+1}{2} \to 1$.

Hence $\frac{x}{x-1}\ln x$ lies between 1 and a quantity that approaches 1.

So $\frac{x}{x-1}\ln x \to 1$ as $x \to 1$.

Note. Other forms of restriction on the range of possible values of $\ln x$ can be derived by choosing other areas related to the graph of $\frac{1}{t}$. For instance a cruder relationship is based on the same areas A and B, and a third area bounded by the t axis and the lines $t = 1$, $t = x$ and $y = 1$.

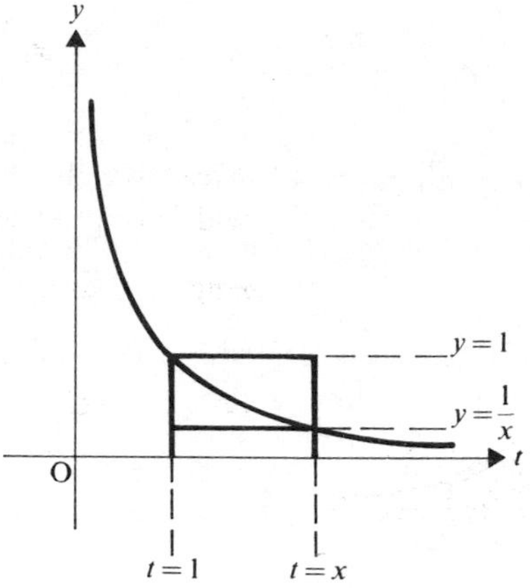

EXERCISE 4b

1) Prove that $\ln a - \ln b = \ln \frac{a}{b}$. (Use the substitution $t \equiv \frac{a}{u}$ where appropriate.)

2) Use the definition $\ln a^n = \int_1^{a^n} \frac{1}{t}\,\mathrm{d}t$ and the substitution $t \equiv u^n$ to prove that $\ln a^n = n \ln a$.

3) By using the substitution $t \equiv \frac{1}{u}$ prove from the definition $\ln x = \int_1^x \frac{1}{t}\,\mathrm{d}t$ that $\ln x = -\ln \frac{1}{x}$. Hence show that the areas bounded by the curve $y = \frac{1}{t}$, the t axis and

(a) the lines $t = \frac{1}{2}$ and $t = 1$,
(b) the lines $t = 1$ and $t = 2$,
are equal.

4) Using the definition $\ln(1 + x) = \int_1^{1+x} \frac{1}{t}\,\mathrm{d}t$ and considering the area under the graph $y = \frac{1}{t}$, prove that, for $x > 0$,

$$x(1 + x) > (1 + x)\ln(1 + x) > x.$$

Hence show that $\ln(1 + x) > x - x^2 + x^3 - x^4 \ldots$ and that, as $x \to 0$, $\frac{1}{x}\ln(1 + x) \to 1$.

5) Use Simpson's Rule with 5 ordinates to evaluate $\int_1^2 \frac{1}{t}\,\mathrm{d}t$ correct to 4 decimal places.

Hence find an approximate value for ln 2.

6) Show that $\displaystyle\int_1^{1.1} \frac{1}{t}\,dt = \int_0^{0.1} \frac{1}{1+u}\,du.$

By using the binomial expansion of $(1+u)^{-1}$ find $\displaystyle\int_0^{0.1} \frac{1}{1+u}\,du$ correct to five decimal places. Hence find $\ln 1.1$ correct to 5 decimal places.

7) Use a method similar to that indicated in Question 6 to find, without using tables or calculator, the value of $\ln \frac{6}{5}$.

8) Show that $\displaystyle\int_2^{3} \frac{1}{t}\,dt = \ln \frac{3}{2}.$

9) Show that $\ln 1 = 0$.

10) By using the definition $\displaystyle\ln(1+x) = \int_1^{1+x} \frac{1}{t}\,dt,$ the substitution $t \equiv 1+u$ and the binomial series, show that, for a certain range of values of x

$$\ln(1+x) = x - \frac{x^2}{2} + \frac{x^3}{3} - \frac{x^4}{4} + \dots.$$

HYPERBOLIC FUNCTIONS

Consider the two exponential functions e^x and e^{-x}.

The function $\frac{1}{2}(e^x + e^{-x})$ is called the *hyperbolic cosine* function which is written *cosh x*
The function $\frac{1}{2}(e^x - e^{-x})$ is called the *hyperbolic sine* function which is written *sinh x*,

i.e. $$\cosh x \equiv \tfrac{1}{2}(e^x + e^{-x})$$

and $$\sinh x \equiv \tfrac{1}{2}(e^x - e^{-x})$$

e^x and e^{-x} are defined for all values of x so it follows that $\cosh x$ and $\sinh x$ are also defined for all real values of x,
i.e. the domain of both $f: x \longmapsto \cosh x$ and $f: x \longmapsto \sinh x$ is $\mathbb{R}$.

HYPERBOLIC RELATIONSHIPS

The names of the two hyperbolic functions we have so far met, i.e. the hyperbolic sine and cosine, suggest that these functions have certain properties that are similar to those of trig functions.
Consider, for instance, the relationship between $\cosh^2 x$ and $\sinh^2 x$

$$\cosh^2 x \equiv \tfrac{1}{4}(e^{2x} + 2 + e^{-2x})$$
$$\sinh^2 x \equiv \tfrac{1}{4}(e^{2x} - 2 + e^{-2x})$$

Hence $\qquad \cosh^2 x - \sinh^2 x \equiv 1$

and $\qquad \cosh^2 x + \sinh^2 x \equiv \tfrac{1}{2}(e^{2x} + e^{-2x}) \equiv \cosh 2x$

These identities can be compared with

$$\cos^2 x + \sin^2 x \equiv 1$$

and $\qquad \cos^2 x - \sin^2 x \equiv \cos 2x$

In each case it can be seen that the term $\sin^2 x$ becomes $-\sinh^2 x$ in the corresponding hyperbolic identity, but that otherwise there is a direct analogy. This is an example of Osborn's Rule which can be used to convert many trig identities into analogous hyperbolic identities.

The rule is:
Change each trig ratio into the comparative hyperbolic function. Whenever a *product of two sines* occurs, change the sign of that term.
Osborn's Rule should be used with care and restraint as there are a number of ways in which a product of two sine ratios can be disguised, e.g. in $\tan^2 x$. Osborn's Rule is justified in Chapter 8, where specific relationships between trig and hyperbolic functions are derived.
Because of these close links, trig terminology is used to define further hyperbolic functions. So we have:

$$\tanh x \equiv \frac{\sinh x}{\cosh x} \equiv \frac{e^x - e^{-x}}{e^x + e^{-x}}$$

$$\operatorname{cosech} x \equiv \frac{1}{\sinh x} \equiv \frac{2}{e^x - e^{-x}}$$

$$\operatorname{sech} x \equiv \frac{1}{\cosh x} \equiv \frac{2}{e^x + e^{-x}}$$

$$\coth x \equiv \frac{1}{\tanh x} \equiv \frac{e^x + e^{-x}}{e^x - e^{-x}}$$

The Graphs of Hyperbolic Functions

The graphs of $\sinh x$ and $\cosh x$ can be obtained by combining the graphs of e^x and e^{-x} as shown below.

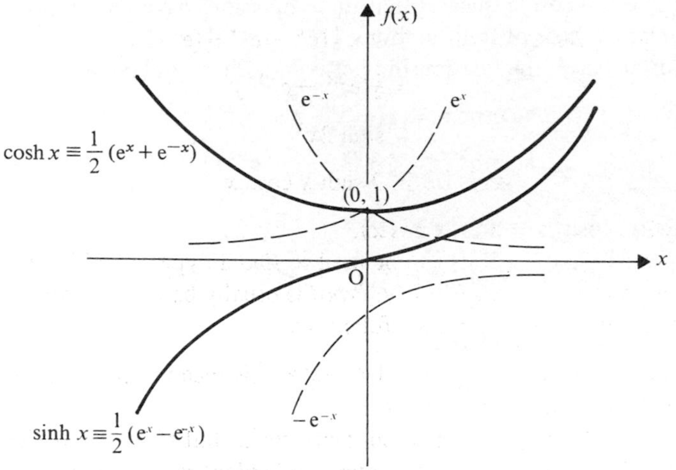

Note that both functions are continuous

$\cosh x \geqslant 1$ for all values of x

$\sinh x \to \cosh x$ as $x \to \infty$

$\sinh x \to -\cosh x$ as $x \to -\infty$

the graph of $\cosh x$ has one turning point (minimum)

the graph of $\sinh x$ has no turning point.

The graph of $\tanh x$ is given by dividing $\sinh x$ by $\cosh x$, i.e.

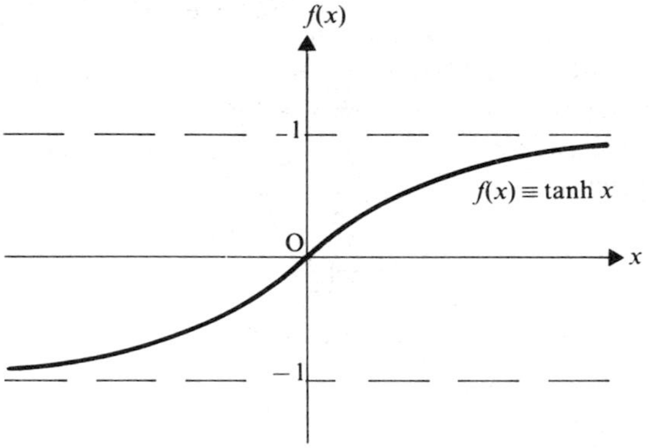

Note that $-1 < \tanh x < 1$.

Note that the similarity between hyperbolic and trig functions is limited to their occurrence in identities. Their graphs have no corresponding relationship because, whilst all the trig functions are periodic, none of the hyperbolic functions is periodic.

EXAMPLES 4c

1) Prove that $\sinh 2x \equiv 2 \sinh x \cosh x$.

Starting on the R.H.S.,

$$2 \sinh x \cosh x \equiv 2 \times \tfrac{1}{2}(e^x - e^{-x})\tfrac{1}{2}(e^x + e^{-x})$$
$$\equiv \tfrac{1}{2}(e^{2x} - e^{-2x})$$
$$\equiv \sinh 2x$$

i.e. $$\sinh 2x \equiv 2 \sinh x \cosh x$$

(compare with $\sin 2x \equiv 2 \sin x \cos x$).
Note that, while Osborn's Rule can be used to *find* a hyperbolic identity, it cannot be used to *prove* that identity. *Proof* is usually based, as in this example, on the *definitions* of the hyperbolic functions.

2) From the trig identity $\tan^2 x + 1 \equiv \sec^2 x$ deduce the corresponding hyperbolic identity and then prove it.
In the trig identity the term $\tan^2 x$ contains $\sin^2 x$ but the term $\sec^2 x$ does not. So we deducc the corresponding hyperbolic identity by converting $\tan^2 x$ to $-\tanh^2 x$ and $\sec^2 x$ to $\operatorname{sech}^2 x$, giving

$$1 - \tanh^2 x \equiv \operatorname{sech}^2 x$$

Considering the L.H.S.

$$1 - \tanh^2 x \equiv 1 - \left(\frac{e^x - e^{-x}}{e^x + e^{-x}}\right)^2$$
$$\equiv \frac{(e^x + e^{-x})^2 - (e^x - e^{-x})^2}{(e^x + e^{-x})^2}$$
$$\equiv \frac{4}{(e^x + e^{-x})^2}$$
$$\equiv \left(\frac{2}{e^x + e^{-x}}\right)^2$$
$$\equiv \operatorname{sech}^2 x$$

This confirms that our deduction was correct.

EXERCISE 4c

1) Prove that $f: x \longmapsto \cosh x$ is an even function.

2) Prove that $\sinh x$ and $\tanh x$ are odd functions of x.

Prove the following identities from the basic definitions of the hyperbolic functions:

3) $\coth^2 x - 1 \equiv \operatorname{cosech}^2 x$

4) $\sinh(x+y) \equiv \sinh x \cosh y + \cosh x \sinh y$

5) $\cosh 3x \equiv 4\cosh^3 x - 3\cosh x$

6) $\tanh 2x \equiv \dfrac{2\tanh x}{1+\tanh^2 x}$.

Use Osborn's Rule to deduce the hyperbolic identities corresponding to the trig identities given in Questions 7–12. In each case prove that your result is correct.

7) $\sin(x+y) \equiv \sin x \cos y + \cos x \sin y$

8) $\cos(x-y) \equiv \cos x \cos y + \sin x \sin y$

9) $\sin 3x \equiv 3\sin x - 4\sin^3 x$

10) $\cos x + \cos y \equiv 2\cos\dfrac{x+y}{2}\cos\dfrac{x-y}{2}$

11) $\sin 2x \equiv \dfrac{2\tan x}{1+\tan^2 x}$

12) $\cos 2x \equiv \dfrac{1-\tan^2 x}{1+\tan^2 x}$

13) If $\sinh x = \frac{3}{4}$ calculate $\operatorname{sech} x$, $\tanh x$, $\cosh 2x$ and $\tanh 2x$.

14) If $\tanh x = \frac{5}{13}$ calculate $\operatorname{cosech} x$, $\cosh x$, $\sinh 2x$ and $\tanh 2x$.

15) Sketch the graphs of $\coth x$, $\operatorname{cosech} x$, $\operatorname{sech} x$, $\cosh x + \sinh x$, $\sinh^2 x$, $\tanh(-x)$ and $\sinh(-x)$.
State which, if any, of these functions are
(a) even (b) odd (c) continuous.

Prove that:

16) $\tanh x \equiv \sqrt{\left(\dfrac{\cosh 2x - 1}{\cosh 2x + 1}\right)}$ $(x > 0)$.

17) $\cosh 3x \cosh^3 x + \sinh 3x \sinh^3 x \equiv \cosh^3 2x$.

18) Show that the point with coordinates $(a\cosh t,\ b\sinh t)$ lies on the curve $\dfrac{x^2}{a^2} - \dfrac{y^2}{b^2} = 1$.

HYPERBOLIC EQUATIONS

To solve equations involving hyperbolic functions, use can be made of the identities proved in the previous section, as well as of the basic definitions of the functions.

EXAMPLES 4d

1) Solve the equation $3\sinh x - \cosh x = 1$.

As no convenient identity relates $\sinh x$ to $\cosh x$ we will use the definitions of these functions, giving

$$\begin{aligned}
&\tfrac{3}{2}(e^x - e^{-x}) - \tfrac{1}{2}(e^x + e^{-x}) = 1\\
\Rightarrow\quad & 2e^x - 4e^{-x} = 2\\
\Rightarrow\quad & e^x - 1 - 2e^{-x} = 0\\
\Rightarrow\quad & (e^x)^2 - e^x - 2 = 0\\
\Rightarrow\quad & (e^x - 2)(e^x + 1) = 0
\end{aligned}$$

So either $e^x - 2 = 0$ or $e^x + 1 = 0$.
But $e^x = -1$ has no real solution.
So $e^x = 2 \Rightarrow x = \ln 2$ is the only solution.

2) Find the values of x for which $12\cosh^2 x + 7\sinh x = 24$.

Using the identity $\cosh^2 x - \sinh^2 x \equiv 1$ gives

$$\begin{aligned}
&12(1 + \sinh^2 x) + 7\sinh x - 24 = 0\\
\Rightarrow\quad & 12\sinh^2 x + 7\sinh x - 12 = 0\\
\Rightarrow\quad & (3\sinh x + 4)(4\sinh x - 3) = 0
\end{aligned}$$

Hence $\sinh x = -\frac{4}{3}$ or $\frac{3}{4}$.
Now using the definition of $\sinh x$, $\frac{1}{2}(e^x - e^{-x}) = -\frac{4}{3}$ or $\frac{3}{4}$,

$$\begin{aligned}
\text{i.e.}\quad & 3e^x + 8 - 3e^{-x} = 0 \quad\text{or}\quad 2e^x - 3 - 2e^{-x} = 0\\
\Rightarrow\quad & 3e^{2x} + 8e^x - 3 = 0 \quad\text{or}\quad 2e^{2x} - 3e^x - 2 = 0\\
\Rightarrow\quad & (3e^x - 1)(e^x + 3) = 0 \quad\text{or}\quad (2e^x + 1)(e^x - 2) = 0\\
\Rightarrow\quad & e^x = \tfrac{1}{3},\ -3,\ -\tfrac{1}{2},\ 2
\end{aligned}$$

But negative values of e^x do not give real values of x, so real solutions are given only by $e^x = \frac{1}{3}, 2$.
Hence $x = \ln\frac{1}{3}$ or $\ln 2$.

EXERCISE 4d

Solve, for real values of x, the equations given in Questions 1–10.

1) $\sinh x + 4 = 4\cosh x$

2) $7 + 2\cosh x = 6\sinh x$

3) $2\sinh x + 6\cosh x = 9$

4) $5\cosh x + \sinh x = 7$

5) $\cosh 2x - 7\cosh x + 7 = 0$

6) $4\tanh^2 x - \operatorname{sech} x = 1$

7) $4\cosh x - e^{-x} = 3$

8) $\sinh^2 x - 5\cosh x + 5 = 0$

9) $4\sinh x + 3e^x + 3 = 0$

10) $20\cosh 2x - 21\sinh x = 200$

11) Express $4\cosh x + 5\sinh x$ in the form $r\sinh(x+y)$ giving the values of r and $\tanh y$.

12) By expressing $13\cosh x + 5\sinh x$ in the form $r\cosh(x+y)$, find its minimum value.

THE CALCULUS OF HYPERBOLIC FUNCTIONS

If $\quad f(x) \equiv \cosh x \equiv \frac{1}{2}(e^x + e^{-x})$

then $\quad f'(x) = \frac{1}{2}(e^x - e^{-x}) \equiv \sinh x$

i.e.
$$\frac{d}{dx}(\cosh x) = \sinh x$$

Also if $\quad f(x) \equiv \sinh x \equiv \frac{1}{2}(e^x - e^{-x})$

then $\quad f'(x) = \frac{1}{2}(e^x + e^{-x}) \equiv \cosh x$

i.e.
$$\frac{d}{dx}(\sinh x) = \cosh x$$

From these results it follows that

$$\int \sinh x\, dx = \cosh x + K$$

and
$$\int \cosh x\, dx = \sinh x + K$$

Other hyperbolic functions can now be differentiated and integrated, using the results proved above.

Note that Osborn's Rule does *not* apply to calculus operations.

EXAMPLES 4e

1) Differentiate (a) cosech x (b) $\tanh^2 x$.

(a) $$f(x) \equiv \operatorname{cosech} x \equiv \frac{1}{\sinh x} \equiv (\sinh x)^{-1}$$

so $$f'(x) = -(\sinh x)^{-2}\frac{\mathrm{d}}{\mathrm{d}x}(\sinh x)$$

$$= -\frac{\cosh x}{\sinh^2 x}$$

(b) $$f(x) \equiv \tanh^2 x$$

so $$f'(x) = 2\tanh x\frac{\mathrm{d}}{\mathrm{d}x}(\tanh x)$$

but $$\tanh x \equiv \frac{\sinh x}{\cosh x}$$

so $$\frac{\mathrm{d}}{\mathrm{d}x}(\tanh x) = \frac{\cosh x\,(\cosh x) - \sinh x\,(\sinh x)}{\cosh^2 x}$$

$$= \frac{1}{\cosh^2 x}$$

$$= \operatorname{sech}^2 x$$

Hence $$\frac{\mathrm{d}}{\mathrm{d}x}(\tanh^2 x) = 2\tanh x \operatorname{sech}^2 x$$

2) Find (a) $\int \tanh x\,\mathrm{d}x$ (b) $\int \sinh^3 x\,\mathrm{d}x$.

(a) $$\int \tanh x\,\mathrm{d}x = \int \frac{\sinh x}{\cosh x}\,\mathrm{d}x$$

$$\equiv \int \frac{f'(x)}{f(x)}\,\mathrm{d}x \quad \text{where} \quad f(x) \equiv \cosh x$$

$$= \ln\,[f(x)] + K$$

So $$\int \tanh x\,\mathrm{d}x = \ln(\cosh x) + K$$

(b) $$\int \sinh^3 x\,\mathrm{d}x \equiv \int \sinh x \sinh^2 x\,\mathrm{d}x$$

$$\equiv \int \sinh x\,(\cosh^2 x - 1)\,\mathrm{d}x$$

$$\equiv \int \sinh x \cosh^2 x\,\mathrm{d}x - \int \sinh x\,\mathrm{d}x$$

$$= \tfrac{1}{3}\cosh^3 x - \cosh x + K$$

EXERCISE 4e

Differentiate the following functions w.r.t x.

1) $\operatorname{sech} x$ 2) $\tanh 2x$ 3) $\coth x$ 4) $\operatorname{sech}^2 x$ 5) $\sinh 4x$

6) $\cosh^3 2x$ 7) $x \sinh x$ 8) $\sinh x \tanh x$ 9) $e^x \sinh x$

10) $\sqrt{(\cosh 5x)}$ 11) $x^2 \tanh^2 3x$ 12) $\dfrac{e^x}{\sinh 2x}$ 13) $e^{\cosh x}$

14) $\ln \sinh x$ 15) $\sqrt{(\tanh x)}$ 16) $e^{\tanh^2 x}$

Find the following integrals.

17) $\int \sinh 5x \, dx$ 18) $\int \cosh 4x \, dx$ 19) $\int \tanh 2x \, dx$

20) $\int \coth x \, dx$ 21) $\int \sinh^2 x \, dx$ 22) $\int \cosh^3 x \, dx$

23) $\int x \sinh 2x \, dx$ 24) $\int \dfrac{\sinh x}{\cosh^2 x} dx$ 25) $\int e^x \cosh x \, dx$

26) $\int \cosh 3x \cosh x \, dx$ 27) $\int \sinh 5x \cosh 3x \, dx$

Evaluate the integrals in Questions 28–31.

28) $\int_0^4 \cosh 3x \, dx$ 29) $\int_1^2 e^x \sinh x \, dx$

30) $\int_0^1 \tanh 4x \, dx$ 31) $\int_1^4 \operatorname{sech}^2 5x \, dx$

32) Find the equation of the tangent and normal to the curve $\dfrac{x^2}{a^2} - \dfrac{y^2}{b^2} = 1$ at the point $(a \cosh u, b \sinh u)$.

33) Investigate the stationary value(s) of $25 \cosh x - 7 \sinh x$.

34) If $y = A \cosh kx + B \sinh kx$, prove that $\dfrac{d^2y}{dx^2} = k^2 y$. Hence find y as a function of x given that $\dfrac{d^2y}{dx^2} = 4y$, and that, when $x = 0$, $y = 2$ and $\dfrac{dy}{dx} = 2$.

INVERSE HYPERBOLIC FUNCTIONS

The function $f: x \longmapsto \sinh x,\ x \in \mathbb{R}$, is a one-one mapping, i.e. one value of $\sinh x$ arises from just one value of x. Therefore the reverse mapping $\sinh x \longmapsto x$ is a function for all real values of x.
It is called the *inverse sinh function* and is denoted by $\sinh^{-1}$ (or arsinh).

So if $f: x \longmapsto \sinh x,\ x \in \mathbb{R}$ then $f^{-1}: x \longmapsto \sinh^{-1} x,\ x \in \mathbb{R}$
and

$$y = \sinh x \iff x = \sinh^{-1} y$$

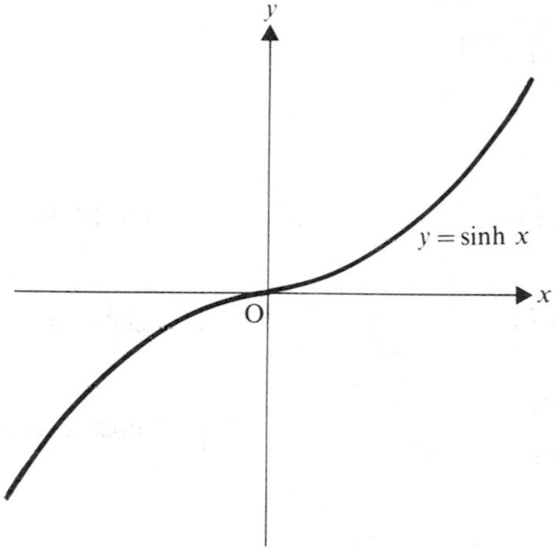

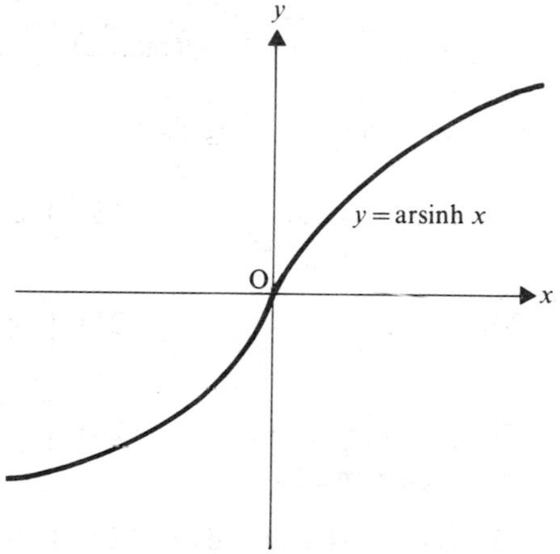

The function $f: x \longmapsto \tanh x,\ x \in \mathbb{R}$, is also a one-one mapping, but its image-set is the open interval $(-1, 1)$. Therefore the reverse relationship $\tanh x \longmapsto x$ is a function with domain $x \in \mathbb{R}, -1 < x < 1$. It is called the *inverse tanh function* and is denoted by $\tanh^{-1}$ (or artanh).
So if $f: x \longmapsto \tanh x,\ x \in \mathbb{R}$
then $f^{-1}: x \longmapsto \tanh^{-1} x,\ x \in \mathbb{R}, -1 < x < 1$

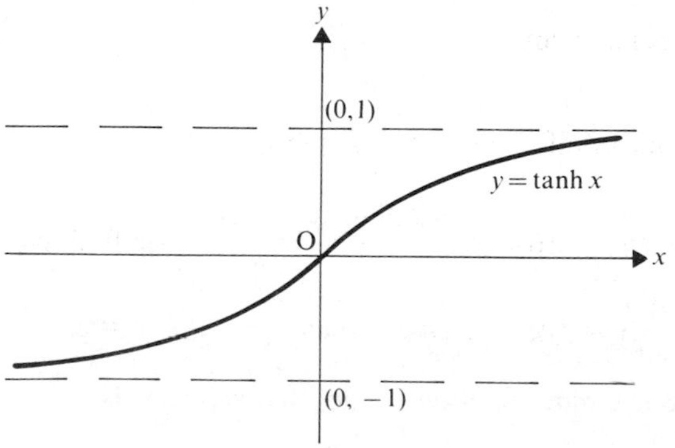

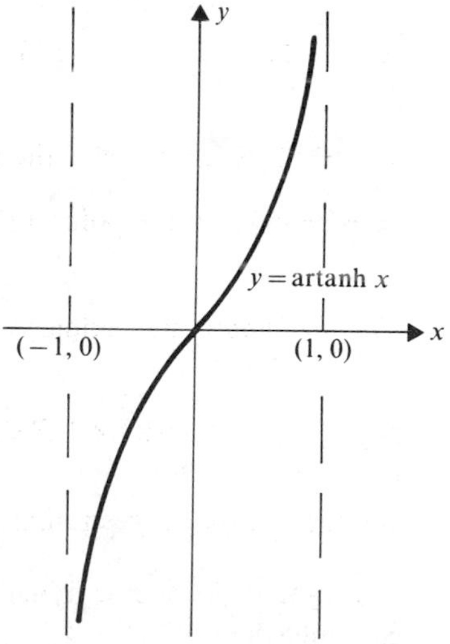

The function $f: x \longmapsto \cosh x, \quad x \in \mathbb{R}$ is not a one-one mapping. A look at the curve $y = \cosh x$ shows that any value of y (except $y = 1$) arises from two, equal and opposite, values of x. So this function does not have an inverse. However, if we change the domain to give the function

$$f: x \longmapsto \cosh x, \quad x \in \mathbb{R}, \quad x \geqslant 0$$

this is a one-one mapping with image-set $[1, \infty)$.
So this function has an inverse with domain $[1, \infty)$ and image-set $[0, \infty)$,
i.e. if $f: x \longmapsto \cosh x, \quad x \in \mathbb{R}, \quad x \geqslant 0$
then $f^{-1}: x \longmapsto \cosh^{-1} x, \quad x \in \mathbb{R}, \quad x \geqslant 1$.

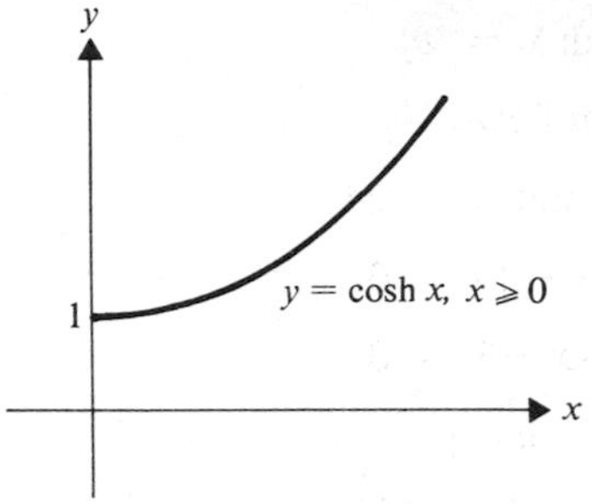

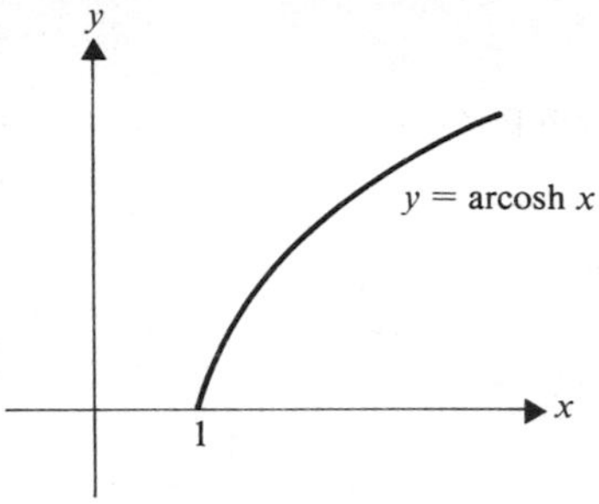

The Logarithmic Form

Now $y = \operatorname{arcosh} x \iff x = \cosh y$ provided that $y \geqslant 0$.

As $$\cosh y \equiv \tfrac{1}{2}(e^y + e^{-y})$$

we have $$x = \tfrac{1}{2}(e^y + e^{-y})$$

$$\Rightarrow \quad e^y - 2x + e^{-y} = 0$$

$$\Rightarrow \quad e^{2y} - 2xe^y + 1 = 0$$

Hence $$e^y = x \pm \sqrt{(x^2 - 1)}$$

i.e. $$y = \ln [x \pm \sqrt{(x^2 - 1)}]$$

But $$x - \sqrt{(x^2 - 1)} \equiv \frac{[x - \sqrt{(x^2 - 1)}][x + \sqrt{(x^2 - 1)}]}{[x + \sqrt{(x^2 - 1)}]}$$

$$\equiv \frac{1}{x + \sqrt{(x^2 - 1)}}$$

So $$y = \ln [x + \sqrt{(x^2 - 1)}]^{\pm 1} = \pm \ln [x + \sqrt{(x^2 - 1)}]$$

But $y \geqslant 0$, so

$$\operatorname{arcosh} x \equiv \ln [x + \sqrt{(x^2 - 1)}]$$

This form of the inverse hyperbolic cosine function verifies that $\operatorname{arcosh} x$ is real only when $x \geqslant 1$.

Similarly $$y = \text{arsinh}\, x \iff x = \sinh y$$

and $$\sinh y \equiv \tfrac{1}{2}(e^{y} - e^{-y})$$

$\Rightarrow$ $$y = \ln\,[x \pm \sqrt{(x^2+1)}]$$

But $x - \sqrt{(x^2+1)} < 0$

So $$\text{arsinh}\, x \equiv \ln\,[x + \sqrt{(x^2+1)}]$$

Note that the logarithmic form of an inverse hyperbolic function is useful when solving some hyperbolic equations.
For instance, the latter part of the solution of the equation

$$12\cosh^2 x + 7\sinh x = 24$$

given on p. 201, can now be abbreviated as follows:

$$12\cosh^2 x + 7\sinh x = 24$$

$\Rightarrow$ $$12\sinh^2 x + 7\sinh x - 12 = 0$$

$\Rightarrow$ $$(3\sinh x + 4)(4\sinh x - 3) = 0$$

$\Rightarrow$ $$\sinh x = -\tfrac{4}{3} \quad\text{or}\quad \tfrac{3}{4}$$

Hence $$x = \text{arsinh}\,(-\tfrac{4}{3}) = \ln\,[-\tfrac{4}{3} + \sqrt{(\tfrac{16}{9}+1)}]$$

or $$x = \text{arsinh}\,\tfrac{3}{4} = \ln\,[\tfrac{3}{4} + \sqrt{(\tfrac{9}{16}+1)}]$$

Thus $$x = \ln\tfrac{1}{3} \quad\text{or}\quad \ln 2$$

EXAMPLES 4f

1) Prove that $\text{artanh}\, x - \text{artanh}\, y \equiv \text{artanh}\,\dfrac{x-y}{1-xy}$.

Let $$\text{artanh}\, x \equiv u \Rightarrow \tanh u \equiv x$$

and $$\text{artanh}\, y \equiv v \Rightarrow \tanh v \equiv y$$

Then the L.H.S. of the required identity is $u - v$.

$$\tanh(u-v) \equiv \frac{\tanh u - \tanh v}{1 - \tanh u \tanh v}$$

$\left(\text{using Osborn's Rule on } \tan(u-v) \equiv \dfrac{\tan u - \tan v}{1 + \tan u \tan v}\right)$

$$\equiv \frac{x-y}{1-xy}$$

Hence $$u - v \equiv \text{artanh}\,\frac{x-y}{1-xy}$$

i.e. $$\text{artanh}\, x - \text{artanh}\, y \equiv \text{artanh}\,\frac{x-y}{1-xy}$$

2) Find x if $\operatorname{arsinh} x = \ln 3$.

If $\qquad \operatorname{arsinh} x = \ln 3$

then $\qquad x = \sinh(\ln 3)$

$$= \tfrac{1}{2}\{e^{\ln 3} - e^{-\ln 3}\}$$

$$= \tfrac{1}{2}\{3 - \tfrac{1}{3}\}$$

Hence $\qquad x = \tfrac{4}{3}$

EXERCISE 4f

1) Evaluate $\operatorname{arcosh} 3$, $\operatorname{arsinh}(-1)$, $\operatorname{artanh} \tfrac{1}{2}$.

2) Sketch the curve with equation $y = \operatorname{arcosh} 3x$ and state the range of values of x for which $\operatorname{arcosh} 3x$ is defined.

3) Sketch the curve with equation $y = \operatorname{artanh} \dfrac{x}{2}$ and give the range of values of x for which y is real.

4) If $y = \operatorname{artanh} x$, express y as a logarithmic function of x.

5) Express $\operatorname{arcosh} 2$ as a logarithm.

6) Prove that

(a) $\operatorname{artanh} x + \operatorname{artanh} y \equiv \operatorname{artanh} \dfrac{x+y}{1+xy}$,

(b) $\operatorname{artanh} \dfrac{2x}{1+x^2} \equiv 2 \operatorname{artanh} x$, $\quad$ (c) $2 \operatorname{artanh} x \equiv \ln \dfrac{1+x}{1-x}$.

7) Express each of the following as logarithms:

(a) $\operatorname{arsech} x$ $\quad$ (b) $\operatorname{arcosh} \dfrac{1}{x}$ $\quad$ (c) $\operatorname{arsinh}(x^2 - 1)$.

8) Solve the equation $2 \operatorname{artanh} x = \ln 3$.

9) Find x if $\operatorname{arcosh} 5x = \operatorname{arsinh} 4x$.

10) Solve the equation $\operatorname{artanh} \left(\dfrac{x^2-1}{x^2+1}\right) = \ln 2$.

Differentiation of Inverse Hyperbolic Functions

If $y = \sinh^{-1} x$ then $x = \sinh y$

$$\Rightarrow \qquad \frac{dx}{dy} = \cosh y$$

$$= +\sqrt{(\sinh^2 y - 1)}$$
$$= \sqrt{(x^2 + 1)}$$

Note that the positive value only of the square root is taken because $\cosh y > 0$ for all real values of y.

Therefore $$\frac{dy}{dx} = \frac{1}{\sqrt{(x^2+1)}}$$

$\Rightarrow$ $$\frac{d}{dx}\sinh^{-1}x = \frac{1}{\sqrt{(x^2+1)}}$$

If $y = \cosh^{-1}x$ then $x = \cosh y$

$\Rightarrow$ $$\frac{dx}{dy} = \sinh y$$
$$= +\sqrt{(\cosh^2 y - 1)}$$
$$= \sqrt{(x^2 - 1)}$$

Note that again the positive square root only is taken, this time because $y = \cosh^{-1}x \Rightarrow y \geqslant 0$ and $\sinh y \geqslant 0$ for these values of y.

Therefore $$\frac{dy}{dx} = \frac{1}{\sqrt{(x^2-1)}}$$

$\Rightarrow$ $$\frac{d}{dx}\cosh^{-1}x = \frac{1}{\sqrt{(x^2-1)}}$$

It can be shown in a similar way that

$$\frac{d}{dx}(\text{artanh}\, x) = \frac{1}{1-x^2}$$

The Use of Hyperbolic Functions in Integration

The reader may recall that trig substitutions played a useful part in integrating certain expressions of the type $\sqrt{(1-x^2)}$, $\frac{1}{\sqrt{(4-x^2)}}$. The trig identity $\cos^2 x \equiv 1 - \sin^2 x$ suggested using the substitutions $x \equiv \sin u$ and $x \equiv 2\sin u$ in these two cases.

Similarly, expressions such as $\sqrt{(x^2 - 1)}$, $\frac{1}{\sqrt{(4+x^2)}}$ suggest using the substitutions $x \equiv \cosh u$ and $x \equiv 2\sinh u$, based on the identity $\cosh^2 x - \sinh^2 x \equiv 1$.

Almost all expressions which are the sum or difference of two squares can be integrated by using a trig or hyperbolic substitution based on the similarity of the given expression to a standard identity.

EXAMPLES 4g

1) Find $\displaystyle\int_0^1 \frac{1}{\sqrt{(1+9x^2)}}\,dx$.

Let $3x \equiv \sinh u$ (using $\cosh^2 u \equiv 1 + \sinh^2 u$)
so that $\ldots 3dx \equiv \ldots \cosh u\, du$

and

x	0	1
u	0	arsinh 3

Then $$\int_0^1 \frac{1}{\sqrt{(1+9x^2)}}\,dx \equiv \int_0^{\text{arsinh } 3} \frac{1}{\sqrt{(1+\sinh^2 u)}}\left(\frac{1}{3}\cosh u\right) du$$

$$\equiv \int_0^{\text{arsinh } 3} \frac{1}{\cosh u}\left(\frac{1}{3}\cosh u\right) du$$

$$= \left[\frac{u}{3}\right]_0^{\text{arsinh } 3}$$

So $$\int_0^1 \frac{1}{\sqrt{(1+9x^2)}}\,dx = \frac{1}{3}\text{arsinh } 3.$$

2) Integrate $\sqrt{(x^2+2x-1)}$ w.r.t. x.

First we convert x^2+2x-1 into a difference of two squares

i.e. $$x^2+2x-1 \equiv (x+1)^2 - 2$$

Then $$\int \sqrt{(x^2+2x-1)}\,dx \equiv \int \sqrt{[(x+1)^2 - (\sqrt{2})^2]}\,dx$$

Let $x+1 \equiv \sqrt{2}\cosh u$ (using $\cosh^2 u - 1 \equiv \sinh^2 u$)
so that $\ldots dx \equiv \ldots \sqrt{2}\sinh u\, du$.

Then $$\int \sqrt{(x^2+2x-1)}\,dx \equiv \sqrt{2}\int \sqrt{(\cosh^2 u - 1)}\sqrt{2}\sinh u\, du$$

$$\equiv 2\int \sinh^2 u\, du$$

$$\equiv \int (\cosh 2u - 1)\, du$$

$$= \tfrac{1}{2}\sinh 2u - u + K$$

But $$\sinh 2u \equiv 2\sinh u\cosh u$$

$$\equiv 2\cosh u\sqrt{(\cosh^2 u - 1)}$$

$$\equiv 2\left(\frac{x+1}{\sqrt{2}}\right)\sqrt{\left(\frac{(x+1)^2}{2} - 1\right)}$$

$$\equiv (x+1)\sqrt{(x^2+2x-1)}$$

and
$$u \equiv \text{arcosh}\,\frac{x+1}{\sqrt{2}}$$
So
$$\int \sqrt{(x^2+2x-1)}\,dx = \tfrac{1}{2}(x+1)\sqrt{(x^2+2x-1)} - \text{arcosh}\left(\frac{x+1}{\sqrt{2}}\right) + K.$$

Certain integrals can be obtained by recognition, thus avoiding the need to make a substitution. The standard results that can be quoted are:

$$\int \frac{1}{\sqrt{(x^2+a^2)}}\,dx = \text{arsinh}\,\frac{x}{a} + K$$
$$\int \frac{1}{\sqrt{(x^2-a^2)}}\,dx = \text{arcosh}\,\frac{x}{a} + K$$

It is left to the reader to derive these results by substitution and thereafter to quote them where needed.

EXERCISE 4g

1) Prove, by using a suitable substitution, that:

(a) $\displaystyle\int \frac{1}{\sqrt{(x^2+a^2)}}\,dx = \text{arsinh}\,\frac{x}{a} + K$

(b) $\displaystyle\int \frac{1}{\sqrt{(x^2-a^2)}}\,dx = \text{arcosh}\,\frac{x}{a} + K.$

2) *Write down* the integral w.r.t. x of:

(a) $\dfrac{1}{\sqrt{(x^2+4)}}$ (b) $\dfrac{1}{\sqrt{(x^2-9)}}$ (c) $\dfrac{1}{\sqrt{(4x^2-16)}}$ (d) $\dfrac{1}{\sqrt{(4x^2+9)}}$.

[*Hint*. In parts (c) and (d) take out a factor 4 from under the square root.]

3) Differentiate:

(a) $\text{arcosh}\,(x+1)$ (b) $\text{arsinh}\,2x$ (c) $\text{artanh}\,x$

(d) $x\,\text{arcosh}\,x$ (e) $\text{arsinh}\,\dfrac{1}{x}$ (f) $\text{arcosh}\,\sqrt{x}$

(g) $\text{arsech}\,x$ (h) $\text{arcosech}\,(x-1)$ (i) $\text{arcosh}\,(\sinh 2x)$

(j) $e^{\text{arcosh}\,x}$.

By making a hyperbolic substitution, find the following integrals.

4) $\displaystyle\int \sqrt{(9x^2+4)}\,dx$ 5) $\displaystyle\int (4x^2-1)^{\frac{3}{2}}\,dx$ 6) $\displaystyle\int \sqrt{(x^2+4x+5)}\,dx$

7) $\displaystyle\int \sqrt{(x^2+4x+3)}\,dx$ 8) $\displaystyle\int \frac{1}{\sqrt{(x^2+2)}}\,dx$ 9) $\displaystyle\int \frac{1}{\sqrt{(2x^2-1)}}\,dx$

10) $\int \frac{1}{\sqrt{(x^2+2x+2)}}\,dx$ 11) $\int \frac{1}{\sqrt{(x^2+2x)}}\,dx$ 12) $\int \sqrt{(4x+x^2)}\,dx$

13) $\int \frac{1}{\sqrt{(4x^2-x)}}\,dx$ 14) $\int \frac{x+1}{\sqrt{(x^2+1)}}\,dx$ 15) $\int \frac{x-1}{\sqrt{(x^2-1)}}\,dx$

Evaluate the following integrals.

16) $\int_1^2 \frac{1}{\sqrt{(9x^2-1)}}\,dx$ 17) $\int_0^1 \frac{1}{\sqrt{(x^2+4)}}\,dx$ 18) $\int_0^1 \frac{1}{\sqrt{(x^2+6x+5)}}\,dx$

19) $\int_2^3 \frac{x+1}{\sqrt{(x^2-4)}}\,dx$ 20) $\int_0^2 \sqrt{(x^2+4)}\,dx$ 21) $\int_{\frac{4}{3}}^2 \sqrt{(9x^2-16)}\,dx$

The following integrals can be found by making a hyperbolic substitution, by making a trig substitution, by using partial functions or by recognition. Choose the appropriate method and hence find each integral.

22) $\int \sqrt{(1-4x^2)}\,dx$ 23) $\int \frac{1}{\sqrt{(4-x^2)}}\,dx$ 24) $\int \frac{1}{\sqrt{(x^2-9)}}\,dx$

25) $\int \frac{1}{9x^2-1}\,dx$ 26) $\int \frac{1}{9x^2+1}\,dx$ 27) $\int \frac{1}{\sqrt{(9x^2+1)}}\,dx$

28) $\int \frac{x+2}{\sqrt{(x^2-4)}}\,dx$ 29) $\int \frac{x+2}{x^2-4}\,dx$ 30) $\int \frac{x}{x^2-4}\,dx$

31) $\int \frac{x}{\sqrt{(x^2-4)}}\,dx$ 32) $\int \sqrt{\left(\frac{x+1}{x^2-1}\right)}\,dx$

33) $\int \frac{\sqrt{(x^2+1)}}{x}\,dx$

By sketching the appropriate graphs, state the number of real roots of the following equations. (Do *not* solve the equations.)

34) $\sin x = \sinh x$ 35) $\operatorname{arcosh} x = e^{-x}$ 36) $\cosh x = \tan x$

37) $\ln x = \operatorname{arsinh} x$

SUMMARY

$f(x)$ is even when $f(-x) = f(x)$

$f(x)$ is odd when $f(-x) = -f(x)$

$f(x)$ is periodic when $f(x) = f(x+a)$ for all values of x in the domain of $f(x)$, and the period is a.

$\ln x$ can be defined by $\ln x = \int_1^x \frac{1}{t}\,dt, \quad x > 0$

$\cosh x = \frac{1}{2}(e^x + e^{-x})$

$\sinh x = \frac{1}{2}(e^x - e^{-x})$

Osborn's Rule states that relationships between hyperbolic functions can be obtained from trig identities by converting trig expressions into the corresponding hyperbolic expressions and changing the sign of any term containing the product of two sines.

$$\frac{d}{dx}(\cosh x) = \sinh x \quad \text{and} \quad \frac{d}{dx}(\sinh x) = \cosh x$$

$$\int \cosh x \, dx = \sinh x + K \quad \text{and} \quad \int \sinh x \, dx = \cosh x + K$$

$$\text{arcosh}\, x \equiv \ln [x + \sqrt{(x^2 - 1)}]$$

$$\text{arsinh}\, x \equiv \ln [x + \sqrt{(x^2 + 1)}]$$

$$\frac{d}{dx}(\text{arcosh}\, x) = \frac{1}{\sqrt{(x^2 - 1)}} \quad \text{and} \quad \int \frac{1}{\sqrt{(x^2 - 1)}} dx = \text{arcosh}\, x + K$$

$$\frac{d}{dx}(\text{arsinh}\, x) = \frac{1}{\sqrt{(x^2 + 1)}} \quad \text{and} \quad \int \frac{1}{\sqrt{(x^2 + 1)}} dx = \text{arsinh}\, x + K$$

$$\frac{d}{dx}(\text{artanh}\, x) = \frac{1}{1 - x^2} \quad \text{and} \quad \int \frac{1}{1 - x^2} dx = \text{artanh}\, x + K.$$

MULTIPLE CHOICE EXERCISE 4

(The instructions for answering these questions are on p. xii.)

TYPE 1

1) $\int \frac{1}{\sqrt{(x^2 - 9)}}$ is

(a) $\sin^{-1}(x/3) + k$ (b) $\sinh^{-1}(x/3) + k$ (c) $\cos^{-1}(x/3) + k$
(d) $\cosh^{-1}(x/3) + k$ (e) none of these.

2) $\frac{d}{dx}(\ln \cosh 2x)$ is

(a) $\frac{1}{\cosh 2x}$ (b) $\frac{2}{\sinh 2x}$ (c) $\frac{1}{2 \sinh 2x}$ (d) $2 \tanh 2x$
(e) $\frac{1}{2} \tanh 2x$.

3) $f(x) = \sinh x$. $f^{-1}(x)$ is

(a) $\operatorname{cosech} x$ (b) $\operatorname{arsinh} x$ (c) $\dfrac{2}{e^x + e^{-x}}$ (d) $\sinh(1/x)$

(e) none of these.

4) $f(x) = \sin x$, $g(x) = x^2 - 1$, $h(x) = |x|$, $hgf(x)$ is
(a) $\sin(x^2 - 1)$ (b) $|\sin(x^2 - 1)|$ (c) $|\sin^2 x - 1|$
(d) $\sin(|x|^2 - 1)$ (e) none of these.

5) $f: x \longmapsto |x^2 - 1|$, $x \in \mathbb{R}$. $f'(1)$ is
(a) 2 (b) -2 (c) 0 (d) 1
(e) meaningless.

6) If $y = \operatorname{arcosh}(x^2 - 1)$ then y is also equal to
(a) $\ln(x^2 - 1)$ (b) $\ln[x^2 - 1 + \sqrt{x}]$ (c) $\frac{1}{2}(e^{x^2-1} - e^{1-x^2})$
(d) $\ln[x^2 - 1 + x\sqrt{(x^2 - 2)}]$ (e) none of these.

7) The curve on the right could have an equation $y = f(x)$ where $f(x)$ is the function
(a) $x^2 + 1$
(b) the inverse of the function $x^2 + 1$
(c) $\cosh^{-1} x$
(d) the equation of the curve cannot be of the form $y = f(x)$.

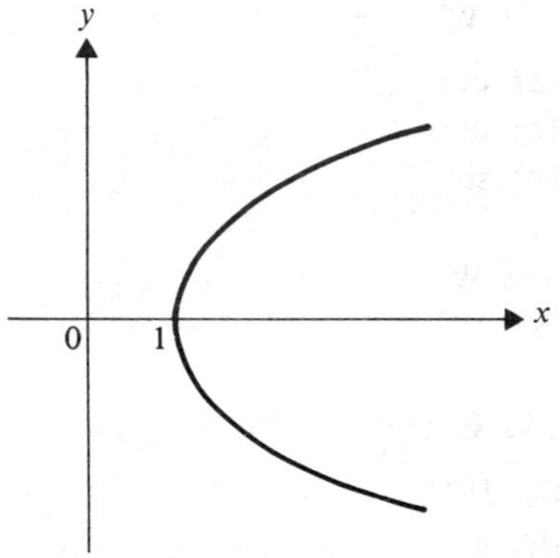

8) If $f(x) = \sin(x - \alpha)\cos(x - \alpha)$, the period of $f(x)$ is
(a) $\pi - \alpha$ (b) 2π (c) $2\pi - 2\alpha$ (d) π
(e) $\frac{1}{2}(\pi - \alpha)$.

9) If f is the function defined by $f: x \longmapsto \dfrac{x}{x^2 - 4}$, the domain of f is

(a) $\mathbb{R}$ (b) $\mathbb{R} - \{0\}$ (c) $\mathbb{R} - \{2, -2\}$ (d) $\mathbb{R} - \{4\}$
(e) $\mathbb{R} - \{0, -2, 2\}$.

10) If f is the function defined by $f: x \longmapsto \dfrac{1}{x}$, $x \in \mathbb{R}$, $x > 0$ the image set of f is

(a) $\mathbb{R}^+$ (b) $\mathbb{R} - \{0\}$ (c) $\mathbb{R}$ (d) $\mathbb{R}^+ + \{0\}$.

TYPE II

11) The function which is represented by this graph

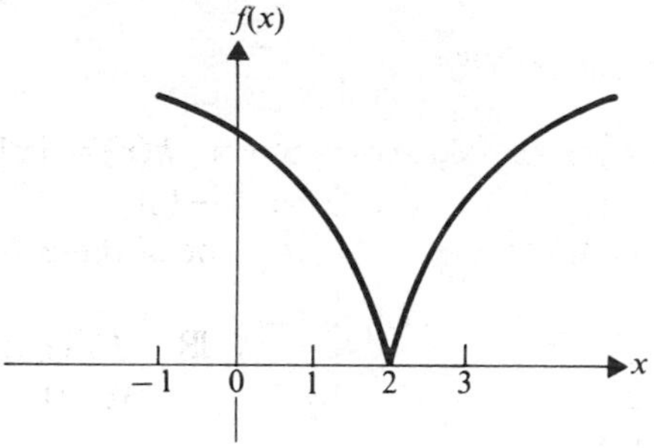

(a) is continuous at $x = 1$
(b) has a discontinuity at $x = 2$
(c) is an even function
(d) is periodic.

12) If $f(x) \equiv \sinh x \cosh x$

(a) $f(x)$ is periodic
(b) $f'(x) = \cosh 2x$
(c) the graph of $f(x)$ is continuous
(d) $f(x) \geqslant 0$ for all real values of x.

13) Which of the following identities is/are correct?

(a) $\cosh^2 x + \sinh^2 x \equiv 1$
(b) $1 - \tanh^2 x \equiv \operatorname{sech}^2 x$
(c) $\cosh(x + y) \equiv \cosh x \cosh y - \sinh x \sinh y$
(d) $\sinh(x - y) \equiv \sinh x \cosh y - \cosh x \sinh y$.

14) Which of the following functions is/are even?

(a) $\cosh x$ (b) $\tanh x$ (c) $\operatorname{arcosh} x$ (d) $\operatorname{arsinh} x$.

15) In which of the following statements is the sign $\Leftrightarrow$ correctly used?

(a) $f(x) = f(-x) \iff f(x)$ is an even function
(b) $y = \cosh x \iff y \geqslant 1$
(c) $f(x) \equiv \sin x \iff f^{-1}(x) \equiv \arcsin x$
(d) $\int f(x)\,dx = \cosh x \iff f(x) \equiv \sinh x$.

16) $\cosh 2x$ is equal to:

(a) $\cosh^2 x - \sinh^2 x$
(b) $\frac{1}{2}(e^{2x} + e^{-2x})$
(c) $\int_0^x \sinh 2x\,dx$
(d) $\frac{d}{dx}(\sinh 2x)$.

17) $f(x) \equiv |2x - 1|$

(a) $f(x)$ is continuous
(b) $f'(x)$ is continuous
(c) $y = f(x)$ undergoes a sudden change in gradient at $x = \frac{1}{2}$
(d) f^{-1} exists.

18) $f(x) \equiv \tanh x$

(a) $f(x)$ is periodic (b) $f(x)$ is continuous

(c) $f'(x)$ is discontinuous (d) $f(x)$ is odd.

19) The value of $\begin{vmatrix} \sinh x & \cosh x \\ \cosh x & \sinh x \end{vmatrix}$ is:

(a) 1 (b) $\sinh^2 x - \cosh^2 x$ (c) -1 (d) $\cosh 2x$.

TYPE III

20) (a) $f(x)$ is an even function of x.
(b) The graph of $f(x)$ is symmetrical about the x axis

21) (a) $y = \cosh x$.
(b) $y \geqslant 1$.

22) (a) $f(4) = f(2)$.
(b) $f(x)$ is a periodic function with period 2.

23) A function $f(x)$ is continuous for $-3 \leqslant x \leqslant 3$.

(a) $\int_{-3}^{3} f(x)\, dx = 0$.

(b) $f(x)$ is an odd function.

24) (a) $e^p = e^q$.
(b) $\cosh p = \cosh q$.

TYPE V

25) The graph of a continuous function contains no sudden change in direction.

26) The graph of $f(x)$ is the reflection of the graph of $f^{-1}(x)$ in the line $y = x$ when f^{-1} is the inverse of f.

27) $f(x) \equiv \sinh x$ is an odd function.

28) If $f(x)$ is continuous in the range $0 \leqslant x \leqslant 1$, then $f'(x)$ is also continuous in the same range.

29) If $f(x)$ is periodic with period $2a$, then $f(x) = f(x + 2a)$ for all values of x.

30) If $y = f(x)$ is continuous for all values of x then so is $y^2 = f(x)$.

31) The point $(a \sinh t, b \cosh t)$ lies on the curve $\dfrac{x^2}{a^2} - \dfrac{y^2}{b^2} = 1$ for all values of t.

32) If $f(x)$ is periodic with period a, then $f'(x)$ is also periodic with period a.

33) If f is an even function and g is any function then fg is an even function.

MISCELLANEOUS EXERCISE 4

1) The function f is defined by

$$f(x) = \begin{cases} \dfrac{3}{x^2}\sin 2x^2, & x < 0 \\[2ex] \dfrac{x^2+2x+c}{1-3x^2}, & x \geqslant 0,\quad x \neq \dfrac{1}{\sqrt{3}} \end{cases}$$

(i) Find the value of c if f is continuous at $x = 0$.
(ii) Evaluate $\lim\limits_{x\to\infty} f(x)$. (JMB)

2) The function f is defined by

$$f(x) = \begin{cases} ax^2+2x-1, & x \leqslant 1 \\[2ex] b-\dfrac{c}{x^2}, & x > 1 \end{cases}$$

where a, b and c are constants. It is given that
(i) f is continuous at $x = 1$,
(ii) f is differentiable at $x = 1$,
(iii) $f'(0) = f'(2)$.
Find the value of b. (JMB)

3) Determine the ranges of the following functions whose domains are as given:
(i) $f: x \longmapsto x^2+3$, x real, $0 \leqslant x \leqslant 3$;
(ii) $g: x \longmapsto x^2-2x+3$, x real, $0 \leqslant x \leqslant 3$;
(iii) $h: x \longmapsto 2\sin x+\cos x$, x real.
State whether f^{-1}, g^{-1}, h^{-1} exist, giving, in each case, brief reasons for your answer. (JMB)

4) The function f is periodic with period π and

$$f(x) = \sin x \quad \text{for} \quad 0 \leqslant x \leqslant \pi/2,$$
$$f(x) = 4(\pi^2-x^2)/(3\pi^2) \quad \text{for} \quad \pi/2 < x < \pi.$$

Sketch the graph of $f(x)$ in the range $-\pi \leqslant x \leqslant 2\pi$. (U of L)

5) The function $f(x)$ is defined for $0 \leqslant x \leqslant 2$ by

$$f(x) = x \quad \text{for} \quad 0 \leqslant x \leqslant 1,$$
$$f(x) = (2-x)^2 \quad \text{for} \quad 1 < x \leqslant 2.$$

Sketch the graph of this function for $0 \leqslant x \leqslant 2$ and find

$$\int_0^2 f(x)\,dx.$$

(U of L)p

6) The function f is defined by

$$f(x) = \sin x \quad \text{for} \quad x \leqslant 0,$$
$$f(x) = x \quad \text{for} \quad x > 0.$$

Sketch the graphs of $f(x)$ and its derivative $f'(x)$ for $-\pi/2 < x < \pi/2$ and decide whether the functions f and f' are continuous at $x = 0$ or not.

(U of L)

7) The function f is periodic with period 3 and

$$f(x) = \sqrt{(9 - 4x^2)} \quad \text{for} \quad 0 < x \leqslant 1.5,$$
$$f(x) = 2x - 3 \quad \text{for} \quad 1.5 < x \leqslant 3.$$

Sketch the graph of $f(x)$ in the range $-3 \leqslant x \leqslant 6$.

(U of L)

8) Defining $\ln x^n$ as $\int_1^{x^n} \frac{1}{t}\,dt$, for $x > 0$, use the substitution $t = u^n$ to prove that $\ln x^n = n \ln x$.

By considering the area under the graph of $y = \frac{1}{t}$ from $t = 1$ to $t = 1 + x$, or otherwise, show that, for $x > 0$,

$$\frac{x}{1+x} < \ln(1+x) < x$$

and deduce that, as x decreases to zero, $\frac{1}{x}\ln(1+x)$ tends to 1.

A periodic function is defined by

$$\begin{cases} f(x) = \dfrac{1}{x}\ln(1+x) & \text{for} \quad 0 < x \leqslant 1, \\ f(x+1) = f(x) & \text{for all } x. \end{cases}$$

Sketch the graph $y = f(x)$ for values of x from -2 to 2.

(U of L)

9) (a) By making the substitution $t = 1/u$ in the integral $\int_1^x \frac{dt}{t}$, prove that, for $x > 0$, $\ln(1/x) = -\ln x$.

(b) The function f is such that $f(x + \pi) = f(x)$ for all values of x. In the interval $0 \leqslant x < \pi$, $f(x) = x - \sin x$. Sketch the curve $y = f(x)$ for $-2\pi \leqslant x \leqslant 2\pi$, and state all the values of x for which the function f is discontinuous. Evaluate the integrals

(i) $\int_{-\pi/2}^{\pi/2} f(x)\,dx$, (ii) $\int_0^{3\pi/2} f(x)\,dx$.

(U of L)

10) Defining $\ln t$ for $t > 1$ as $\int_1^t \frac{dx}{x}$, prove that

$$1 - \frac{1}{t} < \ln t < t - 1.$$

(U of L)p

11) (a) By the substitution $x = u^n$, show that

$$\int_1^{a^n} \frac{1}{x}\,dx = n\int_1^{a} \frac{1}{x}\,dx.$$

(b) Defining e by the equation

$$\int_1^{e} \frac{1}{x}\,dx = 1$$

show that

$$\int_1^{t} \frac{1}{x}\,dx = \log_e t$$

and

$$\int_{t_2}^{t_1} \frac{1}{x}\,dx = \log_e\left(\frac{t_2}{t_1}\right).$$

12) Define $\ln x$ and deduce from the definition that

$$\ln\frac{xy}{z^2} = \ln x + \ln y - 2\ln z.$$ (AEB'71)p

13) Define $\operatorname{cosech} x$ and $\coth x$ in terms of exponential functions and from your definitions prove that

$$\coth^2 x \equiv 1 + \operatorname{cosech}^2 x.$$ (AEB'73)p

14) Find the minimum value of $(5\cosh x + 3\sinh x)$. (U of L)

15) Solve, for real x, the equation

$$5\cosh x - 3\sinh x = 5.$$ (U of L)

16) Find

(a) $\displaystyle\int \frac{dx}{\sqrt{(1+9x^2)}}$, (b) $\displaystyle\int \sinh^2 3x\,dx$. (U of L)

17) If $y = (\operatorname{arcosh} x)^2$ prove that $(x^2-1)\left(\dfrac{dy}{dx}\right)^2 = 4y$. (U of L)

18) Find the possible values of $\sinh x$ if

$$\begin{vmatrix} \cosh x & -\sinh x \\ \sinh x & \cosh x \end{vmatrix} = 2.$$ (U of L)

19) Solve the equation $3\operatorname{sech}^2 x + 4\tanh x + 1 = 0$. (AEB '71)p

20) If $x = \cosh\theta$ and $y = \sinh\theta$, obtain $\dfrac{dy}{dx}$ in terms of the parameter θ. Sketch the graphs of y and $\dfrac{dy}{dx}$ regarded as functions of θ. (U of L)

21) Sketch the graphs of $y = \operatorname{arsinh} x$ and $y = \operatorname{arcosh}(x+2)$.
Find the coordinates of the point of intersection of the two curves.

22) (a) State the value of $\tanh x$ in terms of e^x and e^{-x}.

Prove that (i) $\dfrac{2\tanh x}{1+\tanh^2 x} \equiv \tanh 2x$

(ii) $\operatorname{artanh} x \equiv \dfrac{1}{2}\ln\dfrac{1+x}{1-x}$

If $\tanh 2y = -\frac{4}{5}$, show that $y = -\frac{1}{2}\ln 3$ and find the value of $\tanh y$.

(b) Evaluate $\displaystyle\int_3^5 \frac{1}{\sqrt{(x^2-9)}}\,dx$ and $\displaystyle\int_0^1 \operatorname{arsinh} x\,dx$ (JMB)

23) State the expansions of $\cosh(x+y)$ and $\sinh(x+y)$ in terms of hyperbolic functions of x and of y. Hence, or otherwise, express $\cosh 2x$ and $\cosh 3x$ in terms of $\cosh x$.
The three values of $\cosh x$ given by the equation

$$a\cosh 3x + b\cosh 2x = 0$$

where a and b are non-zero, are y_1, y_2 and y_3.
Show that $y_1y_2 + y_2y_3 + y_3y_1$ is independent of a and b. (JMB)

24) Show that the curve $y = \cosh 2x - 4\sinh x$ has just one stationary point and find its coordinates, giving the x coordinate in logarithmic form. Determine the nature of the stationary point. (JMB)

25) Prove that

$$16\sinh^2 x\cosh^3 x \equiv \cosh 5x + \cosh 3x - 2\cosh x.$$

Hence or otherwise evaluate:

$$\int_0^1 16\sinh^2 x\cosh^3 x\,dx$$

giving your answer in terms of e. (JMB)

26) Show that $\dfrac{1+\tanh^2 x}{1-\tanh^2 x} \equiv \cosh x.$

By means of the substitution $t \equiv \tanh x$, or otherwise, find the indefinite integral $\int \operatorname{sech} 2x\,dx$. (JMB)

27) Prove that $\operatorname{arsinh} x \equiv \ln[x + \sqrt{(1+x^2)}]$, and draw a rough graph of the function.
Prove that $\operatorname{arsinh}\frac{3}{4} + \operatorname{arsinh}\frac{5}{12} = \operatorname{arsinh}\frac{4}{3}$. (O)

28) Express $\operatorname{arcosech} x$ in logarithmic form.
Solve the equation $\operatorname{arcosech} x + \ln x = \ln 3$. (JMB)

29) Solve the equations

$$\cosh x - 3\sinh y = 0,$$
$$2\sinh x + 6\cosh y = 5,$$

giving answers in logarithmic form. (AEB '73)p

30) (a) Find $\displaystyle\int \frac{dx}{\sqrt{(x^2-2x+10)}}$.

(b) Find $\displaystyle\int \frac{dx}{x^2-2x+10}$. (U of L)

31) Prove that $\operatorname{arsinh} x = \ln\{x+\sqrt{(x^2+1)}\}$.

Show that $\dfrac{d}{dx}(\operatorname{arsinh} x) = \dfrac{1}{\sqrt{(x^2+1)}}$.

Evaluate $\displaystyle\int_1^8 \frac{1}{\sqrt{(x^2-2x+2)}}\,dx$ expressing your answer as a natural logarithm.

Show that $\displaystyle\int_1^2 \frac{3}{\sqrt{(x^2-2x+2)}}\,dx = \int_1^8 \frac{1}{\sqrt{(x^2-2x+2)}}\,dx$. (JMB)

32) Define $\cosh x$ and $\sinh x$ and hence prove that

$$\frac{1}{\cosh 2x + \sinh 2x} \equiv \cosh 2x - \sinh 2x$$

Hence, or otherwise, show that

$$\int_0^1 \frac{dx}{\cosh 2x + \sinh 2x} = \tfrac{1}{2}(1 - e^{-2}).$$ (AEB'77)p

33) Find the area enclosed by the curve $y = \cosh x$, the ordinate at $x = \ln 2$, the x axis and the y axis. Find also the volume obtained when this area is rotated completely about the x axis. (AEB'74)p

34) Prove that $\sinh^{-1}x \equiv \ln[x+\sqrt{(1+x^2)}]$ and write down a similar expression for $\cosh^{-1}x$.
If $2\cosh y - 7\sinh x = 3$ and $\cosh y - 3\sinh^2 x = 2$, find the real values of x and y in logarithmic form. (AEB'75)p

35) (a) Show graphically that the equation

$$\operatorname{arsinh} x = \operatorname{arsech} x$$

has only one real root. Prove that this root is $\{(\sqrt{5}-1)/2\}^{\frac{1}{2}}$.

(b) Prove that

$$\int_{\frac{4}{5}}^{1} \operatorname{arsech} x\,dx = 2\arctan 2 - \frac{\pi}{2} - \frac{4}{5}\ln 2.$$

CHAPTER 5

POLAR COORDINATES

POLAR FRAME OF REFERENCE

This system of reference consists of a fixed point O, called *the pole* and a line in a fixed direction from O called *the initial line.*

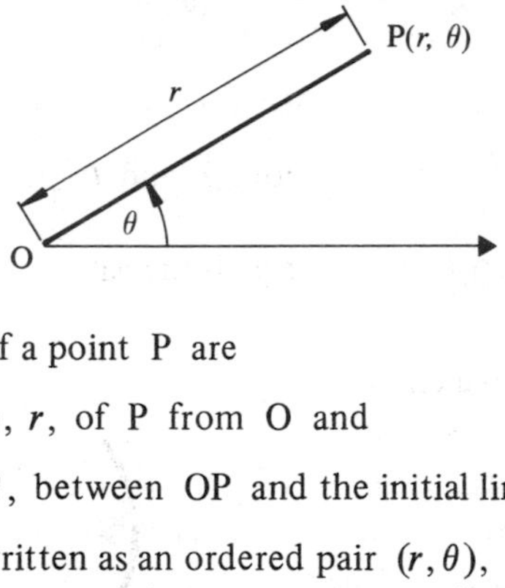

The polar coordinates of a point P are

the distance, r, of P from O and

the angle, θ, between OP and the initial line.

These coordinates are written as an ordered pair (r, θ), i.e. (distance, angle). OP is called the *radius vector* and θ is sometimes called the *vectorial angle.*

Positive values of θ correspond to an anticlockwise sense of rotation from the initial line, and negative values of θ correspond to clockwise rotation.

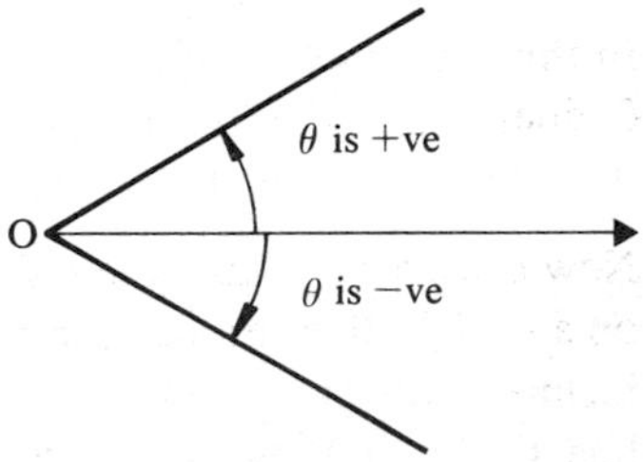

A positive value is given to r when the distance is measured in the direction $\overrightarrow{OP}$ where OP is at an angle θ to the initial line. A negative value of r indicates a distance along $\overrightarrow{PO}$ produced.

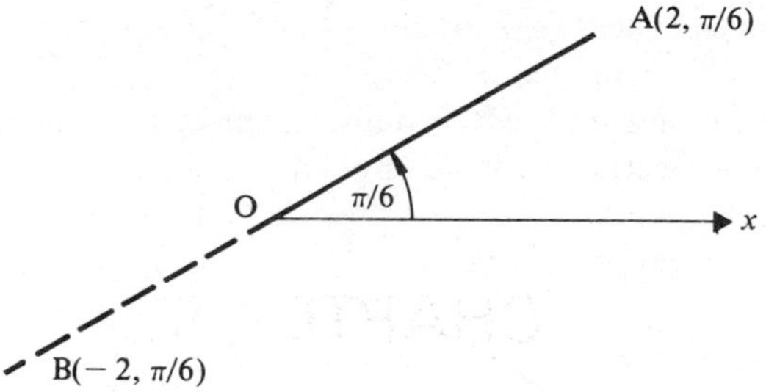

Thus the point with polar coordinates $(2, \pi/6)$ is the point A in the diagram above and the point with polar coordinates $(-2, \pi/6)$ is the point B.
Note, however, that the polar coordinates of a given point are not unique. For example the coordinates of the point B can be given as
$(-2, \pi/6)$ or as $(2, -5\pi/6)$ or as $(2, 7\pi/6)$ or as $(-2, \pi/6 + 2\pi)$ and so on.
In order to avoid this multiplicity of forms, it is conventional to give the polar coordinates of a point in the form where

$$r \geqslant 0 \quad \text{and} \quad -\pi < \theta \leqslant \pi$$

POLAR EQUATIONS

If a point $P(r, \theta)$ is unrestricted, both r and θ can independently take any value.
But if P is constrained to a particular path the values that r and/or θ can take are limited.

For example, if P lies on a circle with centre O and radius a then, for all possible positions of P,

$$r = a$$

Thus $r = a$ is the *polar equation* of the locus of P.

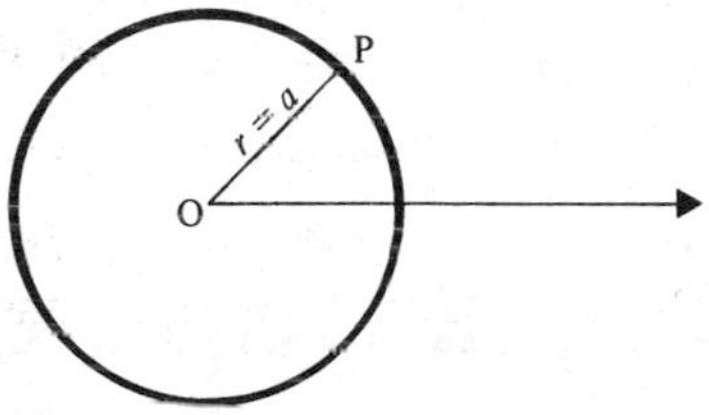

In this example the value of r is fixed and is independent of the value of θ so θ does not appear in the polar equation of the circle.

Now consider the case when P lies on a straight line through the pole, inclined at an angle α to the initial line Ox.

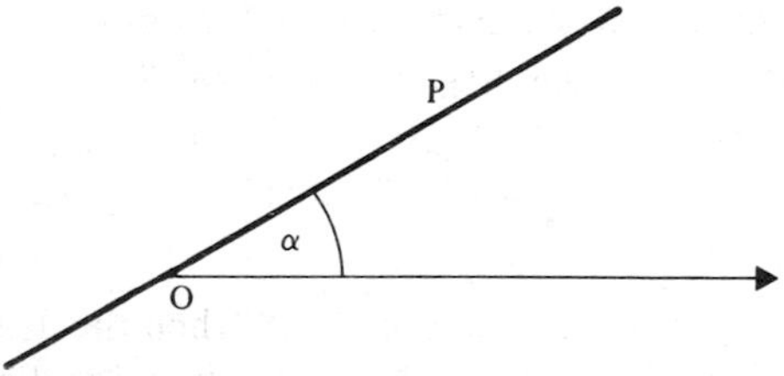

For every position of P, $\theta = \alpha$ and this is the polar equation of the line.

This time θ is independent of r so r does not appear in the equation of the line.

The pole divides this line into two sections, one made up of points for which r is positive, the other section made up of points for which r is negative.

The section corresponding to positive values of r is called the part-line (or half-line) with equation $\theta = \alpha$.

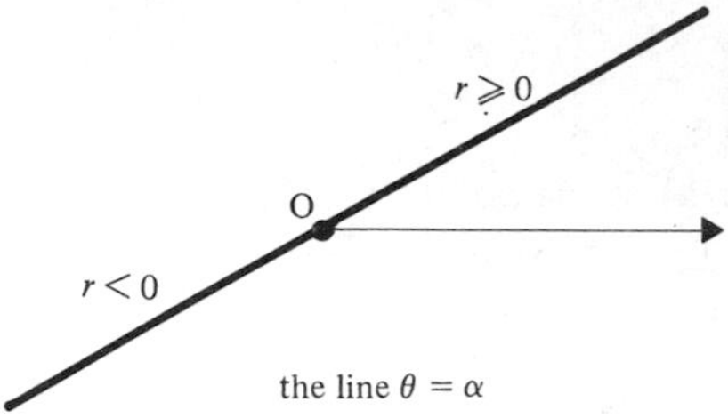

the line $\theta = \alpha$

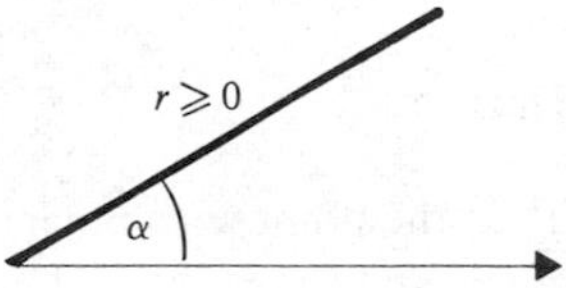

the part-line $\theta = \alpha$

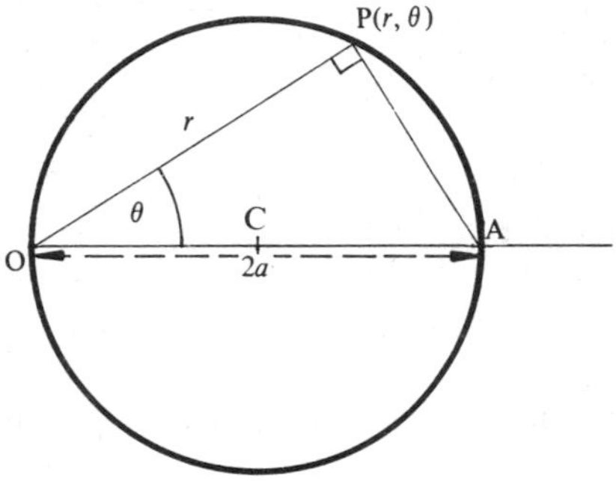

Now consider the set of points P which lie on a circle with centre C on the initial line, with diameter $2a$, and passing through the pole.

If OA is a diameter and P is any point on the circle then angle OPA is a right angle.

Thus, in triangle OPA, $r = 2a\cos\theta$.

This relationship between r and θ is valid for all possible positions of P and for no other points. Hence $r = 2a\cos\theta$ is the polar equation of the specified circle.

Relationship between Polar and Cartesian Coordinates

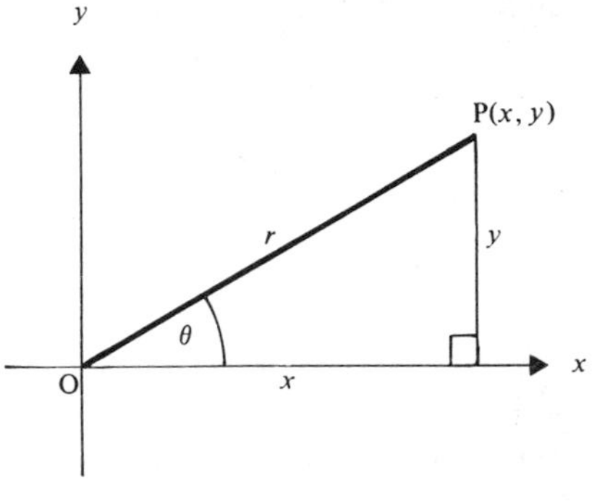

If a point P has Cartesian coordinates (x, y) relative to an origin O and polar coordinates (r, θ) relative to a pole O and initial line Ox, then

$$\begin{cases} x = r\cos\theta \\ y = r\sin\theta \end{cases} \quad \text{or} \quad \begin{cases} r^2 = x^2 + y^2 \\ \tan\theta = \dfrac{y}{x} \end{cases}$$

These relationships can be used to convert a Cartesian equation into polar form and conversely.
For example, to find the Cartesian equation of the circle with polar equation $r = 2a\cos\theta$

we use $$\cos\theta = \frac{x}{r} \quad \text{and} \quad r^2 = x^2 + y^2$$

Thus $$r = 2a\left(\frac{x}{r}\right)$$

$\Rightarrow$ $$r^2 = 2ax$$

i.e. $$x^2 + y^2 - 2ax = 0$$

Conversely, to find the polar equation of the parabola $y^2 = 4x$ we use

$$y = r\sin\theta \quad \text{and} \quad x = r\cos\theta$$

Thus $$r^2\sin^2\theta = 4r\cos\theta$$

$\Rightarrow$ $$r = \frac{4\cos\theta}{\sin^2\theta}$$

Note that the Cartesian x axis corresponds in polar form to the part-lines

$$\theta = 0 \quad \text{and} \quad \theta = \pi$$

and that the Cartesian y axis corresponds to the part-lines

$$\theta = \frac{\pi}{2} \quad \text{and} \quad \theta = -\frac{\pi}{2}\left(\text{or } \frac{3\pi}{2}\right)$$

i.e.

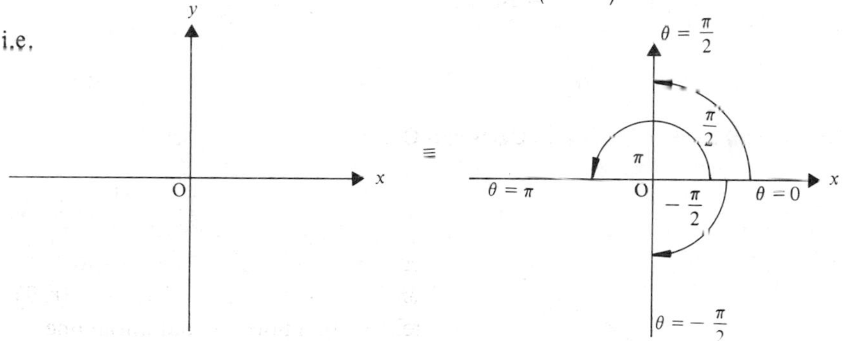

EXERCISE 5a

1) Mark the following points on a diagram:

$$\left(3, \frac{\pi}{4}\right); \quad \left(1, -\frac{\pi}{3}\right); \quad \left(-2, \frac{\pi}{2}\right); \quad \left(-1, -\frac{2\pi}{3}\right); \quad \left(2, \frac{3\pi}{2}\right); \quad \left(-4, \frac{\pi}{4}\right).$$

2) Given the polar coordinates of two points A and B find $(AB)^2$ if A and B are:

(a) $\left(2, \frac{\pi}{4}\right); \left(3, \frac{\pi}{2}\right)$ (b) $\left(-3, -\frac{\pi}{2}\right); \left(4, \frac{5\pi}{6}\right)$

(c) $(4, 0); \left(2, -\frac{\pi}{2}\right)$. (d) $\left(2, \frac{\pi}{3}\right); \left(-2, -\frac{\pi}{3}\right)$

(e) $\left(3, -\frac{\pi}{4}\right); \left(3, \frac{3\pi}{4}\right)$.

3) In a polar frame of reference, where O is the pole, given A(2, 0), B$\left(3, \frac{\pi}{3}\right)$ C$\left(-2, -\frac{\pi}{6}\right)$, D$\left(4, -\frac{\pi}{2}\right)$, E$\left(-3, -\frac{\pi}{3}\right)$, state whether each of the following triangles is isosceles, right-angled or neither of these.

(a) AOD (b) BDE (c) COE (d) COB
(e) AOB (f) DOE (g) BOE

Find the Cartesian equations of the lines or curves whose polar equations are:

4) $r = a$ 5) $r = a \cos 2\theta$ 6) $r^2 = a^2 \sin 2\theta$

7) $r = a(1 + \cos 2\theta)$ 8) $r = d \sec(\theta - \alpha)$.

Find the polar equations of the lines or curves whose Cartesian equations are:

9) $(x - a)^2 + (y - a)^2 = a^2$ 10) $\frac{x^2}{a^2} + \frac{y^2}{b^2} = 1$

11) $y = 2x$ 12) $y = x^2$ 13) $xy = 4$.

POLAR CURVE TRACING

The shape of a curve can be determined from its polar equation by listing corresponding values of θ and r and plotting these coordinates. (Values of θ are usually limited to the range $0 \leqslant \theta \leqslant 2\pi$.) Certain observations can reduce the amount of tabulated work however.

(1) If r is a function of $\cos\theta$, the curve is symmetrical about the line, $\theta = 0$, since $\cos(-\theta) = \cos\theta$.
Such a curve can be traced by calculating the values of r for angles in the range $0 \leqslant \theta \leqslant \pi$, drawing the curve through these points and then reflecting it in the initial line.

(2) If r is a function of $\sin\theta$, the curve is symmetrical about the line $\theta = \frac{\pi}{2}$, since $\sin(\pi - \theta) = \sin\theta$.

Such a curve can be traced by plotting points calculated in the range $-\frac{\pi}{2} \leqslant \theta \leqslant \frac{\pi}{2}$ and reflecting this curve in the line $\theta = \frac{\pi}{2}$.

(3) If the curve passes through O and if in the neighbourhood of O (i.e. where $r \simeq 0$) $\theta \simeq \alpha$, then the line $\theta = \alpha$ is a tangent to the curve at O. Such tangents can be found by finding the values of θ for which $r = 0$.

EXAMPLES 5b

1) Trace the curve with polar equation $r = a(1 + \cos\theta)$. Deduce the shapes of the curves with equations (a) $r = a(1 - \cos\theta)$, (b) $r = a(1 + \sin\theta)$.

(i) There is symmetry about the initial line (because r depends on $\cos\theta$).

(ii) $r = 0$ when $\cos\theta = -1$, i.e. when $\theta = \pi$.
So the line $\theta = \pi$ is a tangent to the curve at O.

(iii) Values of r are now calculated for convenient angles in the range $0 \leqslant \theta \leqslant \pi$.

θ	0	$\frac{\pi}{6}$	$\frac{\pi}{4}$	$\frac{\pi}{3}$	$\frac{\pi}{2}$	$\frac{2\pi}{3}$	$\frac{3\pi}{4}$	$\frac{5\pi}{6}$	π
$\cos\theta$ (to 1 d.p.)	1	0.9	0.7	0.5	0	−0.5	−0.7	−0.9	−1
r	$2a$	$1.9a$	$1.7a$	$1.5a$	a	$0.5a$	$0.3a$	$0.1a$	0

These points are easily located if, first, a framework is prepared of lines at each of the values of θ used above, and a series of circles of radii $\frac{a}{2}$, a, $\frac{3a}{2}$ and $2a$.

Note that it is possible to buy polar graph paper.

Thus

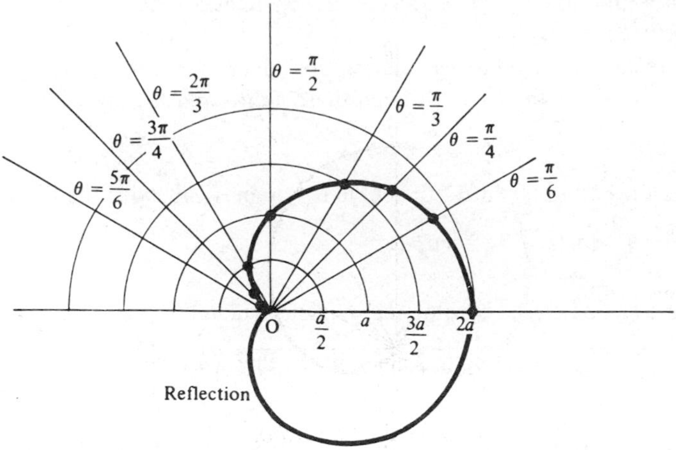

Note. The part-line $\theta = \pi$ is a tangent to the curve at O.
The shape of the curve at O is called a *cusp*.

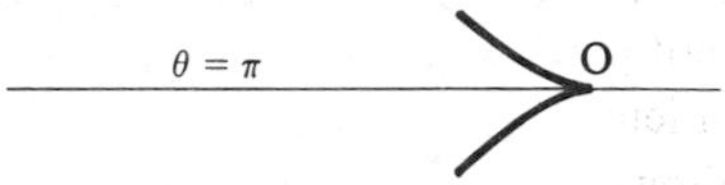

(a) When $r = a(1 - \cos\theta)$, the values of r obtained in the table above are reversed in order giving the following curve trace.

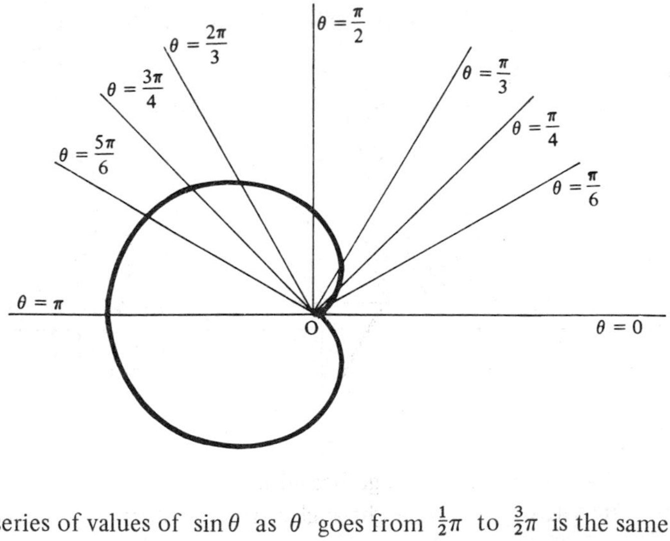

(b) The series of values of $\sin\theta$ as θ goes from $\frac{1}{2}\pi$ to $\frac{3}{2}\pi$ is the same as the series of values of $\cos\theta$ as θ goes from 0 to π.

Thus the curve $r = a(1 + \sin\theta)$ is given by rotating $r = a(1 + \cos\theta)$ through an angle $\frac{1}{2}\pi$,

i.e.

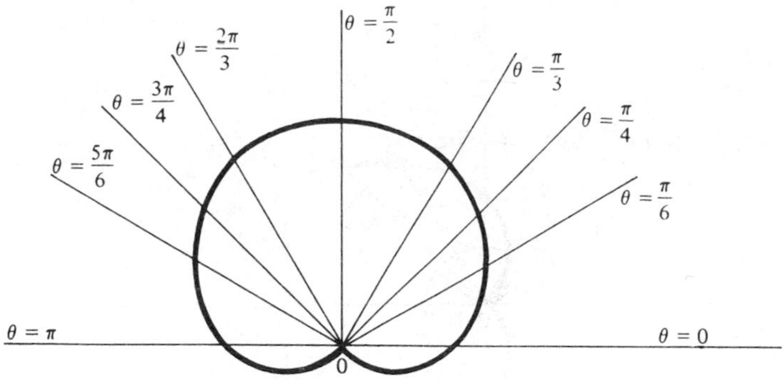

Note. Each of the above curves is called a *cardioid*.

2) N is the foot of the perpendicular from the pole, O, to a given straight line (not passing through O). ON is inclined to the initial line at an angle α and is of length d.
Show that the polar equation of the given line is $r = d\sec(\theta - \alpha)$. Hence sketch the lines (a) $r\cos\theta = 4$, (b) $r\sin\theta = 3$.

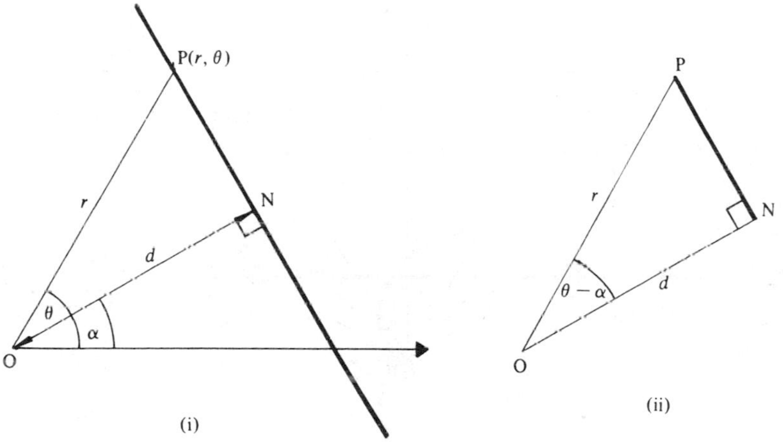

Diagram (i) shows the given data, together with a general point $P(r, \theta)$ on the given line. Diagram (ii) shows the right-angled triangle ONP from which we see that

$$ON = OP\cos PON$$

i.e.
$$d = r\cos(\theta - \alpha)$$

As this condition applies to *any* point P on the given line, the polar equation of the given line is $r = d \sec(\theta - \alpha)$.

(a) If $r \cos\theta = 4$, $r = 4 \sec\theta$.

This is the equation of a line for which $\begin{cases} d = 4 \\ \alpha = 0 \end{cases}$

i.e.

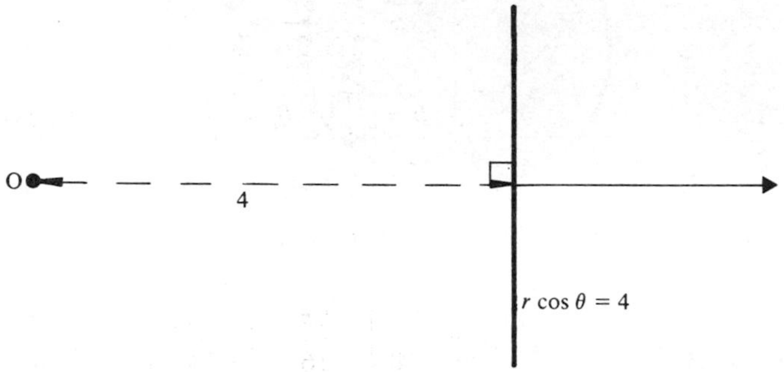

(b) If $r \sin\theta = 3$, $r \cos\left(\theta - \frac{\pi}{2}\right) = 3$

$$\Rightarrow \qquad r = 3 \sec\left(\theta - \frac{\pi}{2}\right)$$

This is the equation of a line for which $\begin{cases} d = 3 \\ \alpha = \frac{\pi}{2} \end{cases}$

i.e.

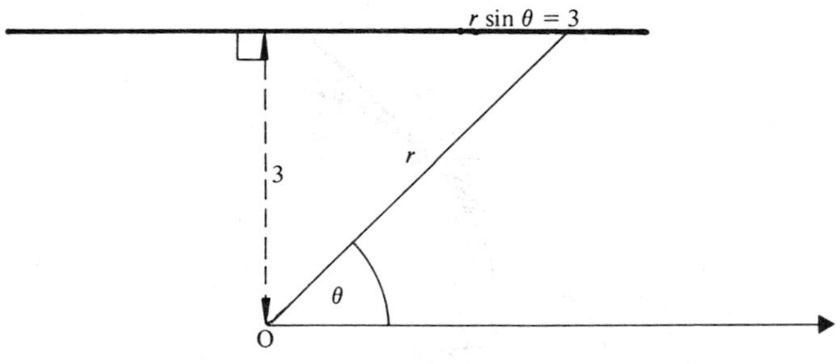

3) Sketch the curve with equation $r = 2a \cos 2\theta$.

(a) The equation can be written as $r = 2a(2\cos^2\theta - 1)$

or as $r = 2a(1 - 2\sin^2\theta)$

Thus there is symmetry about the line $\theta = 0$ (r depends on $\cos\theta$)

and there is also symmetry about the line $\theta = \dfrac{\pi}{2}$ (r depends on $\sin\theta$).

Hence we will tabulate values of θ in the range $0 \leqslant \theta \leqslant \dfrac{\pi}{2}$ only, reflecting the curve so obtained in *both* axes of symmetry.

(b) When $r \simeq 0$, $\cos 2\theta \simeq 0 \Rightarrow 2\theta \simeq \dfrac{\pi}{2}, \dfrac{3\pi}{2}, \dfrac{5\pi}{2}, \dfrac{7\pi}{2}$.

Thus the lines $\theta = \dfrac{\pi}{4}$, $\theta = \dfrac{3\pi}{4}$, $\theta = \dfrac{5\pi}{4}$, $\theta = \dfrac{7\pi}{4}$ are all tangents to the curve at the pole.

(c)

θ	0	$\frac{\pi}{16}$	$\frac{\pi}{8}$	$\frac{3\pi}{16}$	$\frac{\pi}{4}$	$\frac{5\pi}{16}$	$\frac{3\pi}{8}$	$\frac{7\pi}{16}$	$\frac{\pi}{2}$
$\cos 2\theta$	1	0.9	0.7	0.4	0	-0.4	-0.7	-0.9	-1
r	$2a$	$1.8a$	$1.4a$	$0.8a$	0	$-0.8a$	$-1.4a$	$-1.8a$	$-2a$

The polar coordinates in the above table give

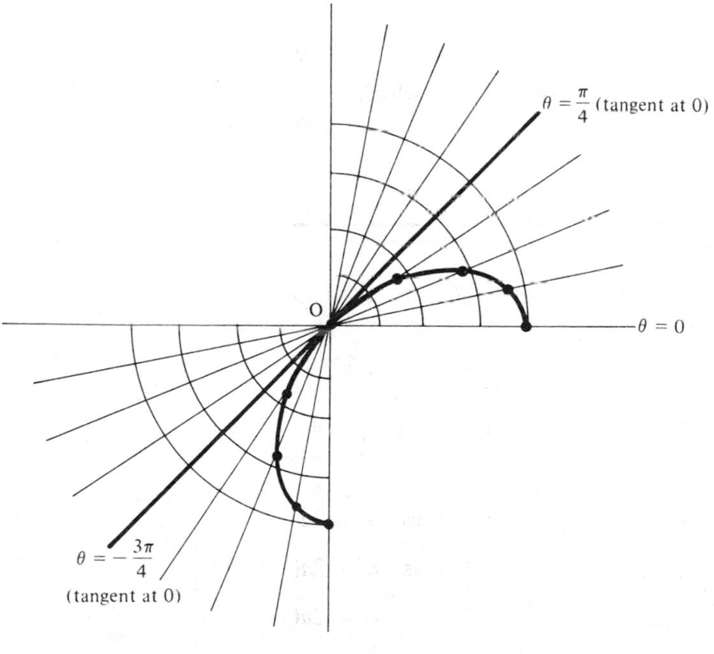

(**Note** that when $\theta = \frac{5\pi}{16}, \frac{3\pi}{8}, \frac{7\pi}{16}, \frac{\pi}{2}$, r is negative.)

Reflecting this curve in both axes of symmetry gives the complete curve with equation $r = 2a \cos 2\theta$.

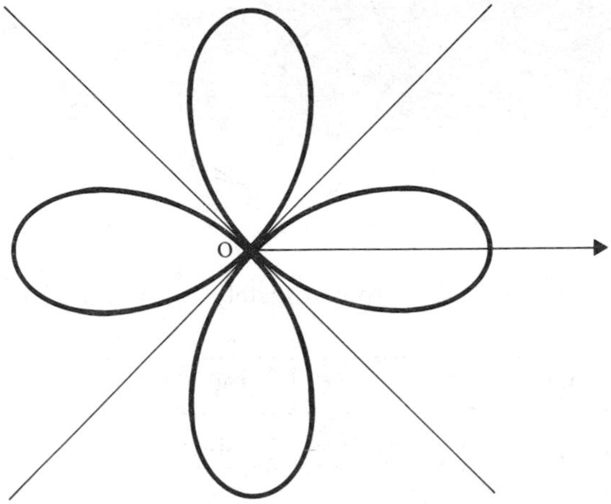

EXERCISE 5b

Trace the curves or lines whose polar equations are given below stating, where appropriate, the equations of the tangents at the pole.

1) $r = 3$ 2) $\theta = -\frac{2\pi}{3}$ 3) $r = 4 \sin \theta$

4) $r = 2 \sin 3\theta$ 5) $r = 2 \cos 3\theta$ 6) $r = a \cos 2\theta$

7) $r = a \cos \theta$ 8) $r^2 = a^2 \cos^2 \theta$

9) $r = k\theta$ when (a) $k > 0$, (b) $k < 0$

10) $r = 2 - 3 \cos \theta$ 11) $r = 3 - 2 \cos \theta$

12) $r^2 = a^2 \cos 2\theta$ (*Hint*. If $\cos 2\theta < 0$, r is not real.)

Note. Because it is useful to be familiar with the sketches of the common polar curves (most of which are included in the exercise above), these curves are given on pages 237 and 238.

AREA BOUNDED BY A POLAR CURVE

Consider two points P and Q on a curve whose polar equation is $r = f(\theta)$. P is the point (r, θ) and Q has coordinates $(r + \delta r, 0 + \delta\theta)$ if Q is close to P.

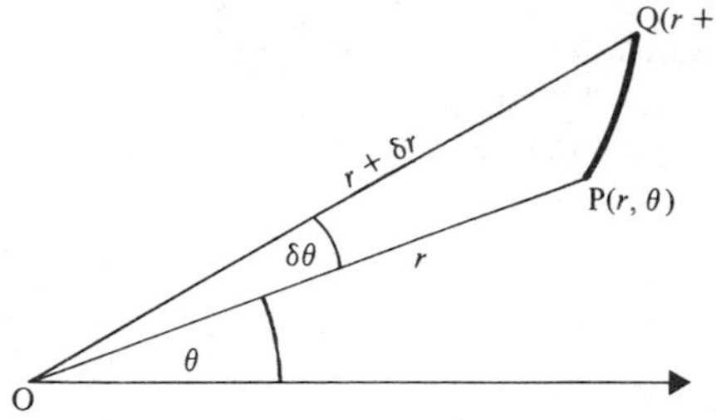

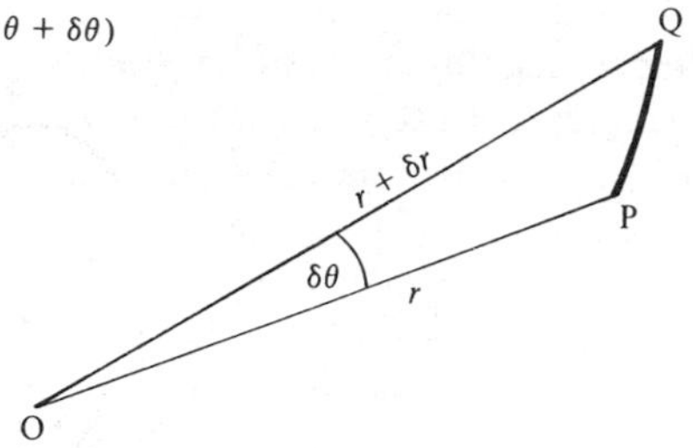

The area, δA, of sector POQ is approximately equal to the area of triangle POQ

i.e.
$$\delta A \simeq \tfrac{1}{2}r(r + \delta r)\sin\delta\theta$$

Hence
$$\frac{\delta A}{\delta\theta} \simeq \tfrac{1}{2}r(r + \delta r)\frac{\sin\delta\theta}{\delta\theta}$$

As $\delta\theta \to 0$: $\quad \delta r \to 0, \quad \dfrac{\sin\delta\theta}{\delta\theta} \to 1 \quad$ and $\quad \dfrac{\delta A}{\delta\theta} \to \dfrac{dA}{d\theta}$

Therefore
$$\frac{dA}{d\theta} = \tfrac{1}{2}r^2$$

Now consider the area bounded by the two part-lines $\theta = \alpha$, $\theta = \beta$ and the part of the curve between these lines.

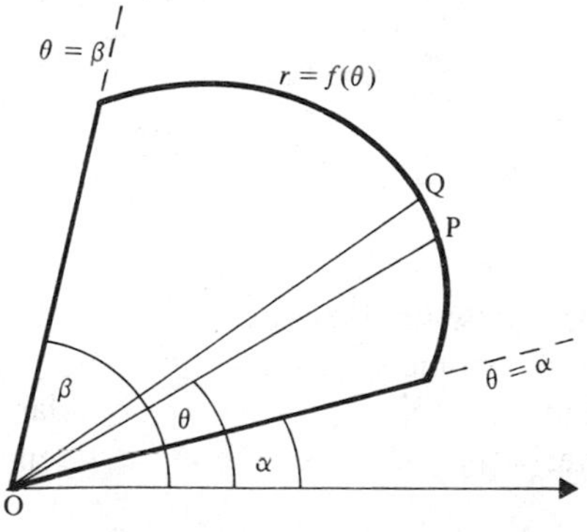

This area comprises an infinite set of elements of the type POQ.

Hence, throughout the whole of the area being considered

$$\frac{dA}{d\theta} = \tfrac{1}{2}r^2$$

so $$A = \tfrac{1}{2}\int_{\alpha}^{\beta} r^2 \, d\theta$$

In deriving this expression it has been assumed that every angle between α and β corresponds to a real value of r.
When using the formula above, care must be taken to ensure that this condition is satisfied.

EXAMPLES 5c

1) Find the area of the cardioid $r = a(1 + \cos\theta)$.

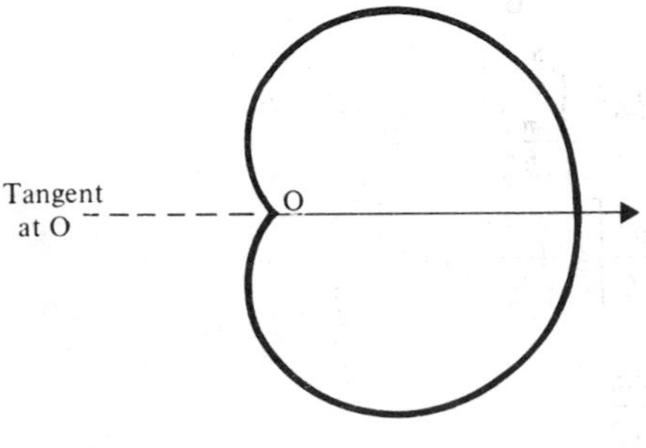

The complete area is traced out by taking values of θ from 0 to 2π, and r is defined for all values of θ.

Thus the area A of the cardioid is given by

$$A = \tfrac{1}{2}\int_0^{2\pi} r^2 \, d\theta = \tfrac{1}{2}\int_0^{2\pi} a^2(1 + \cos\theta)^2 \, d\theta$$

$$A = \tfrac{1}{2}a^2 \int_0^{2\pi} \left\{1 + 2\cos\theta + \tfrac{1}{2}(1 + \cos 2\theta)\right\} d\theta$$

$$= \frac{a^2}{2}\left[\theta + 2\sin\theta + \frac{\theta}{2} + \frac{1}{4}\sin 2\theta\right]_0^{2\pi}$$

$$= \frac{3\pi a^2}{2}$$

Note. The area of the cardioid can also be found by finding the area of the upper half and doubling it (because the curve is symmetrical about $\theta = 0$),

i.e. $$A = 2 \times \tfrac{1}{2}\int_0^{\pi} a^2(1 + \cos\theta)^2 \, d\theta$$

2) Find the area of one loop of the curve $r = a\cos 3\theta$.

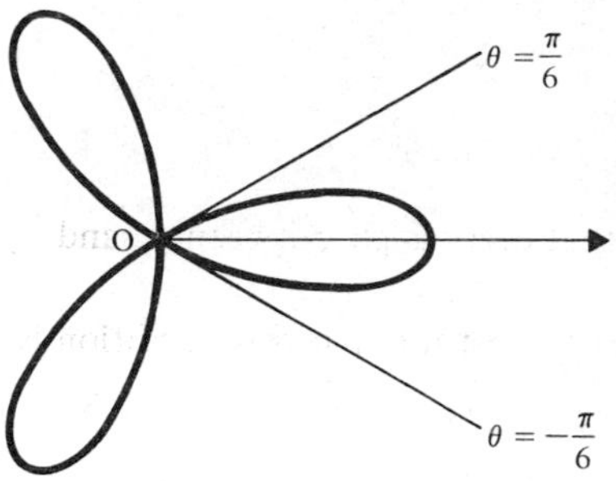

The loops of this curve lie between values of θ for which $r = 0$ i.e. for which $\cos 3\theta = 0$

$$\Rightarrow \quad 3\theta = \pm\frac{\pi}{2}, \pm\frac{3\pi}{2}, \dots$$

or $$\theta = \pm\frac{\pi}{6}, \pm\frac{\pi}{2}, \dots$$

One loop lies between $\theta = -\frac{\pi}{6}$ and $\theta = +\frac{\pi}{6}$

so its area is $$\frac{1}{2}\int_{-\frac{\pi}{6}}^{\frac{\pi}{6}} a^2\cos^2 3\theta\, d\theta = \frac{1}{4}a^2\int_{-\frac{\pi}{6}}^{\frac{\pi}{6}} (1+\cos 6\theta)\, d\theta$$

$$= \frac{a^2}{4}\left[\theta + \frac{1}{6}\sin 6\theta\right]_{-\frac{\pi}{6}}^{\frac{\pi}{6}}$$

$$= \frac{\pi a^2}{12}$$

3) Find the area enclosed by the curve $r^2 = a^2\sin 2\theta$.

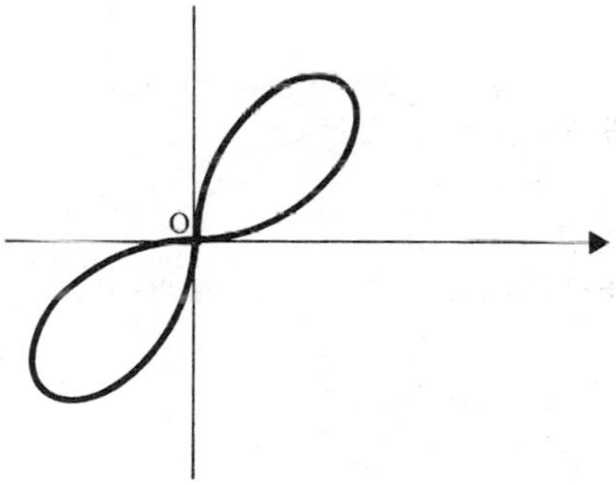

r is real only if $\sin 2\theta \geqslant 0$

i.e.

$0 \leqslant 2\theta \leqslant \pi$ and $2\pi \leqslant 2\theta \leqslant 3\pi$

or

$0 \leqslant \theta \leqslant \frac{\pi}{2}$ and $\pi \leqslant \theta \leqslant \frac{3\pi}{2}$

Using only those values of θ for which r is real, we see that the enclosed area is given by

$$\int_0^{\frac{\pi}{2}} \frac{1}{2}r^2\, d\theta + \int_{\pi}^{\frac{3\pi}{2}} \frac{1}{2}r^2\, d\theta$$

or, since the two loops are identical,

$$\text{area} = 2\int_0^{\frac{\pi}{2}} \tfrac{1}{2}r^2 \, d\theta$$

$$= a^2 \int_0^{\frac{\pi}{2}} \sin 2\theta \, d\theta$$

$$= a^2$$

EXERCISE 5c

In Questions 1–5 find the areas bounded by the given curve and the given radius vectors.
In each case sketch the curve, indicating the required area.

1) $r = a$ between $\theta = 0$ and $\theta = \pi$

2) $r = 2a \sin\theta$ between $\theta = 0$ and $\theta = \dfrac{\pi}{2}$

3) $r = a(1 - \sin\theta)$ between $\theta = 0$ and $\theta = 2\pi$

4) $r = 5\theta$ between $\theta = 0$ and $\theta = \dfrac{\pi}{2}$

5) $r(1 + \cos\theta) = a$ between $\theta = 0$ and $\theta = \dfrac{\pi}{2}$

6) Find the area of one loop of the curve $r = 2a \sin 2\theta$.

7) Sketch the curve $r = 3 - 5\cos\theta$. Find the area of the inner loop of this curve.

8) Sketch the curves $r = 2a\cos\theta$ and $r = a(1 + \cos\theta)$ on the same diagram.
Hence find the area enclosed *between* these curves.

9) The line $r = 4\sec\left(\theta - \dfrac{\pi}{6}\right)$ meets the initial line at G and meets the part-line $\theta = \dfrac{\pi}{2}$ at H. Find the area of triangle GOH.

10) Find the area enclosed by the curve $r = a(4 + 3\cos\theta)$.

11) In answering the question 'find the area enclosed by $r = a\cos 3\theta$' a student calculated

$\tfrac{1}{2}\int_0^{2\pi} r^2 \, d\theta$ and found that this gave double the correct answer. Explain why.

COMMON POLAR LINES AND CURVES

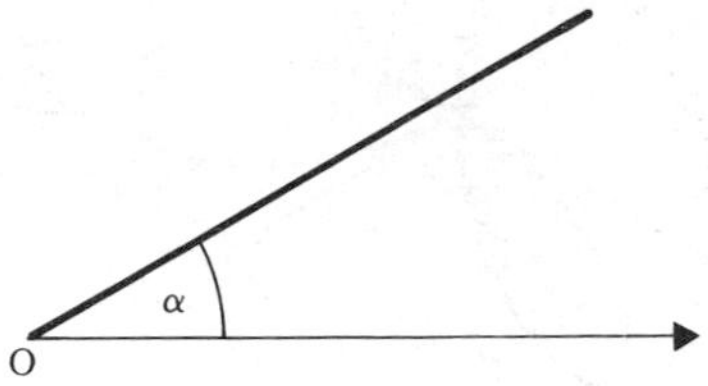

$\theta = \alpha$

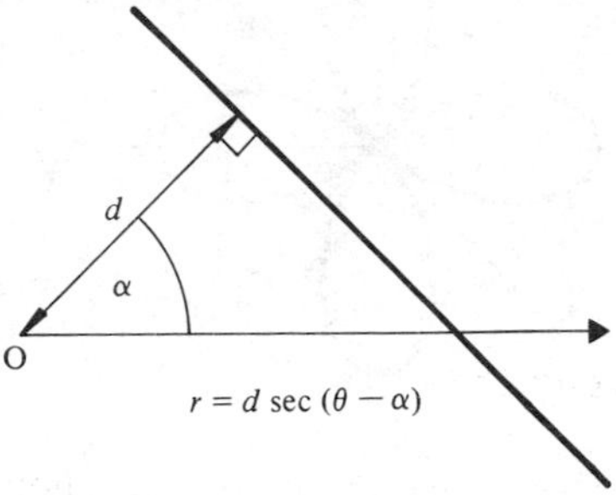

$r = d \sec(\theta - \alpha)$

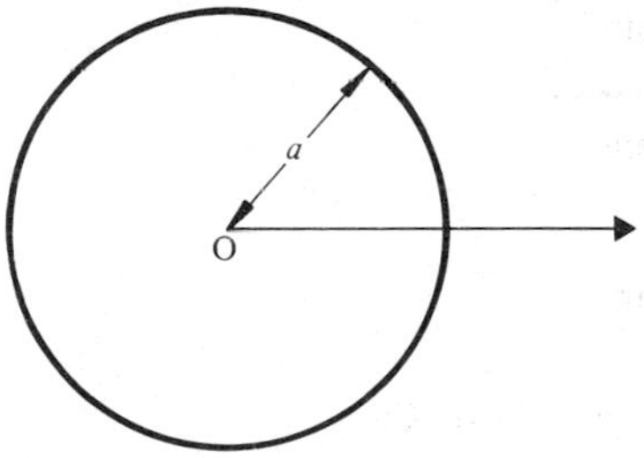

$r = a$

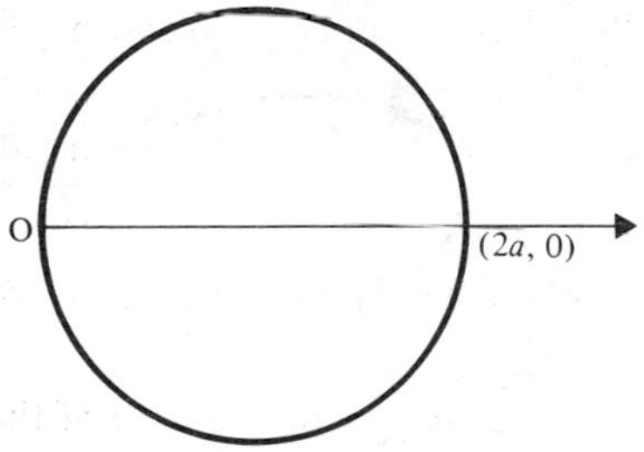

$r = 2a \cos\theta$

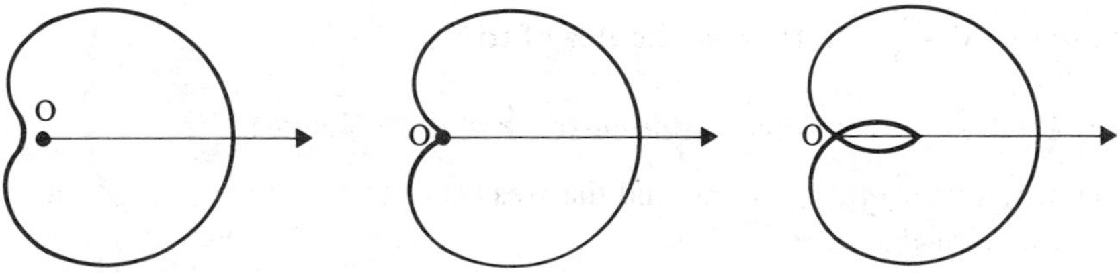

$a > b$ $a = b$ $a < b$

$r = a + b \cos\theta$

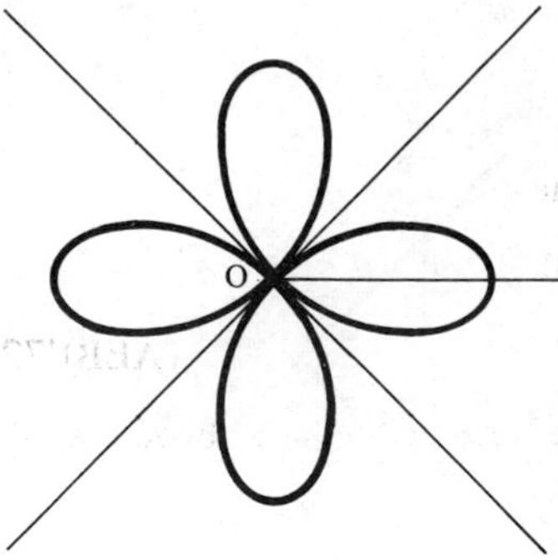

$r = a\cos 2\theta$

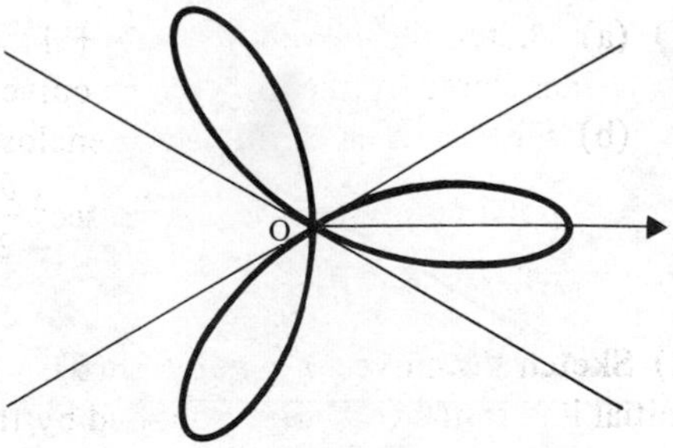

$r = a\cos 3\theta$

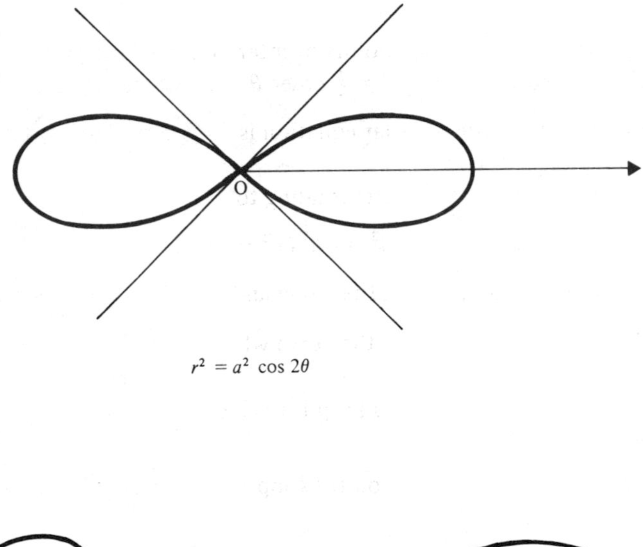

$r^2 = a^2 \cos 2\theta$

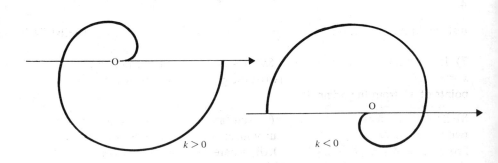

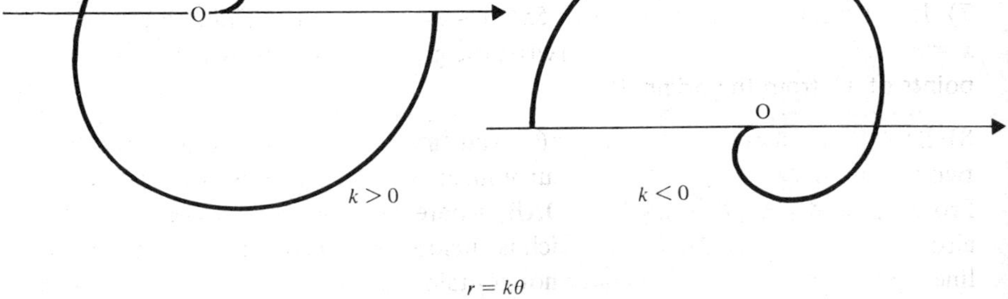

$r = k\theta$

MISCELLANEOUS EXERCISE 5

1) (a) Sketch the curve $y^2 = (x+1)^2(3-x)$ and show that the area enclosed by the loop of the curve is $\frac{256}{15}$.

(b) Find the area of the sector enclosed by the curve whose equation in polar coordinates is $r = a\sec^2\dfrac{\theta}{2}$ and the radii $\theta = 0$, $\theta = \alpha$ where $\alpha < \pi$. (AEB)'72

2) Sketch the curve $r = a(1+\sin\theta)$. Find the ratio of the area above the initial line to the total area bounded by the curve.

3) Sketch the curve whose polar equation is

$$r\cos^2\theta = 1$$

Find the area contained between the curve and the straight lines $\theta = \pm\dfrac{\pi}{4}$.

(AEB)'72

4) Sketch the curve whose polar equation is $r = 1 + 2\cos 2\theta$ for the range $0 \leqslant \theta \leqslant \pi$.
Prove that the values of θ at the points of intersection of the curve $r = 1 + 2\cos 2\theta$ with the line $r = \frac{1}{2}\operatorname{cosec}\theta$ are given by $2\sin 3\theta = 1$.

5) (a) Sketch the curve whose polar equation is $r = a\sin 3\theta$ and find the area of one of its loops.

(b) Find the polar equation corresponding to

$$(x^2+y^2)^2 = a^2(x^2-y^2)$$

Hence, or otherwise, sketch the curve and calculate the enclosed area.

6) Show by means of a sketch that the curve whose polar equation is $r = 2\cos\theta\sin\left(\theta - \dfrac{\pi}{4}\right)$ describes a loop for values of θ in the range $\dfrac{\pi}{4} \leqslant \theta \leqslant \dfrac{\pi}{2}$. Show that the point on the loop farthest from the origin is at a distance from the origin of $1 - \dfrac{1}{\sqrt{2}}$. (AEB)'74

7) If a curve C has the equation $5x^2 + 4xy + 2y^2 = 1$, find, by putting $x = r\cos\theta$, $y = r\sin\theta$ or otherwise, the greatest and least distances of points of C from the origin O. (O)

8) Sketch the curve $r = 5 + 2\cos\theta$ and find the polar coordinates of the two points A and B in which this curve meets the straight line $r\cos\theta = 3$. Prove that the area of the triangle OAB, where O is the pole, is $9\sqrt{3}$. Find also the area of the finite region which is enclosed between the curve and the line $r\cos\theta = 3$ and which does not contain the pole. (U of L)

9) Sketch the curve given in polar coordinates by the equation $r^2 = a^2 \cos 2\theta$, where $r \geqslant 0$ and $a > 0$.
(a) Obtain the angle between the tangents to the curve at the pole.
(b) Calculate the area of the region enclosed by one loop of the curve.
(U of L)p

10) Show that the curve whose polar equation is $r = a + b \sec\theta$, where $a > b > 0$, has a loop.
Prove that the area enclosed by the loop is

$$a^2 \cos^{-1}\frac{b}{a} + b\sqrt{(a^2-b^2)} - 2ab \ln\left(\frac{a+\sqrt{(a^2-b^2)}}{b}\right)$$ (O)

11) Find the points of intersection of the curves $r = 3(1+\cos\theta)$ and $r = 1-\cos\theta$. Sketch the curves.

12) The polar equation of a curve is $r = 2\cos\theta\,(1-\sin\theta)$.
Find the values of θ in the range $0 \leqslant \theta < 2\pi$ for which $r = 0$.
Find also the polar coordinates of the points at which $\dfrac{dr}{d\theta} = 0$,
for $0 \leqslant \theta < 2\pi$. Sketch the curve.
Prove that the area of the region enclosed by the portion of the curve from $\theta = 0$ to $\theta = \pi$ is $\frac{1}{12}(15\pi - 32)$. (C)

13) Sketch the curve with polar equation $r = a \sin 2\theta \sin\theta$.
Find the area of the region enclosed by one loop of the curve. (C)

CHAPTER 6

SERIES

CONVERGENCE OF SERIES

Sequences and series were introduced in Chapter 15 of *The Core Course*. In that chapter we investigated several series and a variety of methods for finding S_n, the sum of the first n terms. We also discussed briefly the concept of convergence of a series and we start this chapter with a more detailed investigation of convergence.

Consider the geometric series

$$1 + \tfrac{1}{2} + (\tfrac{1}{2})^2 + (\tfrac{1}{2})^3 + (\tfrac{1}{2})^4 + \dots$$

The sum of just one term (S_1) is 1
The sum of the first two terms (S_2) is 1.5
The sum of the first three terms (S_3) is 1.75
The sum of the first four terms (S_4) is 1.875
The sum of the first five terms (S_5) is 1.9375

. . . .

The sum of the first n terms (S_n) is $2 - 2(\tfrac{1}{2})^n$.

From the values obtained for S_1, S_2, S_3, S_4, S_5, it appears that, as we add successively more terms of this series, the sum approaches the value 2.
This is confirmed by the general expression $S_n = 2 - 2(\tfrac{1}{2})^n$ in which $(\tfrac{1}{2})^n \to 0$ as $n \to \infty$. So $S_n \to 2$ as $n \to \infty$ and we say that the sum of this series converges to 2.
The series itself is called convergent.
Now consider the arithmetic progression

$$1 + 2 + 3 + 4 + 5 + \dots$$

Here, $S_1 = 1$, $S_2 = 3$, $S_3 = 6$, $\dots$, $S_n = \dfrac{n}{2}(n+1), \dots,$

and it is clear that as $n \to \infty$, $S_n \to \infty$. This series is not convergent.

As a third example, consider the series

$$1-1+1-1+1-1+\dots$$

Now $S_1=1,\ S_2=0,\ S_3=1,\ S_4=0,\ \dots,$ so $S_n=0$ or 1 depending on whether n is even or odd. This series is not convergent either.

In general a convergent series is one for which

$$\lim_{n\to\infty}[S_n] \quad \text{exists and is finite.}$$

A series which does not converge is called *divergent.*

We are now going to consider methods for determining whether a given series is convergent or not. One method is to find S_n in the form $f(n)$ and determine whether $\lim_{n\to\infty}[S_n]$ exists and is finite.

It is not always easy to find S_n, however, so we need other ways to investigate the convergence, or otherwise, of series.
We begin by stating an obvious condition that has to be satisfied by any convergent series, namely that
as $n\to\infty$, *the general term of the series must approach zero.*
This can be seen quite clearly by considering the general series

$$u_1+u_2+u_3+\dots+u_n+\dots.$$

If this series is convergent then, as $r\to\infty$, $S_n\to k$, say.

Now $$u_n = S_n - S_{n-1}$$

and, as $n\to\infty$, both S_n and S_{n-1} approach the value k, so the difference between them approaches zero,
i.e. $\lim_{n\to\infty}[u_n]=0.$

So $$\sum u_n \text{ is convergent} \Rightarrow \lim_{n\to\infty}[u_n]=0$$

This condition can be used *only* to prove that a series is not convergent,

e.g. $$1-2+3-4+5-6+\dots$$

is *not* convergent because u_n does *not* approach zero as $n\to\infty$.
The converse is not true, i.e. the fact that $u_n\to 0$ as $n\to\infty$ does not necessarily mean that the series is convergent,
i.e. $\lim_{n\to\infty}[u_n]=0 \not\Rightarrow \Sigma u_n$ is convergent.
The series examined below illustrates this fact.

The Harmonic Series $\displaystyle\sum_{r=1}^{\infty}\frac{1}{r}$

The series $1+\frac{1}{2}+\frac{1}{3}+\frac{1}{4}+\frac{1}{5}+\dots$ is called the harmonic series
After the first two terms the remaining terms can be grouped in brackets so that

there are 2 terms in the first bracket, 4 terms in the second bracket, 8 terms in the third bracket and so on, as shown below:

$$1 + \tfrac{1}{2} + (\tfrac{1}{3} + \tfrac{1}{4}) + (\tfrac{1}{5} + \tfrac{1}{6} + \tfrac{1}{7} + \tfrac{1}{8}) + (\tfrac{1}{9} + \ldots + \tfrac{1}{16}) + \ldots$$

In any one bracket the last term is the smallest term. Replacing every term in any one bracket by the last term in that bracket gives another series, viz.

$$1 + \tfrac{1}{2} + (\tfrac{1}{4} + \tfrac{1}{4}) + (\tfrac{1}{8} + \tfrac{1}{8} + \tfrac{1}{8} + \tfrac{1}{8}) + (\tfrac{1}{16} + \ldots + \tfrac{1}{16}) + \ldots$$

The sum of the terms in each bracket of this second series is $\frac{1}{2}$, and this is *less* than the sum of the terms in the corresponding bracket of the harmonic series.

So for $n > 2$, the sum of the first n terms of the harmonic series (S_n) is greater than the sum of the first n terms of the second series (S'_n).

The second series can be written

$$1 + \tfrac{1}{2} + (\tfrac{1}{2}) + (\tfrac{1}{2}) + (\tfrac{1}{2}) + \ldots$$

which is clearly divergent.

i.e. $\qquad S'_n \to \infty \quad \text{as} \quad n \to \infty$

But $S_n > S'_n$ for $n > 2$

so $\qquad S_n \to \infty \quad \text{as} \quad n \to \infty$

i.e. $1 + \frac{1}{2} + \frac{1}{3} + \frac{1}{4} + \ldots$ is divergent even though $u_n \to 0$ as $n \to \infty$.

The Comparison Tests

We showed that the harmonic series was divergent by comparing its terms with the (smaller) terms of another series which was clearly divergent. This is an example of a comparison test which we now state formally.

If, after a certain number of terms, all the terms of a given series are greater than the corresponding terms of a known divergent series, then the given series is also divergent.

Similarly if, after a certain number of terms, all the terms of a given series are less than the corresponding terms of a known convergent series, then the given series is also convergent.

It should be noted that these comparison tests apply only to series with positive terms. So to summarize

provided that $u_n > 0$ and $a_n > 0$

if $u_n > a_n$ for all values of $n > N$ where N is fixed,

$$\sum a_r \text{ is divergent} \Rightarrow \sum u_r \text{ is divergent}$$

and if $u_n < a_n$ for all values of $n > N$ where N is fixed, then

$$\sum a_r \text{ is convergent} \Rightarrow \sum u_r \text{ is convergent}.$$

The use of these tests for investigating the convergence, or otherwise, of a particular series requires

(a) an expression for the general term of the series under investigation

(b) a series for comparison which are known to be either convergent or divergent. Useful series for this purpose are

the harmonic series, $\sum \frac{1}{r}$, for divergence

the geometric series, $\sum \left(\frac{1}{k}\right)^r$ when $k > 1$ for convergence.

EXAMPLES 6a

1) Show that the series $1 + \frac{1}{\sqrt{2}} + \frac{1}{\sqrt{3}} + \frac{1}{\sqrt{4}} + \ldots$ is divergent.

The nth term of the given series is $\frac{1}{\sqrt{n}}$.

Comparing with $1 + \frac{1}{2} + \frac{1}{3} + \ldots + \frac{1}{n} + \ldots$ which is divergent,

we have $$u_n = \frac{1}{\sqrt{n}} \quad \text{and} \quad a_n = \frac{1}{n}$$

$$\Rightarrow \qquad \frac{u_n}{a_n} = \frac{1}{\sqrt{n}} \bigg/ \frac{1}{n} = \sqrt{n}$$

Now $\sqrt{n} > 1$ for all values of $n > 1$.
Therefore $u_n > a_n$ for all values of $n > 1$.

So the series $1 + \frac{1}{\sqrt{2}} + \frac{1}{\sqrt{3}} + \ldots$ is divergent.

2) Show that the series $\sum_{r=1}^{\infty} \frac{1}{r!}$ is convergent.

Using the G.P. $\frac{1}{2} + (\frac{1}{2})^2 + (\frac{1}{2})^3 + \ldots$ for comparison we have

$$u_n = \frac{1}{n!} \quad \text{and} \quad a_n = \frac{1}{2^n}$$

So $$\frac{u_n}{a_n} = \frac{2^n}{n!} = \frac{2}{n} \times \frac{2}{n-1} \times \frac{2}{n-2} \times \ldots \times \frac{2}{3} \times \frac{2}{2} \times \frac{2}{1}$$

i.e. $$\frac{u_n}{a_n} < 2 \quad \text{for} \quad n > 2$$

But to get $u_n < a_n$ we need $\frac{u_n}{a_n} < 1$ so we will compare $\sum u_n$ with twice the G.P. used above,
i.e. we will compare $\sum u_n$ with $2(\frac{1}{2}) + 2(\frac{1}{2})^2 + 2(\frac{1}{2})^3 + \ldots$

So we now have $u_n = \frac{1}{n!}$ and $a_n = \frac{2}{2^n}$

$$\Rightarrow \qquad \frac{u_n}{a_n} = \frac{2^n}{(2)(n!)} = \frac{1}{2}\left[\frac{2}{n} \times \frac{2}{n-1} \times \frac{2}{n-2} \times \ldots \times \frac{2}{3} \times \frac{2}{2} \times \frac{2}{1}\right]$$

i.e.
$$\frac{u_n}{a_n} < 1 \quad \text{for} \quad n > 2$$

Therefore $u_n < a_n$ for $n > 2$
and $2(\frac{1}{2}) + 2(\frac{1}{2})^2 + \ldots$ is a G.P. with common ratio $\frac{1}{2}$, so it is convergent.

Therefore $\sum_{r=1}^{\infty} \frac{1}{r!}$ is convergent.

The Series $\sum\left(\frac{1}{r}\right)^k$, where k is a Constant

The series $1 + (\frac{1}{2})^k + (\frac{1}{3})^k + (\frac{1}{4})^k + (\frac{1}{5})^k + \ldots$ is another useful one to use in comparison tests since, as we shall show,

$$\sum_{r=1}^{\infty}\left(\frac{1}{r}\right)^k \quad \text{diverges when} \quad k \leqslant 1$$

and

$$\sum_{r=1}^{\infty}\left(\frac{1}{r}\right)^k \quad \text{converges when} \quad k > 1$$

(a) Proof that $\sum_{r=1}^{\infty}\left(\frac{1}{r}\right)^k$ diverges when $k \leqslant 1$

We have already proved this for two particular values of k, i.e. $k = 1$, which gives the harmonic series $1 + \frac{1}{2} + \frac{1}{3} + \ldots$, and $k = \frac{1}{2}$, which gives the series $1 + \frac{1}{\sqrt{2}} + \frac{1}{\sqrt{3}} + \ldots$.

The general proof for $k < 1$ uses the same method as in Example 6a No. 1) when $k = \frac{1}{2}$.

When $k = p$, say, where $p < 1$, we have the series

$$1 + \frac{1}{2^p} + \frac{1}{3^p} + \frac{1}{4^p} + \ldots, \quad \text{for which} \quad u_n = \frac{1}{n^p}.$$

Comparing with $1+\frac{1}{2}+\frac{1}{3}+\frac{1}{4}+\ldots$, for which $a_n = \dfrac{1}{n}$

we have
$$\frac{u_n}{a_n} = \frac{n}{n^p} = n^{1-p}$$

Now $p<1$ so $1-p>0 \Rightarrow n^{1-p}>1$ for $n>1$.

Therefore $u_n > a_n$ for $n>1$.

So $\displaystyle\sum_{r=1}^{\infty}\left(\frac{1}{r}\right)^k$ is divergent for $k \leqslant 1$.

(b) Proof that $\displaystyle\sum_{r=1}^{\infty}\left(\frac{1}{r}\right)^k$ is convergent for $k>1$

This proof is based on the grouping of terms into brackets, a method similar to that used when investigating the harmonic series.
When $k=q$, say, where $q>1$ we have the series

$$1+\frac{1}{2^q}+\frac{1}{3^q}+\frac{1}{4^q}+\ldots$$

Bracketing first one, then two, then four, then eight, and so on, terms together gives

$$1+\left(\frac{1}{2^q}+\frac{1}{3^q}\right)+\left(\frac{1}{4^q}+\frac{1}{5^q}+\frac{1}{6^q}+\frac{1}{7^q}\right)+\left(\frac{1}{8^q}+\ldots+\frac{1}{15^q}+\ldots\right)$$

In the first bracket, $\dfrac{1}{3^q}<\dfrac{1}{2^q}$

so
$$\frac{1}{2^q}+\frac{1}{3^q}<\frac{2}{2^q}$$

In the second bracket, the last three terms are less than $\dfrac{1}{4^q}$

so
$$\frac{1}{4^q}+\frac{1}{5^q}+\frac{1}{6^q}+\frac{1}{7^q}<\frac{4}{4^q}=\left(\frac{2}{2^q}\right)^2$$

Similarly for the third bracket,

$$\frac{1}{8^q}+\frac{1}{9^q}+\ldots+\frac{1}{15^q}<\frac{8}{8^q}=\left(\frac{2}{2^q}\right)^3$$

and for the fourth bracket,

$$\frac{1}{16^q}+\frac{1}{17^q}+\ldots+\frac{1}{31^q}<\frac{16}{16^q}=\left(\frac{2}{2^q}\right)^4$$

Therefore the sum of the given series is *less* than the sum of the series

$$1+\left(\frac{2}{2^q}\right)+\left(\frac{2}{2^q}\right)^2+\left(\frac{2}{2^q}\right)^3+\ldots$$

This second series is a G.P. whose common ratio is $\frac{2}{2^q}$ or $\frac{1}{2^{q-1}}$.

Now $\frac{1}{2^{q-1}}<1$ because $q>1$, so this series converges.

Therefore the given series $\sum_{r=1}^{\infty}\left(\frac{1}{r}\right)^k$ converges when $k>1$.

EXAMPLES 6a (continued)

3) Using a comparison test, or otherwise, prove that the series $\sum_{r=1}^{\infty}\frac{1}{r(r+1)}$ is convergent.

Using a comparison test, we can compare the given series with $\sum_{r=1}^{\infty}\left(\frac{1}{r}\right)^2$ which we now know is convergent.

For $\sum_{r=1}^{\infty}\frac{1}{r(r+1)}$, $\quad u_n=\frac{1}{n(n+1)}$

and for $\sum_{r=1}^{\infty}\left(\frac{1}{r}\right)^2$, $\quad a_n=\frac{1}{n^2}$.

So $\frac{u_n}{a_n}=\frac{1}{n(n+1)}\div\frac{1}{n^2}=\frac{n}{n+1}$ which is <1 for all values of n.

Therefore $\sum_{r=1}^{\infty}\frac{1}{r(r+1)}$ is convergent.

An alternative method to show that $\sum_{r=1}^{\infty}\frac{1}{r(r+1)}$ is convergent is to find S_n using partial fractions and a difference method,

i.e. $$\frac{1}{n(n+1)}=\frac{1}{n}-\frac{1}{n+1}$$

so $$S_n=(1-\tfrac{1}{2})+(\tfrac{1}{2}-\tfrac{1}{3})+(\tfrac{1}{3}-\tfrac{1}{4})+\ldots+\left(\frac{1}{n}-\frac{1}{n+1}\right)$$

$\Rightarrow$ $$S_n=1-\frac{1}{n+1}$$

As $n\to\infty$, $\frac{1}{n+1}\to 0$ so $S_n\to 1$.

Therefore the series is convergent.

Another Form of Comparison Test

If $\sum u_n$ and $\sum a_n$ are series of positive terms and if $\lim_{n\to\infty}\left[\frac{u_n}{a_n}\right] = k$ where k is finite and non-zero

then either $\sum u_n$ and $\sum a_n$ both converge

or $\sum u_n$ and $\sum a_n$ both diverge.

The proof of this statement uses the fact that
if $\sum a_n$ converges to S then $\sum \mu a_n$ converges to μS,
and if $\sum a_n$ diverges then $\sum \mu a_n$ diverges. (provided that $\mu \neq 0$).

Now, if as $n \to \infty$, $\frac{u_n}{a_n} \to l$ where $l \neq 0$
then after a finite number of terms, $u_n \simeq l a_n$

So
$$\frac{l a_n}{2} \leqslant u_n \leqslant \frac{3 l a_n}{2}$$

Now if $\sum u_n$ converges, then $\sum \frac{l}{2} a_n$ converges so $\sum a_n$ converges

similarly, for $l \neq 0$, if $\sum u_n$ diverges, then $\sum \frac{3l}{2} a_n$ diverges so $\sum a_n$ diverges.

Neat solutions to problems can be produced by using this form of comparison test, but it is very important to remember that if $\lim_{n\to\infty} \frac{u_n}{a_n}$ is zero or infinite, no conclusion can be drawn.

For example, suppose we use this test to investigate the convergence, or otherwise, of the series $\sum_{r=1}^{\infty} \frac{1}{r(r+1)}$, for which $u_n = \frac{1}{n(n+1)}$.

Comparing with $\sum_{r=1}^{\infty} \frac{1}{r^2}$, which is convergent, we have

$$\frac{u_n}{a_n} = \frac{n}{n+1} \Rightarrow \lim_{n\to\infty}\left[\frac{n}{n+1}\right] = \lim_{n\to\infty}\left[1 - \frac{1}{n+1}\right] = 1$$

so $\sum u_n$ is also convergent.

But comparing with $\sum_{r=1}^{\infty} \frac{1}{r^3}$, which is convergent, we have

$$\frac{u_n}{a_n} = \frac{n^2}{n+1} \Rightarrow \lim_{n\to\infty}\left[\frac{n^2}{n+1}\right] = \lim_{n\to\infty}\left[n - 1 + \frac{1}{n+1}\right] = \infty$$

so we cannot draw any conclusion about $\sum u_n$.

Further, comparing with $\sum_{r=1}^{\infty} \frac{1}{r}$ which is divergent, we have

$$\frac{u_n}{a_n} = \frac{n}{n(n+1)} \Rightarrow \lim_{n\to\infty}\left[\frac{1}{n+1}\right] = 0$$

This result is also inconclusive and tells us nothing about the behaviour of $\sum u_n$.

To get a conclusive result from this test it is sensible to choose, for comparison, a series whose general term is of the same order as that of the series to be tested.

We end this section with one more test for convergence which we state but do not prove. This test is not a comparison test, it involves terms of the given series only.

D'Alembert's Ratio Test

If $\sum u_r$ is a series of positive terms and

$$\lim_{n\to\infty} \frac{u_{n+1}}{u_n} = p$$

then if $p < 1$, $\sum u_r$ converges

if $p > 1$, $\sum u_r$ diverges

Note that if $p = 1$ this test is inconclusive.

As an example, consider the series $\sum \frac{2^r}{r!}$, for which

$$u_n = \frac{2^n}{n!} \quad \text{and} \quad u_{n+1} = \frac{2^{n+1}}{(n+1)!}$$

So

$$\frac{u_{n+1}}{u_n} = \frac{2^{n+1}n!}{(n+1)!\,2^n} = \frac{2}{n+1}$$

$$\Rightarrow \lim_{n\to\infty}\left[\frac{u_{n+1}}{u_n}\right] = 0$$

Therefore the series is convergent.

But if we apply the ratio test to the series $\sum \frac{1}{r}$ we have

$$\lim_{n\to\infty}\left[\frac{u_{n+1}}{u_n}\right] = \lim_{n\to\infty}\left[\frac{n}{n+1}\right] = \lim_{n\to\infty}\left[1-\frac{1}{n+1}\right] = 1$$

and if we apply it to the series $\sum \frac{1}{r(r+1)}$ we have

$$\lim_{n \to \infty}\left[\frac{u_{n+1}}{u_n}\right] = \lim_{n \to \infty}\left[\frac{(n+1)(n+2)}{n(n+1)}\right] = \lim_{n \to \infty}\left[1+\frac{2}{n}\right] = 1$$

The application of the ratio test in each of these cases gives $p = 1$ but we know that the first series is divergent and the second series is convergent, showing that this text is inconclusive when $p = 1$.

EXERCISE 6a

Test for convergence the series whose nth term is

1) $\frac{1}{n(n+1)}$ 2) $\frac{1}{n(2^n)}$ 3) $\frac{1}{\sqrt{(n+1)}}$

4) $\frac{n+1}{n(n+2)}$ 5) $\frac{2^n}{n+1}$ 6) $\frac{2^n}{(n+1)(n+2)}$

7) $\frac{n}{2^n}$ 8) $\frac{1}{n^2+4}$ 9) $\frac{n}{\sqrt{(n+1)}}$

10) $\frac{n}{\sqrt{(n^2+1)}}$ 11) $\frac{n}{\sqrt{(n^3+1)}}$ 12) $\sqrt{\left(\frac{n}{n+1}\right)}$

13) $\left(\frac{n+1}{n(n+2)}\right)^{\frac{1}{2}}$ 14) $\frac{n^k}{n!}$ 15) $\frac{n!}{(2n)!}$

16) Use a comparison test to show that if $\sum u_r$ is convergent then so is $\sum (u_r)^2$

17) Show that $\sum_{r=1}^{\infty} \frac{x^r}{r!}$ is convergent for all positive values of x.

POWER SERIES

A series whose terms involve increasing (or decreasing) powers of a variable quantity is called a power series.

In *The Core Course* we have shown how binomial functions of x can be expressed as a series of ascending powers of x, and how the series can be used to find approximate numerical values for certain irrational numbers. There are many other functions of x (such as $e^x, \cos x, \ldots$) which can be expressed as infinite series. These series can be used to find approximate values (to any required degree of accuracy) for quantities such as e^2, $\cos \pi/6$, etc.

MacLAURIN'S SERIES

If $f(x)$ is any function of x then, supposing that $f(x)$ can be expanded as a series of ascending powers of x and that this series can be differentiated term by term, we have

$$f(x) = a_0 + a_1x + a_2x^2 + a_3x^3 + a_4x^4 + a_5x^5 + \ldots + a_rx^r + \ldots \quad [1]$$

where $a_0, a_1, a_2 \ldots$ are constants to be evaluated.

Substituting 0 for x in [1] gives $f(0) = a_0$, i.e. $a_0 = f(0)$.

Differentiating [1] w.r.t x gives

$$f'(x) = a_1 + 2a_2x + 3a_3x^2 + 4a_4x^3 + 5a_5x^4 + \ldots \quad [2]$$

Hence $f'(0) = a_1$ i.e. $a_1 = f'(0)$

Differentiating (2) w.r.t. x gives

$$f''(x) = 2a_2 + (3)(2a_3x) + (4)(3a_4x^2) + (5)(4a_5x^3) + \ldots \quad [3]$$

Hence $f''(0) = 2a_2$ i.e. $a_2 = \dfrac{f''(0)}{2}$

Similarly .

$$f'''(x) = (3)(2a_3) + (4)(3)(2a_4x) + (5)(4)(3a_5x^2) + \ldots$$

so $f'''(0) = (3!)(a_3)$ i.e. $a_3 = \dfrac{f'''(0)}{3!}$

$$f''''(x) = (4)(3)(2a_4) + (5)(4)(3)(2a_5x) + \ldots$$

so $f''''(0) = (4!)(a_4)$ i.e. $a_4 = \dfrac{f''''(0)}{4!}$

Hence we can deduce that, after differentiating $f(x)$ r times,

$$f^{(r)}(x) = r!a_r + (r+1)(r)\ldots(2a_{r+1})x + \ldots$$

so $f^{(r)}(0) = r!a_r$ i.e. $a_r = \dfrac{f^{(r)}(0)}{r!}$

Substituting these values of $a_0, a_1, a_2 \ldots$ in [1] we get

$$f(x) = f(0) + f'(0)x + \frac{f''(0)}{2!}x^2 + \frac{f'''(0)}{3!}x^3 + \ldots + \frac{f^{(r)}(0)}{r!}x^r + \ldots$$

This is known as *Maclaurin's Series* and the series can be found if $f^{(n)}(0)$ exists for all values of n.
For the series obtained to be useful, it must converge to $f(x)$.

A full discussion of the convergence of series is beyond the scope of this book, but it is found that some series converge to $f(x)$ for all values of x and others converge to $f(x)$ for a limited range of values of x. In the following examples the range of values of x for which the expansion is valid will be given but not justified.

We will now use Maclaurin's Series to find the series expansions of some important functions.

EXAMPLES 6b

1) $f(x) = e^x$.

Using $$f(x) = f(0) + f'(0)x + \frac{f''(0)}{2!}x^2 + \ldots + \frac{f^{(r)}(0)}{r!}x^r + \ldots$$

we have
$$f(x) = e^x \qquad f(0) = e^0 = 1$$
$$f'(x) = e^x \qquad f'(0) = e^0 = 1$$
$$f''(x) = e^x \qquad f''(0) = e^0 = 1$$
$$\ldots\ldots\ldots\ldots \qquad \ldots\ldots\ldots\ldots\ldots$$
$$f^{(r)}(x) = e^x \qquad f^{(r)}(0) = e^0 = 1$$

hence $$e^x = 1 + x + \frac{x^2}{2!} + \frac{x^3}{3!} + \frac{x^4}{4!} + \ldots + \frac{x^r}{r!} + \ldots$$

This is valid for all values of x.

2) $f(x) = \ln x$.

Using $f(x) = f(0) + f'(0)x + \ldots$
we have $f(x) = \ln x$, $f(0) = \ln 0$ which is undefined.
Therefore *$\ln x$ cannot be expanded as a series of ascending powers of x.*

3) $f(x) = \ln(1+x)$.

Using $$f(x) = f(0) + f'(0)x + \frac{f''(0)}{2!}x^2 + \ldots + \frac{f^{(r)}(0)}{r!}x^r + \ldots$$

we have
$$f(x) = \ln(1+x) \qquad f(0) = \ln 1 = 0$$
$$f'(x) = \frac{1}{1+x} \qquad f'(0) = 1$$
$$f''(x) = -\frac{1}{(1+x)^2} \qquad f''(0) = -1$$
$$f'''(x) = +\frac{2}{(1+x)^3} \qquad f'''(0) = 2$$

$$f''''(x) = -\frac{2\times 3}{(1+x)^4} \qquad f''''(0) = -3!$$

$$\ldots\ldots\ldots\ldots\ldots\ldots \qquad \ldots\ldots\ldots\ldots\ldots\ldots$$

$$f^{(r)}(x) = (-1)^{r+1}\frac{(r-1)!}{(1+x)^r} \qquad f^{(r)}(0) = (-1)^{r+1}(r-1)!$$

hence

$$\ln(1+x) = 0 + x - \frac{x^2}{2!} + \frac{2!x^3}{3!} - \frac{3!x^4}{4!} + \ldots + \frac{(-1)^{r+1}(r-1)!}{r!}x^r \ldots$$

$$\text{i.e. } \ln(1+x) = x - \frac{x^2}{2} + \frac{x^3}{3} - \frac{x^4}{4} + \ldots + \frac{(-1)^{r+1}x^r}{r} + \ldots$$

This series is valid for $-1 < x \leqslant 1$.

4) $f(x) = \cos x$.

Using $$f(x) = f(0) + f'(0)x + \frac{f''(0)}{2!}x^2 + \ldots + \frac{f^{(r)}(0)}{r!}x^r + \ldots$$

we have

$$f(x) = \cos x \qquad f(0) = \cos 0 = 1$$

$$f'(x) = -\sin x \qquad f'(0) = -\sin 0 = 0$$

$$f''(x) = -\cos x \qquad f''(0) = -\cos 0 = -1$$

$$f'''(x) = \sin x \qquad f'''(0) = \sin 0 = 0$$

$$f''''(x) = \cos x \qquad f''''(0) = \cos 0 = 1$$

Hence $$\cos x = 1 + (0)x - \frac{1}{2!}x^2 + (0)\frac{x^3}{3!} + \frac{1}{4!}x^4 \ldots$$

$$= 1 - \frac{x^2}{2!} + \frac{x^4}{4!} - \frac{x^6}{6!} + \ldots$$

From this we see that the series consists of even powers of x only, and that the terms containing $x^2, x^6, x^{10}, \ldots$ are negative, but the terms containing $x^4, x^8, x^{12}, \ldots$ are positive. A general term of the series is $\pm\dfrac{x^{2r}}{(2r)!}$ and is positive if r is even, negative, if r is odd, so the general term can be written $(-1)^r\dfrac{x^{2r}}{(2r)!}$

Hence $$\cos x = 1 - \frac{x^2}{2!} + \frac{x^4}{4!} - \frac{x^6}{6!} + \ldots + \frac{(-1)^r x^{2r}}{(2r)!} + \ldots$$

This is valid for all values of x.

EXERCISE 6b

Use Maclaurin's Series to expand each of the following as a series of ascending powers of x up to and including the term containing the power of x indicated.

1) e^{-x}, x^4
2) $\sin x$, x^7
3) $\ln(1-x)$, x^4
4) $(1+x)^n$, x^3
5) $\tan x$, x^3
6) $e^x \ln(1-x)$, x^2
7) $(1-x)^{-1}$, x^4
8) $e^x \cos x$, x^3
9) $\tan(e^x)$, x^2

STANDARD EXPANSIONS

Summarizing some of the results from the last section we have:

$$e^x = 1+x+\frac{x^2}{2!}+\frac{x^3}{3!}+\dots+\frac{x^r}{r!}+\dots \text{ for all values of } x$$

$$\ln(1+x) = x-\frac{x^2}{2}+\frac{x^3}{3}-\frac{x^4}{4}+\dots+(-1)^{r+1}\frac{x^r}{r}+\dots \text{ for } -1<x\leqslant 1$$

$$\ln(1-x) = -x-\frac{x^2}{2}-\frac{x^3}{3}-\frac{x^4}{4}-\dots-\frac{x^r}{r}+\dots \text{ for } -1\leqslant x<1$$

$$\cos x = 1-\frac{x^2}{2!}+\frac{x^4}{4!}-\frac{x^6}{6!}+\dots+\frac{(-1)^r x^{2r}}{(2r)!}+\dots \text{ for all values of } x$$

$$\sin x = x-\frac{x^3}{3!}+\frac{x^5}{5!}-\frac{x^7}{7!}+\dots+\frac{(-1)^r x^{2r+1}}{(2r+1)!}+\dots \text{ for all values of } x$$

$$(1+x)^n = 1+nx+\frac{n(n-1)}{2!}x^2+\dots \text{ for } -1<x<1$$

$$(1+x)^{-1} = 1-x+x^2+\dots+(-1)^r x^r+\dots \text{ for } -1<x<1$$

$$(1-x)^{-1} = 1+x+x^2+\dots+x^r+\dots \text{ for } -1<x<1$$

All the series expansions given above may be quoted *unless* their derivation is asked for.

It is interesting to note the Maclaurin Series for $(1+x)^n$, i.e. the binomial expansion, and the series for the special cases $(1+x)^{-1}$ and $(1-x)^{-1}$, which are also geometric series.

It is also interesting to note the Maclaurin Series for even and odd functions. In the case of an even function of x where $f(-x)=f(x)$ we would expect the series to contain only even powers of x as
$x^{2n}=(-x)^{2n}$ but $x^{2n+1}\neq(-x)^{2n+1}$.
This is confirmed in the case of the series obtained for $\cos x$.
Similarly, for an odd function of x (i.e. $f(-x)=-f(x)$) we would expect only odd powers of x in a series expansion. This is confirmed by the series obtained for $\sin x$.

It is particularly important that the reader should *remember* the range of values of x for which a particular expansion is valid. The following examples show how these standard expansions can be used to express some compound functions as series, and the reader is advised to note that *whenever possible* a compound function should be expressed as a sum or difference of functions *before* any analysis of it is carried out.

EXAMPLES 6c

1) Expand (a) $\ln \dfrac{1-2x}{(1+2x)^2}$ (b) $\dfrac{e^{2x}+e^{-2x}}{e^x}$.

as a series of ascending powers of x up to and including the term in x^4. Give the general terms and the range of values of x for which each expansion is valid.

(a)
$$\ln\left[\frac{1-2x}{(1+2x)^2}\right] \equiv \ln(1-2x) - \ln(1+2x)^2$$
$$\equiv \ln(1-2x) - 2\ln(1+2x)$$

Using the expansion for $\ln(1-x)$ and replacing x by $2x$ gives

$$\ln(1-2x) = -(2x) - \frac{(2x)^2}{2} - \frac{(2x)^3}{3} - \frac{(2x)^4}{4} - \dots - \frac{(2x)^r}{r} - \dots$$

$$\text{if } -1 \leqslant 2x < 1$$

Using the expansion for $\ln(1+x)$ and replacing x by $2x$ gives

$$2\ln(1+2x) = 2(2x) - \frac{2(2x)^2}{2} + \frac{2(2x)^3}{3} - \frac{2(2x)^4}{4} + \dots + \frac{2(-1)^{r+1}(2x)^r}{r} + \dots$$

$$\text{if } -1 < 2x \leqslant 1$$

Hence

$$\ln\left[\frac{1-2x}{(1+2x)^2}\right] = \left(-2x - 2x^2 - \tfrac{8}{3}x^3 - 4x^4 - \dots - \frac{2^r x^r}{r} - \dots\right)$$
$$- \left(4x - 4x^2 + \tfrac{16}{3}x^3 - 8x^4 + \dots + \frac{2^{r+1}(-1)^{r+1}x^r}{r} + \dots\right)$$
$$= -6x + 2x^2 - 8x^3 + 4x^4 + \dots - \frac{2^r x^r}{r} + \frac{2^{r+1}(-1)^r x^r}{r} + \dots$$

provided that $-\frac{1}{2} < x \leqslant \frac{1}{2}$ *and* $-\frac{1}{2} \leqslant x < \frac{1}{2}$.

Therefore the series, up to x^4, is $-6x + 2x^2 - 8x^3 + 4x^4$

The general term is $-\dfrac{2^r}{r}\left[1-2(-1)^r\right]x^r$

The expansion is valid for $-\frac{1}{2}<x<\frac{1}{2}$

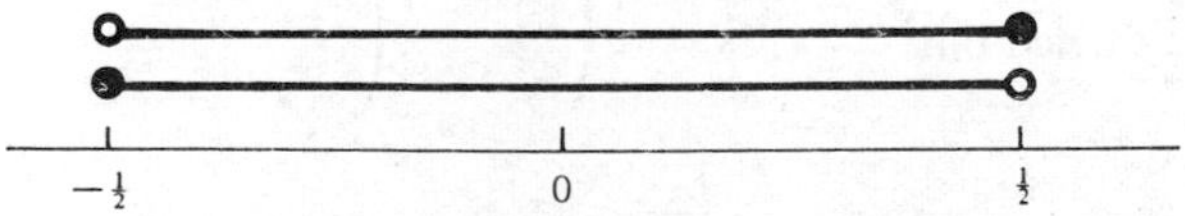

(b)
$$\frac{e^{2x}+e^{-2x}}{e^x} \equiv \frac{e^{2x}}{e^x}+\frac{e^{-2x}}{e^x} \equiv e^x+e^{-3x}$$

Using the expansion of e^x

$$e^x+e^{-3x} = \left(1+x+\frac{x^2}{2!}+\frac{x^3}{3!}+\frac{x^4}{4!}+\ldots+\frac{x^r}{r!}+\ldots\right)$$
$$+\left(1-3x+\frac{(-3x)^2}{2!}+\frac{(-3x)^3}{3!}+\frac{(-3x)^4}{4!}+\ldots+\frac{(-3x)^r}{r!}+\ldots\right)$$
$$= 2-2x+5x^2-\tfrac{13}{3}x^3+\tfrac{41}{12}x^4+\ldots+\frac{1+(-3)^r}{r!}x^r+\ldots$$

As far as the term in x^4, the series is $2-2x-5x^2-\frac{13}{3}x^3+\frac{41}{12}x^4$

The general term is $\dfrac{1+(-3)^r}{r!}x^r$ and the expansion is valid for all values of x.

2) Show that for small values of x, $\sin^2 x \simeq x^2-\frac{1}{3}x^4$.

Using the identity $\sin^2 x \equiv \frac{1}{2}(1-\cos 2x)$ and the expansion of $\cos x$ we have:

$$\sin^2 x \equiv \tfrac{1}{2}-\tfrac{1}{2}\cos 2x = \tfrac{1}{2}-\tfrac{1}{2}\left(1-\frac{(2x)^2}{2!}+\frac{(2x)^4}{4!}-\frac{(2x)^6}{6!}+\ldots\right)$$
$$= x^2-\tfrac{1}{3}x^4+\tfrac{2}{45}x^6+\ldots$$

If x is small enough to neglect x^5 and higher powers of x we have

$$\sin^2 x \simeq x^2-\tfrac{1}{3}x^4$$

3) Expand $\ln\dfrac{x}{x-1}$ as a series of *descending* powers of x as far as the term in x^{-4}. Use your expansion to find $\ln 1.25$ correct to four decimal places.

$$\ln\left(\frac{x}{x-1}\right) = \ln\left(\frac{x-1}{x}\right)^{-1}$$

$$= -\ln\left(1-\frac{1}{x}\right)$$

$$= -\left\{-\left(\frac{1}{x}\right)-\frac{1}{2}\left(\frac{1}{x}\right)^2-\frac{1}{3}\left(\frac{1}{x}\right)^3-\frac{1}{4}\left(\frac{1}{x}\right)^4-\ldots\right\}$$

$$= \frac{1}{x}+\frac{1}{2x^2}+\frac{1}{3x^3}+\frac{1}{4x^4}+\ldots$$

This expansion is valid for $-1 \leqslant \frac{1}{x} < 1$

i.e. for $x \leqslant -1$ and $x > 1$

When $\frac{x}{x-1} = 1.25$, $x = 5$, a value for which the expansion is valid.

Therefore $\ln 1.25 = \frac{1}{5}+\frac{1}{2}(\frac{1}{25})+\frac{1}{3}(\frac{1}{125})+\frac{1}{4}(\frac{1}{625})+\ldots$

$= 0.2 + 0.02 + 0.002\,667 + 0.0004 + \ldots$

$= 0.223\,067 + \ldots$

$= 0.2231$ to 4 d.p.

4) Express $\cosh x$ as a series of ascending powers of x up to and including the term in x^6. Hence show that, in the neighbourhood of $x = 0$, $\cosh x \simeq 1 + \frac{1}{2}x^2$.
Illustrate your result graphically.

By definition $\cosh x \equiv \frac{1}{2}(e^x + e^{-x})$

Now $e^x = 1 + x + \frac{x^2}{2!}+\frac{x^3}{3!}+\frac{x^4}{4!}+\ldots$ for all real values of x

and $e^{-x} = 1 - x + \frac{x^2}{2!}-\frac{x^3}{3!}+\frac{x^4}{4!}-\ldots$ for all real values of x.

So $\frac{1}{2}(e^x + e^{-x}) = \frac{1}{2}\left(2+\frac{2x^2}{2!}+\frac{2x^4}{4!}+\frac{2x^6}{6!}+\ldots\right)$

$\Rightarrow \cosh x = 1 + \frac{x^2}{2!}+\frac{x^4}{4!}+\frac{x^6}{6!}+\ldots$ for all real values of x.

In the neighbourhood of $x = 0$ means values close to zero.
Now if x is small, x^2 is even smaller and x^3 and higher powers of x get progressively more negligible in value.
So if x is small enough for x^4 and higher powers to make a negligible contribution to $\sum x^{2r}/(2r)!$ then

$$\cosh x \simeq 1 + \tfrac{1}{2}x^2$$

By drawing the curves $y = \cosh x$ and $y = 1 + \frac{1}{2}x^2$ for $-1 < x < 1$,

i.e.

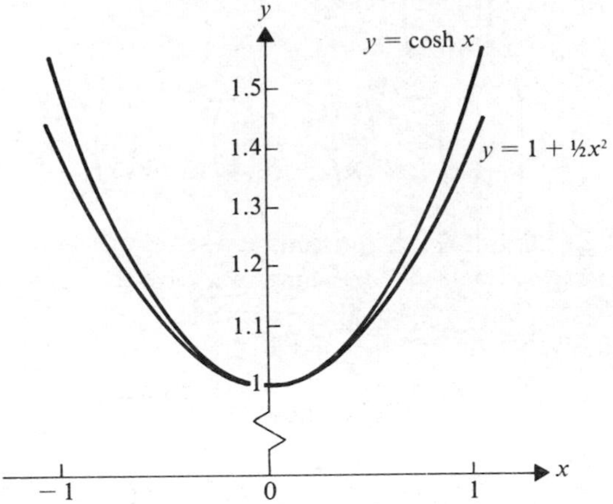

we see that
the closer the value of x is to zero, the better is the approximation.
We say that the function $1 + \frac{1}{2}x^2$ is a quadratic approximation to the function $\cosh x$ for small values of x.

5) By expanding $e^x + e^{2x}$ as a series of ascending powers of x, find a linear approximation to $e^x + e^{2x}$ for small values of x. Interpret the result graphically.

$$e^x + e^{2x} = \left(1 + x + \frac{x^2}{2!} + \dots\right) + \left(1 + 2x + \frac{(2x)^2}{2!} + \dots\right)$$

$$= 2 + 3x + \frac{5x^2}{2!} + \dots$$

A linear function of x is of the form $ax + b$ where a and b are constants. So treating the terms in x^2 and higher powers of x as negligible (which we can do if x is small enough), we have

$$e^x + e^{2x} \simeq 2 + 3x$$

Sketching the curve $y = e^x + e^{2x}$ and the line $y = 2 + 3x$ gives

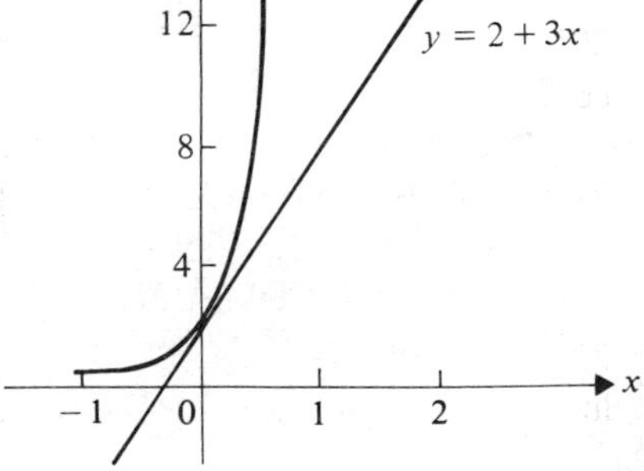

The line $y = 2 + 3x$ is a tangent to the curve $y = e^x + e^{2x}$ at the point $(0, 2)$ because
they both have the same gradient at $x = 0$,
they both pass through $(0, 2)$.

EXERCISE 6c

Expand each of the following functions as a series of ascending powers of x up to and including the term containing x^4. Give the general term and the range of values of x for which the expansion is valid.

1) e^{3x} 2) $\cos \dfrac{x}{2}$ 3) $2 \sin x \cos x$

4) $\ln \sqrt{\left(\dfrac{1-x}{1+x}\right)}$ 5) $\dfrac{e^x + e^{-x}}{e^x}$ 6) $\ln [(1+x)(1-2x)^2]$

7) $\dfrac{x - e^x}{e^x}$ 8) $\cos^2 2x$ 9) $e^{3x} + \ln \sqrt{(1-2x)}$

10) $\sin^3 x$ 11) $\sinh x$ 12) $\cosh x$

Expand each of the following functions as a series of ascending powers of y up to and including the term in y^2.

13) $e^y \ln(1+y)$ 14) $\tan^2 y$ 15) $e^y \cos y$

16) $\sin(a+y)$

Expand each of the following functions as a series of ascending powers of $\dfrac{1}{x}$ up to and including the term in $\left(\dfrac{1}{x}\right)^3$. Give the range of values of x for which the expansion is valid.

17) $\ln\left(\dfrac{x+1}{x}\right)$ 18) $\ln\left(\dfrac{1+2x}{1+x}\right)$ (*Hint*. Divide throughout by x.)

19) Use the expansion of e^x to find the value of e correct to six decimal places.

20) Use the expansion of $\cos x$ to show that when θ is small, $\cos\theta \simeq 1 - \frac{1}{2}\theta^2$

21) Use the expansion of $\ln\left(\dfrac{1+x}{1-x}\right)$ with $x = \frac{1}{3}$ to find $\ln 2$ correct to three decimal places.

22) Use the expansions of $\ln(1+x)$ and $\ln(1-x)$ to find the values of $\ln 1.1$ and $\ln 0.9$ respectively, correct to four decimal places.

23) Find a linear approximation for $f(x) \equiv e^x \cos x$ in the region close to $x = 0$.

24) Use Maclaurin's Series to obtain the expansion of $\arcsin x$ as a series of ascending powers of x as far as the term containing x^3. By substituting $x = \frac{1}{2}$ find the value of π to two significant figures.

25) If x is so small that x^3 and higher powers of x may be neglected, show that $\dfrac{e^x}{1+x} \simeq 1 + \frac{1}{2}x^2$.

26) By using the expansion of $\ln(1+x)$ as a series of ascending powers of x find a quadratic approximation for $\ln(1+x)$ when x is small. Use your approximation to find an approximate value for the smallest root of the equation $\ln(1+x) = x^2$. By sketching the graphs of $f(x) \equiv \ln(1+x)$ and $f(x) \equiv x^2$ illustrate that the value of your root is exact.

27) Expand $\sinh^{-1} x$ as a series of ascending powers of x up to and including the term in x^3.

TAYLOR'S EXPANSION

Maclaurin's expansion cannot be applied to all problems. For instance we have shown that $\ln x$ cannot be expanded as a series of ascending powers of x because $\ln x$ is undefined when $x = 0$. Also the Maclaurin expansion of $\ln(1+x)$ is valid only for $-1 < x \leqslant 1$ and the series so obtained cannot be used to find $\ln(3)$.

Problems such as these can sometimes be overcome by expanding $f(x)$ as a series of ascending powers of $(x-a)$,

i.e. $$f(x) = a_0 + a_1(x-a) + a_2(x-a)^2 + a_3(x-a)^3 + \dots$$

A method similar to that used for obtaining Maclaurin's expansion gives

$$f(x) = f(a) + f'(a)(x-a) + \frac{f''(a)}{2!}(x-a)^2 + \frac{f'''(a)}{3!}(x-a)^3 + \dots$$
$$\dots + \frac{f^r(a)}{r!}(x-a)^r + \dots$$

This is known as Taylor's Expansion and, although we will not discuss the convergence of the series obtained, it can be assumed that any Taylor Series the reader is asked to investigate is convergent for values of x close to a. This series is particularly useful for finding a linear approximation for $f(x)$ when x is close to a.

As $(x-a)$ is small, $(x-a)^2$ and higher powers of $(x-a)$ can be neglected giving

$$f(x) \simeq f(a) + f'(a)(x-a)$$

EXAMPLES 6d

1) Expand $\ln x$ as a series of ascending powers of $(x-a)$ up to the term containing $(x-a)^3$.
Use your expansion to evaluate $\ln 1.01$ correct to five decimal places.

$$f(x) \equiv \ln x \qquad f(a) = \ln a$$

$$f'(x) \equiv \frac{1}{x} \qquad f'(a) = \frac{1}{a}$$

$$f''(x) \equiv -\frac{1}{x^2} \qquad f''(a) = -\frac{1}{a^2}$$

$$f'''(x) \equiv \frac{2}{x^3} \qquad f'''(a) = \frac{2}{a^3}$$

Therefore $\ln x = \ln a + \frac{1}{a}(x-a) - \frac{1}{a^2}\frac{(x-a)^2}{2!} + \frac{2}{a^3}\frac{(x-a)^3}{3!} + \dots$

If $x = 1.01$ and we let $a = 1$, then $(x-a) = 0.01$ which is small.

Hence $\ln 1.01 = \ln 1 + \frac{1}{1}(0.01) - \frac{1}{(1)^2}\frac{(0.01)^2}{2!} + \frac{2}{(1)^3}\frac{(0.01)^3}{3!} + \dots$

$$= 0 + 0.01 - 0.000\,05 + \dots$$

$$= 0.009\,95 \text{ correct to 5 d.p.}$$

2) Show that if $\left(x - \frac{\pi}{4}\right)$ is so small that $\left(x - \frac{\pi}{4}\right)^2$ and higher powers may be neglected

$$\tan x \simeq 1 - \frac{\pi}{2} + 2x$$

Using Taylor's Theorem to expand $\tan x$ as a series of ascending powers of $(x-a)$ we have,

$$f(x) \equiv \tan x \qquad f(a) = \tan a$$

$$f'(x) \equiv \sec^2 x \qquad f'(a) = \sec^2 a$$

So $\tan x = \tan a + (\sec^2 a)(x-a) + \dots$

When $a = \frac{\pi}{4}$ $\quad \tan x \simeq \tan\frac{\pi}{4} + \left(\sec^2\frac{\pi}{4}\right)\left(x - \frac{\pi}{4}\right)$

$$\simeq 1 + (2)\left(x - \frac{\pi}{4}\right)$$

Therefore $\tan x \simeq 1 - \frac{\pi}{2} + 2x$ when $x \simeq \frac{\pi}{4}$

EXERCISE 6d

Use Taylor's Series to expand the following expressions as series of ascending powers of $(x-a)$ up to and including the term in $(x-a)^2$.

1) $\cos x$ 2) e^x 3) $\cosh x$ 4) $\dfrac{1}{x}$

5) $\ln(1+x)$ 6) $\sin x \cos x$

7) Expand $\cos x \ln x$ as a Taylor Series in powers of $(x-1)$ up to and including the term in $(x-1)^2$.

8) Expand $\sin x$ as a series of ascending powers of $(x-\pi/6)$ as far as the third term of the series. Use your series to find an approximate value for $\sin 31°$ given that $1° = 0.017$ rad.

INTEGRATION AND DIFFERENTIATION OF POWER SERIES

If $$f(x) = a_0 + a_1x + a_2x^2 + a_3x^3 + \dots$$

then $$\int_0^x f(x)\,dx = \int_0^x a_0\,dx + \int_0^x a_1x\,dx + \int_0^x a_2x^2\,dx + \dots$$

and $$\frac{d}{dx}f(x) = \frac{d}{dx}a_0 + \frac{d}{dx}(a_1x) + \frac{d}{dx}(a_2x^2) + \dots$$

provided that x lies *inside* the interval for which the series expansion of $f(x)$ is convergent.

Integration can be useful for deriving some series.

For example,

$$\ln(1+x) = \int_0^x (1+x)^{-1}\,dx$$

and $(1+x)^{-1} = 1 - x + x^2 - x^3 + x^4 - \dots$ for $-1 < x < 1$

$$\Rightarrow \quad \ln(1+x) = x - \frac{x^2}{2} + \frac{x^3}{3} - \frac{x^4}{4} + \dots \quad \text{for } -1 < x < 1$$

Note that when this method is used to derive a power series for $\ln(1+x)$, the set of values of x for which we can say that the series converges does not include $x = 1$ although we know, from earlier work, that the series is convergent when $x = 1$.

As a second example consider the expansion of $\cos^{-1}x$.

Now $$\int_0^x \frac{1}{\sqrt{(1-x^2)}}\,dx = \frac{\pi}{2} - \cos^{-1}x$$

and $$(1-x^2)^{-\frac{1}{2}} = 1 + \tfrac{1}{2}x^2 + \tfrac{3}{8}x^4 + \dots \quad \text{for } -1 < x < 1$$

So $$\int_0^x \frac{1}{\sqrt{(1-x^2)}}\,dx = \int_0^x 1\,dx + \int_0^x \tfrac{1}{2}x^2\,dx + \int^x \tfrac{3}{8}x^4\,dx + \ldots$$

$$\Rightarrow \quad \frac{\pi}{2} - \cos^{-1}x = x + \tfrac{1}{6}x^3 + \tfrac{3}{40}x^5 + \ldots \quad \text{for} \quad -1 < x < 1$$

$$\Rightarrow \quad \cos^{-1}x = \frac{\pi}{2} - x - \tfrac{1}{6}x^3 - \tfrac{3}{40}x^5 - \ldots \quad \text{for} \quad -1 < x < 1$$

Recognition of a Number Series as a Power Series

Some number series can be investigated by recognizing their similarity to familiar power series with x replaced by a numerical value.

Consider, for example, the series $\sum_{r=1}^{\infty} \frac{2^r}{r!}$

Now $$\sum_1^{\infty} \frac{2^r}{r!} = 2 + \frac{2^2}{2!} + \frac{2^3}{3!} + \ldots$$

which can be recognized as the expansion of e^x when $x = 2$, but with the first term missing.

Thus $$\sum_{r=1}^{\infty} \frac{2^r}{r!} = e^2 - 1$$

Now consider $\sum_{r=1}^{\infty} rx^r = x + 2x^2 + 3x^3 + \ldots$

We note that the given series is similar to the series

$$1 + x + x^2 + x^3 + x^4 + \ldots$$

and so could, perhaps, be derived from it.

Now $$(1-x)^{-1} = 1 + x + x^2 + x^3 + \ldots \quad \text{for} \quad -1 < x < 1$$

and $$\frac{d}{dx}(1-x)^{-1} = 1 + 2x + 3x^2 + \ldots \quad \text{for} \quad -1 < x < 1$$

$$\Rightarrow \quad x\frac{d}{dx}(1-x)^{-1} = x + 2x^2 + 3x^3 + \ldots \quad \text{for} \quad -1 < x < 1$$

So $\sum_{r=1}^{\infty} rx^r = \frac{x}{(1-x)^2}$ and converges for $-1 < x < 1$.

In a similar way we can investigate a series of the form $\sum \frac{x^r}{r}$ by integrating $1 + x + x^2 + \ldots$.

SUMMARY

The harmonic series, $1+\frac{1}{2}+\frac{1}{3}+\frac{1}{4}+\dots$ is divergent.

The series $\sum_{r=1}^{\infty}\left(\frac{1}{r}\right)^k$ converges when $k>1$ and diverges when $k \leqslant 1$.

The comparison tests for series of positive terms:

1) If $\sum a_r$ converges and
   if $u_n < a_n$ for all values of n after a finite number, then $\sum u_r$ converges

2) If $\sum a_r$ diverges and
   if $u_n > a_n$ for all values of n after a finite number, then $\sum u_r$ diverges.

3) If $\lim_{n\to\infty}\left[\frac{u_n}{a_n}\right] = p$ where p is finite and not zero, then

   either $\sum u_r$ and $\sum a_r$ both converge

   or $\sum u_r$ and $\sum a_r$ both diverge.

Maclaurin's Series:

$$f(x) = f(0)+f'(0)x+\frac{f''(0)x^2}{2!}+\frac{f'''(0)x^3}{3!}+\dots$$

Taylor's Series:

$$f(x) = f(a)+f'(a)(x-a)+\frac{f''(a)}{2!}(x-a)^2+\dots$$

Series expansions of standard functions:

$$(1+x)^n = 1+nx+\frac{n(n-1)}{2!}x^2+\frac{n(n-1)(n-2)}{3!}x^3+\dots$$

for all values of x if $n \in \mathbb{N}$

for $-1<x<1$ if $n \notin \mathbb{N}$.

$$e^x = 1+x+\frac{x^2}{2!}+\frac{x^3}{3!}+\dots+\frac{x^r}{r!}+\dots$$

for all values of x.

$$\ln(1+x) = x-\frac{x^2}{2}+\frac{x^3}{3}-\frac{x^4}{4}+\dots+\frac{(-1)^{r+1}x^r}{r}+\dots$$

for $-1<x\leqslant 1$.

$$\ln(1-x) = -x-\frac{x^2}{2}-\frac{x^3}{3}-\frac{x^4}{4}-\dots-\frac{x^r}{r}-\dots$$

for $-1\leqslant x<1$.

$$\cos x = 1 - \frac{x^2}{2!} + \frac{x^4}{4!} - \frac{x^6}{6!} + \ldots + \frac{(-1)^r x^{2r}}{(2r)!} + \ldots$$

for all values of x.

$$\sin x = x - \frac{x^3}{3!} + \frac{x^5}{5!} - \frac{x^7}{7!} + \ldots + \frac{(-1)^r x^{2r+1}}{(2r+1)!} + \ldots$$

for all values of x.

$$\cosh x = 1 + \frac{x^2}{2!} + \frac{x^4}{4!} + \ldots + \frac{x^{2r}}{(2r)!} + \ldots$$

for all values of x.

$$\sinh x = x + \frac{x^3}{3!} + \frac{x^5}{5!} + \ldots + \frac{x^{2r+1}}{(2r+1)!} + \ldots$$

for all values of x.

MISCELLANEOUS EXERCISE 6

1) Test for convergence the series whose nth term is $\dfrac{2^n}{2^{n-1}+n}$. (U of L)p

2) Show that if $a_n = (n+2)/(n^2+n)$ then a_{n+1} is less than a_n. Show also that, as $n \to \infty$, $a_n \to 0$ and that $\sum a_n$ diverges. (U of L)p

3) Prove that if the series $\sum a_n$ is convergent, a_n must tend to zero as n tends to infinity. Give an example to show that the converse is not true. Find the range of values of x for which the series

$\sum_{r=1}^{\infty} \dfrac{(x-2)^r}{x^r}$ is convergent. (U of L)p

4) Examine for convergence each of the following series.

(a) $\sum_{n=1}^{\infty} \dfrac{n}{2^n}$ (b) $\sum_{n=2}^{\infty} \dfrac{1}{\sqrt{(n^3-1)}}$ (U of L)p

5) Investigate the convergence or divergence of the following series.

(i) $\sum_{n=1}^{\infty} \dfrac{1}{2^n+n}$ (ii) $\sum_{n=1}^{\infty} \dfrac{3^n}{n!}$ (iii) $\sum_{n=1}^{\infty} \dfrac{1+n}{n^2}$ (iv) $\sum_{n=1}^{\infty} \dfrac{1}{ne^n}$

(U of L)p

6) Test for convergence each of the following series.

(a) $\sum_{n=1}^{\infty} \frac{2n-1}{n(n+1)(n+2)}$ (b) $\sum_{n=1}^{\infty} \frac{1+n}{2+n^2}$

(c) $\sum_{n=1}^{\infty} ne^{-n^2}$ (U of L)p

7) (i) Show that the series $\sum_{r=1}^{\infty} \frac{1}{r^k}$ diverges for $0 < k \leqslant 1$, and state the set of real values of k for which the series converges.

(ii) Determine, giving your reasons, whether or not each of the following series converges:

(a) $\sum_{r=1}^{\infty} \frac{\ln r}{r}$, (b) $\sum_{r=1}^{\infty} \frac{\ln r}{r^2}$, (c) $\sum_{r=1}^{\infty} \frac{r}{e^r}$.

(iii) Find the sum of the series:

$$\sum_{r=1}^{\infty} \frac{1}{r(r+2)(r+4)}.$$ (U of L)

8) (i) Show that, for $x > 1$, $\ln x < x$.

Determine whether the series $\sum_{n=1}^{\infty} \frac{\ln n}{2n^3 - 1}$ converges or diverges.

(ii) Find the set of values of p for which the series $\sum_{n=1}^{\infty} n^p x^n$ is convergent

(a) when $0 < x < 1$,
(b) when $x = 1$,
(c) when $x > 1$.

(iii) Show that the series $\sum_{n=1}^{\infty} n \exp(-n^2)$, where $\exp x = e^x$, is convergent. (U of L)

9) In the expansion in ascending powers of x of the function $\frac{1+x}{1+ax} - e^{bx}$ the coefficients of x and x^2 are both zero. Find the two possible pairs of values of the constants a and b. Verify that for one of these pairs of values all coefficients in the expansion are zero. (JMB)

10) (a) Express $\ln\left(\frac{1+x}{1-x}\right)$ as a series of ascending powers of x up to and including the term in x^5.
Prove that $\frac{2}{3}[1 + \frac{1}{3}(\frac{1}{9}) + \frac{1}{5}(\frac{1}{9})^2 + \ldots] = \ln 2$.
(b) Prove that, if x is so small that x^6 and higher powers of x may be neglected, then $\frac{e^{2x}-1}{e^{2x}+1} \simeq x - \frac{x^3}{3} + \frac{2x^5}{15}$. (U of L)

11) If $e^{2y}(1+x) = (1-2x)$ express y as a function of x. Expand y as a series in ascending powers of x up to and including the term in x^3 and state the range of values for which the expansion is valid. (AEB)p '73

12) Write down the first four terms and the rth term u_r in the expansion of e^{-2x} in ascending powers of x.

Find in its simplest form $\frac{u_{r+1}}{u_r}$.

Express $\sum_{r=1}^{\infty} \frac{x^{r-1}u_{r+1}}{u_r}$ in the form $k \ln(1+ax)$ and state the values of a and k. (AEB)p '76

13) Write down the expansions of $\ln(1+x)$ and $\ln(1-x)$ and in each case state the range of values of x for which the expansions are valid.

Deduce that $\ln \frac{y+1}{y-1} = \frac{2}{y} + \frac{2}{3y^3} + \frac{2}{5y^5} + \frac{2}{7y^7} + \ldots$

and state the range of values of y for which the expansion is valid. (AEB)p '75

14) Find the first non-zero term in the expansion of $(1-x)(1-e^x) + \ln(1+x)$ in ascending powers of x. (U of L)

15) If $y = \ln \cos x$, prove that $\frac{d^3y}{dx^3} + 2\frac{d^2y}{dx^2}\frac{dy}{dx} = 0$.

Hence, or otherwise, obtain the Maclaurin expansion of y in terms of x up to and including the term in x^4. Using $x = \frac{\pi}{4}$ show that $\ln 2 \simeq \frac{\pi^2}{16}\left(1 + \frac{\pi^2}{96}\right)$.

16) By expressing x^x as a power of e and using the exponential series, or otherwise, find (without using tables) the value of x^x when $x = 0.01$, correct to three places of decimals. ($\ln 10 = 2.30$ to two decimal places).
(O)

17) If $\sin y = \frac{1}{2}\cos x$, prove that $\dfrac{d^2y}{dx^2} = \tan y\left\{\left(\dfrac{dy}{dx}\right)^2 - 1\right\}$.

If $-\frac{1}{2}\pi < y < \frac{1}{2}\pi$, obtain the expansion of y in ascending powers of x as far as and including the term in x^2, and show that the coefficient of x^3 is zero. (C)

18) If $y = \tan(e^x - 1)$, prove that $\dfrac{d^2y}{dx^2} = \dfrac{dy}{dx}(1 + 2e^x y)$.

Obtain the Maclaurin expansion of y as far as the term in x^4 inclusive. (C)

19) Use Maclaurin's theorem to expand $\ln\cos x$ in a series of ascending powers of x as far as the term in x^4. Check your result by replacing $\cos x$ by $1 - \frac{1}{2}x^2 + \frac{1}{24}x^4$ and using the logarithmic series.
By substituting $x = \frac{1}{4}\pi$, obtain a value for $\ln 2$ to two decimal places, and explain why you might expect a result less than the true value. (C)

20) If $y = \ln\tan x$, find $\dfrac{dy}{dx}$ and $\dfrac{d^2y}{dx^2}$, and prove that

$$\frac{d^3y}{dx^3} = 4(3 + \cos 4x)/\sin^3 2x$$

Obtain the expansion of $\ln\tan(x + \frac{1}{4}\pi)$ in powers of x as far as the term in x^3. (C)

21) Obtain the Taylor expansion of the function $x^2\ln x$ in powers of $(x-1)$ as far as the term in $(x-1)^4$. (U of L)p

22) Find the Taylor expansion of $x\ln x$ about $x = 1$ (i.e. in ascending powers of $(x-1)$), up to and including the term in $(x-1)^5$.
From the expansion, find the value of $x\ln x$ when $x = 1.1$ correct to 5 places of decimals. (O)

23) Prove that, if $-1 < x < 1$, then

$$\tan^{-1}x = x - \tfrac{1}{3}x^3 + \tfrac{1}{5}x^5 - \tfrac{1}{7}x^7 + \dots$$

(You may assume that an infinite series may be integrated term by term, provided that x lies in the range of convergence of the series.)
Find (as a multiple of π) the sum of the infinite series

$$1 - \frac{1}{3\times 3} + \frac{1}{5\times 3^2} - \frac{1}{7\times 3^3} + \dots$$ (O)

CHAPTER 7

DIFFERENTIAL EQUATIONS

There are many physical situations in which different variables increase or decrease at certain rates.
When there is a relationship between these rates of change, it can often be expressed in the form of a differential equation. A direct (i.e. non-differential) relationship between the variables can be found if the differential equation can be solved. So clearly the solution of differential equations is of great importance in many spheres.
There is an endless variety of types of differential equation and the solution of each type has its own special technique.
(There are also some differential equations which cannot be solved to give a direct relationship between the variables.)
Solving a differential equation necessarily involves an integration process and therefore automatically introduces a constant of integration, e.g. a first order differential equation contains $\frac{dy}{dx}$ and therefore is solved by *one* integration operation, thus introducing *one* arbitrary constant.
A second order differential equation contains $\frac{d^2y}{dx^2}$; its solution requires *two* integration processes and therefore contains *two* arbitrary constants.

We saw in *The Core Course* that integration is based on *recognizing* derivatives. Logically then, the solution of some differential equations will also involve recognition.
There are so many different types of differential equation, each with its own solution technique, that the solution of these equations comprises a vast subject in its own right. In this book, however, we are going to study only two further categories of differential equation, one of first order and one of second order (first order linear equations with variables separable were dealt with in *The Core Course*).

FIRST ORDER DIFFERENTIAL EQUATIONS OF THE PRODUCT TYPE

Exact Differential Equations

Knowing that $\dfrac{d}{dx}(uv) = v\dfrac{du}{dx} + u\dfrac{dv}{dx}$, we can recognize the R.H.S. of this formula when it occurs in a differential equation, and quote its integral, uv. For example, consider the differential equation

$$x\frac{dy}{dx} + y = e^x$$

The L.H.S. can be recognized as the derivative w.r.t. x of the product xy. Thus, integrating both sides of the equation with respect to x gives

$$xy = e^x + A$$

A differential equation of this type, in which part of it is the exact derivative of a product, is called an *exact differential equation*.

EXAMPLES 7a

1) Find the general solution of the differential equation

$$x^2 \cos y \frac{dy}{dx} + 2x \sin y = \frac{1}{x^2}$$

The L.H.S. is seen to be the derivative w.r.t. x of $x^2 \sin y$ (the identification of this product is made easier by noting that the term containing $\dfrac{dy}{dx}$ is given by differentiating a term that is a function of y; i.e. $\sin y$ in this case).

Hence, integrating both sides w.r.t. x gives

$$x^2 \sin y = -\frac{1}{x} + A$$

i.e. $$x^3 \sin y = Ax - 1$$

is the general solution of the given differential equation.

2) Find y in terms of x if $y = 0$ when $x = 1$

and $$2ye^x \frac{dy}{dx} + y^2e^x = e^{2x}$$

Rearranging the L.H.S. in the form

$$\left(2y\frac{dy}{dx}\right)(e^x) + (y^2)(e^x)$$

we recognize it as the derivative w.r.t. x of y^2e^x.

Thus
$$y^2e^x = \int e^{2x}\,dx = \tfrac{1}{2}e^{2x} + A$$

$\Rightarrow$
$$y = \pm\sqrt{\left(\frac{e^x}{2} + \frac{A}{e^x}\right)}.$$

But $y = 0$ when $x = 1$, so
$$0 = \frac{e}{2} + \frac{A}{e} \quad \Rightarrow \quad A = -\frac{e^2}{2}$$

Hence
$$y = \pm\sqrt{\left(\frac{e^x}{2} - \frac{e^2}{2e^x}\right)}$$

The reader may have felt tempted to simplify the differential equation in Example 2 above by cancelling e^x, giving
$$2y\frac{dy}{dx} + y^2 = e^x \qquad [1]$$

In this form, however, the L.H.S. is *not* the derivative of a product.
In fact, given equation [1], we would choose to *multiply by* e^x in order to make the L.H.S. exact in form.
The term e^x, which thus makes the L.H.S. of [1] integrable, is called an *integrating factor*. It is often possible to find a suitable integrating factor, either by inspection, or by using the result of the formal method shown below.

The Integrating Factor

Consider a first order differential equation that can be written in the form
$$\frac{dy}{dx} + Fy = G$$

where F and G are both *functions of x only*.
The L.H.S. is not yet exact, but suppose that it becomes exact when it is multiplied by I, a function of x,

then
$$(I)\left(\frac{dy}{dx}\right) + (y)(FI) = GI$$

Comparing the L.H.S. with $v\dfrac{du}{dx} + u\dfrac{dv}{dx}$

we have
$$v = I, \qquad \frac{du}{dx} = \frac{dy}{dx}$$
$$u = y, \qquad \frac{dv}{dx} = FI$$

$$\Rightarrow \qquad \frac{dI}{dx} = FI$$

Now this is a first order linear differential equation with variables separable, so

$$\int \frac{1}{I}\,dI = \int F\,dx$$

$$\Rightarrow \qquad \ln I = \int F\,dx$$

$$\Rightarrow \qquad I = e^{\int F\,dx}$$

Thus we see that $e^{\int F\,dx}$ is an integrating factor for the expression

$$\frac{dy}{dx} + Fy$$

So, *provided that* $e^{\int F\,dx}$ *can be found*, we have:

$$\frac{dy}{dx} + Fy = G \quad \Rightarrow \quad I\frac{dy}{dx} + yIF = IG$$

$$\Rightarrow \quad \int \left(I\frac{dy}{dx} + yIF\right) dx = \int IG\,dx$$

$$\Rightarrow \quad Iy = \int IG\,dx$$

Both I and G are functions of x only, so $\int IG\,dx$ can usually be found.

EXAMPLES 7a (continued)

3) Find a suitable integrating factor and hence solve the differential equation

$$x\frac{dy}{dx} + 3y = \frac{e^x}{x^2}$$

First the equation must be written in the standard form $\dfrac{dy}{dx} + Fy = G$,

i.e. $$\frac{dy}{dx} + \frac{3y}{x} = \frac{e^x}{x^3} \qquad \Rightarrow \qquad F = \frac{3}{x}$$

The integrating factor is I where $I = e^{\int F\,dx}$

$$\int F\,dx = \int \frac{3}{x}\,dx = 3\ln x = \ln x^3$$

$$\Rightarrow \qquad I = e^{\ln x^3} = x^3$$

Multiplying the *standard form* of the given equation by I, we have

$$x^3\frac{dy}{dx} + 3x^2y = e^x$$

Then integrating both sides w.r.t. x gives

$$x^3y = e^x + A$$

Note. A mistake that is easily made, but which must of course be avoided, is to multiply the *original* differential equation, rather than the standard form, by I.

EXERCISE 7a

Find the general solution of each of the following exact differential equations.

1) $x\dfrac{dy}{dx} + y = e^x$

2) $\cos x\dfrac{dy}{dx} - y\sin x = x^2$

3) $\dfrac{x}{y}\dfrac{dy}{dx} + \ln y = x + 1$

4) $\dfrac{1}{x}\dfrac{dy}{dx} - \dfrac{y}{x^2} = \sin x$

5) $e^x y + e^x\dfrac{dy}{dx} = 2$

6) $xe^y\dfrac{dy}{dx} + e^y = e^x$

7) $\ln x\dfrac{dy}{dx} + \dfrac{y}{x} = x\ln x$

8) $(1+x)\dfrac{dy}{dx} + y = x^3$

9) $x\sec^2 y\dfrac{dy}{dx} + \tan y = \tan x$

10) $e^x e^y\dfrac{dy}{dx} + e^x e^y = e^{2x}$

By using integrating factors, solve the following differential equations.

11) $\dfrac{dy}{dx} + 3y = e^{-3x}$

12) $\dfrac{dy}{dx} + y\cot x = \operatorname{cosec} x$

13) $x^2\dfrac{dy}{dx} + xy = x + 1$

14) $\dfrac{dy}{dx} - \dfrac{3y}{x+1} = (x+1)^4$

15) $\tan x\dfrac{dy}{dx} + y = e^x\tan x$

16) $\dfrac{dv}{dt} = t - 2vt$

17) $x\dfrac{dy}{dx} + 2y = \dfrac{\sin x}{x}$

18) $x\dfrac{dy}{dx} = y - x^2e^x$

19) $\dfrac{dr}{d\theta} + r\cot\theta = \sin\theta$

20) $y + x(x-1)\dfrac{dy}{dx} = x^3e^{-x^2}$

21) $x\dfrac{dy}{dx} = y + x^2(\sin x + \cos x)$ and $y = 0$ when $x = \dfrac{\pi}{2}$.

22) $\dfrac{dy}{dx} = x - xy$ and $y = 0$ when $x = 0$.

FORMATION OF SECOND ORDER LINEAR DIFFERENTIAL EQUATIONS

If an equation $y = f(x)$ contains two arbitrary constants, A and B, then by differentiating twice two more equations are produced,

i.e. $$\frac{dy}{dx} = f'(x) \quad \text{and} \quad \frac{d^2y}{dx^2} = f''(x)$$

These two equations, together with the original $y = f(x)$, allow A and B to be eliminated, so forming a second order differential equation. We are now going to consider three types of equation $y = f(x)$, all of which give rise in this way to a linear second order differential equation.

Case (a)

Consider $y = Ae^{2x} + Be^{3x}$

$$\Rightarrow \quad \frac{dy}{dx} = 2Ae^{2x} + 3Be^{3x} = 2y + Be^{3x}$$

$$\Rightarrow \quad \frac{d^2y}{dx^2} = 2\frac{dy}{dx} + 3Be^{3x} = 2\frac{dy}{dx} + 3\left(\frac{dy}{dx} - 2y\right)$$

Thus $\quad y = Ae^{2x} + Be^{3x} \quad \Rightarrow \quad \dfrac{d^2y}{dx^2} - (2+3)\dfrac{dy}{dx} + (2 \times 3)y = 0$

There is a marked similarity between the coefficients of this differential equation and those of the quadratic equation $u^2 - 5u + 6 = 0$
whose roots are 2 and 3.
In order to check that this analogy is general, and not merely coincidence, we now consider the more general equation

$$y = Ae^{\alpha x} + Be^{\beta x}$$

$$\Rightarrow \quad \frac{dy}{dx} = A\alpha e^{\alpha x} + B\beta e^{\beta x}$$

$$\Rightarrow \quad \frac{d^2y}{dx^2} = A\alpha^2 e^{\alpha x} + B\beta^2 e^{\beta x}$$

Eliminating A and B from these three equations gives

$$\frac{d^2y}{dx^2} - (\alpha + \beta)\frac{dy}{dx} + \alpha\beta y = 0$$

which compares with the quadratic equation

$$u^2 - (\alpha + \beta)u + \alpha\beta = 0$$

whose roots are α and β.

This is called the *auxiliary quadratic equation* and it can now be used to *recognize* the general solution of a second order linear differential equation with constant coefficients.

i.e.
$$a\frac{d^2y}{dx^2}+b\frac{dy}{dx}+cy = 0$$

If the auxiliary quadratic equation $au^2+bu+c=0$ has real distinct roots α and β (i.e. $b^2-4ac>0$) then we can quote, by recognition, the solution
$$y = Ae^{\alpha x}+Be^{\beta x}$$

Case (b)

Consider
$$y = e^{2x}(A+Bx)$$

$$\Rightarrow \quad \frac{dy}{dx} = 2y+Be^{2x}$$

$$\Rightarrow \quad \frac{d^2y}{dx^2} = 2\frac{dy}{dx}+2Be^{2x} = 2\frac{dy}{dx}+2\left(\frac{dy}{dx}-2y\right)$$

Thus
$$y = e^{2x}(A+Bx) \Rightarrow \frac{d^2y}{dx^2}-(2+2)\frac{dy}{dx}+(2\times 2)y = 0$$

This time we see that the auxiliary quadratic equation, $u^2-4u+4=0$, has equal roots of value 2.

Again we will check a general example of this type, using
$$y = e^{\alpha x}(A+Bx)$$

$$\Rightarrow \quad \frac{dy}{dx} = \alpha y+Be^{\alpha x}$$

$$\Rightarrow \quad \frac{d^2y}{dx^2} = \alpha\frac{dy}{dx}+B\alpha e^{\alpha x} = \alpha\frac{dy}{dx}+\alpha\left(\frac{dy}{dx}-\alpha y\right)$$

So
$$y = e^{\alpha x}(A+Bx) \Rightarrow \frac{d^2y}{dx^2}-2\alpha\frac{dy}{dx}+\alpha^2y = 0$$

This provides a second form for the solution, by recognition, of the differential equation
$$a\frac{d^2y}{dx^2}+b\frac{dy}{dx}+cy = 0$$

When the auxiliary quadratic equation $au^2+bu+c=0$ has equal roots, α (i.e. $b^2-4ac=0$)

then
$$y = e^{\alpha x}(A+Bx)$$

Case (*c*)

Consider $y = Ae^{2x}\cos(3x+\epsilon)$, where ϵ is an arbitrary constant

$$\Rightarrow \quad \frac{dy}{dx} = 2Ae^{2x}\cos(3x+\epsilon) - 3Ae^{2x}\sin(3x+\epsilon)$$

$$= 2y - 3Ae^{2x}\sin(3x+\epsilon)$$

$$= 2y - 3Ae^{2x}\sin(3x+\epsilon)$$

$$\Rightarrow \quad \frac{d^2y}{dx^2} = 2\frac{dy}{dx} - 6Ae^{2x}\sin(3x+\epsilon) - 9Ae^{2x}\cos(3x+\epsilon)$$

$$= 2\frac{dy}{dx} - 2\left(2y - \frac{dy}{dx}\right) - 9y$$

Thus $\quad y = Ae^{2x}\cos(3x+\epsilon) \;\Rightarrow\; \dfrac{d^2y}{dx^2} - 4\dfrac{dy}{dx} + 13y = 0$

In this case the auxiliary quadratic equation is $u^2 - 4u + 13 = 0$ and it has complex roots $2 \pm 3i$.

Applying the procedure above to the general case when the auxiliary quadratic equation has roots $p \pm qi$ we can show that

if the auxiliary quadratic equation $au^2 + bu + c = 0$ has complex roots $p \pm qi$ (i.e. $b^2 - 4ac < 0$) then we can quote, by recognition, the solution

$$y = Ae^{px}\cos(qx+\epsilon)$$

A reader who is studying mechanics will recognize this form as the equation of damped harmonic motion when $p < 0$, and forced harmonic motion when $p > 0$.

Note that the compound angle identity $A\cos(qx+\epsilon) \equiv B\cos qx + C\sin qx$ provides an alternative form for the solution above,

i.e. $$y = e^{px}(B\cos qx + C\sin qx)$$

SOLUTION OF SECOND ORDER LINEAR DIFFERENTIAL EQUATIONS WITH CONSTANT COEFFICIENTS

From the work done in the preceding section we know that the differential equation

$$a\frac{d^2y}{dx^2} + b\frac{dy}{dx} + cy = 0$$

has a general solution based on the roots of its auxiliary quadratic equation

$$au^2 + bu + c = 0$$

such that

(a) if $b^2 - 4ac > 0$, $y = Ae^{\alpha x} + Be^{\beta x}$

(b) if $b^2 - 4ac = 0$, $y = e^{\alpha x}(A + Bx)$

(c) if $b^2 - 4ac < 0$, $y = Ae^{px}\cos(qx + \epsilon) \equiv e^{px}(B\cos qx + C\sin qx)$

For example,

If $\dfrac{d^2y}{dx^2} + 3\dfrac{dy}{dx} - 4y = 0$, then (using $u^2 + 3u - 4 = 0 \Rightarrow u = -4, 1$)

we have $y = Ae^{-4x} + Be^x$.

If $\dfrac{d^2y}{dx^2} + 6\dfrac{dy}{dx} + 9y = 0$, then (using $u^2 + 6u + 9 = 0 \Rightarrow u = -3$)

we have $y = e^{-3x}(A + Bx)$.

If $\dfrac{d^2y}{dx^2} + 2\dfrac{dy}{dx} + 2y = 0$, then (using $u^2 + 2u + 2 = 0 \Rightarrow -1 \pm i$)

we have $y = Ae^{-x}\cos(x + \epsilon)$.

Note that if the differential equation contains no term in $\dfrac{dy}{dx}$ and the other two terms have the same sign, then the roots of the auxiliary equation are purely imaginary,

e.g. if $$\frac{d^2y}{dx^2} + k^2y = 0$$

then $$u^2 + k^2 = 0 \Rightarrow u = \pm ki$$

The general solution then becomes

$$y = A\cos(kx + \epsilon)$$

(from $y = Ae^{px}\cos(qx + \epsilon)$ when $p = 0$ and $q = k$).
A student of mechanics will notice that this is the equation of simple harmonic motion.

EXERCISE 7b

Write down the general solution for each of the following differential equations.

1) $\dfrac{d^2y}{dx^2} - 3\dfrac{dy}{dx} + 2y = 0$

2) $3\dfrac{d^2y}{dx^2} - 7\dfrac{dy}{dx} + 4y = 0$

3) $\dfrac{d^2y}{dx^2} - 2\dfrac{dy}{dx} + y = 0$

4) $\dfrac{d^2y}{dx^2} + \dfrac{dy}{dx} + y = 0$

5) $\dfrac{d^2y}{dx^2} - 5\dfrac{dy}{dx} + 4y = 0$

6) $\dfrac{d^2y}{dx^2} - 4y = 0$

7) $\dfrac{d^2y}{dx^2} + 4y = 0$ 8) $2\dfrac{d^2y}{dx^2} + \dfrac{dy}{dx} + 2y = 0$

9) $\dfrac{d^2y}{dx^2} - 2\dfrac{dy}{dx} = 0$ 10) $9\dfrac{d^2y}{dx^2} - 6\dfrac{dy}{dx} + y = 0$

The Particular Integral

The second order linear differential equations that we have so far considered have been of the form

$$a\frac{d^2y}{dx^2} + b\frac{dy}{dx} + cy = 0$$

Now we will turn our attention to the form in which the R.H.S. is not zero but is a function of x,

i.e. $$a\frac{d^2y}{dx^2} + b\frac{dy}{dx} + cy = f(x) \qquad [1]$$

Suppose, for example, that we want to solve the equation

$$\frac{d^2y}{dx^2} - 5\frac{dy}{dx} + 6y = e^x$$

The term on the R.H.S. suggests that a solution may be of the form $y = \lambda e^x$. Using this as a *trial solution* we have

$$y = \lambda e^x \quad \Rightarrow \quad \frac{dy}{dx} = \lambda e^x \quad \Rightarrow \quad \frac{d^2y}{dx^2} = \lambda e^x$$

Substituting these expressions into the given equation gives

$$\lambda e^x - 5\lambda e^x + 6\lambda e^x = e^x \quad \Rightarrow \quad \lambda = \tfrac{1}{2}$$

So we see that $y = \frac{1}{2}e^x$ is a solution of the given equation but it cannot be the complete solution because it contains no arbitrary constants.
However it must be part of the complete solution,
and is called a *particular integral* (P.I.).
The remainder of the solution can be found by considering the simpler differential equation

$$\frac{d^2y}{dx^2} - 5\frac{dy}{dx} + 6y = 0$$

whose general solution is $y = Ae^{2x} + Be^{3x}$.

Clearly this solution *alone* does not satisfy the original equation, but, when combining it with the particular integral, we find that

$$\left.\begin{aligned} \text{if} \quad & y = Ae^{2x} + Be^{3x} + \tfrac{1}{2}e^x \\ \text{then} \quad & \frac{dy}{dx} = 2Ae^{2x} + 3Be^{3x} + \tfrac{1}{2}e^x \\ \text{and} \quad & \frac{d^2y}{dx^2} = 4Ae^{2x} + 9Be^{3x} + \tfrac{1}{2}e^x \end{aligned}\right\} \Rightarrow \frac{d^2y}{dx^2} - 5\frac{dy}{dx} + 6y = e^x$$

So the general solution of the given differential equation is

$$y = Ae^{2x} + Be^{3x} + \tfrac{1}{2}e^x$$

which is obtained by adding the *complementary function* $(Ae^{2x} + Be^{3x})$ and the particular integral.

In fact, for all differential equations of the type

$$a\frac{d^2y}{dx^2} + b\frac{dy}{dx} + cy = f(x)$$

the solution is $\quad y = \text{C.F.} + \text{P.I.}$

where the complementary function, C.F., is the solution of the equation

$$a\frac{d^2y}{dx^2} + b\frac{dy}{dx} + cy = 0$$

and the particular integral is *any* solution of the *complete* differential equation.

The Failure Case

If we examine a similar differential equation $\dfrac{d^2y}{dx^2} - 5\dfrac{dy}{dx} + 4y = e^x$, we find that the complementary function is $y = Ae^{4x} + Be^x$ and the trial solution would appear at first sight to be $y = \lambda e^x$.

But λe^x is already included in the term Be^x in the complementary function so $y = \lambda e^x$ satisfies the differential equation when R.H.S. $= 0$ and therefore cannot satisfy the differential equation when R.H.S. $= e^x$.
This fact emerges if $y = \lambda e^x$ is used in the given equation, as we find that $\lambda e^x - 5\lambda e^x + 4\lambda e^x = e^x$ which is inconsistent.
So we need to look for another trial solution.

In this case, when the R.H.S. is $e^{\alpha x}$ (where α is a root of the auxiliary equation)
we use, as a trial solution, $y = \lambda x e^{\alpha x}$.
In the example above,

$$y = \lambda x e^x \quad \Rightarrow \quad \frac{dy}{dx} = \lambda(xe^x + e^x)$$

$$\Rightarrow \quad \frac{d^2y}{dx^2} = \lambda(xe^x + 2e^x)$$

so that $$\lambda(xe^x + 2e^x) - 5\lambda(xe^x + e^x) + 4\lambda xe^x = e^x$$

i.e. $$-3\lambda e^x = e^x \quad \Rightarrow \quad \lambda = -\tfrac{1}{3}$$

Thus the general solution of the given equation is $y = Ae^{4x} + Be^x - \frac{1}{3}xe^x$.

The Complex Index Case

If the R.H.S. of a similar differential equation is of the form $ke^{(p+qi)x}$ then a suitable trial solution is, $y = (\lambda + \mu i)e^{(p+qi)x}$.
For example, to solve the equation

$$\frac{d^2y}{dx^2} + 2\frac{dy}{dx} - 3y = 8e^{(3+2i)x} \qquad [1]$$

we can seek a particular integral by trying

$$y = (\lambda + \mu i)e^{(3+2i)x}$$

$$\Rightarrow \quad \frac{dy}{dx} = (3+2i)(\lambda + \mu i)e^{(3+2i)x}$$

$$\Rightarrow \quad \frac{d^2y}{dx^2} = (3+2i)^2(\lambda + \mu i)e^{(3+2i)x}$$

Substituting these values in [1] gives

$$(3+2i)^2(\lambda + \mu i) + 2(3+2i)(\lambda + \mu i) - 3(\lambda + \mu i) = 8$$

$$\Rightarrow \quad (\lambda - 2\mu) + (2\lambda + \mu)i = 1$$

Hence $$\lambda - 2\mu = 1 \quad \text{and} \quad 2\lambda + \mu = 0$$

$$\Rightarrow \quad \lambda = \tfrac{1}{5} \quad \text{and} \quad \mu = -\tfrac{2}{5}$$

So the P.I. for [1] is $y = \frac{1}{5}(1 - 2i)e^{(3+2i)x}$.
From the auxiliary quadratic equation we find that the C.F. is $Ae^x + Be^{-3x}$.
So the complete solution is

$$y = Ae^x + Be^{-3x} + \tfrac{1}{5}(1 - 2i)e^{(3+2i)x}$$

Note that we assume here that exponential functions with complex indices can be differentiated in the same way as when the index is real.

Further Trial Solutions

When $a\dfrac{d^2y}{dx^2}+b\dfrac{dy}{dx}+cy=f(x)$, certain expressions for $f(x)$ suggest standard trial solutions. If the suggested trial solution is λe^{kx} but e^{kx} is a solution of the differential equation with R.H.S. = 0 then multiply e^{kx} by x, or x^2... if necessary. Other standard trial solutions are given alongside. Their use is demonstrated in the next set of examples.

$f(x)$	Trial solution
p (a constant)	$y=\lambda$
$px+q$	$y=\lambda x+\mu$
px^2+qx+r	$y=\lambda x^2+\mu x+\eta$
$p\sin x$ or $p\cos x$ or $p\sin x+q\cos x$	$y=\lambda\sin x+\mu\cos x$

Calculation of Arbitrary Constants

Because the solution of a second order differential equation contains two arbitrary constants, their evaluation requires two extra facts or initial conditions, e.g. if $\dfrac{d^2y}{dx^2}-3\dfrac{dy}{dx}+2y=0$ and when $x=0$, $y=3$ and $\dfrac{dy}{dx}=5$, the solution is carried out as follows.

From $u^2-3u+2=0$ we have $u=1,2$

so that
$$y = Ae^x+Be^{2x}$$

$\Rightarrow$
$$\frac{dy}{dx} = Ae^x+2Be^{2x}$$

When $x=0$,
$$\begin{cases} y = 3 \Rightarrow 3 = A+B \\ \dfrac{dy}{dx} = 5 \Rightarrow 5 = A+2B \end{cases}$$

Hence $A=1$ and $B=2$.

So
$$y = e^x+2e^{2x}$$

EXAMPLES 7c

1) Solve the equation $\dfrac{d^2y}{dx^2}+3\dfrac{dy}{dx}+2y=10\cos x$

given that $y=1$ and $\dfrac{dy}{dx}=0$ when $x=0$.

First we find the complementary function using

$$u^2+3u+2 = 0 \Rightarrow u = -1,-2$$

i.e. C.F. is
$$y = Ae^{-x}+Be^{-2x}$$

Now for a trial solution we will use

$$y = \lambda \cos x + \mu \sin x$$

$$\Rightarrow \quad \frac{dy}{dx} = -\lambda \sin x + \mu \cos x$$

$$\Rightarrow \quad \frac{d^2y}{dx^2} = -\lambda \cos x - \mu \sin x$$

Then the given equation becomes

$$-\lambda \cos x - \mu \sin x + 3(-\lambda \sin x + \mu \cos x) + 2(\lambda \cos x + \mu \sin x) = 10 \cos x.$$

$$\Rightarrow \quad (\lambda + 3\mu) \cos x - (3\lambda - \mu) \sin x = 10 \cos x$$

Equating coefficients of $\cos x$ and $\sin x$ gives

$$\left.\begin{aligned} \lambda + 3\mu &= 10 \\ 3\lambda - \mu &= 0 \end{aligned}\right\} \Rightarrow \lambda = 1, \quad \mu = 3$$

So the general solution is

$$y = Ae^{-x} + Be^{-2x} + \cos x + 3 \sin x$$

$$\Rightarrow \quad \frac{dy}{dx} = -Ae^{-x} - 2Be^{-2x} - \sin x + 3 \cos x$$

But $y = 1$ and $\dfrac{dy}{dx} = 0$ when $x = 0$, so

$$1 = A + B + 1$$

and

$$\left.\begin{aligned} 1 &= A + B + 1 \\ 0 &= -A - 2B + 3 \end{aligned}\right\} \Rightarrow A = -3, \quad B = 3$$

Hence $y = 3e^{-2x} - 3e^{-x} + \cos x + 3 \sin x$.

2) If $\dfrac{d^2s}{dt^2} + s = t$, $s = 0$ when $t = 0$ and when $t = \dfrac{\pi}{2}$, find s when $t = \dfrac{\pi}{4}$.

The auxiliary equation is $u^2 + 1 = 0 \Rightarrow u = \pm i$.
So the C.F. is $s = A \cos(t + \epsilon)$.
For a trial solution we will use $s = \lambda t$, so that $\dfrac{ds}{dt} = \lambda$ and $\dfrac{d^2s}{dt^2} = 0$.

Then $\qquad 0 + \lambda t = t \Rightarrow \lambda = 1$

So $\qquad s = A \cos(t + \epsilon) + t$

But $\qquad s = 0$ when $t = 0 \Rightarrow A \cos \epsilon = 0$

and $\qquad s = 0$ when $t = \dfrac{\pi}{2} \Rightarrow -A \sin \epsilon + \dfrac{\pi}{2} = 0$

Hence $\epsilon = \frac{\pi}{2}$ and $A = \frac{\pi}{2}$

So $$s = \frac{\pi}{2}\cos\left(t+\frac{\pi}{2}\right)+t$$

When $t = \frac{\pi}{4}$, $$s = \frac{\pi}{2}\cos\left(\frac{3\pi}{4}\right)+\frac{\pi}{4} = \frac{\pi}{4}(1-\sqrt{2})$$

3) Find the general solution of the differential equation

$$\frac{d^2y}{dx^2}-4\frac{dy}{dx}+4y = e^{2x}$$

The auxillary equation is $u^2-4u+4=0 \Rightarrow u=2$.

So the C.F. is $y=Ae^{2x}+Bxe^{2x}$.
Now $y=\lambda e^{2x}$ suggests itself as a trial solution but λe^{2x} is part of the C.F.
Multiplying by x gives $y=\lambda xe^{2x}$ as a trial solution and we find again that λxe^{2x} is part of the C.F.
So λe^{2x} and λxe^{2x} are both solutions of the differential equation with R.H.S. $=0$.

Multiplying by x again gives $y=\lambda x^2e^{2x}$ and using this as a trial solution we have

$$\frac{dy}{dx} = 2\lambda x^2e^{2x}+2\lambda xe^{2x} \quad \text{and} \quad \frac{d^2y}{dx^2} = 4\lambda x^2e^{2x}+8\lambda xe^{2x}+2\lambda e^{2x}$$

then

$$4\lambda x^2e^{2x}+8\lambda xe^{2x}+2\lambda e^{2x}-4(2\lambda x^2e^{2x}+2\lambda xe^{2x})+4\lambda x^2e^{2x} = e^{2x}$$

$$\Rightarrow \qquad 2\lambda = 1 \Rightarrow \lambda = \tfrac{1}{2}$$

So the general solution of the differential equation is

$$y = (A+Bx)e^{2x}+\tfrac{1}{2}x^2e^{2x}$$

Note. In the differential equation

$$a\frac{d^2y}{dx^2}+b\frac{dy}{dx}+cy = f(x)$$

the R.H.S., $f(x)$, can be any function of x. The methods given above for finding the P.I. deal only with certain special forms of $f(x)$ which occur most frequently. Extra help would be needed to find the P.I. when $f(x)$ is not one of those functions whose trial solutions have already been suggested.

EXERCISE 7c

Find the complete solution of the following differential equations.

1) $\dfrac{d^2y}{dx^2}-6\dfrac{dy}{dx}+5y=3$

2) $\dfrac{d^2y}{dx^2}-2\dfrac{dy}{dx}+y=e^{2x}$

3) $\dfrac{d^2y}{dx^2}-2\dfrac{dy}{dx}+y=e^{x}$

4) $\dfrac{d^2y}{dx^2}+3\dfrac{dy}{dx}+2y=\sin x$

5) $\dfrac{d^2y}{dx^2}+\dfrac{dy}{dx}+y=1+x$

6) $\dfrac{d^2y}{dx^2}-4y=3e^{-2x}$

7) $4\dfrac{d^2y}{dx^2}-5\dfrac{dy}{dx}+y=\cos x-\sin x$

8) $9\dfrac{d^2y}{dx^2}+6\dfrac{dy}{dx}+y=x^2+2x+3$

9) $\dfrac{d^2y}{dx^2}+25y=2e^{(1+i)x}$

10) $\dfrac{d^2y}{dx^2}-4\dfrac{dy}{dx}+3y=65\sin 2x$

11) $\dfrac{d^2s}{dt^2}-4\dfrac{ds}{dt}+4s=5$

12) $\dfrac{d^2x}{dt^2}+\dfrac{dx}{dt}+2x=e^{(2-3i)t}$

13) $\dfrac{d^2y}{dx^2}+3\dfrac{dy}{dx}+2y=6e^{x}+\sin x$

14) $\dfrac{d^2\theta}{dt^2}+\theta=5e^{t}\sin t$

(Take $\theta=\lambda e^{t}\sin t+\mu e^{t}\cos t$ as a trial solution in Question 14.)

15) If $y=1$ and $\dfrac{dy}{dx}=12$ when $x=0$, solve the differential equation

$$\frac{d^2y}{dx^2}-3\frac{dy}{dx}-10y=0.$$

16) Find the general solution of the differential equation

$$\frac{d^2y}{dx^2}+8\frac{dy}{dx}+25y=48\cos x-16\sin x.$$

Find also the particular solution for which $y=8$ and $\dfrac{dy}{dx}=-27$ when $x=0$.

17) Find a particular integral $f(x)$ of the differential equation

$$\frac{d^2y}{dx^2}-5\frac{dy}{dx}+6y=(12x-7)e^{-x}$$

in the form $f(x)\equiv(Ax+B)e^{-x}$.
Find the solution of the equation that satisfies the conditions $y=0=\dfrac{dy}{dx}$ when $x=0$.

18) Find the complementary function and a particular integral for the differential equation $\dfrac{d^2y}{dx^2}+4\dfrac{dy}{dx}+4y = 8\sin 2x$.

If, also, $y = 1$ and $\dfrac{dy}{dx} = 0$ when $x = 0$, find the solution of the equation.

SUMMARY

The differential equation $\dfrac{dy}{dx}+Fy = G$ where F and G are functions of x can be solved using an integrating factor I where $I = e^{\int F\,dx}$

The differential equation $a\dfrac{d^2y}{dx^2}+b\dfrac{dy}{dx}+cy = f(x)$ has a complementary function:

(a) $Ae^{\alpha x}+Be^{\beta x}$
if $b^2-4ac > 0$ and the auxiliary equation has roots α, β.

(b) $e^{\alpha x}(A+Bx)$
if $b^2-4ac = 0$ and the auxiliary equation has a repeated root α.

(c) $Ae^{px}\cos(qx+\epsilon)$ or $e^{px}(B\cos qx + C\sin qx)$
if $b^2-4ac < 0$ and $p \pm iq$ are the roots of the auxiliary equation.

MISCELLANEOUS EXERCISE 7

1) Show how the differential equation

$$\frac{dy}{dx}+Py = Q$$

where P and Q are functions of x, can be solved by use of an integrating factor. Find the general solution of the differential equation

$$\frac{dy}{dx}+y\cot x = \cos 3x.$$

(C)

2) Find the general solution of the differential equation

$$\frac{d^2x}{dt^2}-\frac{dx}{dt}-2x = 10\sin t$$

Determine the solution which remains finite as $t\to\infty$ and for which $x = 4$ at $t = 0$. (JMB)

3) (a) Solve the differential equation

$$(1+y^2)\sin 2x + (1+\cos^2 x)\frac{dy}{dx} = 0$$

given that $y = 0$, when $x = \frac{\pi}{2}$.

(b) Solve the differential equation

$$x\frac{dy}{dx} + (1-x)y = x.$$

(AEB '67)

4) If $x = e^t$, show that $x\frac{dy}{dx} = \frac{dy}{dt}$ and $x^2\frac{d^2y}{dx^2} = \frac{d^2y}{dt^2} - \frac{dy}{dt}$.

Use these results to reduce the differential equation

$$x^2\frac{d^2y}{dx^2} + x\frac{dy}{dx} - 4y = 16$$

to a differential equation in y and t, and hence solve it, given that $y = 0$ and $dy/dx = 0$ when $x = 1$. (O)

5) Solve the differential equation

(a) $\sqrt{(3x-x^2)}\frac{dy}{dx} = 1 + \cos 2y$,

(b) $(1-x^2)\frac{dy}{dx} = x(1-2y+x)$. (AEB '69)

6) The three differential equations

$$\frac{d^2x}{dt^2} + x = 0$$

$$\frac{d^2x}{dt^2} - 3\frac{dx}{dt} + 2x = 10\cos t$$

$$\frac{d^2x}{dt^2} + 3\frac{dx}{dt} = -10\cos t$$

are all subject to the conditions that $x = 1$ and $\frac{dx}{dt} = -3$ when $t = 0$.

Find the solutions. (C)

7) Find the curve $y = f(x)$ which passes through the point $(0, 1)$ and satisfies the differential equation

$$\frac{dy}{dx} + y\tan x = 2\sin^2 x\cos x.$$

(JMB)

8) Solve the differential equations

(a) $\dfrac{dy}{dx}+\dfrac{\sqrt{(1+y^2)}}{xy(1+x^2)}=0,$

(b) $\dfrac{dy}{dx}=y+x$ given that $y=3$ when $x=0$. (AEB '66)

9) (a) Find the solution of the differential equation

$$\sin x\frac{dy}{dx}-y\cos x = \sin^2 x\cos x$$

for which $y=2$ when $x=\frac{1}{2}\pi$.

(b) Find the solution of the differential equation

$$\frac{d^2y}{dx^2}+2\frac{dy}{dx}+5y = 2e^{-x}$$

for which $y=0$ when $x=0$ and when $x=\frac{1}{4}\pi$. (O)

10) Solve the equation

$$\frac{dy}{dx}+2y\cot x = \cos x.$$ (AEB'75)p

11) (a) If $(x^2+1)\dfrac{dy}{dx}+4xy=12x^3$, and $y=1$ when $x=1$, express y in terms of x in as simple a form as possible.

(b) If $\dfrac{d^2y}{dx^2}-4\dfrac{dy}{dx}+8y=2e^{2x}$, and $y=0$, $dy/dx=0$ when $x=0$, find y in terms of x. (O)

12) Solve the differential equations

(a) $(1+\cos 2x)\dfrac{dy}{dx}-(1+e^y)\sin 2x=0$ given that $y=0$ when $x=\pi/4$;

(b) $(1+x)e^x\dfrac{dy}{dx}+xe^xy-1=0$. (AEB '73)

13) (i) A particular integral of the differential equation

$$\frac{d^2y}{dx^2}+2\frac{dy}{dx}+5y = 26+15x$$

is given by $y=a+bx$, where a, b are constants. Find the values of a and b and determine the general solution of the differential equation.

(ii) Using the substitution $y=xv$, where v is a function of x, transform the differential equation $x^2\dfrac{dy}{dx} = x^2+xy+y^2$ where $x>0$, into a differential equation relating v and x. Hence find y in terms of x, given that $y=0$ when $x=1$. (U of L)

14) Find the value of the constant a such that $y = axe^{-x}$ is a solution of the differential equation

$$\frac{d^2y}{dx^2} + 3\frac{dy}{dx} + 2y = 2e^{-x}$$

Find the solution of this differential equation for which $y = 1$ and $\frac{dy}{dx} = 3$ when $x = 0$. (U of L)

15) (i) A particular integral of the differential equation

$$3\frac{d^2y}{dx^2} + 8\frac{dy}{dx} + 5y = 9\cos 2x - 23\sin 2x$$

is $P\cos 2x + Q\sin 2x$, where P and Q are constants. Find P and Q and obtain the general solution of the differential equation.

(ii) By differentiating twice, and eliminating the arbitrary constants A and B, obtain the differential equation of which the general solution is

$$y = A\cosh 3x + B\sinh 3x + x^2 + 2$$

(U of L)

16) Find the solution of the differential equation

$$\frac{d^2y}{dx^2} + 9y = 18$$

for which y has a maximum at $(\frac{1}{2}\pi, 6)$. Find the minimum value of y and the values of x for which $y = 0$. (O)

17) At time t the population of the world is x (treated as a continuous variable). The time-rate of increase of x due to births is αx and the time-rate of decrease of x due to deaths is βx. Write down a differential equation relating x and t.
Assuming that $\alpha = 0.060$, $\beta = 0.040$ (t being measured in years), show that the population of the earth will double in 35 years, approximately. (C)p

18) A particle moves on the x axis so that at time t its x coordinate satisfies the differential equation

$$\frac{d^2x}{dt^2} + 3\frac{dx}{dt} + 2x = 6t + 5$$

Given that $x = 11$ and that $\frac{dx}{dt} = -12$ when $t = 0$, prove that the least value of x is $\frac{7}{8} + 3\ln 4$. (C)

19) Find the general solution of the differential equation

$$\frac{d^2x}{dt^2} - \frac{dx}{dt} - 2x = 10 \sin t.$$

Determine the solution which remains finite as $t \to \infty$ and for which $x = 4$ at $t = 0$. (JMB)

CHAPTER 8

COMPLEX NUMBERS

This chapter extends the work on complex numbers started in *The Core Course* and we begin with a brief reminder of some of the results already obtained.

Any complex number $z = x + yi$ can be expressed in the form $r(\cos\theta + i\sin\theta)$ where r is the modulus and θ is the argument of $x + yi$.

Further, $$r = \sqrt{(x^2 + y^2)} \quad \text{and} \quad \tan\theta = \frac{y}{x}$$

where $-\frac{1}{2}\pi < \theta \leqslant \frac{1}{2}\pi$ for the principal argument.

If $\bar{z} = x - yi$, $\bar{z}$ is the conjugate of z

or if $z = r(\cos\theta + i\sin\theta)$ then $\bar{z} = r(\cos\theta - i\sin\theta)$

$$= r[\cos(-\theta) + i\sin(-\theta)]$$

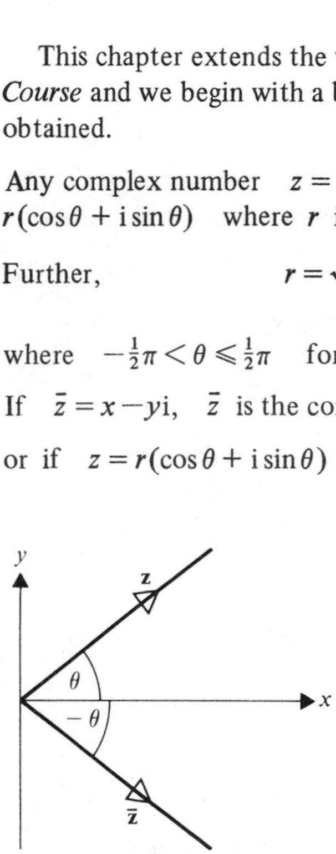

Representing z and $\bar{z}$ on an Argand diagram we see that $\bar{z}$ is the reflection of z in the real axis and z behaves as a vector in the xy plane.

Further, if $z_1 = r_1(\cos\theta_1 + i\sin\theta_1)$ and $z_2 = r_2(\cos\theta_2 + i\sin\theta_2)$

then $|z_1z_2| = r_1r_2$ and $\arg(z_1z_2) = \arg z_1 + \arg z_2$

$|z_1/z_2| = r_1/r_2$ and $\arg(z_1/z_2) = \arg z_1 - \arg z_2$

So $$z_1z_2 = r_1r_2[\cos(\theta_1 + \theta_2) + i\sin(\theta_1 + \theta_2)]$$

and $$\frac{z_1}{z_2} = \frac{r_1}{r_2}[\cos(\theta_1 - \theta_2) + i\sin(\theta_1 - \theta_2)]$$

POWERS OF A COMPLEX NUMBER

If $z = x + yi$ has unit modulus (in which case z behaves as a unit vector in the xy plane) then

$$z = \cos\theta + \mathrm{i}\sin\theta$$

Now consider z^2,

$$|z^2| = |(z)(z)| = 1$$

and $$\arg(z^2) = \arg z + \arg z = 2\theta$$

So $$z^2 = \cos 2\theta + \mathrm{i}\sin 2\theta$$

Similarly considering z^3,

$$|z^3| = |(z^2)(z)| = 1$$

and $$\arg(z^3) = \arg(z^2) + \arg(z) = 2\theta + \theta = 3\theta$$

So $$z^3 = \cos 3\theta + \mathrm{i}\sin 3\theta$$

These two results suggest that, for positive integral values of n,

$$z^n = \cos n\theta + \mathrm{i}\sin n\theta$$

This deduction, known as De Moivre's Theorem, is shown to be correct by using the method of induction.

DE MOIVRE'S THEOREM

De Moivre's Theorem states that, for all real values of n,

$$(\cos\theta + \mathrm{i}\sin\theta)^n \equiv \cos n\theta + \mathrm{i}\sin n\theta$$

Proof when n is a Positive Integer

If p_n is the statement $(\cos\theta + \mathrm{i}\sin\theta)^n \equiv \cos n\theta + \mathrm{i}\sin n\theta$

then p_k is the statement $(\cos\theta + \mathrm{i}\sin\theta)^k \equiv \cos k\theta + \mathrm{i}\sin k\theta$

$$\Rightarrow \quad p_{k+1} \text{ is } (\cos\theta + \mathrm{i}\sin\theta)^{k+1} \equiv (\cos k\theta + \mathrm{i}\sin k\theta)(\cos\theta + \mathrm{i}\sin\theta)$$

$$\equiv (\cos k\theta\cos\theta - \sin k\theta\sin\theta) + \mathrm{i}(\sin k\theta\cos\theta + \cos k\theta\sin\theta)$$

$$\equiv \cos(k+1)\theta + \mathrm{i}\sin(k+1)\theta$$

Hence $$p_k \Rightarrow p_{k+1} \qquad [1]$$

Now p_1 is $(\cos\theta + i\sin\theta)^1 \equiv \cos\theta + i\sin\theta$ which is obviously true. Hence, using [1] above, p_2 is true, p_3 is true, and so on for all positive integral values of n,

e.g. $$z^7 = (\cos\theta + i\sin\theta)^7 = \cos 7\theta + i\sin 7\theta$$

Further, it can be proved that De Moivre's Theorem is valid for any rational value of n.

Proof when n is a Negative Integer

If n is negative, then $n = -m$ where m is a positive integer.

Now $$(\cos\theta + i\sin\theta)^n \equiv (\cos\theta + i\sin\theta)^{-m}$$

$$\equiv \frac{1}{(\cos\theta + i\sin\theta)^m}$$

But m is a positive integer,

so $$(\cos\theta + i\sin\theta)^m \equiv \cos m\theta + i\sin m\theta$$

and $$\frac{1}{\cos m\theta + i\sin m\theta} \equiv \frac{\cos m\theta - i\sin m\theta}{(\cos m\theta + i\sin m\theta)(\cos m\theta - i\sin m\theta)}$$

$$\equiv \cos m\theta - i\sin m\theta$$

$$\equiv \cos(-m\theta) + i\sin(-m\theta)$$

$$\equiv \cos n\theta + i\sin n\theta$$

So, when n is a negative integer,

$$(\cos\theta + i\sin\theta)^n \equiv \cos n\theta + i\sin n\theta$$

In particular when $n = -1$,

$$\frac{1}{z} = z^{-1} = (\cos\theta + i\sin\theta)^{-1}$$

$$= \cos(-\theta) + i\sin(-\theta)$$

$\Rightarrow$ $$\frac{1}{z} = \cos\theta - i\sin\theta = \bar{z}$$

Proof when n is Rational

If n is a rational fraction, $n = \frac{p}{q}$ where p and q are integers.

We wish to prove that

$$(\cos\theta + i\sin\theta)^{p/q} \equiv \cos\frac{p\theta}{q} + i\sin\frac{p\theta}{q}$$

Raising the R.H.S. to power q we have

$$\left(\cos\frac{p\theta}{q} + \mathrm{i}\sin\frac{p\theta}{q}\right)^q \equiv \cos p\theta + \mathrm{i}\sin p\theta \qquad \text{(as } q \text{ is an integer).}$$

Then $$(\cos p\theta + \mathrm{i}\sin p\theta) \equiv (\cos\theta + \mathrm{i}\sin\theta)^p \qquad \text{(as } p \text{ is an integer).}$$

So $$\left(\cos\frac{p\theta}{q} + \mathrm{i}\sin\frac{p\theta}{q}\right)^q \equiv (\cos\theta + \mathrm{i}\sin\theta)^p$$

$\Rightarrow$ $$\cos\frac{p\theta}{q} + \mathrm{i}\sin\frac{p\theta}{q} \equiv (\cos\theta + \mathrm{i}\sin\theta)^{p/q}$$

Hence De Moivre's Theorem applies when n is rational.

Note that $\cos\frac{p\theta}{q} + \mathrm{i}\sin\frac{p\theta}{q}$ is *just one* value of $(\cos\theta + \mathrm{i}\sin\theta)^{p/q}$. There are further values, as subsequent work will show.

Note also that De Moivre's Theorem is, in fact, valid also when n is irrational but we will not attempt a proof at this stage.

Applications of De Moivre's Theorem

Certain trig identities can be derived using De Moivre's Theorem. In particular, expressions such as $\cos n\theta$, $\sin n\theta$, $\tan n\theta$ can be expressed in terms of $\cos\theta$, $\sin\theta$ and $\tan\theta$.
For instance we can find an identity for $\cos 5\theta$ as follows:

$$\begin{aligned}\cos 5\theta &\equiv \mathrm{Re}\,(\cos 5\theta + \mathrm{i}\sin 5\theta)\\ &\equiv \mathrm{Re}\,(\cos\theta + \mathrm{i}\sin\theta)^5 \qquad \text{(De Moivre's Theorem)}\\ &\equiv \mathrm{Re}\,(c^5 + 5c^4\mathrm{i}s + 10c^3\mathrm{i}^2s^2 + 10c^2\mathrm{i}^3s^3 + 5c\mathrm{i}^4s^4 + \mathrm{i}^5s^5)\\ &\qquad\qquad \text{(where } c \equiv \cos\theta \text{ and } s \equiv \sin\theta\text{)}\\ &\equiv c^5 - 10c^3s^2 + 5cs^4\end{aligned}$$

So $$\cos 5\theta \equiv \cos^5\theta - 10\cos^3\theta\sin^2\theta + 5\cos\theta\sin^4\theta$$

If required, the R.H.S. can be expressed entirely in terms of $\cos\theta$ by using $\cos^2\theta + \sin^2\theta \equiv 1$.

Note that the method used above to find an identity for $\cos 5\theta$ provides, at the same time, an identity for $\sin 5\theta$,

i.e. $$\mathrm{i}\sin 5\theta \equiv \mathrm{Im}\,(\cos 5\theta + \mathrm{i}\sin 5\theta)$$

$\Rightarrow$ $$\mathrm{i}\sin 5\theta \equiv 5c^4\mathrm{i}s + 10c^2\mathrm{i}^3s^3 + \mathrm{i}^5s^5$$

$\Rightarrow$ $$\sin 5\theta \equiv 5\cos^4\theta\sin\theta - 10\cos^2\theta\sin^3\theta + \sin^5\theta$$

Further, using the identities so derived for $\cos 5\theta$ and $\sin 5\theta$, $\tan 5\theta$ can also be expressed in terms of $\tan\theta$,

i.e. $$\tan 5\theta \equiv \frac{\sin 5\theta}{\cos 5\theta} \equiv \frac{5\cos^4\theta\sin\theta - 10\cos^2\theta\sin^3\theta + \sin^5\theta}{\cos^5\theta - 10\cos^3\theta\sin^2\theta + 5\cos\theta\sin^4\theta}$$

Then dividing every term by $\cos^5\theta$ gives

$$\tan 5\theta \equiv \frac{5\tan\theta - 10\tan^3\theta + \tan^5\theta}{1 - 10\tan^2\theta + 5\tan^4\theta}$$

The technique demonstrated above can clearly be applied to find similar identities for $\cos n\theta$, $\sin n\theta$ and $\tan n\theta$ for any positive integral value of n.

Properties of z and $\frac{1}{z}$

It has been shown that

if $z = \cos\theta + i\sin\theta$ then $\frac{1}{z} = \cos\theta - i\sin\theta$.

Hence $$z + \frac{1}{z} = 2\cos\theta$$

and $$z - \frac{1}{z} = 2i\sin\theta$$

Further, De Moivre's Theorem shows that

$$\frac{1}{z^n} = (\cos\theta + i\sin\theta)^{-n} \equiv \cos(-n\theta) + i\sin(-n\theta)$$

So $$\frac{1}{z^n} = \cos n\theta - i\sin n\theta$$

Also $$z^n = \cos n\theta + i\sin n\theta$$

Hence $$z^n + \frac{1}{z^n} = 2\cos n\theta$$

and $$z^n - \frac{1}{z^n} = 2i\sin n\theta$$

These relationships are useful for expressing $\cos^n\theta$ or $\sin^n\theta$ in terms of the cosines or sines of multiples of θ.

For example, consider $\cos^4\theta$.

Beginning with the relationship containing $\cos\theta$, i.e. $z+\dfrac{1}{z}=2\cos\theta$,

we have

$$(2\cos\theta)^4 = \left(z+\frac{1}{z}\right)^4$$

$$\Rightarrow \qquad 2^4\cos^4\theta = z^4+4z^3\left(\frac{1}{z}\right)+6z^2\left(\frac{1}{z^2}\right)+4z\left(\frac{1}{z^3}\right)+\frac{1}{z^4}$$

$$= z^4+4z^2+6+\frac{4}{z^2}+\frac{1}{z^4}$$

$$= \left(z^4+\frac{1}{z^4}\right)+4\left(z^2+\frac{1}{z^2}\right)+6$$

Using $z^n+\dfrac{1}{z^n}=2\cos n\theta$ when $n=2$ and $n=4$ on the R.H.S. gives

$$2^4\cos^4\theta \equiv (2\cos 4\theta)+4(2\cos 2\theta)+6$$

So $$\cos^4\theta \equiv \tfrac{1}{8}[\cos 4\theta+4\cos 2\theta+3]$$

Again it is clear that this method can be applied to any positive integral power of $\sin\theta$ or $\cos\theta$. It should be pointed out, however, that when dealing with $\sin^n\theta$ we begin with the relationship $z-\dfrac{1}{z}=2\mathrm{i}\sin\theta$, using

$$(2\mathrm{i}\sin\theta)^n = \left(z-\frac{1}{z}\right)^n$$

In this case extra care must be taken with

(a) the power of i on the L.H.S.,

(b) the signs of the terms on the R.H.S.

EXAMPLES 8a

1) Use De Moivre's Theorem to prove that

$$(\cos p\theta+\mathrm{i}\sin p\theta)(\cos q\theta+\mathrm{i}\sin q\theta) \equiv \cos(p+q)\theta+\mathrm{i}\sin(p+q)\theta.$$

Hence simplify $\dfrac{\cos 3\theta+\mathrm{i}\sin 3\theta}{\cos 5\theta-\mathrm{i}\sin 5\theta}$

De Moivre's Theorem states that

$$(\cos\theta+\mathrm{i}\sin\theta)^n \equiv \cos n\theta+\mathrm{i}\sin n\theta$$

Hence $$\cos p\theta+\mathrm{i}\sin p\theta \equiv (\cos\theta+\mathrm{i}\sin\theta)^p$$

and $$\cos q\theta+\mathrm{i}\sin q\theta \equiv (\cos\theta+\mathrm{i}\sin\theta)^q$$

So $(\cos p\theta + \mathrm{i}\sin p\theta)(\cos q\theta + \mathrm{i}\sin q\theta) \equiv (\cos\theta + \mathrm{i}\sin\theta)^{p+q}$

$$\equiv \cos(p+q)\theta + \mathrm{i}\sin(p+q)\theta.$$

Now
$$\frac{\cos 3\theta + \mathrm{i}\sin 3\theta}{\cos 5\theta - \mathrm{i}\sin 5\theta} \equiv (\cos 3\theta + \mathrm{i}\sin 3\theta)(\cos 5\theta - \mathrm{i}\sin 5\theta)^{-1}$$
$$\equiv (\cos 3\theta + \mathrm{i}\sin 3\theta)(\cos\{-5\theta\} - \mathrm{i}\sin\{-5\theta\})$$
$$\equiv (\cos 3\theta + \mathrm{i}\sin 3\theta)(\cos 5\theta + \mathrm{i}\sin 5\theta)$$
$$\equiv \cos(3+5)\theta + \mathrm{i}\sin(3+5)\theta$$

So
$$\frac{\cos 3\theta + \mathrm{i}\sin 3\theta}{\cos 5\theta - \mathrm{i}\sin 5\theta} \equiv \cos 8\theta + \mathrm{i}\sin 8\theta$$

2) Prove that $\sin^7\theta \equiv \frac{1}{64}(35\sin\theta - 21\sin 3\theta + 7\sin 5\theta - \sin 7\theta)$.

Hence find $\int (35\sin\theta - 64\sin^7\theta)\,d\theta$.

Using $z - \dfrac{1}{z} = 2\mathrm{i}\sin\theta$ gives

$$(2\mathrm{i}\sin\theta)^7 = \left(z - \frac{1}{z}\right)^7$$

$$\Rightarrow \quad 2^7\mathrm{i}^7\sin^7\theta = z^7 - 7z^5 + 21z^3 - 35z + \frac{35}{z} - \frac{21}{z^3} + \frac{7}{z^5} - \frac{1}{z^7}$$

$$\equiv \left(z^7 - \frac{1}{z^7}\right) - 7\left(z^5 - \frac{1}{z^5}\right) + 21\left(z^3 - \frac{1}{z^3}\right) - 35\left(z - \frac{1}{z}\right)$$

But
$$z^n - \frac{1}{z^n} = 2\mathrm{i}\sin n\theta$$

so, using $n = 7, 5, 3$ and 1 we have

$$2^7\mathrm{i}^7\sin^7\theta \equiv (2\mathrm{i}\sin 7\theta) - 7(2\mathrm{i}\sin 5\theta) + 21(2\mathrm{i}\sin 3\theta) - 35(2\mathrm{i}\sin\theta)$$

But $\mathrm{i}^7 = -\mathrm{i}$

So
$$-2^7\mathrm{i}\sin^7\theta \equiv 2\mathrm{i}[\sin 7\theta - 7\sin 5\theta + 21\sin 3\theta - 35\sin\theta]$$

$$\Rightarrow \quad \sin^7\theta \equiv \tfrac{1}{64}[35\sin\theta - 21\sin 3\theta + 7\sin 5\theta - \sin 7\theta]$$

Hence $35\sin\theta - 64\sin^7\theta \equiv 21\sin 3\theta - 7\sin 5\theta + \sin 7\theta$

So

$$\int (35\sin\theta - 64\sin^7\theta)\,d\theta \equiv \int (21\sin 3\theta - 7\sin 5\theta + \sin 7\theta)\,d\theta$$

$$= -7\cos 3\theta + \tfrac{7}{5}\cos 5\theta - \tfrac{1}{7}\cos 7\theta + K$$

3) Show that $\tan 4\theta \equiv \dfrac{4t-4t^3}{1-6t^2+t^4}$ where $t \equiv \tan\theta$.

Use your result to solve the equation

$$t^4+4t^3-6t^2-4t+1 = 0$$

giving your answers correct to three decimal places.

To express $\tan 4\theta$ in terms of $\tan\theta$ we begin by using the complex number

$$\cos 4\theta + \mathrm{i}\sin 4\theta \equiv (c+\mathrm{i}s)^4 \qquad (c \equiv \cos\theta,\quad s \equiv \sin\theta)$$

$$\equiv c^4 + 4c^3\mathrm{i}s + 6c^2\mathrm{i}^2s^2 + 4c\mathrm{i}^3s^3 + \mathrm{i}^4s^4$$

$$\cos 4\theta \equiv \operatorname{Re}(c+\mathrm{i}s)^4 \equiv c^4 - 6c^2s^2 + s^4$$

and $$\mathrm{i}\sin 4\theta \equiv \operatorname{Im}(c+\mathrm{i}s)^4 \equiv \mathrm{i}(4c^3s - 4cs^3)$$

Hence $$\tan 4\theta \equiv \frac{\sin 4\theta}{\cos 4\theta} \equiv \frac{4c^3s-4cs^3}{c^4-6c^2s^2+s^4}$$

Dividing every term on the R.H.S. by c^4 gives

$$\tan 4\theta \equiv \frac{4t-4t^3}{1-6t^2+t^4} \quad \text{where} \quad t \equiv \tan\theta$$

Now if $$t^4+4t^3-6t^2-4t+1 = 0 \qquad [1]$$

then $$\frac{4t-4t^3}{1-6t^2+t^4} = 1$$

$\Rightarrow$ $$\tan 4\theta = 1$$

$\Rightarrow$ $$\theta = \frac{1}{4}\left(n\pi + \frac{\pi}{4}\right)$$

Hence the solutions of equation [1] are the tangents of the angles in the set

$$\theta = \frac{1}{4}\left(n\pi + \frac{\pi}{4}\right)$$

Hence, taking $n = 0, 1, 2, 3,$ we get

$$t = \tan\frac{\pi}{16},\ \tan\frac{5\pi}{16},\ \tan\frac{9\pi}{16},\ \tan\frac{13\pi}{16}$$

i.e. $$t = 0.199,\ 1.497,\ -5.027,\ -0.668$$

Note that further values of n, although they give further values of θ, repeat the values of $\tan\theta$ already found.

EXERCISE 8a

1) Use De Moivre's Theorem to express each of the following complex numbers in the form $\cos n\theta + i \sin n\theta$.

(a) $(\cos\theta + i\sin\theta)^7$ (b) $(\cos\theta + i\sin\theta)^{-3}$

(c) $(\cos\theta + i\sin\theta)^{\frac{1}{2}}$ (d) $\left(\cos\dfrac{\pi}{3} + i\sin\dfrac{\pi}{3}\right)^3$

(e) $\left(\cos\dfrac{\pi}{4} + i\sin\dfrac{\pi}{4}\right)^{-2}$ (f) $(\cos\pi + i\sin\pi)^{\frac{1}{3}}$

2) Express each of the following complex numbers in the form $(\cos\theta + i\sin\theta)^n$.

(a) $\cos 5\theta + i\sin 5\theta$ (b) $\cos 2\theta - i\sin 2\theta$

(c) $\cos\dfrac{\theta}{3} + i\sin\dfrac{\theta}{3}$ (d) $\cos\dfrac{\theta}{2} - i\sin\dfrac{\theta}{2}$

Use De Moivre's Theorem to simplify the following expressions.

3) $(\cos 2\theta + i\sin 2\theta)(\cos 5\theta + i\sin 5\theta)$

4) $\dfrac{\cos 7\theta + i\sin 7\theta}{\cos 2\theta - i\sin 2\theta}$

5) $\dfrac{(\cos 4\theta + i\sin 4\theta)(\cos 3\theta + i\sin 3\theta)}{\cos 5\theta + i\sin 5\theta}$

6) $(\cos 3\theta + i\sin 3\theta)(\cos 6\theta - i\sin 6\theta)$

7) $(\cos 2\theta + i\sin 2\theta)^2(\cos\theta + i\sin\theta)^3$

8) $\dfrac{\cos\theta - i\sin\theta}{\cos 4\theta - i\sin 4\theta}$

9) $\left(\cos\dfrac{\pi}{3} + i\sin\dfrac{\pi}{3}\right)^2\left(\cos\dfrac{2\pi}{3} + i\sin\dfrac{2\pi}{3}\right)^4$

10) $\dfrac{\left(\cos\dfrac{\pi}{4} + i\sin\dfrac{\pi}{4}\right)^5\left(\cos\dfrac{3\pi}{4} + i\sin\dfrac{3\pi}{4}\right)^3}{\left(\cos\dfrac{\pi}{4} - i\sin\dfrac{\pi}{4}\right)^2}$

Write down one value of each of the following expressions.

11) $(\cos 2\theta + i\sin 2\theta)^{\frac{1}{2}}$ 12) $\sqrt[3]{\left(\cos\dfrac{\pi}{2} + i\sin\dfrac{\pi}{2}\right)}$

13) $\sqrt[5]{(\cos\pi + i\sin\pi)}$ 14) $\sqrt[n]{(\cos\theta + i\sin\theta)}$

Prove the following trig identities using methods based on De Moivre's Theorem.

15) $\cos 4\theta \equiv 8\cos^4\theta - 8\cos^2\theta + 1$

16) $\sin^5\theta \equiv \frac{1}{16}(\sin 5\theta - 5\sin 3\theta + 10\sin\theta)$

17) $\tan 6\theta \equiv 2t\left(\dfrac{3 - 10t^2 + 3t^4}{1 - 15t^2 + 15t^4 - t^6}\right)$ where $t \equiv \tan\theta$

18) $\cos^4\theta \equiv \frac{1}{8}(\cos 4\theta + 4\cos 2\theta + 3)$

19) $\sin^6\theta \equiv \frac{1}{32}(10 - 15\cos 2\theta + 6\cos 4\theta - \cos 6\theta)$

20) $\cos^4\theta + \sin^4\theta \equiv \frac{1}{4}(\cos 4\theta + 3)$

21) Prove that $\tan 3\theta \equiv \dfrac{3\tan\theta - \tan^3\theta}{1 - 3\tan^2\theta}$ and *hence* solve the equation $1 - 3t^2 = 3t - t^3$. Give your answers correct to 3 significant figures and verify these results by an algebraic method.

22) Use De Moivre's Theorem to find the following integrals.

(a) $\int 8\cos^4\theta \, d\theta$ (b) $\int (32\cos^6\theta - \cos 6\theta)\, d\theta$

(c) $\int (\cos 4\theta - 8\sin^4\theta)\, d\theta$

CUBE ROOTS OF UNITY

The cube roots of unity are found algebraically in *The Core Course* by solving the equation $z^3 - 1 = 0$. In the same book it is also shown algebraically that each of the complex cube roots of unity is the square of the other, and that the sum of the three cube roots is zero.
We will now use De Moivre's Theorem to find these cube roots and use the Argand diagram to verify the relationship between the roots.

Suppose that $\cos\theta + i\sin\theta$ is one of the cube roots of unity, then

$$(\cos\theta + i\sin\theta)^3 = 1$$

So to find the distinct cube roots of unity we need to solve this equation for values of θ in the range $-\pi < \theta \leqslant \pi$.

Now if $(\cos\theta + i\sin\theta)^3 = 1$ then, using De Moivre's Theorem

$$\cos 3\theta + i\sin 3\theta = 1$$

$$\Rightarrow \quad \cos 3\theta = 1 \quad \text{and} \quad \sin 3\theta = 0$$

$$\Rightarrow \quad 3\theta = 2p\pi$$

$$\Rightarrow \quad \theta = \frac{2p\pi}{3} \quad \text{for integral values of } p.$$

Taking $p = -1, 0, 1$ gives all the possible values of θ in the specified range,
i.e. $\theta = -2\pi/3,\ 0,\ 2\pi/3$.

Therefore there are *three* distinct cube roots of unity,

i.e. $$z_1 = 1$$

$$z_2 = \cos\frac{2\pi}{3} + i\sin\frac{2\pi}{3}$$

$$z_3 = \cos -\frac{2\pi}{3} + i\sin -\frac{2\pi}{3}$$

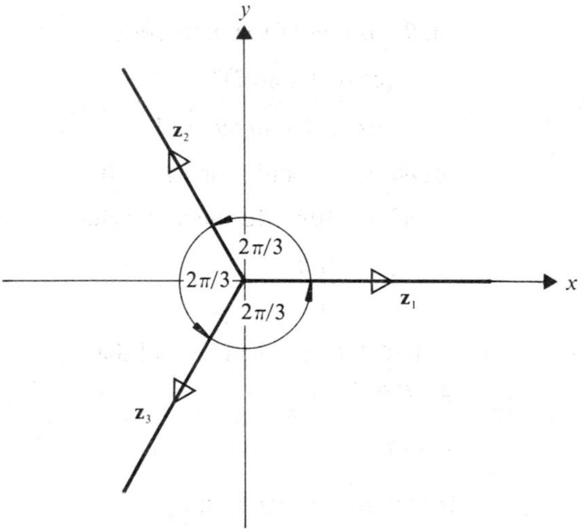

Representing these on an Argand diagram, we note that

(a) the points representing z_1, z_2 and z_3 lie on a circle of radius 1,

(b) the vectors $\mathbf{z}_1, \mathbf{z}_2$ and $\mathbf{z}_3$ are each separated by equal angles of $2\pi/3$,

(c) $2\arg z_2 = 4\pi/3$ or $-2\pi/3$ which is $\arg z_3$
and $2\arg z_3 = -4\pi/3$ or $2\pi/3$ which is $\arg z_2$
i.e. $z_2^2 = z_3$ and $z_3^2 = z_2$.

So if one of the complex roots is denoted by ω, the other is ω^2 i.e. the three cube roots of unity can be denoted by $1, \omega, \omega^2$.

(d) the vectors $\mathbf{z}_1, \mathbf{z}_2, \mathbf{z}_3$ form an equilateral triangle

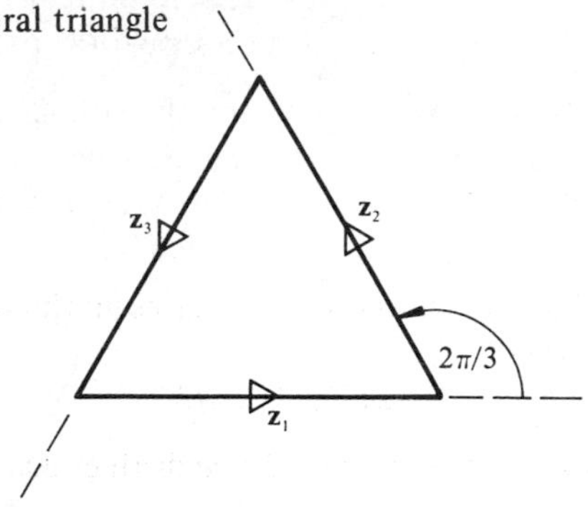

i.e. $\mathbf{z}_1 + \mathbf{z}_2 + \mathbf{z}_3 = 0$

or $1 + \omega + \omega^2 = 0$

(**Note** that simple addition of z_1, z_2 and z_3 also verifies this fact.)

(e) z_3 is the reflection of z_2 in the real axis i.e. ω and ω^2 are conjugate.

The nth Roots of Unity where n is a Positive Integer

The technique used to find the cube roots of unity is now extended to find the nth roots of unity, i.e. the values of $\sqrt[n]{1}$, or the solutions to the equation $z^n - 1 = 0$.

Suppose that $\cos\theta + \mathrm{i}\sin\theta$ is one of the nth roots of unity,

then $$(\cos\theta + \mathrm{i}\sin\theta)^n = 1$$

$\Rightarrow$ $$\cos n\theta + \mathrm{i}\sin n\theta = 1$$

So $$\cos n\theta = 1 \quad \text{and} \quad \sin n\theta = 0$$

$\Rightarrow$ $$n\theta = 2p\pi \quad \text{for integral values of } p.$$

$\Rightarrow$ $$\theta = \frac{2p\pi}{n}$$

Values of θ in the range $0 < \theta \leqslant 2\pi$ will give all the distinct values of $\cos\theta + \mathrm{i}\sin\theta$. So taking $p = 1, 2, \ldots, n$

gives $\theta = \dfrac{2\pi}{n}, \dfrac{4\pi}{n}, \dfrac{6\pi}{n}, \ldots, 2\pi$.

Therefore there are n distinct nth roots of unity,

i.e. $$z_1 = \cos\frac{2\pi}{n} + \mathrm{i}\sin\frac{2\pi}{n}$$

$$z_2 = \cos\frac{4\pi}{n} + \mathrm{i}\sin\frac{4\pi}{n}$$

$$\cdots\cdots\cdots\cdots\cdots\cdots$$

$$z_n = \cos 2\pi + \mathrm{i}\sin 2\pi = 1$$

Now we can see that

each nth root has modulus 1,
one nth root has zero principal argument, i.e. one root is 1,
arguments of successive nth roots differ by $2\pi/n$.

So on an Argand diagram, the points representing the nth roots of unity lie on a circle, radius 1, and are separated by angles $2\pi/n$.

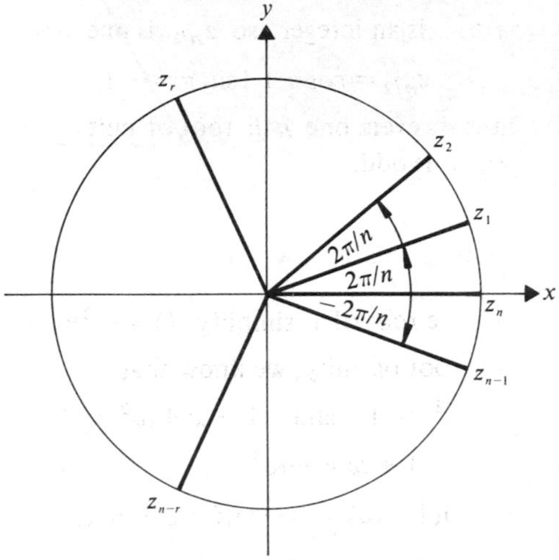

Several interesting relationships between the roots can be observed from the Argand diagram.

(a) $\left.\begin{array}{l}\arg z_2 = 2\arg z_1\\ \arg z_3 = 3\arg z_1\\ \ldots\ldots\ldots\ldots\\ \arg z_r = r\arg z_1\end{array}\right\}$ i.e. $z_2 = z_1^2,\ z_3 = z_1^3,\ \ldots,\ z_r = z_1^r, \ldots$

So if the complex root with smallest positive principal argument is denoted by ω, the other complex roots are $\omega^2, \omega^3, \ldots \omega^{n-1}$ $(\omega^n = 1)$.

(b) The vectors represented by $\mathbf{z}_1, \mathbf{z}_2, \ldots, \mathbf{z}_n$ form a regular n-sided polygon i.e. $\mathbf{z}_1 + \mathbf{z}_2 + \ldots + \mathbf{z}_n = 0$

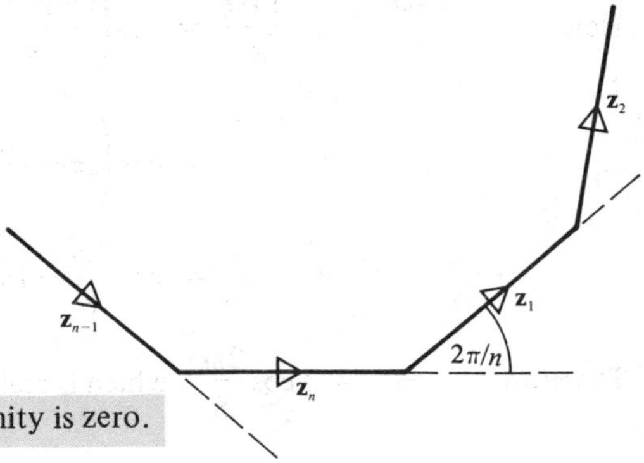

So the sum of the nth roots of unity is zero.

(c) z_{n-1} is the reflection of z_1 in the real axis, i.e. $z_{n-1} = \bar{z}_1$
z_{n-2} is the reflection of z_2 in the real axis, i.e. $z_{n-2} = \bar{z}_2$

....

z_{n-r} is the reflection of z_r in the real axis, i.e. $z_{n-r} = \bar{z}_r$

....

So the nth roots of unity occur in conjugate pairs.

(d) When n is even, $n/2$ is an integer, so $z_{n/2}$ is one nth root of unity where

$$z_{n/2} = \cos\pi + i\sin\pi = -1$$

i.e. when n is even, one nth root of unity is -1
but this is not so when n is odd.

EXAMPLES 8b

1) If ω is a complex cube root of 1, simplify $(1+\omega^2)(1+\omega)$.

If ω is a complex cube root of unity, we know that

$$\omega^3 = 1 \quad \text{and} \quad 1+\omega+\omega^2 = 0$$

So $1+\omega^2 = -\omega$ and $1+\omega = -\omega^2$.

Therefore $\quad (1+\omega^2)(1+\omega) = (-\omega)(-\omega^2) = \omega^3 = 1.$

2) If ω is any complex eighth root of unity show that $\omega + \omega^7$ is real.

There are six complex and two real eighth roots of unity, the real roots being $+1$ and -1.
A general eighth root of 1 has the form

$$\cos\frac{2\pi r}{8} + i\sin\frac{2\pi r}{8}$$

where r is an integer.

If $\omega = \cos\frac{2\pi r}{8} + i\sin\frac{2\pi r}{8}$ then $\omega^7 = \cos\frac{14\pi r}{8} + i\sin\frac{14\pi r}{8}$

$$\text{Now} \quad \cos\frac{14\pi r}{8} + i\sin\frac{14\pi r}{8} = \cos\left(\frac{16\pi r - 2\pi r}{8}\right) + i\sin\left(\frac{16\pi r - 2\pi r}{8}\right)$$

$$= \cos\left(2\pi r - \frac{2\pi r}{8}\right) + i\sin\left(2\pi r - \frac{2\pi r}{8}\right)$$

$$= \cos\frac{2\pi r}{8} - i\sin\frac{2\pi r}{8}$$

Therefore $\omega + \omega^7 = 2\cos\frac{2\pi r}{8}$ which is real.

3) Find the fifth roots of unity. If ω is the root with smallest positive argument and if

$$u = \omega + \omega^4 \quad \text{and} \quad v = \omega^2 + \omega^3,$$

show that

$$u + v = -1 \quad \text{and that} \quad u - v = \sqrt{5}$$

Hence find $\cos 72°$.

Suppose that $\cos\theta + i\sin\theta$ is a fifth root of 1

then $$(\cos\theta + i\sin\theta)^5 = 1$$

$\Rightarrow$ $$\cos 5\theta + i\sin 5\theta = 1$$

$\Rightarrow$ $$\cos 5\theta = 1 \quad \text{and} \quad \sin 5\theta = 0$$

$\Rightarrow$ $$5\theta = 2p\pi \qquad \text{where } p \text{ is an integer.}$$

Solving for θ in the range $0 < \theta \leqslant 2\pi$ to give the distinct values of $\cos\theta + i\sin\theta$, we have

$$\theta = \frac{2\pi}{5}, \frac{4\pi}{5}, \frac{6\pi}{5}, \frac{8\pi}{5}, 2\pi$$

Therefore the fifth roots of unity are

$$\cos\frac{2\pi}{5} + i\sin\frac{2\pi}{5},\ \cos\frac{4\pi}{5} + i\sin\frac{4\pi}{5},\ \cos\frac{6\pi}{5} + i\sin\frac{6\pi}{5},\ \cos\frac{8\pi}{5} + i\sin\frac{8\pi}{5},\ 1$$

The smallest positive argument is $\frac{2\pi}{5}$, so

$$\omega = \cos\frac{2\pi}{5} + i\sin\frac{2\pi}{5}$$

Hence $$\omega^2 = \cos\frac{4\pi}{5} + i\sin\frac{4\pi}{5}$$

$$\omega^3 = \cos\frac{6\pi}{5} + i\sin\frac{6\pi}{5}$$

$$\omega^4 = \cos\frac{8\pi}{5} + i\sin\frac{8\pi}{5}$$

and we know that $$1 + \omega + \omega^2 + \omega^3 + \omega^4 = 0 \qquad [1]$$

Now $$u = \omega + \omega^4 \quad \text{and} \quad v = \omega^2 + \omega^3$$

So $$u + v = \omega + \omega^2 + \omega^3 + \omega^4$$

$$= -1 \qquad \text{from [1]}$$

Also
$$uv = (\omega+\omega^4)(\omega^2+\omega^3)$$
$$= \omega^3(1+\omega^3)(1+\omega)$$
$$= \omega^3(1+\omega+\omega^3+\omega^4)$$
$$= \omega^3(-\omega^2)$$
$$= -1$$

Then
$$(u-v)^2 = (u+v)^2 - 4uv$$
$$= (-1)^2 - 4(-1)$$
$$= 5$$

$\Rightarrow$
$$u - v = \sqrt{5}$$

Now ω and ω^4 are conjugate and so are ω^2 and ω^3.

Hence
$$u = 2\cos\frac{2\pi}{5} \quad\text{and}\quad v = 2\cos\frac{4\pi}{5}$$

$\Rightarrow$
$$u+v = 2\cos\frac{2\pi}{5} + 2\cos\frac{4\pi}{5} = -1$$

and
$$u-v = 2\cos\frac{2\pi}{5} - 2\cos\frac{4\pi}{5} = \sqrt{5}$$

$\Rightarrow$
$$4\cos\frac{2\pi}{5} = \sqrt{5}-1$$

$\Rightarrow$
$$\cos 72^\circ = \tfrac{1}{4}(\sqrt{5}-1)$$

The nth Roots of $r(\cos\alpha + \mathrm{i}\sin\alpha)$

The techniques used to find the nth roots of unity can be extended to find the nth roots of any complex number.

Suppose that $r'(\cos\theta + \mathrm{i}\sin\theta)$ is an nth root of $r(\cos\alpha + \mathrm{i}\sin\alpha)$

then
$$(r')^n(\cos\theta + \mathrm{i}\sin\theta)^n = r(\cos\alpha + \mathrm{i}\sin\alpha)$$

$\Rightarrow$
$$(r')^n(\cos n\theta + \mathrm{i}\sin n\theta) = r(\cos\alpha + \mathrm{i}\sin\alpha)$$

$\Rightarrow$
$$r' = \sqrt[n]{r} \quad\text{and}\quad \cos n\theta + \mathrm{i}\sin n\theta = \cos\alpha + \mathrm{i}\sin\alpha$$

$\Rightarrow$
$$\theta = \frac{\alpha}{n} + \frac{2\pi p}{n}$$

Taking $p = 1, 2, 3, \dots, n$ gives all possible values of θ in the range $\frac{\alpha}{n} < \theta < \frac{\alpha}{n} + 2\pi$ and hence all the distinct values of $\cos\theta + i\sin\theta$.

Hence

> there are n distinct nth roots of $r(\cos\alpha + i\sin\alpha)$
> each nth root has modulus $\sqrt[n]{r}$
> successive arguments differ by $\frac{2\pi}{n}$
> one nth root has argument $\frac{\alpha}{n}$.

So, on an Argand diagram, the points representing the nth roots of $r(\cos\alpha + i\sin\alpha)$ lie on a circle of radius $\sqrt[n]{r}$ and are separated by angles $2\pi/n$.

Further, the vectors representing the nth roots form a regular n-sided polygon, so the sum of the nth roots of z is zero.

However, unless z is real, they do *not* occur in conjugate pairs because they are not symmetrically arranged about the real axis.

The approach in particular cases is demonstrated in the examples that follow.

EXAMPLES 8b (continued)

4) Find the cube roots of $2 + 2i$ and mark this number, and its cube roots, on an Argand diagram. Deduce that the sum of the cube roots is zero.

$|2 + 2i| = \sqrt{8}$ and $\arg(2 + 2i) = \pi/4$

so $$2 + 2i = \sqrt{8}(\cos\pi/4 + i\sin\pi/4)$$

If $r(\cos\theta + i\sin\theta)$ is a cube root of $2 + 2i$

then $$r^3(\cos\theta + i\sin\theta)^3 = \sqrt{8}(\cos\pi/4 + i\sin\pi/4)$$

$$\Rightarrow \quad r^3 = \sqrt{8} \quad \text{and} \quad \cos 3\theta + i\sin 3\theta = \cos\pi/4 + i\sin\pi/4$$

$$\Rightarrow \quad r = \sqrt{2} \quad \text{and} \quad 3\theta = \frac{\pi}{4} + 2p\pi \Rightarrow \theta = \frac{\pi}{12} + \frac{2p\pi}{3}$$

$$\Rightarrow \quad \theta = \frac{\pi}{12}, \frac{3\pi}{4}, \frac{17\pi}{12} \quad \text{for} \quad \frac{\pi}{12} \leqslant \theta < \frac{\pi}{12} + 2\pi$$

So there are three distinct cube roots of $2 + 2i$ given by

$$z_1 = \sqrt{2}\left(\cos\frac{\pi}{12} + i\sin\frac{\pi}{12}\right)$$

$$z_2 = \sqrt{2}\left(\cos\frac{3\pi}{4} + i\sin\frac{3\pi}{4}\right)$$

$$z_3 = \sqrt{2}\left(\cos\frac{17\pi}{12} + i\sin\frac{17\pi}{12}\right) = \sqrt{2}\left[\cos\left(\frac{-7\pi}{12}\right) + i\sin\left(\frac{-7\pi}{12}\right)\right]$$

The modulus of each cube root is $\sqrt{2}$, so the points P_1, P_2, P_3 representing z_1, z_2, z_3 lie on a circle, centre O, radius $\sqrt{2}$.

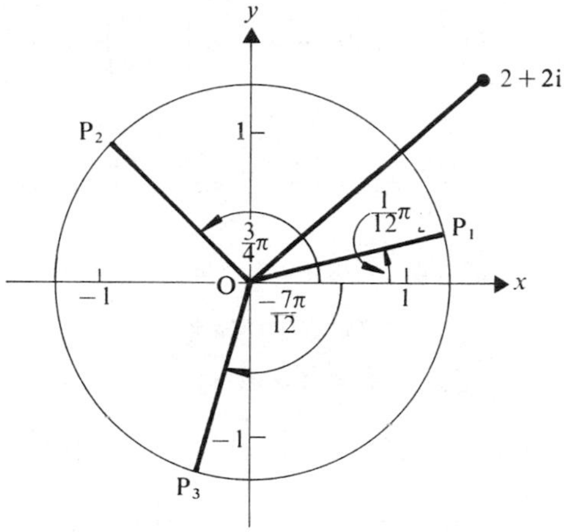

Again $\angle P_1OP_2 = \angle P_2OP_3 = \angle P_3OP_1 = \frac{2}{3}\pi$.
So the vectors represented by $\overrightarrow{OP_1}$, $\overrightarrow{OP_2}$, $\overrightarrow{OP_3}$ form an equilateral triangle and therefore their vector sum is zero

so $$z_1 + z_2 + z_3 = 0$$

5) Find the fifth roots of -1 and hence prove that $\cos\frac{3}{5}\pi + \cos\frac{1}{5}\pi = \frac{1}{2}$.

The modulus and argument of -1, are 1 and π,

so $$-1 = \cos\pi + i\sin\pi$$

If $\cos\theta + i\sin\theta$ is a fifth root of -1

then $$(\cos\theta + i\sin\theta)^5 = \cos\pi + i\sin\pi$$

$\Rightarrow$ $$\cos 5\theta + i\sin 5\theta = \cos\pi + i\sin\pi$$

$\Rightarrow$ $$5\theta = \pi + 2p\pi \quad \Rightarrow \quad \theta = \frac{(2p+1)\pi}{5}$$

Values of θ in the range $\frac{\pi}{5} \leqslant \theta < \frac{11\pi}{5}$ give the distinct values of $\cos\theta + i\sin\theta$.

So $$\theta = \frac{\pi}{5}, \frac{3\pi}{5}, \pi, \frac{7\pi}{5}, \frac{9\pi}{5}$$

Therefore the fifth roots of -1 are given by

$$z_1 = \cos\frac{\pi}{5} + \mathrm{i}\sin\frac{\pi}{5}$$

$$z_2 = \cos\frac{3\pi}{5} + \mathrm{i}\sin\frac{3\pi}{5}$$

$$z_3 = \cos\pi + \mathrm{i}\sin\pi = -1$$

$$z_4 = \cos\frac{7\pi}{5} + \mathrm{i}\sin\frac{7\pi}{5} = \cos\left(-\frac{3\pi}{5}\right) + \mathrm{i}\sin\left(-\frac{3\pi}{5}\right)$$

$$z_5 = \cos\frac{9\pi}{5} + \mathrm{i}\sin\frac{9\pi}{5} = \cos\left(-\frac{\pi}{5}\right) + \mathrm{i}\sin\left(-\frac{\pi}{5}\right)$$

Now the sum of the complex nth roots of any number is zero,

so $$z_1 + z_2 + z_3 + z_4 + z_5 = 0$$

$\Rightarrow$ $$\mathrm{Re}(z_1 + z_2 + z_3 + z_4 + z_5) = 0$$

Note also that $\mathrm{Im}(z_1 + z_2 + z_3 + z_4 + z_5) = 0$ but it is not relevant here.

$\Rightarrow$ $$\cos\frac{\pi}{5} + \cos\frac{3\pi}{5} + (-1) + \cos\left(-\frac{\pi}{5}\right) + \cos\left(-\frac{3\pi}{5}\right) = 0$$

But $\cos\left(-\frac{\pi}{5}\right) = \cos\frac{\pi}{5}$ and $\cos\left(-\frac{3\pi}{5}\right) = \cos\frac{3\pi}{5}$

So $$2\cos\frac{\pi}{5} + 2\cos\frac{3\pi}{5} - 1 = 0$$

i.e. $$\cos\frac{1}{5}\pi + \cos\frac{3}{5}\pi = \frac{1}{2}$$

Note that $z_5 = \bar{z}_1$ and $z_4 = \bar{z}_2$, i.e. the complex fifth roots of -1 occur in conjugate pairs.

EXERCISE 8b

1) Use De Moivre's Theorem to find the square roots of:
(a) $1 - \mathrm{i}$ (b) $3 + 4\mathrm{i}$ (c) $-5 + 12\mathrm{i}$ (d) $\sqrt{3} + \mathrm{i}$
(e) $-2 - 2\mathrm{i}$ (f) i (g) -1.

2) Find the cube roots of each complex number given in Question 1.

3) Without first calculating them, illustrate on an Argand diagram, the nth roots of unity where n is
(a) 2 (b) 3 (c) 4 (d) 5 (e) 6 (f) 7 (g) 8.

4) Without first calculating them, illustrate the fifth roots of z on an Argand diagram, where z is
(a) 1 (b) 32 (c) $9 - 9\sqrt{2}i$ (d) $-1 + \sqrt{3}i$.

5) If ω is a complex cube root of unity, simplify

$$(1 + 6\omega)(1 + 6\omega^2).$$

6) If α is a complex fifth root of unity, simplify

$$(1 + \alpha^4)(1 + \alpha^2)(1 + \alpha).$$

7) If β is a complex fourth root of unity, simplify

$$(1 + \beta)(1 + \beta^3).$$

8) Find the cube roots of -1. Show that they can be denoted by $-1, \lambda, -\lambda^2$ and prove that $\lambda^2 - \lambda + 1 = 0$.

9) If ω is a complex cube root of unity, simplify

$$\begin{vmatrix} a & \omega b & \omega^2 c \\ \omega^2 b & c & \omega a \\ \omega c & \omega^2 a & b \end{vmatrix}.$$

10) By considering the ninth roots of unity show that

$$\cos\frac{2\pi}{9} + \cos\frac{4\pi}{9} + \cos\frac{6\pi}{9} + \cos\frac{8\pi}{9} = -\frac{1}{2}.$$

11) Prove that the fifth roots of 1 can be denoted by $1, \alpha, \alpha^2, \alpha^3, \alpha^4$ and show that $1 + \alpha + \alpha^2 + \alpha^3 + \alpha^4 = 0$.

12) Without using any series expansions, prove that

$$(\sqrt{3} + i)^n + (\sqrt{3} - i)^n \text{ is real.}$$

Find the value of this expression when $n = 12$.

13) By considering the seventh roots of -1 prove that

$$\cos\frac{\pi}{7} + \cos\frac{3\pi}{7} + \cos\frac{5\pi}{7} = \frac{1}{2}.$$

THE EXPONENTIAL FORM FOR A COMPLEX NUMBER

Because any complex number can be given in the form $r(\cos\theta + i\sin\theta)$, other ways of writing a complex number can be found by expressing $\cos\theta$ and/or $\sin\theta$ in alternative forms. A particularly interesting form is found from the series expansions of $\sin\theta$ and $\cos\theta$.

$$\cos\theta = 1 - \frac{\theta^2}{2!} + \frac{\theta^4}{4!} - \frac{\theta^6}{6!} + \ldots$$

$$\sin\theta = \theta - \frac{\theta^3}{3!} + \frac{\theta^5}{5!} - \ldots$$

Adding these series together gives

$$\cos\theta + \sin\theta = 1 + \theta - \frac{\theta^2}{2!} - \frac{\theta^3}{3!} + \frac{\theta^4}{4!} + \frac{\theta^5}{5!} - \ldots$$

This series is very similar to the expansion of e^θ,

i.e.
$$e^\theta = 1 + \theta + \frac{\theta^2}{2!} + \frac{\theta^3}{3!} + \frac{\theta^4}{4!} + \frac{\theta^5}{5!} + \ldots$$

The discrepancy is that the signs of *pairs of consecutive terms* alternate. As this type of sign change is a property of powers of i,

i.e. $\quad i, i^2, i^3, i^4, i^5, i^6, i^7 \ldots = i, -1, -i, 1, i, -1, -i \ldots$

it suggests examining the expansion of $e^{i\theta}$.

$$\begin{aligned} e^{i\theta} &= 1 + i\theta + \frac{(i\theta)^2}{2!} + \frac{(i\theta)^3}{3!} + \frac{(i\theta)^4}{4!} + \frac{(i\theta)^5}{5!} + \ldots \\ &= 1 + i\theta - \frac{\theta^2}{2!} - \frac{i\theta^3}{3!} + \frac{\theta^4}{4!} + \frac{i\theta^5}{5!} - \ldots \\ &= 1 - \frac{\theta^2}{2!} + \frac{\theta^4}{4!} - \ldots \\ &\quad + i\left(\theta - \frac{\theta^3}{3!} + \frac{\theta^5}{5!} - \ldots\right) \end{aligned}$$

Comparing this series with those for $\cos\theta$ and $\sin\theta$ we see that

$$e^{i\theta} \equiv \cos\theta + i\sin\theta$$

Thus any complex number can be written in any of the forms

$$z = x + yi = r(\cos\theta + i\sin\theta) = re^{i\theta}$$

Most of the work carried out earlier in this chapter using $r(\cos\theta + i\sin\theta)$ can now be done using $re^{i\theta}$, and some readers may prefer the brevity that this form offers.

For instance, to find the cube roots of $4+3i$, whose modulus is 5 and whose argument is α where $\tan\alpha = 3/4$, we can use

$$4+3i = 5e^{i\alpha}$$

Then, if $re^{i\theta}$ is any cube root of $5e^{i\alpha}$,

$$r^3e^{3i\theta} = 5e^{i\alpha}$$

$$\Rightarrow \qquad r = \sqrt[3]{5} \quad \text{and} \quad 3\theta = \alpha + 2p\pi$$

$$\Rightarrow \qquad \theta = \frac{\alpha}{3} + \frac{2p\pi}{3}$$

$$= \frac{\alpha}{3}, \frac{\alpha+2\pi}{3}, \frac{\alpha+4\pi}{3}$$

for values of θ in the range $\frac{\alpha}{3} \leqslant \theta \leqslant \frac{\alpha}{3} + 2\pi$.

So the cube roots of $5e^{i\alpha}$ are $\sqrt[3]{5}e^{i\alpha/3}$, $\sqrt[3]{5}e^{i(\alpha+2\pi)/3}$, $\sqrt[3]{5}e^{i(\alpha+4\pi)/3}$.

Note that when using $re^{i\theta}$, θ *must* be given in radians, because the series expansions for $\sin\theta$ and $\cos\theta$ are valid only for an angle measured in radians.

EXAMPLES 8c

1) Express in the form $re^{i\theta}$, each of the complex numbers

$$z_1 = 1+i, \quad z_2 = \sqrt{3}-i, \quad z_3 = \frac{1+i}{\sqrt{3}-i}, \quad z_4 = (1+i)(\sqrt{3}-i)$$

and represent them on an Argand diagram.

$$|z_1| = \sqrt{2} \quad \text{and} \quad \arg z_1 = \frac{\pi}{4} \quad \Rightarrow \quad z_1 = \sqrt{2}e^{i\frac{\pi}{4}}$$

$$|z_2| = 2 \quad \text{and} \quad \arg z_2 = -\frac{\pi}{6} \quad \Rightarrow \quad z_2 = 2e^{-i\frac{\pi}{6}}$$

$$\left.\begin{aligned} |z_3| &= \left|\frac{z_1}{z_2}\right| = \frac{\sqrt{2}}{2} \\ \arg z_3 &= \arg\left(\frac{z_1}{z_2}\right) = \arg z_1 - \arg z_2 = \frac{5\pi}{12} \end{aligned}\right\} \Rightarrow \quad z_3 = \frac{\sqrt{2}}{2}e^{i\frac{5}{12}\pi}$$

$$\left.\begin{aligned} |z_4| &= |z_1|\,|z_2| = 2\sqrt{2} \\ \arg z_4 &= \arg z_1 + \arg z_2 = \frac{\pi}{12} \end{aligned}\right\} \Rightarrow \quad z_4 = 2\sqrt{2}e^{i\frac{\pi}{12}}$$

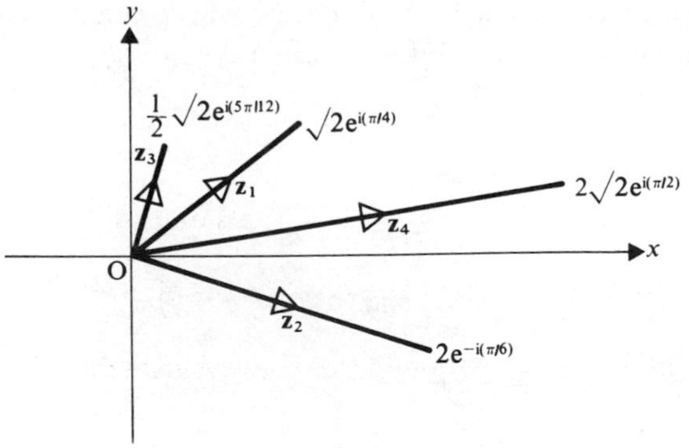

2) Without integrating by parts, find $\int_0^{\frac{\pi}{2}} e^\theta \cos\theta \, d\theta$.

$$\cos\theta \equiv \text{Re}(\cos\theta + i\sin\theta) \equiv \text{Re}(e^{i\theta})$$

Then
$$I = \text{Re}\int_0^{\frac{\pi}{2}} e^\theta e^{i\theta}\, d\theta = \text{Re}\int_0^{\frac{\pi}{2}} e^{(1+i)\theta}\, d\theta$$

$$= \text{Re}\left[\frac{1}{1+i}e^{(1+i)\theta}\right]_0^{\frac{\pi}{2}}$$

$$= \text{Re}\left[\frac{e^\theta}{1+i}(\cos\theta + i\sin\theta)\right]_0^{\frac{\pi}{2}}$$

$$= \text{Re}\left[\frac{(1-i)}{2}e^\theta(\cos\theta + i\sin\theta)\right]_0^{\frac{\pi}{2}}$$

$$= \left[\tfrac{1}{2}e^\theta \cos\theta + \tfrac{1}{2}e^\theta \sin\theta\right]_0^{\frac{\pi}{2}}$$

$$= \tfrac{1}{2}e^{\frac{\pi}{2}} - \tfrac{1}{2}$$

i.e.
$$I = \tfrac{1}{2}(e^{\frac{\pi}{2}} - 1)$$

Note. Assumptions are made in this solution that integration of complex functions is carried out in the same way as for real functions.

EXERCISE 8c

1) Express in the form $re^{i\theta}$

(a) $1 + i$ (b) i (c) $2 - 2\sqrt{3}i$ (d) $-1 + i$

(e) 4 (f) $3 + 4i$ (g) $\dfrac{1+\sqrt{3}i}{1-\sqrt{3}i}$.

2) Express in the form $a + bi$,
(a) $e^{-i\frac{\pi}{3}}$ (b) $2e^{i\frac{5\pi}{6}}$ (c) $5e^{i\pi}$ (d) $e^{-i\frac{\pi}{2}}$ (e) $4e^{-i\pi}$.

Find, in the form $re^{i\theta}$,

3) the cube roots of -1, 4) the fifth roots of $-4-4i$,

5) the seventh roots of 1, 6) the fourth roots of $3-\sqrt{7}i$.

Use the relationship $e^{i\theta} = \cos\theta + i\sin\theta$ to evaluate the following integrals.

7) $\int_0^{\frac{\pi}{2}} e^{-\theta}\sin\theta\, d\theta$ 8) $\int_0^{\frac{\pi}{4}} e^{\theta}\cos 2\theta\, d\theta$

9) Express $\cos\theta$ and $\sin\theta$ each in terms of $e^{i\theta}$ and $e^{-i\theta}$.
Use your results to prove that

$$16\cos^3\theta\sin^2\theta \equiv 2\cos\theta - \cos 3\theta - \cos 5\theta.$$

RELATIONSHIPS BETWEEN HYPERBOLIC AND TRIGONOMETRIC FUNCTIONS

Examination of the series expansions of $\cos x$, $\sin x$, $\cosh x$ and $\sinh x$ discloses a number of interesting relationships connecting these functions and the exponential function.

$$\cos x = 1 - \frac{x^2}{2!} + \frac{x^4}{4!} - \frac{x^6}{6!} + \ldots \qquad [1]$$

$$\cosh x = 1 + \frac{x^2}{2!} + \frac{x^4}{4!} + \frac{x^6}{6!} + \ldots \qquad [2]$$

$$\sin x = x - \frac{x^3}{3!} + \frac{x^5}{5!} - \frac{x^7}{7!} + \ldots \qquad [3]$$

$$\sinh x = x + \frac{x^3}{3!} + \frac{x^5}{5!} + \frac{x^7}{7!} + \ldots. \qquad [4]$$

Remembering that $i^{4n} = 1$ and $i^{(4n-2)} = -1$
it is clear that replacing x by ix in [1] gives

$$\cos ix = 1 + \frac{x^2}{2!} + \frac{x^4}{4!} + \frac{x^6}{6!} + \ldots$$

$\Rightarrow$ $$\cos ix = \cosh x$$

Replacing x by ix in [3] gives

$$\sin ix = i\left\{x + \frac{x^3}{3!} + \frac{x^5}{5!} + \frac{x^7}{7!} + \dots\right\}$$

$$\Rightarrow \qquad \sin ix = i \sinh x$$

If, however, x is replaced by ix in equations [2] and [4] we find that

$$\cosh ix = 1 - \frac{x^2}{2!} + \frac{x^4}{4!} - \frac{x^6}{6!} + \dots$$

$$= \cos x$$

$$\Rightarrow \qquad \cosh ix = \cos x$$

$$\sinh ix = i\left\{x - \frac{x^3}{3!} + \frac{x^5}{5!} - \frac{x^7}{7!} + \dots\right\}$$

$$\Rightarrow \qquad \sinh ix = i \sin x$$

The two latter relationships present yet another way of denoting a complex number,

i.e. $$\cos\theta + i\sin\theta = \cosh i\theta + \sinh i\theta.$$

OSBORN'S RULE

The reader will recall that, when applying Osborn's Rule to convert a trig identity into a hyperbolic identity, every term containing $\sin^2\theta$ becomes $-\sinh^2 x$.

The reason for this change of sign can now be appreciated by considering the relationships

$$\begin{cases} \cos ix = \cosh x \\ \sin ix = i \sinh x \end{cases}$$

Replacing ix by θ gives

$$\begin{cases} \cos\theta = \cosh x \\ \sin\theta = i\sinh x \end{cases}$$

Now any term containing $\sin^2\theta$ becomes $(i \sinh x)^2$ and any term containing $\cos^2\theta$ becomes $(\cosh x)^2$,

i.e. $$\begin{cases} \sin^2\theta = -\sinh^2 x \\ \cos^2\theta = \cosh^2 x, \end{cases}$$

e.g. $$\cos^2\theta + \sin^2\theta \equiv 1 \quad \Rightarrow \quad \cosh^2 x - \sinh^2 x \equiv 1$$

Exponential Form for Trigonometric Functions

Using the relationships

$$\cosh ix = \cos x$$

$$\sinh ix = i \sin x$$

and the definitions

$$\cosh ix \equiv \tfrac{1}{2}(e^{ix} + e^{-ix})$$

$$\sinh ix \equiv \tfrac{1}{2}(e^{ix} - e^{-ix})$$

we have

$$\cos x = \tfrac{1}{2}(e^{ix} + e^{-ix})$$

$$\sin x = \frac{1}{2i}(e^{ix} - e^{-ix})$$

All the relationships derived in this section, and others that the reader may produce, are based on the assumption that the expansions of trig, hyperbolic and exponential functions have the same form for an imaginary variable as for a real variable. In fact at this stage we are actually *defining* trig, hyperbolic and exponential functions of an imaginary variable.

LOCI

So far in this Chapter we have concentrated on the algebraic aspects of complex numbers. But remembering that in the Argand diagram, $z = x + yi$ behaves as the position vector, **z**, of the point $P(x, y)$, we see that complex numbers can be used to describe lines and curves in a plane.
When $z = x + yi$, the point $P(x, y)$ can be anywhere on the Argand diagram. But if a special condition is imposed on z, this affects the possible positions of P.
For instance, if $|z| = 4$ the line OP is of constant length 4 units.

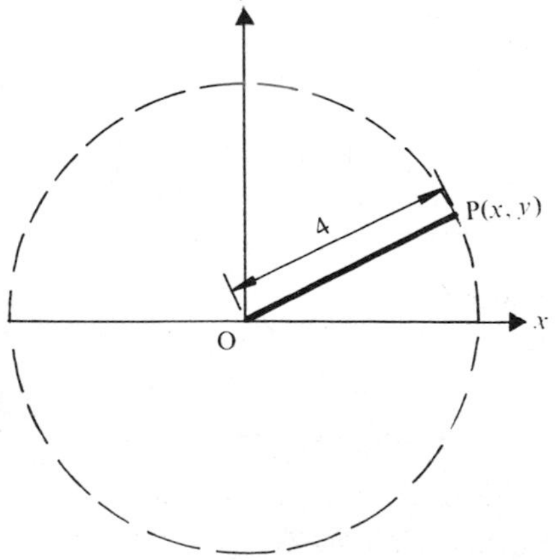

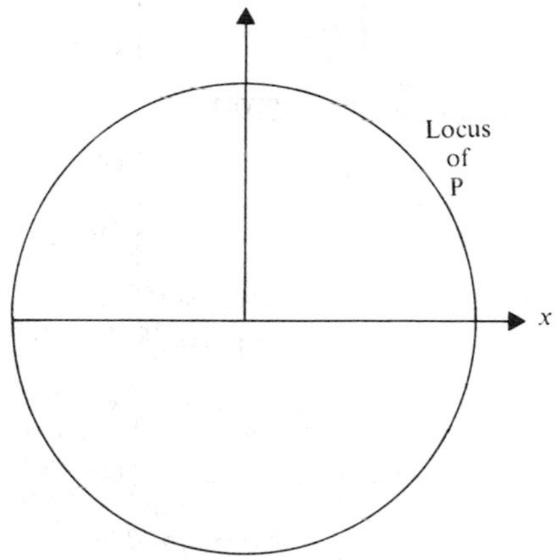

Thus P is restricted to any point on the circumference of a circle with centre O and radius 4, i.e. $|z| = 4$ defines the circle, centre O and radius 4.
This circle, which is the locus of P, has an equation

$$x^2 + y^2 = 4^2 \quad \text{in Cartesian form}$$

$$r = 4 \quad \text{in polar form}$$

and $$|z| = 4 \quad \text{in complex form}$$

Alternatively the Cartesian equation of the locus of P can be found directly from the complex equation, without reference to a diagram, as follows:

$$z = x + yi \quad \text{so} \quad |x + yi| = 4$$

But $$|x + yi| = \sqrt{(x^2 + y^2)}$$

Hence $$\sqrt{(x^2 + y^2)} = 4$$

or $$x^2 + y^2 = 16$$

which is the Cartesian equation of a circle, centre O and radius 4.

(Each of these methods has advantages in different problems. It is unwise to use one of them exclusively.)

Now consider the case when $|z - z_1| = 4$ where z_1 is a fixed point representing $x_1 + y_1 i$.
If $P(x, y)$ represents $x + yi$ and $A(x_1, y_1)$ represents $x_1 + y_1 i$ on an Argand diagram, then $z - z_1$ is represented by the line joining A to P.

So $|z - z_1|$ means 'the length of the line AP'.

Therefore $|z - z_1| = 4$ means that the length of AP is always 4 units.

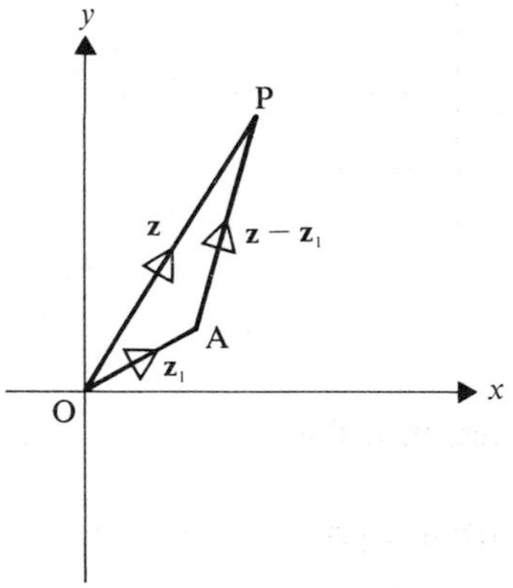

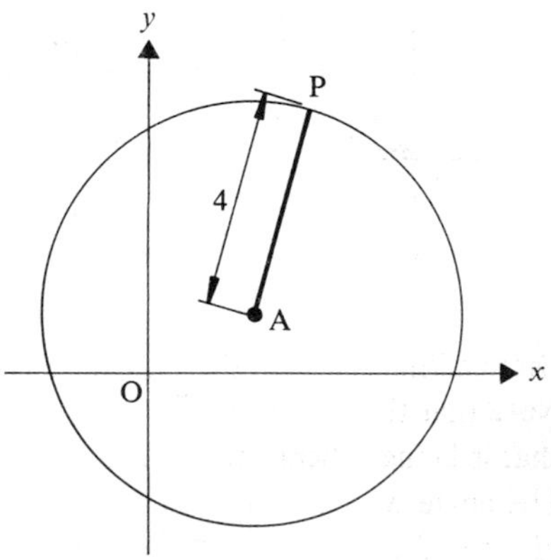

So the locus of P, as z varies, is a circle, centre A and radius 4 units.
The Cartesian equation of this circle is

$$(x - x_1)^2 + (y - y_1)^2 = 4^2$$

Before dealing with a locus problem involving an argument, a careful appraisal of the meaning of $\arg(z - z_1)$ is necessary.

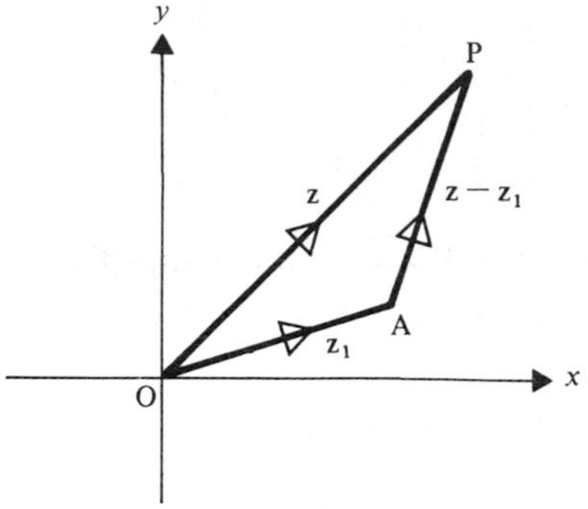

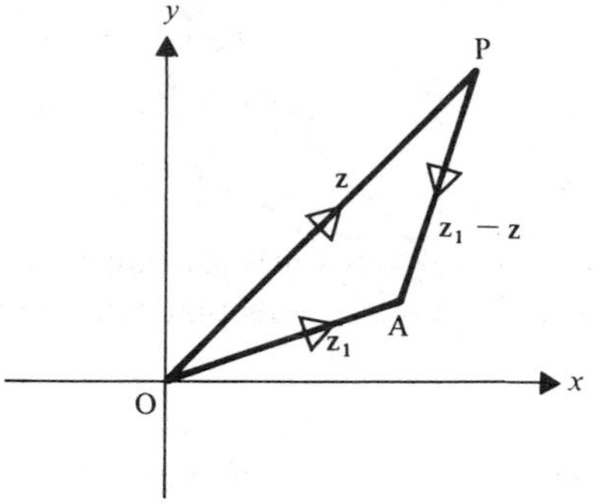

$z - z_1$ is represented by the line joining A to P.
$z_1 - z$ is represented by the line joining P to A.
These lines have opposite directions so

$$\arg(z - z_1) \neq \arg(z_1 - z)$$

One method for determining the argument of a line that does not pass through O is given below.

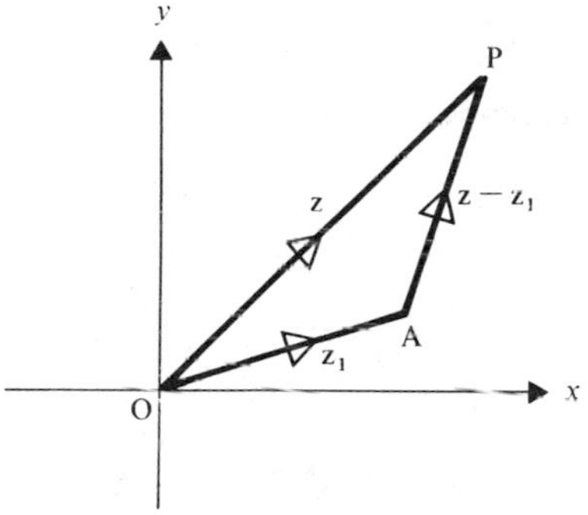

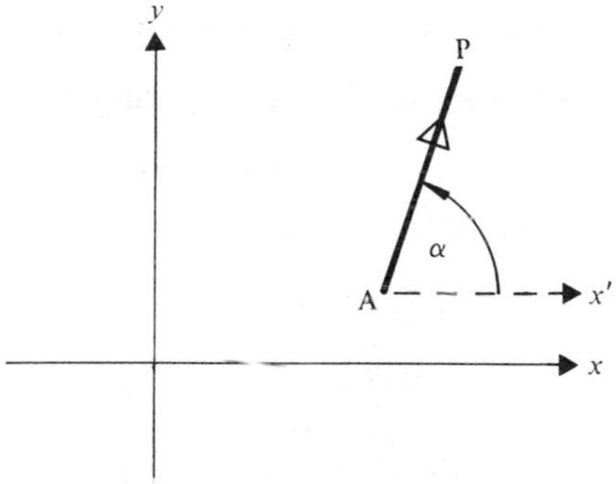

Draw a line Ax', parallel to the *positive* x axis.
Note that the line is drawn through the point representing the complex number that is being subtracted.
The angle $x'AP$, shown as α in the diagram, is then the principal argument of $z - z_1$.
Now if $\arg(z_1 - z)$ is required, a line parallel to Ox is drawn through P (because P represents z which is the complex number that is being subtracted this time).

Then $\arg(z_1 - z) = \beta$ as shown in the diagrams below.

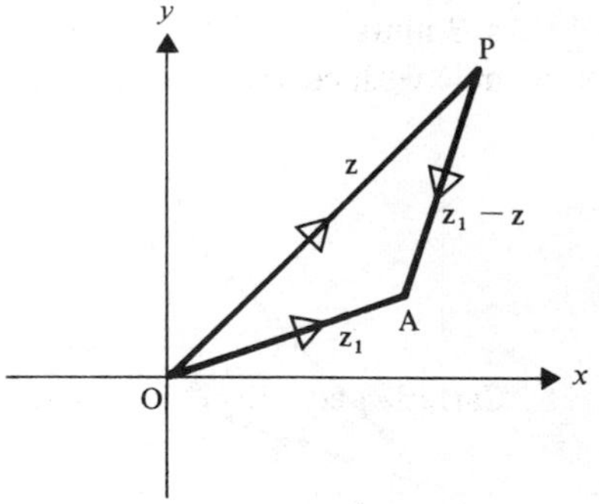

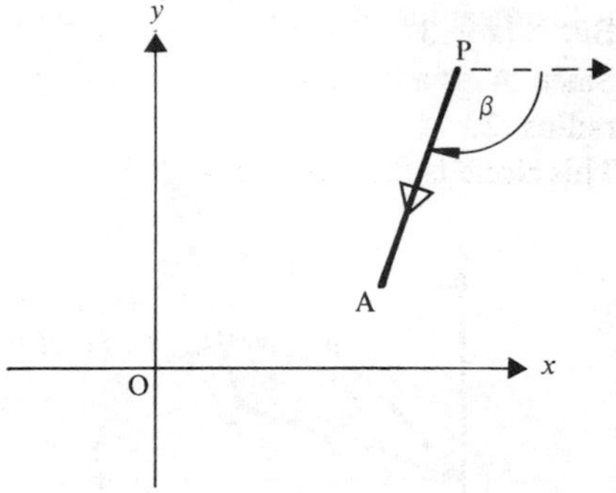

Consider now the locus defined by the equation $\arg(z - z_1) = \frac{\pi}{4}$.

If z is represented by $P(x, y)$ and z_1 is represented by $A(x_1, y_1)$ where $z = x + yi$ and $z_1 = x_1 + y_1i$, then one possible position of P is shown in the diagram

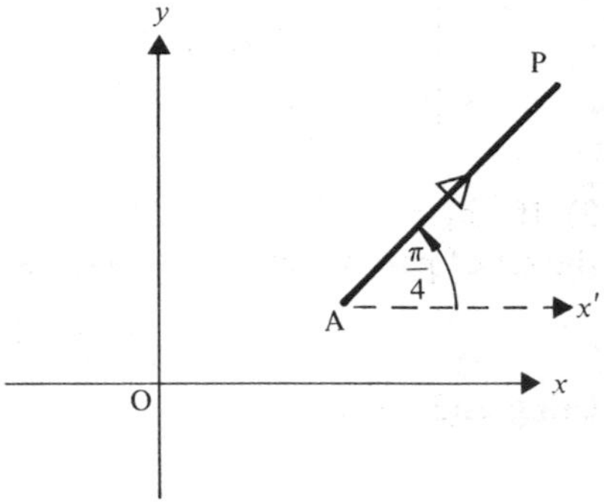

But the length of AP is unspecified so the line AP can be produced indefinitely and every point on it satisfies the condition $\arg(z - z_1) = \frac{\pi}{4}$.

The line PA must *not* be produced however, since $\arg(z - z_1) \neq \frac{\pi}{4}$ when P is below A as shown.

So the locus of P in a case like this is described as a *part-line* or *half-line*

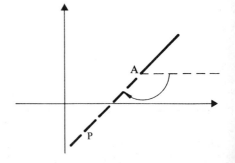

EXAMPLES 8d

1) If $|z - 3| = 2$, sketch the locus of $P(x, y)$ which represents z on an Argand diagram. Write down the Cartesian equation of this locus.

If $$z - 3 = z - z_1$$

then $z_1 = 3 + 0i$ which can be represented by A(3, 0).

The line AP represents $z-3$.
But $|z-3|=2$ so the length of AP is always 2 units.
Since A is a fixed point, P lies anywhere on a circle with centre A and radius 2.
This circle is thus the locus of P.

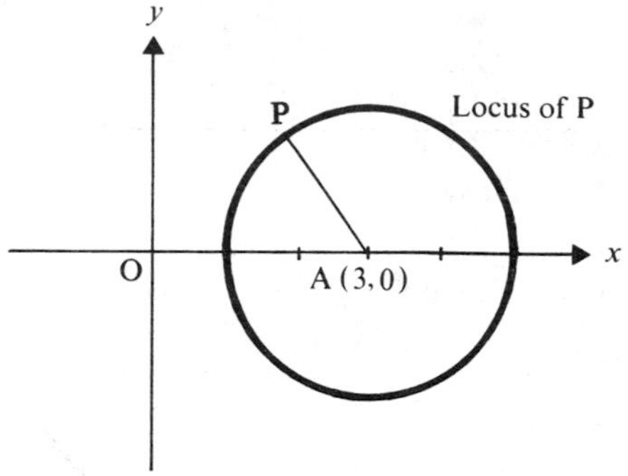

The Cartesian equation of the circle is

$$(x-3)^2+(y-0)^2 = 2^2$$

$$\Rightarrow \quad x^2+y^2-6x+5 = 0$$

2) If $z_1=3+i$, $z_2=-3-i$ and $z=x+yi$ determine the locus of the set of points $P(x, y)$ on the Argand diagram for which

$$|z-z_1| = |z-z_2|$$

Using A(3, 1) and B(−3, −1) the given condition becomes

$$AP = BP$$

i.e. P is always equidistant from the two fixed points A and B,
i.e. P always lies on the perpendicular bisector of the line AB.

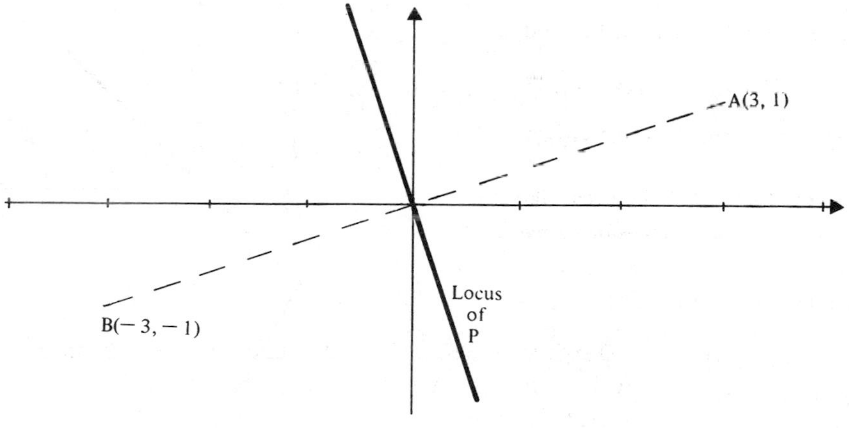

The locus of P is therefore the perpendicular bisector of AB and its Cartesian equation is $y=-3x$.

3) $P(x, y)$ is the point on an Argand diagram representing $z = x + yi$.
If $|z+2| = 3|z-2-4i|$ show that the Cartesian equation of the locus of P is $x^2 + y^2 - 5x - 9y + 22 = 0$.

To use the geometric approach in this problem we would first write $z+2$ in the form $z - z_1$ so that it corresponds to a line joining two points,

i.e. $$|z-(-2+0i)| = 3|z-(2+4i)|$$

Then, using $A(-2, 0)$, $B(2, 4)$ and $P(x, y)$ we find that $AP = 3BP$. It is not obvious from this property however, where P can lie, so the method based on the algebraic definition of a modulus is used instead.
Writing $x + yi$ for z we have

$$z+2 = x+yi+2 = (x+2)+yi$$

$\Rightarrow$ $$|z+2| = \sqrt{(x+2)^2+(y)^2}$$

and $$z-2-4i = x+yi-2-4i = (x-2)+(y-4)i$$

$\Rightarrow$ $$|z-2-4i| = \sqrt{(x-2)^2+(y-4)^2}$$

Thus the given condition can be written

$$\sqrt{[(x+2)^2+y^2]} = 3\sqrt{[(x-2)^2+(y-4)^2]}$$

i.e. $$x^2+4x+4+y^2 = 9(x^2-4x+4+y^2-8y+16)$$

$\Rightarrow$ $$8x^2+8y^2-40x-72y+176 = 0$$

$\Rightarrow$ $$x^2+y^2-5x-9y+22 = 0$$

Because this is the condition which the coordinates of P must satisfy, it is the equation of the locus of P, which is seen to be a circle.

4) Sketch on an Argand diagram the locus of the set of points $P(x, y)$ where $z = x + yi$, for which $\arg(z-2+3i) = \frac{2}{3}\pi$

$$z-2+3i = z-(2-3i)$$

So $z-2+3i$ is represented by the line joining $A(2, -3)$ to $P(x, y)$.

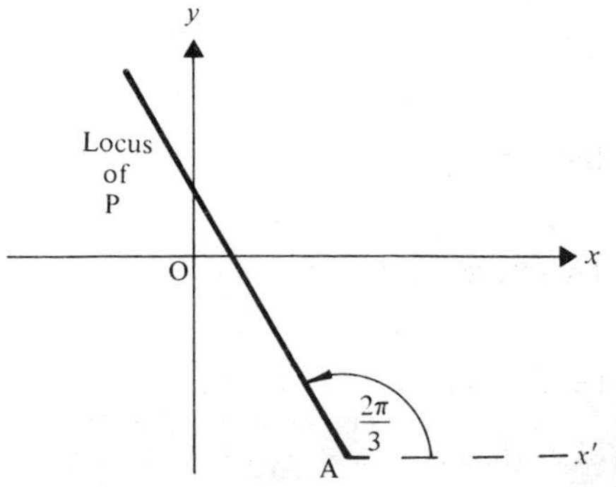

Since A is fixed, the line Ax' can be drawn and, taking an angle $\frac{2}{3}\pi$ in the anticlockwise sense from Ax', a line AP can be drawn. Any point on this part-line represents P and the line is therefore the locus of P.

5) If $\arg(z-1) - \arg(z+1) = \dfrac{\pi}{4}$, show that the point $P(x, y)$ representing z on an Argand diagram lies on an arc of a circle. Give the coordinates of the centre of this circle.

If A is $(1, 0)$ and B is $(-1, 0)$, $\arg(z-1) = \alpha$ and $\arg(z+1) = \beta$ we have:

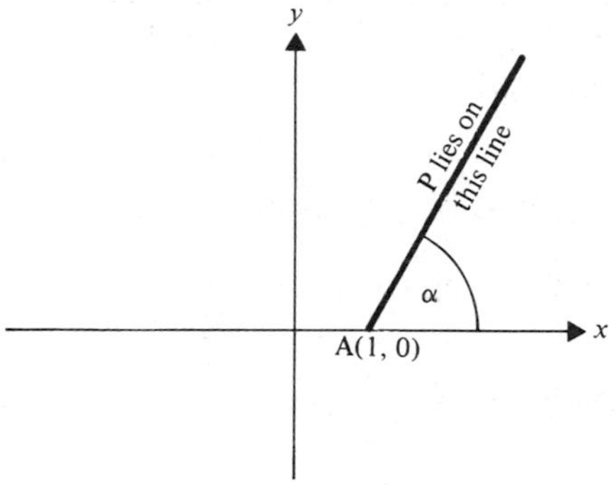

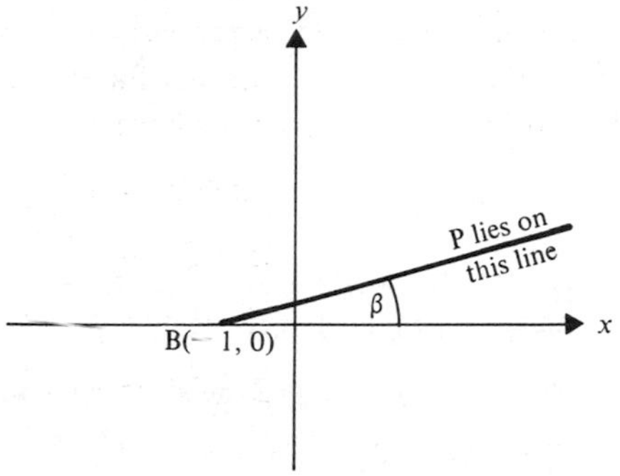

Thus

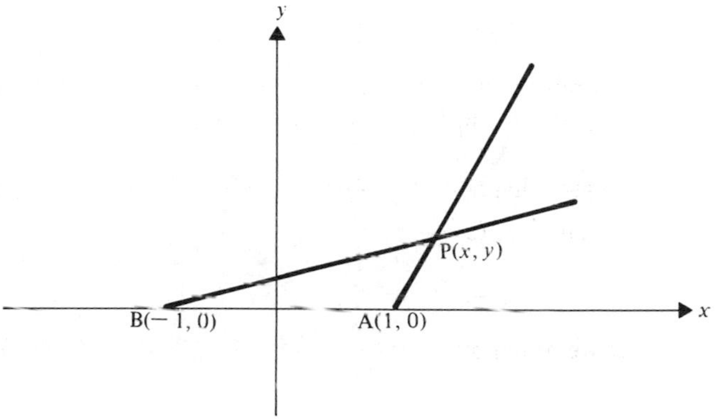

But $\alpha - \beta = \dfrac{\pi}{4}$ and angle APB $= \alpha - \beta$.

Therefore angle APB $= \dfrac{\pi}{4}$,

i.e. the line AB subtends a constant angle $\dfrac{\pi}{4}$ at P, so P must lie on an arc of a circle cut off by the chord AB.

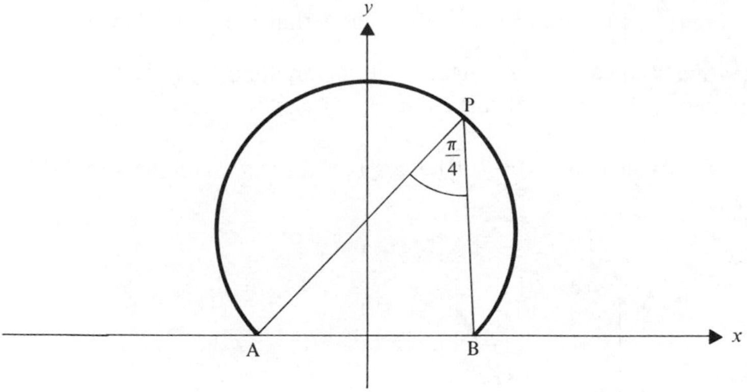

If C is the centre of this circle, then angle $ACB = \frac{\pi}{2}$ (the angle at the centre is twice the angle at the circumference) and C lies on the y axis (perpendicular bisector of chord AB).

The locus of P can now be drawn more clearly:

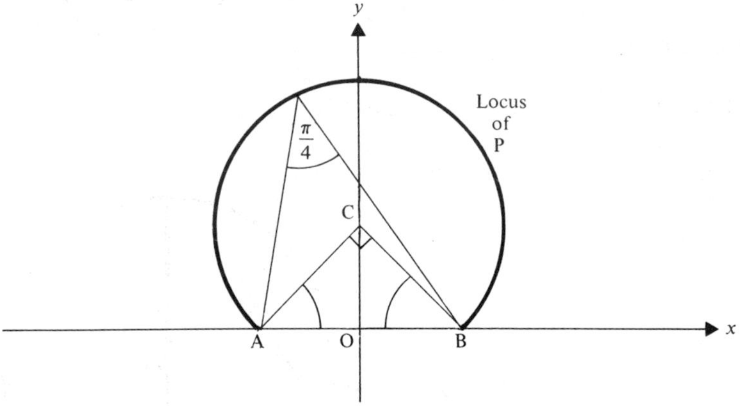

Since OB = OC $\left(\angle CAO = \angle CBO = \frac{\pi}{4}\right)$ the coordinates of C are (0, 1).

Note. The Cartesian equation of a circle with centre (0, 1) and radius $\sqrt{2}$ (i.e. AC) is $x^2 + y^2 - 2y = 1$. This equation represents a *complete* circle however and not just the major arc which is the locus of P.
Thus the Cartesian equation of the locus of P is

$$x^2 + y^2 - 2y = 1 \quad \textit{and} \quad y > 0$$

6) If $\text{Re}\left(z - \frac{1}{z}\right)$ is zero, find the polar equation of the locus of $P(r, \theta)$ which represents z on an Argand diagram. Sketch this locus.

If $$z = r(\cos\theta + \mathrm{i}\sin\theta)$$

then $$\frac{1}{z} = \frac{1}{r(\cos\theta + \mathrm{i}\sin\theta)}$$

$$= \frac{\cos\theta - \mathrm{i}\sin\theta}{r}$$

Hence $$\mathrm{Re}\left(z - \frac{1}{z}\right) = r\cos\theta - \frac{\cos\theta}{r}$$

$$= \frac{1}{r}\cos\theta\,(r^2 - 1)$$

So $$\mathrm{Re}\left(z - \frac{1}{z}\right) = 0 \quad\Rightarrow\quad (r^2 - 1)\cos\theta = 0, \quad r \neq 0$$

and this is the polar equation of the locus of z when $\mathrm{Re}\left(z - \frac{1}{z}\right) = 0$

As $$(r^2 - 1)\cos\theta = 0 \quad\Rightarrow\quad \cos\theta = 0 \quad\text{or}\quad r = \pm 1$$

the locus is seen to be in two parts:

the line $\theta = \dfrac{\pi}{2}$, excluding the origin

and the circle, centre O, of radius 1. i.e.

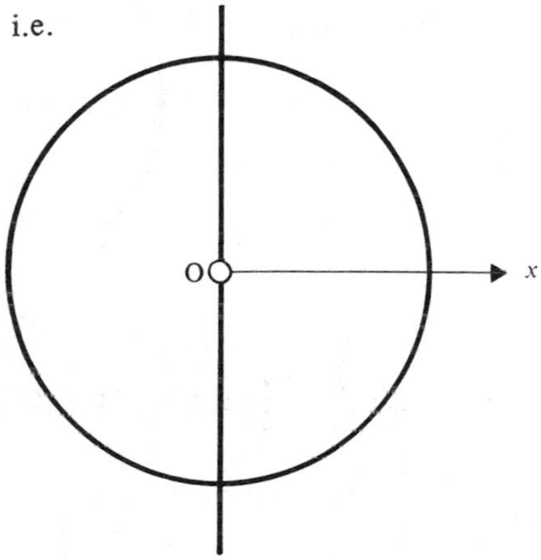

EXERCISE 8d

Sketch the locus of the point $P(x, y)$, where P represents the complex number $z = x + y\mathrm{i}$ on an Argand diagram, if:

1) $|z| = 1$

2) $|z - 1| = 3$

3) $|z - 2\mathrm{i}| = 3$

4) $|z + 2| = 2$

5) $|z-1+i|=4$

6) $\arg z=\frac{\pi}{3}$

7) $\arg(z-1)=-\frac{3\pi}{4}$

8) $\arg(z+3-4i)=\frac{\pi}{4}$

9) $|z-2-3i|=|z+4-5i|$

10) $|z|=|z+4i|$

11) $\arg(z-4)-\arg z=\frac{\pi}{3}$

12) $\arg(z+2)=\arg(z-2)-\frac{\pi}{6}$

13) $\arg(z-i)+\frac{\pi}{4}=\arg(z+i)$

14) $\arg(z-3)=\frac{2\pi}{3}+\arg z$.

Express, in complex form, the equations of the following loci:

15) A circle with centre (h, k) and radius r when
(a) $(h,k)=(3,4)$; $r=5$ (b) $(h,k)=(5,0)$; $r=2$
(c) $(h,k)=(0,4)$; $r=4$.

16) Write down the polar equation of the locus of $P(r,\theta)$ where $z=r(\cos\theta+i\sin\theta)$ if:

(a) $|z|=1$ (b) $|z|=4$ (c) $\arg z=\frac{\pi}{4}$ (d) $\arg z=-\frac{2\pi}{3}$.

17) Write down the Cartesian equation of each locus given in Question 16.

18) Find the equation (i) in complex form, (ii) in Cartesian form, of the perpendicular bisector of the line joining the points A and B where the coordinates of A and B are:
(a) $(4,0), (-8,0)$ (b) $(1,2), (7,-4)$ (c) $(0,6), (6,0)$.

19) Write down the equation, in argument form, of the following part-lines:

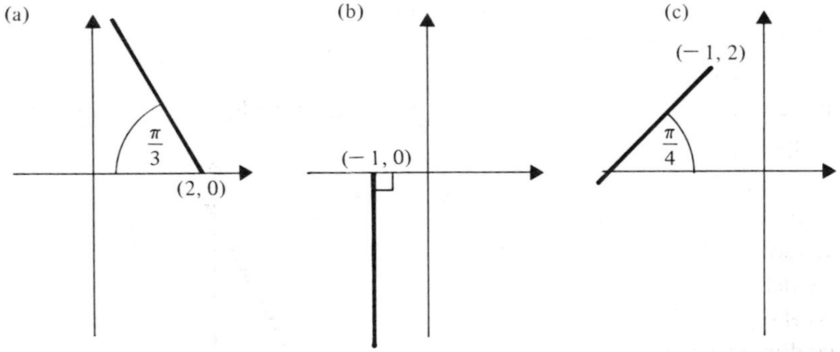

Find the Cartesian equations equivalent to the following equations, in which $z=x+yi$. Sketch the locus of $P(x,y)$ on an Argand diagram.

20) $2|z-2|=|z+4i|$

21) $\operatorname{Im}(z^2)=0$

Find the polar equations of the loci in Questions 22 to 24 using $z=r(\cos\theta+i\sin\theta)$. Sketch the locus of z on an Argand diagram.

22) $\operatorname{Im}\left(z-1+\frac{4}{z}\right)=0$ 23) $\operatorname{Re}\left(\frac{z-2}{z}\right)=0$

24) $\operatorname{Re}\left(z-\frac{1}{\bar{z}}\right)=0$ where $\bar{z}$ denotes the conjugate of z.

25) If $|z_1-z_2|=|z_1+z_2|$ show that the arguments of z_1 and z_1 differ by $\frac{\pi}{2}$.

26) Sketch the loci defined by the following conditions, determining first the equivalent Cartesian equation where necessary.

(a) $\left|\frac{z-2}{z+6}\right|=1$ (b) $\left|\frac{z-3i}{z-1+i}\right|=3$

(c) $\arg\left(\frac{z}{z+5}\right)=\frac{\pi}{2}$ (d) $\arg\left(\frac{z-3-i}{z+5-3i}\right)=\frac{\pi}{3}$

FURTHER USE OF THE ARGAND DIAGRAM

Many interesting and varied problems involving complex numbers can be solved very simply by using an Argand diagram. This approach is illustrated by the following examples, in all of which $z=x+yi$, P is a point (x,y) and $\overrightarrow{OP}$ represents z on the Argand diagram.

EXAMPLES 8e

1) If $\arg(z+3)=\frac{\pi}{3}$ find the least value of $|z|$.

Since $\arg(z+3)=\arg[z-(-3)]=\frac{\pi}{3}$, P lies anywhere on the line shown in the diagram.

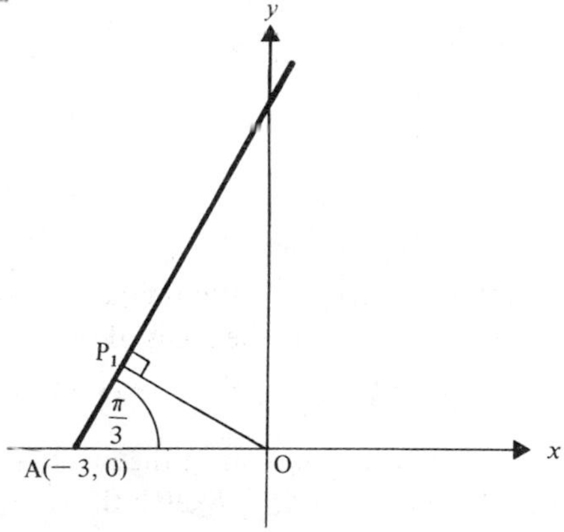

Now $|z|$ means 'the length of OP'. So we require the point P on the line, which is nearest to O. This is the point P_1 where OP_1 is perpendicular to AP_1.

Then $OP_1=OA\sin\frac{\pi}{3}=\frac{3\sqrt{3}}{2}$

Thus the least value of $|z|$ is $\frac{3\sqrt{3}}{2}$

2) Shade the area represented on an Argand diagram by $|z + i| < 4$.

As $z + i = z - (-i)$ the equation can be expressed in the form:

'the length of AP < 4 units' where A is $(0, -1)$.

When AP $= 4$ units, P lies on a circle with centre A and radius 4.
When AP < 4, P lies inside this circle.
Therefore the area is as shown below.

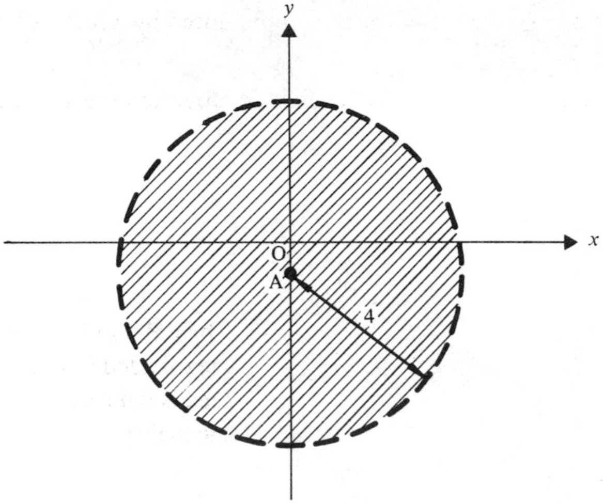

3) If $|z - 3 + 3i| = 2$, find the greatest and least values of $|z + 1|$.

Given $|z - (3 - 3i)| = 2$, we see that $P(x, y)$ lies on a circle of radius 2 and centre $(3, -3)$.
Also $|z + 1| = |z - (-1)|$ means 'the length of AP' where A is the point $(-1, 0)$.

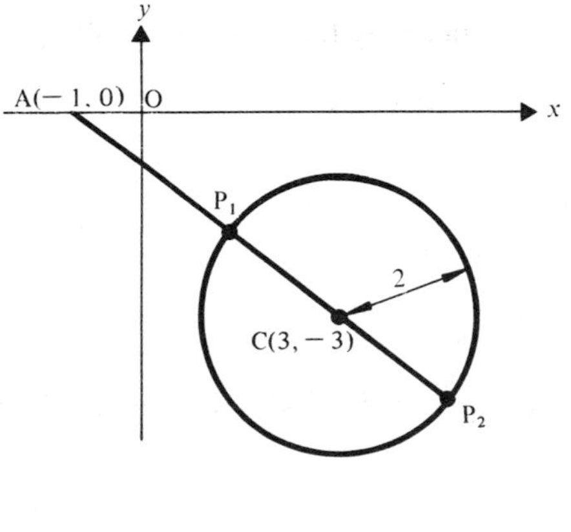

Therefore we require the least and greatest distances from A to any point on the circle.
These are the distances AP_1 and AP_2 where AP_1P_2 passes through the centre C of the circle.

The length of AC $= \sqrt{[3-(-1)]^2+[-3]^2} = 5$

Therefore $AP_1 = AC - CP_1 = 5-2 = 3$

and $AP_2 = AC + CP_2 = 5+2 = 7$

Thus the greatest value of $|z+1|$ is 7 and the least value of $|z+1|$ is 3.

(**Note.** If the coordinates of P_1 and P_2 are required, the Cartesian equations of the circle and the line AC can be solved simultaneously.)

4) Indicate on an Argand diagram what is meant by $0 < \arg(z+2+3i) \leqslant \frac{\pi}{6}$

$z+2+3i = z-(-2-3i)$ which is represented by a line AP where A is the point $(-2,-3)$.

The inclination of this line to the positive x direction must be between 0 and $\frac{\pi}{6}$.

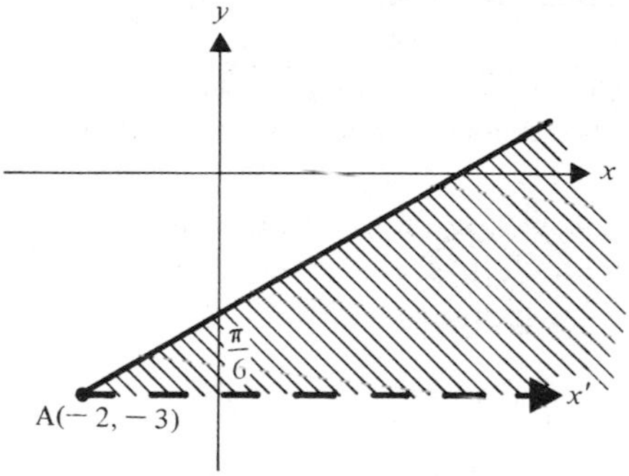

Thus P can lie anywhere in the area shaded in the diagram, so this area is indicated by the given inequality.

EXERCISE 8e

1) Shade on an Argand diagram the areas represented by the following inequalities.

(a) $|z-1| < 4$ (b) $|z+3i| > 2$

(c) $|z+1-i| < 1$ (d) $\frac{\pi}{3} < \arg z < \frac{2\pi}{3}$

2) Shade on an Argand diagram the region occupied by the set of points $P(x, y)$ for which $|z| < 5$ *and* $-\frac{\pi}{6} < \arg z < \frac{\pi}{6}$, where $z = x + yi$.

3) Find the least value of $|z-1|$ if:

(a) $\arg(z+1) = -\frac{\pi}{4}$ (b) $|z+3-i| = 2$.

4) Find the value(s) of z for which:

(a) $|z| = 4$ and $\arg z = \frac{\pi}{4}$ (b) $|z+2+i| = 5$ and $\text{Re}(z-1) = 0$.

5) Find the points of intersection of the loci on an Argand diagram defined by:
(a) $|z-1+2i| = |z+1|$ and $|z-1| = \sqrt{2}$

(b) $\arg z = -\dfrac{\pi}{4}$ and $|z| = 2$.

6) Indicate on an Argand diagram the set of points $P(x, y)$ for which:

(a) $0 \leqslant \arg(z+1) \leqslant \dfrac{\pi}{3}$ and $|z+i| = 3$

(b) $|z+3-2i| < 4$ and $\arg(z+1) = \dfrac{5\pi}{6}$

(c) $|z| > 1$, $|z| < 4$ and $\arg z = -\dfrac{3\pi}{4}$.

TRANSFORMATIONS OF THE ARGAND DIAGRAM

We have seen, in Chapter 1, that matrices can be used to describe transformations of a plane. We can also use complex numbers to describe transformations of a plane.
If $z = x + yi$ is represented by the point $P(x, y)$ in the z plane and $w = u + vi$ is represented by the point $Q(u, v)$ in the w plane, then a relationship between z and w defines the mapping of the point P to the point Q.
For example, if $w = 2z$

i.e. $$u + vi = 2(x + yi) = 2x + 2yi$$

then a point $P(x, y)$ is mapped to the point $Q(u, v)$
where $u = 2x$ and $v = 2y$.
So under this transformation, the image of P will have coordinates that are twice those of P,
i.e. this transformation is an enlargement (centred on O) by a factor of 2,

i.e. if $w = 2z$:

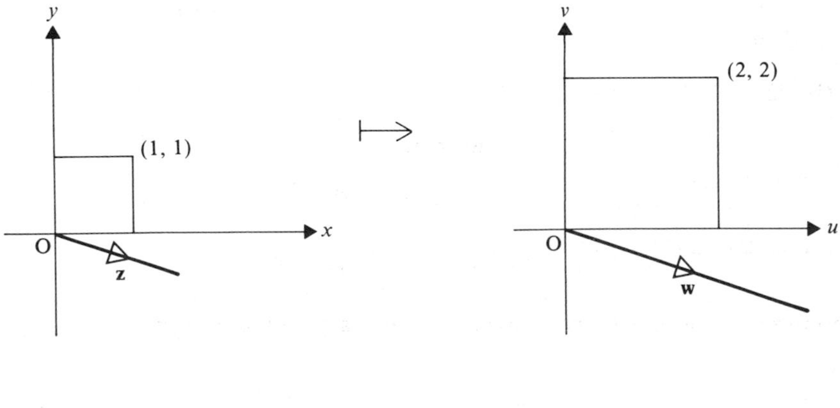

Any relationship of the form $w = kz$, where k is real and positive, represents

an enlargement for $k > 1$

a reduction for $k < 1$

the identity transformation for $k = 1$.

We will now interpret some other relationships between z and w as transformations.

Consider $w = \bar{z}$.

Under this transformation, z is mapped to its conjugate, which is the reflection of z in the real axis,

i.e. if $w = \bar{z}$:

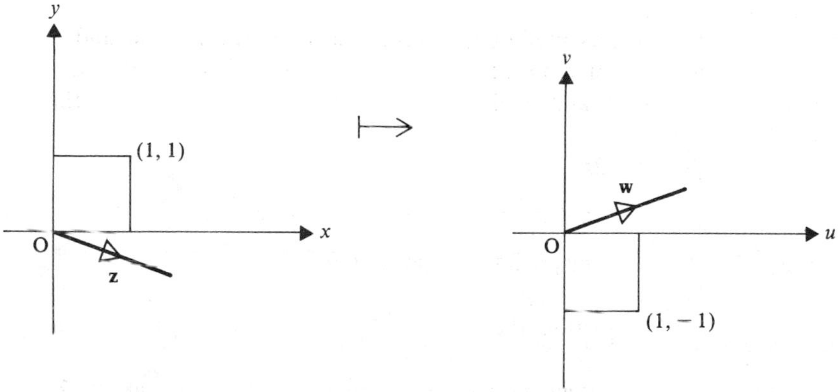

So $w = \bar{z}$ represents a reflection in the real axis.

Consider $w = ze^{i\alpha}$

Writing z as $re^{i\theta}$ gives

$$w = re^{i\theta}e^{i\alpha}$$

$$= re^{i(\theta + \alpha)}$$

$$\Rightarrow \quad |w| = |z| \quad \text{but} \quad \arg w = (\arg z) + \alpha$$

So this transformation maps $|z|$ to $|z|$ and $\arg z$ to $\alpha + \arg z$

i.e. if $w = ze^{i\alpha}$:

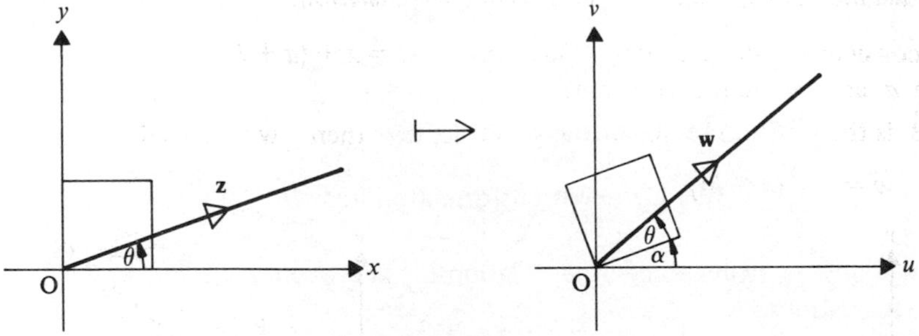

So $w = ze^{i\alpha}$ rotates z through an angle α about O.

Consider $w = z + a$ where a is real and constant.
On an Argand diagram, a can be represented by the vector $\overrightarrow{OA}$ where A is the point $(a, 0)$.
So $w = z + a$ maps **z** to $\mathbf{z} + \overrightarrow{OA}$

i.e. if $w = z + a$

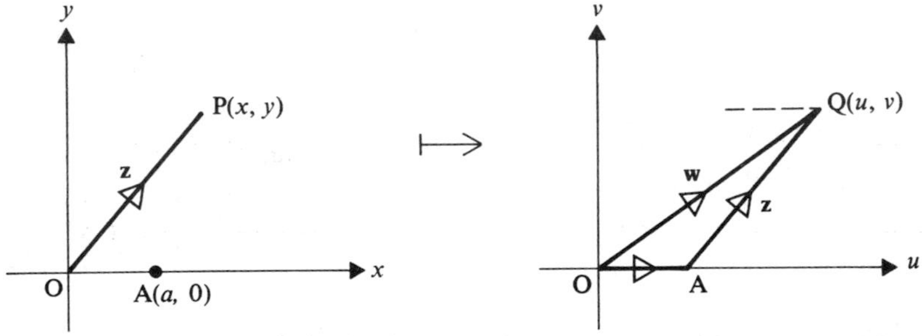

This becomes clearer if we consider the coordinates of P and Q:

$$w = z + a \Rightarrow u + vi = x + yi + a = (x + a) + yi$$

So $$u = x + a \quad \text{and} \quad v = y$$

The effect of this on a unit square is shown in the diagrams below.

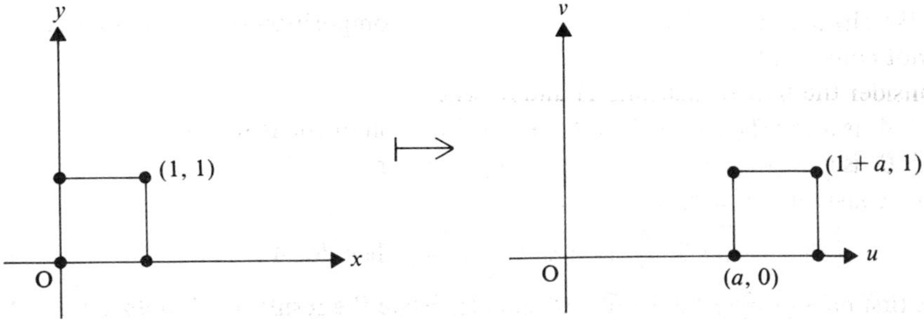

A transformation such as this one, where all points in the z plane are moved a fixed distance in a given direction, is called a *translation.*

Now consider the more general relationship $w = z + (a + bi)$ where a and b are real constants.
If $\overrightarrow{OB}$ is the position vector of the point (a, b) then $\mathbf{w} = \mathbf{z} + \overrightarrow{OB}$

i.e. if $w = z + (a + bi)$:

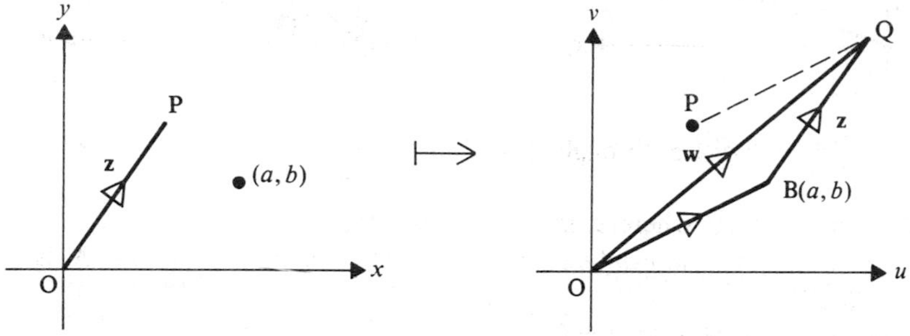

and

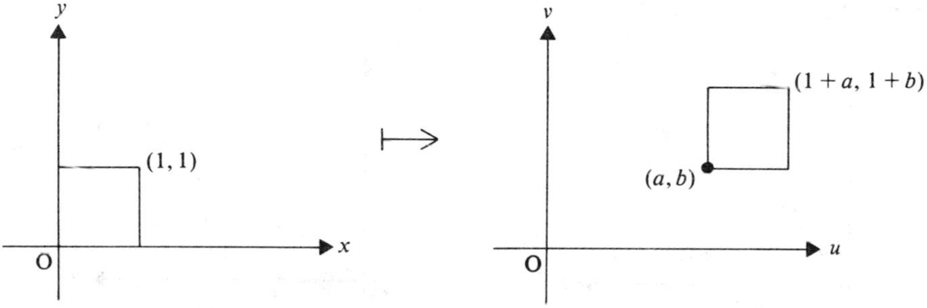

So the point Q is obtained by moving P a distance $|a + bi|$ in the direction of the vector representing $a + bi$,

i.e. $w = z + (a + bi)$ represents a translation by a distance $|a + bi|$ in the direction of the vector representing $a + bi$.

Composition of Transformations

We already know, from Chapter 1, that the composition of transformations is not commutative.
Consider the transformations A and B where
A is a translation of 3 units in the direction of the real axis
and B is an enlargement (centre O) by a factor 2.
Now consider the composition $A \circ B$ where

$A \circ B$ means 'do B first then do A'

i.e. first enlarge by a factor 2 and then translate the result by 3 units along Ox.

So $A \circ B$ can be defined by $w = 2z + 3$ because this means that z is first multiplied by 2 and then 3 is added to the result.

Next consider the relationship $w = 2(z + 3)$.
This time, first 3 is added to z and then the result is multiplied by 2.
So $w = 2(z + 3)$ defines the composition $B \circ A$, i.e. first translate, then enlarge.
Now $2z + 3 \neq 2(z + 3)$ and this confirms that, in general, composition of transformations is not commutative.

EXAMPLES 8f

1) The transformations A and B of the z plane to the w plane are defined as follows:

A is a rotation of $\pi/3$ radians about O

B is a translation of 2 units in the direction of the negative real axis.

Write down the relationship between w and z defining $A \circ B$ and $B \circ A$.

A is defined by $w = ze^{i\pi/3}$
and B is defined by $w = z - 2$.

So $A \circ B$ is defined by $w = (z - 2)e^{i\pi/3}$
and $B \circ A$ is defined by $w = ze^{i\pi/3} - 2$

Finding the Image of a Curve

Either the direct relationship between w and z, or the relationship between the coordinates of P and Q, can be used to find the image of a curve under a given transformation.

For example, to find the image of the circle $|z| = 4$ under the transformation $w = 2z$, we can either use the fact that

$$w = 2z \Rightarrow |w| = 2|z|$$

$$\Rightarrow \text{ the circle } |z| = 4 \text{ maps to the circle } |w| = 8,$$

or we can use the fact that

$$w = 2z \Rightarrow u = 2x \text{ and } v = 2y$$

and $|z| = 4$ has Cartesian equation $x^2 + y^2 = 4^2$

so $x^2 + y^2 = 16$ maps to $\left(\dfrac{u}{2}\right)^2 + \left(\dfrac{v}{2}\right)^2 = 4$

$$\Rightarrow \qquad u^2 + v^2 = 64$$

EXAMPLES 8f (continued)

2) The relationship $w = z - 3$ maps $P(x, y)$ to $Q(u, v)$.
If P describes the line $x - 2y = 5$, find the equation of the locus of Q.

$w = z - 3$ represents a translation of 3 units in the direction of the negative real axis.
So the line $x - 2y = 5$ in the z plane has, as its image in the w plane, the line with the same gradient but with its intercept on the real axis moved 3 units to the left.

i.e.

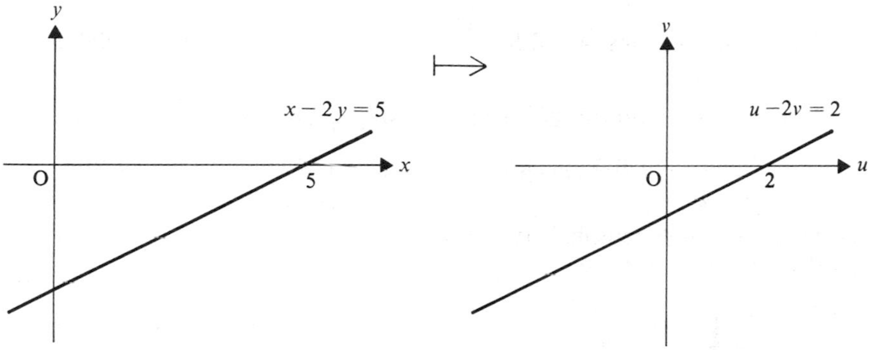

So $x - 2y = 5$ maps to $u - 2v = 2$.

Alternatively, working from the relationship between the coordinates of $P(x, y)$ and $Q(u, v)$, we have

$$w = z - 3 \Rightarrow u + vi = x + yi - 3$$

$$\Rightarrow u + 3 = x \quad \text{and} \quad v = y$$

i.e. $w = z - 3$ maps $\begin{cases} x \text{ to } u + 3 \\ y \text{ to } v \end{cases}$

So $x - 2y = 5$ maps to $u + 3 - 2v = 5$

$$\Rightarrow \qquad u - 2v = 2$$

So far we have investigated relationships between w and z that have all had obvious geometric interpretations as transformations. However, some relationships are not easy to interpret geometrically. When finding the image of a curve in such a case, knowledge of the form of the transformation is not necessary.

3) The relationship $w = z^2$, where $z = x + yi$ and $w = u + vi$, maps $P(x, y)$ to $Q(u, v)$. Find the locus of Q if P describes
(a) the circle $x^2 + y^2 = 25$,
(b) the curve $x^2 - y^2 = 4$.

(a) The equation $x^2+y^2=25$ can be written in the form $|z|=5$.

Now $w=z^2 \Rightarrow |w|=|z^2|=|z||z|$
Hence $|z|=5$ maps to $|w|=25$.

(b) $x^2-y^2=4$ cannot easily be expressed in complex form, so we use the relationships between x, y and u, v.

Now $w=z^2 \Rightarrow u+v\text{i}=(x+y\text{i})^2=x^2-y^2+2xy\text{i}$

$$\Rightarrow \quad \begin{cases} u = x^2-y^2 \\ v = 2xy \end{cases}$$

So $x^2-y^2=4$ maps to $u=4$.

4) If $z=x+y\text{i}$ and $w=u+v\text{i}$ represent points $\text{P}(x, y)$ and $\text{Q}(u, v)$ and $w=\dfrac{z+1}{z-1}$, find the locus of Q when P describes

(a) the circle $x^2+y^2=4$
(b) the parabola $y^2=4x$

(a) Although $x^2+y^2=4$ can be expressed as $|z|=2$, this form cannot be used directly from the relationship $w=\dfrac{z+1}{z-1}$

Rearranging this, however, gives $z=\dfrac{w+1}{w-1}$

Then $$|z| = 2 \Rightarrow \left|\frac{w+1}{w-1}\right| = 2$$

i.e. $$|w+1| = 2|w-1|$$

$$\Rightarrow \quad (u+1)^2+v^2 = 4[(u-1)^2+v^2]$$

$$\Rightarrow \quad 3u^2+3v^2-10u+3 = 0$$

Therefore the image of the circle $x^2+y^2=4$ is another circle with equation $3u^2+3v^2-10u+3=0$.

(b) To find the image of $y^2=4x$ we need x and y separately in terms of u and v.

$$z = \frac{w+1}{w-1} \Rightarrow x+yi = \frac{u+1+vi}{(u-1)+vi} = \frac{(u^2+v^2-1)-2vi}{(u-1)^2+v^2}$$

$$\Rightarrow \quad x = \frac{u^2+v^2-1}{(u-1)^2+v^2} \quad \text{and} \quad y = \frac{-2v}{(u-1)^2+v^2}$$

So $y^2 = 4x$ maps to $\dfrac{4v^2}{[(u-1)^2+v^2]^2} = \dfrac{4(u^2+v^2-1)}{(u-1)^2+v^2}$

$$\Rightarrow \qquad v^2 = [(u-1)^2+v^2]\,[u^2+v^2-1]$$

provided that $(u-1)^2+v^2 \neq 0$.

Therefore the image of $y^2 = 4x$ is the curve

$$(u^2+v^2-1)(u^2+v^2-2u+1) = v^2$$

EXERCISE 8f

In this exercise x, y, u, v are real numbers and z, w are complex numbers such that $z = x + yi$, $w = u + vi$. Describe the locus of the point (u, v) in the Argand diagram, in Questions 1-5.

1) $w = z + 4$ and
(a) $|z| = 3$ (b) $\arg(z) = \dfrac{\pi}{3}$ (c) $|z+4| = 5$
(d) $x^2 + y^2 = 1$ (e) $y = 2x$ (f) $y^2 = 4x$.

2) $w = -2z$ and
(a) $x^2 - y^2 = 1$ (b) $\arg(z) = \dfrac{\pi}{2}$ (c) $|z| = 4$
(d) $y = 2x + 3$.

3) $w = \dfrac{1}{z}$ and
(a) $|z| = 6$ (b) $\arg z = -\dfrac{\pi}{4}$ (c) $\dfrac{x^2}{4} + y^2 = 1$
(d) $xy = 4$ (e) $y = 3x + 4$.

4) $w^2 = z$ and $P(x, y)$ describes
(a) the circle $x^2 + (y-1)^2 = 1$,
(b) the positive x axis,
(c) the negative y axis.

5) $w = \dfrac{1}{z+1}$ and
(a) $2x + y + 1 = 0$ (b) $x = 0$ (c) $x + 1 = 0$.

6) Given that $z = w^2$, prove that
(a) if u varies but v is constant, then the locus of $P(x, y)$ is a parabola,
(b) if v varies but u is constant, the locus of P is again a parabola.
Show that these two parabolas have the same axis and focus.

SUMMARY

$(\cos\theta + i\sin\theta)^n \equiv \cos n\theta + i\sin n\theta$ (De Moivre's Theorem)

$$\left.\begin{aligned} z^n + \frac{1}{z^n} &= 2\cos n\theta \\ z^n - \frac{1}{z^n} &= 2i\sin n\theta \end{aligned}\right\} \quad \text{where} \quad z = \cos\theta + i\sin\theta.$$

If $z = r(\cos\theta + i\sin\theta) = re^{i\theta}$, the nth roots of z are given by

$$\sqrt[n]{r}e^{i(\theta + 2p\pi)/n}$$

The cube roots of 1 can be expressed as $1, \omega, \omega^2$, and $1 + \omega + \omega^2 = 0$.

MISCELLANEOUS EXERCISE 8

1) Simplify, without the use of tables,

$$\frac{(\cos\frac{1}{7}\pi - i\sin\frac{1}{7}\pi)^3}{(\cos\frac{1}{7}\pi + i\sin\frac{1}{7}\pi)^4}.$$ (U of L)p

2) Prove that, when n is a positive integer,

$$(\cos\theta + i\sin\theta)^n = \cos n\theta + i\sin n\theta.$$

Find the modulus and argument of

$$\frac{[(\sqrt{3})(\cos\theta + i\sin\theta)]^4}{\cos 2\theta - i\sin 2\theta}.$$ (JMB)

3) If $z = \cos\theta + i\sin\theta$, show that

(a) $z + \dfrac{1}{z} = 2\cos\theta$, (b) $z^n + \dfrac{1}{z^n} = 2\cos n\theta$.

Hence, or otherwise, show that

$$\cos^4\theta = \tfrac{1}{8}(\cos 4\theta + 4\cos 2\theta + 3).$$ (U of L)p

4) Use De Moivre's theorem to prove that, if θ is not a multiple of π,

$$\frac{\sin 5\theta}{\sin\theta} = 16\cos^4\theta - 12\cos^2\theta + 1.$$ (U of L)p

5) Prove by induction that if n is a positive integer

$$(\cos\theta + i\sin\theta)^n = \cos n\theta + i\sin n\theta.$$

Evaluate $(1 + i)^n + (1 - i)^n$ when $n = 20$. (U of L)p

6) Assuming De Moivre's theorem for an index which is a positive integer, prove that one value of $(\cos\theta + i\sin\theta)^{p/q}$ is

$$\cos(p\theta/q) + i\sin(p\theta/q)$$

when p and q are positive integers and p/q is in its lowest terms. What are the other values?
Find, in the form $a + ib$, the six roots of the equation $z^6 + 1 = 0$, and represent them on the Argand diagram. (O)

7) Expand $\left(z + \dfrac{1}{z}\right)^4$ and $\left(z - \dfrac{1}{z}\right)^4$.

By putting $z = \cos\theta + i\sin\theta$, deduce that

$$\cos^4\theta + \sin^4\theta = \tfrac{1}{4}(\cos 4\theta + 3).$$ (U of L)p

8) By using De Moivre's theorem, or otherwise, express $\sin 4\theta$ and $\cos 4\theta$ in terms of powers of $\sin\theta$ and $\cos\theta$ and show that

$$\tan 4\theta = \frac{4\tan\theta(1 - \tan^2\theta)}{1 - 6\tan^2\theta + \tan^4\theta}.$$ (U of L)p

9) By comparing the expressions for $(\cos\theta + i\sin\theta)^5$ given by De Moivre's theorem and by the binomial theorem, prove that

$$\cos 5\theta = 16\cos^5\theta - 20\cos^3\theta + 5\cos\theta.$$

By considering the equation $\cos 5\theta = 0$, prove that

$$\cos(\pi/10).\cos(3\pi/10) = \tfrac{1}{4}\sqrt{5}.$$ (U of L)p

10) Draw the locus $|z| = 1$ on an Argand diagram. Mark also a point $z = a + bi$, for which $a > b > 0$ and $a^2 + b^2 > 1$.
On the same diagram join the points representing

$$z^2,\ \frac{1}{z}\ \text{and}\ z + \frac{1}{z},$$

to the origin indicating any equal angles.
(a) If $u = x + yi$, find the complete set of values of u such that $u + \dfrac{1}{u}$ is real.
(b) If $\omega = \tfrac{1}{2}(-1 + i\sqrt{3})$, express in the form $p + qi$ the complex numbers ω^4 and ω^5. (U of L)

11) (a) If $\omega = \cos 120° + i\sin 120°$, represent the numbers ω, ω^2 and ω^3 in an Argand diagram, and find all the possible values of $(\omega^n + \omega^{2n} + \omega^{3n})$, where n is an integer.
(b) If a^2 is greater than $(b^2 + c^2)$ in the triangle ABC, show that the value of the expression

$$\frac{(1 + \cos 2B + i\sin 2B)(1 + \cos 2C + i\sin 2C)}{(1 + \cos 2A - i\sin 2A)}$$

is real and positive. (U of L)

12) Obtain the three cube roots of unity in the form $a + ib$ where a and b are real.
These three roots are denoted by $\omega_1, \omega_2, \omega_3$. Show that the real parts of $\dfrac{1}{1+\omega_1}, \dfrac{1}{1+\omega_2}$ and $\dfrac{1}{1+\omega_3}$ are all equal to $\frac{1}{2}$, and interpret this result geometrically. (JMB)

13) (a) If $z = 3 - 4i$, express z^2 and $1/z$ in the form $a + bi$ where a and b are real and represent them in an Argand diagram.
If $w^2 = z$, express the two values of w in the form $a + bi$.
(b) State De Moivre's theorem for a positive integral power and use it to express $\sin 3\theta$ and $\cos 3\theta$ in terms of $\sin\theta$ and $\cos\theta$ respectively. Hence show that, if $\sin 3\theta + \cos 3\theta = 0$, either $\tan\theta = 1$ or $\sin 2\theta = -\frac{1}{2}$. (U of L)

14) Prove De Moivre's Theorem,

$$(\cos\theta + i\sin\theta)^n = \cos n\theta + i\sin n\theta$$

in the case where n is a positive integer.
Express $\cos 6\theta$ as a polynomial in $\cos\theta$.

If $z = \cos\theta + i\sin\theta$, show by expanding $\left(z + \dfrac{1}{z}\right)^5\left(z - \dfrac{1}{z}\right)^5$, or otherwise, that

$$\sin^5\theta\cos^5\theta = \frac{1}{2^9}(\sin 10\theta - 5\sin 6\theta + 10\sin 2\theta).$$

Evaluate

$$\int_0^{\frac{\pi}{2}} \sin^5\theta\cos^5\theta\, d\theta.$$

15) (a) Find the cube roots of -1, either by using De Moivre's theorem or by factorizing $z^3 + 1$ and hence solving algebraically the equation $z^3 + 1 = 0$.
Show that if either complex root is denoted by λ, the other is $-\lambda^2$, and that

$$(X + \lambda Y - \lambda^2 Z)(X - \lambda^2 Y + \lambda Z) = X^2 + Y^2 + Z^2 - YZ + ZX + XY.$$

(b) Express the complex number $z = 8(1 + i)/\sqrt{2}$ in the form $r(\cos\theta + i\sin\theta)$ and hence show that the three values of $z^{\frac{2}{3}}$ are $-4i$, $2(\sqrt{3} + i)$, $2(-\sqrt{3} + i)$. (U of L)

16) (a) If $1, \omega$ and ω^2 are the three cube roots of unity, find the value of
(i) $1 + \omega + \omega^2$,
(ii) $(1 + 2\omega + 3\omega^2)(1 + 2\omega^2 + 3\omega)$.
Also show that, if the equations $x^3 - 1 = 0$ and $px^5 + qx + r = 0$ have a common root, then

$$(p + q + r)(p\omega^5 + q\omega + r)(p\omega^{10} + q\omega^2 + r) = 0.$$

(b) If $|z-1| = 3|z+1|$, prove that the locus of z in an Argand diagram is a circle and find its centre and radius. (U of L)

17) Prove that $\tan \frac{\pi}{15}$ is a root of the equation

$$t^4 - 6\sqrt{3}t^3 + 8t^2 + 2\sqrt{3}t - 1 = 0.$$

Give the other roots in the form $\tan \frac{r\pi}{15}$. (O)

18) Express in the form $\cos\theta + i\sin\theta$ each of the cube roots of unity. If $\alpha^3 = \beta^3 = 1$ and $\alpha \neq \beta$, use the Argand diagram to find the value of $|\alpha - \beta|$. On the same diagram plot the points representing the three possible values of $\alpha + \beta$, and evaluate $(\alpha+\beta)^3$. (U of L)p

19) Express $\sqrt{3} - i$ in the form $re^{i\theta}$, where $r > 0$ and $-\pi < \theta \leqslant \pi$. Hence show that, when n is a positive integer,

$$(\sqrt{3}-i)^n + (\sqrt{3}+i)^n = 2^{n+1}\cos(n\pi/6).$$ (U of L)p

20) (a) Show on an Argand diagram,
(i) the three roots of the equation $z^3 - 1 = 0$,
(ii) the four roots of the equation $z^4 - 16 = 0$.
(b) If $z = e^{i\theta}$, show that
(i) $z + z^{-1} = 2\cos\theta$,
(ii) $z^n + z^{-n} = 2\cos n\theta$.
Show also that

$$\cos^6\theta = \tfrac{1}{32}(\cos 6\theta + 6\cos 4\theta + 15\cos 2\theta + 10).$$ (U of L)

21) Given that z is one of the three cube roots of unity, find the two possible values of the expression $z^2 + z + 1$.
Given that ω is a complex cube root of unity, simplify each of the expressions

$$(1 + 3\omega + \omega^2)^2 \quad \text{and} \quad (1 + \omega + 3\omega^2)^2,$$

and show that their product is equal to 16 and that their sum is -4. (JMB)

22) Solve the equation

$$z^3 = 8i,$$

giving the roots in the form $re^{i\theta}$ where $r > 0$ and $0 \leqslant \theta < 2\pi$. (U of L)

23) When $z = 4\sqrt{3}e^{i\pi/3} - 4e^{i5\pi/6}$ express z in the form $re^{i\theta}$
Hence

(a) show that $\dfrac{z}{8} + i\left(\dfrac{z}{8}\right)^2 + \left(\dfrac{z}{8}\right)^3 = 2e^{i\pi/2}$,

(b) find the cube roots of z in the (r, θ) form. (AEB '77)p

24) (a) Solve the equation $z^5 = 1$ and represent the roots on an Argand diagram.
If ω denotes any one of the non-real roots of $z^5 = 1$, show that $1 + \omega + \omega^2 + \omega^3 + \omega^4 = 0$.

(b) By expressing $\sin\theta$ and $\cos\theta$ in terms of $e^{i\theta}$ and $e^{-i\theta}$, or otherwise, prove that

$$2^5 \sin^4\theta \cos^2\theta = \cos 6\theta - 2\cos 4\theta - \cos 2\theta + 2.$$

(U of L)

25) Points P and Q represent complex numbers w and z respectively in an Argand diagram. If $w = u + iv$, $z = x + iy$ and $w = \dfrac{1 + zi}{z + i}$, express u and v in terms of x and y.
Prove that when P describes the portion of the imaginary axis between the points representing $-i$ and i, Q describes the whole of the positive half of the imaginary axis. (U of L)p

26) If $(1 + 3i)z_1 = 5(1 + i)$, express z_1 and z_1^2 in the form $x + yi$, where x and y are real.
Sketch in an Argand diagram the circle $|z - z_1| = |z_1|$ giving the coordinates of its centre. (U of L)p

27) Sketch on an Argand diagram the locus of a point P representing the complex number z, where

$$|z - 1| = |z - 3i|$$

and find z when $|z|$ has its least value on this locus. (U of L)p

28) Sketch on an Argand diagram the curve described by the equation $|z - 3 + 6i| = 2|z|$ and express the equation of this curve in Cartesian form. (U of L)p

29) Find the ratio of the greatest value of $|z + 1|$ to its least value when $|z - i| = 1$. (U of L)p

30) If P is the point on an Argand diagram representing the complex number z and $|z - 1| = 3|z + i|$, sketch the locus of P and express the equation of this locus in Cartesian form.
Find the points on the locus which satisfy the equation

$$|z| = |z - 1 + i|$$

(U of L)p

31) z is the point $x + yi$ in the Argand diagram and $\left|\dfrac{z}{z-3}\right| = \dfrac{1}{2}$.

Find the Cartesian equation of the locus of z

32) If z is any complex number such that $|z| = 1$ prove, using an Argand diagram or otherwise, that $1 \leqslant |2 + z| \leqslant 3$ and that $-\frac{1}{6}\pi \leqslant \arg(2 + z) \leqslant \frac{1}{6}\pi$. (C)p

33) (i) Given that the complex numbers w_1 and w_2 are the roots of the equation

$$z^2 - 5 - 12i = 0,$$

express w_1 and w_2 in the form $a + ib$, where a and b are real.

(ii) Indicate the point sets in an Argand diagram corresponding to the sets of complex numbers

$$A = z : |z| = 3, \ z \in \mathbb{C}$$

$$B = z : |z| = 2, \ z \in \mathbb{C}$$

Shade the region corresponding to values of z for which the inequalities

$$2 < |z| < 3,$$

$$\pi/6 < \arg z < \pi/3$$

are simultaneously satisfied. (U of L)

34) The transformation $w = \dfrac{z + i}{iz + 2}$, $z \neq 2i$, maps the complex number $z = x + iy$ on to the complex number $w = u + iv$. Find the two points in the complex plane which are invariant under the transformation. Show that if z lies on the imaginary axis, then w also lies on the imaginary axis, and that if z lies on the real axis, then w lies on the circle

$$2u^2 + 2v^2 + v - 1 = 0.$$

Sketch this circle and indicate clearly on to which part the *positive* real axis of the z plane is mapped. (JMB)

35) The transformation $T: z \to w$ in the complex plane is defined by

$$w = \frac{az + b}{z + c}, \quad (a, b, c \in \mathbb{R})$$

Given that $w = 3i$ when $z = -3i$, and $w = 1 - 4i$ when $z = 1 + 4i$, find the values of a, b and c.

(i) Show that the points for which z is transformed to $\bar{z}$ lie on a circle and give the centre and radius of this circle.

(ii) Show that the line through the point $z = 4$ and perpendicular to the real axis is invariant under T. (C)

CHAPTER 9

POLYNOMIAL FUNCTIONS AND EQUATIONS

POLYNOMIAL FUNCTIONS OF ONE VARIABLE

Remainders and Factors

In *The Core Course* we saw that a polynomial of degree n has the form

$$f(x) \equiv p_n x^n + p_{n-1}x^{n-1} + \ldots + p_0$$

where n is a positive integer and $p_n, p_{n-1}, \ldots$ are constants, of which at least p_n is non-zero.
If $f(x)$ is divided by the linear function $(x-a)$, the quotient, $Q(x)$, is a polynomial of degree $n-1$ and the remainder, R, is a constant,

i.e. $$f(x) \equiv (x-a)Q(x) + R.$$

Substituting a for x gives $f(a) = R$. This result, which is known as the Remainder Theorem, is introduced in *The Core Course* and is restated below.

When a polynomial $f(x)$ is divided by $(x-a)$ the remainder is $f(a)$.

Division of a Polynomial by a Quadratic Function

Consider the polynomial $f(x) \equiv x^6 - 2x^4 + x^2 - 2$.
Dividing $f(x)$ by $x^2 - x - 2$ by long division gives

$$
\begin{array}{r l}
 & x^4 + x^3 + x^2 + 3x + 6 \\
x^2 - x - 2 \,\big) & x^6 \quad - 2x^4 \quad + x^2 \quad - 2 \\
 & x^6 - x^5 - 2x^4 \\
\hline
 & x^5 \quad + x^2 \quad - 2 \\
 & x^5 - x^4 - 2x^3 \\
\hline
 & x^4 + 2x^3 + x^2 \quad - 2 \\
 & x^4 - x^3 - 2x^2 \\
\hline
 & 3x^3 + 3x^2 \quad - 2 \\
 & 3x^3 - 3x^2 - 6x \\
\hline
 & 6x^2 + 6x - 2 \\
 & 6x^2 - 6x - 12 \\
\hline
 & 12x + 10
\end{array}
$$

So $x^6 - 2x^4 + x^2 - 2$, when divided by $x^2 - x - 2$, gives a quotient $x^4 + x^3 + x^2 + 3x + 6$ and a remainder $12x + 10$.
If the divisor factorizes, the remainder can be found by adapting the Remainder Theorem as follows:
remainder is a linear function $Ax + B$,

i.e. $$f(x) \equiv (ax^2 + bx + c)Q(x) + Ax + B$$

Similarly we can deduce that, when a polynomial is divided by a cubic function, the remainder is quadratic.
If the divisor factorizes, the remainder can be found by adapting the remainder theorem as follows:
Consider again $f(x) \equiv x^6 - 2x^4 + x^2 - 2$ when divided by $x^2 - x - 2$.

As $$x^2 - x - 2 \equiv (x-2)(x+1)$$

we have $$f(x) \equiv (x-2)(x+1)Q(x) + Ax + B$$

$$\Rightarrow \quad \begin{cases} f(2) = 2A + B \\ f(-1) = -A + B \end{cases}$$

$$\Rightarrow \quad \begin{cases} 34 = 2A + B \\ -2 = -A + B \end{cases} \Rightarrow A = 12, \quad B = 10$$

In general, if dividing $f(x)$ by $(x-\alpha)(x-\beta)$ gives a remainder $Ax + B$ then

$$f(x) \equiv (x-\alpha)(x-\beta)Q(x) + Ax + B$$

$$\Rightarrow \quad f(\alpha) = A\alpha + B \quad \text{and} \quad f(\beta) = A\beta + B$$

This method for finding the remainder may be extended to division of a polynomial by a cubic function.

For example, if $f(x) \equiv 2x^6 - x^5 - 2x^3 - 2$ is divided by $(x-1)(x+1)(2x-1)$, the form of the remainder is $Ax^2 + Bx + C$,

i.e. $$f(x) \equiv (x-1)(x+1)(2x-1)Q(x) + Ax^2 + Bx + C$$

$$\left.\begin{aligned} f(1) &= A+B+C \\ f(-1) &= A-B+C \\ f(\tfrac{1}{2}) &= \frac{A}{4}+\frac{B}{2}+C \end{aligned}\right\} \Rightarrow \left\{\begin{aligned} -3 &= A+B+C \\ 3 &= A-B+C \\ -\frac{9}{4} &= \frac{A}{4}+\frac{B}{2}+C \end{aligned}\right.$$

$\Rightarrow$ $$A = 1, \quad C = -1, \quad B = -3.$$

Note that if the divisor does *not* factorize, and/or the quotient is required, then long division must be used.

Repeated Factors

The Factor Theorem, which follows from the Remainder Theorem, is introduced in *The Core Course* and states that

if, for a polynomial $f(x)$, $f(a) = 0$, then $(x-a)$ is a factor of $f(x)$.

If $f(x)$ has a repeated factor $(x-a)$,

i.e. $$f(x) \equiv (x-a)^2 g(x)$$

then $$f'(x) = \frac{d}{dx}[(x-a)^2 g(x)]$$

$$= (x-a)^2 g'(x) + 2(x-a)g(x) \qquad \left[\text{using } \frac{d}{dx}(uv)\right]$$

$$= (x-a)[(x-a)g'(x) + 2g(x)].$$

i.e. if $f(x)$ has a repeated factor $(x-a)$, then $f'(x)$ has a factor $(x-a)$.

This property can be verified from the graph of $f(x)$ for, if $(x-a)$ is a repeated factor of $f(x)$, then $x = a$ is a repeated root of the equation $f(x) = 0$. So the graph of $y = f(x)$ touches the x axis at $x = a$ and hence $f'(a) = 0$.

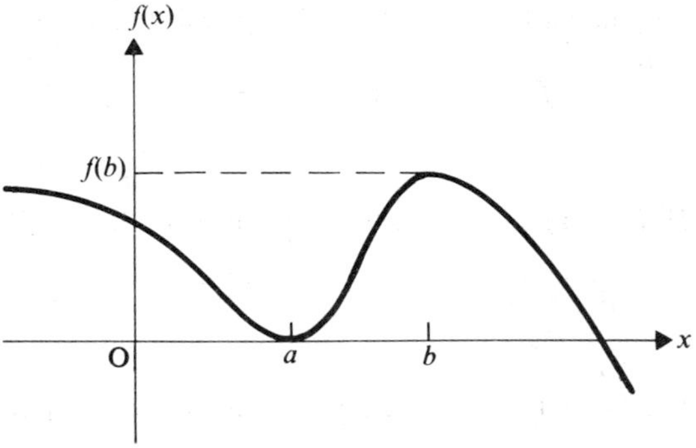

Note. A linear factor of $f'(x)$ is *not necessarily* a repeated factor of $f(x)$. For example, in the diagram above

$$f'(b) = 0 \Rightarrow (x-b) \text{ is a factor of } f'(x)$$

But $\quad f(b) \neq 0$ so $(x-b)$ is *not* a factor of $f(x)$

So the necessary and sufficient condition for $f(x)$ to have a repeated factor $(x-a)$ is that *both* $f(a)=0$ *and* $f'(a)=0$.

EXAMPLES 9a

1) Determine whether $f(x) \equiv 3x^4 - 8x^3 - 6x^2 + 24x - 13$ has any repeated factors, and, if so, find them.

$$\begin{aligned} f'(x) &\equiv 12x^3 - 24x^2 - 12x + 24 \\ &\equiv 12(x^3 - 2x^2 - x + 2) \\ &\equiv 12(x-1)(x^2 - x - 2) \\ &\equiv 12(x-1)(x+1)(x-2) \end{aligned}$$

Now $f'(x) = 0$ when $x = 1, -1$ or 2.
Checking the value of $f(x)$ for these values of x we have

$$\begin{aligned} f(1) &= 3 - 8 - 6 + 24 - 13 = 0 \\ f(-1) &= 3 + 8 - 6 - 24 - 13 \neq 0 \\ f(2) &= 48 - 64 - 24 + 48 - 13 \neq 0 \end{aligned}$$

So $(x+1)$ and $(x-2)$ are not factors of $f(x)$.
Hence $(x-1)$ is the only repeated factor of $f(x)$

2) If the equation $3x^4 + 2x^3 - 6x^2 - 6x + p = 0$ has two equal roots, find the possible values of p.

If $f(x) \equiv 3x^4 + 2x^3 - 6x^2 - 6x + p$, the equation $f(x) = 0$ has two equal roots if $f(x)$ has two equal factors, i.e. a repeated factor.
Any linear factor of $f'(x)$ is a possible repeated factor of $f(x)$.
Now

$$\begin{aligned} f'(x) \equiv 12x^3 + 6x^2 - 12x - 6 &\equiv 6(2x^3 + x^2 - 2x - 1) \\ &\equiv 6(x-1)(x+1)(2x+1) \end{aligned}$$

If $(x-1)$ is a repeated factor of $f(x)$ then $f(1) = 0$

$$\Rightarrow \qquad 3 + 2 - 6 - 6 + p = 0$$

$$\Rightarrow \qquad p = 7$$

Similarly $(x+1)$ is a repeated factor of $f(x)$ if $f(-1) = 0$.

$$\Rightarrow \qquad 3 - 2 - 6 + 6 + p = 0 \Rightarrow p = -1$$

and $(2x+1)$ is a repeated factor of $f(x)$ if $f(-\frac{1}{2}) = 0$

$$\Rightarrow \qquad \tfrac{3}{16}-\tfrac{2}{8}-\tfrac{6}{4}+\tfrac{6}{2}+p = 0 \quad \Rightarrow \quad p = -\tfrac{23}{16}$$

So the possible values of p are $7, -1, -\frac{23}{16}$.

3) Without performing long division, find the remainder when x^3-5x^2+6x-2 is divided by $(x-2)^2$.

$f(x) \equiv x^3-5x^2+6x-2$ has a remainder $Ax+B$ when divided by $(x-2)^2$,

i.e. $$f(x) \equiv (x-2)^2Q(x)+Ax+B \qquad [1]$$

Now $$f(2) = 2A+B$$

$$\Rightarrow \qquad 2A+B = -2$$

No other equation linking A and B can be obtained from [1] without involving $Q(x)$. But differentiating [1] w.r.t. x gives

$$f'(x) = \frac{\mathrm{d}}{\mathrm{d}x}[(x-2)^2Q(x)]+A$$

$$\Rightarrow \qquad 3x^2-10x+6 \equiv (x-2)^2Q'(x)+2(x-2)Q(x)+A$$

$$\Rightarrow \qquad f'(2) = 12-20+6 = A \quad \Rightarrow \quad A = -2$$

So $$B = 2$$

Therefore the remainder is $-2x+2$.

4) Find the constant p such that x^2+2 is a factor of x^4-6x^2+p. Hence factorize x^4-6x^2+p.

If x^2+2 is a factor of x^4-6x^2+p

then $$x^4-6x^2+p \equiv (x^2+2)f(x)$$

where $f(x)$ is of degree 2, i.e. $f(x) \equiv ax^2+bx+c$.

So $$x^4-6x^2+p \equiv (x^2+2)(ax^2+bx+c)$$

Comparing coefficients of x^4 gives $1 = a$.
Comparing coefficients of x^3 gives $0 = b$.
Comparing coefficients of x^2 gives $-6 = 2a+c \Rightarrow c = -8$.
Comparing constants gives $p = 2c = -16$.

Therefore $$x^4-6x^2+p \equiv x^4-6x^2-16 \equiv (x^2+2)(x^2-8)$$
$$\equiv (x^2+2)(x-2\sqrt{2})(x+2\sqrt{2}).$$

Common Factors

If two polynomials $f(x)$ and $g(x)$ have a common factor $(x-a)$ then

$$f(x) \equiv (x-a)h(x) \qquad [1]$$

$$g(x) \equiv (x-a)j(x) \qquad [2]$$

For any constant K

$$f(x)+Kg(x) \equiv (x-a)h(x)+K(x-a)j(x)$$

$$\equiv (x-a)[h(x)+Kj(x)]$$

i.e. if $f(x)$ and $g(x)$ have a common factor $(x-a)$ then, for any constant K, $(x-a)$ is a factor of $f(x)+Kg(x)$.

This property is sometimes useful for solving problems concerning two polynomials with a common factor.
Also, from identities [1] and [2], it follows that

$$f(a) = 0 \quad \text{and} \quad g(a) = 0$$

giving a pair of simultaneous equations which provides another approach to problems involving a common factor. Both of these approaches are illustrated in the following examples.

EXAMPLES 9a (continued)

5) Find the constants p and q such that $x-2$ is a common factor of

$$x^3-x^2-2px+3q \quad \text{and} \quad qx^3-px^2+x+2.$$

If $(x-2)$ is a factor of $f(x) \equiv x^3-x^2-2px+3q$, then

$$f(2) \equiv 0 \Rightarrow 4-4p+3q = 0 \qquad [1]$$

If, also $(x-2)$ is a factor of $g(x) \equiv qx^3-px^2+x+2$ then

$$g(2) = 0 \Rightarrow 8q-4p+4 = 0 \qquad [2]$$

Solving equations [1] and [2] simultaneously gives

$$p = 1, \qquad q = 0$$

6) Show that if $(x+1)$ is a common factor of x^3-ax^2+b and $x^4-ax^3+bx^2+c$ then $2a = 2b-2 = -2-c$.

Let $f(x) \equiv x^3-ax^2+b$ and $g(x) \equiv x^4-ax^3+bx^2+c$.
For *any* value of K, $(x+1)$ is factor of $f(x)+Kg(x)$

therefore $$f(-1)+Kg(-1) = 0$$

i.e. for any value of K.

$$(-1-a+b)+K(1+a+b+c) = 0$$

When $K=1$, $2b+c=0$.
When $K=0$, $a-b=-1$.

Hence
$$a = b-1 = -\frac{c}{2}-1$$

$\Rightarrow$
$$2a = 2b-2 = -2-c$$

7) Find the relationship between a, b and c such that the equations $x^2-ax+b=0$ and $ax^2+x-c=0$ have a common root.
If the common root is α, then α satisfies both given equations,

i.e.
$$\begin{cases} \alpha^2 - a\alpha + b = 0 & [1] \\ a\alpha^2 + \alpha - c = 0 & [2] \end{cases}$$

To find a relationship between a, b and c, we must eliminate α and α^2 from equations [1] and [2].

[1] $+a$[2] $\Rightarrow$
$$(1+a^2)\alpha^2 + (b-ac) = 0$$

$\Rightarrow$
$$\alpha^2 = \frac{ac-b}{1+a^2} \quad [3]$$

[2] $-a$[1] $\Rightarrow$
$$(1+a^2)\alpha - (c+ab) = 0$$

$\Rightarrow$
$$\alpha = \frac{c+ab}{1+a^2} \quad [4]$$

From equations [3] and [4] we have
$$\frac{ac-b}{1+a^2} = \left(\frac{c+ab}{1+a^2}\right)^2$$

$\Rightarrow$
$$(ac-b)(1+a^2) = (c+ab)^2$$

EXERCISE 9a

1) Find the remainder when $x^4-5x^3+6x^2-7$ is divided by $(x-1)(x-3)$.

2) Find the remainder when x^4+x^2-7 is divided by x^2-4.

3) Find the constants p and q such that when x^3-px+q is divided by x^2-3x+2, the remainder is $4x-1$.

4) Find the constants a, b and c such that when x^5-7x^3+4x-2 is divided by $(x-1)(x+1)(x-3)$ the remainder is ax^2+bx+c.

5) Find the remainder when $x^3 - 5x^2 + 7$ is divided by $(x-1)^2$.

6) Find the constants m and n such that when $x^4 - mx^2 + n$ is divided by $(x+1)^2$ the remainder is $5x - 2$.

7) Determine whether the given functions have any repeated factors and, if they have, find them.
(a) $x^4 - 16$ (b) $x^4 - 18x^2 + 81$ (c) $2x^3 - 3x^2 + 1$ (d) $x(x^2 - 4)$.

8) If the equation $2x^3 - 9x^2 + 12x + p = 0$ has two equal roots, find the possible values of p.

9) Find the value of a for which the function $2x^3 - ax^2 - 12x - 7$ has a repeated factor.

10) Find the constant m for which $x^2 + 1$ is a factor of $mx^4 + x^2 - 1$.

11) Show that $x^2 + 3$ is a factor of $x^3 - x^2 + 3x - 3$.

12) Show that if $(x - a)$ is a factor of $p_n x^n + p_{n-1}x^{n-1} + \ldots + p_0$ then $|a|$ is a factor of $|p_0|$.

13) Find the constant a for which the functions $f(x) \equiv ax^2 + 2x - 1$ and $g(x) \equiv x^2 + 4x + a$ have a common factor.

14) Find the constants p and q such that $(x - 1)$ is a common factor of $x^4 - 2px^2 + 2$ and $x^4 + x^2 - q$.

15) Show that if the cubic functions $x^3 + ax^2 + b$ and $ax^3 + bx^2 + x - a$ have a common factor, it is also a factor of the quadratic function $(b - a^2)x^2 + x - a(1 + b)$.

16) Determine the condition that the equations

$$px^2 + qx + 1 = 0 \quad \text{and} \quad x^2 + px + q = 0$$

have a common root.

17) Determine the value of m such that the equations

$$x^3 + mx - 1 = 0 \quad \text{and} \quad x^3 - 3x + m = 0$$

have a common root.

POLYNOMIAL FUNCTIONS OF MORE THAN ONE VARIABLE

A polynomial in three variables x, y, z consists of the sum of terms such as $px^l y^m z^n$ where p is a constant and l, m, n are positive integers. This definition can be extended to any number of variables. The *degree* of such a term is the *sum* of the powers of the variables. For example $2x^2y^3$ is a term of degree five, $5xy^3z^5$ is of degree nine, $4a^3b^3$ is of degree six. The term of highest degree determines the degree of the polynomial.

Homogeneous Functions

A homogeneous function is a polynomial, each of whose terms is of the same degree. For example

x^2+xy-y^2 is a homogeneous polynomial in x and y of degree 2.

$\alpha^2\beta+\beta^2\gamma+\gamma^2\alpha$ is a homogeneous polynomial in α, β, γ of degree 3.

Cyclic Functions

Consider the polynomial $(x-y)(y-z)(z-x)$.
If x and y (or any other *pair* of x, y, z) are interchanged, the polynomial changes.
But if x is replaced by y, y is replaced by z and z is replaced by x (i.e. the variables are interchanged in the cyclic order  the polynomial becomes $(y-z)(z-x)(x-y)$ which is identical to the original function.
Any function that remains the same when *all* the variables are interchanged in *cyclic order* is said to be a *cyclic function*.

For example

$$\left.\begin{aligned}&\alpha\beta+\beta\gamma+\gamma\alpha\\&xy^2+yz^2+zx^2\\&(a^2-b^2)(b^2-c^2)(c^2-d^2)(d^2-a^2)\end{aligned}\right\}\text{ are cyclic functions}$$

but

$$\left.\begin{aligned}&x^2+xy-y^2\\&(a^2+b^2)(b^2-c^2)(c^2+a^2)\end{aligned}\right\}\text{ are not cyclic.}$$

The Sigma Notation for Cyclic Functions

The sigma notation provides a useful shorthand form for a cyclic function which is a *sum* of terms.
For example $f(xyz) \equiv xy+yz+zx$ can be written as $f(xyz) \equiv \sum xy$ where $\sum xy$ means the *sum* of all *different* terms found by interchanging x, y and z in cyclic order. Similarly $f(xyz) \equiv \sum x(y+z)$ means the sum of all different terms found by interchanging x, y and z in cyclic order,

i.e. $$\sum x(y+z) \equiv x(y+z)+y(z+x)+z(x+y).$$

Note that $\sum x(y+z)$ is *not* the same as $\sum xy$, in fact $\sum x(y+z) = 2\sum xy$.

EXERCISE 9b

State the degree of the following functions and state also whether they are homogeneous and/or cyclic.

1) $x+y+z$

2) $x^2-y^2+z^2$

3) $ab+a^2+b^2$

4) $a(b-c)+b(a-c)$

5) $\alpha^2(\beta+\gamma)+\beta^2(\gamma+\alpha)+\gamma^2(\beta-\alpha)$

6) $\alpha^2(\beta+1)+\beta^2(\gamma+1)+\gamma^2(\alpha+1)$

7) $(p-q)(q-r)(r-p)$

8) $a^2-b^2-c^2$

9) $ab+bc+ca+a^2+b^2+c^2$

10) $2(x^2+y^2+z^2)-3(xy+yz+zx)$

11) Write out the following cyclic functions in full.

(a) $f(\alpha\beta\gamma)\equiv\sum\alpha(\alpha^2-\beta^2)$

(b) $f(xyz)\equiv\sum xy^2$

(c) $f(\alpha\beta\gamma)\equiv\sum\alpha(\beta^2-\gamma^2)$

(d) $f(abcd)\equiv\sum ab^2$

12) Write the following cyclic functions in the sigma notation.

(a) $x^2(y^2+z^2)+y^2(z^2+x^2)+z^2(x^2+y^2)$

(b) $\alpha^2(\beta+\gamma)+\beta^2(\gamma+\alpha)+\gamma^2(\alpha+\beta)$

(c) $\alpha+\beta+\gamma$

(d) $a^2b+b^2c+c^2d+d^2a$

FACTORIZATION OF HOMOGENEOUS AND CYCLIC POLYNOMIALS

When attempting to factorize a polynomial in several variables, the following considerations should be noted.

(a) If the polynomial is homogeneous, its factors also are homogeneous.

(b) If the polynomial is cyclic, the product of its factors is a cyclic function.

These points help in selecting likely linear factors of a particular polynomial. The Factor Theorem can then be used to determine whether they are, or are not, factors.

EXAMPLES 9c

1) Factorize a^3-b^3.

$f(ab)\equiv a^3-b^3$ is homogeneous, so any linear factors will be of the form $a\pm b$.

When $a=-b$, $\quad f(-bb) = -b^3-b^3 \neq 0$

so $a+b$ is not a factor.

When $a=b$, $\quad f(bb) = b^3-b^3 = 0$

so $a-b$ is a factor.

Therefore $\quad a^3-b^3 \equiv (a-b)g(ab)$

Now $g(ab)$ is homogeneous, and of degree 2.

So the general form of $g(ab)$ is $Aa^2 + Bab + Cb^2$ where A, B, and C are constants,

i.e. $$a^3 - b^3 \equiv (a-b)(Aa^2 + Bab + Cb^2)$$

Comparing coefficients of a^3, b^3, a^2b gives $A = B = C = 1$

Therefore $$a^3 - b^3 \equiv (a-b)(a^2 + ab + b^2)$$

As $a^2 + ab + b^2$ has no real linear factors, $a^3 - b^3$ cannot be factorized further.

Note that in this and the following examples, what would normally be 'thought processes' are written down to clarify the working. In practice, the factorization of $a^3 - b^3$ would be written down simply as:

When $a = b$, $$a^3 - b^3 = 0$$

Therefore $$a^3 - b^3 \equiv (a-b)(a^2 + kab + b^2)$$

Comparing coefficients of a^2b gives $k = 1$.

Hence $$a^3 - b^3 \equiv (a-b)(a^2 + ab + b^2)$$

The result in the example above is a special case of the following more general results.
Consider the function $x^n - a^n$ where n is any positive integer.

When $x = a$, $$x^n - a^n = a^n - a^n = 0$$

So $(x - a)$ is a factor of $x^n - a^n$.
Now consider the function $x^n + a^n$.
It is clear that $(x - a)$ is not a factor of this polynomial.
But if n is odd, i.e. $n = 2m + 1$
then when $x = -a$,

$$x^n + a^n = (-a)^{2m+1} + a^{2m+1} = 0$$

So $(x + a)$ is a factor of $x^n + a^n$ when n is odd.
These results are quotable and are summarized below:

If n is a positive integer
$(x - a)$ is a factor of $x^n - a^n$ for all values of n
$(x + a)$ is a factor of $x^n + a^n$ for odd values of n.

In particular

$$x^3 - y^3 \equiv (x-y)(x^2 + xy + y^2)$$
$$x^3 + y^3 \equiv (x+y)(x^2 - xy + y^2)$$

EXAMPLES 9c (continued)

2) Factorize $a(b^2-c^2)+b(c^2-a^2)+c(a^2-b^2)$.

Let $$f(abc) \equiv a(b^2-c^2)+b(c^2-a^2)+c(a^2-b^2)$$

$f(abc)$ is homogeneous so any linear factors are of the form
$$a \pm b, \quad b \pm c, \quad c \pm a.$$

When $a=b$, $f(abc) = b(b^2-c^2)+(c^2-b^2)+0 = 0$

so $(a-b)$ is a factor.
As $f(abc)$ is cyclic, it follows that $(a-b)$ is one of a set of factors that, as a whole, is cyclic, so $(b-c)$ and $(c-a)$ also are factors.
Now $f(abc)$ is of degree 3 and we have found three linear factors. Hence the only other possible factor is a constant, k,

i.e. $$f(abc) \equiv k(a-b)(b-c)(c-a)$$

Comparing coefficients of ab^2 gives $k=1$.
Therefore
$$a(b^2-c^2)+b(c^2-a^2)+c(a^2-b^2) \equiv (a-b)(b-c)(c-a)$$

3) Factorize $(x+y)^3(x-y)+(y+z)^3(y-z)+(z+x)^3(z-x)$.

Let $$f(xyz) \equiv (x+y)^3(x-y)+(y+z)^3(y-z)+(z+x)^3(z-x)$$

$f(xyz)$ is homogeneous so likely linear factors are of the form
$$x \pm y, \quad x \pm z, \quad y \pm z.$$

When $x=y$,
$$f(yyz) = 0+(y+z)^3(y-z)+(z+y)^3(z-y) = 0$$

Hence $(x-y)$ is a factor.
Similarly $(y-z)$ and $(z-x)$ are factors.
Therefore $(x-y)(y-z)(z-x)$ is a factor of $f(xyz)$.
Now $f(xyz)$ is cyclic and of degree 4 and the product of the factors that we have found so far is also cyclic and of degree 3. Hence the remaining factor is linear and must be cyclic in x, y *and* z. So it can only be of the form $x+y+z$. The only other possible factor is a constant k.

Therefore $$f(xyz) \equiv k(x+y+z)(x-y)(y-z)(z-x)$$

Comparing coefficients of x^3y gives $k=1$.

So $$f(xyz) \equiv (x+y+z)(x-y)(y-z)(z-x)$$

4) Factorize $a^3+b^3+c^3-3abc$.

Let $$f(abc) \equiv a^3+b^3+c^3-3abc$$

$f(abc)$ is homogeneous, so any linear factors are of the form

$$a \pm b, \quad b \pm c, \quad c \pm a, \quad a \pm b \pm c$$

when $a = \pm b$, $f(\pm bbc) \neq 0$ so $(a+b)$ and $(a-b)$ are not factors. Similarly $b \pm c$, $c \pm a$ are not factors.
When $a = -(b+c)$

$$\begin{aligned} f(-\{b+c\}bc) &= -(b+c)^3 + b^3 + c^3 + 3(b+c)bc \\ &= -b^3 - 3b^2c - 3bc^2 - c^3 + b^3 + c^3 + 3b^2c + 3bc^2 \\ &= 0 \end{aligned}$$

Hence $(a+b+c)$ is a factor.
Note that $f(abc)$ is cyclic, so its factors *as a whole* must make a cyclic function. The one factor found so far, $a+b+c$, is cyclic so the remaining factor(s) must form a cyclic group. Hence $a-b-c$ and $a+b-c$ cannot be factors.
Now $f(abc)$ is cyclic and of degree 3, and the factor that we have found is also cyclic. So the other factor must be cyclic and of degree 2. The most general form for such an expression is $k_1(a^2+b^2+c^2) + k_2(ab+bc+ca)$ where k_1 and k_2 are constants.
Therefore

$$a^3 + b^3 + c^3 - 3abc \equiv (a+b+c)[k_1(a^2+b^2+c^2) + k_2(ab+bc+ca)]$$

Comparing coefficients of a^3 and of abc gives $k_1 = 1$ and $k_2 = -1$.
Hence

$$a^3 + b^3 + c^3 - 3abc \equiv (a+b+c)(a^2+b^2+c^2-ab-bc-ca)$$

EXERCISE 9c

Factorize the following functions.

1) $(a-b)^3 + (a+b)^3$

2) $x^2 + y^2 + z^2 + 2xy + 2yz + 2zx$

3) $x^2(y-z) + y^2(z-x) + z^2(x-y)$

4) $a(b^2-c^2) + b(c^2-a^2) + c(a^2-b^2)$

5) $(a-b)^3 + (b-c)^3 + (c-a)^3$

6) $x^4(y-z) + y^4(z-x) + z^4(x-y)$

7) $a^6 - b^6$

8) $x^6 - 64$

9) $pq(p-q) + qr(q-r) + rp(r-p)$

10) $a^3 + b^3 + c^3 + 3bc(b+c)$

11) Find the sum of the n terms of the geometric progression

$$x^{n-1} + ax^{n-2} + a^2x^{n-3} + \ldots + a^{n-1}.$$

Hence show that

$$x^n - a^n \equiv (x-a)(x^{n-1} + ax^{n-2} + \ldots + a^{n-1}).$$

Use this result to write down the factors of

$$x^5 - 32 \quad \text{and} \quad a^5 - b^5.$$

12) If m and n are integers, show that $(x-y)$ is a factor of

$$x^n(y^m - z^m) + y^n(z^m - x^m) + z^n(x^m - y^m).$$

13) Show that $(x-a)^2$ is a factor of $x^3 - ax^2 - a^2x + a^3$.
Hence factorize $p^3 - p^2q - pq^2 + q^3$.

14) Factorize $a^3 + 8b^3 + 27c^3 - 18abc$.

POLYNOMIAL EQUATIONS

The Nature of the Roots

It was seen in *The Core Course*, that a quadratic equation has either two real roots (distinct or equal) or two conjugate complex roots.
Now consider a cubic equation $f(x) = 0$ which, as we already know, *must* have at least one real root. Therefore the cubic function $f(x)$ has at least one linear factor, $x - \alpha$.

So $$f(x) \equiv (x-\alpha)(ax^2 + bx + c)$$

Now $$f(x) = 0 \Rightarrow x = \alpha \quad \text{or} \quad ax^2 + bx + c = 0$$

But $ax^2 + bx + c = 0$ has either two real roots $(b^2 - 4ac \geqslant 0)$ or two conjugate complex roots $(b^2 - 4ac < 0)$.
Thus a cubic equation has

either three real roots (not necessarily distinct)

or one real root and a pair of conjugate complex roots.

Polynomial equations of degree higher than three have a similar property,

i.e. if a polynomial equation has any complex roots, they occur in conjugate pairs.

(A proof of this property is given later in this chapter.)
It therefore follows that no polynomial equation can have an odd number of complex roots.
Assuming that a polynomial equation of degree n has n roots, it also follows that such an equation has

at least one real root if n is odd, but *may* have no real roots if n is even.

RELATIONSHIPS BETWEEN ROOTS AND COEFFICIENTS

It has already been established that if a quadratic equation $ax^2 + bx + c = 0$ has roots α and β, then

$$\alpha + \beta = -\frac{b}{a} \quad \text{and} \quad \alpha\beta = \frac{c}{a}$$

Similar relationships between the roots and the coefficients of polynomial equations of higher degree can be found as follows.

Cubic Equations

The general cubic equation can be written

$$ax^3 + bx^2 + cx + d = 0 \qquad [1]$$

and if its roots are α, β, γ, then the equation can also be written in the form

$$(x-\alpha)(x-\beta)(x-\gamma) = 0 \qquad [2]$$

Dividing equation [1] by a
and comparing it with the expansion of equation [2], we have

$$\begin{cases} x^3 + \dfrac{b}{a}x^2 + \dfrac{c}{a}x + \dfrac{d}{a} = 0 \\ x^3 - (\alpha+\beta+\gamma)x^2 + (\alpha\beta+\beta\gamma+\gamma\alpha)x - \alpha\beta\gamma = 0 \end{cases}$$

As these two forms represent the same equation *and* the terms in x^3 are identical, the terms in x^2, x and the constant, must also be identical,

i.e.
$$\frac{b}{a} = -(\alpha+\beta+\gamma) \equiv -\sum\alpha$$

$$\frac{c}{a} = \alpha\beta+\beta\gamma+\gamma\alpha \equiv \sum\alpha\beta$$

$$\frac{d}{a} = -\alpha\beta\gamma$$

or
$$\sum\alpha = -\frac{b}{a}$$

$$\sum\alpha\beta = \frac{c}{a}$$

$$\alpha\beta\gamma = -\frac{d}{a}$$

Quartic Equations

Carrying out a similar investigation of the general quartic equation

$$ax^4 + bx^3 + cx^2 + dx + e = 0$$

with roots $\alpha, \beta, \gamma, \delta$, leads to comparing

$$x^4 + \frac{b}{a}x^3 + \frac{c}{a}x^2 + \frac{d}{a}x + \frac{e}{a} = 0$$

with $$x^4 - \left(\sum\alpha\right)x^3 + \left(\sum\alpha\beta\right)x^2 - \left(\sum\alpha\beta\gamma\right)x + \alpha\beta\gamma\delta = 0$$

Equating coefficients of corresponding terms gives

$$\sum\alpha = -\frac{b}{a}$$

$$\sum\alpha\beta = \frac{c}{a}$$

$$\sum\alpha\beta\gamma = -\frac{d}{a}$$

$$\alpha\beta\gamma\delta = \frac{e}{a}$$

The results obtained for the relationships between the roots and the coefficients of quadratic, cubic and quartic equations, establish a pattern which suggests further relationships for higher degree equations.

Quadratic	Cubic	Quartic	Quintic
$\sum\alpha = -\frac{b}{a}$	$\sum\alpha = -\frac{b}{a}$	$\sum\alpha = -\frac{b}{a}$	$\sum\alpha = -\frac{b}{a}$
$\alpha\beta = \frac{c}{a}$	$\sum\alpha\beta = \frac{c}{a}$	$\sum\alpha\beta = \frac{c}{a}$	$\sum\alpha\beta = \frac{c}{a}$
	$\alpha\beta\gamma = -\frac{d}{a}$	$\sum\alpha\beta\gamma = -\frac{d}{a}$	$\sum\alpha\beta\gamma = -\frac{d}{a}$
		$\alpha\beta\gamma\delta = \frac{e}{a}$	$\sum\alpha\beta\gamma\delta = \frac{e}{a}$
			$\alpha\beta\gamma\delta\epsilon = -\frac{f}{a}$

The reader can verify the truth of the results quoted above, by extrapolation, for a fifth degree (quintic) equation, by adopting the method already used for the lower powers.

Note that even if some of the roots are complex, the sum and the product of the roots are both real, verifying that complex roots occur in conjugate pairs.

APPLICATIONS OF THE RELATIONSHIPS BETWEEN ROOTS AND COEFFICIENTS

Many problems that are based on relationships between roots and coefficients can be approached by using the standard relationships derived in this chapter (but the *solution* of cubic or quartic equations is *not*, in general, assisted by these relationships).
The following examples illustrate some of the methods that can be adopted.

EXAMPLES 9d

1) If the equation $x^3+px^2+qx+r=0$ has roots α, β, γ express in terms of p, q and r,

(a) $\sum \alpha^2$, (b) $\sum \alpha\beta(\alpha+\beta)$.

(a)
$$\sum \alpha^2 \equiv \alpha^2+\beta^2+\gamma^2$$
$$\equiv (\alpha+\beta+\gamma)^2-(2\alpha\beta+2\beta\gamma+2\gamma\alpha)$$
$$\equiv \left(\sum \alpha\right)^2 - 2\left(\sum \alpha\beta\right)$$

But
$$\sum \alpha = -\frac{b}{a} = -p \quad \text{and} \quad \sum \alpha\beta = \frac{c}{a} = q$$

So
$$\sum \alpha^2 = p^2-2q$$

(b) Terms such as $\alpha^2\beta$ occur, amongst other terms, in the product of $(\alpha+\beta+\gamma)$ and $(\alpha\beta+\beta\gamma+\gamma\alpha)$, so we will consider this product in full.

$$(\alpha+\beta+\gamma)(\alpha\beta+\beta\gamma+\gamma\alpha) \equiv \alpha^2\beta+\beta^2\alpha+\beta^2\gamma+\gamma^2\beta+\gamma^2\alpha+\alpha^2\gamma+3\alpha\beta\gamma$$
$$\equiv \sum \alpha\beta(\alpha+\beta)+3\alpha\beta\gamma$$

So
$$\sum \alpha\beta(\alpha+\beta) \equiv \left(\sum \alpha\right)\left(\sum \alpha\beta\right) - 3\alpha\beta\gamma$$
$$= (-p)(q)-3(-r)$$
$$= 3r-pq$$

Note. The expression $\sum \alpha\beta(\alpha+\beta)$ is an unambiguous representation of the *six* terms

$$\alpha^2\beta+\alpha\beta^2+\beta^2\gamma+\beta\gamma^2+\gamma^2\alpha+\gamma\alpha^2$$

whereas $\sum \alpha^2\beta$ means only the *three* terms in cyclic order

$$\alpha^2\beta+\beta^2\gamma+\gamma^2\alpha$$

However the reader may well encounter $\sum \alpha^2\beta$ being used to represent the full set of six terms above, and is therefore warned to interpret the meaning of $\sum \alpha^2\beta$ with caution and by taking account of the context.

2) If the equation $x^3 + px^2 + qx + r = 0$ has roots that are in arithmetic progression, show that $2p^3 - 9pq + 27r = 0$.

If the roots form an arithmetic progression with common difference λ, and α is the middle root, then the three roots are

$$\alpha - \lambda,\ \alpha,\ \alpha + \lambda.$$

From the given equation we see that

$$a = 1, \quad b = p, \quad c = q, \quad d = r$$

so $$\sum \alpha = (\alpha - \lambda) + \alpha + (\alpha + \lambda) = -p$$

$\Rightarrow$ $$3\alpha = -p \qquad [1]$$

But α satisfies the given equation so we have

$$\left(-\frac{p}{3}\right)^3 + p\left(-\frac{p}{3}\right)^2 + q\left(-\frac{p}{3}\right) + r = 0$$

$\Rightarrow$ $$2p^3 - 9pq + 27r = 0$$

Equations With Related Roots

Suppose that a cubic equation $ax^3 + bx^2 + cx + d = 0$ has roots α, β, γ and that a second cubic equation $a_1x^3 + b_1x^2 + c_1x + d_1 = 0$ has roots $\alpha_1, \beta_1, \gamma_1$ where α_1, β_1 and γ_1 are functions of α, β and γ, then a_1, b_1, c_1 and d_1, must be related to a, b, c and d. The following examples show how this relationship can be applied in problems.

EXAMPLES 9d (continued)

3) If the roots of the equation $4x^3 + 7x^2 - 5x - 1 = 0$ are α, β and γ, find the equation whose roots are

(a) $\alpha + 1$, $\beta + 1$ and $\gamma + 1$, (b) $\alpha^2, \beta^2, \gamma^2$.

(a) If $4x^3 + 7x^2 - 5x - 1 = 0$
then $x = \alpha, \beta, \gamma$.
If $f(X) = 0$ is the required equation
then $X = \alpha + 1, \beta + 1, \gamma + 1$.
So, for each of the roots,

$$X = x + 1 \quad \Rightarrow \quad x = X - 1$$

But x satisfies the given equation, so

$$4(X-1)^3 + 7(X-1)^2 - 5(X-1) - 1 = 0$$

i.e. the required equation is $$4X^3 - 5X^2 - 7X + 7 = 0$$

(b) This time $$x = \alpha, \beta, \gamma$$

and $$X = \alpha^2, \beta^2, \gamma^2$$

So $$X = x^2 \quad \Rightarrow \quad x = \pm X^{\frac{1}{2}}$$

The given equation thus becomes

$$\pm 4X^{\frac{3}{2}} + 7X \mp 5X^{\frac{1}{2}} - 1 = 0$$

This is not in a satisfactory form so, first, we rearrange it to isolate $X^{\frac{1}{2}}$, then square both sides, as follows:

$$\pm (4X^{\frac{3}{2}} - 5X^{\frac{1}{2}}) = 1 - 7X$$

$$\Rightarrow \quad [\pm X^{\frac{1}{2}}(4X - 5)]^2 = (1 - 7X)^2$$

$$\Rightarrow \quad X(16X^2 - 40X + 25) = 1 - 14X + 49X^2$$

Thus the required equation is

$$16X^3 - 89X^2 + 39X - 1 = 0$$

Note. A relationship between X and x can be found in some examples where there does not, initially, appear to be a simple connection.
For example, if the required roots are $\beta\gamma, \gamma\alpha$ and $\alpha\beta$ we can convert them into

$$\frac{\beta\gamma\alpha}{\alpha}, \frac{\gamma\alpha\beta}{\beta} \text{ and } \frac{\alpha\beta\gamma}{\gamma}$$

Then, as $\alpha\beta\gamma = -\frac{d}{a}$, the required roots are $-\frac{d}{a\alpha}, -\frac{d}{a\beta}$ and $-\frac{d}{a\gamma}$

Thus $$X = -\frac{d}{ax} \quad \Rightarrow \quad x = -\frac{d}{aX}$$

Note also that it is not *always* possible to find a simple relationship between X and x. An alternative method for such cases is given in the following example.

4) If α, β and γ are the roots of the equation $x^3 + 7x + 5 = 0$, find the equation whose roots are $\alpha^2 + 1$, $\beta^2 + 1$ and $\gamma^2 + 1$.

From the given equation we have

$$\sum \alpha = -\frac{b}{a} = 0$$

$$\sum \alpha\beta = \frac{c}{a} = 7$$

$$\alpha\beta\gamma = -\frac{d}{a} = -5$$

Now if the required equation is

$$X^3 + \frac{B}{A}X^2 + \frac{C}{A}X + \frac{D}{A} = 0$$

we have

$$-\frac{B}{A} = \sum(\alpha^2 + 1) \equiv \left(\sum \alpha^2\right) + 3$$

$$\equiv \left(\sum \alpha\right)^2 - 2\sum \alpha\beta + 3$$

$$= 0 - 2(7) + 3 = -11$$

$$\Rightarrow \frac{B}{A} = 11$$

$$\frac{C}{A} = \sum (\alpha^2 + 1)(\beta^2 + 1) \equiv \sum \alpha^2\beta^2 + 2\sum\alpha^2 + 3$$

But $\sum \alpha^2\beta^2 \equiv (\alpha\beta + \beta\gamma + \gamma\alpha)^2 - 2(\alpha\beta^2\gamma + \beta\gamma^2\alpha + \gamma\alpha^2\beta)$

$$\equiv \left(\sum \alpha\beta\right)^2 - 2\alpha\beta\gamma \sum \alpha$$

$$= 49 - 0$$

and $\sum\alpha^2 = \left(\sum \alpha\right)^2 - 2\sum \alpha\beta$

$$= 0 - 14$$

So $\frac{C}{A} = 49 - 28 + 3$

$$\Rightarrow \frac{C}{A} = 24$$

$$-\frac{D}{A} = (\alpha^2 + 1)(\beta^2 + 1)(\gamma^2 + 1)$$

$$\equiv \alpha^2\beta^2\gamma^2 + \sum \alpha^2\beta^2 + \sum \alpha^2 + 1$$

$$= (-5)^2 + 49 - 14 + 1 = 61$$

$$\Rightarrow \frac{D}{A} = -61.$$

So the required equation is

$$X^3 + 11X^2 + 24X - 61 = 0$$

EXERCISE 9d

1) If α, β, γ are the roots of the following equations, write down the values of $\sum\alpha$, $\sum\alpha\beta$ and $\alpha\beta\gamma$.

(a) $4x^3 - x^2 + 2x = 7$ (b) $x^3 - 3x + 1 = 0$ (c) $8x^3 = 1$
(d) $x^3 - x = 0$ (e) $x^3 + 4x^2 = 5$

2) Find the values of $\sum \alpha^2$, $\sum \alpha^2\beta^2$, $\sum \alpha\beta(\alpha+\beta)$ and $\sum\dfrac{1}{\alpha}$ in each of the following cases.

(a) $x^3 - 3x^2 + x + 5 = 0$ (b) $3x^3 + x^2 - 4x + 1 = 0$
(c) $4x^3 + 3x + 7 = 0$ (d) $x^4 - x^3 + 2x + 3 = 0$
(e) $x^3 + 1 = 0$ (f) $x^4 + x = 1$
(g) $x^4 + x^3 = 0$

3) If the equation $x^3 + 2x^2 - 5x + 1 = 0$ has roots α, β, γ, find the equation with roots

(a) $\alpha - 2,\ \beta - 2,\ \gamma - 2$ (b) $\dfrac{1}{\alpha}, \dfrac{1}{\beta}, \dfrac{1}{\gamma}$
(c) $2\alpha,\ 2\beta,\ 2\gamma$ (d) $\alpha^2, \beta^2, \gamma^2$
(e) $\beta\gamma,\ \gamma\alpha,\ \alpha\beta$ (f) $\alpha+\beta,\ \beta+\gamma,\ \gamma+\alpha$
(g) $\alpha^2 + \alpha,\ \beta^2 + \beta,\ \gamma^2 + \gamma$.

4) The equation $7x^3 - 4x - 11 = 0$ has roots α, β, γ. Prove that

(a) $7\sum\alpha^3 = 33 + 4\sum\alpha$ (b) $7\sum\alpha^4 = 11\sum\alpha + 4\sum\alpha^2$.

Evaluate $\sum\alpha^3$ and $\sum\alpha^4$ and write down a similar expression for $\sum\alpha^5$.

5) Find the relationship between a, b, c and d if the roots of the equation $ax^3 + bx^2 + cx + d = 0$ are:

(a) in geometric progression,
(b) such that one root is equal to the sum of the other two,
(c) all equal.

6) If $\alpha + \beta + \gamma = 2$, $\alpha\beta + \beta\gamma + \gamma\alpha = -5$ and $\alpha\beta\gamma = -6$, write down the equation with roots α, β, γ and hence evaluate α, β and γ.

7) If $\alpha, \beta, \gamma, \delta$ are the roots of the equation

$$3x^4 + 4x^3 - 7x^2 + 5x - 3 = 0$$

write down the values of $\sum\alpha$, $\sum\alpha\beta$, $\sum\alpha\beta\gamma$ and $\alpha\beta\gamma\delta$.

Find the values of $\sum\alpha^2$ and $\sum\frac{1}{\alpha}$.

Find, also, the equation whose roots are $\frac{\alpha}{2}, \frac{\beta}{2}, \frac{\gamma}{2}$ and $\frac{\delta}{2}$.

8) If $f(x)$ is a quartic function of x with a repeated factor, $(x-a)$, prove that $(x-a)$ is also a factor of $f'(x)$.
Hence solve the equation

$$3x^4-8x^3-6x^2+24x = 8.$$

9) If the roots of the equation

$$x^4-px^3+qx^2-pqx+1 = 0$$

are α, β, γ and δ, show that

$$(\alpha+\beta+\gamma)(\alpha+\beta+\delta)(\alpha+\gamma+\delta)(\beta+\gamma+\delta) = 1.$$

10) Find the equation whose roots are given by adding 2 to the roots of the equation

$$x^4+3x^3-13x^2-51x-36 = 0$$

Hence solve the given equation.

Complex Roots of Polynomial Equations

We have seen that $z^n-1=0$ has complex roots (i.e. the nth roots of unity) and that they occur in conjugate pairs.
We shall now prove the more general case, that

if a polynomial equation with real coefficients has complex roots, they occur in conjugate pairs.

Consider the polynomial function $f(x)$, where $f(x)$ has real coefficients.
If $x=a+bi$ then $f(a+bi)=c+di$ where c and d are real.
Now even powers of (bi) are real and odd powers of (bi) are imaginary,

i.e. c contains only even powers of b

while d contains only odd powers of b.

So if the sign of b is changed, c does not change sign but d does change sign.

i.e. if $f(a+bi) = c+di$ then $f(a-bi) = c-di$

Now, if $a+bi$ is a root of the equation $f(x)=0$ then

$$f(a+bi) = 0 \Rightarrow c = 0 \text{ and } d = 0$$

$$\Rightarrow c-di = 0 \qquad \text{so } f(a-bi) = 0$$

therefore $a-bi$ is also a root of the equation $f(x)=0$.

(This fact is not true of equations with non-real coefficients. For example, the roots of $z^3-(2+2i)=0$ are found in Examples 8b, No 4, and they do not occur in conjugate pairs.)
Knowing that the complex roots occur in conjugate pairs can help reduce the work when solving polynomial equations, provided that the equations can be solved. In the examples that follow, we look at two types of equation: those that reduce to a quadratic and those that can be expressed in the form $x^n-1=0$.

EXAMPLES 9e

1) Solve the equation $x^6-2x^3+4=0$.

First, we can regard the equation as quadratic in x^3,

i.e. $$(x^3)^2-2(x^3)+4 = 0$$

Using the formula for solving a quadratic equation gives

$$x^3 = 1\pm i\sqrt{3}$$

Taking $x^3=1+i\sqrt{3}=2e^{i\pi/3}$ and supposing that $re^{i\theta}$ is a value of x gives

$$r^3e^{i3\theta} = 2e^{i\pi/3}$$

$\Rightarrow$ $r = 2^{\frac{1}{3}}$ and $\theta = \pi/9,\ 7\pi/9,\ -5\pi/9$ for $-\pi<\theta\leqslant\pi$

So $x_1=2^{\frac{1}{3}}(\cos\pi/9+i\sin\pi/9)$
$x_2=2^{\frac{1}{3}}(\cos 7\pi/9+i\sin 7\pi/9)$
$x_3=2^{\frac{1}{3}}(\cos 5\pi/9-i\sin 5\pi/9)$

We now have three roots of the given equation and we could repeat the procedure for the other value of x^3 to find the remaining three roots. But, we know that the roots occur in conjugate pairs and that there are no conjugate pairs contained in x_1, x_2 and x_3.
So the other roots are $\bar{x}_1$, $\bar{x}_2$ and $\bar{x}_3$
i.e. $2^{\frac{1}{3}}(\cos\pi/9-i\sin\pi/9)$, $2^{\frac{1}{3}}(\cos 7\pi/9-i\sin 7\pi/9)$
and $2^{\frac{1}{3}}(\cos 5\pi/9+i\sin 5\pi/9)$.

2) Find the four complex roots of the equation

$$x^4+x^3+x^2+x+1 = 0.$$

From Question 11 in Exercise 9c, we know that

$$x^n-a^n \equiv (x-a)(x^{n-1}+ax^{n-2}+\ldots+a^{n-1})$$

when $a=1$ this becomes

$$x^n-1 \equiv (x-1)(x^{n-1}+x^{n-2}+\ldots+1)$$

So multiplying the given equation by $x-1$ and using the result above we have

$$(x-1)(x^4+x^3+x^2+x+1) = x^5-1 = 0$$

Now the roots of the equation $x^5-1=0$ are the fifth roots of unity *including* the real root, $x=1$.
Hence the roots of $x^4+x^3+x^2+x+1=0$ are the four *complex* fifth roots of unity,

i.e. $\cos 2\pi/5 + i\sin 2\pi/5, \quad \cos 2\pi/5 - i\sin 2\pi/5$

$\cos 4\pi/5 + i\sin 4\pi/5, \quad \cos 4\pi/5 - i\sin 4\pi/5$

Real Quadratic Factors

If $x_1 = a+bi$ is a root of the equation $f(x)=0$, where $f(x)$ is a polynomial with real coefficients, then $x_2 = a-bi$ is also a root.
So $x-(a+bi)$ and $x-(a-bi)$ are both factors of $f(x)$.
Now $[x-(a+bi)][x-(a-bi)]$ is *real* and equal to

$$x^2-2ax+(a^2+b^2)$$

i.e. we have found a real quadratic factor of $f(x)$.
This means that a polynomial, $f(x)$, with *real* coefficients can be expressed as a product of real factors, none of which is of degree greater than 2 (there may be some linear factors as well).
Whether we can find these factors is another story – it usually depends on being able to solve the equation $f(x)=0$.

EXAMPLES 9e (continued)

3) Express x^6-2x^3+4 as the product of three quadratic factors with real coefficients.

The roots of the equation $x^6-2x^3+4=0$ are found in Example 1 above.
Taking these roots in conjugate pairs gives the real quadratic factors as follows,

$$(x-x_1)(x-\bar{x}_1) = x^2-(2^{\frac{4}{3}}\cos\pi/9)x+2^{\frac{2}{3}}$$

$$(x-x_2)(x-\bar{x}_2) = x^2-(2^{\frac{4}{3}}\cos 7\pi/9)x+2^{\frac{2}{3}}$$

$$(x-x_3)(x-\bar{x}_3) = x^2-(2^{\frac{4}{3}}\cos 5\pi/9)x+2^{\frac{2}{3}}$$

4) Express $f(x)=x^4-x^3+2x^2+x+3$ as the product of two quadratic factors with real coefficients.

There is no obvious way of solving the equation $x^4-x^3+2x^2+x+3=0$ so we try to find the factors by observation and undetermined coefficients.
Now $x^4-x^3+2x^2+x+3$

$$\equiv \begin{cases} \text{either} & (x^2+ax+3)(x^2+bx+1) \qquad [1] \\ \text{or} & (x^2+px-3)(x^2+qx-1) \qquad [2] \end{cases}$$

Comparing the coefficients of x^3, x^2 and x

[1] gives

$$\begin{aligned} a+b &= -1 \\ 4+ab &= 2 \\ a+3b &= 1 \end{aligned}$$

Solving the first and last equations for a and b gives $b=1$ and $a=-2$ and these values are consistent with the middle equation.

[2] gives

$$\begin{aligned} p+q &= -1 \\ pq-4 &= 2 \\ -p-3q &= 1 \end{aligned}$$

This time the first and last equations give $q=0$ and $p=-1$.
But these values do not satisfy the remaining equation, so there are no values of p and q for which $f(x)$ can have factors of type [2].

Hence $x^4-x^3+2x^2+x+3 \equiv (x^2-2x+3)(x^2+x+1)$.

EXERCISE 9e

Solve the following equations, giving any complex roots in the form $re^{i\theta}$.

1) $x^6-4x^3+8=0$

2) $x^4-6x^2+25=0$

3) $x^6-1=0$

4) $(x-2)^3=1$ (*Hint.* Solve for $x-2$.)

5) $x^5+x^4+x^3+x^2+x+1=0$

6) $x^3+x^2+x+1=0$

7) $x^3-x^2+x-1=0$ (*Hint.* Multiply by $x+1$.)

8) $x^4-x^3+x^2-x+1=0$

9) Express the L.H.S. of each equation given above as the product of linear and/or quadratic factors with real coefficients.

SUMMARY

When a polynomial function $f(x)$ is divided by $(x-a)(x-b)$ the remainder is of the form $\mathrm{P}x+\mathrm{Q}$, and

$$\begin{cases} f(a) = \mathrm{P}a+\mathrm{Q} \\ f(b) = \mathrm{P}b+\mathrm{Q} \end{cases}$$

When a polynomial function $f(x)$ has a repeated factor $(x-a)$, then $(x-a)$ is also a factor of $f'(x)$, i.e. $$f(a) = 0 \quad and \quad f'(a) = 0$$

If $\alpha, \beta, \gamma, \delta \ldots$ are the roots of a polynomial equation

$$ax^n + bx^{n-1} + cx^{n-2} + \ldots = 0$$

then

$$\sum\alpha = -\frac{b}{a}, \quad \sum\alpha\beta = \frac{c}{a}, \quad \sum\alpha\beta\gamma = -\frac{d}{a}, \quad \sum\alpha\beta\gamma\delta = \frac{e}{a}\ldots$$

A polynomial equation of degree n with real coefficients has
(a) n roots (real and/or complex),
(b) complex roots, if any, in conjugate pairs,
(c) at least one real root if n is odd.

MULTIPLE CHOICE EXERCISE 9

(The instructions for answering these questions are on p. xii.)

TYPE I

1) When $x^3 + 5x - 2$ is divided by $(x-1)(x-2)$ the remainder is:
(a) 4 (b) $12x - 8$ (c) $12x$ (d) 16 (e) 0.

2) The function $p^3 + q^3 + r^3 - 3pqr$ has a factor:
(a) $p - q$ (b) $p + q + r$ (c) $p + r$ (d) $p - q - r$ (e) $q - p$.

3) The function $x^4 - 8x^3 + 22x^2 - 24x + 9$ has a repeated factor:
(a) $x - 1$ (b) $x^2 + 1$ (c) $x - 2$ (d) $x^2 - 1$ (e) $x + 1$.

4) The value of a for which $x^2 - ax - 2$ and $x^3 + x^2 - x - 1$ have a common factor is:
(a) -2 (b) -1 (c) $\frac{1}{2}$ (d) 2 (e) 0.

5) The sum of the squares of the roots of the equation $4x^3 + 3x^2 - 2x + 1 = 0$ is:
(a) $\frac{9}{16}$ (b) $-\frac{9}{16}$ (c) $-\frac{7}{16}$ (d) $\frac{25}{16}$ (e) none of these.

6) If α, β, γ are the roots of the equation $x^3 - 2x^2 + 3x - 4 = 0$, the equation whose roots are $\frac{1}{\alpha}, \frac{1}{\beta}, \frac{1}{\gamma}$ is:

(a) $x^3 - \frac{1}{2}x^2 + \frac{1}{3}x - \frac{1}{4} = 0$ (b) $4x^3 - 3x^2 + 2x - 1 = 0$

(c) $\frac{x^3}{4} - \frac{x^2}{3} + \frac{x}{2} - 1 = 0$ (d) $\frac{1}{x^3 - 2x^2 + 3x - 4} = 0$

(e) none of these.

TYPE II

7) $f(ab) \equiv a^n - b^n$.
(a) $f(ab)$ is cyclic.
(b) $f(ab)$ is homogeneous.
(c) $a-b$ is a factor of $f(ab)$.

8) $f(x) \equiv x^2 - 2x + 1$ and $g(x) \equiv x^3 - x^2 - x + 1$.
(a) $f(x)$ and $g(x)$ have a common factor.
(b) $g(x)$ has a repeated factor $x - 1$.
(c) $f(x) + Kg(x)$ has a real linear factor for all real values of K.

9) The equation $5x^3 - 9x^2 + 12x + 4 = 0$:
(a) has a root between 0 and -1,
(b) has three real roots,
(c) has roots α, β, γ where $\alpha\beta\gamma = \frac{4}{5}$,
(d) has a repeated root.

10) If $x^3 + px + q = 0$ has a repeated root, α, then
(a) $3\alpha^2 = p$,
(b) $\alpha^3 + p\alpha + q = 0$,
(c) the third root is -2α.

TYPE III

11) (a) $f(abc)$ is a cyclic polynomial of degree 3 with 3 real linear factors.
(b) $f(abc) \equiv (a+b)(b+c)(c+a)$.

12) (a) $f(abc)$ is a cyclic polynomial of degree 2.
(b) $f(abc) \equiv \pm(a+b+c)^2$.

13) (a) $f(xy)$ is a homogeneous polynomial.
(b) $f(xy)$ is a cyclic polynomial.

14) (a) $f'(x)$ has a factor $(x - \alpha)$.
(b) $f(x)$ has a repeated factor $(x - \alpha)$.

15) (a) $f(x) \equiv x^n - a^n$.
(b) $(x - a)$ is a factor of $f(x)$.

16) (a) $f(x) + Kg(x)$ has a factor $x - 1$ for all real values of K.
(b) $x - 1$ is a common factor of $f(x)$ and $g(x)$.

17) The roots of the equation $x^3 + qx^2 + rx + s = 0$ are α, β, γ.
(a) α, β and γ are all real.
(b) q, r and s are all real. (U of L)

18) $f(x)$ is a polynomial function.
(a) The equation $f'(x) = 0$ has two equal roots.
(b) The equation $f(x) = 0$ has three equal roots.

19) $f(x) = 0$ is a polynomial equation.
(a) $f(x) = 0$ has exactly one real root.
(b) $f'(x) = 0$ has no real roots.

TYPE V

20) $f(x)$ is a polynomial function of x of degree n. When $f(x)$ is divided by x^m $(m < n)$, the remainder is a polynomial of degree $n - m$.

21) If $f(x)$ has a stationary value of zero when $x = a$, then $x - a$ is a repeated factor of $f(x)$.

22) If $f(xyz)$ is cyclic then each of its linear factors (if it has any) is also cyclic.

23) If $f(xyz)$ is homogeneous, any factor of $f(xyz)$ will also be homogeneous.

24) If $f(x)$ is a polynomial of degree n, $f(x) = 0$ has n roots.

25) One root of the equation $x^3 + 3x - 5 = 0$ is approximately equal to the solution of the equation $3x - 5 = 0$.

MISCELLANEOUS EXERCISE 9

1) Let $f(x) \equiv 2x^4 + ax^2 + bx - 60$. The remainder when $f(x)$ is divided by $(x - 1)$ is -94. One factor of $f(x)$ is $(x - 3)$. Determine the constants a and b. (U of L)

2) Find integers m and n such that $(x + 1)^2$ is a factor of $x^5 + 2x^2 + mx + n$. (U of L)p

3) Determine the quadratic function $f(x)$ which is exactly divisible by $(2x + 1)$ and has remainders -6 and -5 when divided by $(x - 1)$ and $(x - 2)$ respectively. Determine $g(x) \equiv (px + q)^2 f(x)$, where p, q are constants, given that, on division by $(x - 2)^2$, the remainder is $-39 - 3x$. (AEB '72)

4) When a polynomial in x is divided by $(x - a)$ the remainder is R_1 and when it is divided by $(x - b)$ the remainder is R_2. Find the remainder when the polynomial is divided by $(x - a)(x - b)$. (U of L)p

5) Given that $x^2 + 1$ is a factor of $x^4 + px^3 + 3x + q$, find the values of p and q. Hence find the real roots of the equation

$$x^4 + px^3 + x^2 + 3x + q + 1 = 0.$$ (U of L)p

6) When the polynomial $f(x)$ is divided by

$$(x - 1)(x - 2)(x - 3)$$

the remainder equals

$$a(x - 2)(x - 3) + b(x - 3)(x - 1) + c(x - 1)(x - 2)$$

Express the constants a, b, c in terms of $f(1), f(2)$ and $f(3)$.

Without performing the division, find the value of the constant k for which the remainder when $(x^5 + kx^2)$ is divided by $(x-1)(x-2)(x-3)$ contains no term in x^2. (U of L)p

7) Show that the remainder when the polynomial $f(x)$ is divided by $(x-a)$ is $f(a)$. Show further that, if $f(x)$ is divided by $(x-a)(x-b)$, where $a \neq b$, then the remainder is

$$\left[\frac{f(a)-f(b)}{a-b}\right]x + \frac{af(b)-bf(a)}{a-b}.$$

(C)p

8) A polynomial $f(x)$ is divided by $x^2 - a^2$, where $a \neq 0$, and the remainder is $px + q$. Prove that

$$p = \frac{1}{2a}[f(a)-f(-a)]$$

$$q = \tfrac{1}{2}[f(a)+f(-a)]$$

Find the remainder when $x^n - a^n$ is divided by $x^2 - a^2$ for the cases when (a) n is even, (b) n is odd. (JMB)

9) Show that if $(x+t)$ is a common factor of $x^3 + px^2 + q$ and $ax^3 + bx + c$, then it is also a factor of $apx^2 - bx + aq - c$. Show that $x^3 + \sqrt{7}x^2 - 14\sqrt{7}$ and $2x^3 - 13x - \sqrt{7}$ have a common factor and hence find all the roots of the equation $2x^3 - 13x - \sqrt{7} = 0$. (U of L)

10) Show that $(x-y)$ is a factor of

$$x(y-z)^3 + y(z-x)^3 + z(x-y)^3$$

and hence factorize the expression completely. (U of L)p

11) Show, by putting $a = b + c$ or otherwise, that $a - b - c$ is a factor of

$$a^4 + b^4 + c^4 - 2b^2c^2 - 2c^2a^2 - 2a^2b^2$$

Factorize this expression completely. (C)

12) Given that m and n are positive integers, prove that

$$x^m(b^n - c^n) + b^m(c^n - x^n) + c^m(x^n - b^n)$$

is divisible by $x^2 - x(b+c) + bc$. (U of L)

13) Express $4b^2c^2 - (b^2 + c^2 - a^2)^2$ as the product of four factors and hence determine the sign of the expression when a, b and c denote the lengths of the sides of a triangle. (U of L)p

14) Factorize $\quad a^2(b-c) + b^2(c-a) + c^2(a-b)$

and $\quad a^4(b-c) + b^4(c-a) + c^4(a-b)$

and show that, if a, b, c are real quantities, no two of which are equal, $a^4(b-c) + b^4(c-a) + c^4(a-b)$ cannot be zero. (U of L)

15) Prove that, if the roots of the equation

$$ax^3 + bx^2 + cx + d = 0, \quad (a \neq 0)$$

are α, β, γ, then $\alpha + \beta + \gamma = -b/a$, $\alpha\beta\gamma = -d/a$.
Solve the equation

$$32x^3 - 14x + 3 = 0$$

given that one root is twice another. (U of L)p

16) The roots of the equation $x^3 + 3x + 2 = 0$ are α, β and γ. Find the equation whose roots are $\alpha + \dfrac{1}{\alpha}$, $\beta + \dfrac{1}{\beta}$, $\gamma + \dfrac{1}{\gamma}$. (AEB '73)

17) If the roots of the equation

$$ax^3 + bx^2 + cx + d = 0$$

are in arithmetic progression (that is, one root is half the sum of the other two), prove that

$$2b^3 - 9abc + 27a^2d = 0$$

Solve the equation

$$18x^3 + 27x^2 + x - 4 = 0.$$

(O)

18) If α, β, γ are the roots of the equation

$$x^3 - x^2 - 4x + 5 = 0$$

find cubic equations whose roots are
(a) 2α, 2β and 2γ,
(b) $1/\alpha$, $1/\beta$ and $1/\gamma$,
(c) $\alpha + \beta$, $\beta + \gamma$ and $\gamma + \alpha$.

Evaluate $\sum (\alpha + \beta)^2$. (U of L)

19) The roots of the equation $x^3 - 26x^2 + 156x + p = 0$ are in geometric progression. Find p. (U of L)

20) The roots of the equation $x^3 + px^2 + qx + 30 = 0$ are in the ratios 2:3:5. Find the values of p and q. (U of L)

21) If $\alpha, \beta, \gamma, \delta$ are the roots of the equation

$$x^4 + ax^3 + bx^2 + cx + d = 0$$

and $\alpha + \beta = \gamma + \delta$, show that the roots of the equation

$$ay^2 - 2cy + ad = 0$$

are $\alpha\beta$ and $\gamma\delta$. Prove, also, that

$$a^3 - 4ab + 8c = 0$$

Solve the equation

$$4x^4 - 8x^3 - 23x^2 + 27x + 18 = 0$$

(O)

22) If the roots of the equation $x^3 - 9x^2 + 3x - 39 = 0$ are α, β, γ, show that an equation whose roots are $\alpha - 3$, $\beta - 3$, and $\gamma - 3$ is

$$x^3 - 24x - 84 = 0.$$

Show also that the equation $x^3 - 24x - 84 = 0$ has only one real root, and show that this root lies between 6 and 7.
Sketch the two curves $y = x^3 - 9x^2 + 3x - 39$ and $y = x^3 - 24x - 84$ on the same diagram. (U of L)

23) If α, β, γ are the roots of the equation $x^3 - px^2 - q = 0$, prove that $\alpha^2 + \beta^2 + \gamma^2 = p^2$ and express

$$\beta^2\gamma^2 + \gamma^2\alpha^2 + \alpha^2\beta^2$$

in terms of p and q.
Use the Remainder Theorem to find the remainder when $x^3 - 7x^2 + 36$ is divided by $x + 2$. Hence solve the equation $x^3 - 7x^2 + 36 = 0$ and verify that your expression for

$$\beta^2\gamma^2 + \gamma^2\alpha^2 + \alpha^2\beta^2$$

is correct in this case. (U of L)

24) Sketch the curve $y = x^4 + 3x^3 + x^2$, giving the coordinates and nature of its turning points.
For the equation $x^4 + 3x^3 + x^2 - k = 0$
(a) find the complex roots in the form $a + ib$, a and b being real, when $k = -4$,
(b) find the integer root when $k = 9$, and find two consecutive integers between which the other real root must lie in this case,
(c) calculate the sum of the squares of the roots, showing that it is independent of k. (JMB)

25) Given that $p + iq$, where p and q are real and $q \neq 0$, is a root of the equation

$$a_0z^n + a_1z^{n-1} + \ldots + a_n = 0$$

where $a_0, a_1, \ldots, a_n$ are all real, prove that $p - iq$ is also a root.

Given that $1 + 3i$ is a root of the equation

$$z^4 - 6z^3 + az^2 + bz + 70 = 0$$

where a and b are real, find a, b and the other three roots of the equation. (JMB)

26) Find the six (complex) roots of the equation

$$z^6 - z^3 + 1 = 0$$

Express $z^6 - z^3 + 1$ as the product of three quadratic factors, all with real coefficients. (O)

27) Show that the roots of $z^6+z^3+1=0$ are included in the roots of $z^9-1=0$. Find the solutions of $z^6+z^3+1=0$ in the form $\cos\phi+\mathrm{i}\sin\phi$ and hence find the values of θ between 0 and 2π which satisfy both the equations

$$\cos 6\theta+\cos 3\theta+1 = 0 \quad \text{and} \quad \sin 6\theta+\sin 3\theta = 0$$

28) Find the modulus and the argument of each of the roots of the equation

$$z^5+32 = 0$$

Hence express

$$z^4-2z^3+4z^2-8z+16$$

as the product of two quadratic factors of the form $z^2-az\cos\theta+b$, where a, b and θ are real. (JMB)

29) Find the two complex numbers whose squares are equal to i.
Hence, or otherwise, find the four (complex) roots of the equation $z^4+4=0$, giving the modulus and argument of each.
Hence express z^4+4 as the product of two real quadratic polynomials. (O)

30) Solve the equation $z^5-1=0$, giving your answers in the form $re^{\mathrm{i}\theta}$.
Show the solutions of the equation as points on an Argand diagram.
Hence, or otherwise, solve the equation $z^9+z^5-z^4=1$, giving your answers in the form $re^{\mathrm{i}\theta}$. (C)p

CHAPTER 10

FURTHER INTEGRATION AND SOME APPLICATIONS

REDUCTION METHOD OF INTEGRATION

Certain methods of integration which are applicable to functions involving a power n where n is a positive integer, are viable only when n is relatively small. For instance $\int \cos^n x \, dx$ can be found when $n = 4$ by using the identity $\cos^2\theta \equiv \frac{1}{2}(1 + \cos 2\theta)$ as often as necessary, but the same method applied to $\int \cos^{20} x \, dx$ would be extremely unwieldy. In cases like this, a means of systematically reducing the value of n is useful, and is called a reduction method. Usually a reduction method is based on the technique of integration by parts, but it should not be assumed that this is the only approach.

EXAMPLES 10a

1) If $I_n \equiv \int \cos^n x \, dx$, show that

$$I_n = \frac{1}{n} \sin x \cos^{n-1} x + \frac{n-1}{n} I_{n-2}$$

Hence find $\int \cos^5 x \, dx$.

To use integration by parts, $\cos^n x$ is written in the product form $\cos x \cos^{n-1} x$,

so that

$$I_n = \int \cos x \cos^{n-1} x \, dx$$

If $v = \cos^{n-1}x$; $\dfrac{du}{dx} = \cos x$

then $\dfrac{dv}{dx} = (n-1)\cos^{n-2}x(-\sin x)$; $u = \sin x$

So
$$I_n = \sin x \cos^{n-1}x + \int (n-1)\cos^{n-2}x \sin^2 x \, dx$$
$$= \sin x \cos^{n-1}x + (n-1)\int \cos^{n-2}x(1-\cos^2 x)\, dx$$
$$= \sin x \cos^{n-1}x + (n-1)\int \cos^{n-2}x \, dx - (n-1)\int \cos^n x \, dx$$

Hence

$$I_n = \sin x \cos^{n-1}x + (n-1)I_{n-2} - (n-1)I_n$$

or
$$nI_n = \sin x \cos^{n-1}x + (n-1)I_{n-2}$$

So
$$I_n = \frac{1}{n}\sin x \cos^{n-1}x + \frac{n-1}{n}I_{n-2}$$

Now we can apply this reduction formula to find $\int \cos^5 x \, dx$, first using $n = 5$,

i.e.
$$I_5 = \tfrac{1}{5}\sin x \cos^4 x + \tfrac{4}{5}I_3 \qquad [1]$$

Further, using $n = 3$ we have,

$$I_3 = \tfrac{1}{3}\sin x \cos^2 x + \tfrac{2}{3}I_1 \qquad [2]$$

Now

$$I_1 = \int \cos x \, dx = \sin x + K \qquad [3]$$

Combining [1], [2] and [3] gives,

$$I_5 = \tfrac{1}{5}\sin x \cos^4 x + \tfrac{4}{5}(\tfrac{1}{3}\sin x \cos^2 x + \tfrac{2}{3}\sin x) + K$$

2) Establish a reduction formula that could be used to find $\int x^n e^x \, dx$ and use it when $n = 4$.

Let $I_n = \int x^n e^x \, dx$.

If $v = x^n$ $\qquad \dfrac{du}{dx} = e^x$

then $\dfrac{dv}{dx} = nx^{n-1}$ $\qquad u = e^x$

Then $$I_n = x^n e^x - \int nx^{n-1}e^x \,dx$$

i.e. $$I_n = x^n e^x - nI_{n-1}$$

This is a suitable reduction formula.

Using
$$\begin{cases} n = 4 \quad \text{gives} \quad I_4 = x^4e^x - 4I_3 \\ n = 3 \quad \text{gives} \quad I_3 = x^3e^x - 3I_2 \\ n = 2 \quad \text{gives} \quad I_2 = x^2e^x - 2I_1 \\ n = 1 \quad \text{gives} \quad I_1 = xe^x - I_0 \end{cases}$$

Now $$I_0 = \int x^0 e^x \,dx = e^x + K$$

So $$I_4 = x^4e^x - 4[x^3e^x - 3\{x^2e^x - 2(xe^x - e^x)\}] + K$$

Note that some reduction formulae cause the value of n to fall by 2 in each step while in other cases n falls only by 1 per step.

3) If $I_n = \int \tan^n\theta \,d\theta$, find a reduction formula for I_n and use it to evaluate $\int_0^{\frac{\pi}{4}} \tan^6\theta \,d\theta$.

If $I_n = \int \tan^n\theta \,d\theta$, writing $\tan^n\theta$ as $\tan^2\theta \tan^{n-2}\theta$ gives

$$I_n = \int (\sec^2\theta - 1)\tan^{n-2}\theta \,d\theta$$

$$= \int \sec^2\theta \tan^{n-2}\theta \,d\theta - \int \tan^{n-2}\theta \,d\theta$$

$$\Rightarrow \qquad I_n = \frac{1}{n-1}\tan^{n-1}\theta - I_{n-2}$$

When $n = 6$ we have

$$I_6 = \left[\tfrac{1}{5}\tan^5\theta\right]_0^{\frac{\pi}{4}} - I_4$$

$$= \left[\tfrac{1}{5}\tan^5\theta\right]_0^{\frac{\pi}{4}} - \left[\tfrac{1}{3}\tan^3\theta\right]_0^{\frac{\pi}{4}} + I_2$$

$$= \left[\tfrac{1}{5}\tan^5\theta\right]_0^{\frac{\pi}{4}} - \left[\tfrac{1}{3}\tan^3\theta\right]_0^{\frac{\pi}{4}} + \left[\tan\theta\right]_0^{\frac{\pi}{4}} - I_0$$

$$= \tfrac{1}{5} - \tfrac{1}{3} + 1 - \int_0^{\frac{\pi}{4}} d\theta$$

$$= \tfrac{13}{15} - \left[\theta\right]_0^{\frac{\pi}{4}}$$

$$= \tfrac{13}{15} - \tfrac{1}{4}\pi$$

EXERCISE 10a

Establish a reduction formula for each of the following integrals.

1) $\int \sin^n x \, dx$ 2) $\int \tan^n 2x \, dx$

3) $\int (x+1)^n e^{2x} \, dx$ 4) $\int \cos^n 2\theta \, d\theta$

5) $\int x^n e^{ax} \, dx$ 6) $\int x(\ln x)^n \, dx$

7) $\int \cosh^n x \, dx$ 8) $\int \sec^n x \, dx$

Use a reduction method to find each of the following integrals.

9) $\int \cos^6 x \, dx$ 10) $\int \sin^7 x \, dx$

11) $\int (1-x)^3 e^{4x} \, dx$ 12) $\int \cos^4 (ax+b) \, dx$

13) $\int \sin^4 \left(\frac{\pi}{4} - 3\theta\right) d\theta$ 14) $\int \tan^5 x \, dx$

15) $\int x^3 \sin x \, dx$ 16) $\int \sinh^5 x \, dx$

17) $\int \sec^4 x \, dx$ 18) $\int x^6 e^{-x} \, dx$

19) $\int x(\ln x)^3 \, dx$ 20) $\int \cosh^4 3x \, dx$

21) Prove that, if $I_n = \int x^n (1+x^3)^7 \, dx$ then

$$I_n = \frac{1}{n+22}\left\{x^{n-2}(1+x^3)^8 - (n-2)I_{n-3}\right\}$$

(*Hint*. Use $x^n \equiv x^{n-2}x^2$.)

Hence or otherwise determine $\int x^5 (1+x^3)^7 \, dx$

22) If $I_n = \int \frac{\cos^{2n} x}{\sin x} \, dx$, write down a similar expression for I_{n+1}.

Hence, by using the identity $\cos^2 x + \sin^2 x \equiv 1$, prove that

$$(2n+1)I_{n+1} = (2n+1)I_n + \cos^{2n+1} x$$

Hence or otherwise determine $\int \frac{\cos^6 x}{\sin x} \, dx$.

DEFINITE INTEGRATION BY REDUCTION METHODS

Again we will consider integrating $\cos^n x$ but this time as a definite integral with boundaries at $x = 0$ and $x = \dfrac{\pi}{2}$, so that

$$I_n = \int_0^{\frac{\pi}{2}} \cos^n x \, dx$$

Integrating by parts as before gives

$$I_n = \left[\frac{1}{n}\sin x \cos^{n-1} x\right]_0^{\frac{\pi}{2}} + \frac{n-1}{n} I_{n-2}$$

$$= 0 + \frac{n-1}{n} I_{n-2}$$

i.e. $$I_n = \int_0^{\frac{\pi}{2}} \cos^n x \, dx \Rightarrow I_n = \frac{n-1}{n} I_{n-2}$$

If we now consider $\int_0^{\frac{\pi}{2}} \sin^n x \, dx$, it can be shown that

$$I_n = \int_0^{\frac{\pi}{2}} \sin^n x \, dx \Rightarrow I_n = \frac{n-1}{n} I_{n-2}$$

When these reduction formulae are used successively, we find that

$$I_n = \left(\frac{n-1}{n}\right) I_{n-2}$$

$$= \left(\frac{n-1}{n}\right)\left(\frac{n-3}{n-2}\right) I_{n-4}$$

$$= \left(\frac{n-1}{n}\right)\left(\frac{n-3}{n-2}\right)\left(\frac{n-5}{n-4}\right) I_{n-6} \quad \text{etc.}$$

The final term in the formula depends on whether n is odd or even.

(a) If n is even we have

$$I_n = \left(\frac{n-1}{n}\right)\left(\frac{n-3}{n-2}\right)\cdots\left(\frac{3}{4}\right)\left(\frac{1}{2}\right) I_0$$

But $$I_0 = \int_0^{\frac{\pi}{2}} \cos^0 x \, dx \quad \text{or} \quad \int_0^{\frac{\pi}{2}} \sin^0 x \, dx$$

i.e. $$I_0 = \frac{\pi}{2} \quad \text{in both cases.}$$

So, when n is even,

$$I_n = \left(\frac{n-1}{n}\right)\left(\frac{n-3}{n-2}\right)\ldots\left(\frac{3}{4}\right)\left(\frac{1}{2}\right)\left(\frac{\pi}{2}\right)$$

(b) If n is odd we have

$$I_n = \left(\frac{n-1}{n}\right)\left(\frac{n-3}{n-2}\right)\ldots\left(\frac{4}{5}\right)\left(\frac{2}{3}\right)I_1$$

But $$I_1 = \int_0^{\frac{\pi}{2}} \cos x \, dx \quad \text{or} \quad \int_0^{\frac{\pi}{2}} \sin x \, dx$$

i.e. $I_1 = 1$ in both cases.

So, when n is odd

$$I_n = \left(\frac{n-1}{n}\right)\left(\frac{n-3}{n-2}\right)\ldots\left(\frac{4}{5}\right)\left(\frac{2}{3}\right)$$

These very simple and convenient reduction formulae can be quoted but their use is, of course, valid only when the limits of integration are 0 and $\frac{\pi}{2}$,

e.g. $$\int_0^{\frac{\pi}{2}} \cos^{10}x \, dx = \left(\frac{9}{10}\right)\left(\frac{7}{8}\right)\left(\frac{5}{6}\right)\left(\frac{3}{4}\right)\left(\frac{1}{2}\right)\left(\frac{\pi}{2}\right)$$

and $$\int_0^{\frac{\pi}{2}} \sin^9 x \, dx = \left(\frac{8}{9}\right)\left(\frac{6}{7}\right)\left(\frac{4}{5}\right)\left(\frac{2}{3}\right)$$

but $$\int_0^{\pi} \cos^8 x \, dx \text{ is } \textit{not } \left(\frac{7}{8}\right)\left(\frac{5}{6}\right)\left(\frac{3}{4}\right)\left(\frac{1}{2}\right)\left(\frac{\pi}{2}\right)$$

A Graphical Extension to the use of Reduction Formulae

If a definite integral of $\sin^n x$ or $\cos^n x$ is required between limits 0 and $k\pi/2$ (where k is an integer) a graphical approach makes it possible to use the simplified reduction formula which is valid only for the limits 0 to $\frac{\pi}{2}$.

Consider, for instance, the graphs of $\cos x$, $\cos^2 x$ and $\cos^3 x$.

(a)

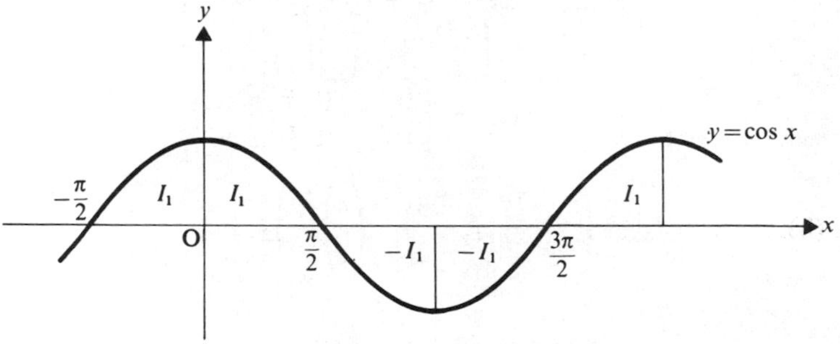

(b)

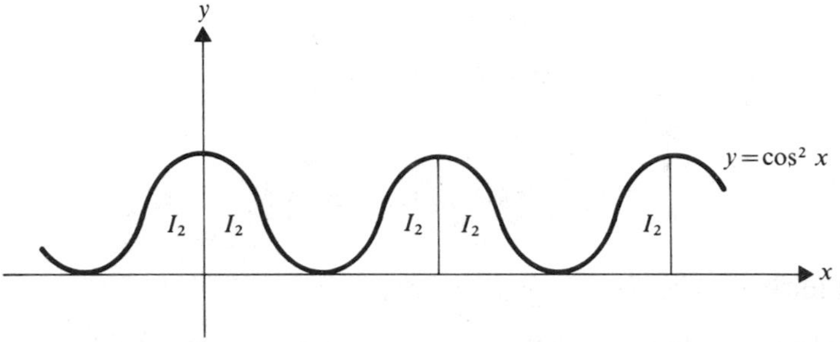

Note in graph (b) that, since $y = (\cos x)^2$, $y \geqslant 0$ for all values of x.

(c)

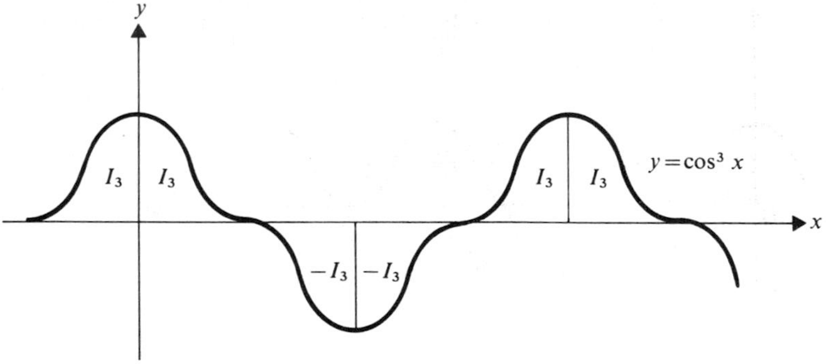

In case (a), if $$\int_0^{\frac{\pi}{2}} \cos x \, dx = I_1$$

then $$\int_{\frac{\pi}{2}}^{\pi} \cos x \, dx = -I_1 \quad \text{etc.}$$

So $$\int_0^{\pi} \cos x \, dx = I_1 - I_1 = 0$$

and $$\int_0^{\frac{3\pi}{2}} \cos x \, dx = I_1 - I_1 - I_1 = -I_1$$

$\Rightarrow$ $$\int_0^{\frac{k\pi}{2}} \cos x \, dx = \begin{cases} 0 & \text{if } k \text{ is even} \\ \pm I_1 & \text{if } k \text{ is odd.} \end{cases}$$

In case (b), if $$\int_0^{\frac{\pi}{2}} \cos^2 x \, dx = I_2$$

then $$\int_0^{\pi} \cos^2 x \, dx = I_2 + I_2 = 2I_2$$

and $$\int_0^{\frac{3\pi}{2}} \cos^2 x \, dx = I_2 + I_2 + I_2 = 3I_2 \quad \text{etc.}$$

$\Rightarrow$ $$\int_0^{\frac{k\pi}{2}} \cos^2 x \, dx = kI_2 \quad \text{for any integer } k.$$

Case (c) follows the same pattern as case (a).

In general the graph of $\cos^n x$ is

either

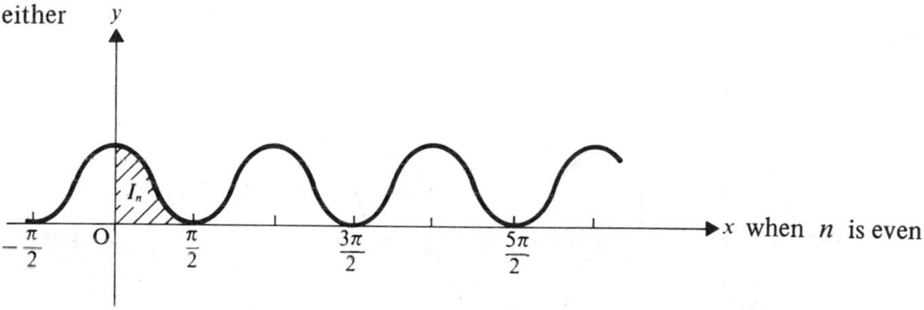

or

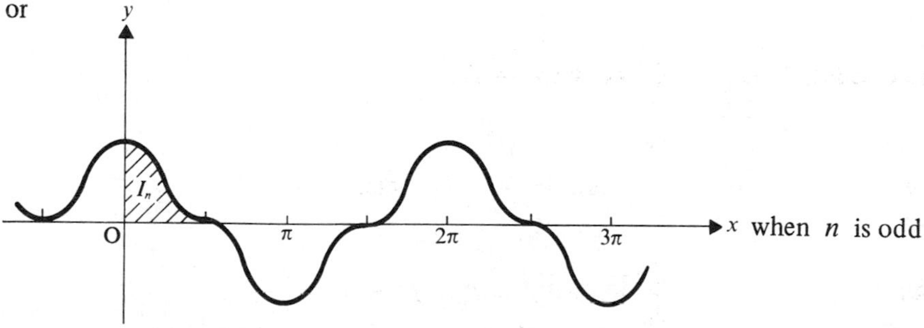

From the graphs it can be seen that

$$\int_0^{\frac{k\pi}{2}} \cos^n x \, dx = kI_n \qquad \text{when } n \text{ is even}$$

and $$\int_0^{\frac{k\pi}{2}} \cos^n x \, dx \quad \text{is either zero or } \pm I_n \qquad \text{if } n \text{ is odd.}$$

These conclusions should not be regarded as quotable but simply as examples of the graphical approach to such integrals. In all cases the required integral is expressed in terms of I_n where $I_n = \int_0^{\frac{\pi}{2}} \cos^n x \, dx$ and which can be calculated using the simple reduction formula, $I_n = \frac{n-1}{n} I_{n-2}$.

In a similar way, $\int_0^{\frac{k\pi}{2}} \sin^n x \, dx$ can be expressed in terms of $\int_0^{\frac{\pi}{2}} \sin^n x \, dx$.

Note that this approach can be used when the lower limit also is a multiple of $\frac{\pi}{2}$, e.g. to find $\int_{-\frac{\pi}{2}}^{2\pi} \sin^3 x \, dx$, we use the graph of $\sin^3 x$ as follows.

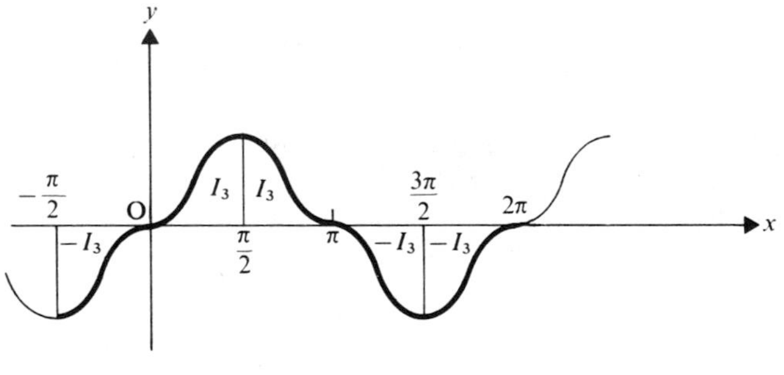

If $$I_3 = \int_0^{\frac{\pi}{2}} \sin^3 x \, dx$$

then $$\int_{-\frac{\pi}{2}}^{2\pi} \sin^3 x \, dx = -I_3 + I_3 + I_3 - I_3 - I_3 = -I_3$$

But $$I_3 = \tfrac{2}{3} I_1 = \tfrac{2}{3}$$

So $$\int_{-\frac{\pi}{2}}^{2\pi} \sin^3 x \, dx = -\tfrac{2}{3}$$

Note that the graph of $\sin^n x$, when n is even, is

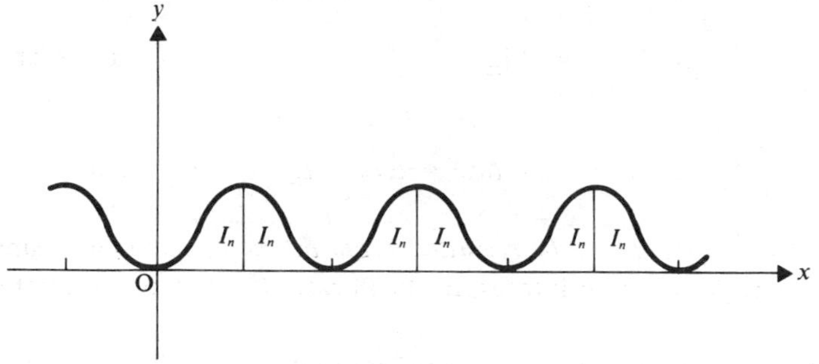

Reduction Formula with Change of Variable

The periodic property of the integrand cannot always be used to carry out the integration and when this is the case a change of variable may *sometimes* help.

For example, to evaluate $\int_0^{\frac{\pi}{3}} \cos^5 3x \, dx$, we can use the substitution

$$u \equiv 3x \quad \Rightarrow \quad \ldots du \equiv \ldots 3\, dx$$

and

x	0	$\frac{\pi}{3}$
u	0	π

Then $$\int_0^{\frac{\pi}{3}} \cos^5 3x \, dx \equiv \tfrac{1}{3} \int_0^{\pi} \cos^5 u \, du$$

Now $\int_0^{\pi} \cos^5 u \, du$ can be found using the graph of $\cos^5 u$, and in this way $\int_0^{\frac{\pi}{3}} \cos^5 3x \, dx$ can be evaluated.

EXAMPLES 10b

1) Use a suitable change of variable to evaluate $\int_{-\frac{\pi}{4}}^{\frac{\pi}{4}} \cos^8 2\theta \, d\theta$.

Let $$u \equiv 2\theta \quad \Rightarrow \quad \ldots du \equiv \ldots 2\, d\theta$$

and

θ	$-\frac{\pi}{4}$	$\frac{\pi}{4}$
u	$-\frac{\pi}{2}$	$\frac{\pi}{2}$

Then $$\int_{-\frac{\pi}{4}}^{\frac{\pi}{4}} \cos^8 2\theta \, d\theta \equiv \tfrac{1}{2} \int_{-\frac{\pi}{2}}^{\frac{\pi}{2}} \cos^8 u \, du$$

As the limits are not 0 to $\frac{\pi}{2}$, we now refer to the graph of $\cos^8 u$,

i.e.

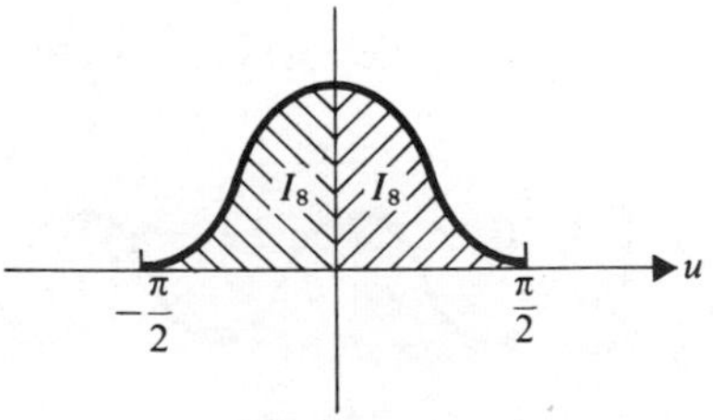

From the graph it is seen that

$$\int_{-\frac{\pi}{2}}^{\frac{\pi}{2}} \cos^8 u \, du = 2I_8$$

$$\Rightarrow \quad \int_{-\frac{\pi}{4}}^{\frac{\pi}{4}} \cos^8 2\theta \, d\theta = I_8$$

$$= \left(\frac{7}{8}\right)\left(\frac{5}{6}\right)\left(\frac{3}{4}\right)\left(\frac{1}{2}\right)\left(\frac{\pi}{2}\right)$$

2) If $I_{p,q} = \int_0^{\frac{\pi}{2}} \sin^p x \cos^q x \, dx$, prove that

$$I_{p,q} = \frac{p-1}{p+q} I_{p-2,q}$$

Use your result to evaluate $\int_0^{\frac{\pi}{2}} \sin^{10} x \cos^2 x \, dx$.

In order to integrate $\sin^p x \cos^q x$ by parts, the integral is arranged in the form

$$I_{p,q} = \int_0^{\frac{\pi}{2}} (\sin^{p-1} x)(\sin x \cos^q x) \, dx$$

Then we can use $\quad v = \sin^{p-1} x \qquad \frac{du}{dx} = \sin x \cos^q x$

$$\frac{dv}{dx} = (p-1)\sin^{p-2} x \cos x \qquad u = -\frac{1}{q+1}\cos^{q+1} x$$

$$\Rightarrow \quad I_{p,q} = \left[-\frac{1}{q+1}\cos^{q+1} x \sin^{p-1} x\right]_0^{\frac{\pi}{2}} + \frac{p-1}{q+1}\int_0^{\frac{\pi}{2}} \sin^{p-2} x \cos^{q+2} x \, dx$$

$$\Rightarrow \quad I_{p,q} = \frac{p-1}{q+1}\int_0^{\frac{\pi}{2}} \sin^{p-2} x \cos^{q+2} x \, dx$$

The R.H.S. is not yet in the form $I_{p-2,q}$, so we use $\cos^2 x \equiv 1-\sin^2 x$ to reduce $\cos^{q+2}x$ to $\cos^q x$.

Then
$$I_{p,q} = \frac{p-1}{q+1}\int_0^{\frac{\pi}{2}} \sin^{p-2}x \cos^q x(1-\sin^2 x)\,dx$$
$$= \frac{p-1}{q+1}\int_0^{\frac{\pi}{2}} (\sin^{p-2}x\cos^q x - \sin^p x \cos^q x)\,dx$$
$$= \frac{p-1}{q+1}(I_{p-2,q} - I_{p,q})$$

$$\Rightarrow \qquad I_{p,q} = \frac{p-1}{p+q} I_{p-2,q}$$

Using $p = 10$, $q = 2$, we have

$$\int_0^{\frac{\pi}{2}} \sin^{10}x \cos^2 x\,dx = I_{10,2} = \tfrac{9}{12} I_{8,2}$$
$$= (\tfrac{9}{12})(\tfrac{7}{10}) I_{6,2}$$
$$\cdots\cdots\cdots\cdots$$
$$= (\tfrac{9}{12})(\tfrac{7}{10})(\tfrac{5}{8})(\tfrac{3}{6})(\tfrac{1}{4}) I_{0,2}$$
$$= \frac{9.7.5.3.1}{6.5.4.3.2(2^5)} I_{0,2}$$

But
$$I_{0,2} = \int_0^{\frac{\pi}{2}} \cos^2 x\,dx$$
$$= \left(\frac{1}{2}\right)\left(\frac{\pi}{2}\right) = \left(\frac{\pi}{4}\right)$$

So
$$\int_0^{\frac{\pi}{2}} \sin^{10}x \cos^2 x\,dx = \frac{9.7.5.3\pi}{6!(2^7)} = \frac{21\pi}{2^{11}}$$

EXERCISE 10b

Evaluate the following definite integrals:

1) $\int_0^{\frac{\pi}{2}} \sin^{11}x\,dx$

2) $\int_{-\frac{\pi}{2}}^{0} \cos^{10}x\,dx$

3) $\int_0^{\pi} \cos^7 x\,dx$

4) $\int_{-\frac{\pi}{2}}^{\frac{\pi}{2}} \sin^6\theta\,d\theta$

5) $\int_0^{\frac{3\pi}{2}} \sin^8 x \, dx$

6) $\int_{-\frac{\pi}{2}}^{\pi} \cos^5 \theta \, d\theta$

7) $\int_0^{2\pi} \sin^4 x \, dx$

8) $\int_{-\pi}^{\pi} \cos^{12} x \, dx$

9) $\int_0^{\infty} x^8 e^{-x} \, dx$

10) $\int_0^{\frac{\pi}{4}} \tan^6 \theta \, d\theta$

11) $\int_0^{\frac{\pi}{2}} \sin^8 x \cos^2 x \, dx$

12) $\int_0^{\frac{\pi}{4}} \sin^7 2x \, dx$

13) $\int_0^{\frac{\pi}{3}} \cos^5 3\theta \, d\theta$

14) $\int_0^{\frac{\pi}{8}} \sin^4 4x \, dx.$

15) If $I_n = \int_0^1 x^n e^x \, dx$, express I_n in terms of I_{n-1}. Find I_5 in terms of e.

16) If $I_n = \int_0^1 x(1-x^3)^n \, dx$ prove that

$$(3n+2)I_n = 3nI_{n-1}$$

and find I_n in terms of n.

17) Given $I_n = \int_1^0 x^n \cosh x \, dx$, prove that, for $n \geqslant 2$

$$I_n = \sinh 1 - n \cosh 1 + n(n-1)I_{n-2}$$

Evaluate (a) $\int_1^0 x^4 \cosh x \, dx$ (b) $\int_1^0 x^3 \cosh x \, dx$.

18) If $I_n = \int_0^1 x^n \sqrt{(1-x)} \, dx$ prove that $(2n+3)I_n = 2nI_{n-1}$.

Hence find I_9.

19) If $I_n = \int_0^{\frac{\pi}{2}} x^n \cos x \, dx$ prove that

$$I_n = \left(\frac{\pi}{2}\right)^n - n(n-1)I_{n-2} \quad \text{for} \quad n \geqslant 2$$

Hence find I_6.

20) If $I_n = \int^1 \frac{\sinh^{2n}x}{\cosh x}\,dx$, prove that

$$I_{n+1} + I_n = \frac{\sinh^{2n+1}(1)}{2n+1}$$

Show that $I_0 = -\frac{\pi}{2} + 2\arctan e$ and evaluate I_3.

21) If $I_{m,n} = \int_{-1}^{1} (1+x)^m(1-x)^n\,dx$ where m and n are positive integers, prove that $(n+1)I_{m,n} = mI_{m-1,n+1}$.
Hence, or otherwise, evaluate $I_{5,6}$.

IMPROPER INTEGRALS

Consider the definite integral $\int_a^b f(x)\,dx$.

We have already shown that this exists if $f(x)$ is continuous for all values of x in the interval $[a, b]$ and that we can represent the integral as the area of the region enclosed by $y = f(x)$, the x axis and the ordinates $x = a$ and $x = b$,

i.e.

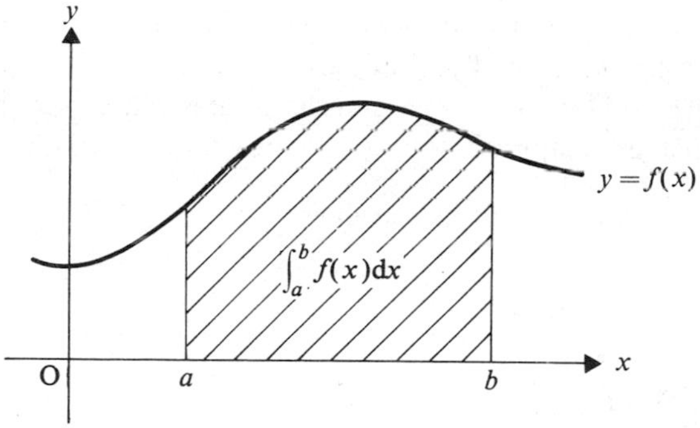

If one or both of the limits of integration are infinite, the region whose area is represented by the integral is unbounded so it is not clear what value we should give to the integral.

For example,

the area represented by $\int_0^\infty x\,dx$ is shown in diagram (i),

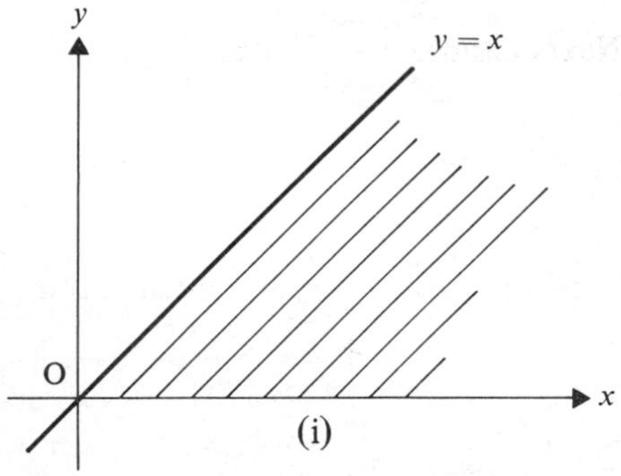

(i)

and the area represented by $\int_0^\infty e^{-x}\,dx$ is shown in diagram (ii).

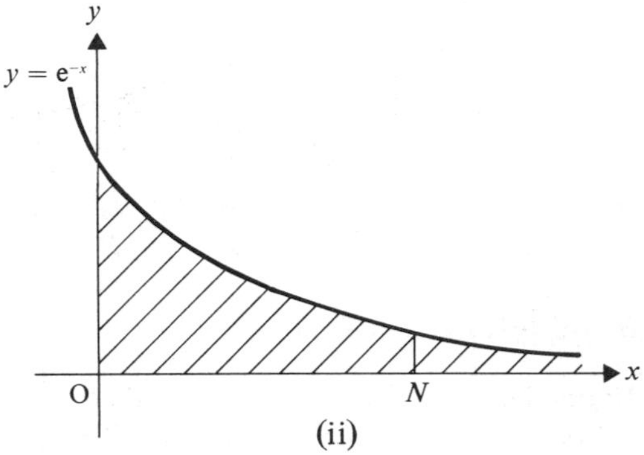

(ii)

In both cases the integral is called *improper.*

Considering the first example, $\int_0^\infty x\,dx$, it is obvious from the diagram that the area is infinite and so $\int_0^\infty x\,dx$ is not defined.

However, Example (ii) is not so straightforward because, as x approaches ∞, e^{-x} approaches zero very rapidly. Thus, after a certain point where $x = N$ say, the area enclosed by $y = e^{-x}$ and the x axis will be very small. This suggests that, as x approaches ∞, the area *may* approach a finite value.

Now $\int_0^x e^{-x}\,dx$ exists for all finite values of x

and $\int_0^x e^{-x}\,dx = [-e^{-x}]_0^x = -e^{-x} + 1$

As $x \to \infty$, $e^{-x} \to 0$ so $\int_0^x e^{-x}\,dx \to 1$

Thus the area *does* converge to a finite value and

$\int_0^\infty e^{-x}\,dx$ is called convergent and its value is 1.

Next consider $\displaystyle\int_2^\infty \frac{1}{x}\,dx.$

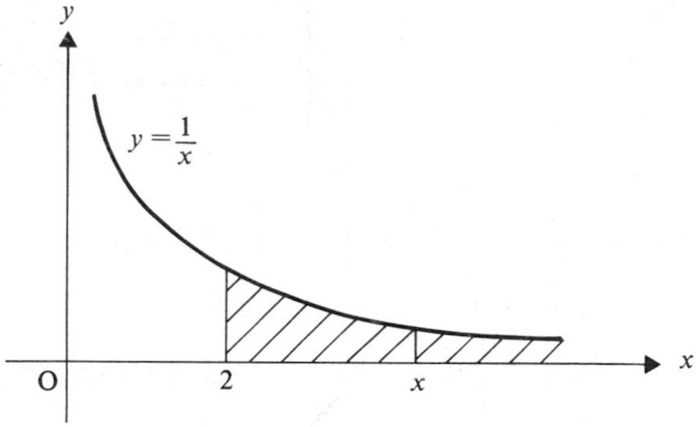

This is similar to the last example in that $\frac{1}{x} \to 0$ as $x \to \infty$ again suggesting that the area may approach a finite value as x approaches ∞.

Now $\displaystyle\int_2^x \frac{1}{x}\,dx = [\ln x]_2^x = \ln x - \ln 2.$

As $x \to \infty$, $\ln x \to \infty$ so $\displaystyle\int_2^x \frac{1}{x}\,dx \to \infty,$

i.e. this time the area diverges as $x \to \infty$,

so $\displaystyle\int_2^\infty \frac{1}{x}\,dx$ is not convergent and has no value.

Using the idea of a limit, we can now give a general definition of an improper integral.

$$\int_a^\infty f(x)\,dx = \lim_{x \to \infty}\left[\int_a^x f(x)\,dx\right]$$

and

$$\int_{-\infty}^a f(x)\,dx = \lim_{x \to -\infty}\left[\int_x^a f(x)\,dx\right]$$

A similar situation arises when one of the limits of integration is where $f(x)$ has a singularity, and such integrals also are called improper.

Consider, for example, $\int_0^2 \frac{1}{x}\,dx$.

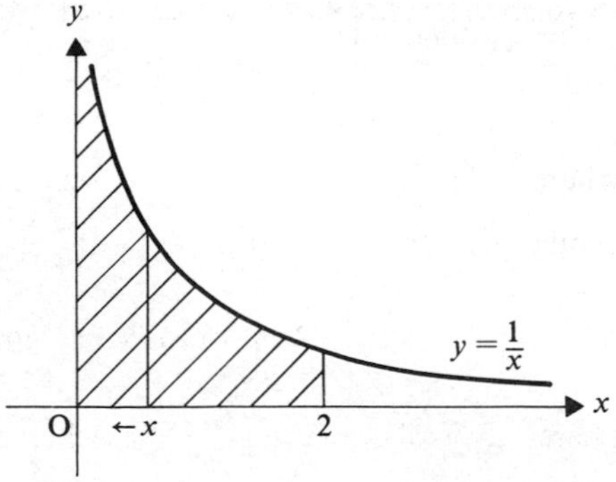

Now $\frac{1}{x}$ has a singularity at $x = 0$, but, provided x is positive, $\int_x^2 \frac{1}{x}\,dx$ exists and is equal to $\ln 2 - \ln x$. As x approaches zero from positive values, $\ln x \to -\infty$, so $\int_x^2 \frac{1}{x}\,dx \to \infty$ and the area does not converge to a finite value,

so $\int_0^2 \frac{1}{x}\,dx$ does not have a value.

Now consider $\int_0^2 \frac{1}{\sqrt{x}}\,dx$.

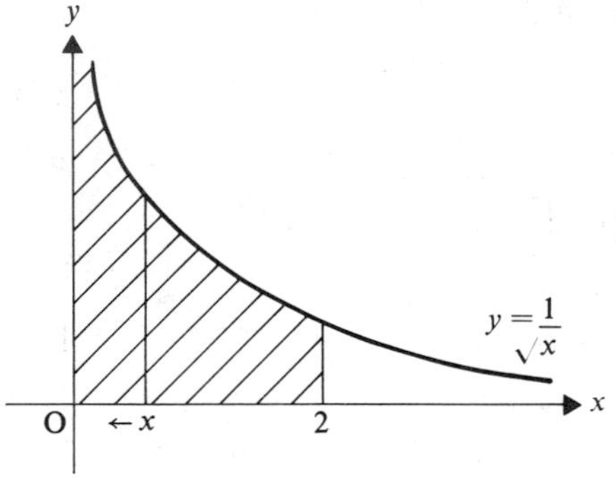

Again, $\frac{1}{\sqrt{x}}$ has a singularity at $x = 0$ while, provided $x > 0$,

$\int_x^2 \frac{1}{\sqrt{x}}\,dx$ exists and is equal to $[2\sqrt{x}]_x^2 = 2\sqrt{2} - 2\sqrt{x}$.

As x approaches zero from positive values, $2\sqrt{x} \to 0$

i.e. $\int_x^2 \frac{1}{\sqrt{x}}\,dx \to 2\sqrt{2}$.

So $\int_0^2 \frac{1}{\sqrt{x}}\,dx$ is convergent and is equal to $2\sqrt{2}$.

In general, if $f(x)$ has a singularity at $x = b$ then

$$\int_a^b f(x)\,dx = \lim_{x \to b-} \left[\int_a^x f(x)\,dx\right]$$

where $x \to b-$ means 'x approaches b from values less than b'

Similarly, if $f(x)$ has a singularity at $x = a$,

$$\int_a^b f(x)\,dx = \lim_{x \to a+} \left[\int_x^b f(x)\,dx\right]$$

where $x \to a+$ means 'x approaches a from values greater than a'.

Further, if $f(x)$ has a singularity at $x = c$, where $a < c < b$ then the integral can be considered in two parts,

i.e.
$$\int_a^b f(x)\,dx = \int_a^c f(x)\,dx + \int_c^b f(x)\,dx$$

$$= \lim_{x \to c-} \left[\int_a^x f(x)\,dx\right] + \lim_{x \to c+} \left[\int_x^b f(x)\,dx\right]$$

EXAMPLE 10c

Find the value, if its exists, of

(a) $\displaystyle\int_2^\infty \cos x\,dx$ (b) $\displaystyle\int_1^\infty \frac{1}{1+x^2}\,dx$ (c) $\displaystyle\int_{-1}^0 \frac{1}{1-x^2}\,dx.$

(a) $$\int_2^\infty \cos x\,dx = \lim_{x \to \infty} \left[\int_2^x \cos x\,dx\right]$$

$$= \lim_{x \to \infty} [\sin x - \sin 2]$$

As $x \to \infty$, $\sin x$ oscillates between -1 and 1

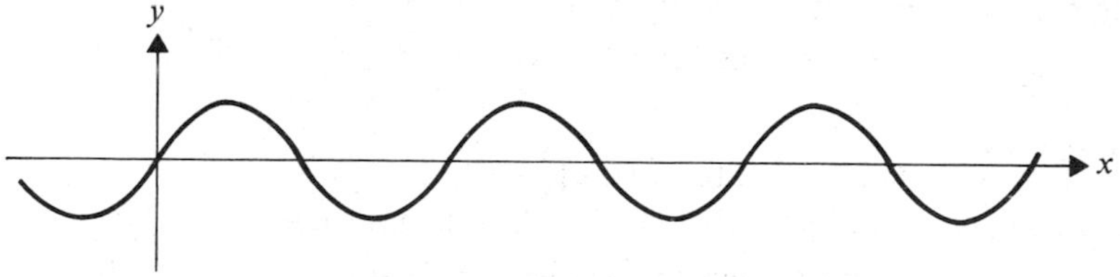

so $\displaystyle\lim_{x \to \infty} [\sin x]$ does not exist.

Therefore $\displaystyle\int_2^\infty \cos x\,dx$ has no value.

(b) $$\int_1^\infty \frac{1}{1+x^2}\,dx = \lim_{x\to\infty}\left[\int_1^x \frac{1}{1+x^2}\,dx\right]$$
$$= \lim_{x\to\infty}\,[\arctan x - \arctan 1]$$

As $x \to \infty$, $\arctan x \to \frac{\pi}{2}$

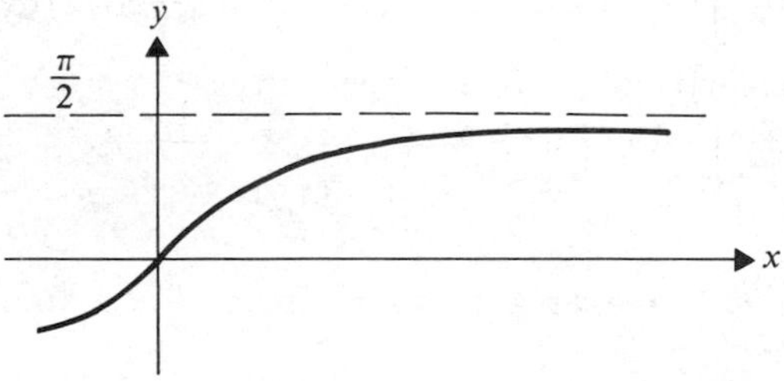

So $$\int_1^\infty \frac{1}{1+x^2}\,dx = \tfrac{1}{2}\pi - \tfrac{1}{4}\pi = \tfrac{1}{4}\pi$$

(c) $$\int_{-1}^0 \frac{1}{1-x^2}\,dx = \int_{-1}^0 \left(\frac{\frac{1}{2}}{1+x} + \frac{\frac{1}{2}}{1-x}\right)dx$$
$$= \int_{-1}^0 \frac{1}{2(1+x)}\,dx + \int_{-1}^0 \frac{1}{2(1-x)}\,dx$$

Now $\dfrac{1}{2(1+x)}$ has a singularity at $x = -1$ but

$\dfrac{1}{2(1-x)}$ is continuous in the interval $[-1, 0]$.

So $$\int_{-1}^0 \frac{1}{1-x^2}\,dx = \lim_{x\to(-1)+}\left[\int_x^0 \frac{1}{2(1+x)}\,dx\right] + \int_{-1}^0 \frac{1}{2(1-x)}\,dx$$
$$= \lim_{x\to(-1)+}\,[-\tfrac{1}{2}\ln(1+x)] + \int_{-1}^0 \frac{1}{2(1-x)}\,dx$$

As $x \to (-1)+$, $\ln(1+x) \to -\infty$.

Therefore $\displaystyle\int_{-1}^0 \frac{1}{1-x^2}\,dx$ has no value.

EXERCISE 10c

Find the values, when they exist, of the following integrals.

1) $\displaystyle\int_1^\infty \frac{1}{x}\,dx$

2) $\displaystyle\int_1^\infty \frac{1}{x^{\frac{1}{3}}}\,dx$

3) $\displaystyle\int_1^\infty \frac{1}{x^3}\,dx$

4) $\displaystyle\int_0^2 \frac{1}{x^3}\,dx$

5) $\displaystyle\int_0^2 \frac{1}{x^{\frac{1}{3}}}\,dx$

6) $\displaystyle\int_1^\infty \frac{x}{1+x^2}\,dx$

7) $\displaystyle\int_1^\infty \frac{x^2}{1+x^2}\,dx$ 8) $\displaystyle\int_0^\infty e^{-x}\cos x\,dx$ 9) $\displaystyle\int_1^2 \frac{1}{1-x^3}\,dx$

10) $\displaystyle\int_0^{\frac{\pi}{2}} \tan x\,dx$ 11) $\displaystyle\int_0^{\frac{\pi}{2}} (1-\cos x)\,dx$ 12) $\displaystyle\int_0^\pi \frac{1}{\sin^2 x}\,dx$

13) $\displaystyle\int_0^1 x\ln x\,dx$ 14) $\displaystyle\int_0^a \frac{1}{x-a}\,dx$ 15) $\displaystyle\int_{2a}^\infty \frac{1}{x^2-a^2}\,dx$

THE MEAN VALUE OF A FUNCTION

Consider the curve $y = f(x)$ for values of x in the interval $[a, b]$. The mean value of y in this interval is the arithmetic average value of y.

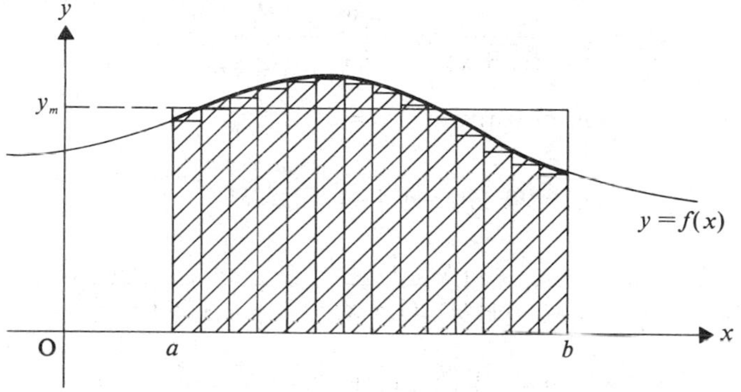

The shaded area in the diagram can be thought of as approximately equal to a sum of ordinates (i.e. values of y), so the average value of these ordinates multiplied by the length of the interval also gives an approximation to the area. Hence we define the mean value of y in the interval $[a, b]$ to be that value, y_m, which, when multiplied by the length of the interval, gives the exact integral $\int_a^b f(x)\,dx$

i.e.
$$y_m(b-a) = \int_a^b f(x)\,dx$$

$\Rightarrow$
$$y_m = \frac{1}{b-a}\int_a^b f(x)\,dx$$

EXAMPLES 10d

1) Find correct to 3 d.p. the mean value of the function $\sin\theta$ for $0 \leqslant \theta \leqslant \pi$.

If $y = \sin\theta$, then for $0 \leqslant \theta \leqslant \pi$

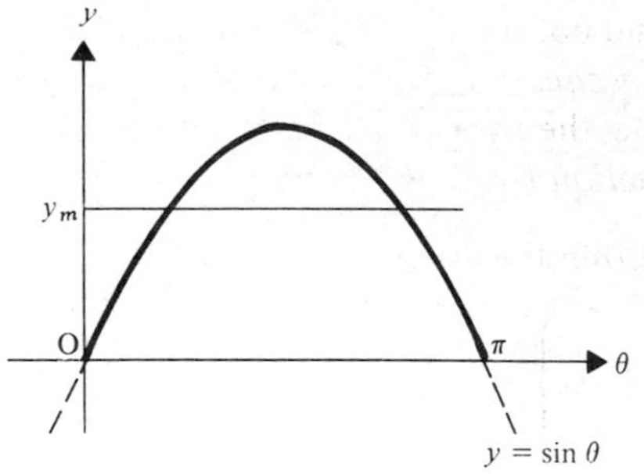

$$y_m = \frac{1}{\pi - 0}\int_0^{\pi} \sin\theta \, d\theta$$

$$= \frac{1}{\pi}\Big[-\cos\theta\Big]_0^{\pi}$$

$$= \frac{1}{\pi}[1+1] = \frac{2}{\pi}$$

$$= 0.637 \text{ correct to 3 d.p.}$$

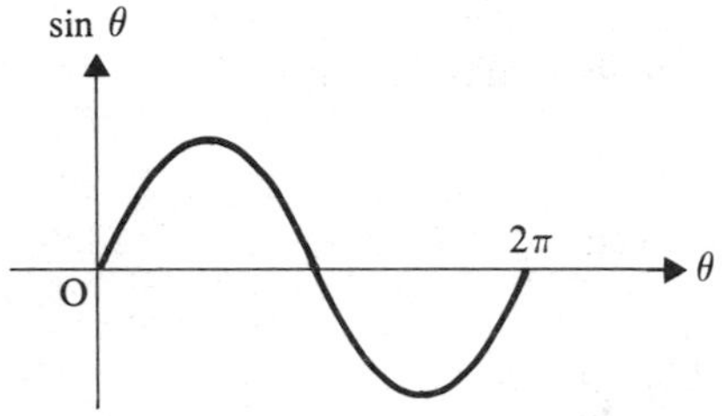

Note that for $0 \leqslant \theta \leqslant 2\pi$, the mean value of $\sin\theta$ is zero as the part of the curve for $\pi \leqslant \theta \leqslant 2\pi$ is the reflection in the x axis of the part of the curve for $0 \leqslant \theta \leqslant \pi$.

2) Find the mean value of the function $f(x) \equiv 4x(x-1)(x-2)$ for the range $-1 \leqslant x \leqslant 1$.

If $y = 4x(x-1)(x-2) = 4x^3 - 12x^2 + 8x$
then for the range $-1 \leqslant x \leqslant 1$

$$y_m = \frac{1}{1-(-1)}\int_{-1}^{1} (4x^3 - 12x^2 + 8x)\,dx$$

$$= \frac{1}{2}\Big[x^4 - 4x^3 + 4x^2\Big]_{-1}^{1}$$

$$= -4$$

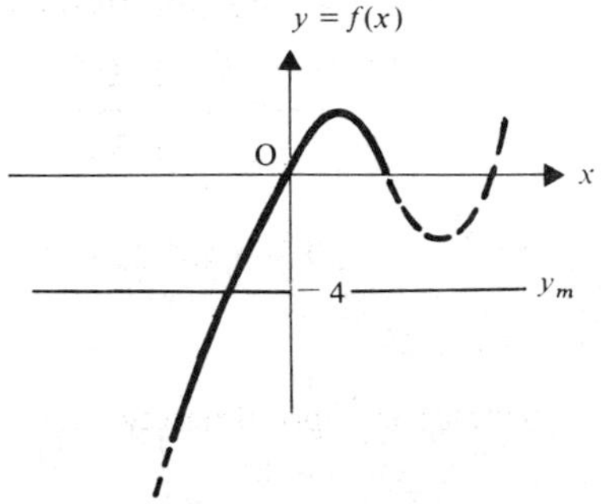

THE ROOT MEAN SQUARE VALUE OF A FUNCTION

The mean value of a function $f(x)$ is the value of the average ordinate of the graph of $y = f(x)$. Thus the mean value of a function takes into account the sign of the function. It is sometimes desirable to find an average that considers only the numerical value, and not the sign, of the function within the specified range. This can be achieved by ***squaring*** $f(x)$ and finding the mean of these squared values, subsequently taking the square root of this mean, to give the *root mean square value of the function* (or R.M.S. value)

i.e. the R.M.S. value of a function $f(x)$ within the range $a \leqslant x \leqslant b$

is given by
$$\left\{\frac{1}{b-a}\int_a^b [f(x)]^2\,dx\right\}^{\frac{1}{2}}$$

If $[f(x)]^2$ is integrable, the R.M.S. value can be determined exactly. Otherwise an approximate value for $\sum_{x=a}^{x=b} [f(x)]^2\,\delta x$ can be found using Simpson's Rule or some other numerical method.

For example, the R.M.S. value of $\sin x$ between $x = 0$ and $x = \frac{\pi}{2}$ is found from

$$\left\{\frac{1}{(\pi/2-0)}\int_0^{\frac{\pi}{2}} \sin^2 x\,dx\right\}^{\frac{1}{2}} = \left\{\frac{1}{\pi}\int_0^{\frac{\pi}{2}} (1-\cos 2x)\,dx\right\}^{\frac{1}{2}}$$
$$= \left\{\frac{1}{\pi}\left[\frac{\pi}{2}\right]\right\}^{\frac{1}{2}}$$
$$= \tfrac{1}{2}\sqrt{2}.$$

Similarly the R.M.S. value of e^{x^2} between $x = 0$ and $x = 1$ is given by

$$\left\{\frac{1}{1-0}\int_0^1 e^{2x^2}\,dx\right\}^{\frac{1}{2}}$$

but $\int e^{2x^2}\,dx$ cannot be found. So an approximate value for this integral is obtained by using Simpson's Rule with, say, five ordinates,

i.e.

x	0	0.25	0.5	0.75	1.0
e^{2x^2}	1	1.133	1.649	3.080	7.389

$$\Rightarrow \quad \int_0^1 e^{2x^2}\,dx \simeq \frac{0.25}{3}\{8.389 + 4(4.213) + 2(1.649)\}$$
$$\simeq 2.378$$

Hence the required R.M.S. value is approximately $\left\{\frac{1}{(1-0)}(2.378)\right\}^{\frac{1}{2}} = 1.542.$

EXERCISE 10d

Find the mean value of the given function over the interval indicated.

1) $(x-1)(x-2)$, $[1, 2]$

2) $\dfrac{1}{t}$, $[\frac{1}{2}, 1]$

3) e^{-x}, $[1, 5]$

4) $\tan\theta$, $\left[0, \dfrac{\pi}{4}\right]$

5) $\sin^2\theta$, $[0, 2\pi]$

6) $\dfrac{1}{(x-1)(x-2)}$, $[-1, 0]$

7) $\cos^3\theta$, $\left[0, \dfrac{\pi}{2}\right]$

8) xe^x, $[1, 2]$

9) $\sqrt{1-x^2}$, $[0, 1]$

10) $\dfrac{1}{\sqrt{(1-x^2)}}$, $[0, \frac{1}{2}]$

Find the root mean square value of the following functions over the specified ranges.
Where necessary use Simpson's Rule using five ordinates giving these results to three significant figures.

11) $f(x) \equiv \dfrac{1}{x}$ for $1 \leqslant x \leqslant 4$

12) $f(x) \equiv \tan x$ for $0 \leqslant x \leqslant \frac{1}{4}\pi$

13) $f(x) \equiv 1 + \sin x$ for $0 \leqslant x \leqslant \frac{1}{2}\pi$

14) $f(x) \equiv e^x$ for $1 \leqslant x \leqslant 2$

15) $f(x) \equiv \sqrt{(4-x^2)}$ for $0 \leqslant x \leqslant 2$

16) $f(x) \equiv \sinh x$ for $1 \leqslant x \leqslant 2$

17) $f(x) \equiv x \sin x$ for $0 \leqslant x \leqslant \frac{1}{2}\pi$

18) $f(x) \equiv \ln x$ for $1 \leqslant x \leqslant 2$

19) $f(x) \equiv 1 + x^2$ for $0 \leqslant x \leqslant 4$

THE LENGTH OF AN ARC OF A CURVE

Cartesian Coordinates

There is no simple formula for calculating the length of any portion of a curve other than a circle. So if the length of a particular arc is required, we use the method of summing small elements of arc length.
Suppose that the arc PQ, of length δs, is such an element, then the length, s, of the curve AB is given by $s = \displaystyle\sum_{x=x_1}^{x=x_2} \delta s.$

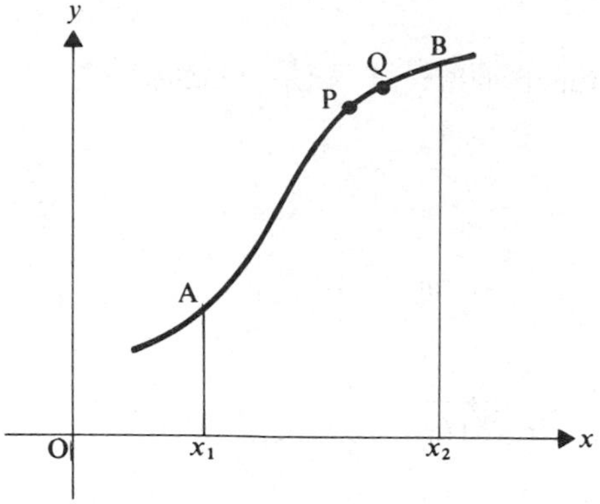

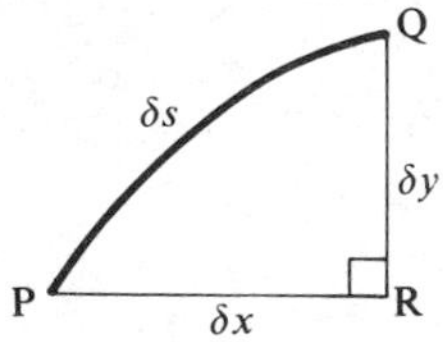

Now between P and Q, x increases by δx, y increases by δy and, as δs is small, δs is approximately equal to the hypotenuse of ΔPQR,

i.e. $$(\delta s)^2 \simeq (\delta x)^2 + (\delta y)^2$$

$$\Rightarrow \qquad \left(\frac{\delta s}{\delta x}\right)^2 \simeq 1 + \left(\frac{\delta y}{\delta x}\right)^2$$

$$\Rightarrow \qquad \delta s \simeq \left\{1 + \left(\frac{\delta y}{\delta x}\right)^2\right\}^{\frac{1}{2}} \delta x$$

Now $$s = \sum_{x=x_1}^{x=x_2} \delta s$$

$$\simeq \sum_{x=x_1}^{x=x_2} \left\{1 + \left(\frac{\delta y}{\delta x}\right)^2\right\}^{\frac{1}{2}} \delta x$$

As $$\delta x \to 0, \qquad \frac{\delta y}{\delta x} \to \frac{dy}{dx}$$

So $$s = \lim_{\delta x \to 0} \sum_{x=x_1}^{x=x_2} \left\{1 + \left(\frac{\delta y}{\delta x}\right)^2\right\}^{\frac{1}{2}} \delta x$$

i.e. $$s = \int_{x_1}^{x_2} \left\{1 + \left(\frac{dy}{dx}\right)^2\right\}^{\frac{1}{2}} dx$$

The length of any particular section of a curve whose Cartesian equation is known can now be evaluated provided that the necessary integration can be performed (otherwise an approximate value for $\Sigma\, \delta s$ can be found by a numerical method, e.g. Simpson's Rule).

EXAMPLES 10e

1) Find the length of the portion of the curve $y = x^2$ between $x = 0$ and $x = 1$.

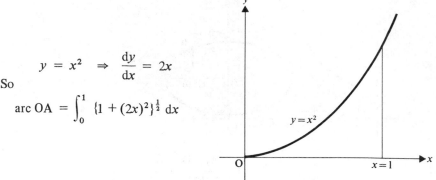

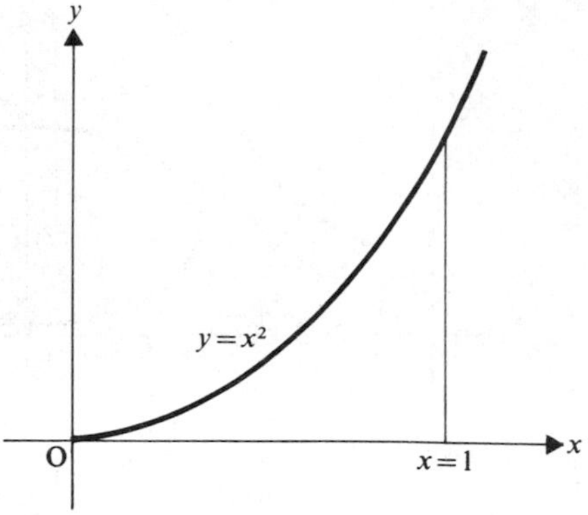

$$y = x^2 \quad \Rightarrow \quad \frac{dy}{dx} = 2x$$

So

$$\text{arc OA} = \int_0^1 \{1 + (2x)^2\}^{\frac{1}{2}}\,dx$$

To carry out the integration we can use the substitution

$$2x \equiv \sinh u \quad \Rightarrow \quad \ldots 2\,dx \equiv \ldots \cosh u\,du$$

and

x	0	1
u	0	arsinh 2

Then
$$\int (1+4x^2)^{\frac{1}{2}}\,dx \equiv \tfrac{1}{2}\int \cosh^2 u\,du$$

$$\equiv \tfrac{1}{4}\int (1 + \cosh 2u)\,du$$

$$= \tfrac{1}{4}(u + \tfrac{1}{2}\sinh 2u).$$

Hence
$$\int_0^1 (1+4x^2)^{\frac{1}{2}}\,dx = \tfrac{1}{4}[u + \sinh u \cosh u]_0^{\text{arsinh } 2}$$

$$= \tfrac{1}{4}\text{ arsinh } 2 + \tfrac{1}{2}\sqrt{(1+4)}$$

So the length of the arc OA is $\frac{1}{2}\sqrt{5} + \frac{1}{4}$ arsinh 2
or $\frac{1}{2}\sqrt{5} + \frac{1}{4}\ln(2+\sqrt{5})$

2) Find an approximate value for the length of the portion of the ellipse $\dfrac{x^2}{4} + y^2 = 1$ that lies in the first quadrant between $x = 0$ and $x = 1$.

The length, s, of the arc AB is given by

$$s = \int_0^1 \left\{1 + \left(\frac{dy}{dx}\right)^2\right\}^{\frac{1}{2}} dx$$

where $$\frac{dy}{dx} = -\frac{x}{4y}$$

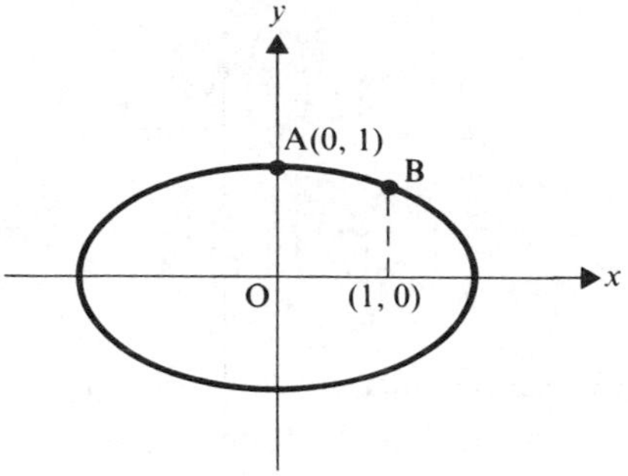

So $$s = \int_0^1 \left\{1 + \frac{x^2}{16y^2}\right\}^{\frac{1}{2}} dx$$

As the indefinite integral $\int \left\{1 + \frac{x^2}{16y^2}\right\}^{\frac{1}{2}} dx$ cannot easily be found for the given ellipse, we find an approximate value for s as follows.

Using Simpson's Rule with five ordinates, we have

x	0	0.25	0.5	0.75	1.00
$\left\{1 + \frac{x^2}{16y^2}\right\}^{\frac{1}{2}}$	1.0000	1.0020	1.0083	1.0202	1.0408

Then $s \simeq \frac{0.25}{3}[1.0000 + 1.0408 + 4(1.0020 + 1.0202) + 2(1.0083)]$

$\Rightarrow \quad s \simeq 1.0122$

Parametric Coordinates

Let us now consider a curve whose equations are given in terms of a parameter t say.

The approximate relationship

$$(\delta s)^2 \simeq (\delta x)^2 + (\delta y)^2$$

can be used again, but this time we will divide each term by $(\delta t)^2$, giving

$$\left(\frac{\delta s}{\delta t}\right)^2 \simeq \left(\frac{\delta x}{\delta t}\right)^2 + \left(\frac{\delta y}{\delta t}\right)^2$$

or $$\delta s \simeq \left\{\left(\frac{\delta x}{\delta t}\right)^2 + \left(\frac{\delta y}{\delta t}\right)^2\right\}^{\frac{1}{2}} \delta t$$

Then, as $\delta t \to 0$, $\dfrac{\delta x}{\delta t} \to \dfrac{dx}{dt}$ and $\dfrac{\delta y}{\delta t} \to \dfrac{dy}{dt}$

So $$s = \lim_{\delta t \to 0} \sum \left\{\left(\frac{\delta x}{\delta t}\right)^2 + \left(\frac{\delta y}{\delta t}\right)^2\right\}^{\frac{1}{2}} \delta t$$

$\Rightarrow$ $$s = \int \left\{\left(\frac{dx}{dt}\right)^2 + \left(\frac{dy}{dt}\right)^2\right\}^{\frac{1}{2}} dt$$

In this way the length of a particular section of a curve bounded by points where $t = t_1$ and $t = t_2$, can be found either by integration or by an approximate method of summation.

EXAMPLES 10e (continued)

3) Find the length of the circumference of the astroid $x = a\cos^3\theta$, $y = a\sin^3\theta$.

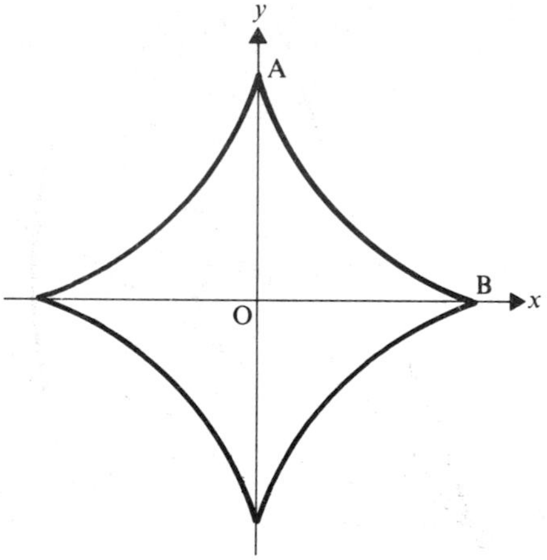

As the astroid is symmetrical about both Ox and Oy, the length of the arc AB is one quarter of the circumference.

Now at A, $x = 0$ so $\cos\theta = 0$

$\Rightarrow$ $$\theta = \frac{\pi}{2}$$

and at B, $y = 0$ so $\sin\theta = 0$

$\Rightarrow$ $$\theta = 0$$

Thus $$\text{arc AB} = \int_0^{\frac{\pi}{2}} \left\{\left(\frac{dx}{d\theta}\right)^2 + \left(\frac{dy}{d\theta}\right)^2\right\}^{\frac{1}{2}} d\theta$$

$$= \int_0^{\frac{\pi}{2}} \{(-3a\cos^2\theta\sin\theta)^2 + (3a\sin^2\theta\cos\theta)^2\}^{\frac{1}{2}}\, d\theta$$

$$= 3a\int_0^{\frac{\pi}{2}} \cos\theta\sin\theta\{\cos^2\theta + \sin^2\theta\}^{\frac{1}{2}}\, d\theta$$

$$= \frac{3a}{2}\int_0^{\frac{\pi}{2}} \sin 2\theta\, d\theta$$

$$= \frac{3a}{4}\Big[-\cos 2\theta\Big]_0^{\frac{\pi}{2}} = \frac{3a}{2}$$

So the length of the circumference of the astroid is $6a$.

Note. The reader may find that, occasionally, the integral used to calculate an arc length gives a negative result. If the limits of the integration are reversed, the result is of equal magnitude but is positive. So a negative result is a property only of the sense in which δs is summed and the arc length can therefore be taken as $|\Sigma\, \delta s|$.

Polar Coordinates

In this case we use a right-angled triangle with sides along and perpendicular to the radius vector, as shown in the diagram, so that the hypotenuse PQ is approximately equal to δs.

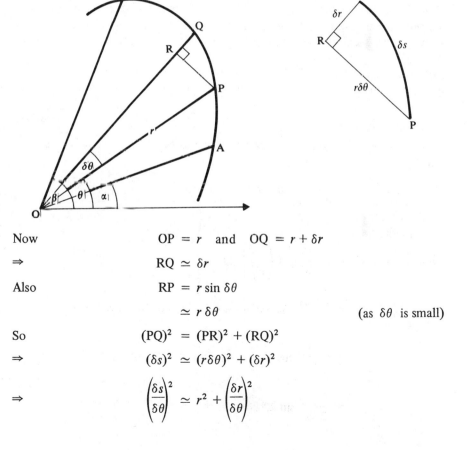

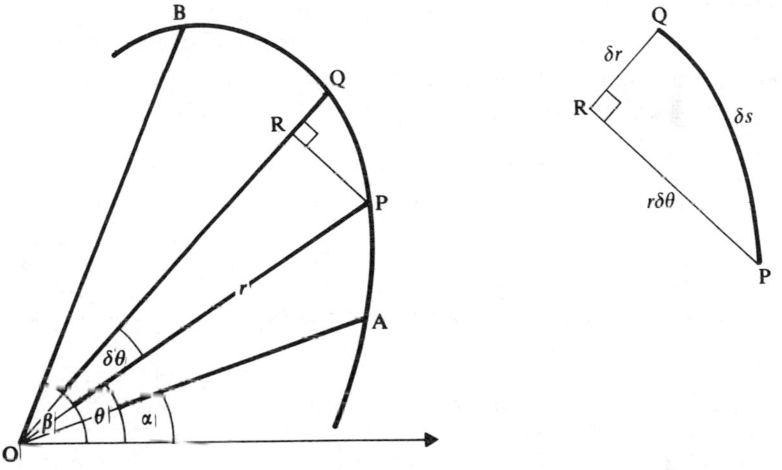

Now $\quad \text{OP} = r \quad \text{and} \quad \text{OQ} = r + \delta r$

$\Rightarrow \quad \text{RQ} \simeq \delta r$

Also $\quad \text{RP} = r \sin \delta\theta$

$\simeq r\,\delta\theta \quad$ (as $\delta\theta$ is small)

So $\quad (\text{PQ})^2 = (\text{PR})^2 + (\text{RQ})^2$

$\Rightarrow \quad (\delta s)^2 \simeq (r\delta\theta)^2 + (\delta r)^2$

$\Rightarrow \quad \left(\frac{\delta s}{\delta\theta}\right)^2 \simeq r^2 + \left(\frac{\delta r}{\delta\theta}\right)^2$

i.e. $$\delta s \simeq \left\{r^2 + \left(\frac{\delta r}{\delta \theta}\right)^2\right\}^{\frac{1}{2}} \delta\theta$$

Now the length, s, of the arc AB is given by

$$s = \sum_{\theta=\alpha}^{\theta=\beta} \delta s$$

$$\simeq \sum_{\theta=\alpha}^{\theta=\beta} \left\{r^2 + \left(\frac{\delta r}{\delta \theta}\right)^2\right\}^{\frac{1}{2}} \delta\theta$$

In the limiting case when $\delta\theta \to 0$ and $\dfrac{\delta r}{\delta\theta} \to \dfrac{dr}{d\theta}$,

$$s = \int_\alpha^\beta \left\{r^2 + \left(\frac{dr}{d\theta}\right)^2\right\}^{\frac{1}{2}} d\theta$$

Note that each of the expressions for finding the length of an arc can easily be derived from the appropriate right angled triangle and so *need* not be memorised. However there is no reason why they should not be quoted unless their proof is required.

Note also that, in each of the preceding cases, the arc length can be found either by integration or by an appropriate numerical method of summation.

EXAMPLES 10e (continued)

4) Find the length of the circumference of the cardioid $r = a(1 + \cos\theta)$

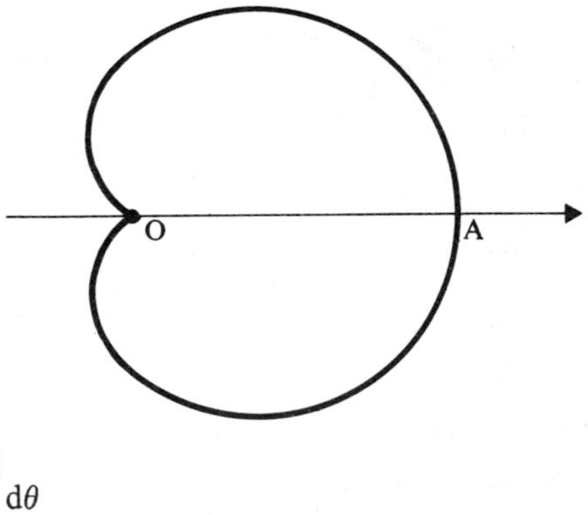

As OA is an axis of symmetry we need calculate only the upper half-circumference

At A, $\theta = 0$ and at O, $\theta = \pi$.

So the length, s, of the curve AO (in the anticlockwise sense) is given by

$$s = \int_0^\pi \left\{r^2 + \left(\frac{dr}{d\theta}\right)^2\right\}^{\frac{1}{2}} d\theta$$

$$= \int_0^\pi \{a^2(1+\cos\theta)^2 + (-a\sin\theta)^2\}^{\frac{1}{2}} d\theta$$

$$= a\int_0^{\pi} \{2(1+\cos\theta)\}^{\frac{1}{2}}\, d\theta$$

$$= a\int_0^{\pi} 2\cos\frac{\theta}{2}\, d\theta$$

$$= \left[4a\sin\frac{\theta}{2}\right]_0^{\pi}$$

$$= 4a$$

So the length of the circumference of the cardioid $r = a(1+\cos\theta)$ is $8a$.

EXERCISE 10e

(**Note** that the problems set throughout this chapter are basically exercises in integration. A reader who finds the integration in a particular problem difficult is recommended to omit it and continue with the next one.)

In each of the following questions, the equation(s) of a curve and a specific section of the curve, are given. In each case find the length of the specified portion.

1) The circle $x^2 + y^2 = 1$; between A$(-1, 0)$ and B$(1, 0)$.

2) $y = c\cosh\dfrac{x}{c}$ (c is a constant); between A$(0, c)$ and B(x, y).

3) The semi-cubical parabola $x^3 = y^2$; between A$(0, 0)$ and B$(1, 1)$.

4) The parabola $2y = x^2$; between A$(0, 0)$ and B$(1, \frac{1}{2})$.

5) $y = \ln\sec x$; between A$(x = 0)$ and **B**$\left(x = \dfrac{\pi}{6}\right)$.

6) The cycloid $x = a(\theta - \sin\theta)$, $y = a(1-\cos\theta)$; between A$\left(\theta = \dfrac{\pi}{2}\right)$ and B$(\theta = \pi)$.

7) The parabola $x = at^2$, $y = 2at$; between the vertex A$(t = 0)$ and B$(at^2, 2at)$.

8) The circle $x = a\cos\theta$, $y = a\sin\theta$; *find* the length of the circumference.

9) $x = \tanh t$, $y = \operatorname{sech} t$; between A$(t = 0)$ and B$(t = 1)$. What is this curve called?

10) The spiral $r = k\theta$ $(k > 0)$; between A$(0, 0)$ and **B**$\left(k\dfrac{\pi}{2}, \dfrac{\pi}{2}\right)$.

11) The cardioid $r = a(1-\sin\theta)$; find the length of its circumference.

12) The circle $r = a$; between $A(a, 0)$ and $B\left(a, \dfrac{\pi}{2}\right)$.

13) The equiangular spiral $r = ae^{\theta}$ between $A(\theta = 0)$ and $B(\theta = \pi)$.

AREA OF SURFACE OF REVOLUTION

When a section of a curve rotates through one revolution about the x axis, a three dimensional surface is formed. The surface area of such an object can be found by summing the areas of small elements of the total surface. A suitable element is formed by the rotation of an elemental arc of length δs tracing out a 'ring' of radius y and width δs.
The surface area, δA, of this ring element is given by

$$\delta A \simeq 2\pi y\, \delta s$$

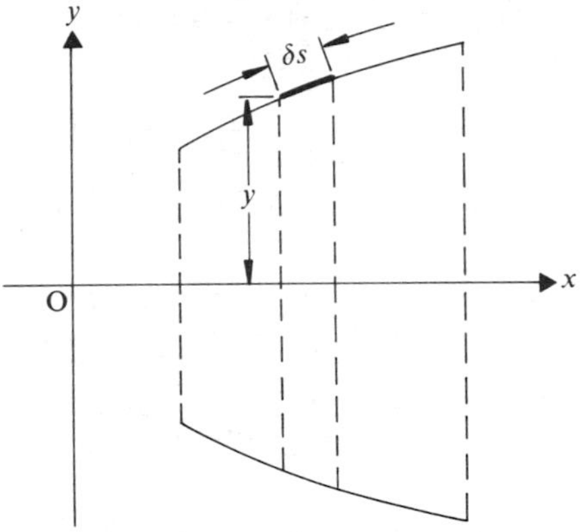

Now the equation of the curve whose rotation has formed the surface may be in Cartesian, parametric or polar form. In each case the total surface area is obtained by evaluating $\Sigma\, \delta A$, but the method of performing this summation depends upon the type of equation.

Cartesian Equation

Using the relationship

$$\delta s \simeq \left\{1 + \left(\frac{\delta y}{\delta x}\right)^2\right\}^{\frac{1}{2}} \delta x$$

gives

$$\delta A \simeq 2\pi y \left\{1 + \left(\frac{\delta y}{\delta x}\right)^2\right\}^{\frac{1}{2}} \delta x$$

$$\Rightarrow \qquad \sum \delta A \simeq \sum 2\pi y \left\{1 + \left(\frac{\delta y}{\delta x}\right)^2\right\}^{\frac{1}{2}} \delta x$$

Thus, when $\delta x \to 0$, the total surface area, A, is given by

$$A = \lim_{\delta x \to 0} \sum 2\pi y \left\{1 + \left(\frac{\delta y}{\delta x}\right)^2\right\}^{\frac{1}{2}} \delta x$$

$$\Rightarrow \qquad A = 2\pi \int y \left\{1 + \left(\frac{dy}{dx}\right)^2\right\}^{\frac{1}{2}} dx$$

Parametric Equations

This time we use

$$\delta s \simeq \left\{\left(\frac{\delta x}{\delta t}\right)^2 + \left(\frac{\delta y}{\delta t}\right)^2\right\}^{\frac{1}{2}} \delta t$$

so that

$$\delta A \simeq 2\pi y \left\{\left(\frac{\delta x}{\delta t}\right)^2 + \left(\frac{\delta y}{\delta t}\right)^2\right\}^{\frac{1}{2}} \delta t$$

Again taking the limit, as $\delta x \to 0$, of $\Sigma\, \delta A$, we get

$$A = 2\pi \int y \left\{\left(\frac{dx}{dt}\right)^2 + \left(\frac{dy}{dt}\right)^2\right\}^{\frac{1}{2}} dt$$

Polar Equation

Using the third of the forms we derived earlier for δs, we have

$$\delta s \simeq \left\{r^2 + \left(\frac{\delta r}{\delta \theta}\right)^2\right\}^{\frac{1}{2}} \delta \theta$$

$$\Rightarrow \qquad \delta A \simeq 2\pi y \left\{r^2 + \left(\frac{\delta r}{\delta \theta}\right)^2\right\}^{\frac{1}{2}} \delta \theta$$

But

$$y = r \sin \theta$$

so

$$\delta A \simeq 2\pi r \sin \theta \left\{r^2 + \left(\frac{\delta r}{\delta \theta}\right)^2\right\}^{\frac{1}{2}} \delta \theta$$

Taking the limit, as $\delta\theta \to 0$, of $\Sigma\, \delta A$, the total surface area becomes

$$A = 2\pi \int r \sin\theta \left\{ r^2 + \left(\frac{dr}{d\theta}\right)^2 \right\}^{\frac{1}{2}} d\theta$$

Note. The reader is recommended to remember *how these formulae are derived* rather than to memorise all three individual results.

EXAMPLE 10f

Find the area of the surface generated when each of the following curves is rotated through an angle 2π about the x axis:
(a) the arc of the parabola $y^2 = 4x$ between the origin and the point $(4, 4)$,
(b) the cycloid $x = a(\theta - \sin\theta)$, $y = a(1 - \cos\theta)$ between $(0, 0)$ and $(2\pi a, 0)$,
(c) the cardioid $r = a(1 + \cos\theta)$ between $(2a, 0)$ and $(0, \pi)$.

(a)
$$y^2 = 4x \quad \Rightarrow \quad \frac{dy}{dx} = \frac{2}{y}$$

So
$$\left\{1 + \left(\frac{dy}{dx}\right)^2\right\}^{\frac{1}{2}} = \left\{1 + \frac{4}{y^2}\right\}^{\frac{1}{2}} = \left\{\frac{x+1}{x}\right\}^{\frac{1}{2}}$$

Then
$$A = 2\pi \int y \left\{1 + \left(\frac{dy}{dx}\right)^2\right\}^{\frac{1}{2}} dx$$

$\Rightarrow$
$$A = 2\pi \int_0^4 2x^{\frac{1}{2}} \left\{\frac{x+1}{x}\right\}^{\frac{1}{2}} dx$$

$$= 4\pi \int_0^4 (x+1)^{\frac{1}{2}}\, dx$$

$$= 4\pi \left[\tfrac{2}{3}(x+1)^{\frac{3}{2}}\right]_0^4$$

$\Rightarrow$
$$A = \frac{8\pi}{3}(5\sqrt{5} - 1)$$

(b)
$$x = a(\theta - \sin\theta) \quad \Rightarrow \quad \frac{dx}{d\theta} = a(1 - \cos\theta)$$

$$y = a(1 - \cos\theta) \quad \Rightarrow \quad \frac{dy}{d\theta} = a \sin\theta$$

So
$$\left\{\left(\frac{dx}{d\theta}\right)^2 + \left(\frac{dy}{d\theta}\right)^2\right\}^{\frac{1}{2}} = \{a^2(1 - 2\cos\theta + \cos^2\theta) + a^2 \sin^2\theta\}^{\frac{1}{2}}$$

$$= a\{2 - 2\cos\theta\}^{\frac{1}{2}}$$

$$= a\left\{4\sin^2\frac{\theta}{2}\right\}^{\frac{1}{2}}$$

$$= 2a\sin\frac{\theta}{2}$$

Then A is given by

$$2\pi\int y\left\{\left(\frac{dx}{d\theta}\right)^2 + \left(\frac{dy}{d\theta}\right)^2\right\}^{\frac{1}{2}} d\theta = 2\pi\int a(1-\cos\theta)2a\sin\frac{\theta}{2}\,d\theta$$

As x goes from 0 to $2\pi a$, θ goes from 0 to 2π.

So $$A = 4\pi a^2\int_0^{2\pi} 2\sin^3\frac{\theta}{2}\,d\theta$$

$$= 8\pi a^2\int_0^{2\pi}\sin\frac{\theta}{2}\left(1-\cos^2\frac{\theta}{2}\right)d\theta$$

$$= 8\pi a^2\left[-2\cos\frac{\theta}{2} + \frac{2}{3}\cos^3\frac{\theta}{2}\right]_0^{2\pi}$$

$\Rightarrow$ $$A = \tfrac{64}{3}\pi a^2$$

(c) $$r = a(1+\cos\theta) \Rightarrow \frac{dr}{d\theta} = -a\sin\theta$$

So $$\left\{r^2 + \left(\frac{dr}{d\theta}\right)^2\right\}^{\frac{1}{2}} = \{a^2(1 + 2\cos\theta + \cos^2\theta) + a^2\sin^2\theta\}^{\frac{1}{2}}$$

$$= a\{2 + 2\cos\theta\}^{\frac{1}{2}}$$

$$= 2a\cos\frac{\theta}{2}$$

Then $$A = 2\pi\int_0^{\pi} r\sin\theta\left\{r^2 + \left(\frac{dr}{d\theta}\right)^2\right\}^{\frac{1}{2}} d\theta$$

$\Rightarrow$ $$A = 2\pi\int_0^{\pi} a(1+\cos\theta)\sin\theta\left(2a\cos\frac{\theta}{2}\right)d\theta$$

$$= 4\pi a^2\int_0^{\pi}\left(2\cos^2\frac{\theta}{2}\right)\left(2\sin\frac{\theta}{2}\cos\frac{\theta}{2}\right)\left(\cos\frac{\theta}{2}\right)d\theta$$

$$= 16\pi a^2 \int_0^{\pi} \cos^4 \frac{\theta}{2} \sin \frac{\theta}{2} \, d\theta$$

$$= 16\pi a^2 \left[-\frac{2}{5} \cos^5 \frac{\theta}{2} \right]_0^{\pi}$$

$$\Rightarrow \quad A = \tfrac{32}{5}\pi a^2$$

VOLUME OF REVOLUTION

When an area rotates completely about a fixed line, a solid is formed with circular cross-section. The volume of such a solid of revolution can be found by summing the volumes of small elements of the solid.
The detailed application of this method, when the area is bounded by a curve with a Cartesian equation, was explained in *The Core Course*. We shall now adapt the method to deal with the rotation of an area bounded by a curve with parametric equations.
Consider, for example, the volume generated when the area in the first quadrant bounded by the ellipse $x = 4\cos\theta$, $y = 3\sin\theta$ rotates completely about the x axis.

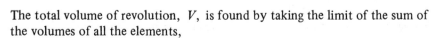

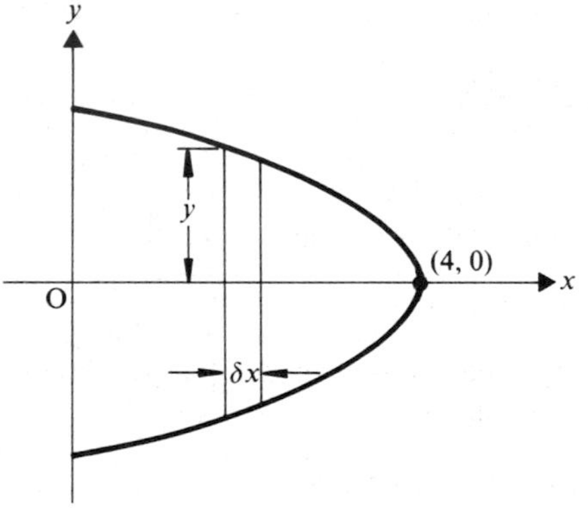

Dividing the volume into thin slices perpendicular to the x axis, we see that each slice is approximately a disc of radius y and thickness δx.
So the volume, δV, of one element is given by

$$\delta V \simeq \pi y^2 \, \delta x$$

The total volume of revolution, V, is found by taking the limit of the sum of the volumes of all the elements,

i.e. $$V = \lim_{\delta x \to 0} \sum_{x=0}^{x=4} \delta V = \lim_{\delta x \to 0} \sum_{x=0}^{x=4} \pi y^2 \, \delta x$$

$$= \int_0^4 \pi y^2 \, dx$$

But $x = 4\cos\theta$ $\Rightarrow$ $\ldots dx \equiv \ldots -4\sin\theta \, d\theta$

and $y = 3\sin\theta$

Also, when $x = 0$, $\theta = \dfrac{\pi}{2}$

and when $x = 4$, $\theta = 0$.

So
$$V = \pi \int_{\frac{\pi}{2}}^{0} (3 \sin \theta)^2(-4 \sin \theta)\, d\theta$$
$$= 36\pi \int_0^{\frac{\pi}{2}} \sin^3\theta \, d\theta$$
$$= 36\pi(\tfrac{2}{3}) \qquad \text{(using a reduction formula).}$$

Hence the volume generated by rotating a quadrant of the given ellipse is 24π.

Note that the degree of rotation necessary for an area to generate a complete solid of revolution, is not always a full revolution. For example, a sphere can be generated either when a *circle* rotates about a diameter through an angle π, or when a *semi-circle* rotates about its diameter through an angle 2π.

EXERCISE 10f

Find the areas of the surfaces generated by the complete rotation about the x axis (or the initial line) of each of the following arcs.

1) The quarter circle in the first quadrant whose equation is $x^2 + y^2 = a^2$.

2) The arc of the astroid $x = a \cos^3 t$, $y = a \sin^3 t$ that lies above the x axis.

3) The line $2y = x$ between the origin and the point $(4, 2)$. (Do not assume any formulae for a cone.)

4) The arc of the curve $r = 2a \cos \theta$ bounded by $\theta = 0$ and $\theta = \dfrac{\pi}{2}$.

5) The complete curve $r = a$. (Do not assume the formula for surface area of a sphere.)

6) The arc between $t = 0$ and $t = 4$ of the parabola $x = t^2$, $y = 2t$.

7) The section of the spiral $r = e^{\theta}$ between $\theta = 0$ and $\theta = \dfrac{\pi}{2}$.

8) The part of the rectangular hyperbola $x^2 - y^2 = a^2$ that lies in the first quadrant between $x = a$ and $x = 2a$.

9) The portion of the curve $y = \cosh x$ between $x = 0$ and $x = 1$.

10) The section between $x = 0$ and $x = 1$ of the curve $y = e^x$.

11) The arc of the curve $x = t^3$, $y = 3t^2$ between the points where $t = 0$ and $t = 3$.

Find the volume generated by the complete rotation of the following areas about the specified lines.

12) The area bounded by the x axis, the curve $x = 2t^2$, $y = 8t$ and the line $x = 8$, rotated about
(a) the x axis, (b) the y axis.

13) The area in the first quadrant, of the astroid $x = \cos^3 t$, $y = \sin^3 t$, rotated about the x axis.

14) The area bounded by the x axis and the arc of the cycloid $x = a(\theta - \sin\theta)$, $y = a(1 - \cos\theta)$ between $\theta = 0$ and $\theta = 2\pi$, rotated about the x axis.

15) The area between the parabola $x = t^2$, $y = 2t$ and the line $x = 2$, rotated about the line $x = 2$.

16) The area bounded by the lines $x = 1$, $x = 2$, $y = 1$ and the curve $x = 4t$, $y = \dfrac{4}{t}$, rotated about
(a) the line $x = 1$ (b) the line $y = 1$
(c) the x axis (d) the y axis.

CENTROID AND FIRST MOMENT

In *The Core Course*, we saw that the *centroid* of an object is its geometric centre. For a symmetrical object therefore, the centroid lies on the axis of symmetry. For an object made of uniform material, the centroid coincides with the point at which the body can be supported in a perfectly balanced state.
Relative to a specified line, an object has a *first moment* which depends upon
(a) the distance, h, of its centroid from the specified line,
(b) $\begin{cases} \text{the volume of the object if it is a solid,} \\ \text{the area of the object if it is a surface,} \\ \text{the length of the object if it is an arc of a curve.} \end{cases}$
In *The Core Course* the first moments of a volume and an area were defined,

i.e. first moment of a volume = volume $\times h$
first moment of an area = area $\times h$.

We now define the first moment of an arc in a similar way,

i.e. first moment of an arc = length $\times h$.

In cases when h is not known, the first moment of an object can be found by summing the first moments of all the elements into which it can be divided. Thus, if h_e is the distance from the specified line of the centroid of one typical element of a solid object, we have
first moment of volume $= \Sigma\, V_e h_e$ where V_e is the volume of the element.

But first moment of volume is also given by Vh.

So

$$Vh = \sum V_e h_e \qquad [1]$$

Similarly, for an object with an area of magnitude A we have

$$Ah = \sum A_e h_e \qquad [2]$$

And, for an object that is an arc of a curve of length S and which therefore has a first moment of arc length,

$$Sh = \sum S_e h_e \qquad [3]$$

The application of equations [1] and [2] above to the locating of centroids of volumes and areas is dealt with in *The Core Course*. The centroid of an arc length, however, was not found at that time, so we will now consider an example of this type.

EXAMPLES 10g

1) Find the first moment, relative to the x axis, of a semi-circular arc of radius a, bounded by the x axis. Hence find the position of the centroid of the arc.

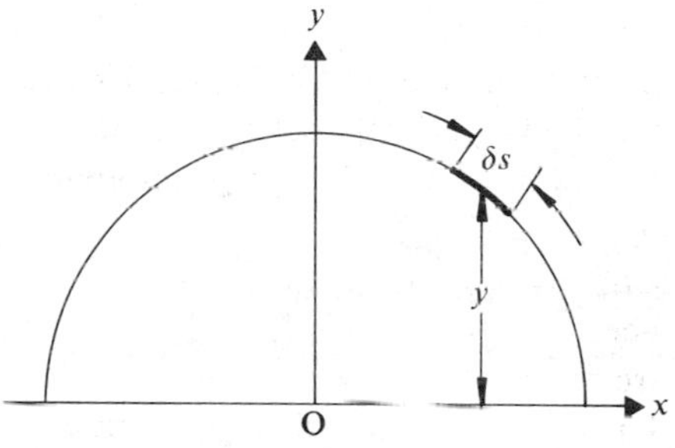

Taking an element of the arc of length δs, whose centroid is at a distance y from Ox we have:

First moment of arc

$$\simeq \sum y\,\delta s$$

But $\quad \delta s = a\delta\theta$

and $\quad y \simeq a \sin\theta$

So $\quad \sum y\,\delta s \simeq \sum a^2 \sin\theta\,\delta\theta$

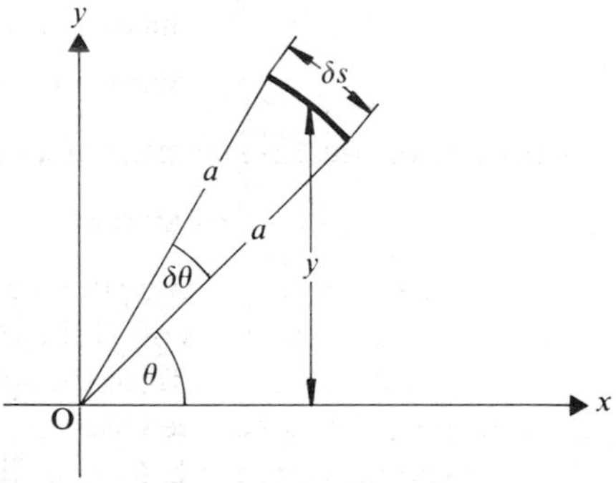

Taking the limit of the summation, and working from $\theta = 0$ to $\theta = \pi$, we find that the first moment, M, of the semicircular arc is given by

$$M = \int_0^\pi a^2 \sin\theta \, d\theta$$

$$= a^2 [-\cos\theta]_0^\pi$$

i.e. $$M = 2a^2$$

Now the total length of the arc is πa and, from symmetry, its centroid lies on the y axis. If the distance of its centroid from Ox is $\bar{y}$, then the first moment of the arc length is also given by

$$M = \pi a \bar{y}$$

$\Rightarrow$ $$\pi a \bar{y} = 2a^2$$

$\Rightarrow$ $$\bar{y} = \frac{2a}{\pi}$$

So the centroid of the arc is the point $\left(0, \dfrac{2a}{\pi}\right)$.

Note that the centroid is *not* on the arc, a result common to most arcs.

Note also that the relationship $(\delta s)^2 \simeq (\delta x)^2 + (\delta y)^2$ can be used where appropriate.

THE THEOREMS OF PAPPUS

Theorem 1

Consider a section of a curve of perimeter s, that does not cross the x axis, and whose centroid is G. If this curve rotates completely about the x axis, then Pappus' First Theorem states that the surface area, A, of the surface so formed, is given by

$$A = s \times \text{distance travelled by } G$$

Proof

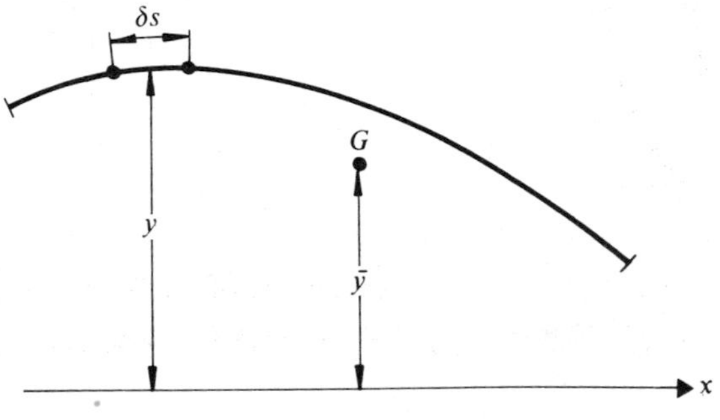

First we find the ordinate, $\bar{y}$, of the centroid, G, using

first moment of whole curve = $\sum$first moment of element,

i.e. $$s\bar{y} \simeq \sum y\,\delta s$$

$\Rightarrow$ $$\bar{y} \simeq \frac{1}{s}\sum y\,\delta s$$

Now the distance travelled by G when the curve rotates about Ox through one revolution is $2\pi\bar{y}$

where $$2\pi\bar{y} \simeq \frac{2\pi}{s}\sum y\,\delta s = \frac{1}{s}\sum 2\pi y\,\delta s$$

But $2\pi y\,\delta s$ is the approximate area of an elemental 'ring' formed when δs rotates completely.
So the total surface area of revolution is given approximately by $\sum 2\pi y\,\delta s$,

i.e. $$A \simeq \sum 2\pi y\,\delta s$$

$$\simeq s\left[\frac{1}{s}\sum 2\pi y\,\delta s\right]$$

Taking the limit, as $\delta x \to 0$, of this summation we have

$$A = s \times (\text{distance travelled by } G).$$

EXAMPLES 10g (continued)

2) Find the surface area of a lifebelt whose cross-section is a circle of radius 8 cm and whose inner radius is 20 cm.

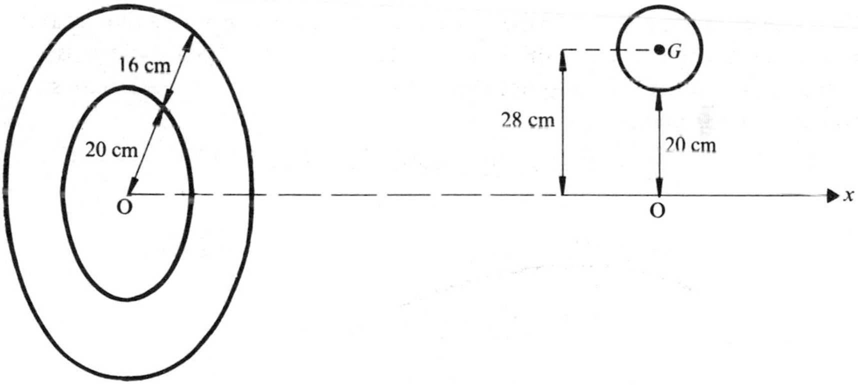

The surface of the lifebelt is formed when a circle of radius 8 cm, with its centre at a distance 28 cm from the x axis, rotates completely about the x axis.

The centroid of the circle is its centre, G, so using Pappus' Theorem, we have

$$\begin{aligned}\text{surface area} &= (\text{perimeter of circle}) \times (\text{distance travelled by } G)\\ &= 2\pi(8) \times 2\pi(28)\end{aligned}$$

Hence, taking $\pi^2 \simeq 10$, we find that the surface area of the lifebelt is approximately 8960 square centimetres.

Theorem 2

Pappus' Second Theorem refers to the area, A, enclosed by a curve that does not cross the x axis, and whose centroid (of area) is G. If this area rotates completely about the x axis, then this theorem states that the volume V of the solid of revolution is given by

$$V = A \times \text{distance travelled by } G$$

Proof

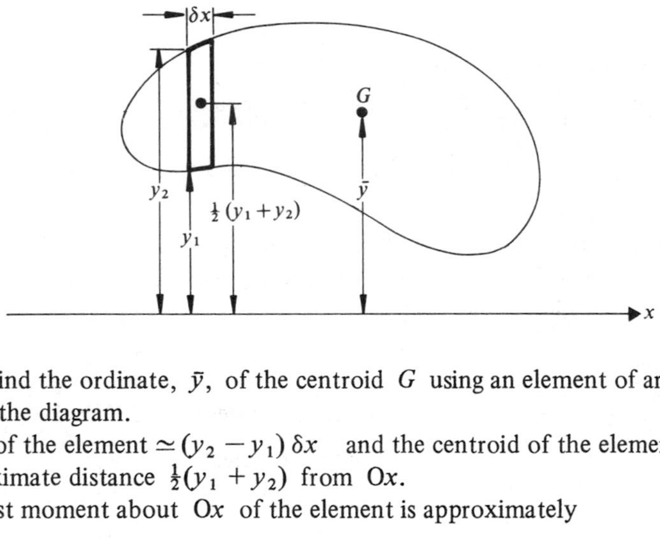

First we find the ordinate, $\bar{y}$, of the centroid G using an element of area as shown in the diagram.
The area of the element $\simeq (y_2 - y_1)\,\delta x$ and the centroid of the element is at an approximate distance $\frac{1}{2}(y_1 + y_2)$ from Ox.
So the first moment about Ox of the element is approximately

$$\tfrac{1}{2}(y_1 + y_2)(y_2 - y_1)\,\delta x$$

But first moment of whole area = Σ first moment of element,

i.e.
$$A\bar{y} \simeq \Sigma \frac{1}{2}(y_1 + y_2)(y_2 - y_1)\,\delta x$$

$\Rightarrow$
$$\bar{y} \simeq \frac{1}{2A}\Sigma(y_2^2 - y_1^2)\,\delta x$$

Now the distance travelled by $\bar{y}$ when the area rotates about Ox through one revolution is $2\pi\bar{y}$,

where
$$2\pi\bar{y} \simeq 2\pi\left\{\frac{1}{2A}\sum(y_2^2 - y_1^2)\right\}\delta x$$
$$= \frac{1}{A}\sum(\pi y_2^2 - \pi y_1^2)\,\delta x$$

But $(\pi y_2^2 - \pi y_1^2)\,\delta x$ is the approximate volume of the elemental annulus formed when the element of area rotates about Ox.
So the total volume of revolution, V, is given by
$$V \simeq \sum(\pi y_2^2 - \pi y_1^2)\,\delta x$$
i.e.
$$V \simeq A\left[\frac{1}{A}\sum(\pi y_2^2 - \pi y_1^2)\,\delta x\right]$$

Taking the limit, as $\delta x \to 0$, of this summation, we have
$$V = A \times \text{distance travelled by } G$$

EXAMPLES 10g (continued)

3) Find the volume of the lifebelt described in Example 2 (page 413).

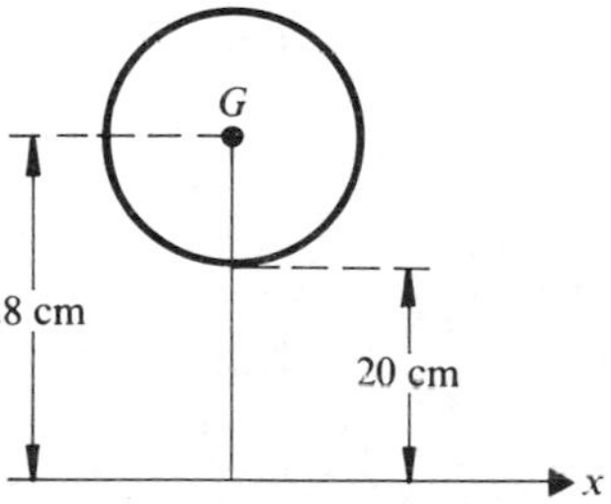

The area enclosed by the circular cross-section is $\pi(8^2) = 64\pi\ \text{cm}^2$.

The distance travelled by G when the circle rotates about Ox is $2\pi \times 28$ cm.

Using Pappus' Second Theorem to find the volume, V, of the lifebelt, we have
$$V = (64\pi)(56\pi)\ \text{cm}^3$$
i.e.
$$V \simeq 35\,840\ \text{cm}^3 \qquad (\pi^2 \simeq 10)$$

4) Find the position of the centroid of a semicircle by applying Pappus' Second Theorem to the volume of a sphere.

Consider a semicircle of radius a, with its diameter on the x axis.
When this semicircle rotates completely about Ox, a sphere is generated whose volume is $\frac{4}{3}\pi a^3$.

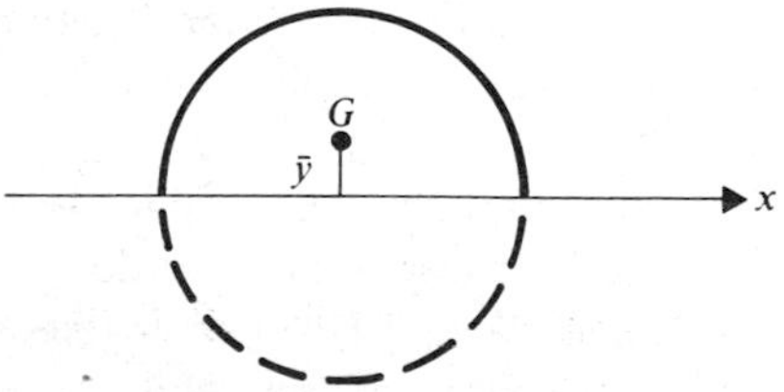

The closed area of the semicircle is $\frac{1}{2}\pi a^2$ and the distance travelled by G during rotation is $2\pi\bar{y}$.

Then Pappus' Second Theorem states that

$$\text{volume generated} = \text{area} \times \text{distance travelled by } G$$

i.e.
$$\tfrac{4}{3}\pi a^3 = \tfrac{1}{2}\pi a^2 \times 2\pi\bar{y}$$

$\Rightarrow$
$$\bar{y} = \frac{4a}{3\pi}$$

EXERCISE 10g

1) Find the first moment about Ox of the arc of the curve $y = \sin x$ between $x = 0$ and $x = \pi$.

2) Find the first moment about Oy of the curve $y = x^2$ between $(1, 1)$ and $(2, 4)$.

3) Find the first moment about Ox of the part of the parabola $y = 1 - x^2$ that lies above the x axis.

4) An area is defined by $0 \leqslant x \leqslant 4$, $0 \leqslant y \leqslant 2\sqrt{x}$. Find its first moment (a) about the x axis, (b) about the y axis.

5) Find the first moments about the x axis of

(a) the area under the curve $y = \cos x$ from $x = -\frac{\pi}{2}$ to $x = \frac{\pi}{2}$;

(b) the area bounded by the curve $y = e^x$ and the lines $x = 0$, $y = 0$, $x = 2$;

(c) the area in the first quadrant enclosed by the astroid $x = a\cos^3\theta$, $y = a\sin^3\theta$;

(d) the area defined by $1 \leqslant x \leqslant 2$ and $\frac{1}{x} \leqslant y \leqslant 1$.

6) Evaluate each of the areas defined in Question 5.

7) Use the results of Questions 5 and 6 to find the y coordinate of the centroid of each of the areas described.

8) Each of the areas described in Question 5 rotates completely about the x axis to form a solid of revolution.
Use Pappus' Second Theorem, together with your answers to Questions 6 and 7 to find the volume of each solid.

9) Find the first moment, about the y axis, of the arc of the parabola $y = (1 - x^2)$ for which $x \geqslant 0$. Find also the length of this arc and hence find the centroid of the arc. Use Pappus' First Theorem to find the area of the surface generated when the arc rotates completely about the y axis.

10) Find the volume of the solid generated when the part of the ellipse $x^2 + 4y^2 = 4$ that lies above the x axis, rotates completely about the x axis. Use Pappus' Second Theorem to find the centroid of this area

$$\left(\text{the area of the ellipse } \frac{x^2}{a^2} + \frac{y^2}{b^2} = 1 \text{ is } \pi ab\right).$$

11) The line $4y = 3x$ between the origin and the point $(4, 3)$ rotates completely about the x axis generating the surface of a cone. Write down the coordinates of the centroid of the line. Use Pappus' First Theorem to find the surface area of the cone.

12) Apply Pappus' Second Theorem to the rotation of a rectangular area about one of its sides, to derive the formula for the volume of a cylinder.

CURVATURE

The curvature of a curve is a measure of its rate of turning. Suppose that, at a point P on a curve, the tangent makes an angle ψ with the x axis, and at an adjacent point Q, the angle between the tangent and x axis is $\psi + \delta\psi$.

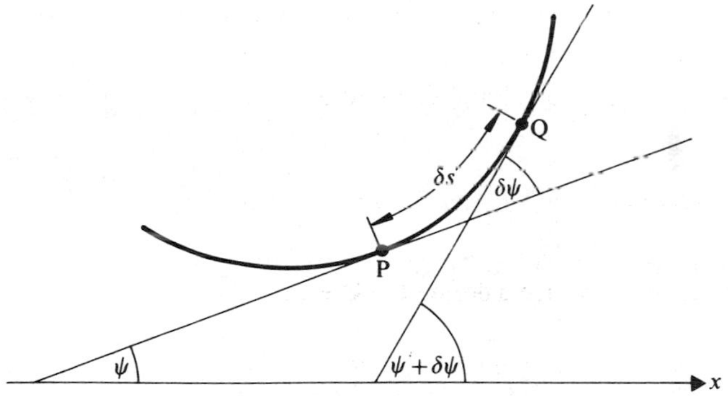

The curvature, κ, at P is defined by

$$\kappa = \lim_{\delta s \to 0} \frac{\delta \psi}{\delta s} = \frac{d\psi}{ds}$$

RADIUS OF CURVATURE

If, at the points P and Q described above, lines are drawn at right angles to the tangents at these points, to meet at C, then C is the centre of a circle passing through P and Q.

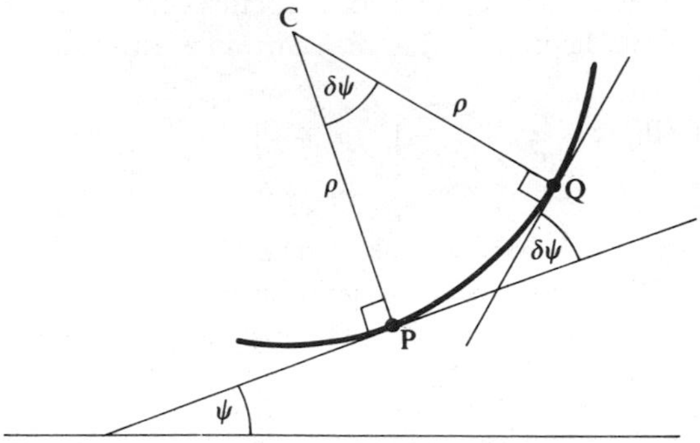

The radius of this circle is denoted by ρ (where $\rho = \text{CP} = \text{CQ}$), so the length of the *circular* arc between P and Q is $\rho\,\delta\psi$.
Now as Q approaches P, the length of the circular arc PQ approaches δs,

i.e. $$\rho\,\delta\psi \simeq \delta s$$

$\Rightarrow$ $$\rho \simeq \frac{\delta s}{\delta\psi}$$

In the limiting case, when $\delta s \to 0$, we have

$$\rho = \frac{ds}{d\psi} = \frac{1}{\kappa}$$

and ρ is called the *radius of curvature* at P of the given curve.

Thus the radius of curvature at a point on a curve is the reciprocal of the curvature at that point.

CALCULATION OF RADIUS OF CURVATURE

(a) Cartesian Form

The gradient of the tangent at P to the given curve is $\tan\psi$,

i.e. $$\frac{dy}{dx} = \tan\psi$$

$\Rightarrow$ $$\frac{d^2y}{dx^2} = \frac{d}{dx}(\tan\psi) = \sec^2\psi\,\frac{d\psi}{dx} \qquad [1]$$

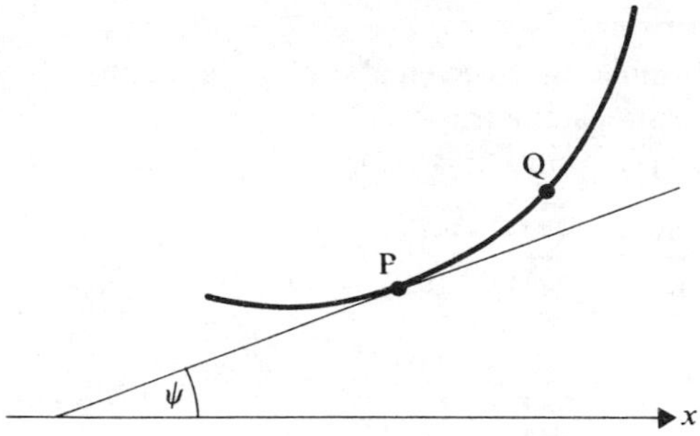

If we consider again the arc PQ of the curve, we see that, when δs is small,

$$\delta x \simeq \delta s \cos \psi$$

or
$$\frac{\delta x}{\delta s} \simeq \cos \psi$$

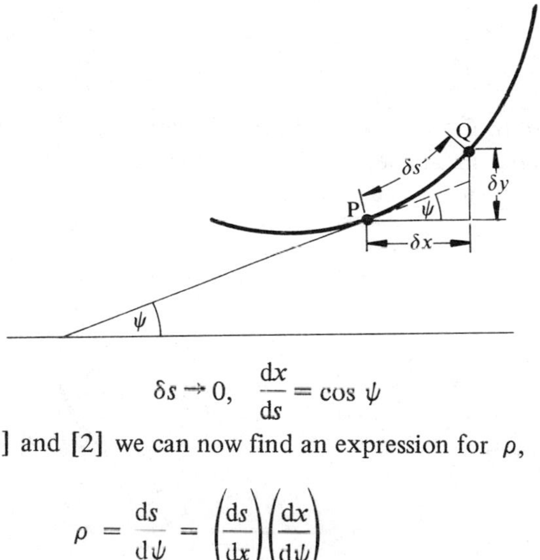

So, when
$$\delta s \to 0, \quad \frac{dx}{ds} = \cos \psi \qquad [2]$$

Using results [1] and [2] we can now find an expression for ρ,

i.e.
$$\rho = \frac{ds}{d\psi} = \left(\frac{ds}{dx}\right)\left(\frac{dx}{d\psi}\right)$$

$$= (\sec \psi)\left(\sec^2 \psi \bigg/ \frac{d^2y}{dx^2}\right).$$

But
$$\sec^2 \psi \equiv 1 + \tan^2 \psi = 1 + \left(\frac{dy}{dx}\right)^2$$

So
$$\rho = \left[1 + \left(\frac{dy}{dx}\right)^2\right]^{\frac{3}{2}} \bigg/ \frac{d^2y}{dx^2}$$

(b) Parametric Form

The Cartesian formula for ρ given above can be adapted for use when the equations of a curve are parametric.
Suppose that $y = f(t)$ and $x = g(t)$,

then $$\frac{dy}{dx} = \frac{f'(t)}{g'(t)}$$

and $$\frac{d^2y}{dx^2} = \frac{1}{f'(t)}\frac{d}{dt}\left(\frac{f'(t)}{g'(t)}\right)$$ (*The Core Course*, p. 286.)

Hence, as both $\frac{dy}{dx}$ and $\frac{d^2y}{dx^2}$ can be found (in terms of t), the Cartesian formula can be used.

(c) Polar Form

If P is a point (r, θ) on a curve with a polar equation, the tangent at P is inclined at ψ to the initial line, and ϕ is the angle between this tangent and the radius vector OP, then $\psi = \theta + \phi$.

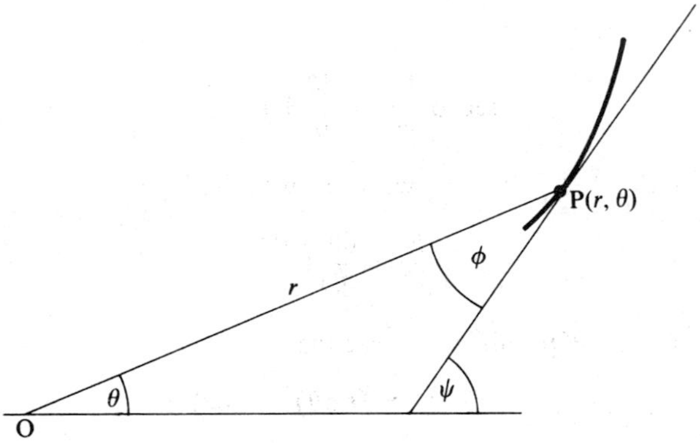

$Q(r + \delta r, \theta + \delta\theta)$ is a point close to P and PR is perpendicular to OQ.

Hence $$RQ \simeq \delta r \quad \text{and} \quad PR \simeq r\,\delta\theta$$

$\Rightarrow$ $$\tan OQP \simeq \frac{r\,\delta\theta}{\delta r}$$

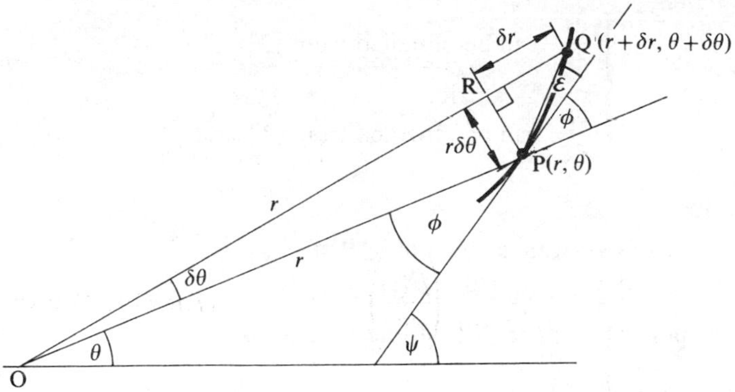

Also, if ϵ is the angle between the tangent at P and the chord PQ, then

$$\phi + \epsilon = \text{OQP} + \delta\theta$$

So $$\tan(\phi + \epsilon - \delta\theta) \simeq \frac{r\,\delta\theta}{\delta r}$$

But, as $\delta\theta \to 0$, $\dfrac{\delta\theta}{\delta r} \to \dfrac{d\theta}{dr}$ and $\epsilon \to 0$ (the chord PQ merges with the tangent at P when $\delta\theta \to 0$).

So $$\tan\phi = r\frac{d\theta}{dr}$$

Differentiating w.r.t. r, gives

$$\sec^2\phi\,\frac{d\phi}{dr} = \frac{d\theta}{dr} + r\frac{d^2\theta}{dr^2} \qquad [1]$$

Now differentiating $\psi = \theta + \phi$ w.r.t. r, we have

$$\frac{d\psi}{dr} = \frac{d\theta}{dr} + \frac{d\phi}{dr} \qquad [2]$$

Also, from 'triangle' PQR above, we see that

$$(\delta s)^2 \simeq (r\,\delta\theta)^2 + (\delta r)^2$$

$$\Rightarrow \qquad \left(\frac{ds}{dr}\right)^2 = \left(r\frac{d\theta}{dr}\right)^2 + 1 \qquad [3]$$

These three expressions can now be used to find ρ for a curve with a polar equation, as

$$\rho = \frac{ds}{d\psi} = \left(\frac{ds}{dr}\right)\left(\frac{dr}{d\psi}\right)$$

and $\begin{cases} \dfrac{ds}{dr} \text{ can be obtained from [3],} \\ \dfrac{dr}{d\psi} \text{ can be obtained from [1] and [2].} \end{cases}$

EXAMPLE 10h

Find the radius of curvature at the point P if

(a) P is the point (x, y) on the curve $y = x^2$,

(b) P is the point $(\sin^3\theta, \cos^3\theta)$ on the astroid $x = \sin^3\theta$, $y = \cos^3\theta$,

(c) P is the point $\left(3, \dfrac{\pi}{3}\right)$ on the cardioid $r = 2(1 + \cos\theta)$.

(a) Using the quotable Cartesian formula for ρ we have,

$$\rho = \left[1 + \left(\frac{dy}{dx}\right)^2\right]^{\frac{3}{2}} \Bigg/ \frac{d^2y}{dx^2}$$

where $$y = x^2 \Rightarrow \frac{dy}{dx} = 2x \Rightarrow \frac{d^2y}{dx^2} = 2$$

So $\rho = \frac{1}{2}(1 + 4x^2)^{\frac{3}{2}}$ at any point on $y = x^2$.

(b) If $x = \sin^3\theta$ and $y = \cos^3\theta$ then

$$\frac{dy}{dx} = \frac{3\cos^2\theta(-\sin\theta)}{3\sin^2\theta(\cos\theta)} = -\cot\theta$$

and
$$\frac{d^2y}{dx^2} = \frac{d}{dx}(-\cot\theta) = \frac{d}{d\theta}(-\cot\theta)\frac{d\theta}{dx}$$

$$= \frac{\operatorname{cosec}^2\theta}{3\sin^2\theta\cos\theta}$$

$$= \frac{1}{3\sin^4\theta\cos\theta}$$

Then
$$\rho = [1 + (-\cot\theta)^2]^{\frac{3}{2}}(3\sin^4\theta\cos\theta)$$

$$= (\operatorname{cosec}^3\theta)(3\sin^4\theta\cos\theta)$$

So $\rho = 3\sin\theta\cos\theta$ at any point on the given astroid.

(c) For the cardioid $r = 2(1 + \cos\theta)$ we have:

$$\tan\phi = r\frac{d\theta}{dr} = r\Bigg/\frac{dr}{d\theta}$$

$$= 2(1 + \cos\theta)/(-2\sin\theta)$$

$$= -\cot\frac{\theta}{2}$$

Hence $$\phi = \frac{\pi}{2} + \frac{\theta}{2}$$

$\Rightarrow$ $$\frac{d\phi}{dr} = \frac{1}{2}\frac{d\theta}{dr} = \frac{1}{2(-2\sin\theta)}$$

When $$\theta = \frac{\pi}{3}, \quad \frac{d\theta}{dr} = -\frac{1}{2\sqrt{3}}$$

Now using $\psi = \theta + \phi$, we have

$$\frac{d\psi}{dr} = \frac{d\theta}{dr} + \frac{d\phi}{dr}$$

$$= \frac{-1}{2\sin\theta} + \frac{d\phi}{dr}$$

So, when $\theta = \frac{\pi}{3}$, $$\frac{d\psi}{dr} = -\frac{1}{\sqrt{3}} - \frac{1}{2\sqrt{3}} = -\frac{\sqrt{3}}{2}.$$

Also $$\left(\frac{ds}{dr}\right)^2 = \left(r\frac{d\theta}{dr}\right)^2 + 1$$

$$= \left(\frac{r}{-2\sin\theta}\right)^2 + 1$$

When $\theta = \frac{\pi}{3}$ and $r = 3$, $\left(\frac{ds}{dr}\right)^2 = 4 \Rightarrow \frac{ds}{dr} = 2$

Now we are in a position to find ρ at the point $\left(3, \frac{\pi}{3}\right)$ using

$$\rho = \frac{ds}{d\psi} = \left(\frac{ds}{dr}\right)\left(\frac{dr}{d\psi}\right) = (2)\left(-\frac{2}{\sqrt{3}}\right)$$

The sign obtained for ρ serves only to indicate whether the curve is concave or convex at the given point.

So the radius of curvature of the cardioid at the point $\left(3, \frac{\pi}{3}\right)$ is $\frac{4\sqrt{3}}{3}$

Note. A rigorous explanation of the significance of the sign of ρ is beyond the scope of this book.

EXERCISE 10h

Find the radius of curvature of each of the following curves at the specified point.

1) $y = 2x^2$; $x = 1$

2) $y = \cos x$; $x = \dfrac{\pi}{3}$

3) $y = e^{-x}$; $x = 0$

4) $y = \ln x$; $x = 2$

5) $y^2 = 4ax$; $(a, 2a)$

6) $xy = 1$; $(1, 1)$

7) $x^2 - y^2 = 1$; $(1, 0)$

8) $y = \cosh x$; $x = 1$

9) $x = t^2$, $y = 2t$; $t = 1$

10) $x = a\cos\theta$, $y = b\sin\theta$; $\theta = \dfrac{\pi}{6}$

11) $x = ct$, $y = \dfrac{c}{t}$; $t = 2$

12) $x = a\cos^3\theta$, $y = a\sin^3\theta$; $\theta = \dfrac{\pi}{4}$

13) $x = 2\cosh u$, $y = \sinh u$, $u = 0$

14) $x = a\cos\theta$, $y = a\sin\theta$; $\theta = \dfrac{\pi}{2}$

Find the curvature of each of the following curves at the specified point.

15) $y^2 = x^3$; $(1, 1)$

16) $x = 2t^2$, $y = 4t$; $(8, 8)$

17) $r = a\cos\theta$; $\theta = \dfrac{\pi}{4}$

18) $r = a\sin 3\theta$; $\theta = \dfrac{\pi}{12}$

19) $r = a(1 + \cos\theta)$; $\theta = 0$

20) $r^2 = a^2\cos 2\theta$; (r, θ)

Find the angle ϕ between the radius vector OP and the tangent at $P(r, \theta)$ for each of the following curves.

21) $r = a\cos 2\theta$

22) $r = a(1 - \sin\theta)$

23) $r = ae^{\theta}$

24) $r = a\sin 3\theta$

25) Prove that the tangents to the cardioid $r = a(1 + \cos\theta)$ at the points $\theta = \alpha$, $\theta = \alpha + \frac{2}{3}\pi$ and $\theta = \alpha + \frac{4}{3}\pi$, are all parallel.

MISCELLANEOUS EXERCISE 10

1) Given that $I_n = \displaystyle\int_0^1 x^p(1 - x)^n \, dx$, where p and n are positive, show that

$$(n + p + 1)I_n = nI_{n-1}$$

Evaluate I_2 for $p = \frac{1}{2}$. (JMB)

2) If $I_n = \int_0^{\frac{\pi}{2}} e^{2x} \sin^n x \, dx, \quad n > 1,$ show that

$$(n^2+4)I_n = n(n-1)I_{n-2} + 2e^{\pi}$$

Hence, or otherwise, find

$$\int_0^{\frac{\pi}{2}} e^{2x} \sin^3 x \, dx.$$

(AEB '71)

3) Given that

$$I_{p,n} = \int_0^1 (1-x)^p x^n \, dx, \quad (p \geqslant 0, \quad n \geqslant 0)$$

prove that, for $p \geqslant 1$,

$$(n+1)I_{p,n} = pI_{p-1,n+1}$$

and also that

$$(p+n+1)I_{p,n} = pI_{p-1,n}$$

Hence prove that, if p and n are positive integers,

$$I_{p,n} = \frac{p!n!}{(p+n+1)!}.$$

(JMB)

4) Express $\dfrac{1}{1-x^2}$ in partial fractions and show that

$$\int_0^{\frac{1}{2}} \frac{dx}{1-x^2} = \frac{1}{2}\ln 3$$

Given that

$$I_n = \int_0^{\frac{1}{2}} \frac{dx}{(1-x^2)^n}$$

show that

$$I_{n-1} = I_n - \int_0^{\frac{1}{2}} \frac{x^2}{(1-x^2)^n} \, dx$$

Hence, or otherwise, prove that, for $n > 1$

$$2(n-1)I_n = (2n-3)I_{n-1} + \tfrac{1}{2}(\tfrac{4}{3})^{n-1}$$

Evaluate I_2.

(JMB)

5) Derive a reduction formula for I_m in terms of I_{m-1} when

$$I_m = \int x^3 (\log_e x)^m \, dx$$

Hence find $\int x^3 (\log_e x)^3 \, dx$.

(AEB'75)p

6) If $I_n = \int \tan^n x \, dx$, obtain a reduction formula for I_n.

Hence, or otherwise, show that $\int_0^{\frac{\pi}{4}} \tan^4 x \, dx = \dfrac{3\pi - 8}{12}$. (AEB'76)p

7) Given that $I_n = \int_0^1 x^n \cosh x \, dx$ prove that, for $n \geqslant 2$,

$$I_n = \sinh 1 - n \cosh 1 + n(n-1)I_{n-2}$$

Evaluate: (a) $\int_0^1 x^4 \cosh x \, dx$ (b) $\int_0^1 x^3 \cosh x \, dx$

expressing each answer in terms of e.

8) Find a reduction formula for

$$I_n = \int_0^1 (1+x^2)^n \, dx.$$

9) If $I_n = \int \dfrac{\sin 2n\theta}{\sin \theta} \, d\theta$, prove that

$$I_n = I_{n-1} + \frac{2}{2n-1} \sin (2n-1)\theta$$

Hence, or otherwise, evaluate

$$\int_0^{\frac{\pi}{2}} \frac{\sin 5\theta}{\sin \theta} \, d\theta.$$ (AEB '76)

10) Given that

$$I_n = \int \frac{\sin nx}{\sin x} \, dx$$

prove that, for $n > 2$

$$I_n - I_{n-2} = \frac{2 \sin (n-1)x}{n-1} + \text{a constant}.$$

Hence find the general solution of the differential equation

$$\frac{dy}{dx} - y \cot x = \sin 5x.$$

11) Evaluate the integrals

(i) $\int_0^1 \dfrac{dx}{\sqrt{[x(1-x)]}}$ (ii) $\int_1^\infty \dfrac{dx}{x\sqrt{(1+x^2)}}$ (iii) $\int_1^\infty \dfrac{1}{x^2} \ln x \, dx$ (O)

12) Evaluate the integrals

(a) $\displaystyle\int_0^{\frac{1}{2}} \frac{dx}{\sqrt{(2x-x^2)}}$ (b) $\displaystyle\int_1^2 \frac{dx}{3x^2-6x+4}$ (c) $\displaystyle\int_3^4 \frac{dx}{\sqrt{(x^2-4x+3)}}$

13) Find the value of a for which the integral

$$\int_1^{\infty} \frac{a}{x(2x+a)}\,dx$$

converges to the value 1. (U of L)p

14) The region bounded by the x axis, the ordinates $x=2$ and $x=6$ and the arc of the parabola $y=\sqrt{x}$ between these ordinates, is rotated through 360° about the x axis. Find the area of the curved surface of this solid. (JMB)p

15) The tangent at a point P on the curve whose parametric equations are

$x=a\left(t-\dfrac{t^3}{3}\right)$, $y=at^2$ cuts the x axis at T. Prove that the distance of the point T from the origin O is one half of the length of the arc OP.

(AEB '67)

16) Find the area enclosed by the curve $y=\cosh x$, the ordinate at $x=\ln 2$, the x axis and the y axis. Find also
(a) the length of the perimeter enclosing this area,
(b) the volume obtained when this area is rotated completely about the x-axis.

(AEB '74)

17) Show that the length of the arc of the curve

$$y = x^2$$

between the origin and the point $(1,1)$ is equal to $\frac{1}{4}(2\sqrt{5}+\sinh^{-1}2)$. (JMB)

18) The curve $y=\frac{1}{4}(e^{2x}+e^{-2x})$ meets the y axis at A, and the tangent at a variable point P on the curve is inclined to the x axis at an angle ψ. Show that the length of the curve between the points A and P is $\frac{1}{2}\tan\psi$. (AEB '67)

19) The parametric equations of a curve are

$$x=a(2\cos t-\cos 2t), \qquad y=a(2\sin t-\sin 2t)$$

where a is a constant.
Show that the angle between the tangent at the point whose parameter is t and the positive direction of the x axis is $\dfrac{3t}{2}$. Find the length of the arc of the curve for which $0\leqslant t\leqslant\pi$. Find also the area of the surface formed when this arc is revolved once about the x axis. (JMB)

20) (a) The curve given parametrically by the equations $x = \sin^3 t$, $y = \cos^3 t$ is rotated through $180°$ about the x axis. Calculate the area of the surface generated.
(b) Find the radius of curvature of the curve $y = 2\cosh\frac{1}{2}x$ at the point where $x = 2$. Find also the length of the curve from the point where $x = -2$ to the point where $x = 2$. (AEB '78)

21) A curve is given by the parametric equations

$$x = a\cos^3 t, \quad y = a\sin^3 t \quad (-\pi < t \leqslant \pi)$$

Show that near $t = 0$ the equations reduce approximately to

$$a - x \simeq 3at^2/2, \quad y \simeq at^3$$

Sketch the curve.
Show that the total length of the curve is $6a$.
Find the area of the surface generated when the part of the curve which is in the first quadrant is rotated once about the x axis. (JMB)

22) A surface of revolution is formed by rotating completely about the x axis the arc of the parabola $x = at^2$, $y = 2at$ from $t = 0$ to $t = \sqrt{3}$.
Show that its surface area, S, is $56\pi a^2/3$.
Show also that the x coordinate, $\bar{x}$, of the centroid of this surface is given by

$$S\bar{x} = 8\pi a^3 \int_0^3 t^3\sqrt{(1+t^2)}\,dt$$

Hence find $\bar{x}$, using the substitution $u^2 = 1 + t^2$ or otherwise. (U of L)

23) Find the length of the arc of the curve $ay^2 = x^3$ between the origin and the point on the curve where $x = 5a$.
Show that, when this arc is rotated about the axis of x, the volume enclosed by the surface so formed and the plane $x = 5a$ is $625\pi a^3/4$. (JMB)p

24) A curve is given by the parametric equations

$$x = 4\cos t + \cos 2t,$$
$$y = \sin 2t + 4\sin t + 2t$$

Find the length of the curve between the points $t = 0$ and $t = \pi/4$. Find the smallest positive value of t for which the radius of curvature of the curve has the absolute value 6. (AEB '72)

25) Find the root mean square value of $\sin x$ with respect to x for $0 \leqslant x \leqslant \pi$. (U of L)

26) Use Pappus' theorem for areas to find, for a semi-circular arc of radius r, the distance of the centroid from the bounding diameter. (U of L)

27) Given that the area of the ellipse $\frac{x^2}{a^2}+\frac{y^2}{b^2}=1$ is πab and that the volume generated when this area rotates through an angle π about the x axis is $\frac{4}{3}\pi ab^2$, use Pappus' theorem to find
(a) the centroid,
(b) the first moment about Ox of that part of the area for which $y \geqslant 0$.

28) A metal ring is cast in the shape of the solid obtained when a square ABCD of side 1 cm rotates completely about a line parallel to AB and distant 10 cm from AB. Find
(a) the volume,
(b) the total surface area of the ring.

29) Find the coordinates of all the points of inflexion of the graph of the function

$$y = x - \sin x$$

in the range $0 \leqslant x \leqslant 4\pi$. Sketch the graph in this range.
The region bounded by the x axis, the line $x = 4\pi$ and the curve whose equation is

$$y = x - \sin x$$

is rotated once about the x axis. Prove that the volume swept out is $\frac{2}{3}\pi^2(32\pi^2 + 15)$.
Find the mean value of y^2 with respect to x over the range $0 \leqslant x \leqslant 4\pi$.
(JMB)

CHAPTER 11

NUMERICAL METHODS FOR THE SOLUTION OF DIFFERENTIAL EQUATIONS

Differential equations were introduced in *The Core Course* and in Chapter 7 of this book, where we investigated analytical methods for solving a few specific types of differential equation. In practical situations, however, where problems involving rates of change give rise to differential equations, these are rarely of a type that can be solved analytically. So the need arises for numerical methods which provide approximate solutions.

The use of numerical methods for the solution of differential equations is a vast topic and in this chapter only a brief introduction to the subject is given.
Now a solution of a differential equation gives a direct relationship between the dependent variable (y) and the independent variable (x). In a practical context the required results are often just some modestly accurate values of y over a limited range of values of x. In this chapter we examine a few methods by which this can be achieved.

It is important to remember that to get a unique solution (as opposed to a general solution involving arbitrary constants) we need more information than the differential equation alone. For a first order differential equation we need a value of y at a given value of x; for a second order differential equation we need two pairs of corresponding values (e.g. values for both y and $\dfrac{dy}{dx}$ at a specified value of x). Such information is often referred to as initial conditions or initial values.

There are basically two approaches to the problem of finding an approximate solution to a differential equation. The first is to find a function that approximates to the actual solution. The second is to find approximate numerical values for y at regularly spaced values of x using what is called a step-by-step method.
We begin with a method for finding a function that approximates to the actual solution.

Method 1. Polynomial Approximations using Taylor Series

In Chapter 5 we saw that most reasonable functions of x can be expanded as a series of ascending powers of $(x-x_0)$, where x_0 is a constant,

i.e. $$f(x) = f(x_0)+f'(x_0)(x-x_0)+\frac{f''(x_0)}{2!}(x-x_0)^2+\dots$$

This is called the Taylor expansion of $f(x)$ about $x=x_0$. By truncating the series after a certain number of terms we get a polynomial function that approximates to $f(x)$. How good this approximation is depends on a number of factors such as, the number of terms included, how close x is to x_0, and so on. It is not necessary to know $f(x)$ in order to derive the series. Provided that initial conditions are given, the series can be derived from a differential equation.

Differential equations are usually formulated in terms of $y, x, \frac{dy}{dx}, \frac{d^2y}{dx^2}, \dots,$ where $y=f(x)$, so it is convenient to write Taylor's expansion of $f(x)$ in the form

$$y = y_0+\left(\frac{dy}{dx}\right)_0(x-x_0)+\left(\frac{d^2y}{dx^2}\right)_0\frac{(x-x_0)^2}{2!}+\dots$$

where $y_0, \left(\frac{dy}{dx}\right)_0, \left(\frac{d^2y}{dx^2}\right)_0, \dots$ are the values of $y, \frac{dy}{dx}, \frac{d^2y}{dx^2}, \dots$ when $x=x_0$.

To illustrate the use of this expansion in order to find an approximate solution to a differential equation, consider the first order equation

$$\frac{dy}{dx} = xy \qquad [1]$$

given that $y=1$ when $x=0$.

Now $y=1$ when $x=0$ so $y_0=1$ and $x_0=0$.

From [1] $\left(\frac{dy}{dx}\right)_0 = (0)(1) = 0$

Differentiating [1] gives $$\frac{d^2y}{dx^2} = y+x\frac{dy}{dx} \qquad [2]$$

From [2] $\left(\frac{d^2y}{dx^2}\right)_0 = 1+(0)(0) = 1$

Differentiating [2] gives $$\frac{d^3y}{dx^3} = 2\frac{dy}{dx}+x\frac{d^2y}{dx^2} \qquad [3]$$

From [3] $\left(\dfrac{d^3y}{dx^3}\right)_0 = 2(0) + (0)(1) = 0$

Differentiating [3] gives $$\frac{d^4y}{dx^4} = 3\frac{d^2y}{dx^2} + \frac{d^3y}{dx^3} \qquad [4]$$

From [4] $\left(\dfrac{d^4y}{dx^4}\right)_0 = 3(1) + (0) = 3$

Theoretically this process can be continued indefinitely but, in practice, only a few terms of the expansion are used to give an approximation for y in terms of x.

Choosing to stop, as we did, at $\dfrac{d^4y}{dx^4}$, gives the terms of the series as far as the term in x^4 and hence y as a quartic function of x,

i.e. $$y \simeq 1 + \tfrac{1}{2}x^2 + \tfrac{1}{8}x^4 \qquad [5]$$

Assuming that this series converges, the error between this approximation and the true solution is the sum of the remaining terms of the series. Obviously a better approximation can be found by calculating further terms.

As we shall see later, [5] gives good accuracy for y for values of x close to zero and modest accuracy for y (to two significant figures) for values of x in the interval $[-1, 1]$.

Adding on the next two terms of the series gives

$$y \simeq 1 + \tfrac{1}{2}x^2 + \tfrac{1}{8}x^4 + \tfrac{1}{48}x^6 + \tfrac{1}{384}x^8$$

This is a better approximation which, if required, can be used either to increase the accuracy of y for values of x close to zero, or to extend the interval over which modest accuracy is obtained for y.
The example chosen to illustrate this method is fairly simple and happens to have an exact solution, namely

$$y = \exp \tfrac{1}{2}x^2$$

Tabulated opposite are values of y obtained from the two approximations and from the true solution, to illustrate this discussion on accuracy.

This table shows that accuracy deteriorates rapidly as x moves further away from its initial value and, in order to get even very modest accuracy for y over the interval $[0, 2]$, yet more terms of the series are needed. This illustrates one of the disadvantages of this method: to obtain results that are accurate enough, even over a small interval, a polynomial of high degree may be required and this involves many differentiations which is a tedious and sometimes complicated process. Bearing in mind that numerical solutions will generally be used only

when an exact solution cannot be found, it is clear that, for a method to be of practical use, a way of estimating the accuracy of results is required.
A general discussion of error estimation is beyond the scope of this book, but another disadvantage of the series solution is that it is often difficult to estimate the accuracy of results.

x	Approx. y $(1+\frac{1}{2}x^2+\frac{1}{8}x^4)$	Approx. y $(1+\frac{1}{2}x^2+\frac{1}{8}x^4+\frac{1}{48}x^6+\frac{1}{384}x^8)$	True y $(\exp\frac{1}{2}x^2)$
0	1	1	1
0.1	1.005 01	1.005 01	1.005 01
0.2	1.020 20	1.020 20	1.020 20
0.5	1.132 81	1.133 15	1.133 15
1.0	1.625 00	1.648 43	1.648 72
1.5	2.757 81	3.061 86	3.080 22
2.0	5.000 00	7.000 00	7.389 06

EXAMPLE 11a

Find, as a series of ascending powers of x up to and including the term in x^4, an approximate solution to the differential equation

$$\frac{d^2y}{dx^2}+2x\frac{dy}{dx}+y^2 = 0$$

where $y=1$ and $\frac{dy}{dx}=2$ when $x=0$.

In this problem $x_0=0$, $y_0=1$, $\left(\frac{dy}{dx}\right)_0=2$

With $x_0=0$, the Taylor Series up to the term in x^4 becomes

$$y \simeq y_0+\left(\frac{dy}{dx}\right)_0 x+\left(\frac{d^2y}{dx^2}\right)_0\frac{x^2}{2!}+\left(\frac{d^3y}{dx^3}\right)_0\frac{x^3}{3!}+\left(\frac{d^4y}{dx^4}\right)_0\frac{x^4}{4!}$$

From the given differential equation

$$\left(\frac{d^2y}{dx^2}\right)_0+(0)(2)+(1)^2 = 0 \quad\Rightarrow\quad \left(\frac{d^2y}{dx^2}\right)_0 = -1$$

Differentiating the given equation twice gives

$$\frac{d^3y}{dx^3}+2\frac{dy}{dx}+2x\frac{d^2y}{dx^2}+2y\frac{dy}{dx} = 0 \qquad [1]$$

and

$$\frac{d^4y}{dx^4}+4\frac{d^2y}{dx^2}+2x\frac{d^3y}{dx^3}+2y\frac{d^2y}{dx^2}+2\left(\frac{dy}{dx}\right)^2 = 0 \qquad [2]$$

Then, by substitution into [1] and then into [2], we have

$$\left(\frac{d^3y}{dx^3}\right)_0 = -8 \quad \text{and} \quad \left(\frac{d^4y}{dx^4}\right)_0 = -4$$

Hence $y \simeq 1 + 2x - \frac{1}{2}x^2 - \frac{4}{3}x^3 - \frac{1}{6}x^4$ when x is reasonably small.

EXERCISE 11a

1) Find a polynomial function of x, of degree 3, which approximates to y when x is close to zero given that

(a) $\dfrac{dy}{dx} + 2xy^2 = 0$ and $y = 1$ when $x = 0$,

(b) $\dfrac{dy}{dx} = \dfrac{x}{y}$ and $y = 2$ when $x = 0$,

(c) $(x+1)\dfrac{dy}{dx} = y^2 + 1$ and $y = -1$ when $x = 0$.

In each case find an approximate value for y when $x = 0.5$.

2) Find the exact solutions to the differential equations given in Question 1 and hence state the accuracy of the results obtained for y when $x = 0.5$.

3) Find, as a series of ascending powers of x up to and including the term in x^3, the solution of the differential equation

$$\frac{d^2y}{dx^2} + 4\frac{dy}{dx} - 3 = 0$$

given that $y = 3$ and $\dfrac{dy}{dx} = 1$ when $x = 0$.

Use your solution to find approximate values for y when $x = 0.5$ and when $x = 1$.

4) Find, as a series of ascending powers of x up to and including the term in x^4, the solution of the differential equation

$$\frac{d^2y}{dx^2} + y\frac{dy}{dx} + x = 0$$

given that $y = 1$ and $\dfrac{dy}{dx} = 1$ when $x = 0$

Use your solution to find values of y when $x = 0.1$ and when $x = 0.5$.

5) Given that $x\dfrac{dy}{dx} + y = 0$ and $y = 1$ when $x = 1$, find y as a series of ascending powers of $(x - 1)$ up to and including the term in $(x - 1)^3$. Hence find a cubic function of x that approximates to y for values of x close to 1 and use it to find approximate values for y when $x = 1.1, 1.5$ and 1.9.

6) Find the exact solution to the differential equation in Question 5 and use it to state the accuracy of the values obtained for y.

7) By finding y as a series of ascending powers of $(x + 1)$ up to and including the term in $(x + 1)^2$, obtain a quadratic function of x that approximates to y given that

$$\frac{dy}{dx} = x^2 + y^2$$

and $y = 0$ when $x = -1$.

In Questions 8–13 find an approximate value for y at the given value of x by obtaining a quadratic function of x that approximates to y.

8) Find y when $x = 1.1$ given that $\dfrac{dy}{dx} = x + y$ and $y = 2$ when $x = 1$.

9) Find y when $x = 0.9$ given that $y\dfrac{dy}{dx} = x + 1$ and $y = 2$ when $x = 1$.

10) Find y when $x = 1.5$ given that $x\dfrac{dy}{dx} = x^2 - y^2$ and $y = 0$ when $x = 2$.

11) Find y when $x = \pi/3$ given that $\dfrac{dy}{dx} = \cos x + \cos y$ and $y = 2$ when $x = \pi/2$.

12) Find y when $x = -0.1$ given that $\dfrac{d^2y}{dx^2} = \left(\dfrac{dy}{dx}\right)^2$ and $y = 1$, $\dfrac{dy}{dx} = 4$ when $x = 0$.

13) Find y when $x = 1.1$ given that $\frac{d^2y}{dx^2} + \frac{dy}{dx} + y = \cos x$ and $y = 1$, $\frac{dy}{dx} = 2$ when $x = 1$.

14) Find, as a series of ascending powers of x up to and including the term in x^7, the solution of the differential equation $\frac{d^2y}{dx^2} + y = x$ given that $y = 0$ and $\frac{dy}{dx} = 1 - \frac{\pi}{2}$ when $x = 0$.
Use your series to find an approximate value for y when $x = \pi/4$. By finding the exact solution of the differential equation comment on the accuracy of your approximate value for y.

STEP-BY-STEP METHODS

Step-by-step methods for solving differential equations do not give a function that approximates to the correct solution; instead they provide values of y at regular steps along the x axis.

Thus if a given differential equation has the initial values $y = y_0$ when $x = x_0$, then a step-by-step solution will give a sequence of values for y,

$$y_1,\ y_2,\ y_3,\ y_4,\ \dots$$

corresponding to the values of x,

$$x_0 + h,\ x_0 + 2h,\ x_0 + 3h,\ x_0 + 4h,\ \dots$$

so giving a set of points $(x_0, y_0), (x_1, y_1), (x_2, y_2), \dots$ which can, if required, be joined up to get an approximation to the solution curve $y = f(x)$.

i.e.

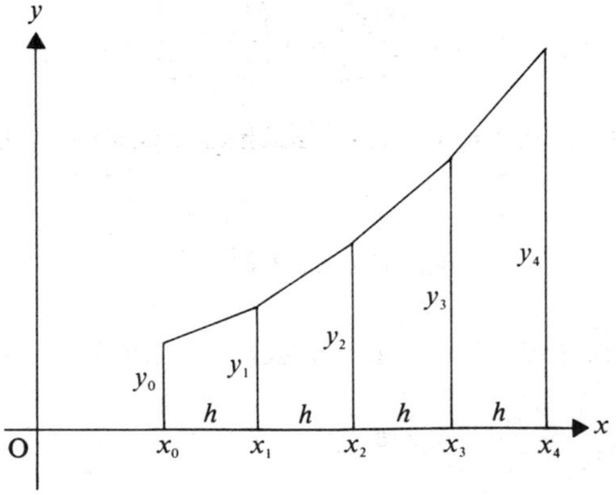

Method 2. Linear Approximation (for First Order Differential Equations)

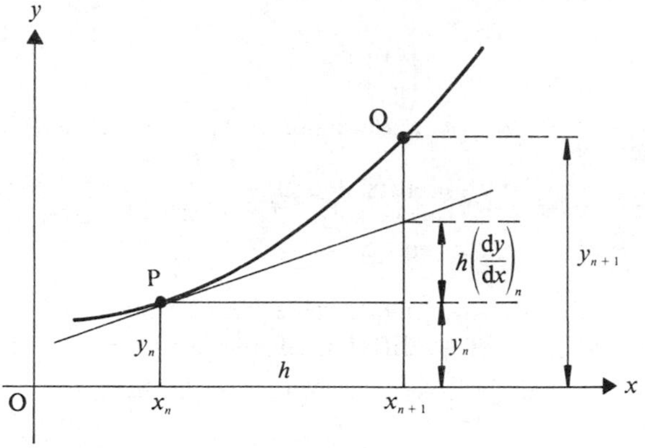

If $P(x_n, y_n)$ is a point on the curve $y = f(x)$ and $Q(x_{n+1}, y_{n+1})$ is another point on the curve where $x_{n+1} - x_n = h$, then from the diagram we see that, if h is fairly small,

$$y_{n+1} \simeq y_n + h\left(\frac{dy}{dx}\right)_n$$

This formula can also be obtained from the Taylor Series expansion of y about $x = x_n$,

i.e. $$y = y_n + \left(\frac{dy}{dx}\right)_n (x - x_n) + \left(\frac{d^2y}{dx^2}\right)_n \frac{(x - x_n)^2}{2!} + \ldots$$

If $y = y_{n+1}$ when $x = x_n + h$, this becomes

$$y_{n+1} = y_n + \left(\frac{dy}{dx}\right)_n h + \left(\frac{d^2y}{dx^2}\right)_n \frac{h^2}{2!} + \ldots$$

If h is fairly small, so that terms in h^2 and higher powers of h can be neglected, we have

$$y_{n+1} \simeq y_n + h\left(\frac{dy}{dx}\right)_n \qquad [1]$$

To illustrate the use of this formula, consider again the differential equation

$$\frac{dy}{dx} = xy \quad \text{where} \quad y = 1 \quad \text{when} \quad x = 0$$

We will calculate approximate values of y corresponding to values of x from 0 to 1 in steps of 0.1.
From the given differential equation,

$$\left(\frac{dy}{dx}\right)_n = x_n y_n$$

So [1] becomes

$$y_{n+1} \simeq y_n + hx_n y_n \qquad [2]$$

First step ($n = 0$)

Using $x_0 = 0$, $y_0 = 1$ and $h = 0.1$ in [2] gives

$$y_1 \simeq 1 + (0.1)(0)(1) \quad \Rightarrow \quad y_1 \simeq 1$$

Second step ($n = 1$)

Using $x_1 = 0.1$, $y_1 = 1$ and $h = 0.1$ in [2] gives

$$y_2 \simeq 1 + (0.1)(0.1)(1) \quad \Rightarrow \quad y_2 \simeq 1.01$$

Third step ($n = 2$)

Using $x_2 = 0.2$, $y_2 = 1.01$ and $h = 0.1$ in [2] gives

$$y_3 \simeq 1.01 + (0.1)(0.2)(1.01) \quad \Rightarrow \quad y_3 \simeq 1.0302$$

This procedure is then repeated for the required number of steps. The calculation can be arranged conveniently in tabular form as shown opposite, where the arrows indicate the order of working. We also include, for comparison, the true values of y obtained from the exact solution $y = \exp \frac{1}{2}x^2$.
(Calculations are corrected to 4 decimal places.)

Comparing the approximate values of y with the true values shows that Method 2 is not very accurate. Accuracy can be improved slightly by halving the step length, but this doubles the number of calculations.

This method is included because it is simple and because it illustrates the basic principle underlying all step-by-step methods: each new value of y is found from an approximation based on the previous value obtained for y.
In Method 1, the approximation used to find all values of y is based only on the initial values x_0 and y_0. So, in theory, the step-by-step method should give reasonable results over a larger interval. The reason that this is not so in the example above is that a comparatively poor approximation (a linear one) was used as opposed to the much better quartic approximation used in Method 1. Even so, if Method 2 was used for further steps, it would become clear that the error involved increases at a slower rate than does the error in Method 1.

n	x_n	y_n		$y_n + hx_ny_n$	$y_n = \exp \frac{1}{2}x_n^2$
0	0	1	→	1	1
1	0.1	1	→	1.01	1.0050
2	0.2	1.01	→	1.0302	1.0202
3	0.3	1.0302	→	1.0611	1.0460
4	0.4	1.0611	→	1.1036	1.0833
5	0.5	1.1036	→	1.1587	1.1331
6	0.6	1.1587	→	1.2283	1.1972
7	0.7	1.2283	→	1.3143	1.2776
8	0.8	1.3143	→	1.4194	1.3771
9	0.9	1.4194	→	1.5471	1.4973
10	1.0	1.5471			1.6487

The next method uses an improved linear approximation.

Method 3. A Better Linear Approximation

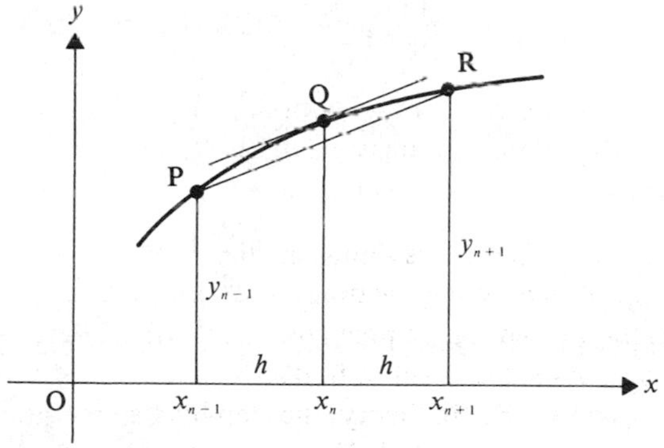

If $P(x_{n-1}, y_{n-1})$, $Q(x_n, y_n)$ and $R(x_{n+1}, y_{n+1})$ are three points on the curve $y = f(x)$, where $x_{n+1} - x_n = x_n - x_{n-1} = h$ then, from the diagram we see that if h is fairly small

the gradient at Q $\simeq$ the gradient of PR

i.e.
$$\left(\frac{dy}{dx}\right)_n \simeq \frac{y_{n+1}-y_{n-1}}{2h}$$

or
$$y_{n+1} \simeq y_{n-1} + 2h\left(\frac{dy}{dx}\right)_n$$

This approximation (although fairly obvious from the diagram) is based on the Mean-value Theorem, which states that if two points P and R are on a curve $y = f(x)$, which is continuous between P and R, then at some point between P and R, the gradient of the curve is equal to the gradient of the line PR.

The approximation can also be obtained from the Taylor expansion of y about x_n

i.e.
$$y = y_n + \left(\frac{dy}{dx}\right)_n (x - x_n) + \ldots \qquad [1]$$

When $x = x_n + h$, [1] becomes
$$y_{n+1} = y_n + \left(\frac{dy}{dx}\right)_n h + \ldots \qquad [2]$$

When $x = x_n - h$, [1] becomes
$$y_{n-1} = y_n - \left(\frac{dy}{dx}\right)_n h + \ldots \qquad [3]$$

[2] − [3] $\Rightarrow$
$$y_{n+1} - y_{n-1} = 2h\left(\frac{dy}{dx}\right)_n + \ldots$$

If h is small enough to neglect terms in h^2 and higher powers of h we have
$$y_{n+1} - y_{n-1} \simeq 2h\left(\frac{dy}{dx}\right)_n$$

$\Rightarrow$
$$y_{n+1} \simeq y_{n-1} + 2h\left(\frac{dy}{dx}\right)_n \qquad [4]$$

This approximation will now be used to solve, again, the differential equation $\frac{dy}{dx} = xy$ where $y = 1$ when $x = 0$.

In this problem, $\left(\frac{dy}{dx}\right)_n = x_n y_n$ so [4] becomes
$$y_{n+1} \simeq y_{n-1} + 2hx_n y_n \qquad [5]$$

First step ($n = 0$)

Using $x_0 = 0$, $y_0 = 1$ and $h = 0.1$ in [5] gives

$$y_1 \simeq y_{-1} + 2(0.1)(0)(1)$$

But the value of y_{-1} is not given so, in order to get started, y_{-1} has to be calculated by a different method.
An approximate value can be obtained using the Taylor quartic approximation for y, found in Method 1, for values of x close to zero, i.e. $y \simeq 1 + \frac{1}{2}x^2 + \frac{1}{8}x^4$.
Hence, when $x = -0.1$, $y_{-1} \simeq 1.005\,05$.
Now, returning to the first step approximation we have

$$y_1 \simeq (1.005\,05) + 2(0.1)(0)(1) \quad \Rightarrow \quad y_1 \simeq 1.005\,05$$

Second step ($n = 1$)

Using $x_1 = 0.1$, $y_1 = 1.005\,05$, $y_0 = 1$ and $h = 0.1$ in [5] gives

$$y_2 \simeq (1) + 2(0.1)(0.1)(1.005\,05) \quad \Rightarrow \quad y_2 \simeq 1.020\,101$$

Third step ($n = 2$)

Using $x_2 = 0.1$, $y_2 = 1.020\,101$, $y_1 = 1.005\,05$ and $h = 0.1$ in [5] gives

$$y_3 \simeq 1.005\,05 + 2(0.1)(0.2)(1.020\,101) \quad \Rightarrow \quad y_3 \simeq 1.045\,854$$

Repeating this procedure up to $x_{10} = 1.0$ and arranging the calculations in a table, we have

$$y_{-1} = 1.005\,05$$

n	x_n	y_n	$y_{n-1} + 2hx_ny_n$
0	0	1	1.0051
1	0.1	1.0051	1.0201
2	0.2	1.0201	1.0459
3	0.3	1.0459	1.0829
4	0.4	1.0829	1.1325
5	0.5	1.1325	1.1961
6	0.6	1.1961	1.2760
7	0.7	1.2760	1.3747
8	0.8	1.3747	1.4960
9	0.9	1.4960	1.6440
10	1.0	1.6440	

Comparing the approximate values obtained for y with the true values (page 439) we see that this method is a considerable improvement on Method 2.

EXAMPLE 11b

Use the approximation

$$y_{n+1} \simeq y_{n-1} + 2h\left(\frac{dy}{dx}\right)_n$$

with $h = 0.2$ to find an approximation to the value of y when $x = 1.6$ given that $(x+1)\frac{dy}{dx} = 2x - y$ and $y = 0.5$ when $x = 1$.

From the given differential equation

$$\left(\frac{dy}{dx}\right)_n = \frac{2x_n - y_n}{x_n + 1} \qquad [1]$$

so the given approximation becomes

$$y_{n+1} \simeq y_{n-1} + (0.4)\left(\frac{2x_n - y_n}{x_n + 1}\right)$$

We are also given $x_0 = 1$ and $y_0 = 0.5$, but to get started we need y_{-1}.

So we use the first two terms of the Taylor expansion of y about x_0,

$$y \simeq y_0 + h\left(\frac{dy}{dx}\right)_0 \qquad [2]$$

From [1], $\left(\frac{dy}{dx}\right)_0 = 0.75$

From [2], when $h = -0.2$, we get $y_{-1} \simeq 0.35$.

n	x_n	y_n	$y_{n-1} + (0.4)\left(\frac{2x_n - y_n}{x_n + 1}\right)$
0	1	0.5	0.65
1	1.2	0.65	0.8182
2	1.4	0.8182	0.9803
3	1.6	0.9803	

Therefore $y \simeq 0.9803$ when $x = 1.6$.

EXERCISE 11b

In Questions 1–4 use the approximation $y_{n+1} \simeq y_n + h\left(\frac{dy}{dx}\right)_n$.

1) Find y when $x = 0.5$, using a step length of 0.1, given that $\frac{dy}{dx} = 2xy^2$ and $y = -1$ when $x = 0$.

2) Using a step length of 0.2, find y when $x = 2$, given that $\frac{dy}{dx} = x + y$ and $y = 0$ when $x = 1$.

3) Using a step length of 0.05, find y when $x = 0.2$ given that $\frac{dy}{dx} = x - y$ and $y = 1$ when $x = 0$.

4) Find y when $x = 1$, using a step length of 0.2 given that $\frac{dy}{dx} = x - xy$ and $y = 0$ when $x = 0$.

In Questions 5–8 use the approximation $y_{n+1} \simeq y_{n-1} + 2h\left(\frac{dy}{dx}\right)_n$.

5) Using a step length of 0.1, find y when $x = 0.5$ given that $\frac{dy}{dx} = 2xy^2$ and $y = -1$ when $x = 0$.

6) Using a step length of 0.5 find y when $x = 3$ given that $\frac{dy}{dx} = xy + 2$ and $y = 0$ when $x = 1$.

7) With a step length of 0.1, find y when $x = 1$ given that $x\frac{dy}{dx} = x + y$ and $y = 1$ when $x = 0.5$.

8) With a step length of 0.1, find y when $x = 0.2$ and when $x = 0.4$ given that $\frac{dy}{dx} = x^2 + y^2$ and $y = 1$ when $x = 0$.

9) Find the exact solution to $\frac{dy}{dx} = 2xy^2$ and $y = -1$ when $x = 0$.
Use your result to compare with the approximate values found in Questions 1 and 5.

10) Given that $\dfrac{dy}{dx} = x + y$ where $y = 1$ when $x = 0$, use a step length of 0.1 in the approximation

$$y_{n+1} \simeq y_n + hf\left[x_n + \tfrac{1}{2}h,\ y_n + \tfrac{1}{2}h\left(\frac{dy}{dx}\right)_n\right] \quad \text{where} \quad \frac{dy}{dx} = f(x,y),$$

to find the approximate values of y when $x = 0.1$ and $x = 0.2$.

11) Use the approximation given in Question 10 to find y when $x = 0.5$ and 1.0 given that $\dfrac{dy}{dx} = x + xy$ and $y = 0$ when $x = 0$.

This chapter ends with one more step-by-step method, using a quadratic approximation, which is useful for second order linear differential equations.

Method 4. For Second Order Equations of the Form $\dfrac{d^2y}{dx^2} = f(x, y)$

For equations of this type an approximation involving $\dfrac{d^2y}{dx^2}$, but not $\dfrac{dy}{dx}$, is useful.

The Taylor expansion of $y = f(x)$ about $x = x_n$ is

$$y = y_n + \left(\frac{dy}{dx}\right)_n (x - x_n) + \left(\frac{d^2y}{dx^2}\right)_n \frac{(x - x_n)^2}{2!} + \dots \qquad [1]$$

when $x = x_n + h$, [1] becomes

$$y_{n+1} = y_n + h\left(\frac{dy}{dx}\right)_n + \frac{h^2}{2}\left(\frac{d^2y}{dx^2}\right)_n + \dots \qquad [2]$$

when $x = x_n - h$, [1] becomes

$$y_{n-1} = y_n - h\left(\frac{dy}{dx}\right)_n + \frac{h^2}{2}\left(\frac{d^2y}{dx^2}\right)_n + \dots \qquad [3]$$

Adding [2] and [3] gives

$$y_{n+1} + y_{n-1} = 2y_n + h^2\left(\frac{d^2y}{dx^2}\right)_n + \dots$$

If h is small, terms in h^4 and higher powers can be neglected giving

$$y_{n+1} \simeq 2y_n - y_{n-1} + h^2\left(\frac{d^2y}{dx^2}\right)_n \qquad [4]$$

We will illustrate how this approximation can be used to find values of y from

$$\frac{d^2y}{dx^2} = x^2 + y^2$$

where $y = 1$ and $\frac{dy}{dx} = 2$ when $x = 0$, using steps of 0.2.

In this problem, [4] becomes

$$y_{n+1} \simeq 2y_n - y_{n-1} + h^2(x_n^2 + y_n^2) \qquad [5]$$

First step

Using $x_0 = 0$, $y_0 = 1$, $h = 0.2$ in [5] gives

$$y_1 \simeq 2 - y_{-1} + (0.04)(0 + 1)$$

So, to get started, we need a value for y_{-1}.

Because $\left(\frac{dy}{dx}\right)_0 = 2$ is given, we can find y_{-1} from the Taylor approximation

$$y \simeq y_0 + \left(\frac{dy}{dx}\right)_0 (x - x_0)$$

which, when $x = -0.2$, gives

$$y_{-1} \simeq 1 + (2)(-0.2) = 0.6$$

Returning to the approximation for y_1 we have

$$y_1 \simeq 2 - 0.6 + (0.04)(1) = 1.44$$

Subsequent steps contain no complications and further calculations are tabulated below.

$$\frac{d^2y}{dx^2} = x^2 + y^2: \qquad y_{-1} = 0.6, \quad x_0 = 0, \quad y_0 = 1, \quad h = 0.2$$

n	x_n	y_n	$x_n^2 + y_n^2$	$2y_n - y_{n-1} + (h)^2(x_n^2 + y_n^2)$
0	0	1	1	1.44
1	0.2	1.44	2.1136	1.5245
2	0.4	1.5245	2.4841	1.7084
3	0.6	1.7084		
				and so on

In the example that follows we illustrate how a more general second order equation can be solved numerically.

EXAMPLE 11c

Using the approximations

$$h^2\left(\frac{d^2y}{dx^2}\right)_n \simeq y_{n+1} - 2y_n + y_{n-1}$$

and

$$2h\left(\frac{dy}{dx}\right)_n \simeq y_{n+1} - y_{n-1}$$

find, with $h = 0.2$, an approximate value for y when $x = 2$ given that

$$\frac{d^2y}{dx^2} + 2x\frac{dy}{dx} + y = 0$$

and $y = 1$ when $x = 1$, $y = 1.4$ when $x = 1.2$.

From the differential equation

$$\left(\frac{d^2y}{dx^2}\right)_n + 2x_n\left(\frac{dy}{dx}\right)_n + y_n = 0 \qquad [1]$$

Using the given approximations in [1] gives

$$\frac{(y_{n+1} - 2y_n + y_{n-1})}{h^2} + \frac{2x_n(y_{n+1} - y_{n-1})}{h} + y_n \simeq 0$$

$$\Rightarrow \qquad y_{n+1} \simeq \frac{y_n(2 - h^2) + y_{n-1}(hx_{n-1})}{hx_n + 1}$$

With $h = 0.2$ this formula simplifies to

$$y_{n+1} \simeq \frac{9.8y_n + (x_n - 5)y_{n-1}}{x_n + 5} \qquad [2]$$

$x_0 = 1$, $y_0 = 1$, $x_1 = 1.2$, $y_1 = 1.4$ are given so there is enough information to calculate y_2 directly from [2].
The calculations are tabulated opposite.

n	x_n	y_n	$\dfrac{9.8y_n + (x_n - 5)y_{n-1}}{x_n + 5}$
0	1	1	
1	1.2	1.4	1.6
2	1.4	1.6	1.6625
3	1.6	1.6625	1.6443
4	1.8	1.6443	1.5874
5	2.0	1.5874	

EXERCISE 11c

In Questions 1–5 use the approximation

$$y_{n+1} \simeq 2y_n - y_{n-1} + h^2\left(\frac{d^2y}{dx^2}\right)_n$$

1) Using a step length of 0.1 find y when $x = 1.5$ given that $\dfrac{d^2y}{dx^2} = y$ and $y = 1$, $\dfrac{dy}{dx} = 1$ when $x = 1$.

2) Using a step length of 0.2 find y when $x = 0.6$ and when $x = 0.8$ given that $\dfrac{d^2y}{dx^2} = x + y$ and $y = 1$, $\dfrac{dy}{dx} = 0$ when $x = 0$.

3) Using a step length of 0.1 find y when $x = 0.5$ given that $\dfrac{d^2y}{dx^2} = x - y$ and $y = 1$ when $x = 0$ and $y = 1.1$ when $x = 0.1$.

4) Using a step length of 0.5 find y when $x = 2$ given that $\dfrac{d^2y}{dx^2} = x^2 - y^2$ and $y = 1$, $\dfrac{dy}{dx} = 1$ when $x = 0$.

5) Using a step length of 0.2 find y when $x = 0.8$ given that $\dfrac{d^2y}{dx^2} = xy$ and $y = 1$, $\dfrac{dy}{dx} = 0$ when $x = 0$.

In Questions 6–10 use the approximations

$$h^2\left(\frac{d^2y}{dx^2}\right)_n \simeq y_{n+1} - 2y_n + y_{n-1} \quad \text{and} \quad 2h\left(\frac{dy}{dx}\right)_n \simeq y_{n+1} - y_{n-1}$$

6) Using a step length of 0.2 find y when $x = 0.6$ given that $x\dfrac{d^2y}{dx^2} + \dfrac{dy}{dx} + xy = 0$ and $y = 1$, $\dfrac{dy}{dx} = 0$ when $x = 0$.

7) Using a step length of 0.1 find y when $x = 1.5$ given that $\dfrac{d^2y}{dx^2} + y\dfrac{dy}{dx} + x = 0$ and $y = 1$ when $x = 1$ and $y = 1.2$ when $x = 1.1$.

8) Using a step length of 0.1 find y when $x = 0.5$ given that $\dfrac{d^2y}{dx^2} + x\dfrac{dy}{dx} + y = 2x$ and $y = 1$ at $x = 0$ and $y = 0.9$ at $x = 0.1$.

9) Using a step length of 0.1 find y when $x = 0.5$ given that $\dfrac{d^2y}{dx^2} + 2\dfrac{dy}{dx} + y = 0$ and $y = 1$ at $x = 0$ and at $x = 0.1$.

Find the exact solution of this equation and compare results.

10) Using a step length of 0.2 find y when $x = 1$ given that $\dfrac{d^2y}{dx^2} + 6\dfrac{dy}{dx} + 8y = x$ and $y = 1$, $\dfrac{dy}{dx} = 0$ at $x = 0$.

Find the exact solution of this equation and compare results.

CHAPTER 12

CURVE SKETCHING AND INEQUALITIES

The general approach to curve sketching is discussed in Chapter 11 of *The Core Course* and we recommend that the reader should revise that work before studying this chapter.

RATIONAL FUNCTIONS WITH A QUADRATIC DENOMINATOR

In *The Core Course* we sketched some curves with an equation of the form

$$y = \frac{ax^2+bx+c}{px^2+qx+r}$$

but only those where either

the denominator had real linear factors so that the curve had vertical asymptotes,

or the numerator was a constant so the curve could be deduced by considering the reciprocal of a quadratic function.

Range of Values

In general a function of the form $f(x) = \dfrac{ax^2+bx+c}{px^2+qx+r}$ is such that any real value of x (except those, if any, for which $px^2+qx+r=0$) gives a real value of $f(x)$. But the converse is not necessarily true, because it does not follow that $f(x)$ can take all real values. Finding the range of values of $f(x)$ is often helpful when sketching its graph.

For example, if $y = \dfrac{x^2-x+2}{x^2+x+1}$ then for all real values of x,

$$y(x^2+x+1) = x^2-x+2$$

Rearranging as a quadratic equation in x gives

$$(y-1)x^2+(y+1)x+(y-2) = 0 \qquad [1]$$

and we can say that x is real provided that

$$(y+1)^2-4(y-1)(y-2) \geqslant 0$$

$\Rightarrow$ $$-(y+1)(3y+7) \geqslant 0$$

$\Rightarrow$ $$-\tfrac{7}{3} \leqslant y \leqslant -1$$

So this function is such that, whatever the value of x, $-\tfrac{7}{3}\leqslant f(x)\leqslant -1$.

EXAMPLES 12a

1) Prove that $\dfrac{2}{27} \leqslant \dfrac{x^2-2x+2}{x^2+3x+9} \leqslant 2.$

This can be proved in two quite different ways.

The first method involves finding the range of possible values of the function

$$y = \frac{x^2-2x+2}{x^2+3x+9}$$

i.e. $$(y-1)x^2+(3y+2)x+9y-2 = 0$$

Now x is real provided that

$$(3y+2)^2-4(y-1)(9y-2) \geqslant 0$$

$\Rightarrow$ $$(27y-2)(y-2) \geqslant 0$$

$\Rightarrow$ $$\tfrac{2}{27} \leqslant y \leqslant 2$$

In the second method, we prove separately that

$$\frac{x^2-2x+2}{x^2+3x+9} \geqslant \frac{2}{27} \qquad [1]$$

and $$\frac{x^2-2x+2}{x^2+3x+9} \leqslant 2 \qquad [2]$$

For [1] we consider

$$\frac{x^2-2x+2}{x^2+3x+9}-\frac{2}{27} \equiv \frac{25x^2-60x+36}{27(x^2+3x+9)}$$

$$\equiv \frac{(5x-6)^2}{27[(x+\tfrac{3}{2})^2+\tfrac{27}{4}]}$$

But $$(5x-6)^2 \geqslant 0$$

and $(x+\frac{3}{2})^2 \geqslant 0$ so $(x+\frac{3}{2})^2+\frac{27}{4} > 0.$

Hence $$\frac{x^2-2x+2}{x^2+3x+9}-\frac{2}{27} \geqslant 0$$

Now for [2] we consider

$$\frac{x^2-2x+2}{x^2+3x+9}-2 \equiv -\frac{x^2+8x+16}{x^2+3x+9}$$

$$\equiv -\frac{(x+4)^2}{(x+\frac{3}{2})^2+\frac{27}{4}}$$

But $(x+4)^2 \geqslant 0$ and $(x+\frac{3}{2})^2+\frac{27}{4} > 0.$

So $$\frac{x^2-2x+2}{x^2+3x+9}-2 \leqslant 0$$

Hence $$\frac{x^2-2x+2}{x^2+3x+9} \leqslant 2$$

Horizontal Asymptotes

An equation of the form

$$y = \frac{ax^2+bx+c}{px^2+qx+r} \qquad [1]$$

can be written in proper fraction form to give

$$y = \frac{a}{p}+\frac{dx+e}{px^2+qx+r} \qquad [2]$$

from which we see that as $x \to \infty$, $y = a/p$
So there is one horizontal asymptote, $y = a/p$.
Alternatively, the horizontal asymptote can be found by observing [1] when rearranged as a quadratic equation in x. (This has often already been done in order to find the range of values of y.)

For example, if $y = \dfrac{x^2-2x+2}{x^2+3x+9}$

then $$(y-1)x^2+(3y+2)x+9y-2 = 0 \qquad [3]$$

We saw in *The Core Course* (p. 11) that if, in a quadratic equation, the x^2 term vanishes then one solution is $x = \infty$.
In [3], the x^2 term vanishes when $y = 1$

$\Rightarrow$ $x = \infty$ when $y = 1$

$\Rightarrow$ $y = 1$ is a horizontal asymptote.

Maximum and Minimum Values

From [2], $\frac{dy}{dx} = \frac{\text{(polynomial of degree 2 or less)}}{(px^2+qx+r)^2}$

So there are, at most, two distinct values of x for which $\frac{dy}{dx} = 0$

i.e. there are, at most, two turning points on the curve $y = \frac{ax^2+bx+c}{px^2+qx+r}$.

It is not necessary to differentiate in order to find these turning points; they can be deduced from knowledge of the range of values of y. This is illustrated in the following examples.

EXAMPLES 12a (continued)

2) Sketch the curve $y = \frac{x^2-x-1}{\frac{1}{2}x^2+x+1}$.

(a) To find the values that y can take we rearrange the equation as a quadratic in x to give

$$(\tfrac{1}{2}y-1)x^2+(y+1)x+(y+1) = 0 \quad [1]$$

x is real provided that

$$(y+1)^2-4(\tfrac{1}{2}y-1)(y+1) \geqslant 0$$

$$\Rightarrow \quad (y+1)(5-y) \geqslant 0$$

$$\Rightarrow \quad -1 \leqslant y \leqslant 5$$

(b) The curve is continuous (there are no vertical asymptotes).

From [1], x^2 vanishes when $y=1$, so $y=1$ is a horizontal asymptote.

(c) From (a), the least value of y is -1 and the greatest value of y is 5. Since the curve is continuous we deduce that these are minimum and maximum values of y.
At a minimum or maximum value of y, [1] has equal roots given by

$$x = \frac{-(y+1)}{2(\frac{1}{2}y-1)}$$

So when $y=-1$, $x=0$
and when $y=5$, $x=-2$

We now have enough information to sketch the curve.

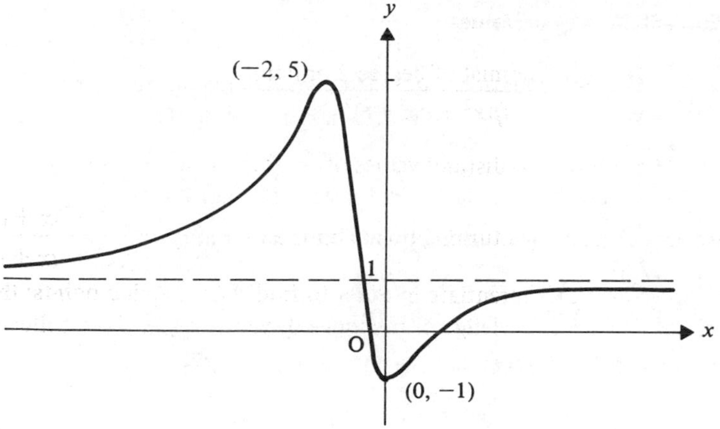

3) Sketch the curve $y = \dfrac{x^2 - x + 2}{x^2 - x + 1}$

(a) Rearranging as a quadratic equation in x gives

$$(y-1)x^2 - (y-1)x + (y-2) = 0 \qquad [1]$$

x is real provided that

$$(y-1)^2 - 4(y-1)(y-2) \geqslant 0$$

$$\Rightarrow \qquad (y-1)(7-3y) \geqslant 0 \qquad [2]$$

Before proceeding, note that $y = 1$ is a horizontal asymptote (because the x^2 term in [1] vanishes when $y = 1$). Note also that when $y = 1$, [1] becomes a contradiction, so $y = 1$ is not a possible value of y.

Returning to [2] we have

$$1 < y \leqslant \tfrac{7}{3}.$$

(b) The curve is continuous (there are no vertical asymptotes).

From (a), $y = 1$ is a horizontal asymptote and the curve does not cross this asymptote. Thus we deduce that y has a maximum value of $\frac{7}{3}$ and no minimum value. At a maximum value of y, [1] has equal roots.

So when $y = \frac{7}{3}$, $\quad x = \dfrac{(y-1)}{2(y-1)} = \frac{1}{2}$.

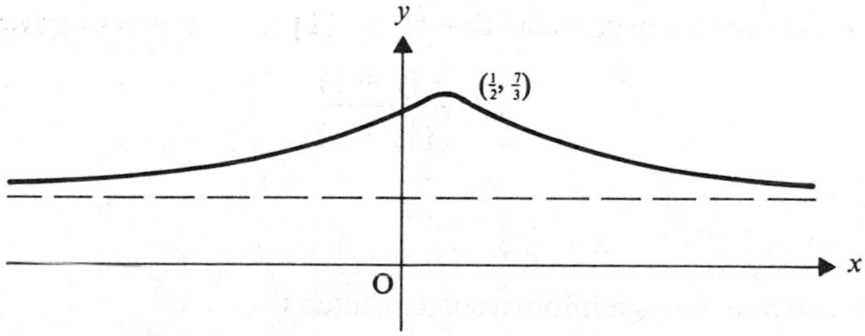

THE EQUATION $y^2 = f(x)$

The curve C_1 with equation $y^2 = f(x)$, can be deduced from the curve C_2 with equation $y = f(x)$, by noting that,

(a) because $y^2 \geqslant 0$,
C_1 is limited to those values of x for which $f(x) \geqslant 0$.

(b) $y = \pm\sqrt{f(x)}$ so C_1 is symmetrical about the x axis.

(c) When $f(x) > 1$, $\sqrt{f(x)} < f(x)$ but
when $0 < f(x) < 1$, $\sqrt{f(x)} > f(x)$.

Consider, for example, $f(x) \equiv (x+1)(x-1)(x-2)$

so that C_1 is $y^2 = (x+1)(x-1)(x-2)$

and C_2 is $y = (x+1)(x-1)(x-2)$

The graph of C_2 is shown in diagram (i).

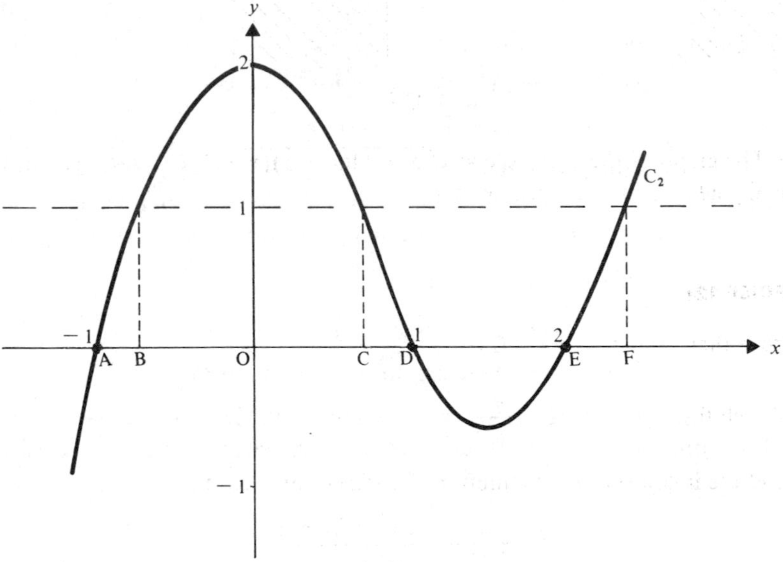

(i)

Now $f(x) \geqslant 0$ for $-1 \leqslant x \leqslant 1$ and $x \geqslant 2$ so C_2 exists only within these ranges.
Between A and B, C and D, E and F, $f(x) < 1$ so $\sqrt{f(x)} > f(x)$
Between B and C and right of F, $f(x) > 1$ so $\sqrt{f(x)} < f(x)$
Thus the graph of C_1 is as shown in diagram (ii).

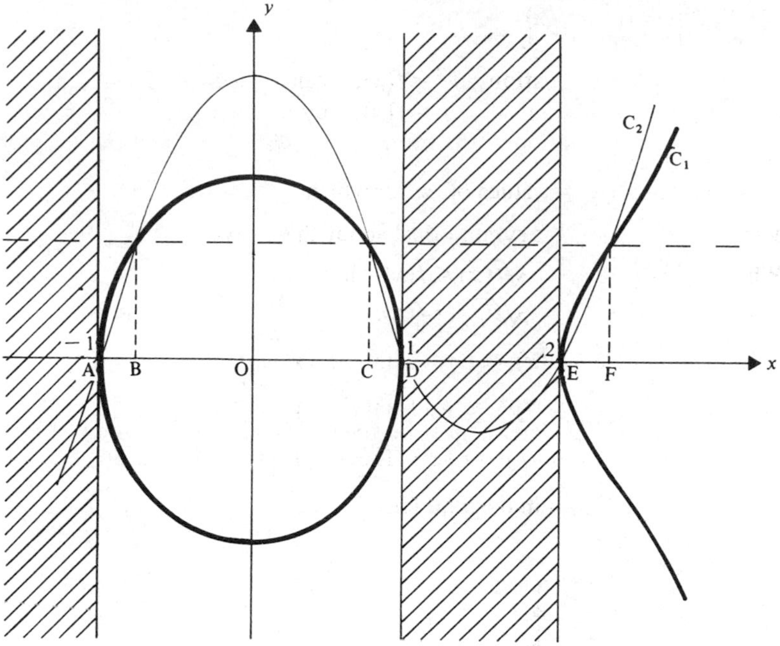

(ii)

Note. The graph of the curve $y = \sqrt{(x+1)(x-1)(x-2)}$ is that part of the curve C_1 *which lies above the x axis.*

EXERCISE 12a

1) Prove that $-1 \leqslant \dfrac{4x}{x^2+4} \leqslant 1$

and sketch the curve $y = \dfrac{4x}{x^2+4}$.

2) Find the image set of the function f defined by

$$f: x \longmapsto \frac{2x^2+2x+1}{x^2+x-1}.$$

3) Sketch the curve $y = \dfrac{2x^2+2x+1}{x^2+x-1}$.

4) Give the domain and image set of the function f defined by

$$f: x \longmapsto \frac{x^2-2x}{x^2+1}$$

Sketch the curve $y = f(x)$.

5) Sketch the graph of $y=\dfrac{2+x^2}{1+x^2}$ by

(a) considering the range of values that y can take and the turning points,

(b) considering the equation in the form $y=1+\dfrac{1}{1+x^2}$ and the reciprocal of the curve $y=1+x^2$.

6) Sketch the curve $y=\dfrac{x^2+x+3}{x^2-x+1}$.

7) Sketch the curve $y=\dfrac{x^2-2x+4}{x^2-2x+3}$.

Sketch the graph of $f(x)$ and hence the graph of $y^2=f(x)$ when $f(x)$ is:

8) x^2-3x+2. 9) x^3-3x^2+2x. 10) $\dfrac{x}{x-1}$. 11) $\dfrac{x}{x^2-4}$.

12) xe^x. 13) $\sin x$. 14) $\ln x$. 15) $\tan x$.

16) e^{x^2}. 17) $e^x \sin x$ (for $-2\pi \leqslant x \leqslant 2\pi$).

THE CYCLOID

Some curves arise naturally in physical situations. Consider, for instance, the curve known as a cycloid. This is the path traced out by a marked point on the rim of a disc when the disc is rolled along a plane. The parametric equations of a cycloid can be found as follows.

Take P as the marked point on the rim of a disc of radius a and let O be the fixed point on the plane which is initially in contact with P.

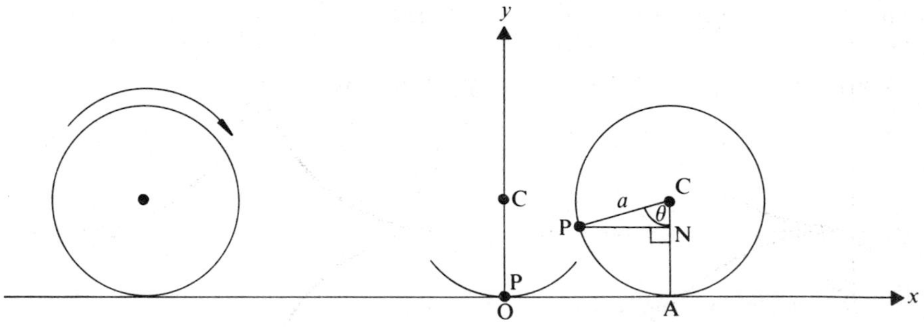

When the disc has rotated through an angle θ, the point of contact is A and the distance OA is equal to the length of the arc AP, i.e. $a\theta$.

Now $PN = a\sin\theta$ and $CN = a\cos\theta$

so the coordinates of P relative to axes along and perpendicular to the plane as shown are

$$x = OA - PN = a\theta - a\sin\theta$$

$$y = AC - CN = a - a\cos\theta$$

Thus the parametric equations of the cycloid are

$$x = a(\theta - \sin\theta), \qquad y = a(1 - \cos\theta)$$

It is not easy to eliminate θ from these equations so a cycloid cannot be represented simply in Cartesian form. Consequently its graph is drawn by taking a set of values of θ and plotting the corresponding values of x and y. The resulting curve is shown below and it is seen to be periodic with a period $2\pi a$.

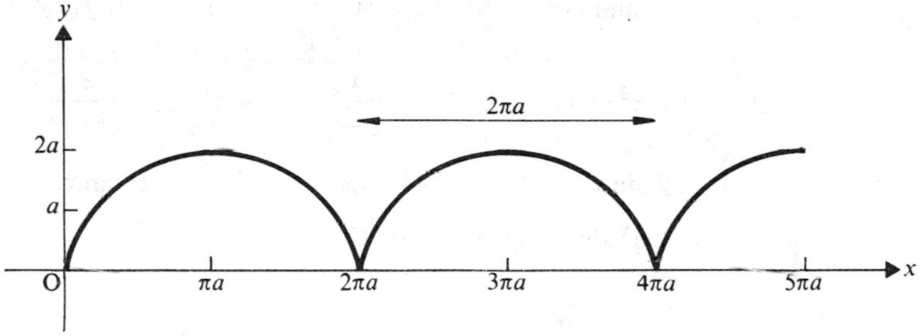

Other curves whose sketches the reader may find useful to remember are,
(a) the semi-cubical parabola,

$$y^2 = x^3;$$

(b) the astroid,

$$x = a\cos^3\theta, \quad y = a\sin^3\theta.$$

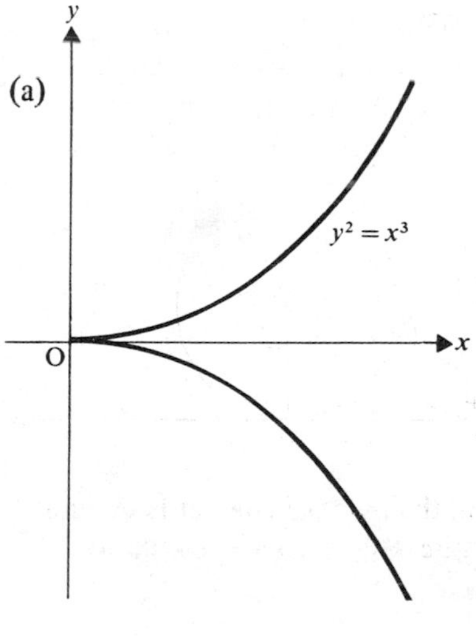

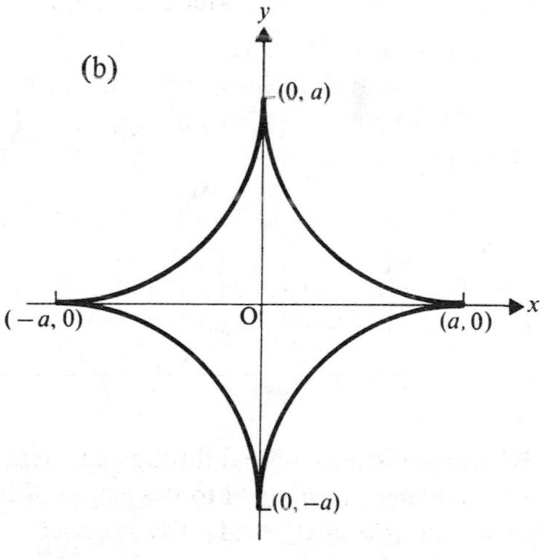

Note that the part of the curve in the first quadrant corresponds to values of θ in the range $0 < \theta \leqslant \frac{\pi}{2}$.

These curves will feature in further work in this chapter.

THE TANGENT TO A CURVE AT THE ORIGIN

Consider first a curve that passes through the origin and which has a second order Cartesian equation. The general equation of such a curve is

$$ax^2 + 2hxy + by^2 + 2gx + 2fy = 0$$

The gradient of the curve at any point can be found by differentiating this equation with respect to x,

i.e. $$2ax + 2h\left(y + x\frac{dy}{dx}\right) + 2by\frac{dy}{dx} + 2g + 2f\frac{dy}{dx} = 0$$

so $$\frac{dy}{dx} = -\frac{ax + hy + g}{hx + by + f}$$

The gradient at the origin, where x and y are both zero, is therefore $-\frac{g}{f}$, and the equation of the tangent to the given curve at the origin

$$y = -\frac{g}{f}x \quad \Rightarrow \quad gx + fy = 0$$

This result can be obtained more simply by using the linear approximation to the given equation when x and y are both very small. In these circumstances, x^2, y^2 and xy are negligible compared with x and y, so the equation of the curve approximates to

$$2gx + 2fy = 0$$

for points very near to the origin.
Thus the equation of the tangent at the origin is

$$gx + fy = 0$$

For example, the equation of the parabola $y^2 = 4ax$ approximates to $4ax = 0$ near to the origin (as y^2 is insignificant compared with $4ax$ when x and y are both very small).

So the tangent at the origin has equation

$$4ax = 0$$

i.e.
$$x = 0$$

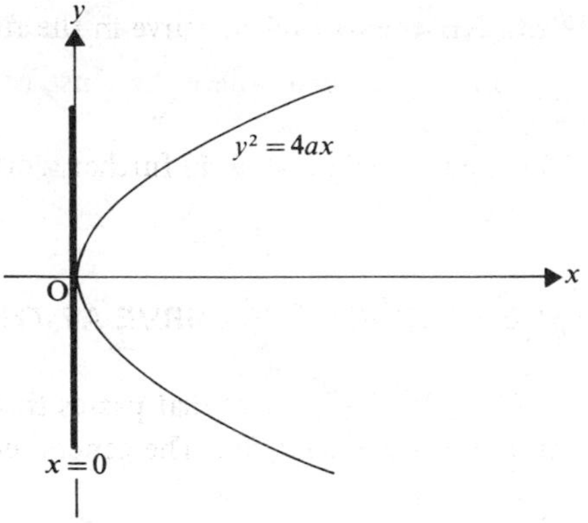

Similarly, to find the equation of the tangent at the origin to the circle

$$x^2 + y^2 - 2x - 4y = 0$$

we delete x^2 and y^2 giving

$$x + 2y = 0$$

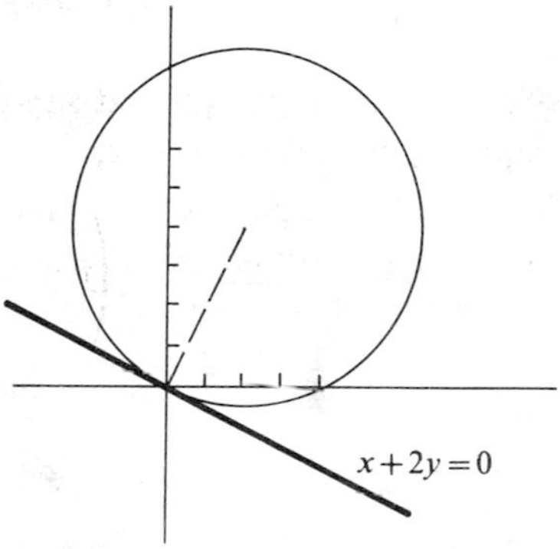

Curves whose equations are of higher degree in x and/or y can be treated in the same way:
e.g., the tangent at O to the curve $x^2 + y^3 - 2y^2 + 3y - 2x = 0$
is $3y - 2x = 0$ (deleting all non-linear terms).
If the equation of a particular curve contains no linear terms, the equation(s) of the tangent(s) at the origin is/are found by deleting all terms except those whose power is least.
For example, the curve with equation $2x^3 + y^3 + 4x^2 = 9y^2$ has two tangents at the origin, whose equations are

$$4x^2 = 9y^2 \Rightarrow 3y = \pm 2x$$

Note. The equation of the tangent at the origin, to a curve whose equation is not algebraic, can sometimes be found by using a series expansion,

e.g.
$$y = x + \sin x \Rightarrow y = x + \left(x - \frac{x^3}{3!} + \ldots\right)$$

which approximates to $y = 2x$ when x is small, so $y = 2x$ is the equation of the tangent at the origin. Otherwise the gradient of the tangent at O can be found from the value of $\frac{dy}{dx}$ when $x = 0$ and $y = 0$.

INFLEXION

A point on a curve where the sense of the curvature changes, is called a *point of inflexion*,
e.g.

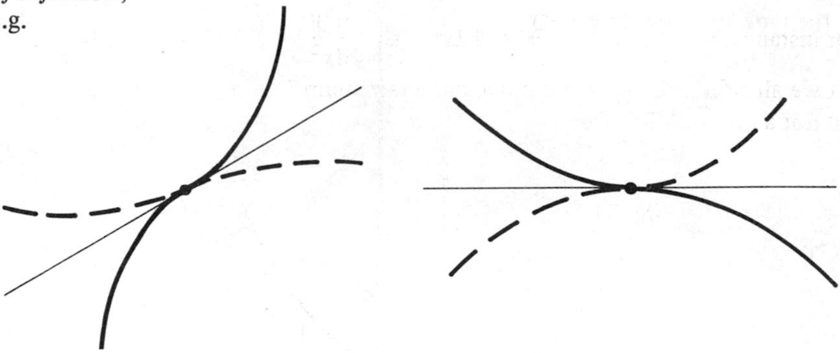

Note that, at any point of inflexion, the curve crosses the tangent drawn at that point.
On one side of a point of inflexion, $P(x_i, y_i)$ a curve is concave and on the other side it is convex. So on one side of P the gradient is increasing as x increases, while on the other side of P the gradient decreases as x increases,

e.g.

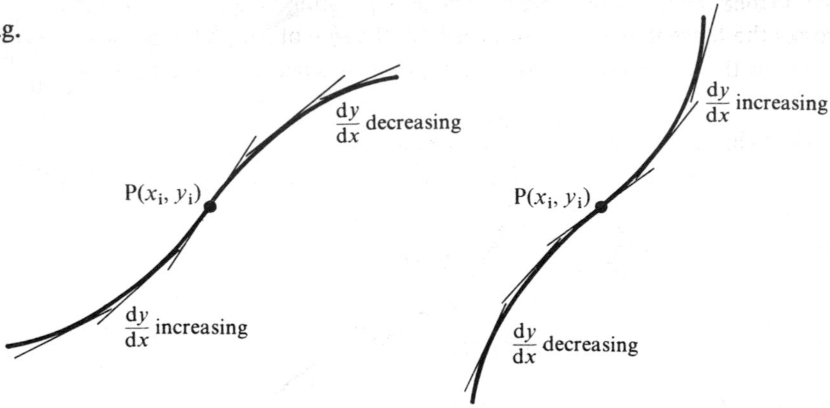

Now if the gradient is increasing, $\frac{d}{dx}\left(\frac{dy}{dx}\right) > 0$

and if the gradient is decreasing, $\frac{d}{dx}\left(\frac{dy}{dx}\right) < 0.$

Therefore, as a curve passes through a point of inflexion, $\frac{d}{dx}\left(\frac{dy}{dx}\right)$ changes sign. It follows that, unless $\frac{d^2y}{dx^2}$ has a singularity where $x = x_i$,

$$P \text{ is a point of inflexion} \Rightarrow \frac{d^2y}{dx^2} = 0 \text{ at P.}$$

The converse of this statement is not always true however so we *cannot* assume that we have an inflexion whenever $\dfrac{d^2y}{dx^2} = 0$.

For instance, if $y = x^4$, $\dfrac{d^2y}{dx^2} = 12x^2$ so $\dfrac{d^2y}{dx^2} = 0$ when $x = 0$.

But we already know that this curve has a minimum turning point when $x = 0$ and not a point of inflexion.

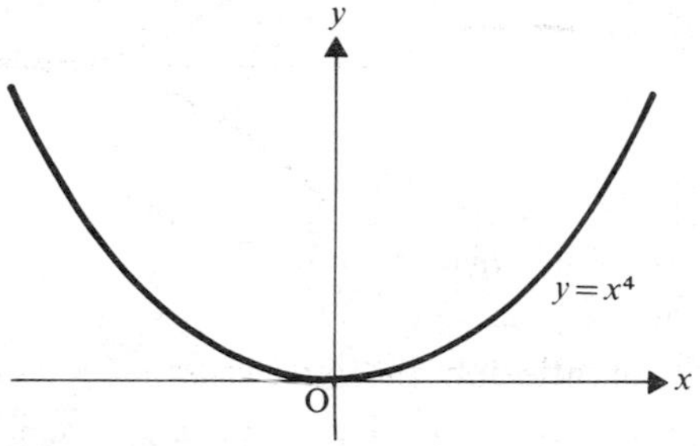

Thus the condition $\dfrac{d^2y}{dx^2} = 0$, is not *sufficient* to define a point of inflexion.

The further condition we need is obtained by noting that, because a curve crosses the tangent at a point of inflexion, the *sign* of the gradient *does not change* as the curve passes through this point (whereas at a turning point $\dfrac{dy}{dx}$ *does* change sign) and that $\dfrac{d^2y}{dx^2}$ changes sign.

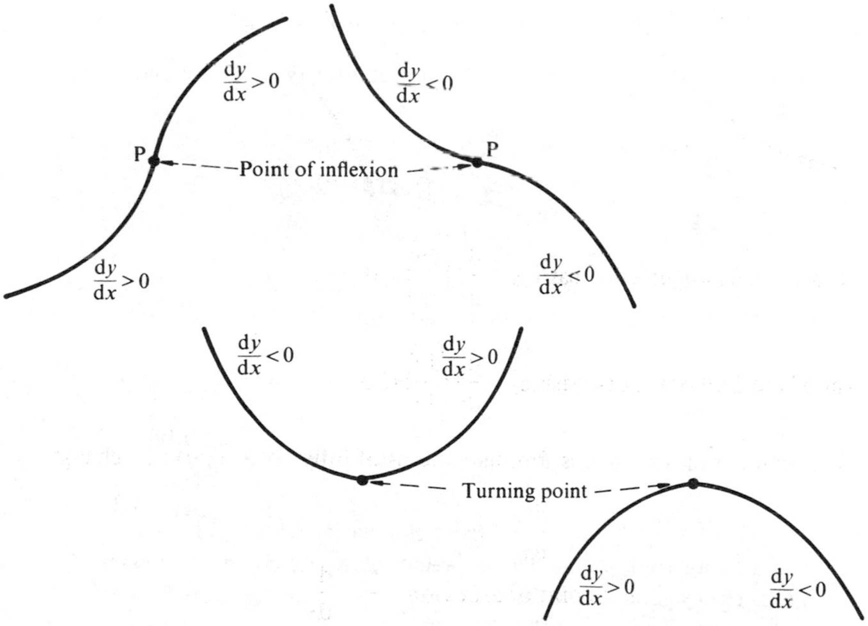

So the *necessary and sufficient* conditions for a point P to be a point of inflexion are that, as the curve passes through P,

$\frac{d^2y}{dx^2}$ changes sign

$\frac{dy}{dx}$ does not change sign.

EXAMPLES 12b

1) Prove that the cycloid $x = a(\theta - \sin\theta)$, $y = a(1 - \cos\theta)$
(a) is periodic with period p where $p = 2\pi a$,
(b) is such that y is an even function of x.
Prove that the cycloid has no point of inflexion.

(a) A curve $y = f(x)$ is periodic if $f(x) = f(x + np)$ where p is the period and n is an integer, i.e. if values of x separated by regular intervals give the same value of y.
In this example the value suggested for p is $2\pi a$.
When the parameter is θ,

$$x = a(\theta - \sin\theta)$$

When the parameter is $\theta + 2n\pi$,

$$\begin{aligned} x &= a\{(\theta + 2n\pi) - \sin(\theta + 2n\pi)\} \\ &= a(\theta - \sin\theta) + 2\pi na. \end{aligned}$$

So when θ increases by regular integral multiples of 2π, x increases by regular intervals of $2\pi a$.
When the parameter is $\theta + 2n\pi$,

$$\begin{aligned} y &= a\{1 - \cos(\theta + 2n\pi)\} \\ &= a(1 - \cos\theta) \end{aligned}$$

Hence y takes the value $a(1 - \cos\theta)$ for values of x at regular intervals of $2\pi a$ showing that the cycloid is periodic with a period $2\pi a$.

(b) y is an even function of x if $y = f(x)$ and $f(x) = f(-x)$ for all values of x.

For the cycloid, $x = a(\theta - \sin\theta) \Rightarrow -x = a(-\theta + \sin\theta)$

$$= a\{(-\theta) - \sin(-\theta)\}$$

So for parameters θ and $-\theta$ the values of x are equal in value and opposite in sign.

At the point with parameter θ, $y = a(1 - \cos\theta)$.
At the point with parameter $-\theta$, $y = a\{1 - \cos(-\theta)\}$

$$= a(1 - \cos\theta),$$

i.e. the values of y are equal when x is replaced by $-x$, showing that, for the cycloid, y is an even function of x.

If there is a point of inflexion, then at that point $\dfrac{d^2y}{dx^2} = 0$

$$\frac{dy}{dx} = \frac{a\sin\theta}{a(1-\cos\theta)} = \cot\frac{\theta}{2}$$

$$\frac{d^2y}{dx^2} = \frac{d}{dx}\left(\cot\frac{\theta}{2}\right) = \left(-\frac{1}{2}\operatorname{cosec}^2\frac{\theta}{2}\right)\frac{d\theta}{dx}$$

$$= -\frac{1}{2}\operatorname{cosec}^2\frac{\theta}{2} \Big/ a(1-\cos\theta)$$

$$= -\frac{1}{4a}\operatorname{cosec}^4\frac{\theta}{2}$$

Now $\operatorname{cosec}\dfrac{\theta}{2}$ is never zero so $\dfrac{d^2y}{dx^2}$ cannot be zero

Thus there is no point of inflexion on the cycloid.

2) Show that the cubic curve $y = \dfrac{x^3}{6} - \dfrac{x^2}{2} + x$ has one point of inflexion and give the coordinates of this point.
Write down the equation of the tangent to this curve at the origin and sketch the curve.

At any point of inflexion, $\dfrac{d^2y}{dx^2} = 0.$

For the given curve $\qquad \dfrac{dy}{dx} = \dfrac{x^2}{2} - x + 1$

and $\qquad \dfrac{d^2y}{dx^2} = x - 1$

As $\dfrac{d^2y}{dx^2} = 0$ when $x = 1$, there *may* be an inflexion at the point $P(1, \frac{2}{3})$.
Taking values of x close to, and on either side of, this point we can check whether $\dfrac{dy}{dx}$ changes sign at P.

When $x = 0.9$, $\dfrac{dy}{dx} > 0$ and when $x = 1.1$, $\dfrac{dy}{dx} > 0$.

So $\dfrac{dy}{dx}$ does not change sign as the curve passes through $P(1, \frac{2}{3})$ and P is therefore a point of inflexion and at this point $\dfrac{dy}{dx} = \frac{1}{2}$.

Near to the origin, x^3 and x^2 are insignificant compared with x and y, so the equation of the tangent at the origin is

$$y = x$$

Noting also that $y = 0$ for only one real value of x, i.e. $x = 0$, the curve can now be sketched.

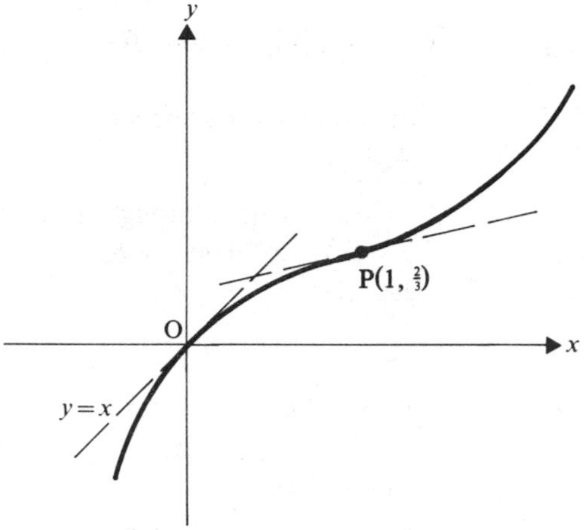

EXERCISE 12b

Write down the equation of the tangent(s) at the origin to each of the following curves, illustrating your result on a diagram.

1) $\dfrac{(x-2)^2}{4} + y^2 = 1$ 2) $y = x(x-3)$

3) $(y+1)^2 + (x-1)^2 = 2$ 4) $(x-3)^2 - \dfrac{(y-4)^2}{2} = 1$

5) $y^2 = 8x$ 6) $x^2 + 2y = 0$

7) $y^3 = x^2 - y^2$ 8) $2x^4 + y^4 - 3x^3 = 4y^2 - x^2$

Determine which of the following six curves contain a point of inflexion, finding the coordinates of this point when it exists.

9) $y = x(x-1)(x-2)$ 10) $y = x^4 - 1$

11) $y = x^4 + x^2 + 1$ 12) $y = x^3 - 1$

13) $y = x^3$ 14) $x = at^2, \quad y = a(t-1)^3$

15) Determine whether either of the curves $y^3 = x$ and $y^3 = x^2$ is such that
(a) y is an odd function of x,
(b) y is an even function of x,
(c) it has a point of inflexion.
Find the equation of the tangent at the origin to each curve.
Sketch both curves.

16) A cubic curve, $y = f(x)$, has a point of inflexion at the point $(2, -22)$.
The equation of the tangent to this curve at the origin is $y + 3x = 0$.
Determine the equation of the curve and sketch it.

17) The equation of a curve is $y = f(x)$, where $f(x)$ is a quadratic function of x.
Find the equation of the curve if it touches the line $4y = 5x$ at the origin and passes through the point $(4, 21)$.

18) A curve with equation $y = f(x)$ passes through the origin and has a point of inflexion at $(1, \frac{2}{3})$. If $f(x) \equiv x^3 + ax^2 + bx + c$, find the equation of the tangent to the curve at the origin.

INEQUALITIES

For any real number, n, we know that

$$n^2 \geqslant 0 \quad \text{and} \quad -n^2 \leqslant 0.$$

These simple properties can be used in proving the validity of many inequalities, some of which are standard.

The Sum of a Positive Real Number and its Reciprocal Cannot be Less than Two

i.e. $$p + \frac{1}{p} \geqslant 2 \quad \text{if} \quad p > 0.$$

The proof of this quotable property begins by considering the expression

$$p + \frac{1}{p} - 2 \equiv \frac{p^2 + 1 - 2p}{p}$$

$$\equiv \frac{(p-1)^2}{p}$$

But $p > 0$ and $(p-1)^2 \geqslant 0$.

So $$\frac{(p-1)^2}{p} \geqslant 0 \Rightarrow p + \frac{1}{p} - 2 \geqslant 0$$

i.e. $$p+\frac{1}{p} \geqslant 2 \quad \text{if} \quad p > 0$$

The Geometric Mean of Two Positive Real Numbers Cannot Exceed their Arithmetic Mean

If p and q are positive real numbers we can say

$$p \equiv m^2 \quad \text{and} \quad q \equiv n^2.$$

The geometric mean of p and q is $\sqrt{(pq)} \equiv mn$.
The arithmetic mean of p and q is $\frac{1}{2}(p+q) \equiv \frac{1}{2}(m^2+n^2)$.
We are required to prove 'hat

$$\sqrt{(pq)} \leqslant \tfrac{1}{2}(p+q)$$

i.e. $$mn \leqslant \tfrac{1}{2}(m^2+n^2)$$

To do this we begin by considering

$$\tfrac{1}{2}(m^2+n^2)-mn \equiv \tfrac{1}{2}(m^2+n^2-2mn)$$
$$\equiv \tfrac{1}{2}(m-n)^2$$

But $$(m-n)^2 \geqslant 0$$

So $$\tfrac{1}{2}(m^2+n^2)-mn \geqslant 0$$

i.e. $$\sqrt{(pq)} \leqslant \tfrac{1}{2}(p+q)$$

Note that each of the above proofs began with consideration of an expression obtained by rearranging the required inequality in a form with zero on one side. This practice should be adopted generally when attempting to prove any inequality.

EXAMPLES 12c

1) If x, y and z are unequal real numbers prove that

$$x^2+y^2+z^2 > xy+yz+zx$$

Consider $$x^2+y^2+z^2-xy-yz-zx$$
$$\equiv \tfrac{1}{2}[2x^2+2y^2+2z^2-2xy-2yz-2zx]$$

This step is introduced because the aim is to convert our expression into perfect squares, and the product term (e.g. xy) in a perfect square contains a factor of 2.

Then $$\tfrac{1}{2}[2x^2+2y^2+2z^2-2xy-2yz-2zx]$$
$$\equiv \tfrac{1}{2}[x^2-2xy+y^2+y^2-2yz+z^2+z^2-2zx+x^2]$$
$$\equiv \tfrac{1}{2}[(x-y)^2+(y-z)^2+(z-x)^2]$$

But, as x, y and z are unequal, $(x-y)^2>0$, $(y-z)^2>0$ and $(z-x)^2>0$.
So $x^2+y^2+z^2-xy-yz-zx>0$.

GRAPHICAL REPRESENTATION OF INEQUALITIES

A Cartesian equation relating x and y corresponds to a line or a curve in the xy plane such that the coordinates of every point on this locus satisfy the given equation.

We saw in *The Core Course* that a Cartesian inequality relating x and y corresponds to an area in the xy plane such that the coordinates of every point in this area satisfy the inequality. The boundary of the area is the line or curve whose equation is given by replacing the inequality sign by an equality sign.

For example, $y^2>4x$ corresponds to an area in the xy plane bounded by the parabola $y^2=4x$, but *not* including points on the parabola.

In diagrams we use a broken line when points on the line are *not* included in the inequality and a continuous line for points on the boundary that are included in the inequality.

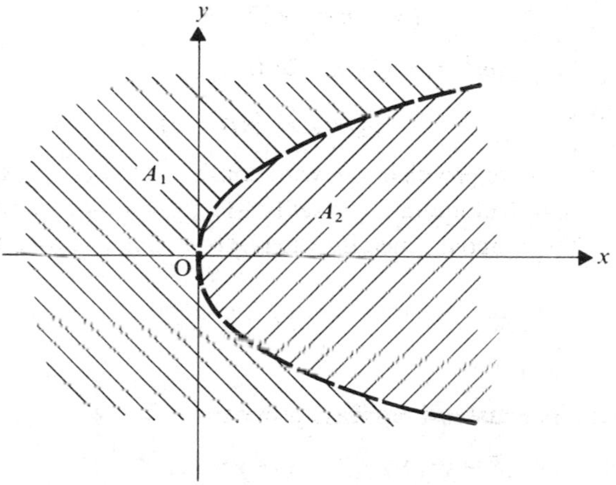

Now the parabola divides the plane into two areas A_1 and A_2, *one* of which represents $y^2>4x$. A quick way to decide which is the required area is to take the coordinates of some simple point within one of the two areas and check whether it satisfies the given inequality. In this example, for instance, we could choose the point $(1, 0)$, which is obviously in A_2, and for which

$$y^2 = 0$$

$$4x = 4$$

So for the point $(1, 0)$, $y^2 \ngtr 4x$, and we deduce that A_2 is *not* the correct area and therefore that A_1 *is* the required area.

Shading the area that is *not* required we have

$y^2 > 4x \quad \Rightarrow$

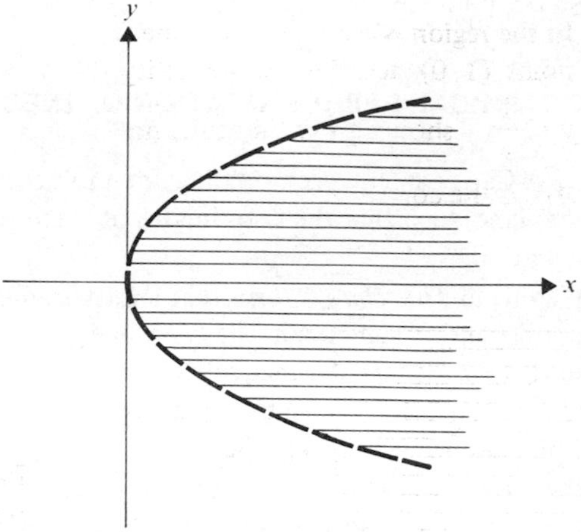

Note. When choosing a particular point within one of the possible areas bounded by a line or curve, the origin is the best choice unless it lies *on* the line or curve.

Note also that in some problems it may be preferable to shade the area that *is* required. As there is no hard and fast rule, it is essential to state, in all cases, what the shading represents.

A different situation arises when the curve forming the boundary of an area has a singularity.

In this case separate investigation must be carried out, on either side of the singularity, to determine the appropriate area.

Consider, for example, the area representing the inequality $y < \dfrac{1}{x}$.

$\dfrac{1}{x}$ is undefined when $x = 0$ so the inequality makes sense only in the two regions

where $x > 0$ (i.e. A_1 and A_2)
and where $x < 0$ (i.e. B_1 and B_2).

Note also that $x = 0$ is *not* part of the solution set as $\dfrac{1}{x}$ is undefined when $x = 0$.

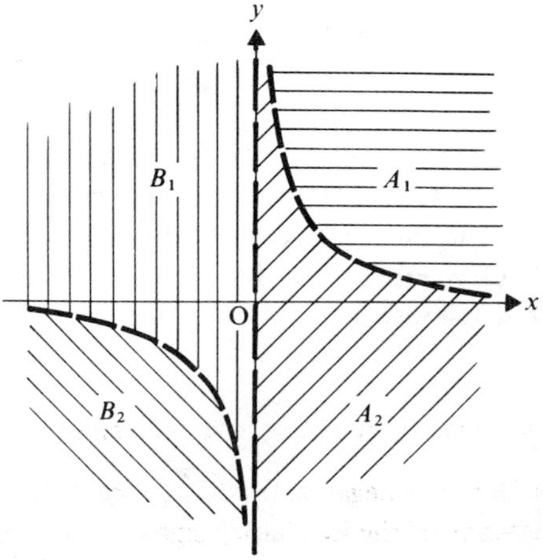

In the region where $x > 0$, the point $(1, 0)$ satisfies the inequality $y < \frac{1}{x}$, showing that A_2 (and not A_1) is the correct area.

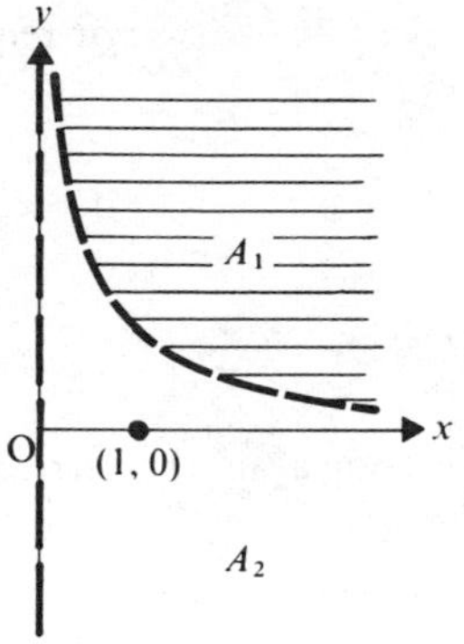

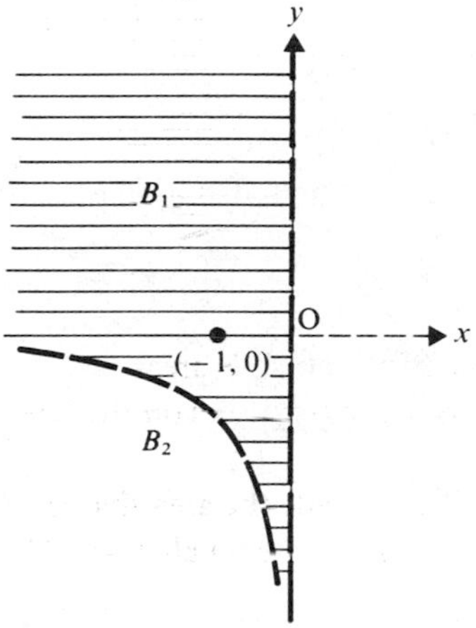

In the region where $x < 0$, the point $(-1, 0)$ does not satisfy the inequality, showing that B_2 (and not B_1) is the correct area.

So, shading the areas *not* required, we have

$$y < \frac{1}{x} \quad \Rightarrow$$

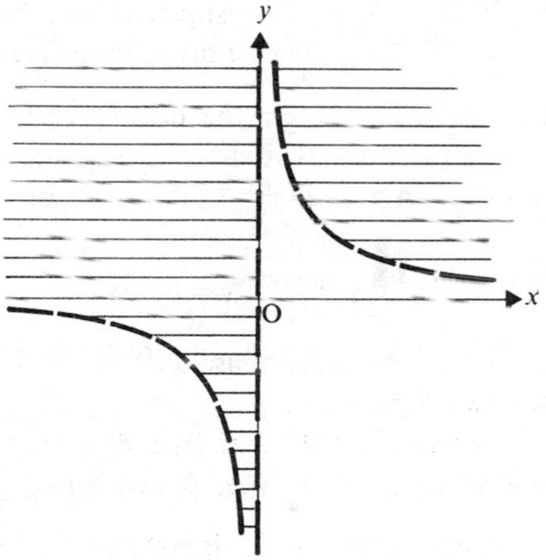

Note that $xy < 1$ is *not* the same inequality as $y < \frac{1}{x}$. This is because, when x is negative, dividing $xy < 1$ by x must be accompanied by a reversal of the inequality sign, i.e. $\qquad$ if $x>0$, $xy<1 \Rightarrow y<\frac{1}{x}$

but $\qquad$ if $x<0$, $xy<1 \Rightarrow y>\frac{1}{x}$

THE SIGN OF *f(xy)*

Considering again the inequality $y^2>4x$ we see that it can be written in the form

$$y^2-4x>0$$

or $\qquad f(xy)>0$ where $f(xy)\equiv y^2-4x$.

But $y^2>4x$ is represented in the xy plane by an area bounded by the parabola $y^2=4x$.
So we see that an inequality of the form

$$f(xy)>0 \quad \text{(i.e. } f(xy) \text{ has a positive sign)}$$

or

$$f(xy)<0 \quad \text{(i.e. } f(xy) \text{ has a negative sign)}$$

is represented by an area bounded by the curve $f(xy)=0$.

Simultaneous Inequalities

The solution of a number of simultaneous inequalities is the set of coordinates (x, y) that satisfy all of the inequalities. The graphical representation of the solution set is therefore obtained by sketching separately the areas that represent each of the individual inequalities and then locating the common area.
This method was used in *The Core Course* for simultaneous linear inequalities and we shall now apply it to those of higher degree.

EXAMPLES 12c (continued)

2) Indicate in a diagram, the area represented by $\frac{x^2}{4}+y^2-1<0$.

If, also, $x^3-y^2>0$, indicate the appropriate area.

The equation $\frac{x^2}{4}+y^2-1=0$ represents an ellipse, so the inequality $\frac{x^2}{4}+y^2-1<0$ represents an area bounded by the ellipse, but not including points on the ellipse.
As the origin $(0, 0)$ satisfies the inequality we see that the required area is *inside* the ellipse as shown.

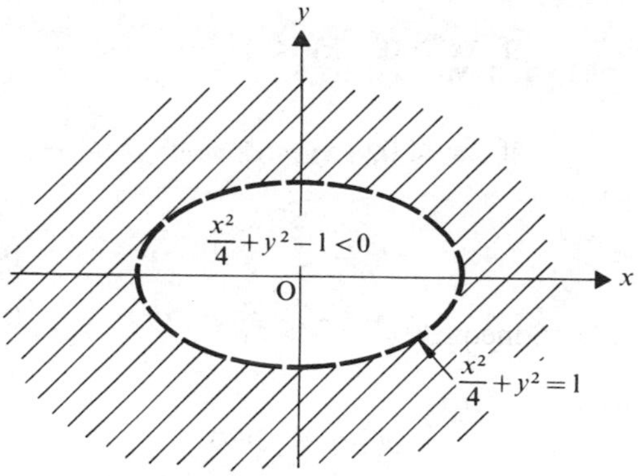

The points that satisfy $x^3 - y^2 > 0$ lie in an area which is bounded by, but does not include points on, the semi-cubical parabola $y^2 = x^3$. As $(1, 0)$ satisfies $x^3 - y^2 > 0$, the area that represents this inequality is on the right of the curve $y^2 = x^3$ as shown.

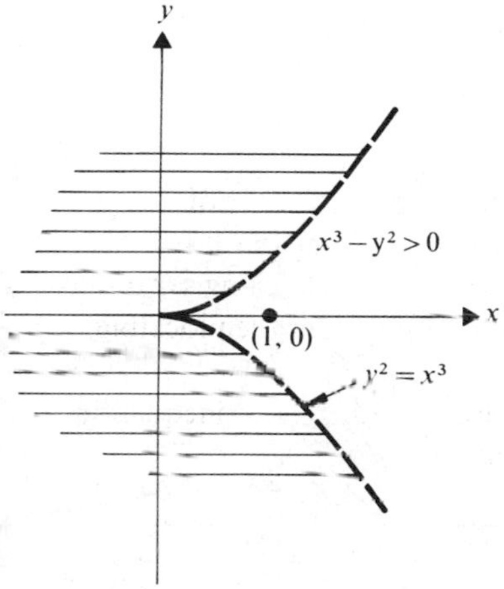

Hence the area common to the two given inequalities is the unshaded region in the diagram opposite.

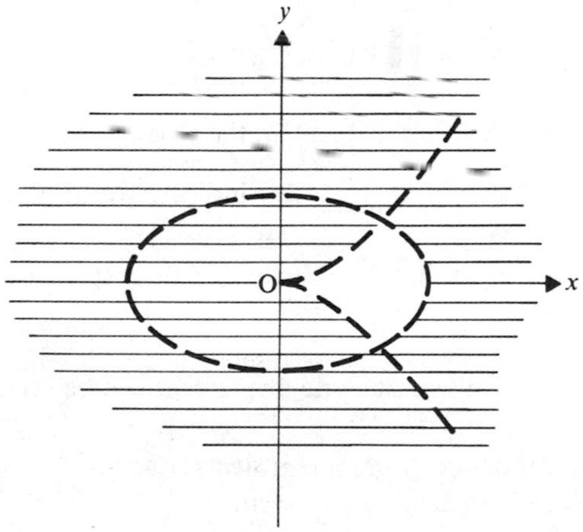

3) Interpret the solution of the inequality $(x^2+y^2-4)(xy-1) \geqslant 0$ and illustrate your explanation with a diagram.

The product $(x^2+y^2-4)(xy-1)$ is positive if x^2+y^2-4 and $xy-1$ have the same sign,

i.e. if $x^2+y^2-4 > 0$ *and* $xy-1 > 0$

or if $x^2+y^2-4 < 0$ *and* $xy-1 < 0$.

Also the given product is zero if

$$\text{either} \quad x^2+y^2-4 = 0 \quad \text{or} \quad xy-1 = 0.$$

So the solution of the given inequality is the set of points for which

$$x^2+y^2-4 \geqslant 0 \quad and \quad xy-1 \geqslant 0$$

and $\quad x^2+y^2-4 \leqslant 0 \quad and \quad xy-1 \leqslant 0$

The relevant regions in the xy plane are bounded by

the circle $\quad x^2+y^2 = 4$

and the rectangular hyperbola $\quad xy = 1$

and do include points on these curves.

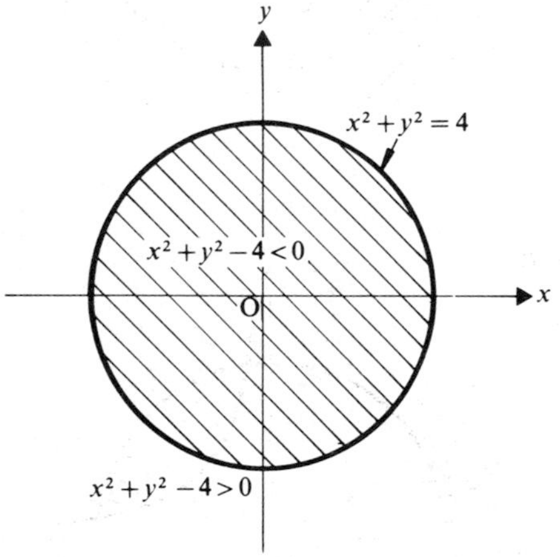

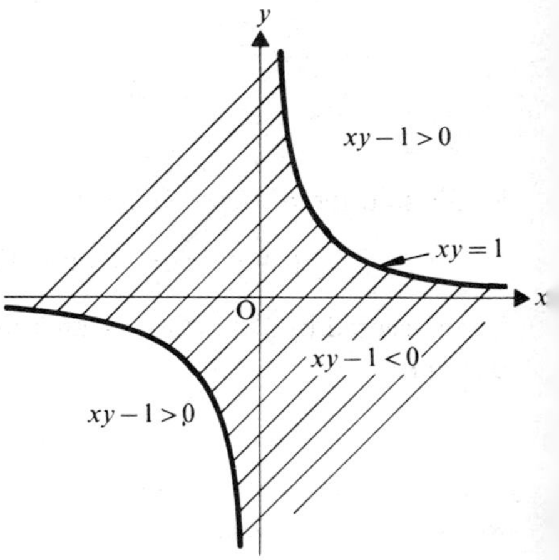

From these separate diagrams it can be seen that the unshaded area in the following diagram represents $\begin{cases} x^2+y^2-4 \geqslant 0 \\ xy-1 \geqslant 0 \end{cases}$

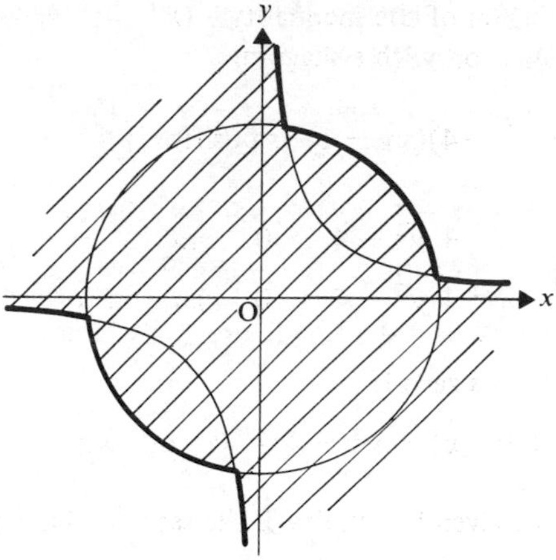

similarly $\begin{cases} x^2 + y^2 - 4 \leqslant 0 \\ xy - 1 \leqslant 0 \end{cases} \Rightarrow$

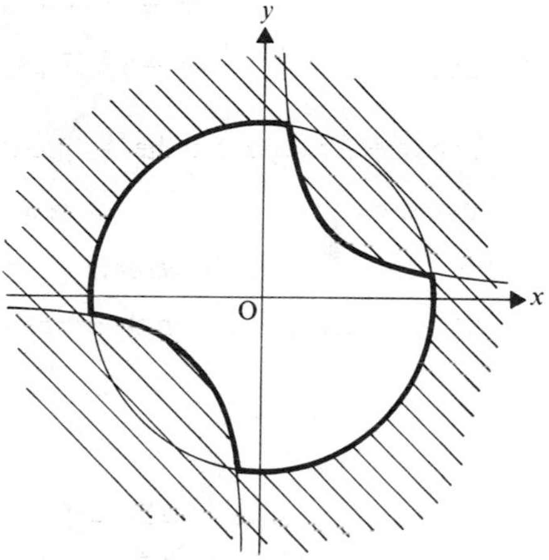

So the solution of the inequality

$$(x^2 + y^2 - 4)(xy - 1) \geqslant 0$$

is represented by the unshaded area shown opposite.

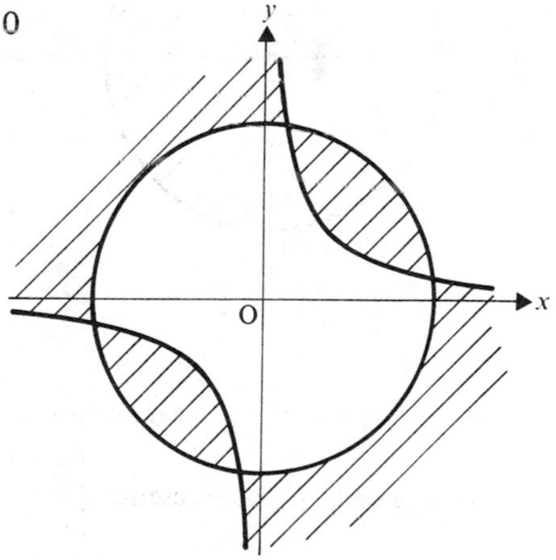

4) If $f(xy) \equiv (x^2 - y^2 - 1)(y^2 - 4x)$ indicate the region in the xy plane where the sign of $f(xy)$ is negative.

If the sign of $f(xy)$ is negative, then $(x^2 - y^2 - 1)(y^2 - 4x) < 0$.

In this case

either $\qquad x^2 - y^2 - 1 < 0$ *and* $y^2 - 4x > 0$

or $\qquad x^2 - y^2 - 1 > 0$ *and* $y^2 - 4x < 0$.

Considering $x^2 - y^2 - 1$ and shading the region where $x^2 - y^2 - 1 < 0$ we have

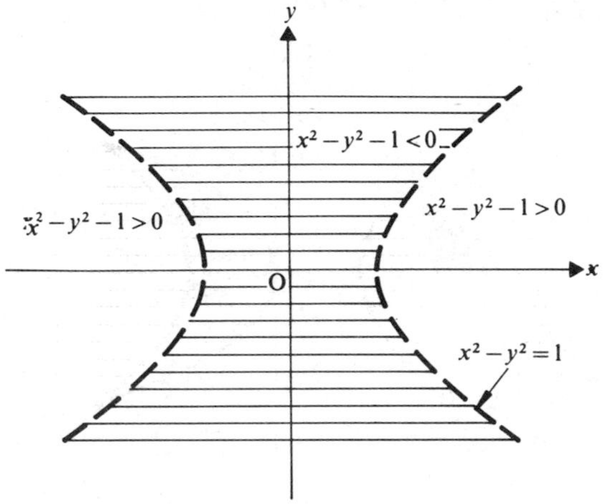

Similarly considering $y^2 - 4x$, we have

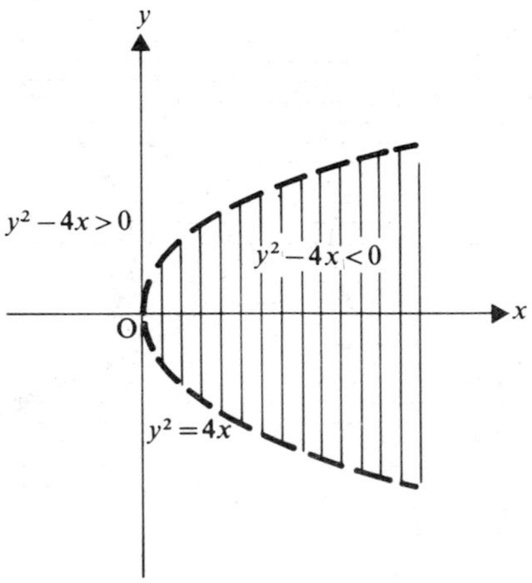

Combining these diagrams shows that

$$\left.\begin{aligned} x^2 - y^2 - 1 &< 0 \\ \text{and} \qquad y^2 - 4x &> 0 \end{aligned}\right\} \text{ corresponds to the unshaded area in diagram (i).}$$

$$\left.\begin{aligned} x^2 - y^2 - 1 &> 0 \\ \text{and} \qquad y^2 - 4x &< 0 \end{aligned}\right\} \text{ corresponds to the unshaded area in diagram (ii).}$$

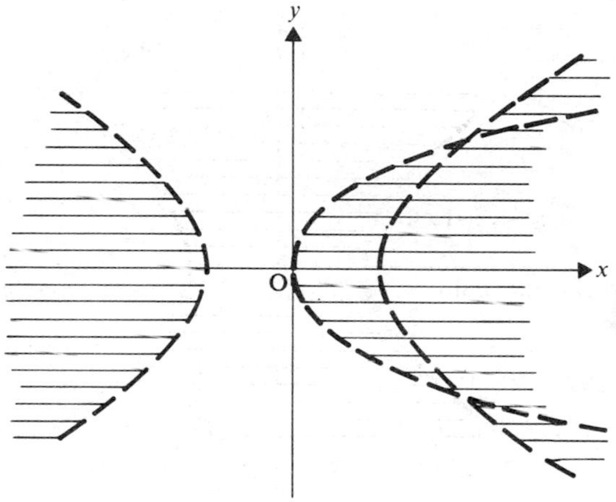

(i)

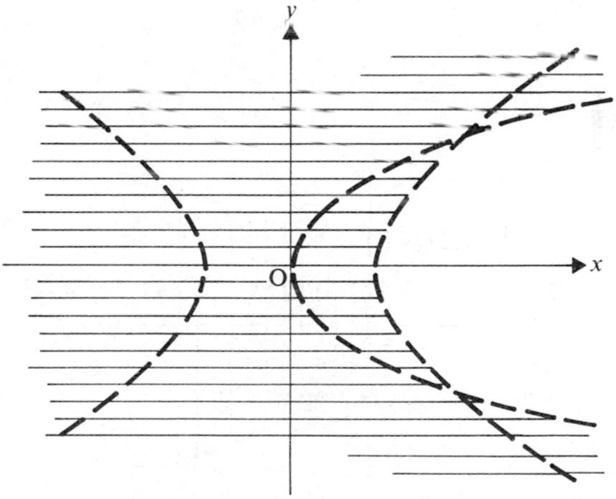

(ii)

So the region where $f(xy) \equiv (x^2 - y^2 - 1)(y^2 - 4x)$ has a negative sign corresponds to the unshaded area shown below.

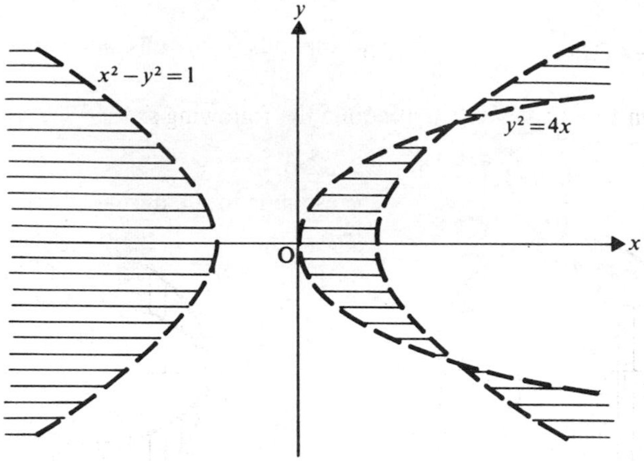

EXERCISE 12c

Prove the following inequalities, for all real values of the variables unless otherwise stated.

1) $x^2 + y^2 - 6y + 9 \geqslant 0$ 2) $x^2 + 10x + y^2 + 26 > 0$

3) $2x^3 + 2y^3 \geqslant (x^2 + y^2)(x + y)$ given that $x > 0$ and $y > 0$.

4) $\dfrac{1}{x^2} + \dfrac{1}{y^2} \geqslant \dfrac{8}{(x + y)^2}$ given that $x > 0$ and $y > 0$.

5) $a^2b^2 + b^2c^2 + c^2a^2 \leqslant a^4 + b^4 + c^4$

6) $3x^2 + 3y^2 + 3z^2 \geqslant (x + y + z)^2$ 7) $x^2 + y^2 \geqslant \frac{1}{2}(x + y)^2$

8) $-1 \leqslant \dfrac{4x}{x^2 + 4} \leqslant 1$ 9) $(a + b)^2 \geqslant 4ab$

10) $a^4 + b^4 + c^4 + d^4 \geqslant 4abcd$

Interpret in the xy plane the meaning of $f(xy) > 0$ if $f(xy)$ is:

11) $y - x$ 12) $y - |x|$ 13) $|y| - x$ 14) $1 - x^2 + y^2$

15) $y^2 + 2x$.

Illustrate the following inequalities graphically:

16) $x^2 + y^2 \leqslant 9$ *and* $y^2 > 4x$

17) $\dfrac{(x - 1)^2}{9} + \dfrac{y^2}{4} < 1$ *and* $xy > 1$

18) $(x^2 + y^2 - 4)(x + y) \geqslant 0$

19) $(y - x)(y^2 - x^3) \geqslant 0$

20) $x(y^2 - x) < 0$

21) $(x^2 - 4y^2 - 4)(4x^2 + 9y^2 - 36) \leqslant 0$

22) $(y^2 - 4x)(xy - 1)(y - 1) \leqslant 0.$

Write down the inequalities that define the following shaded areas.

23)

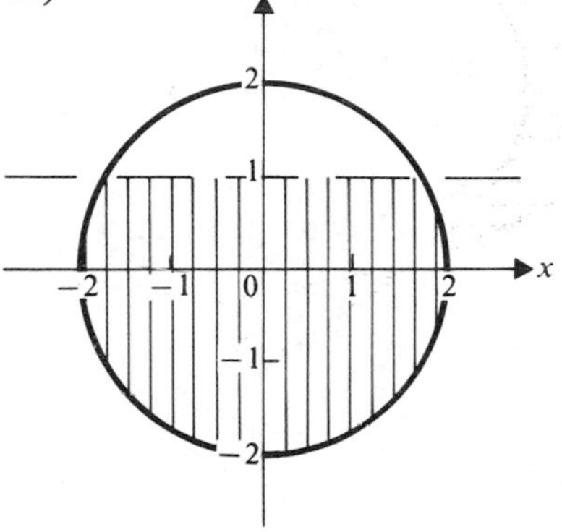

24)

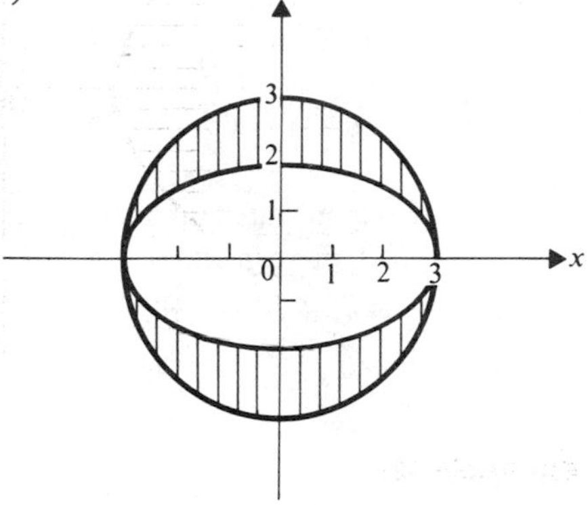

25)

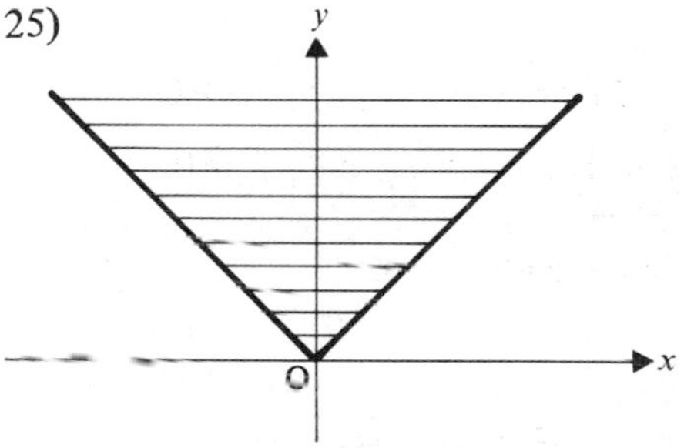

26)

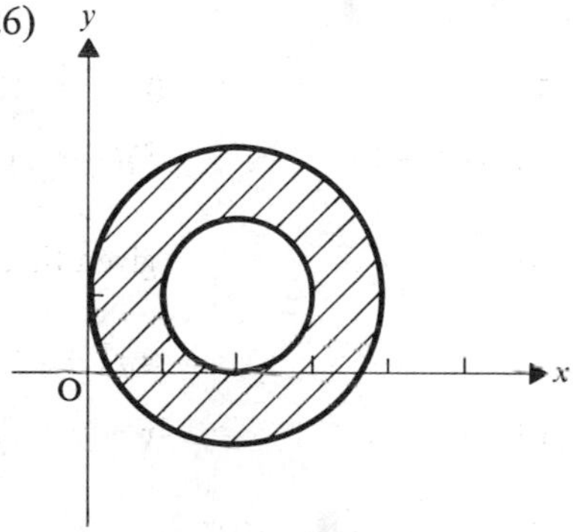

27)

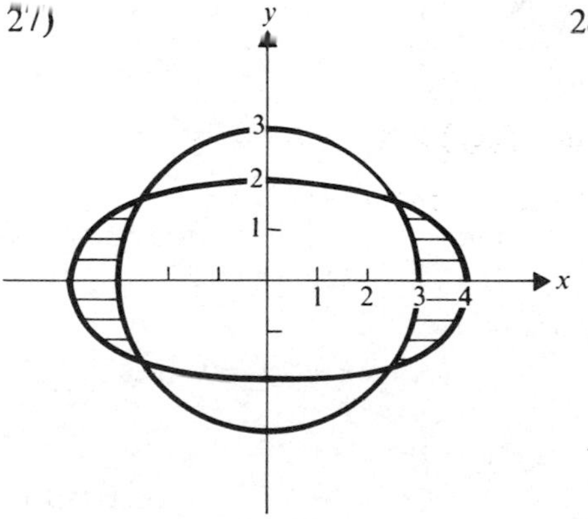

28)

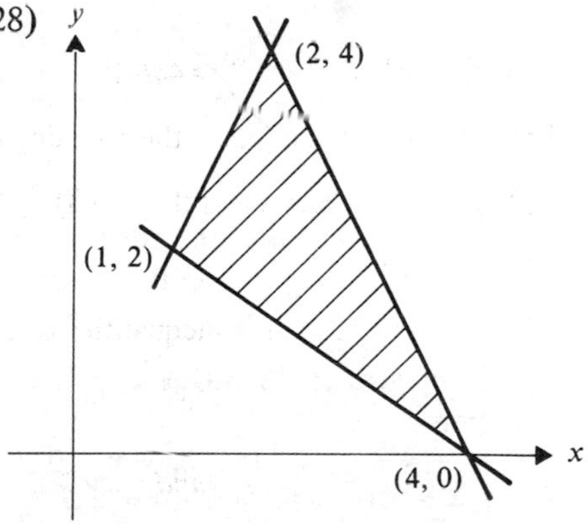

MULTIPLE CHOICE EXERCISE 12

(*Instructions for answering these questions are given on p. xii.*)

TYPE I

1) If x is a real number, $|x+1|<2$ and $2|x-2|>3$ are together equivalent to:

(a) $-3<x<\frac{1}{2}$ (b) $-3<x<-1$ (c) $-1<x<\frac{1}{2}$
(d) $-1<x<1$ (e) $x>3\frac{1}{2}$ or $x<-3$. (U of L)

2) State the line in which an error is first made in the following proof that $a^2+b^2>ab$ where a and b are real.

(a) $(a-b)^2 \equiv a^2+b^2-2ab$
(b) $(a-b)^2 \geqslant 0$
(c) $a^2+b^2 \geqslant 2ab$
(d) $2ab>ab$
(e) $a^2+b^2>ab$.

3)

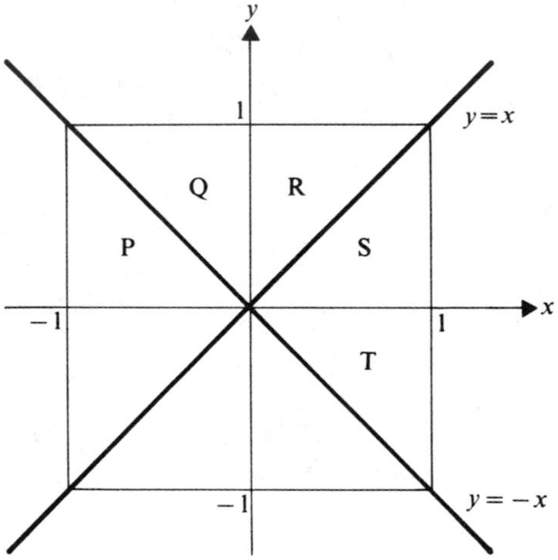

All points (x, y) satisfying both $y>|x|$ and $|y|<1$ lie in the region:
(a) P and S (b) Q and R (c) R (d) S (e) S and T. (U of L)

4) The tangent at the origin to the curve $x^2-y^2+4x+2y=0$ has equation
(a) $y+2x=0$ (b) $2y+x=0$ (c) $y=2x$
(d) $x+y=0$ (e) $x-y=0$. (U of L)

5)

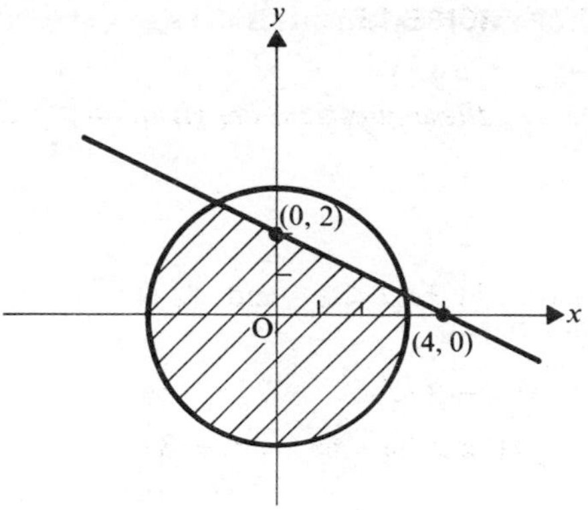

The diagram shows the circle $x^2 + y^2 = 9$ and the straight line $2y + x = 4$. The shaded region consists of those points (x, y) for which:

(a) $x^2 + y^2 - 9 < 2y + x - 4$

(b) $\begin{cases} x^2 + y^2 < 9 \\ 2y + x > 4 \end{cases}$ (c) $\begin{cases} x^2 + y^2 > 9 \\ 2y + x < 4 \end{cases}$

(d) $\begin{cases} x^2 + y^2 > 9 \\ 2y + x > 4 \end{cases}$ (e) $\begin{cases} x^2 + y^2 < 9 \\ 2y + x < 4 \end{cases}$. (U of L)

6) The necessary and sufficient conditions for a point of inflexion are:

(a) $\frac{dy}{dx} = 0$ and $\frac{d^2y}{dx^2} = 0$

(b) $\frac{d^2y}{dx^2} = 0$ and $\frac{dy}{dx} \neq 0$

(c) $\frac{dy}{dx} = 0$ and $\frac{d^2y}{dx^2}$ does not change sign

(d) $\frac{d^2y}{dx^2} = 0$ and $\frac{dy}{dx}$ does not change sign.

TYPE II

7) The equation of the tangent at the origin to a curve $f(xy) = 0$ can be found if:

(a) $f(xy) \equiv x^2 + y^2 - x + 2y$ (b) $f(xy) \equiv 2x^2 + y^2 + 4x$

(c) $f(xy) \equiv x^2 - y^2 + 4x + 1$ (d) $f(xy) \equiv 2x^2 + 3y^2$.

8) In which of the following statements is the sign $\Leftrightarrow$ correctly used?
(a) $(|x| > |y|) \Leftrightarrow (x^2 > y^2)$
(b) $(x^2 + y^2 > 2xy) \Leftrightarrow (x > y)$
(c) $(x^2 = 4) \Leftrightarrow (x = 2)$. (U of L)

TYPE III

9) a, b, c are real numbers.
(a) $ab < ac$.
(b) $b < c$. (U of L)

10) (a) $0 < x < 1$.
(b) $|2 - x| < 2$.

11) (a) P is a point of inflexion on a curve $y = f(x)$.
(b) At P, $\dfrac{d^2y}{dx^2} = 0$.

MISCELLANEOUS EXERCISE 12

1) Sketch the cycloid $x = t - \sin t$, $y = 1 - \cos t$ for $0 \leqslant t \leqslant 2\pi$, showing that it is symmetrical about the line $x = \pi$. Find the centroid of the area contained between this curve and the x axis. (AEB'73)p

2) Write down the equation of the tangent at the origin to the curve $y = 4x - x^3$. Sketch the curve and this tangent for $-2 < x < 2$. (U of L)

3) Find the points of inflexion of the curve $y = \dfrac{x}{x^2+1}$, and show that they lie on a straight line. (O)p

4) Find the (finite) points of inflexion of the curve $y = \dfrac{1}{1+x+x^2}$. Find also the point of intersection of the tangents at these points. (O)

5) Find the coordinates of all the points of inflexion of the graph of the function

$$y = x - \sin x$$

in the range $0 \leqslant x \leqslant 4\pi$. Sketch the graph in this range. (JMB)

6) Find the points of inflexion of the curve

$$y = \frac{4(x+3)}{x^2+6x+12}$$

and show that they lie on a straight line. (O)

7) Find the values of x for which the function

$$f(x) = e^x(2x^2 - 3x + 2)$$

has (a) a maximum, (b) a minimum, (c) an inflexion.
Draw a rough sketch of the graph of the function. (O)

8) If $y^2 = x^2(x-2)$, obtain an expression for dy/dx in terms of x, and hence show that on the graph of y against x there are no turning points. Show that when $x = 2\frac{2}{3}$, $dy/dx = \pm\sqrt{6}$ and $d^2y/dx^2 = 0$.
Sketch the form of the graph of y against x, paying special attention to the point $(2, 0)$ and to the points where $d^2y/dx^2 = 0$. (C)

9) A region R of the plane is defined by $y^2 - 4ax \leqslant 0$, $x^2 - 4ay \leqslant 0$, $x + y - 3a \leqslant 0$. Find the area of R. (O)p

10) Sketch the region in the xy plane within which all three of the following inequalities are satisfied:
(a) $y < x + 1$, (b) $y > (x-1)^2$, (c) $xy < 2$.
Determine the area of this region. (JMB)

11) Show that, for real x,

$$-1 \leqslant \frac{2x}{x^2+1} \leqslant 1.$$ (U of L)

12) Sketch the three lines whose equation are

$$y - x - 6 = 0, \quad 2y + x - 18 = 0, \quad 2y - x = 0.$$

Shade on your diagram the domain defined by

$$y - x - 6 < 0, \quad 2y + x - 18 < 0, \quad 2y - x > 0.$$ (U of L)

13) Shade the region or regions of the xy plane within which $(y^2 - 8x)(x^2 + y^2 - 9)$ is negative. (U of L)

14) Show, by shading on a sketch of the xy plane, the region for which $x^2 + y^2 \leqslant 1$, $y \geqslant x$ and $y \leqslant x + 1$.
Hence find
(a) the greatest value of y,
(b) the least value of $x + y$
for which these inequalities hold.

15) The equation of a curve is $x^2y^2 - x^2 + y^2 = 0$.
(a) Find the equations of the tangents at the origin.
(b) Find the equations of the real asymptotes.
(c) Show that the numerical value of y is never greater than the corresponding value of x.
(d) Show that the numerical value of y is always less than unity.
(e) Sketch the curve. (AEB '71)

16) Sketch on the same diagram the line $x + y = 1$ and the parabola $y^2 = 4x$.
Shade on your sketch the regions of the xy plane for which

$$(x + y - 1)(y^2 - 4x) > 0.$$ (U of L)

17) Sketch the curve whose equation is $y = f(x)$, where

$$f(x) = \frac{x(x-2)}{(x-1)(x-3)}$$

obtaining the equation of each asymptote and showing how the curve approaches its asymptotes. Indicate also the coordinates of the points where the curve intersects the axes.
Find the sets of values of x for which
(a) $f(x) > 0$, (b) $f(x) > 1$.
Sketch also, on a separate diagram, the curve whose equation is $y = |f(x)|$.
(U of L)

18) A curve has the equation

$$x^2 - y^2 + 10x - 2y = 0$$

Write down the equation of the tangent to the curve at the origin and find the angle between the asymptotes of the curve. (U of L)

19) Show that, for all positive values of a and b,

$$a^3 + 2b^3 \geqslant 3ab^2.$$ (U of L)

20) Determine the set of values of λ for which the equation

$$\lambda(x^2 - x + 1) = (x^2 + x + 1)$$

has real roots.
Sketch the graph of $y = \dfrac{x^2 + x + 1}{x^2 - x + 1}$ indicating clearly the asymptote and how the curve approaches its asymptote.
Sketch, on a separate diagram, the curve whose equation is

$$y = \frac{x^2 - x + 1}{x^2 + x + 1}.$$ (U of L)

CHAPTER 13

COORDINATE GEOMETRY

CONIC SECTIONS

The circle and the parabola were introduced in *The Core Course*. These curves, together with the ellipse, hyperbola and a pair of straight lines, are collectively known as conic sections. The reason for this is seen by considering a double cone (obtained by rotating a straight line through one revolution about an axis that intersects the line at an angle α say). When a plane cuts this double cone, the shape of the section formed depends upon the inclination θ of the plane to the axis.

If $\theta = \frac{\pi}{2}$, the plane cuts only one half of the double cone and the cross-section is a circle.

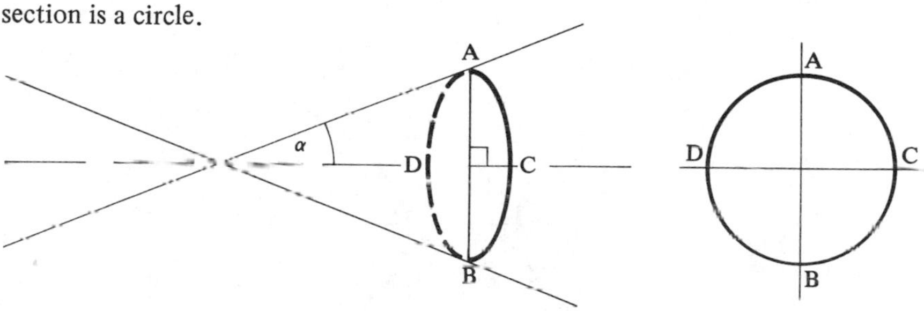

If $\alpha < \theta < \frac{\pi}{2}$, the plane again cuts only one half of the cone and the cross-section is an ellipse.

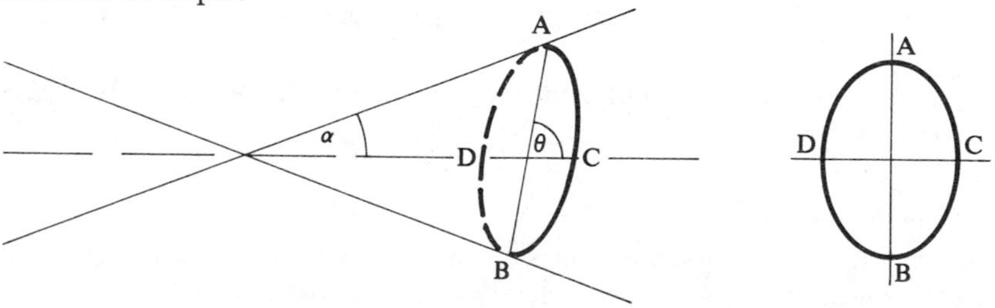

If $\theta = \alpha$, the plane is parallel to a generator of the cone and hence cuts only one half of the double cone in a section which is open-ended. This section is a parabola.

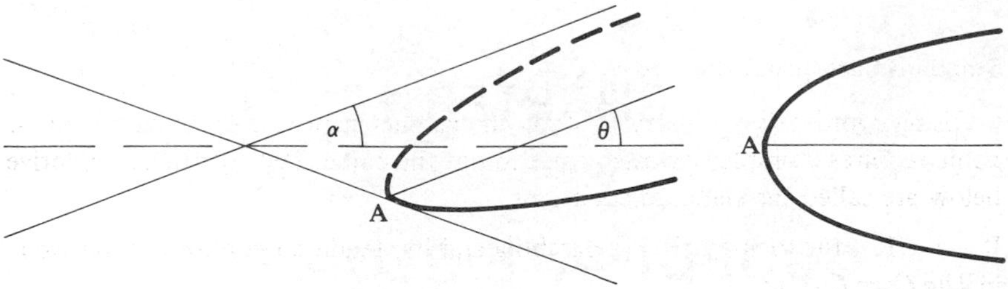

If $\theta < \alpha$, the plane cuts into both halves of the double cone, producing (unless the plane passes through the vertex) a section comprising two open-ended curves. This section is called a hyperbola.

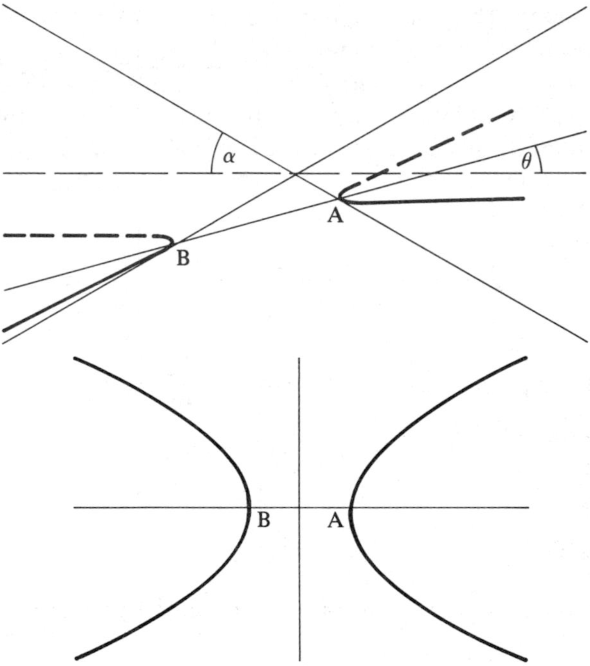

If $\theta < \alpha$ and the plane *does* pass through the vertex, the section is a pair of straight lines.
We now have the complete set of conic sections (often referred to simply as *the conics*).

Conic Sections as Loci

Another factor linking the members of this group of curves is that each is the locus of a set of points, P, obeying the same basic restriction, which is that:

the distances of P from a fixed point (the focus) and a fixed straight line (the directrix) are in a constant ratio e (the eccentricity).

Different values of e correspond to different conics.

Standard Cartesian Equations

Using coordinate geometry to study the geometric properties of each type of conic requires a simple Cartesian equation of the curve. The equations we derive below are called the standard equations.

If $e = 1$, the locus of P is a parabola and its standard equation was derived in *The Core Course*.

If $e \neq 1$, and if the point $S(ae, 0)$ is the focus, the line $x = a/e$ is the directrix and $P(x, y)$ is any point

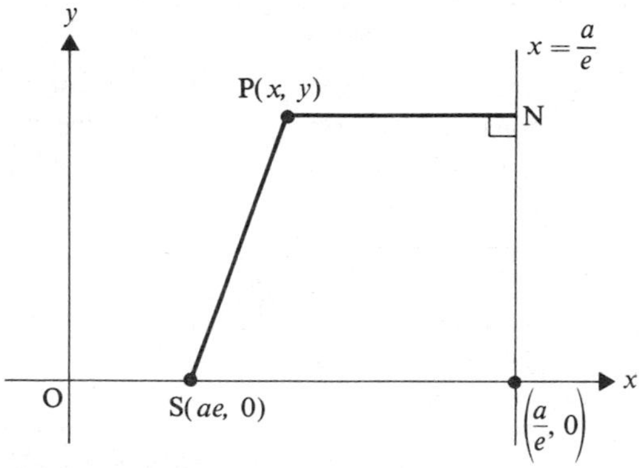

then $P(x, y)$ is a point on a conic $\iff$ $\dfrac{PS}{PN} = e.$

Now $PS^2 = (x - ae)^2 + (y - 0)^2$

and $PN = \dfrac{a}{e} - x.$

But $PS^2 = e^2 PN^2$

so $P(x, y)$ is on the conic $\iff$ $(x - ae)^2 + y^2 = e^2\left(\dfrac{a}{e} - x\right)^2$

Hence $$x^2(1 - e^2) + y^2 = a^2(1 - e^2)$$

$$\Rightarrow \qquad \frac{x^2}{a^2} + \frac{y^2}{a^2(1 - e^2)} = 1 \qquad [1]$$

If $0 < e < 1$, $a^2(1 - e^2) > 0$ and so can be replaced by b^2 to give

$$\frac{x^2}{a^2} + \frac{y^2}{b^2} = 1$$

which is the standard equation of an ellipse.

If $e = 0$, equation [1] becomes $x^2 + y^2 = a^2$ which is the equation of a circle.
If $e > 1$, $a^2(1 - e^2) < 0$ and so can be replaced by $-b^2$ to give

$$\frac{x^2}{a^2} - \frac{y^2}{b^2} = 1$$

which is the standard equation of a hyperbola.
So there is both a geometric and a definitive relationship linking the conic sections. But each member of the set has certain unique characteristics and properties. Some of these were dealt with in *The Core Course* and these the reader is recommended to revise at this stage.
We will now investigate individual conics.

THE PARABOLA

As all parabolas have the same geometric properties, we usually study one with a simple Cartesian equation. This standard parabola has its vertex at the origin, its focus, S, on the x axis at a point $(a, 0)$ and the line $x = -a$ as its directrix. We saw in *The Core Course* that:

(a) its Cartesian equation is $y^2 = 4ax$,
(b) its parametric equations are $x = at^2$, $y = 2at$,
(c) the gradient at the point $(at^2, 2at)$ is $\dfrac{1}{t}$,
(d) the parametric equation of the tangent at the point $(at^2, 2at)$ is $ty = x + at^2$.

Optical Property of a Parabola

Any ray of light parallel to the axis of a parabolic mirror is reflected through the focus. This property, which is of considerable practical use in optics, can be proved by showing that the normal at any point P on a parabola, bisects the angle between PS and the line PQ that is parallel to the axis of the parabola (the angle of incidence and the angle of reflection are equal).

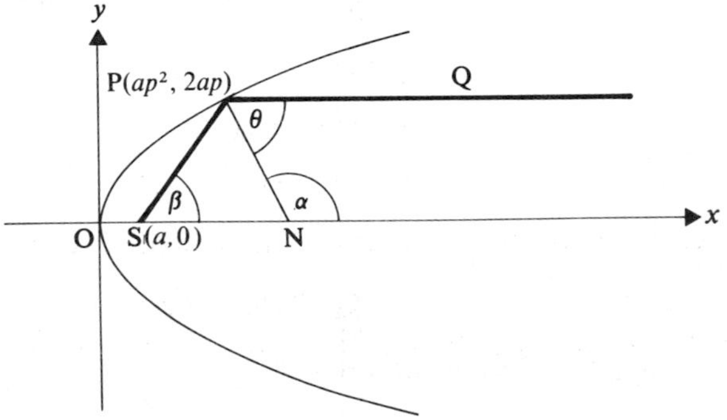

In the diagram PN is the normal at $P(ap^2, 2ap)$,

so $\tan\alpha = -p$

and $\tan\theta = \tan(\pi-\alpha) = p$

Also $\tan\beta = \dfrac{2ap}{ap^2-a} = \dfrac{2p}{p^2-1}$

so $\tan \text{QPS} = \tan(\pi-\beta) = \dfrac{2p}{1-p^2} = \tan 2\theta$

Hence $\angle\text{QPS} = 2(\angle\text{QPN})$

i.e. PN bisects $\angle$QPS.

Focal Chord Properties

A line joining two points P and Q on a parabola is a focal chord if PQ passes through the focus S. The focal chord that is perpendicular to the axis of the parabola is called the *latus rectum*.

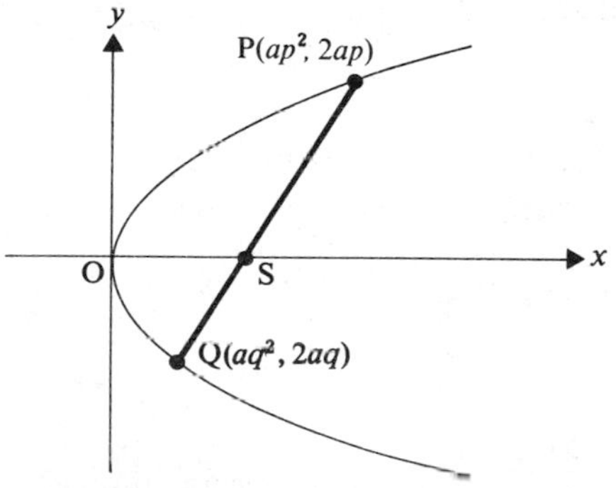

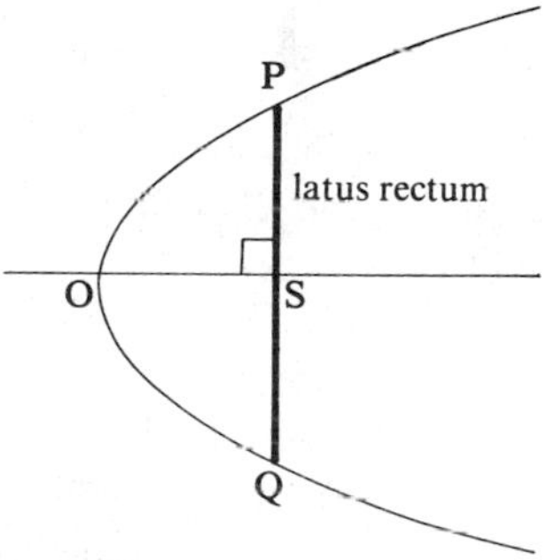

If PSQ is a straight line, the gradients of QS and PS are equal. So, taking $(ap^2, 2ap)$ and $(aq^2, 2aq)$ as the coordinates of P and Q, we have

$$\frac{2aq}{aq^2-a} = \frac{2ap}{ap^2-a}$$

$$\Rightarrow \quad q(p^2-1) = p(q^2-1)$$

$$\Rightarrow \quad pq(p-q) = -(p-q)$$

$$\Rightarrow \quad pq = -1 \quad (\text{as } p \neq q)$$

This property of the parametric values at the ends of a focal chord is quotable.

Now consider the tangents at P and at Q. Their gradients are $\dfrac{1}{p}$ and $\dfrac{1}{q}$ respectively.

But if $pq = -1$, then $\left(\dfrac{1}{p}\right)\left(\dfrac{1}{q}\right) = -1$ showing that the tangents at P and Q are perpendicular to each other.
Further, the point R where these tangents intersect can be found by solving simultaneously the equations

$$\begin{cases} py = x + ap^2 & \text{(tangent at P)} \\ qy = x + aq^2 & \text{(tangent at Q)} \end{cases}$$

giving $\{apq, a(p+q)\}$ as the coordinates of R.

But if $pq = -1$, $apq = -a$ showing that R is a point on the directrix $(x = -a)$.
So we see that, if PQ is a focal chord,

(1) $pq = -1$,

(2) the tangents at P and Q are perpendicular,

(3) the tangents at P and Q meet on the directrix.

Note. The tangents at *any* two points P and Q on the parabola meet where

$$x = apq \quad \text{and} \quad y = a(p+q).$$

The reader may observe that these coordinates are the geometric and arithmetic mean respectively of the coordinates of P and Q, which provides an interesting aid to memory.

Diameter of a Parabola

We usually think of a diameter as existing only in a closed curve (e.g. circle or ellipse) but, defining a diameter as 'the locus of midpoints of a set of parallel chords when this locus is a straight line' we find that a parabola has diameters.

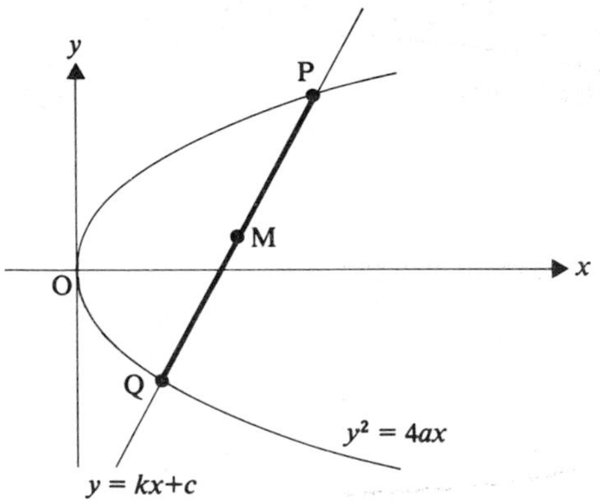

Consider any chord of fixed gradient k and thus with equation

$$y = kx + c$$

This chord meets the parabola $y^2 = 4ax$ at points P and Q whose y coordinates are the roots of the equation

$$y = k\left(\frac{y^2}{4a}\right) + c$$

or $$ky^2 - 4ay + 4ac = 0 \qquad [1]$$

If these roots are y_P and y_Q then, at the midpoint, M, of PQ,

$$y_M = \tfrac{1}{2}(y_P + y_Q),$$

i.e.

$$y_M = \tfrac{1}{2}(\text{sum of roots of equation [1]})$$

$$= \tfrac{1}{2}\left(\frac{4a}{k}\right)$$

$$= \frac{2a}{k}$$

But k is constant, so the y coordinate of M is fixed.
That is, as P and Q move on the parabola, the gradient of PQ remaining constant, the locus of the midpoint of PQ is a fixed horizontal line with equation $y = \dfrac{2a}{k}$.

Thus we see that diameters of a parabola are straight lines parallel to the axis of the parabola.

The following exercise makes use of the work on tangents and normals covered in *The Core Course* together with the methods used and properties derived in this chapter.

EXAMPLES 13a

1) The points $P(ap^2, 2ap)$, $Q(aq^2, 2aq)$ on the parabola $y^2 = 4ax$ are such that PQ passes through the focus of the parabola. Show that the locus of the midpoint of PQ is another parabola and find the coordinates of its focus.

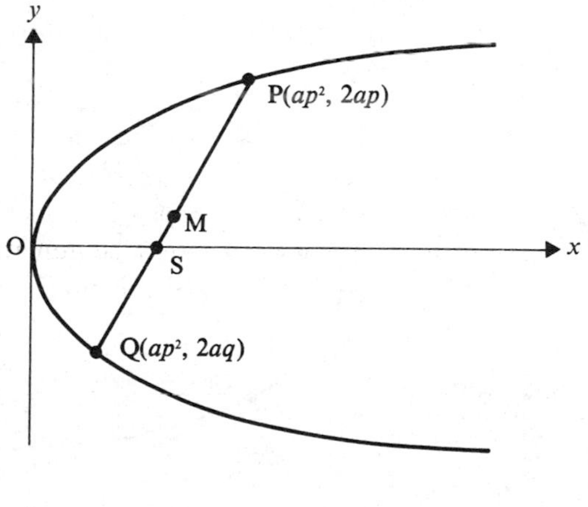

As PQ is a focal chord, $pq = -1$ [1]

At M
$$\begin{cases} x = \tfrac{1}{2}(ap^2 + aq^2) \;\Rightarrow\; \dfrac{2x}{a} = p^2 + q^2 & [2] \\ y = \tfrac{1}{2}(2ap + 2aq) \;\Rightarrow\; \dfrac{y}{a} = p + q & [3] \end{cases}$$

To identify the locus of M we need its Cartesian equation so we must eliminate p and q from equations [2] and [3].

[3] $\Rightarrow$
$$\left(\frac{y}{a}\right)^2 = p^2 + 2pq + q^2$$

Using [1] and [2] we have
$$\left(\frac{y}{a}\right)^2 = \frac{2x}{a} - 2$$

$\Rightarrow$
$$y^2 = 2a(x - a) \quad [4]$$

and this is the Cartesian equation of M.

Using the transformation $Y = y$, $X = x - a$ on [4] gives
$$Y^2 = 2aX$$

which is a parabola whose focus is the point where
$$X = \frac{a}{2}, \quad Y = 0$$

$\Rightarrow$
$$x = \frac{3a}{2}, \quad y = 0$$

2) P and Q are two points on a parabola whose focus is S. The tangents to the parabola at P and Q meet at T. Prove that
$$TS^2 = PS \cdot QS$$

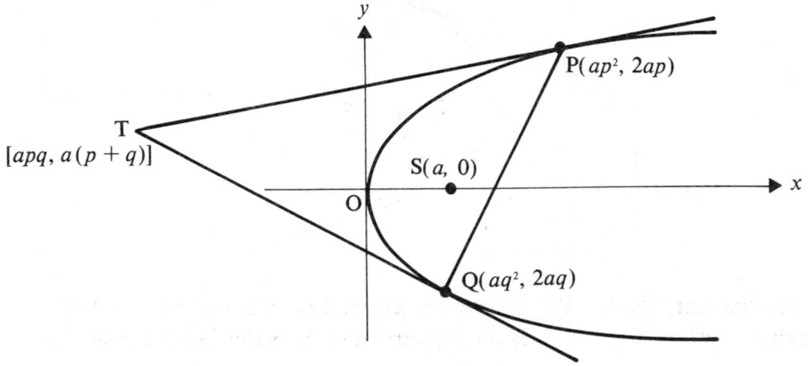

Using the standard equation, $y^2 = 4ax$, for the parabola and taking P as the point $(ap^2, 2ap)$ and Q as the point $(aq^2, 2aq)$ then as S is the point $(a, 0)$ and T is the point $[apq, a(p+q)]$

we have

$$\begin{aligned} TS^2 &= (apq-a)^2 + [a(p+q)-0]^2 \\ &= a^2[p^2q^2 + 1 + p^2 + q^2] \\ &= a^2(p^2+1)(q^2+1) \\ PS^2 &= (ap^2-a)^2 + (2ap-0)^2 \\ &= a^2[p^4 + 2p^2 + 1] \\ &= a^2(p^2+1)^2 \end{aligned}$$

Similarly $QS^2 = a^2(q^2+1)^2$.

So $$PS^2.QS^2 = a^4(p^2+1)^2(q^2+1)^2$$

$\Rightarrow$ $$PS.QS = a^2(p^2+1)(q^2+1) = TS^2$$

EXERCISE 13a

1) The tangent at a point P on the parabola $y^2 = 4ax$ meets the directrix at Q. The line through Q parallel to the axis of the parabola meets the normal at P at the point R. Find the equation of the locus of R. Show that the locus is another parabola and find its vertex.

2) Prove that the line $x - ty + at^2 = 0$ touches the parabola $y^2 = 4ax$ for all values of t. Find the coordinates of the point of contact.

3) The tangent at P to the parabola $y^2 = 4ax$ meets the x axis at T. Prove that $PS = TS$ where S is the focus.

4)

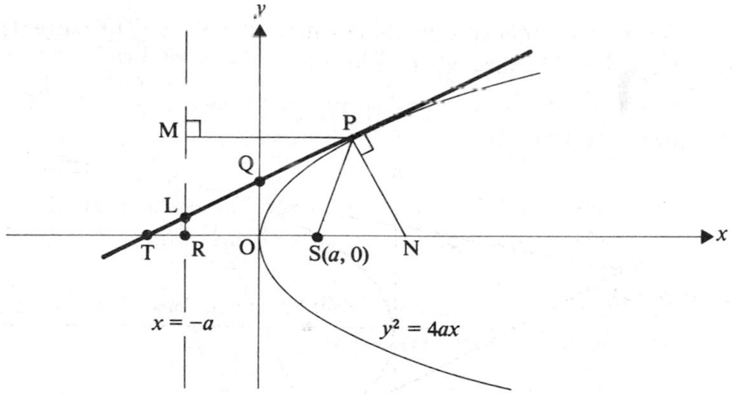

In the diagram, PT and PN are the tangent and normal respectively to the parabola $y^2 = 4ax$. MR is the directrix and S is the focus. Prove that

(a) PT bisects $\angle$MPS,
(b) SPMT is a parallelogram,
(c) the perpendicular from S to the tangent PT meets PT at Q,
(d) the line through L, perpendicular to PL, touches the parabola,
(e) $(SQ)^2 = (SO)(ST)$,
(f) if a line PV is drawn so that $\angle NPS = \angle SPV$, then MPV is a straight line.

5) PQ is a focal chord of the parabola $y^2 = 4ax$, and O is the origin. Using parametric coordinates $(ap^2, 2ap)$ for P, find the coordinates of the centroid, G, of triangle OPQ and hence find the equation of the locus of G as PQ varies.

6) If $A(h, k)$ lies on the normal at $P(at^2, 2at)$ to the parabola $y^2 = 4ax$, show that $at^3 + (2a - h)t - k = 0$.
By considering the possible number of real roots of this cubic equation for t, show that, for certain points A, three normals can be drawn to the parabola, while from a different set of points A, only one normal can be drawn. Illustrate your deduction on a diagram.

7) If the normal at $P(at^2, 2at)$ to the parabola $y^2 = 4ax$ meets the curve again at $Q(aq^2, 2aq)$, prove that $p^2 + pq + 2 = 0$. The tangents at P and Q intersect at R. If the line through R parallel to the axis of the parabola meets the parabola at T, show that PT is a focal chord.

8) PQ is a focal chord of the parabola with parametric equations $x = at^2$, $y = 2at$.
PR is another chord that meets the x axis at the point $(ka, 0)$. If RQ produced meets the axis at T, prove that $RT = kQT$.

9) $P(ap^2, 2ap)$ and $Q(aq^2, 2aq)$ are two variable points on the parabola $y^2 = 4ax$. If PQ subtends a right angle at the origin prove that $pq = -4$.

(a) Show that PQ passes through a fixed point on the axis of the parabola.
(b) The tangents at P and Q meet at R. Find the equation of the locus of R.

10) A variable tangent to the parabola $y^2 = 4x$ meets the parabola $y^2 = 8x$ at P and Q. Find the equation of the locus of the midpoint of PQ.

11) Find the equation of the locus of the midpoints of focal chords of the parabola with parametric equations $x = at^2$, $y = 2at$.

12) P and Q are the points of contact of the tangents drawn from a point R to a parabola with focus S and M is the midpoint of PQ.
Prove that $RP.RQ = 2RS.RM$.

THE ELLIPSE

We saw on page 485 that the standard Cartesian equation of an ellipse is

$$\frac{x^2}{a^2}+\frac{y^2}{b^2} = 1$$

where $\quad b^2 = a^2(1-e^2)$

The shape and position of the locus can be deduced by rearranging the equation as follows:

(1) $$x^2 = \frac{a^2}{b^2}(b^2-y^2)$$

$$x^2 \geqslant 0 \quad \text{so} \quad b^2-y^2 \geqslant 0$$

$$\Rightarrow \quad (b-y)(b+y) \geqslant 0$$

Hence $\quad -b \leqslant y \leqslant b.$

Also, since $\quad x = \pm\frac{a}{b}\sqrt{b^2-y^2},\quad$ the curve is symmetrical about the y axis.

(2) $$y^2 = \frac{b^2}{a^2}(a^2-x^2)$$

$$y^2 \geqslant 0 \quad \text{so} \quad a^2-x^2 \geqslant 0$$

$$\Rightarrow \quad (a-x)(a+x) \geqslant 0$$

Hence $\quad -a \leqslant x \leqslant a.$

Also, since $\quad y = \pm\frac{b}{a}\sqrt{a^2-x^2},\quad$ the curve is symmetrical about the x axis.

(3) When $\begin{cases} x = 0, & y = \pm b \\ y = 0, & x = \pm a \end{cases}$

(4) $\dfrac{dy}{dx} = -\dfrac{b^2x}{a^2y}$ (implicit differentiation).

Thus $\begin{cases} \text{when } x = 0 & \dfrac{dy}{dx} = 0 \\ \text{when } y = 0 & \dfrac{dy}{dx} = \infty \end{cases}$

It is now possible to sketch the ellipse.

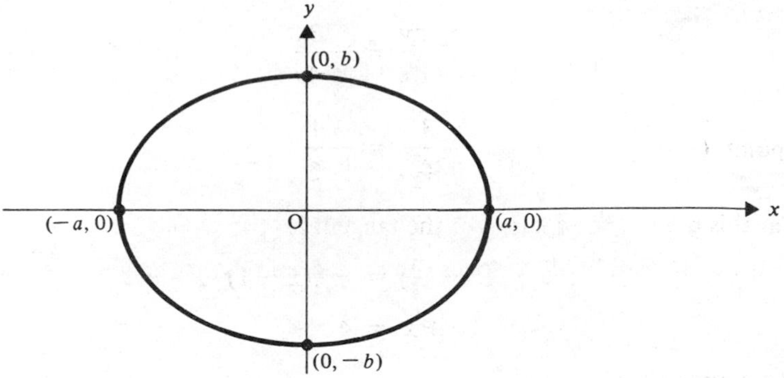

The ellipse is symmetrical about both the x axis and the y axis and so has two symmetrical foci and directrices.

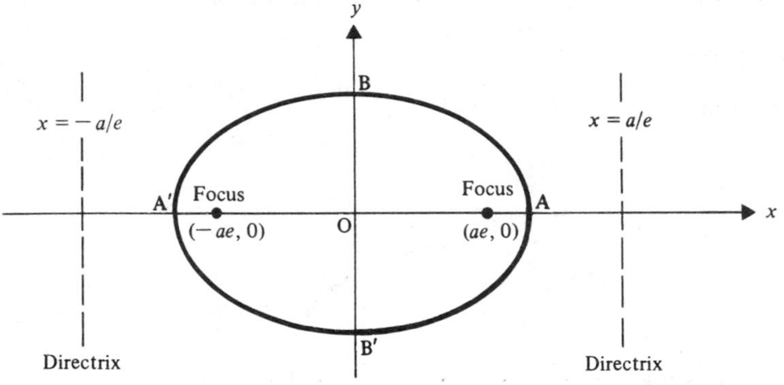

The line A′A is called the *major axis* of the ellipse and is of length $2a$ and B′B is called the *minor axis* of the ellipse and is of length $2b$.

The midpoint of AA′ (or BB′) is called the *centre* of the ellipse. (In this case the centre of the ellipse is the origin.)

Tangent and Normal

The gradient of an ellipse can be found by differentiating its equation. For example the equations of the tangent and normal at the point $(1, \frac{3}{2})$ to the ellipse $\frac{x^2}{4} + \frac{y^2}{3} = 1$ can be found as follows.

Differentiating we have $$\frac{2x}{4} + \frac{2y}{3}\frac{dy}{dx} = 0$$

or
$$\frac{dy}{dx} = -\frac{3x}{4y}$$

At the point $(1, \frac{3}{2})$, $\quad \frac{dy}{dx} = -\frac{3}{4(\frac{3}{2})} = -\frac{1}{2}$

Hence, at this point, the equation of the tangent is
$$(y - \tfrac{3}{2}) = -\tfrac{1}{2}(x - 1)$$
$$\Rightarrow \quad 2y + x = 4$$

and the equation of the normal is
$$(y - \tfrac{3}{2}) = 2(x - 1)$$
$$\Rightarrow \quad 2y = 4x - 1$$

Note. Comparing the given equation with the standard form $\frac{x^2}{a^2} + \frac{y^2}{b^2} = 1$ we see that $a^2 = 4$ and $b^2 = 3$.

Thus the given ellipse can be sketched.

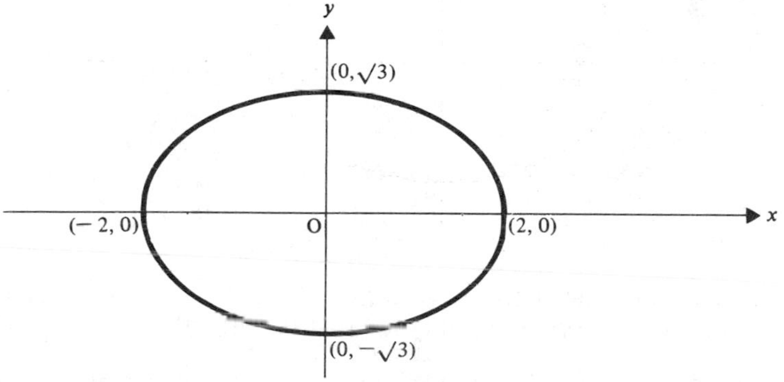

General Ellipse

Although all ellipses have geometric properties in common and we would usually use the standard equation of the ellipse to study these, it is useful to be able to recognize the equation of an ellipse when the major and minor axes are not in their standard positions.
If the two foci lie on the y axis then the major axis of the ellipse is vertical and the Cartesian equation is then
$$\frac{x^2}{b^2} + \frac{y^2}{a^2} = 1$$

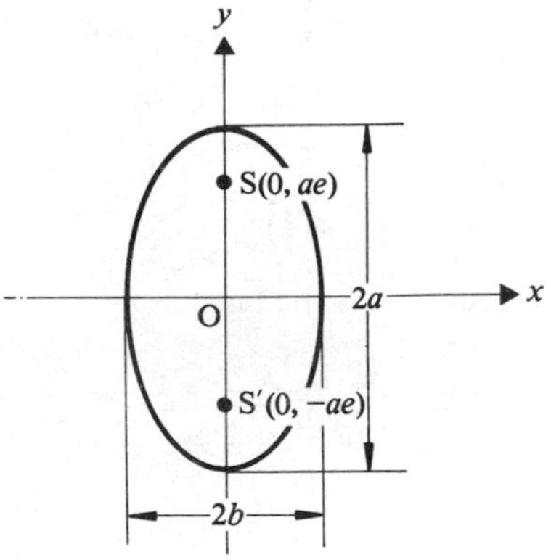

Note. If $a = b$ the equation of the standard ellipse becomes $x^2 + y^2 = a^2$ which is the equation of a circle.
In this case the eccentricity is given by $a^2 = a^2(1 - e^2) \Rightarrow e = 0$.
This confirms that the conic section with zero eccentricity is a circle.
The centre, foci and axes lengths of any ellipse with horizontal and vertical axes can be identified by comparing its equation with that of a standard ellipse

$$\frac{X^2}{a^2} + \frac{Y^2}{b^2} = 1$$

whose centre is given by $X = 0$, $Y = 0$
and whose axes lengths are $2a$ and $2b$.
For example, given an ellipse with equation

$$\frac{(x-1)^2}{25} + \frac{y^2}{16} = 1$$

comparison with $$\frac{X^2}{a^2} + \frac{Y^2}{b^2} = 1$$

shows that $X = x - 1$, $Y = y$, $a = 5$, $b = 4$.

So the centre $(X = 0, Y = 0)$ is $(1, 0)$.

The major axis is horizontal and of length 10 and the minor axis is of length 8. From $b^2 = a^2(1 - e^2)$ we see that $e = \frac{3}{5}$. The foci, which are always on the major axis, are each distant ae $(= 3)$ from the centre, i.e. at $(4, 0)$ and $(-2, 0)$.

The directrices are perpendicular to the major axis and distant $\frac{a}{e}\left(= \frac{25}{3}\right)$ from the centre. So their equations are $x = \frac{28}{3}$ and $x = -\frac{22}{3}$.

The given ellipse can now be sketched.

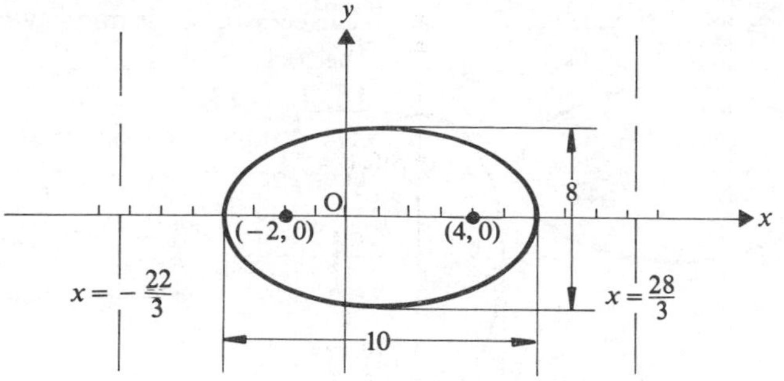

Note that $\dfrac{(x-1)^2}{25}+\dfrac{y^2}{16}=1$ is transformed into $\dfrac{X^2}{25}+\dfrac{Y^2}{16}=1$ by moving the origin to the point $(1, 0)$ and using a Y axis moved to the position $x = 1$.

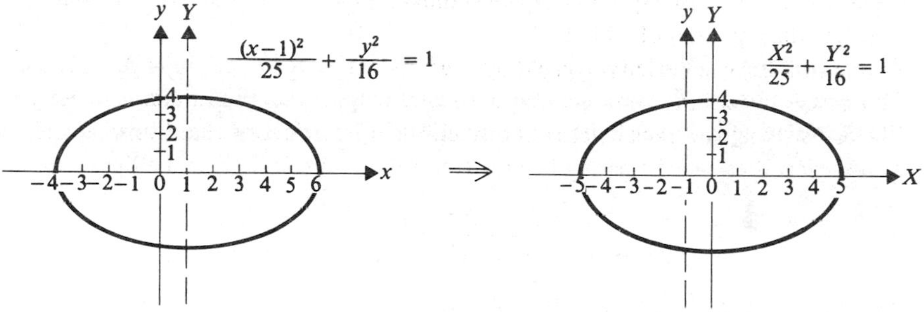

In general, the equation $\dfrac{(x-p)^2}{a^2}+\dfrac{(y-q)^2}{b^2}=1$ can be transformed into the form $\dfrac{X^2}{a^2}+\dfrac{Y^2}{b^2}=1$ by moving the axes to $x = p$ and $y = q$.

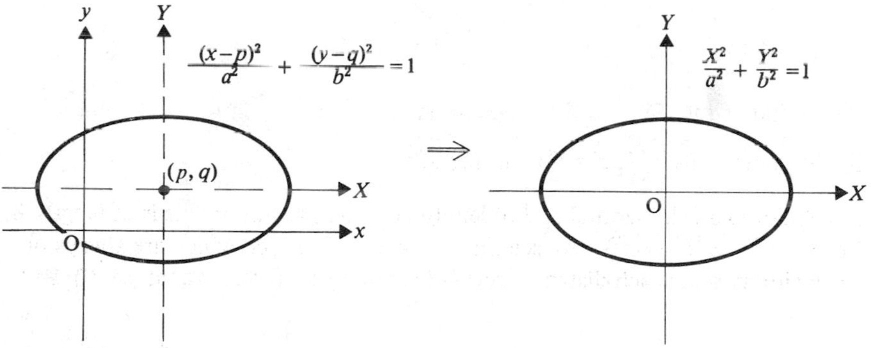

EXAMPLES 13b

1) An ellipse has its foci at the points $(-1, 0)$ and $(7, 0)$ and its eccentricity is $\frac{1}{2}$. Find its Cartesian equation.

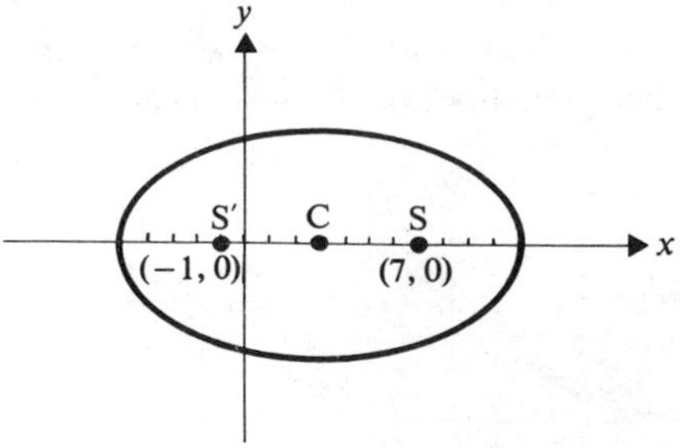

The centre, C, is midway between the foci,
i.e. C is $(3, 0)$.
The distance between the centre and each focus is ae,

i.e. $\quad CS = 4 = \frac{1}{2}a$

$\Rightarrow \quad a = 8$

Now using $b^2 = a^2(1-e^2)$, we have

$$b^2 = 64(1-\tfrac{1}{4}) = 48$$

so the equation of the ellipse is

$$\frac{(x-3)^2}{64} + \frac{y^2}{48} = 1$$

2) An ellipse has its foci on the y axis and its centre at the origin. The distance between the foci is 6 and the major axis is of length 10. Find the eccentricity, the Cartesian equation and the equations of the directrices of this ellipse.

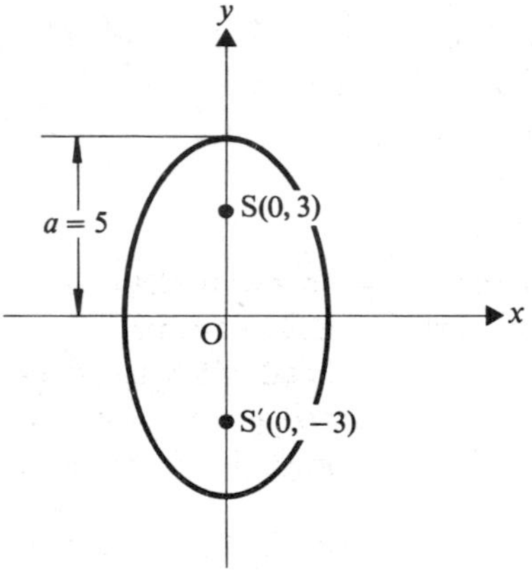

The centre of the ellipse is the origin so the foci are the points $(0, 3)$ and $(0, -3)$ and $ae = 3$.
But $a = 5 \Rightarrow e = \frac{3}{5}$.

Using $b^2 = a^2(1-e^2)$ gives $b^2 = 25(1-\frac{9}{25}) = 16$.
As the major axis is vertical, the Cartesian equation is of the form $\dfrac{x^2}{b^2} + \dfrac{y^2}{a^2} = 1$,

i.e. $$\frac{x^2}{16} + \frac{y^2}{25} = 1$$

The directrices are horizontal and distant $\dfrac{a}{e}$ from O.

So their equations are $y = \pm\frac{25}{3}$.

EXERCISE 13b

Write down the values of a and b and hence sketch the ellipse in each of the following cases.

1) $\dfrac{x^2}{9}+\dfrac{y^2}{4}=1$

2) $\dfrac{x^2}{16}+\dfrac{y^2}{9}=1$

3) $x^2+4y^2=4$

4) $4x^2+9y^2=36$

5) $x^2+25y^2=1$

Find the equation of the tangent and the normal to the following ellipses at the points specified.

6) $\dfrac{x^2}{9}+\dfrac{y^2}{4}=1;\qquad \left(2,\ \frac{2}{3}\sqrt{5}\right)$

7) $\dfrac{x^2}{16}+\dfrac{y^2}{9}=1;\qquad \left(\sqrt{7},\ \frac{9}{4}\right)$

8) $x^2+4y^2=4;\qquad \left(\sqrt{3},\ \frac{1}{2}\right)$

9) Write down the equations of the following ellipses.
(a) Centre 0, major axis horizontal and of length 6, minor axis of length 4.
(b) Centre $(0, 1)$, major axis vertical and of length 8, minor axis of length 4.
(c) Centre $(2, 0)$, major axis horizontal and of length $4a$, eccentricity $\frac{1}{2}$.
(d) Centre $(3, 4)$, minor axis horizontal and of length 6, eccentricity $\frac{1}{4}\sqrt{7}$.

10) Find the coordinates of the centre, the lengths of the axes and the eccentricity of the following ellipses.

(a) $\dfrac{x^2}{9}+\dfrac{y^2}{16}=1$ (b) $\dfrac{x^2}{4}+(y-1)^2=1$

(c) $16x^2+25y^2=400$ (d) $16x^2+25y^2=16$

(e) $x^2-4x+4y^2=12$ (*Hint*: Complete the square on the x terms.)

(f) $b^2(x-a)^2+a^2(y-b)^2=a^2b^2,\quad a>b$

(g) $\dfrac{(x+3)^2}{3}+\dfrac{(y-4)^2}{2}=1.$

11) Find the coordinates of the foci and the equations of the directrices of each of the ellipses in Question 1. Sketch each ellipse.

12) An ellipse has its centre at the origin and its foci on the x axis. The distance between the foci is 3 cm and the distance between the directrices is 12 cm. Find the eccentricity and the length of the major axis. Hence find the Cartesian equation of the ellipse.

13) Use the locus definition of an ellipse to find the equations of the following ellipses.
(a) A focus is at $(3, 4)$, the corresponding directrix is the line $x+y=1$ and the eccentricity is $\frac{1}{2}$.
(b) The set of points that are three times as far from the x axis as they are from the point $(0, 2)$.

The Focal Distance Property

If S and S' are the foci of an ellipse whose major axis is of length $2a$ and P is any point on the ellipse, then

$$PS + PS' = 2a$$

This property provides the following simple method of drawing an ellipse. Take a piece of string of length $2a$ and fix its ends to two points (S and S′) whose distance apart is less than $2a$. A pencil (P) placed in the loop of string, keeping the string taut, now satisfies the condition $PS + PS' = 2a$. So, as the pencil moves, it traces out an ellipse.

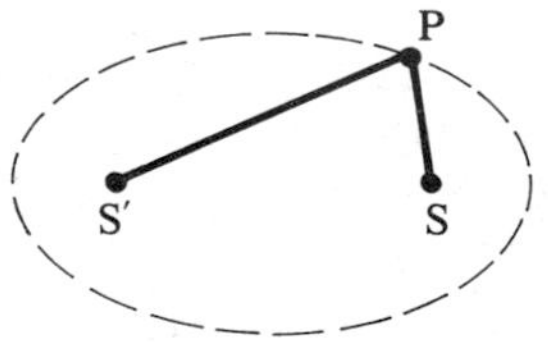

Note that it is not necessary to use this method to *sketch* an ellipse.
This property can be proved from the locus definition of an ellipse as follows:

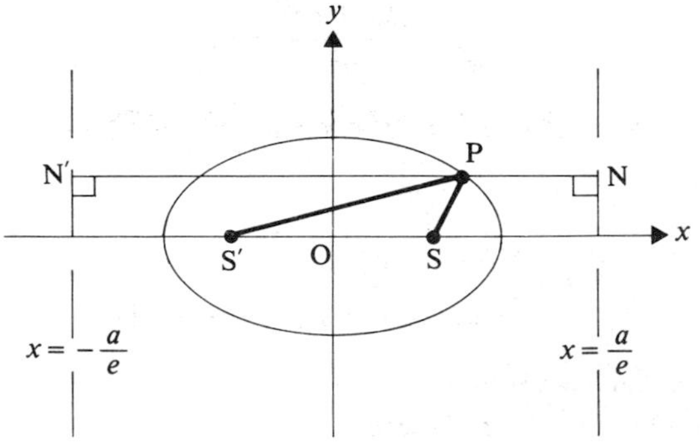

$$PS = ePN$$

$$PS' = ePN'$$

$$PS + PS' = e(PN + PN')$$

$$= e(NN')$$

But NN′ is the distance between the directrices,

i.e. $$NN' = \frac{2a}{e}$$

Hence $$PS + PS' = 2a$$

Note that the locus definition of an ellipse can be applied to derive a variety of its geometric properties.

Diameters

A diameter of any conic is the locus of the midpoints of a set of parallel chords.
In the case of an ellipse we find that each of its diameters passes through the centre of the ellipse.

Consider the ellipse with equation $\dfrac{x^2}{a^2}+\dfrac{y^2}{b^2}=1$ and a set of chords with constant gradient k.
One such chord with equation $y = kx + c$ cuts the ellipse at A and B.

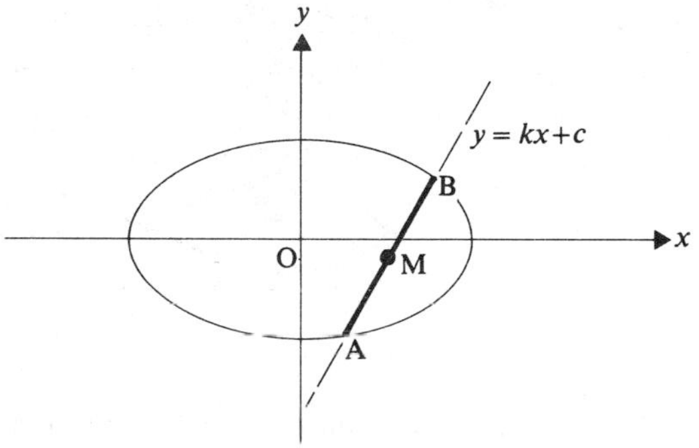

The x coordinates of A and B are the roots of the equation

$$\frac{x^2}{a^2}+\frac{(kx+c)^2}{b^2} = 1$$

or $$(b^2+a^2k^2)x^2 + 2a^2kcx + a^2(c^2-b^2) = 0 \qquad [1]$$

At the midpoint, M(X, Y), of AB,

$$X = \tfrac{1}{2}(x_A + x_B)$$

$$= \tfrac{1}{2}(\text{sum of roots of equation [1]})$$

$$= -\frac{a^2kc}{b^2+a^2k^2} \qquad [2]$$

Also, as M is a point on the line with equation $y = kx + c$, we have

$$Y = kX + c \qquad [3]$$

Equations [2] and [3] are valid for all chords with gradient k, so the Cartesian equation of the locus of M is given by eliminating c (which varies as A and B move on the ellipse) from these two equations;

i.e.
$$Y = kX - \left(\frac{b^2 + a^2k^2}{a^2k}\right)X$$

$\Rightarrow$
$$a^2kY + b^2X = 0$$

This line passes through O for all values of k, so all diameters of an ellipse pass through its centre.

Note that for a set of chords of the ellipse $\dfrac{x^2}{a^2} + \dfrac{y^2}{b^2} = 1$ with gradient k the diameter has gradient $-\dfrac{b^2}{a^2k}$.

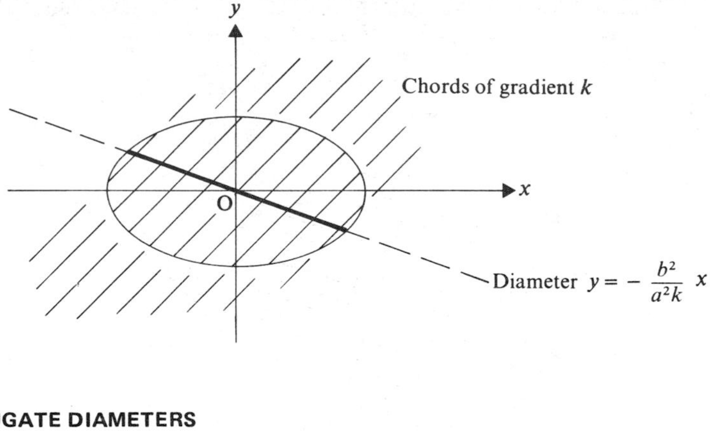

CONJUGATE DIAMETERS

If AB is a diameter with gradient m then the midpoints of all chords parallel to AB lie on a diameter CD.

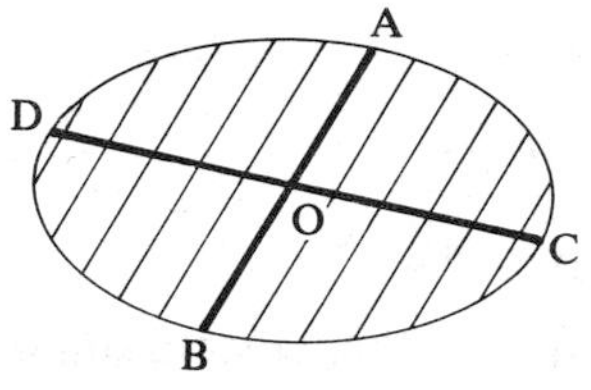

If CD has gradient m' then $m' = -\dfrac{b^2}{a^2m}$.

The product of the gradients of AB and CD is mm' where

$$mm' = -\frac{b^2}{a^2}$$

This expression is symmetric in m and m' (i.e. m and m' can be interchanged without altering its value), so the midpoints of all chords parallel to CD lie on AB. Such a pair of diameters are called conjugate. Thus
a pair of diameters are conjugate

$\iff$ the midpoints of chords parallel to one diameter lie on the other diameter.

$\iff$ the product of their gradients is $-\frac{b^2}{a^2}$.

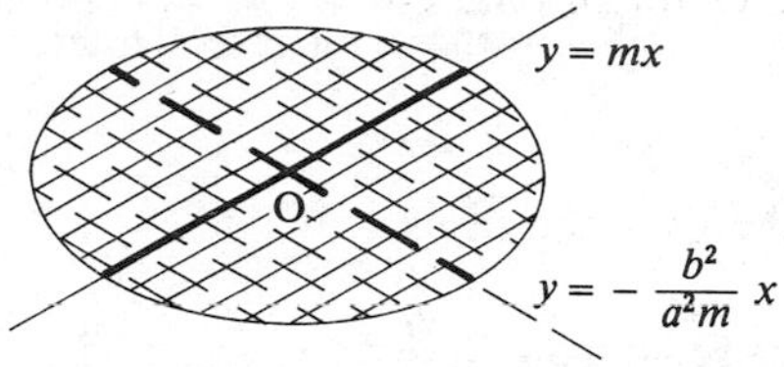

Another interesting way to look at conjugate diameters is to consider the circle

$$X^2 + Y^2 = b^2$$

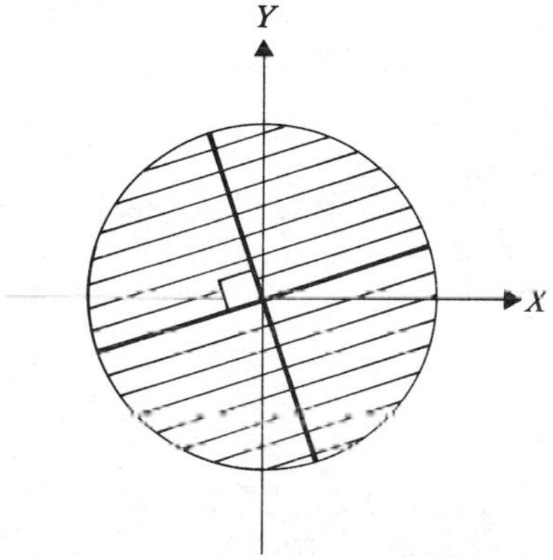

For a circle, a pair of conjugate diameters is defined as a pair of perpendicular diameters.
Now the transformation

$$x = \frac{aX}{b}, \qquad y = Y$$

takes the circle $X^2 + Y^2 = b^2$ to the standard ellipse

$$\frac{x^2}{a^2} + \frac{y^2}{b^2} = 1$$

As this transformation is just a change of scale on the x axis, it obviously preserves parallels and straight lines (but does change gradient).
So the conjugate diameters in the circle are transformed to a pair of diameters in the ellipse such that the midpoints of chords parallel to one of them lie on the other.

But note that, as the product of the gradients of conjugate diameters is $-\frac{b^2}{a^2}$, they are *not* perpendicular (unless $m = 0$ or ∞).
The tangent at the end of a diameter is the limiting member of the set of parallel chords. So the tangents at the ends of the diameter $y = -\frac{b^2}{a^2 m}x$ have gradient m, i.e. they are parallel to the conjugate diameter.

Condition for a Line to Touch an Ellipse

In *The Core Course*, Chapter 10, the problem of tangency was dealt with by first considering the possible intersection of a line and a curve. We shall now apply this method when the curve is an ellipse.

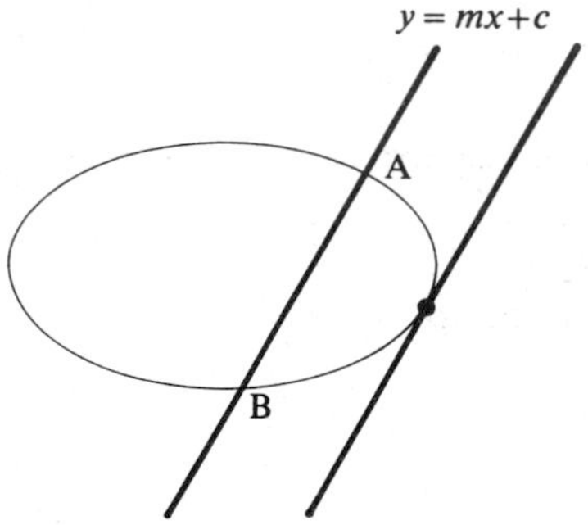

Consider a line $y = mx + c$ and the ellipse $\frac{x^2}{a^2} + \frac{y^2}{b^2} = 1$.
If the line meets the curve at A and B, the x coordinates of A and B are given by the roots of the equation

$$\frac{x^2}{a^2} + \frac{(mx+c)^2}{b^2} = 1$$

or

$$(b^2 + a^2m^2)x^2 + 2a^2mcx + a^2(c^2 - b^2) = 0$$

If the line touches the ellipse, A and B coincide, so this equation has equal roots, i.e.

$$(2a^2mc)^2 = 4a^2(b^2 + a^2m^2)(c^2 - b^2)$$

which simplifies to give

$$c^2 = b^2 + a^2m^2$$

This condition should not be regarded as quotable. Instead, the method used to derive it should be adapted to each individual problem on tangency.

However, it is useful to remember that, from this condition we can deduce that the line

$$y = mx \pm \sqrt{(a^2m^2 + b^2)}$$

always touches the ellipse $\dfrac{x^2}{a^2} + \dfrac{y^2}{b^2} = 1$.

This provides a useful approach to problems involving two tangents to an ellipse from a given point, a technique that is illustrated in the following paragraph.

The Director Circle

If two perpendicular tangents are drawn from a point $P(x, y)$ to an ellipse, $\dfrac{x^2}{a^2} + \dfrac{y^2}{b^2} = 1$, the locus of P as the points of contact vary, is a circle called the *director circle* and its equation is

$$x^2 + y^2 = a^2 + b^2$$

This can be proved by considering the equations of the two tangents from $P(x, y)$ to the ellipse. These equations can be written in the form

$$y = mx \pm \sqrt{(a^2m^2 + b^2)}$$

or $$(x^2 - a^2)m^2 - 2xym + (y^2 - b^2) = 0$$

The roots of this equation, m_1 and m_2, are the gradients of the two tangents from P.

So when the tangents are at right angles, $m_1m_2 = -1$.

In this case the coordinates of P satisfy the relationship

$$\frac{y^2 - b^2}{x^2 - a^2} = -1$$

$\Rightarrow$ $$x^2 + y^2 = a^2 + b^2$$

EXAMPLES 2c

1) The normal at a point $P(x_1, y_1)$ on an ellipse of eccentricity e, meets the major axis at G. Prove that $GS = ePS$ where S is a focus.

Using the standard ellipse with equation

$$\frac{x^2}{a^2} + \frac{y^2}{b^2} = 1$$

differentiating gives $$\frac{2x}{a^2} + \frac{2y}{b^2}\frac{dy}{dx} = 0$$

i.e. $$\frac{dy}{dx} = -\frac{b^2x_1}{a^2y_1} \quad \text{at P.}$$

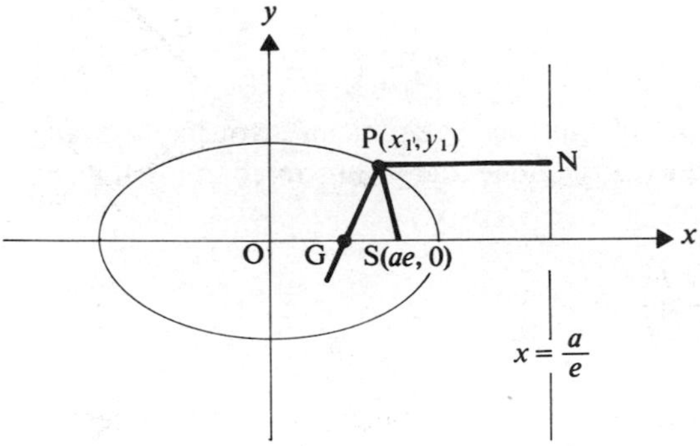

The normal at P has equation

$$y - y_1 = \frac{a^2y_1}{b^2x_1}(x - x_1)$$

At G, $y = 0$, so $$a^2y_1x = (a^2 - b^2)x_1y_1$$

$\Rightarrow$ $$x = \frac{(a^2 - b^2)}{a^2}x_1$$

But $$b^2 = a^2(1 - e^2)$$

or $$a^2 - b^2 = a^2e^2$$

So at G, $$x = e^2x_1$$

Hence $$\text{SG} = ae - e^2x_1$$

Now by definition $$\text{PS} = e\text{PN}$$

$$= e\left(\frac{a}{e} - x_1\right)$$

$$= a - ex_1$$

So $$e\text{PS} = ae - e^2x_1$$

Hence $$\text{SG} = ePS$$

2) The line $y = 2x$ is a diameter of the ellipse $4x^2 + 9y^2 = 36$. Find the equation of the tangents at the ends of this diameter. Show that the area enclosed by the tangent at the ends of this diameter and its conjugate, is 24 square units.

Writing the equation of the ellipse as

$$\frac{x^2}{9}+\frac{y^2}{4} = 1$$

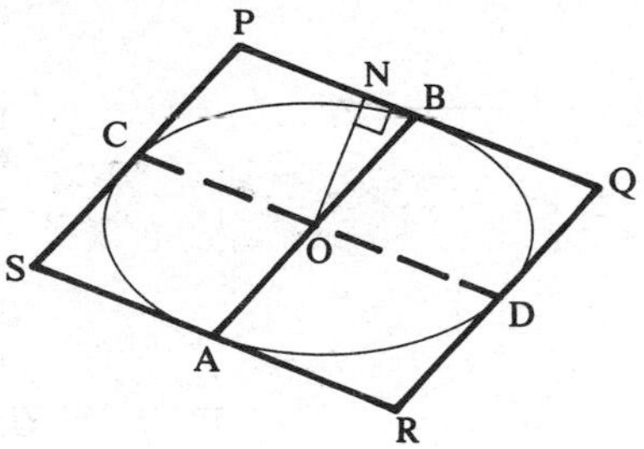

we see that

$a^2 = 9$ and $b^2 = 4$.

The conjugate of the diameter $y = 2x$ has gradient

$$-\frac{b^2}{2a^2} = -\frac{2}{9}$$

The tangents at the ends of the diameter $y = 2x$ therefore have a gradient $-\frac{2}{9}$ because they are parallel to the conjugate diameter.

At A and B, $4x^2 + 9(4x^2) = 36 \Rightarrow x = \pm\frac{3}{\sqrt{10}}$

So the equations of the tangents at A and B are

$$\begin{cases} y - \dfrac{6}{\sqrt{10}} = -\dfrac{2}{9}\left(x - \dfrac{3}{\sqrt{10}}\right) \\[2ex] y + \dfrac{6}{\sqrt{10}} = -\dfrac{2}{9}\left(x + \dfrac{3}{\sqrt{10}}\right) \end{cases}$$

i.e. $9y + 2x = \pm 6\sqrt{10}$

To find the area of the parallelogram PQRS we will find the length of CD (the conjugate diameter) and the perpendicular distance from O to PQ.
The equation of CD is $y = -\frac{2}{9}x$.
Therefore at C and D

$$4x^2 + 9(-\tfrac{2}{9}x)^2 = 36$$

$$\Rightarrow \qquad x = \pm\frac{9}{\sqrt{10}}$$

Therefore at D, $x = \dfrac{9}{\sqrt{10}}$ and $y = -\dfrac{2}{\sqrt{10}}$

$$\Rightarrow \qquad \text{OD} = \sqrt{\frac{85}{10}} = \text{OC}$$

Also, ON is the perpendicular distance from $(0,0)$ to the line $9y+2x-6\sqrt{10}=0$,

i.e. $$\text{ON} = \left|\frac{-6\sqrt{10}}{\sqrt{(81+4)}}\right| = 6\sqrt{\frac{10}{85}}$$

Then
$$\begin{aligned}\text{area PQRS} &= 4 \times \text{area OCPB}\\ &= 4 \times \text{OC} \times \text{ON}\\ &= 4 \times \sqrt{\frac{85}{10}} \times 6 \times \sqrt{\frac{10}{85}}\\ &= 24 \text{ square units}\end{aligned}$$

Note. In general such a parallelogram is of area $4ab$

EXERCISE 13c

1) Find the values of c if $y = 3x + c$ is a tangent to the ellipse $x^2 + 4y^2 = 4$.
Find the coordinates of the points of contact.

2) Prove that the line $3y = 2x + 5$ touches the ellipse $\dfrac{x^2}{4} + y^2 = 1$.

3) Determine whether the line $y = x + 4$ meets the ellipse $\dfrac{x^2}{n^2} + y^2 = 1$ in two distinct points, touches the ellipse or misses it completely, when (a) $n = 2$, (b) $n = 4$, (c) $n = \sqrt{15}$.

4) Find the equation of the locus of the midpoints of chords with gradient $\frac{1}{2}$ of the ellipse $4x^2 + 9y^2 = 36$. What is the equation of the conjugate diameter?
Find the equations of the tangents at the ends of this conjugate diameter.

5) Prove that the line $y = mx \pm \sqrt{(1 + 4m^2)}$ is a tangent to the ellipse $x^2 + 4y^2 = 4$, for all values of m.

6) Find the distances from the foci of the point $P(x_1, y_1)$ on the ellipse $\dfrac{x^2}{9} + \dfrac{y^2}{25} = 1$. (*Hint*: Use the focal distance property.)

7) $P(x, y)$ is a point on the ellipse $\dfrac{x^2}{36} + \dfrac{y^2}{25} = 1$. If P is distant 5 units from one focus, how far is it from the other?

8) The foci of an ellipse are S, S′ and P is any point on the curve. The normal at P meets SS′ at G. Prove that

$$PG^2 = (1-e^2)PS\,.\,PS'$$

where e is the eccentricity of the ellipse.

9) An ellipse with major axis horizontal and centre at the origin has foci that are 8 units apart and directrices that are 18 units apart. Find its Cartesian equation. A tangent to this ellipse is equally inclined to the axes. If the tangent cuts the axes at A and B prove that $AB = 4\sqrt{7}$.

10) Prove that the line $lx+my+n=0$ is a tangent to the ellipse $\dfrac{x^2}{a^2}+\dfrac{y^2}{b^2}=1$ if $a^2l^2+b^2m^2=n^2$.
Tangents are drawn from a point P to the ellipse. Lines from the origin are drawn perpendicular to these two tangents. If these lines are conjugate diameters prove that the equation of the locus of P is

$$a^2x^2+b^2y^2 = a^4+b^4.$$

11) Two conjugate diameters of the ellipse $\dfrac{x^2}{a^2}+\dfrac{y^2}{b^2}=1$ are drawn to meet a directrix at P and Q. PM is perpendicular to OQ and QN is perpendicular to OP.
Prove that PM and QN meet at a fixed point and state its coordinates.

12) Find (i.e. do not quote) the equation of the director circle of the ellipse

$$\frac{x^2}{9}+\frac{y^2}{16} = 1.$$

PARAMETRIC COORDINATES

For the standard ellipse with Cartesian equation

$$\frac{x^2}{a^2}+\frac{y^2}{b^2} = 1$$

suitable parametric equations are

$$x = a\cos\theta, \quad y = b\sin\theta$$

Thus the coordinates of a point P on this ellipse can be given in the form $(a\cos\theta, b\sin\theta)$ where each value of θ corresponds to one and only one point on the ellipse.

The Eccentric Angle

The geometric significance of the parameter θ is explained in the following diagram.

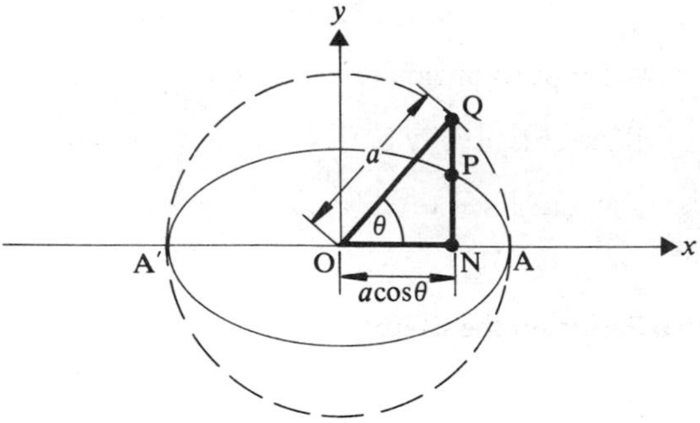

A circle is drawn on AA' as diameter. For a point P on the ellipse, the ordinate NP is produced to cut the circle at Q.

Now since $OQ = a$ and $ON = OQ\cos\angle QON$ we see that

$$\text{the } x \text{ coordinate of P} = ON = a\cos\angle QON.$$

But $$\text{the } x \text{ coordinate of P} = a\cos\theta.$$

So θ is the angle QON and it is sometimes called the *eccentric angle* for the point P on the ellipse. The circle on AA' as diameter is referred to as the *eccentric circle or auxiliary circle.*

Parametric Analysis of the Ellipse

For each value of the parameter θ, there is one and only one point P on the ellipse, so the behaviour of the curve at P is uniquely related to θ. The equations of the tangent and normal at P can therefore be found in terms of θ.

The Parametric Equation of the Tangent

At P, $x = a\cos\theta$ and $y = b\sin\theta$ so the gradient of the tangent at P is the value of $\dfrac{dy}{dx}$ at P.

$$\frac{dy}{dx} = \frac{dy}{d\theta}\bigg/\frac{dx}{d\theta} = -\frac{b\cos\theta}{a\sin\theta}$$

The equation of the tangent at P is therefore

$$y - b\sin\theta = -\frac{b\cos\theta}{a\sin\theta}(x - a\cos\theta)$$

$$\Rightarrow \qquad bx\cos\theta + ay\sin\theta = ab(\sin^2\theta + \cos^2\theta)$$

$$\Rightarrow \qquad bx\cos\theta + ay\sin\theta = ab$$

or
$$\frac{x}{a}\cos\theta + \frac{y}{b}\sin\theta = 1$$

Note. The trig simplification used above is not so obvious if $\frac{dy}{dx}$ is simplified to $-\frac{b}{a}\cot\theta$. In parametric work on an ellipse it is usually better to use $\cos\theta$ and $\sin\theta$ rather than $\tan\theta$ or $\cot\theta$.

The Equation of a Chord Joining Two Points on the Ellipse

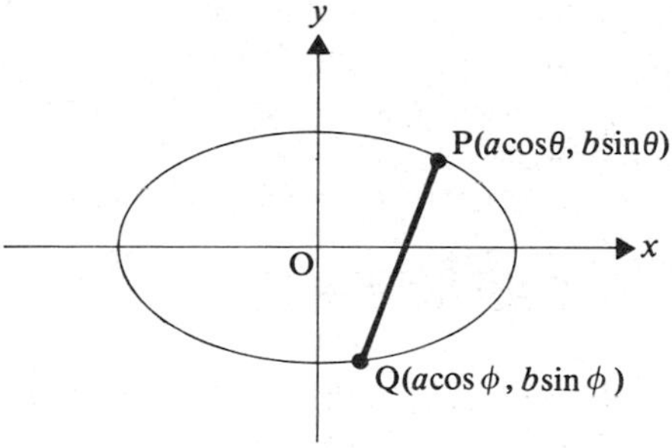

For two different points on the ellipse we use two different eccentric angles, θ and ϕ in this case.
The equation of PQ is

$$\frac{y - b\sin\theta}{x - a\cos\theta} = \frac{b(\sin\theta - \sin\phi)}{a(\cos 0 - \cos\phi)}$$

$$\Rightarrow \qquad ay(\cos\theta - \cos\phi) - ab\sin\theta(\cos\theta - \cos\phi)$$
$$= bx(\sin\theta - \sin\phi) - ab\cos\theta(\sin\theta - \sin\phi)$$

$$\Rightarrow \qquad ay(\cos\theta - \cos\phi) - bx(\sin\theta - \sin\phi)$$
$$= ab(\sin\phi\cos\theta - \sin\theta\cos\phi)$$

But
$$\cos\theta - \cos\phi \equiv 2\sin\frac{\theta+\phi}{2}\sin\frac{\phi-\theta}{2}$$

$$\sin\theta - \sin\phi \equiv 2\cos\frac{\theta+\phi}{2}\sin\frac{\theta-\phi}{2}$$

and
$$\sin\phi\cos\theta - \sin\theta\cos\phi \equiv \sin(\phi - \theta)$$
$$\equiv 2\sin\frac{\phi-\theta}{2}\cos\frac{\phi-\theta}{2}$$

So the equation of the chord becomes

$$ay \sin \frac{\theta + \phi}{2} + bx \cos \frac{\theta + \phi}{2} = ab \cos \frac{\theta - \phi}{2}$$

Note. This particular piece of work is included primarily to indicate that the parametric analysis of an ellipse requires familiarity with standard trig identities which should be revised at this stage.

From the equation of a chord we can deduce the equation of a tangent by considering the result when Q→P, i.e. when $\phi \to \theta$
The equation of the chord then becomes

$$ay \sin \theta + bx \cos \theta = ab$$

Clearly this is not a good method for *finding* the equation of the tangent at P but it is an interesting deduction from the equation of a chord.
Note that a chord through a focus that is perpendicular to the major axis is called the *latus rectum* (see page 487).

Conjugate Diameters and Related Parameters

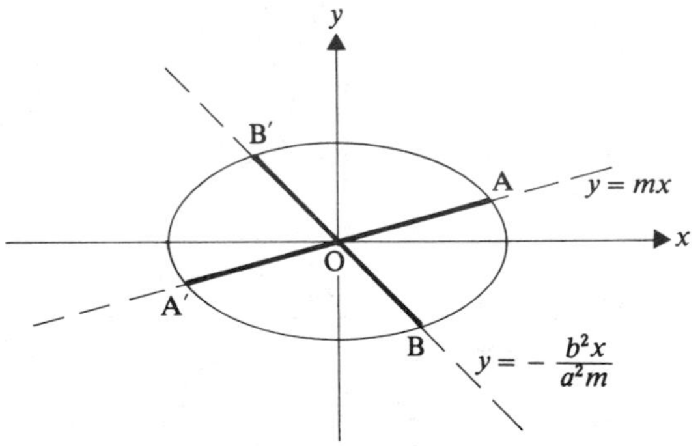

It has already been proved that the product of the gradients of conjugate diameters of the ellipse $\frac{x^2}{a^2} + \frac{y^2}{b^2} = 1$ is $-\frac{b^2}{a^2}$.
AA′ and BB′ are a pair of conjugate diameters where A is the point $(a \cos \theta, b \sin \theta)$ and B is the point $(a \cos \phi, b \sin \phi)$.
The gradient of OA is $\frac{b \sin \theta}{a \cos \theta}$, i.e. $\frac{b}{a} \tan \theta$

and the gradient of OB is $\frac{b}{a} \tan \phi$.

Hence $$\left(\frac{b}{a} \tan \theta\right)\left(\frac{b}{a} \tan \phi\right) = -\frac{b^2}{a^2}$$

$$\Rightarrow \qquad \tan\theta\tan\phi = -1$$

$$\Rightarrow \qquad \tan\theta = -\cot\phi$$

$$= \tan\left(\phi + \frac{\pi}{2}\right)$$

$$\Rightarrow \qquad \theta = \phi + \frac{\pi}{2}$$

So we see that the eccentric angles at the ends of a pair of conjugate diameters differ by $\frac{\pi}{2}$.

This result can also be obtained by using the transformation

$$X = x, \qquad Y = \frac{ay}{b}$$

which takes the point $P(a\cos\theta, b\sin\theta)$ on the ellipse
to the point $Q(a\cos\theta, a\sin\theta)$
and Q is on the auxiliary circle $X^2 + Y^2 = a^2$.

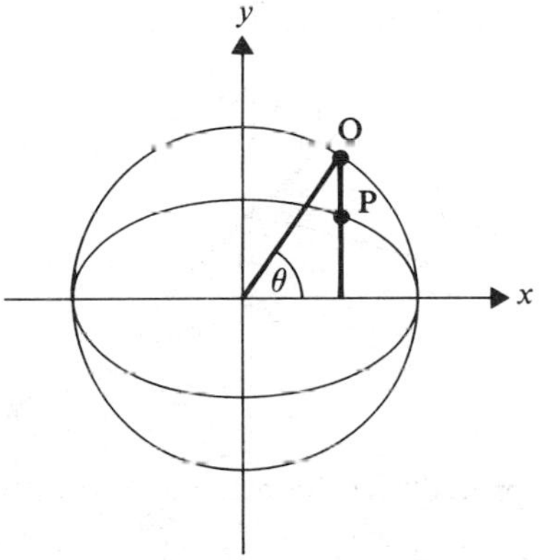

Now conjugate diameters of a circle are perpendicular so it follows that the eccentric angles at the ends of a pair of conjugate diameters of an ellipse differ by $\pi/2$.

EXAMPLES 13d

1) Show that the curve with parametric equations $x = 1 + 4\cos\theta$, $y = 2 + 3\sin\theta$ is an ellipse. State the coordinates of the centre and the lengths of the semi-axes. Find the equation of the tangent to the ellipse at the point $(1 + 4\cos\theta, 2 + 3\sin\theta)$.

Rearranging the parametric equations in the form $\cos\theta = \dfrac{x-1}{4}$, $\sin\theta = \dfrac{y-2}{3}$ and using $\cos^2\theta + \sin^2\theta \equiv 1$, we have

$$\left(\frac{x-1}{4}\right)^2 + \left(\frac{y-2}{3}\right)^2 = 1$$

which is the Cartesian equation of an ellipse with centre $(1, 2)$ and semi-axes of lengths 4 and 3.
The gradient at any point is given by

$$\frac{dy}{dx} = \frac{dy}{d\theta}\Big/\frac{dx}{d\theta} = -\frac{3\cos\theta}{4\sin\theta}$$

So the equation of the tangent at the point $(1+4\cos\theta, 2+3\sin\theta)$ is

$$y-(2+3\sin\theta) = -\frac{3\cos\theta}{4\sin\theta}\{x-(1+4\cos\theta)\}$$

i.e. $\quad 4y\sin\theta + 3x\cos\theta = 3\cos\theta(1+4\cos\theta) + 4\sin\theta(2+3\sin\theta)$

i.e. $\quad 4y\sin\theta + 3x\cos\theta = 3\cos\theta + 8\sin\theta + 12$

2) Show that the tangent at the point $A(a\cos\theta, b\sin\theta)$ on the ellipse $b^2x^2 + a^2y^2 = a^2b^2$ meets the tangent at $B(a\cos\theta, a\sin\theta)$ on the circle $x^2+y^2 = a^2$ at a point on the x axis.

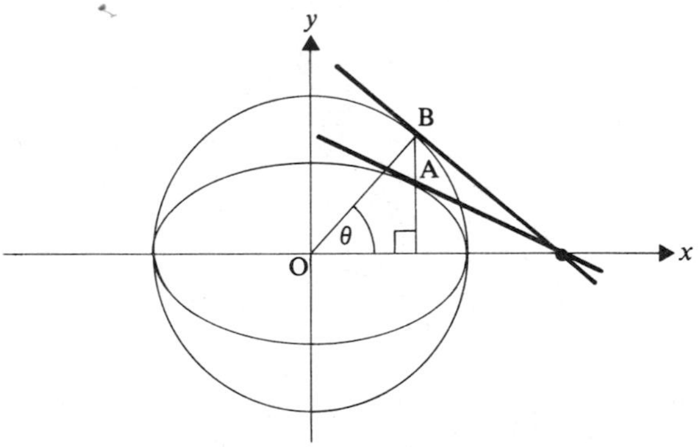

The equation of the tangent at A to the ellipse is

$$y - b\sin\theta = -\frac{b\cos\theta}{a\sin\theta}(x - a\cos\theta) \qquad [1]$$

The equation of the tangent at B to the circle is

$$y - a\sin\theta = -\frac{a\cos\theta}{a\sin\theta}(x - a\cos\theta) \qquad [2]$$

[1] becomes $$\frac{ay\sin\theta}{b} - a\sin^2\theta = -x\cos\theta + a\cos^2\theta$$

[2] becomes $$y\sin\theta - a\sin^2\theta = -x\cos\theta + a\cos^2\theta$$

$$(1)-(2) \quad\Rightarrow\quad \left(\frac{a}{b}-1\right)y\sin\theta = 0 \quad\Rightarrow\quad y = 0$$

So the two tangents meet on the x axis.

3) The normal at a point $P(4\cos\theta, 3\sin\theta)$ on the ellipse $\frac{x^2}{16}+\frac{y^2}{9}=1$ meets the x and y axes at A and B. Show that the locus of M, the midpoint of AB, is an ellipse with the same eccentricity as the given ellipse. Sketch the two ellipses on the same diagram.

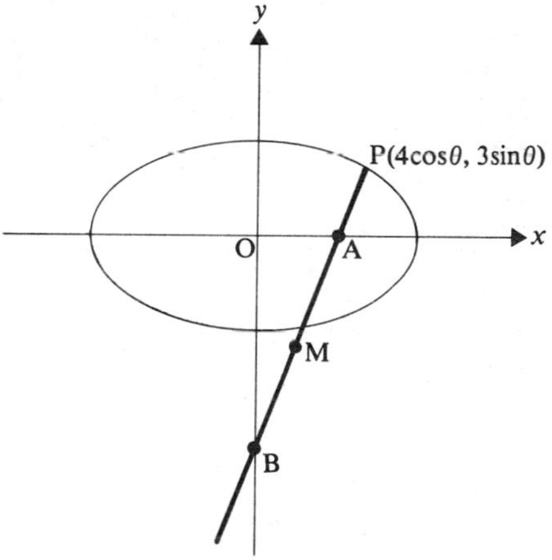

For the given ellipse, $a=4$ and $b=3$, so the major axis is horizontal and

$$e^2 = 1-\frac{b^2}{a^2} = 1-\frac{9}{16},$$

i.e. $\quad e = \frac{1}{4}\sqrt{7}.$

The gradient of the tangent at P is given by

$$\frac{dy}{dx} = -\frac{3\cos\theta}{4\sin\theta}$$

so the equation of the normal at P is

$$y-3\sin\theta = \frac{4\sin\theta}{3\cos\theta}(x-4\cos\theta)$$

i.e. $$4x\sin\theta - 3y\cos\theta = 7\sin\theta\cos\theta$$

Therefore at A, $\quad y = 0$ and $x = \frac{7}{4}\cos\theta$

and at B, $\quad x = 0$ and $y = -\frac{7}{3}\sin\theta$

The coordinates of M, the midpoint of AB, are therefore

$$x = \tfrac{7}{8}\cos\theta,\quad y = -\tfrac{7}{6}\sin\theta$$

and, using $\cos^2\theta+\sin^2\theta\equiv 1$, the equation of the locus of M becomes

$$\left(\frac{8x}{7}\right)^2+\left(-\frac{6y}{7}\right)^2 = 1$$

i.e. $$\frac{x^2}{(\frac{7}{8})^2}+\frac{y^2}{(\frac{7}{6})^2} = 1$$

This is the equation of an ellipse with semi-axes of lengths $\frac{7}{8}$ and $\frac{7}{6}$, the major axis being vertical $(\frac{7}{6} > \frac{7}{8})$. So the eccentricity of this ellipse is given by

$$e^2 = 1-(\tfrac{7}{8})^2/(\tfrac{7}{6})^2 = \tfrac{7}{16}$$

i.e. $e = \frac{1}{4}\sqrt{7}$ which is equal to the eccentricity of the given ellipse. The two ellipses can now be sketched as shown

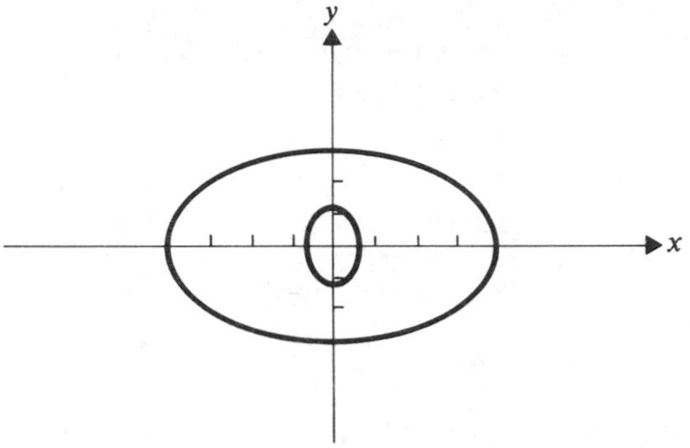

EXERCISE 13d

1) Sketch the loci of the following points as θ varies from 0 to 2π:

(a) $(4\cos\theta, 3\sin\theta)$ (b) $(\cos\theta, 2\sin\theta)$

(c) $(5+4\cos\theta, 2-3\sin\theta)$ (d) $(\cos\theta-1, 2\sin\theta)$.

2) Find the eccentricity, the coordinates of the centre and the coordinates of the foci of the ellipses in Question 1.

3) Obtain the Cartesian equations of the ellipses in Question 1.

4) Obtain parametric equations for the following ellipses:

(a) $\dfrac{(x-1)^2}{9}+\dfrac{(y-3)^2}{4} = 1$ (b) $3(x+2)^2+2(y+1)^2 = 6$

(c) $16x^2+9y^2 = 144$ (d) $(x+1)^2+4(y-1)^2 = 4$.

5) Find the equations of the tangents to the following ellipses at the specified points:

(a) $x = 4\cos\theta$, $y = 3\sin\theta$, at the point where $\theta = \dfrac{\pi}{3}$,

(b) $x = \cos\theta$, $y = 2\sin\theta$, at the point where $\theta = -\dfrac{\pi}{6}$,

(c) $x = 4\cos\theta$, $y = \sin\theta$, at the point where $\theta = \dfrac{3\pi}{4}$,

(d) $x = 5\cos\theta$, $y = 3\sin\theta$, at the point where $\theta = -\dfrac{2\pi}{3}$.

6) Find the equation of the tangent and the normal at the point $P(2\cos\theta, \sin\theta)$ on the ellipse $\dfrac{x^2}{4} + y^2 = 1$.

7) Find an equation of the chord joining the points A and B if OA and OB are conjugate semi-diameters of the ellipse $\dfrac{x^2}{9} + \dfrac{y^2}{4} = 1$. (A and B both have positive ordinates.)

8) Find the coordinates of the points where
(a) the tangent (b) the normal
at the point $P(a\cos\theta, b\sin\theta)$ on the ellipse $\dfrac{x^2}{a^2} + \dfrac{y^2}{b^2} = 1$ meet:
(i) the x axis, (ii) the y axis.

9) (a) Show that the point with parametric coordinates $(1 - 3\cos\theta, 2 + \sin\theta)$ always lies on the ellipse $\dfrac{(x-1)^2}{9} + (y-2)^2 = 1$.

(b) Find a suitable pair of parametric equations for the ellipse $4(x+2)^2 + (y-5)^2 = 4$.

10) S is the focus, on the positive x axis, of the ellipse $\dfrac{x^2}{9} + \dfrac{y^2}{4} = 1$ and $P(3\cos\theta, 2\sin\theta)$ is a variable point on the ellipse. If SP is produced to Q so that PQ = 2PS, find the equation of the locus of Q as P moves on the ellipse.

11) The tangent at P on the ellipse $\dfrac{x^2}{a^2} + \dfrac{y^2}{b^2} = 1$ meets the x and y axes at A and B.
Find, in terms of the eccentric angle of P, the ratio of the lengths of AP and BP.

12) Repeat Question 6 using the normal at P.

13) The tangent at $P(4\cos\theta, 3\sin\theta)$ on the ellipse $9x^2 + 16y^2 = 144$ meets the tangent at the positive end of the major axis at Q and meets the tangent at the positive end of the minor axis at R. Find
(a) the ratio of the lengths PQ and PR,
(b) the parametric equations of the locus of the midpoint of QR.

14) The tangents at the ends of a pair of conjugate diameters of the ellipse $x^2+4y^2=16$ are inclined to each other at an angle $\arctan\frac{5}{3}$. Find the coordinates of the extremities of the diameters.

15) AB is a diameter of the ellipse $x^2+9y^2=25$ and the eccentric angle of A is $\frac{\pi}{6}$. Find
(a) the eccentric angle of B,
(b) the equations of the tangents at A and B,
(c) the equation of the conjugate diameter.

16) Prove that the line $3x\cos\phi+4y\sin\phi=12$ is a tangent to the ellipse $\frac{x^2}{16}+\frac{y^2}{9}=1$ for all values of ϕ.

17) P is a point on an ellipse whose major axis is AB and the tangent at P meets the minor axis at Q. PA and PB cut the minor axis at R and T. Prove that Q bisects RT.

THE HYPERBOLA

The definition of a hyperbola, in common with the other conic sections, is the locus of points whose distances from a focus and a directrix are in a fixed ratio e (>1).

We saw on page 486 that the standard Cartesian equation of the hyperbola is

$$\frac{x^2}{a^2}-\frac{y^2}{b^2}=1 \quad \text{where} \quad b^2=a^2(e^2-1)$$

The general shape of this curve can be deduced by noting that

(a) There is symmetry about both the x axis and the y axis (as the equation contains only even powers of both x and y).

(b) $\frac{y^2}{b^2}=\frac{x^2}{a^2}-1 \quad \Rightarrow \quad \frac{x^2}{a^2}-1\geqslant 0 \quad \Rightarrow \quad |x|\geqslant a.$

(c) $\begin{cases}\text{As } x\to\infty, & y\to\pm\infty\\ \text{As } x\to-\infty, & y\to\pm\infty.\end{cases}$

(d) $\frac{dy}{dx}=\frac{b^2x}{a^2y}$ so $\frac{dy}{dx}$ is infinite when $y=0$.

From these properties the possible shape of the hyperbola is deduced.

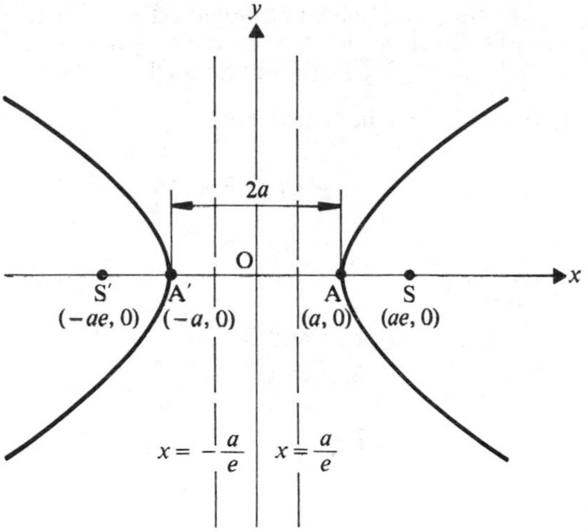

The hyperbola has two foci and two directrices (from symmetry).
AA′ is called the major axis, the points A, A′ are the vertices, and the midpoint of AA′ is the centre. The foci are always on the major axis produced.

If the major axis is vertical the equation of the hyperbola becomes

$$\frac{y^2}{a^2} - \frac{x^2}{b^2} = 1$$

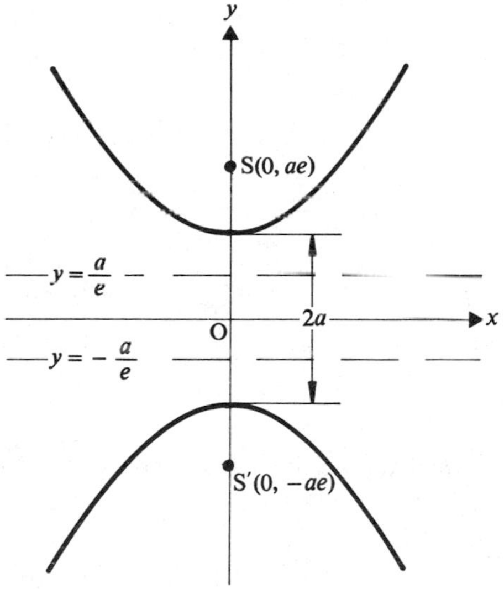

Note that the major axis is not necessarily given by the larger denominator in the Cartesian equation. It is the *positive term* whose denominator gives this axis.

For example, $\frac{x^2}{9} - \frac{y^2}{64} = 1$ has a horizontal major axis of length 6,

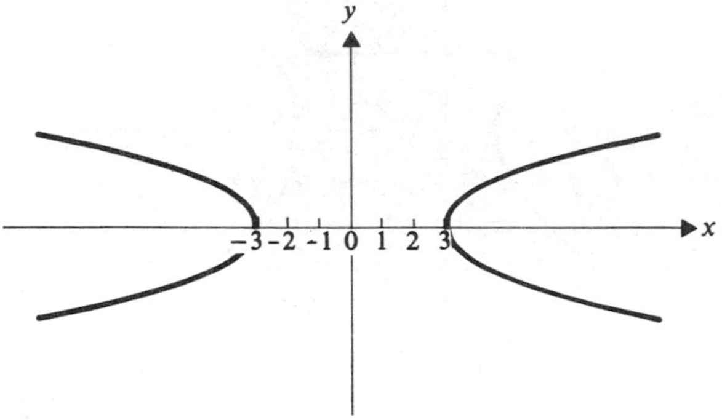

while $\frac{y^2}{25} - \frac{x^2}{16} = 1$ has a vertical major axis of length 10.

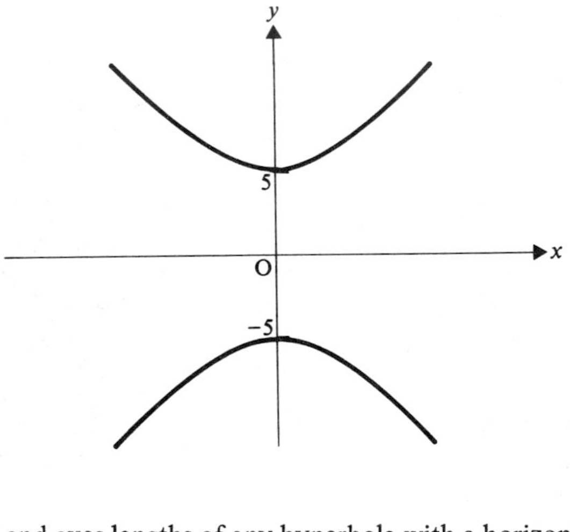

The centre, foci and axes lengths of any hyperbola with a horizontal or vertical major axis can be identified by comparing its equation with that of a standard hyperbola

$$\frac{X^2}{a^2} - \frac{Y^2}{b^2} = 1$$

whose centre is given by $X = 0, \quad Y = 0$
and whose major axis is of length $2a$.

For example, comparing the hyperbola

$$\frac{(x-3)^2}{16}-\frac{(y+2)^2}{9}=1$$

with $$\frac{X^2}{a^2}-\frac{Y^2}{b^2}=1$$

(i.e. by moving the axes to $x = 3$, $y = -2$)
we see that $x - 3 = X$, $y + 2 = Y$, $a = 4$, $b = 3$.
So the centre $(X = 0, \quad Y = 0)$ is the point $(3, -2)$.
The major axis is horizontal (as the x^2 term is positive) and of length 8.
From $b^2 = a^2(e^2 - 1)$ we see that $e = \frac{5}{4}$.
The foci, which are on the major axis and distant ae from the centre, are the points $(3 \pm 5, -2)$, i.e. $(8, -2)$ and $(-2, -2)$.

The directrices (distant $\dfrac{a}{e}$ from the centre) are the lines $x = 3 \pm \frac{16}{5}$, i.e. $x = \frac{31}{5}$ and $x = -\frac{1}{5}$.

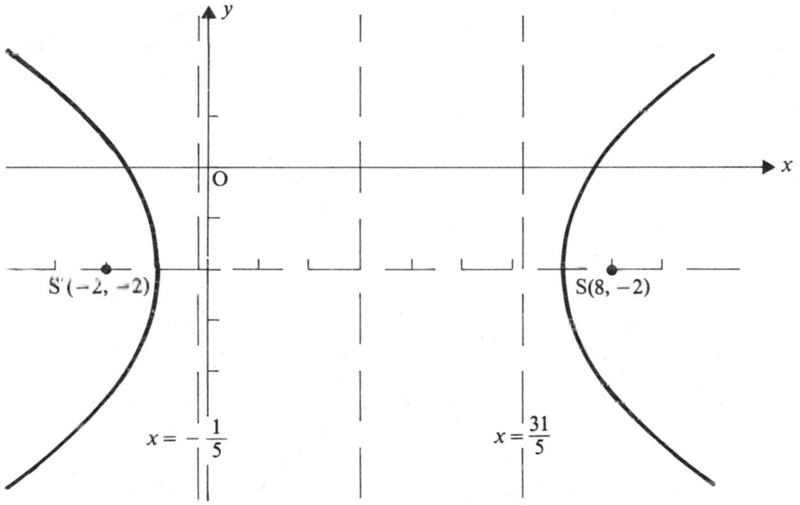

EXERCISE 13e

1) Find the equations of the following hyperbolas:
(a) foci at $(\pm 6, 0)$ and eccentricity 2;
(b) centre at the origin, one directrix $x = 2$, distance between vertices 12;
(c) centre $(2, 0)$, eccentricity 3, directrices vertical, one focus at $(4, 0)$;
(d) foci at $(0, 4)$ and $(0, -2)$, eccentricity $\frac{3}{2}$.

2) Find the centre and the distance between the vertices of each of the following hyperbolas:

(a) $\frac{x^2}{4} - y^2 = 1$ (b) $\frac{x^2}{16} - \frac{y^2}{25} = 1$

(c) $9(x-1)^2 - y^2 = 9$ (d) $\frac{(y+1)^2}{4} - \frac{(x+2)^2}{9} = 1$

(e) $x^2 - y^2 = 1$ (f) $y^2 - x^2 = 4$.

3) Find the coordinates of the foci and the equations of the directrices of each of the hyperbolas in Question 2.

4) Find, from the first principles, the equation of a hyperbola with a focus at $(2, 1)$ a directrix $y = x$, and eccentricity 2.

5) A set of points is such that each point is three times as far from the y axis as it is from the point $(4, 0)$. Find the equation of the locus of P and sketch the locus.

The Focal Distance Property

The distances of any point P on a hyperbola from the foci S and S′ are such that

$$|\text{PS} - \text{PS}'| = 2a$$

This property can be proved by using the definition of a hyperbola as a locus, as follows:

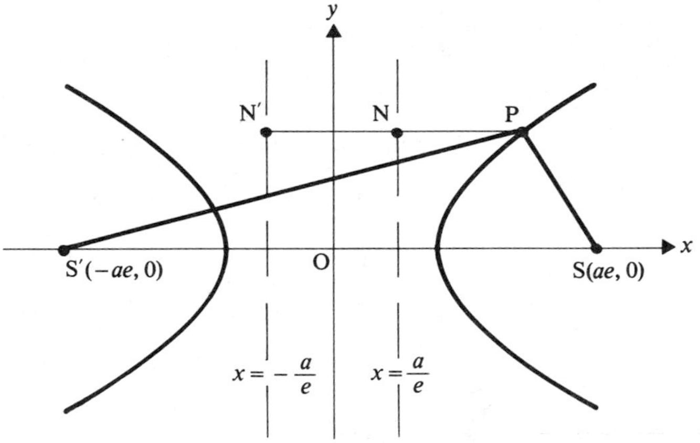

For any point P on the hyperbola,

$$\text{PS} = e\text{PN}$$

$$\text{PS}' = e\text{PN}'$$

so $$\text{PS}' - \text{PS} = e(\text{PN}' - \text{PN})$$

But $$\text{PN}' - \text{PN} = \text{NN}'$$
$$= 2\left(\frac{a}{e}\right)$$

Hence $$\text{PS}' - \text{PS} = 2a$$

If P is on the branch of the hyperbola with focus S′, we find that

$$\text{PS} - \text{PS}' = 2a$$

So, in general, $|\text{PS} - \text{PS}'| = 2a$.

ASYMPTOTES

When x and y both become very large, the equation $\frac{x^2}{a^2} - \frac{y^2}{b^2} = 1$ approximates to $\frac{x^2}{a^2} - \frac{y^2}{b^2} = 0$ (since 1 is negligible compared with x^2 and y^2).

But $\frac{x^2}{a^2} - \frac{y^2}{b^2} = 0$ is the equation of two straight lines $\frac{x}{a} - \frac{y}{b} = 0$ and $\frac{x}{a} + \frac{y}{b} = 0$.

So, for large values of x and y, the hyperbola $\frac{x^2}{a^2} - \frac{y^2}{b^2} = 1$ approximates to a pair of straight lines.

Hence the pair of lines $y = \frac{b}{a}x$ and $y = -\frac{b}{a}x$ are the asymptotes of the hyperbola.

The hyperbola $\frac{x^2}{a^2} - \frac{y^2}{b^2} = 1$ can now be sketched more accurately,

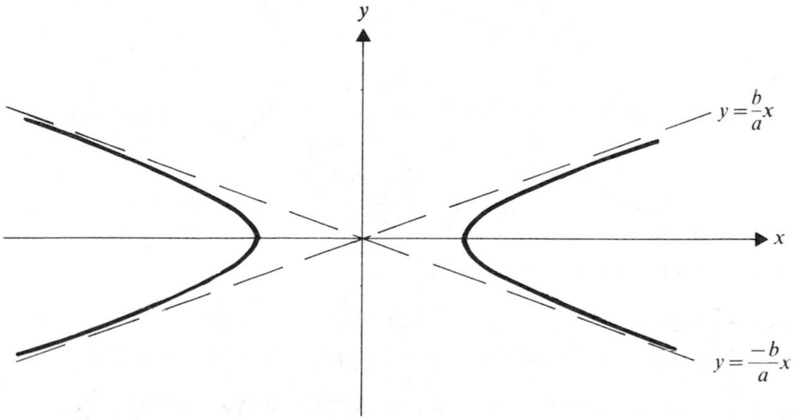

The ratio $b:a$ determines the shape of the hyperbola.

If b is small compared with a the hyperbola is sharp.

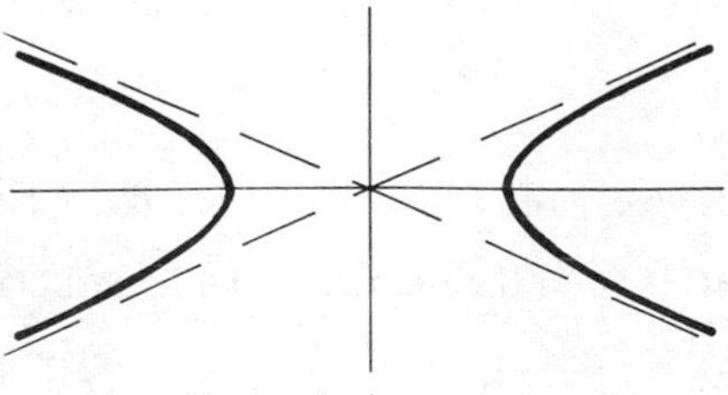

whereas the curve is more rounded when b is large compared with a.

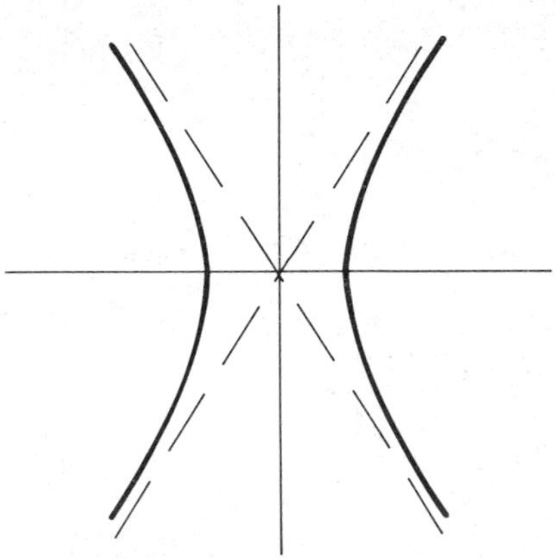

The special case when a and b are equal gives a hyperbola whose asymptotes are perpendicular to each other.

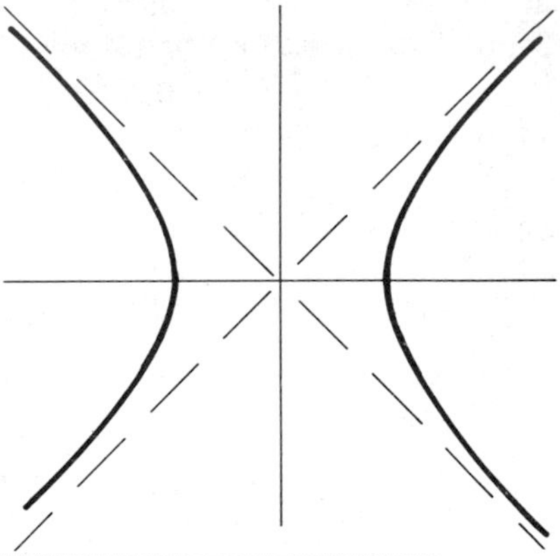

This curve is called a *rectangular hyperbola* and its standard equation is

$$x^2 - y^2 = a^2$$

The eccentricity of a rectangular hyperbola $(a = b)$ is given by
$a^2 = a^2(e^2 - 1)$

i.e. $$e = \sqrt{2}$$

We will investigate the rectangular hyperbola later, but first we will complete an analysis of the general hyperbola in its standard position. As most of the techniques and methods used for a curve with Cartesian equation $\dfrac{x^2}{a^2} - \dfrac{y^2}{b^2} = 1$

are clearly similar to those used for the ellipse with equation $\dfrac{x^2}{a^2}+\dfrac{y^2}{b^2}=1$, no worked examples involving Cartesian methods are given but problems are set in Exercise 13f. Problems solved by parametric analysis, although based on principles used already in the case of the ellipse, involve different trigonometric detail, so this section is now covered more fully.

PARAMETRIC ANALYSIS

The similarity of the equation $\dfrac{x^2}{a^2}-\dfrac{y^2}{b^2}=1$ to the trig identity $\sec^2\theta-\tan^2\theta\equiv 1$ suggests the parametric equations that are usually used,

i.e. $$x = a\sec\theta,\qquad y = b\tan\theta$$

Note that the parameter, θ, has no *obvious* geometric significance in this case.

Gradient

At any point $P(a\sec\theta, b\tan\theta)$, the gradient is given by

$$\frac{dy}{dx}=\frac{dy}{d\theta}\bigg/\frac{dx}{d\theta}=\frac{b\sec^2\theta}{a\sec\theta\tan\theta}=\frac{b\sec\theta}{a\tan\theta}$$

(although a further simplification gives $\dfrac{dy}{dx}=\dfrac{b}{a\sin\theta}$, this form is not recommended if the gradient is going to be used to find the equations of the tangent and the normal).

The Equation of the Tangent at $P(a\sec\theta, b\tan\theta)$

Using the gradient in the form $\dfrac{b\sec\theta}{a\tan\theta}$, the equation of the tangent is

$$y-b\tan\theta = \frac{b\sec\theta}{a\tan\theta}(x-a\sec\theta)$$

i.e. $$bx\sec\theta-ay\tan\theta = ab(\sec^2\theta-\tan^2\theta) = ab$$

The Equation of the Normal at $P(a\sec\theta, b\tan\theta)$

The gradient of the normal is $-\dfrac{a\tan\theta}{b\sec\theta}$ so the equation of the normal is

$$y-b\tan\theta = -\frac{a\tan\theta}{b\sec\theta}(x-a\sec\theta)$$

i.e. $$by\sec\theta+ax\tan\theta = (a^2+b^2)\sec\theta\tan\theta$$

The Equation of the Chord Joining P(a sec θ, b tan θ) and Q(a sec ϕ, b tan ϕ)

The equation of PQ is

$$y - b\tan\theta = \left(\frac{b\tan\theta - b\tan\phi}{a\sec\theta - a\sec\phi}\right)(x - a\sec\theta) \quad or$$

$$\frac{x}{a}(\tan\theta - \tan\phi) - \frac{y}{b}(\sec\theta - \sec\phi) = \sec\theta(\tan\theta - \tan\phi) - \tan\theta(\sec\theta - \sec\phi)$$

$$= \tan\theta\sec\phi - \sec\theta\tan\phi$$

Multiplying throughout by $\cos\theta\cos\phi$ gives

$$\frac{x}{a}(\sin\theta\cos\phi - \cos\theta\sin\phi) - \frac{y}{b}(\cos\phi - \cos\theta) = \sin\theta - \sin\phi$$

Then, using compound angle and factor formulae, we get

$$\frac{x}{a}\left(2\sin\frac{\theta-\phi}{2}\cos\frac{\theta-\phi}{2}\right) - \frac{y}{b}\left(2\sin\frac{\theta+\phi}{2}\sin\frac{\theta-\phi}{2}\right) = 2\cos\frac{\theta+\phi}{2}\sin\frac{\theta-\phi}{2}$$

and finally the equation of the chord PQ can be written

$$\frac{x}{a}\cos\frac{\theta-\phi}{2} - \frac{y}{b}\sin\frac{\theta+\phi}{2} = \cos\frac{\theta+\phi}{2}$$

Note that any problem involving two points on a hyperbola given in parametric form will necessarily contain some demanding trig manipulation.
Note that, if Q $\rightarrow$ P
so that $\phi \rightarrow \theta$ and the chord PQ $\rightarrow$ the tangent at P,
the equation of the chord becomes

$$\frac{x}{a} - \frac{y}{b}\sin\theta = \cos\theta$$

or $\dfrac{x}{a}\sec\theta - \dfrac{y}{b}\tan\theta = 1$, which is the equation of the tangent at P.

The Tangent-Intercept Property

If the tangent at a point P on a hyperbola cuts the asymptotes at A and B, then

$$PA = PB$$

This property can be proved as follows.

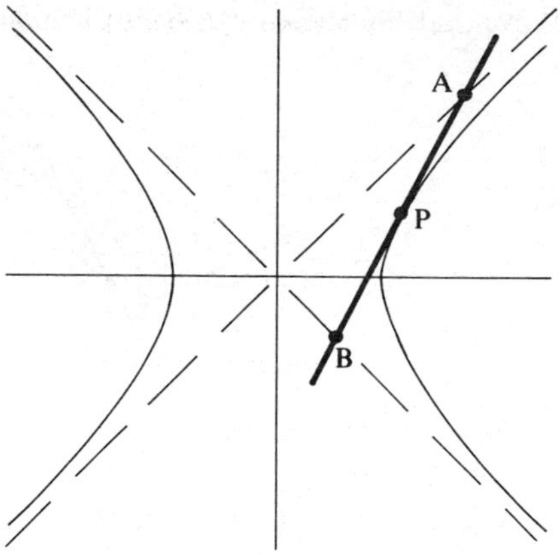

The tangent at P has equation

$$bx\sec\theta - ay\tan\theta = ab$$

and the equations of the asymptotes are

$$y = \pm\frac{b}{a}x$$

So, at A and B

$$bx\sec\theta - (\pm bx)\tan\theta = ab$$

$\Rightarrow$
$$x = \frac{a}{\sec\theta \mp \tan\theta}$$

At the midpoint of AB

$$x = \frac{1}{2}\left\{\frac{a}{\sec\theta + \tan\theta} + \frac{a}{\sec\theta - \tan\theta}\right\}$$

$$= \frac{1}{2}\left\{\frac{2a\sec\theta}{\sec^2\theta - \tan^2\theta}\right\}$$

$$= a\sec\theta$$

= the x coordinate of P

So P is the midpoint of the segment of the tangent at P cut off by the asymptotes.

EXAMPLES 13f

1) The tangent and normal at a point on the hyperbola $\dfrac{x^2}{a^2} - \dfrac{y^2}{b^2} = 1$ cut the

y axis at A and B. Prove that the circle on AB as diameter passes through the foci of the hyperbola.

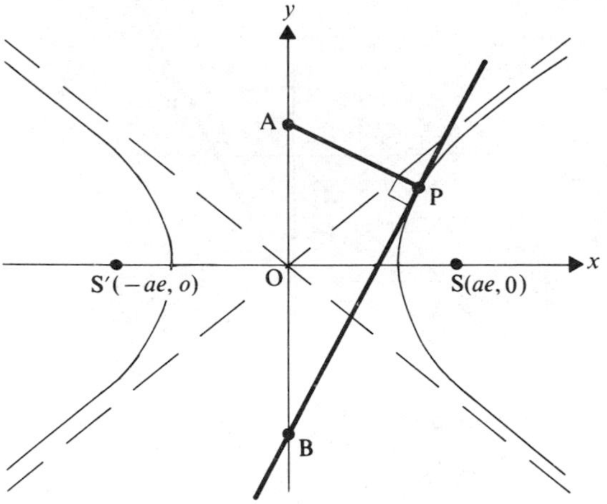

If P is the point $(a \sec \theta, b \tan \theta)$ then the equation of the tangent is

$$\frac{x \sec \theta}{a} - \frac{y \tan \theta}{b} = 1$$

And the equation of the normal (treating it as a line through P perpendicular to the tangent) is

$$\frac{x}{b} \tan \theta + \frac{y}{a} \sec \theta = \frac{a \sec \theta}{b} \tan \theta + \frac{b \tan \theta}{a} \sec \theta$$

$$= \left(\frac{a^2 + b^2}{ab}\right) \sec \theta \tan \theta$$

Hence at A $\qquad y = \left(\dfrac{a^2 + b^2}{b}\right) \tan \theta = \dfrac{a^2 e^2}{b} \tan \theta$

and at B $\qquad y = -\dfrac{b}{\tan \theta}$

The gradient of AS is

$$\left(\frac{a^2 e^2}{b} \tan \theta\right) \Big/ (-ae) = -\frac{ae}{b} \tan \theta$$

and the gradient of BS is

$$\left(\frac{b}{\tan \theta}\right) \Big/ ae = \frac{b}{ae \tan \theta}$$

The product of these gradients is -1, showing that $\angle$ASB is a right angle.

So S is on the circle with AB as diameter.
From symmetry, S′ also lies on this circle.

2) Prove that all diameters of a hyperbola pass through the centre.

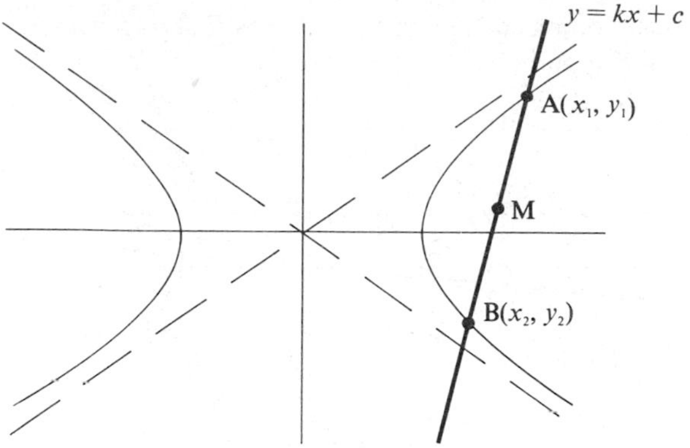

Consider a set of chords of the hyperbola $\dfrac{x^2}{a^2}-\dfrac{y^2}{b^2}=1$, that have a constant gradient k.
One such chord, with equation $y = kx + c$, meets the hyperbola at A and B where

$$\frac{x^2}{a^2}-\frac{(kx+c)^2}{b^2} = 1$$

i.e. $$(b^2-a^2k^2)x^2-2kca^2x-a^2(b^2+c^2) = 0 \qquad [1]$$

The x coordinate of M, the midpoint of AB, is

$$\tfrac{1}{2}(x_1+x_2) = \tfrac{1}{2} \qquad \text{(sum of roots of equation [1])}$$

$$= \frac{1}{2}\left(\frac{2kca^2}{b^2-a^2k^2}\right)$$

So, at M $$x = \frac{kca^2}{b^2-a^2k^2}$$

and $$y = kx+c$$

Eliminating c, which varies as A and B move on the hyperbola while k remains constant, gives

$$y = kx+\frac{(b^2-a^2k^2)}{ka^2}x$$

$\Rightarrow$ $$y = \frac{b^2x}{a^2k}$$

This is the equation of the locus of the midpoints of a set of chords with gradient k, so it is the equation of a diameter of the hyperbola. As it is satisfied by $x=0$, $y=0$, it passes through the origin which is the centre of the hyperbola, for any value of k.
Note that this proof is equally true for chords that join two points on different branches of the hyperbola.

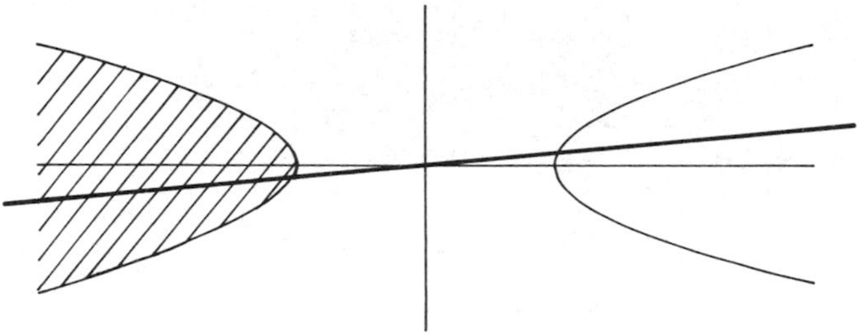

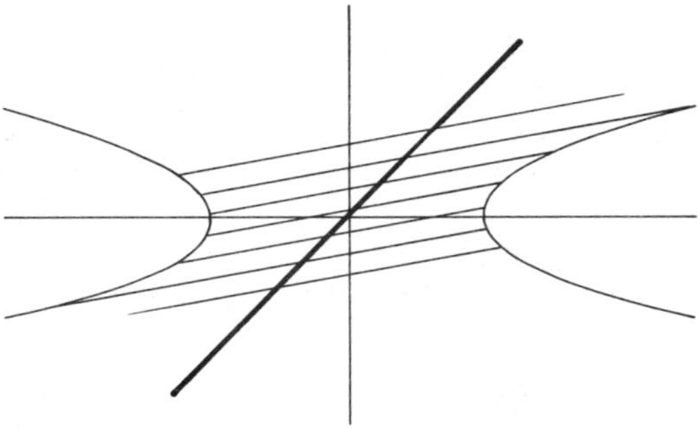

Conjugate Diameters of a Hyperbola

In the last example we saw that for a set of parallel chords with gradient k, the diameter through their midpoints has gradient $\dfrac{b^2}{a^2k}$.
So if AB is a diameter with gradient k, then the midpoints of all chords parallel to AB lie on a diameter CD of gradient k', where $k' = \dfrac{b^2}{a^2k}$.

The product of the gradients of AB and CD is $\dfrac{b^2}{a^2}$ and this is independent of the value of k and of k'.
Such a pair of diameters are called conjugate. Thus

> a pair of diameters of a hyperbola are called conjugate
> $\Longleftrightarrow$ the product of their gradients is b^2/a^2
> $\Longleftrightarrow$ the midpoints of chords parallel to one lie on the other.

EXAMPLES 13f (continued)

3) Find the equations of the tangents to the hyperbola $x^2 - 9y^2 = 9$ that are drawn from the point $(3, 2)$ and find the points of contact.
Find the area of the triangle that these tangents form with their chord of contact

Consider any line with equation $y = mx + c$. This line meets the hyperbola $x^2 - 9y^2 = 9$ at points whose x coordinates are given by

$$x^2 - 9(mx + c)^2 = 9$$

i.e.

$$(9m^2 - 1)x^2 + 18mcx + 9(1 + c^2) = 0 \qquad [1]$$

If the line touches the curve, the roots of this equation are equal.
In this case

$$(18mc)^2 = 36(9m^2 - 1)(1 + c^2)$$

$$\Rightarrow \qquad 9m^2 = 1 + c^2 \qquad [2]$$

But the point $(3, 2)$ is on the line $y = mx + c$

so

$$2 = 3m + c \qquad [3]$$

From [2] and [3],

$$9m^2 = 1 + (2 - 3m)^2$$

$$\Rightarrow \qquad m = \tfrac{5}{12} \quad \text{or } m \text{ is infinte,}$$

i.e.

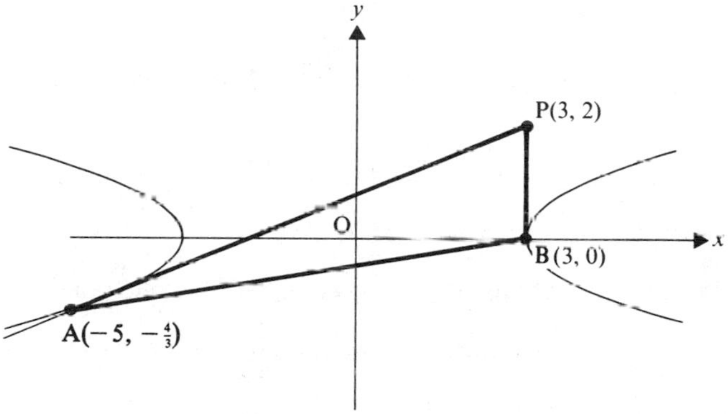

If $m = \frac{5}{12}$, then $c = 2 - 3m = \frac{3}{4}$, so the equation of one of the tangents is

$$12y = 5x + 9$$

If $m \to \infty$, the tangent is parallel to the y axis, and as it passes through the point $(3, 2)$ its equation is

$$x = 3$$

Now when the roots of equation [1] are equal, $x = -\dfrac{18mc}{2(9m^2 - 1)}$, and this is the x coordinate of the point of contact.

So, when $m = \frac{5}{12}$, $x = -5$, showing that the tangent $12y = 5x + 9$ touches the hyperbola at the point $(-5, -\frac{4}{3})$.
By observation, the other tangent touches the hyperbola at the vertex $(3, 0)$.
The triangle formed by the two tangents, PA and PB, and the chord of contact, AB, has vertices $(3, 2), (-5, -\frac{4}{3}), (3, 0)$.

The area, Δ, of the triangle APB is given by

$$2\Delta = \begin{vmatrix} 1 & 1 & 1 \\ x_A & x_P & x_B \\ y_A & y_P & y_B \end{vmatrix} = \begin{vmatrix} 1 & 1 & 1 \\ 3 & -5 & 3 \\ 2 & -\frac{4}{3} & 0 \end{vmatrix}$$

$\Rightarrow$ $$\Delta = 8 \text{ square units.}$$

EXERCISE 13f

1) *Write down* the equation of the following tangents:

(a) to $\dfrac{x^2}{4} - \dfrac{y^2}{3} = 1$ at the point $\left(\dfrac{4}{\sqrt{3}}, 1\right)$,

(b) to $x^2 - y^2 = 9$ at the point $(4, 5)$,

(c) to $x^2 - \dfrac{y^2}{3} = 1$ at the point $(2, 3)$.

2) State the equations of the asymptotes of each of the hyperbolas in Question 1.

3) If the line $y = 2x + 1$ is a tangent to the hyperbola $x^2 - ky^2 = 1$, find the value of k.

4) Find the condition that the line $y = mx + c$ shall touch the hyperbola $\dfrac{x^2}{a^2} - \dfrac{y^2}{b^2} = 1$. Hence find the equations of the tangents from the origin to the hyperbola.

5) Prove that the line $y = x + \sqrt{3}$ is a tangent to the hyperbola $\dfrac{x^2}{4} - y^2 = 1$ and find the coordinates of the point of contact.

6) For what value of m does the line $y = mx + 1$ touch the hyperbola $9x^2 - 4y^2 = 36$?

7) Prove that the lines $y = mx \pm a\sqrt{(m^2 - 1)}$ are tangents to the hyperbola $x^2 - y^2 = a^2$ for all values of m.

8) Find the equation of the locus of midpoints of the chords of the hyperbola $x^2 - 4y^2 = 4$ that have a gradient of 2.

9) Prove that the product of the distances from the foci of the hyperbola $\frac{x^2}{a^2} - \frac{y^2}{b^2} = 1$ to any tangent is b^2.

10) The asymptotes to the hyperbola $b^2x^2 - a^2y^2 = a^2b^2$ are inclined to one another at an angle α.

Prove that $\tan \alpha = \dfrac{2ab}{a^2 - b^2}$.

11) $P(a \sec \theta, b \tan \theta)$ is a point on the hyperbola $\frac{x^2}{a^2} - \frac{y^2}{b^2} = 1$. The ordinate at P and the tangent at P meet one asymptote at Q and R respectively. If the normal at P meets the x axis at N prove that RQ and QN are perpendicular.

12) PQ is a diameter of the hyperbola $\frac{x^2}{4} - y^2 = 1$, whose foci are S and S′. If ∠PSQ and ∠PS′Q are both right angles find the area of PSQS′.

13) $P(a \sec \theta, b \tan \theta)$ is a point on the hyperbola $x^2 - y^2 = a^2$, and A is the fixed point $(2a, 0)$. Show that the locus of M, the midpoint of AP, is another hyperbola. Sketch this locus, indicating the coordinates of the centre and the vertices, and the gradients of its asymptotes.

14) Find the equations of the tangents to the hyperbola $\frac{x^2}{3} - \frac{y^2}{2} = 1$, that are parallel to the line $y = x$. State the coordinates of the points of contact and find the area of the triangle formed by one of these tangents and the asymptotes.

THE RECTANGULAR HYPERBOLA

A hyperbola is said to be rectangular if its asymptotes are perpendicular to each other.

We saw on page 524 that the eccentricity of a rectangular hyperbola is $\sqrt{2}$ and the standard equation is $x^2 - y^2 = a^2$.

If this hyperbola is rotated through an angle $\frac{\pi}{4}$ about the origin, its asymptotes become a suitable pair of Cartesian axes.

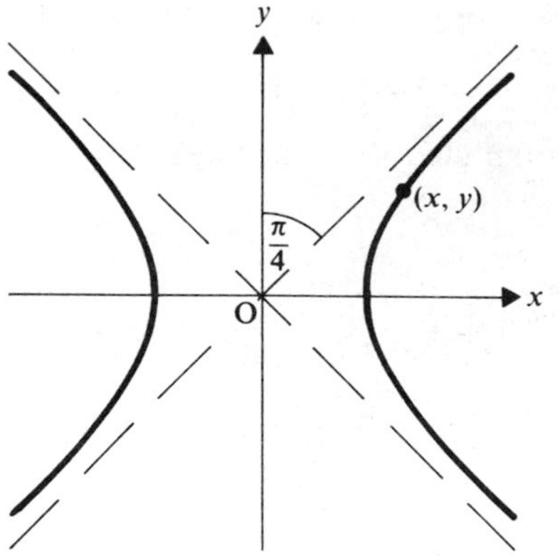

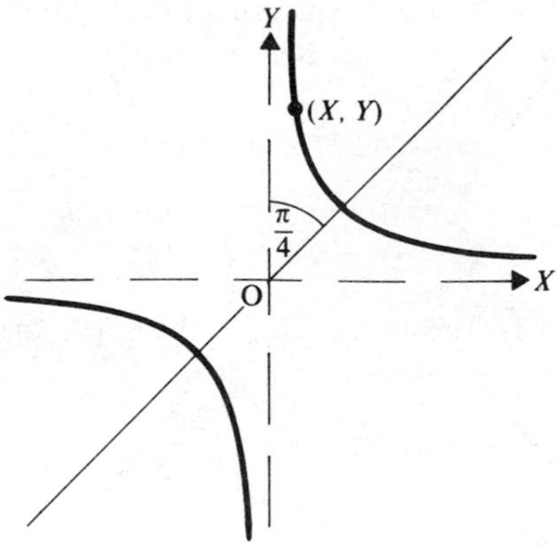

To carry out this transformation, a point (x, y) is mapped to a point (X, Y) by the matrix

$$\begin{pmatrix} \cos\frac{\pi}{4} & -\sin\frac{\pi}{4} \\ \sin\frac{\pi}{4} & \cos\frac{\pi}{4} \end{pmatrix}, \text{ i.e. } \begin{pmatrix} \frac{1}{\sqrt{2}} & -\frac{1}{\sqrt{2}} \\ \frac{1}{\sqrt{2}} & \frac{1}{\sqrt{2}} \end{pmatrix}$$

So $$\begin{pmatrix} X \\ Y \end{pmatrix} = \begin{pmatrix} \frac{1}{\sqrt{2}} & -\frac{1}{\sqrt{2}} \\ \frac{1}{\sqrt{2}} & \frac{1}{\sqrt{2}} \end{pmatrix}\begin{pmatrix} x \\ y \end{pmatrix} = \begin{pmatrix} \frac{x}{\sqrt{2}} - \frac{y}{\sqrt{2}} \\ \frac{x}{\sqrt{2}} + \frac{y}{\sqrt{2}} \end{pmatrix}$$

Hence $$\begin{cases} X = \frac{1}{\sqrt{2}}(x-y) \\ Y = \frac{1}{\sqrt{2}}(x+y). \end{cases}$$

Multiplying these equations together gives

$$XY = \tfrac{1}{2}(x^2 - y^2)$$

But $$x^2 - y^2 = a^2$$

So $$XY = \frac{a^2}{2} = c^2 \quad \text{where} \quad a = \sqrt{2}c.$$

Thus the Cartesian equation of a rectangular hyperbola referred to its asymptotes as axes is $xy = c^2$.

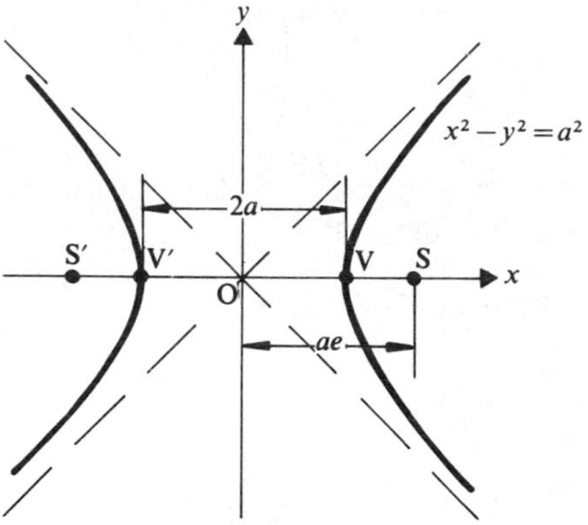

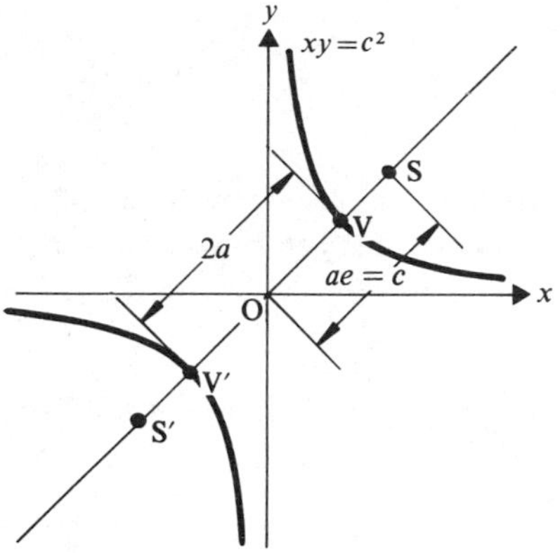

The distance between the centre and each vertex is a, i.e. $\sqrt{2}c$. So the coordinates of the vertices of $xy = c^2$ are (c, c) and $(-c, -c)$. The distance OS between the centre and a focus is ae. But $e = \sqrt{2}$ and $a = \sqrt{2}c$, so $OS = OS' = 2c$. Thus the coordinates of the foci of $xy = c^2$ are $(\sqrt{2}c, \sqrt{2}c)$ and $(-\sqrt{2}c, -\sqrt{2}c)$.

Note that the equation of a rectangular hyperbola located in the second and fourth quadrants is $xy = -c^2$.

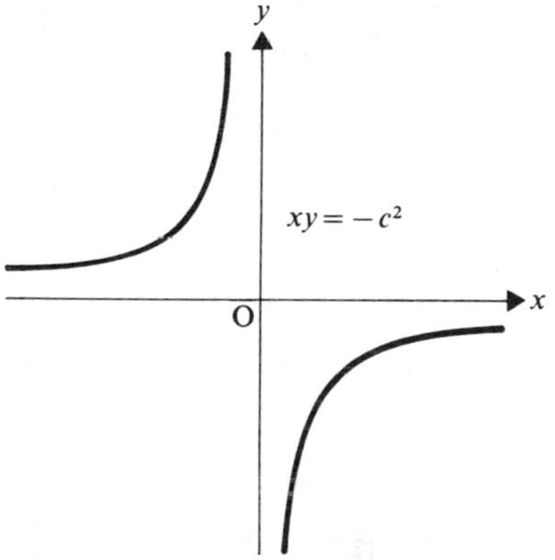

PARAMETRIC ANALYSIS

There are several suitable pairs of parametric equations that can be used for the hyperbola $xy = c^2$ but the commonest are

$$x = ct, \qquad y = \frac{c}{t}$$

Gradient

At a point $P\left(ct, \frac{c}{t}\right)$ the gradient is given by

$$\frac{dy}{dx} = \frac{dy}{dt} \Big/ \frac{dx}{dt} = -\frac{c}{t^2} \Big/ c$$

i.e. $$\frac{dy}{dx} = -\frac{1}{t^2}$$

Note that this result confirms that the gradient of $xy = c^2$ is always negative, a property that can be seen from the shape of the curve.

The Parametric Equation of the Tangent

is $$y - \frac{c}{t} = -\frac{1}{t^2}(x - ct)$$

i.e. $$t^2y + x = 2ct$$

The Parametric Equation of the Normal

is $$y - \frac{c}{t} = t^2(x - ct)$$

i.e. $$ty - t^3x = c(1 - t^4)$$

The Equation of a Chord Joining P $\left(cp, \frac{c}{p}\right)$ and Q $\left(cq, \frac{c}{q}\right)$

The gradient of PQ is $$\frac{\frac{c}{p} - \frac{c}{q}}{cp - cq} = \frac{(q - p)}{pq(p - q)} = -\frac{1}{pq}$$

So the equation of PQ is

$$y - \frac{c}{p} = -\frac{1}{pq}(x - cp)$$

i.e. $$pqy + x = c(p + q)$$

As we have seen before, the equation of the tangent at P can be deduced from the equation of the chord through P and Q by allowing Q to approach P so that $q \to p$ and the equation of PQ $\to$ the equation of the tangent at P.

Note that a chord may join either two points on one branch of the curve or two points, one on each branch of the curve.

The Condition for a Line to Touch a Rectangular Hyperbola

A general line $y = mx + k$ meets the rectangular hyperbola $xy = c^2$ at points whose x coordinates are given by

$$\frac{c^2}{x} = mx + k \qquad [1]$$

so the line is a tangent if the roots of this equation are equal,

i.e. if $$k^2 + 4mc^2 = 0$$

(**Note** that, as c is a standard symbol in the equation of the rectangular hyperbola, it cannot be used for the constant term in the equation of the line, so k is used instead.)

Alternatively, if a line *with a specified gradient* is to touch the rectangular hyperbola $xy = c^2$, the problem can be dealt with parametrically using the gradient of a tangent in the form $-\dfrac{1}{t^2}$.

For example, if a line with gradient -4 is to touch $xy = c^2$, then the value of t at the point(s) of contact must satisfy

$$-\frac{1}{t^2} = -4 \quad \Rightarrow \quad t = \pm\frac{1}{2}$$

Thus the points of contact are $\left(\dfrac{c}{2}, 2c\right), \left(-\dfrac{c}{2}, -2c\right)$

and the equations of the required lines are

$$y - c = -4\left(x - \frac{c}{2}\right)$$

and $$y + c = -4\left(x + \frac{c}{2}\right)$$

The Number of Normals from a Given Point

If a normal to the rectangular hyperbola is drawn from a point $A(h, k)$, not on the hyperbola, then the equation of the normal is of the form

$$ty - t^3x = c(1 - t^4)$$

where t is the value of the parameter of the point, P, where the normal meets the hyperbola. Also this equation must be satisfied by $x = h$, $y = k$,

i.e. $$tk - t^3h = c(1 - t^4) \qquad [1]$$

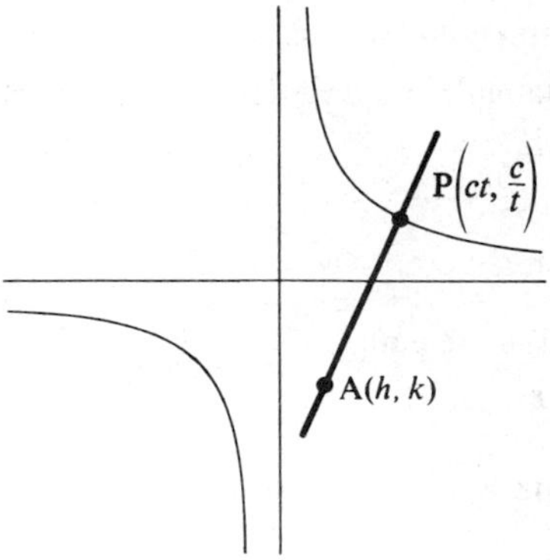

So the position of P is determined by finding the value of t that satisfies equation [1].

But equation [1] is a quartic equation for t,

i.e. $$ct^4 - ht^3 + kt - c = 0$$

and so may have either 4 real roots
or 2 real roots
or no real roots

(since complex roots always occur in conjugate pairs).

Now when $t = 0$, $ct^4 - ht^3 + kt - c < 0$

but when $t \to \infty$ $ct^4 - ht^3 + kt - c > 0$

when $t \to -\infty$ $ct^4 - ht^3 + kt - c > 0$

So equation [1] must have at least one real positive root and at least one real negative root.

Thus we deduce that P is not unique and that from certain points it is possible to draw four normals to the hyperbola while from other points only two normals can be drawn.

e.g.

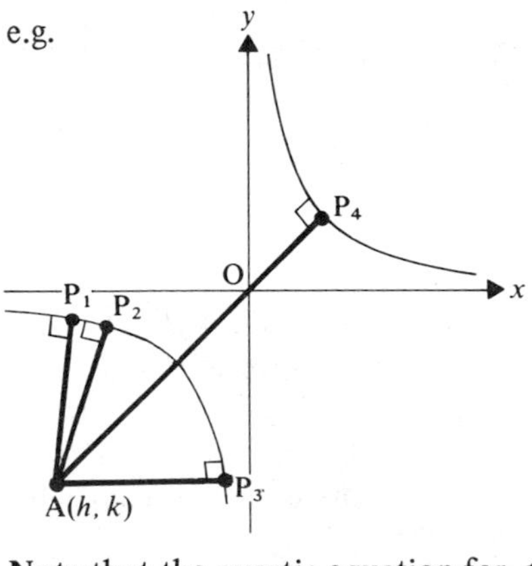

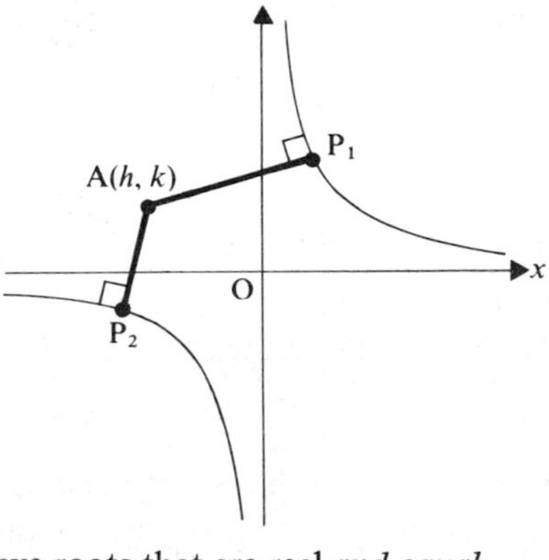

Note that the quartic equation for t may have roots that are real *and equal* and, if so, the corresponding normals coincide.

EXAMPLES 13g

1) A chord PQ of the rectangular hyperbola $xy = c^2$ subtends a right angle at another point R on the hyperbola. Prove that the normal at R is parallel to PQ.

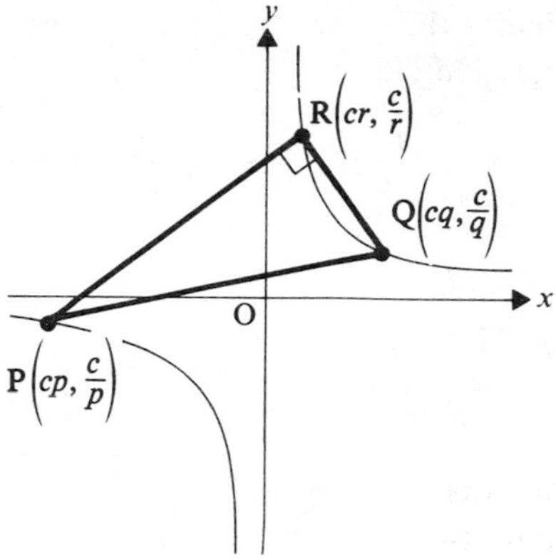

The gradient of PR is

$$\left(\frac{c}{r}-\frac{c}{p}\right)\Big/(cr-cp) = -\frac{1}{pr}$$

Similarly the gradient of QR is $-\dfrac{1}{qr}$.

But PR is perpendicular to QR

so $$\left(-\frac{1}{pr}\right)\left(-\frac{1}{qr}\right) = -1$$

i.e. $$\frac{1}{pqr^2} = -1 \qquad [1]$$

The gradient of the normal at R is r^2

and the gradient of PQ is $-\dfrac{1}{pq}$.

But, from [1], $\dfrac{1}{pq} = -r^2$, so the gradient of PQ is r^2.

Thus the normal at R is parallel to PQ.

2) A chord PQ of the rectangular hyperbola $xy = c^2$ meets the asymptotes at L and M.
Prove that PL = QM.
A tangent to the parabola $y^2 = 4x$ meets the hyperbola $xy = 2$ at two points P and Q.
Find, and sketch, the equation of the locus of the midpoint of PQ.

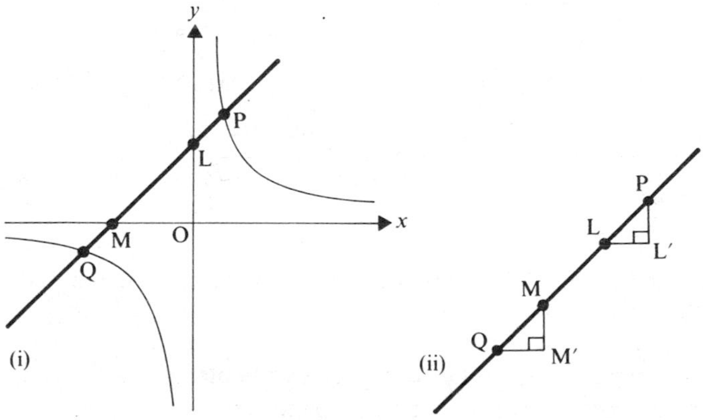

The chord joining $P\left(cp, \dfrac{c}{p}\right)$ and $Q\left(cq, \dfrac{c}{q}\right)$ has equation

$$y - \frac{c}{p} = -\frac{1}{pq}(x - cp)$$

i.e.
$$pqy + x = c(p+q)$$

The asymptotes of the rectangular hyperbola $xy = c^2$ are the x and y axes. Therefore

at L:
$$x = 0, \quad y = \frac{c(p+q)}{pq}$$

at M:
$$y = 0, \quad x = c(p+q)$$

So, in diagram (ii),

$$\begin{aligned} LL' &= x_P - x_L &&= cp - 0 \\ QM' &= x_{M'} - x_Q &&= c(p+q) - cq \\ & &&= cp \end{aligned}$$

Therefore, since triangles $QM'M$ and $LL'P$ are similar and have equal bases,

$$QM = LP.$$

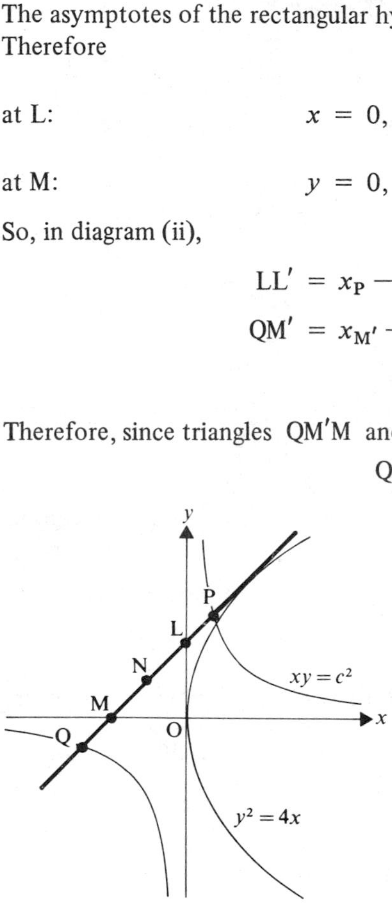

For the hyperbola $xy = 9$, $c = 3$.
So the equation of PQ is

$$pqy + x = 3(p+q)$$

If PQ touches the parabola $y^2 = 4x$, then

$$pqy + \frac{y^2}{4} = 3(p+q)$$

gives equal values for y.

i.e. $y^2 + 4pqy - 12(p+q) = 0$

has equal roots.

So
$$(4pq)^2 + 48(p+q) = 0$$

i.e.
$$(pq)^2 + 3(p+q) = 0 \qquad [1]$$

The coordinates of N, the midpoint of PQ, and therefore the midpoint of LM (since $QM = LP$), are

$$x = \frac{3}{2}(p+q) \qquad [2]$$

and $$y = \frac{3}{2pq}(p+q) \qquad [3]$$

The equation of the locus of N is given by eliminating p and q from equations [1], [2] and [3].

From [1] and [2]

$$(pq)^2 + 2x = 0$$

Then, squaring [3], we have

$$4(pq)^2y^2 = 9(p+q)^2$$

$$\Rightarrow \quad 4(-2x)y^2 = 9\left(\frac{2x}{3}\right)^2$$

So the equation of the locus of N is

$$-2y^2 = x$$

which is a parabola.

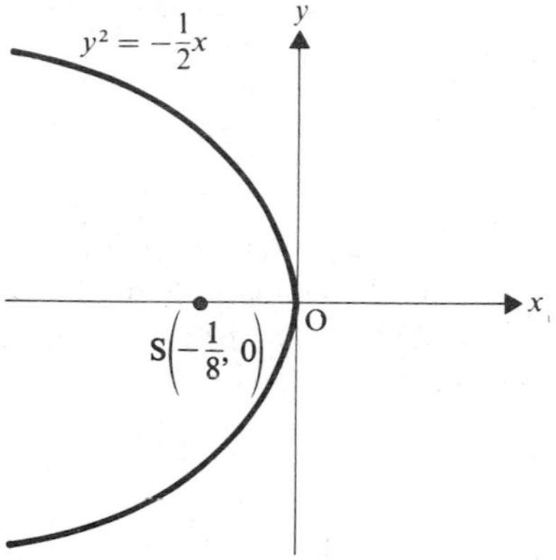

3) The normal at $P\left(ct, \frac{c}{t}\right)$ on the rectangular hyperbola $xy = c^2$ meets the curve again at Q. Find the coordinates of Q.
Hence find the equation of the locus of midpoints of normal chords of the given hyperbola. (A normal chord is normal to the hyperbola at one end.)

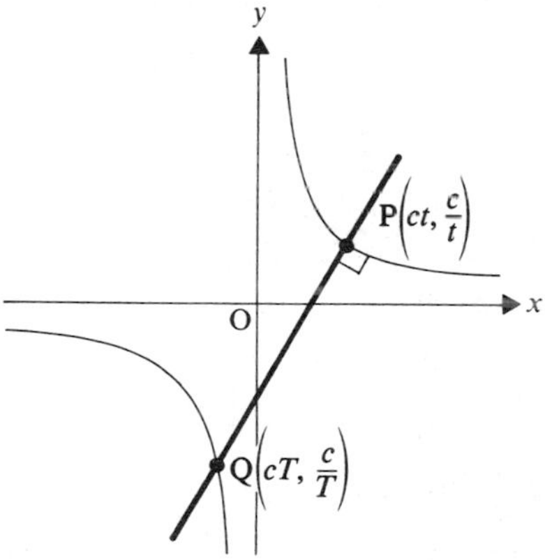

The equation of the normal at P is

$$ty - t^3x = c(1-t^4)$$

Q is on the rectangular hyperbola, so its coordinates can be used in the form $\left(cT, \frac{c}{T}\right)$.

Q is also on the normal at P

so $$\frac{tc}{T} - ct^3T = c(1-t^4)$$

i.e. $$t^3T^2 + (1-t^4)T - t = 0.$$

One solution of this equation is $T = t$ (since P is on both the hyperbola and the normal).

The other solution is therefore $T = -\frac{t}{t^3} \div t$ $\left(\text{using } \alpha\beta = \frac{c}{a}\right)$.

So at Q, $T = -\frac{1}{t^3}$ and therefore the coordinates of Q are $\left(-\frac{c}{t^3}, -ct^3\right)$.

As PQ is a normal chord, we require the locus of its midpoint as P varies. The midpoint, M, of PQ has coordinates

$$x = \frac{c}{2}\left(t - \frac{1}{t^3}\right), \qquad y = \frac{c}{2}\left(\frac{1}{t} - t^3\right)$$

i.e.

$$2x = \frac{c}{t^3}(t^4 - 1), \quad 2y = \frac{c}{t}(1 - t^4)$$

To find the equation of the locus of M, t must be eliminated from these two equations. It is impossible to eliminate t in one step (as t appears in the forms t, t^3 and t^4), so we first eliminate $(t^4 - 1)$, which is a factor in both equations,

i.e.

$$t^4 - 1 = \frac{2xt^3}{c} = -\frac{2yt}{c} \quad \Rightarrow \quad t^2 = -\frac{y}{x}$$

Now we use $t^2 = -\frac{y}{x}$ in one of the parametric equations,

e.g.

$$2y = \frac{c}{t}(1 - t^4) \quad \Rightarrow \quad 4y^2 = \frac{c^2}{t^2}(1 - t^4)^2$$

$$\Rightarrow \quad 4y^2 = c^2\left(-\frac{x}{y}\right)\left(1 - \frac{y^2}{x^2}\right)^2$$

In this way t is completely eliminated and the equation of the locus of M becomes

$$4x^3y^3 + c^2(x^2 - y^2)^2 = 0$$

EXERCISE 13g

1) Write down parametric equations for the following rectangular hyperbolas:

(a) $xy = 16$ (b) $xy + 25 = 0$ (c) $(x - 2)y = 1$

(d) $y = \frac{9}{x}$ (e) $x - 1 = \frac{4}{y}$ (f) $4xy = 1$.

2) Find the Cartesian equation of each of the following loci:

(a) $\left(2t, \frac{2}{t}\right)$ (b) $\left(3t, -\frac{3}{t}\right)$ (c) $\left(\frac{1}{t}, t\right)$

(d) $\left(1 + t, \frac{1}{t}\right)$ (e) $\left(4t, 1 - \frac{4}{t}\right)$ (f) $\left(-2t, \frac{2}{t}\right)$.

3) Find the coordinates of the centre, the foci and the vertices of each of the rectangular hyperbolas given in Questions 1 and 2 and sketch each curve.

4) Find the equation of the tangent and the normal to each of the following hyperbolas at the specified points:

(a) $xy = 16; \left(4t, \dfrac{4}{t}\right)$ (b) $xy = 1; (2, \frac{1}{2})$

(c) $xy = -9; \left(-3t, \dfrac{3}{t}\right)$ (d) $xy = 4$; each vertex.

5) Find the equations of the tangents to the rectangular hyperbola $xy = 4$ that
(a) have a gradient of $-\frac{1}{2}$,
(b) pass through the point $(2, 0)$.

6) The normal at the point $A(6, \frac{3}{2})$ on the rectangular hyperbola $x = 3t$, $y = \dfrac{3}{t}$ meets the curve again at B. Find the coordinates of B and the length of AB.

7) The tangent at $P(5, 2)$ on the curve $xy = 10$ meets the asymptotes at Q and R. Find the length of QR.

8) Repeat Question 7 for the normal at P.

9) Prove that the line $y + m^2x = 2cm$ touches the rectangular hyperbola $xy = c^2$ for all values of m. Find the coordinates of the point of contact.

10) A normal chord AB is drawn at the point $A(4, 1)$ on the rectangular hyperbola $xy = 4$. Find its length.

11) Prove that the area of the triangle bounded by the asymptotes and a tangent to the rectangular hyperbola $xy = c^2$ is constant.

12) The tangent at a point on the curve $x = ct$, $y = \dfrac{c}{t}$ meets the x and y axes at P and Q respectively. The normal at the same point meets the lines $y = x$ and $y = -x$ at R and S respectively. Prove that PRQS is a rhombus unless $t^2 = 1$.

13) P and Q are variable points on the hyperbola $xy = 9$. The tangents at P and Q meet at R. If PQ passes through the point $(6, 2)$, find the equation of the locus of R.

14) Find the condition that the line $lx + my + n = 0$ shall be a tangent to the hyperbola $xy = c^2$.

15) The tangent at a point P on the hyperbola $xy = c^2$ meets the asymptotes at A and B. Prove that P bisects AB.

16) $A(x_1, y_1)$ and $B(x_2, y_2)$ are the ends of a diameter of a circle. Show that the equation of the circle is

$$(x-x_1)(x-x_2)+(y-y_1)(y-y_2) = 0$$

If A and B are also points on opposite branches of the hyperbola $xy = c^2$ and the circle on AB as diameter cuts the hyperbola again at C and D, prove that CD is a diameter of the hyperbola.

17) A line drawn from the centre of a rectangular hyperbola, perpendicular to the tangent at $P\left(ct, \dfrac{c}{t}\right)$, meets the tangent at T. As P moves on the hyperbola, find the equation of the locus of T.

18) The tangents at two points P and Q on a rectangular hyperbola, cut one asymptote at A and B and the other asymptote at C and D. Prove that PQ bisects both AB and CD.

THE LINE PAIR

Each of the conic sections we have analysed so far has had a Cartesian equation of the second degree. The last conic section, a pair of straight lines, also has a second order equation as can be seen by considering two lines with equations

$$m_1x - y + c_1 = 0 \quad \text{and} \quad m_2x - y + c_2 = 0.$$

Separately, these equations are linear, but they can be combined to form a single equation

$$(m_1x - y + c_1)(m_2x - y + c_2) = 0$$

This second order equation is satisfied by any point on either line and therefore represents the pair of lines.
Conversely, an equation of order two represents a line pair if it is made up of two linear factors,
e.g. $x^2 + 3y^2 + 4xy + x + 3y = 0$ can be factorized to give $(x + y + 1)(x + 3y) = 0$ and therefore is the equation of the pair of lines

$$x + y + 1 = 0 \quad \text{and} \quad x + 3y = 0.$$

THE LINE PAIR THROUGH THE ORIGIN

The equations of any two lines through the origin can be given in the form

$$m_1x - y = 0 \quad \text{and} \quad m_2x - y = 0.$$

When these equations are combined, the corresponding line pair becomes

$$(m_1x - y)(m_2x - y) = 0$$

i.e.
$$m_1m_2x^2 - (m_1 + m_2)xy + y^2 = 0 \qquad [1]$$

As *every term* in this equation is of the same degree, two, it is called a *homogeneous equation of the second order*.
In general, such a homogeneous equation can be expressed in the form

$$ax^2 + 2hxy + by^2 = 0$$

or
$$\frac{a}{b}x^2 + \frac{2h}{b}xy + y^2 = 0 \qquad [2]$$

and represents a pair of lines through the origin provided that it has real linear factors,

i.e. if
$$\left(\frac{2h}{b}\right)^2 - 4\left(\frac{a}{b}\right) \geqslant 0$$

$\Rightarrow$
$$h^2 - ab \geqslant 0$$

Comparing equations [1] and [2] it can be seen that, when $ax^2 + 2hxy + by^2 = 0$ represents a line pair through O, then the gradients of these lines, m_1 and m_2, are such that

$$m_1 + m_2 = -\frac{2h}{b} \quad \text{and} \quad m_1m_2 = \frac{a}{b}$$

These relationships can be used to find various properties of pairs of lines. For example, the two lines are

(a) perpendicular if m_1 and m_2 are real and $m_1m_2 = -1$,

i.e. if $\quad h^2 - ab \geqslant 0 \quad$ and $\quad \frac{a}{b} = -1 \quad \Rightarrow \quad a + b = 0.$

In this case equation [2] is of the form

$$x^2 + kxy - y^2 = 0 \qquad \left(k = \frac{2h}{a}\right)$$

(b) identical if m_1 and m_2 are real and equal,
i.e. if $h^2 = ab$.
In this case the LHS of equation [2] is a perfect square,

i.e.
$$ax^2 + 2\sqrt{(ab)}\,xy + by^2 = 0$$

$\Rightarrow$
$$(x\sqrt{a} + y\sqrt{b})^2 = 0$$

The Angle Contained by the Line Pair

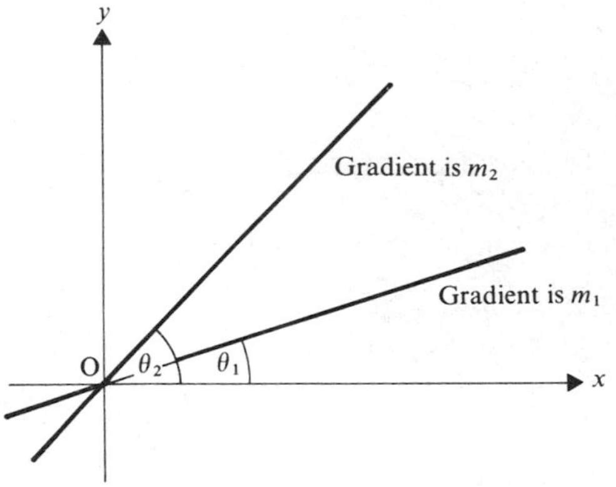

If the two lines are inclined at angles θ_1, θ_2 to the x axis, then

$$\tan\theta_1 = m_1$$

$$\tan\theta_2 = m_2$$

and $\alpha = \theta_2 - \theta_1$.

Thus $$\tan\alpha = \tan(\theta_2 - \theta_1)$$

i.e. $$\tan\alpha = \frac{m_2 - m_1}{1 + m_1 m_2}$$

which can be expressed in terms of a, b and h.

ASYMPTOTES AS LINE PAIRS

Consider the hyperbola $4x^2 - 9y^2 = 36$.
As x and y both become very large, the terms $4x^2$ and $9y^2$ are so large that the term 36 is negligible and the equation of the hyperbola approximates to

$$4x^2 - 9y^2 = 0$$

$$\Rightarrow \quad (2x - 3y)(2x + 3y) = 0$$

so the hyperbola approximates to the pair of lines $2x - 3y = 0$ and $2x + 3y = 0$ for large values of both x and y. These lines are therefore the asymptotes of the hyperbola.
This method can be applied to any second order equation that approximates to a homogeneous equation of the second degree when x and y are both very large, in order to determine whether it has real asymptotes.
A general second order equation can be written in the form

$$ax^2 + 2hxy + by^2 + 2gx + 2fy + c = 0$$

Then if $x \to \infty$ and $y \to \infty$. the terms $2gx, 2fy$ and c are negligible compared with $ax^2, 2hxy$ and by^2.
So the general equation approximates to $ax^2 + 2hxy + by^2 = 0$.
If the curve has real asymptotes, this equation must represent a line pair. Therefore it must have real roots. Thus the general second order curve has real asymptotes if $h^2 - ab \geqslant 0$.

These asymptotes pass through the origin if $g = f = c = 0$. If g, f and c are not all zero the equation $ax^2 + 2hxy + by^2 = 0$ represents a pair of lines parallel to the asymptotes.

EXAMPLES 13h

1) Find the equations of the lines with equation $2x^2 + 5xy - 12y^2 = 0$, and find the angle between these lines.

$$2x^2 + 5xy - 12y^2 \equiv (2x - 3y)(x + 4y)$$

so $2x^2 + 5xy - 12y^2 = 0$ is the combined equation of the lines

$$2x - 3y = 0 \quad \text{and} \quad x + 4y = 0$$

The gradients of these lines are $\frac{2}{3}$ and $-\frac{1}{4}$.

The angle, α, between them is given by

$$\tan\alpha = \frac{\frac{2}{3} - (-\frac{1}{4})}{1 + (\frac{2}{3})(-\frac{1}{4})} = \frac{11}{10} \Rightarrow \alpha = \arctan\frac{11}{10}$$

2) Find the value(s) of λ for which $6x^2 - xy + \lambda y^2 = 0$ represents
(a) two perpendicular lines,
(b) two distinct lines,
(c) two lines inclined at an angle $\dfrac{\pi}{4}$.

Comparing $\quad 6x^2 - xy + \lambda y^2 = 0$

with $\quad ax^2 + 2hxy + by^2 = 0$

we see that $a = 6$, $h = -\frac{1}{2}$ and $b = \lambda$.
The equation represents a pair of real lines if

$$h^2 - ab \geqslant 0$$

i.e. $$\tfrac{1}{4} - 6\lambda \geqslant 0$$

$\Rightarrow$ $$\lambda \leqslant \tfrac{1}{24}$$

(a) The two lines are perpendicular if *also* the product of their gradients $\left(\dfrac{a}{b}\right)$ is -1,

i.e. if $$\frac{6}{\lambda} = -1 \Rightarrow \lambda = -6$$

This value of λ is within the range for which the given equation represents real lines, so $\lambda = -6$ gives a perpendicular line pair.

(b) The lines are distinct if the given equation has real different factors,

i.e. if $$h^2 - ab > 0 \Rightarrow \lambda < \tfrac{1}{24}$$

(c) If the lines are inclined at $\frac{\pi}{4}$, then their gradients, m_1 and m_2, satisfy

$$\tan\frac{\pi}{4} = 1 = \left|\frac{m_1 - m_2}{1 + m_1 m_2}\right|$$

But
$$\begin{aligned}(m_1 - m_2)^2 &\equiv (m_1 + m_2)^2 - 4m_1 m_2\\ &= \left(-\frac{2h}{b}\right)^2 - 4\left(\frac{a}{b}\right)\\ &= \left(\frac{1}{\lambda}\right)^2 - 4\left(\frac{6}{\lambda}\right)\\ &= \frac{1}{\lambda^2}(1 - 24\lambda)\end{aligned}$$

So
$$\left|\frac{m_1 - m_2}{1 + m_1 m_2}\right| = \left|\frac{\sqrt{(1 - 24\lambda)}}{\lambda\left(1 + \frac{6}{\lambda}\right)}\right| = 1$$

i.e. $$1 - 24\lambda = \lambda^2 + 12\lambda + 36$$

$\Rightarrow$ $$\lambda^2 + 36\lambda + 35 = 0$$

$\Rightarrow$ $$(\lambda + 35)(\lambda + 1) = 0$$

$\Rightarrow$ $$\lambda = -35 \quad \text{or} \quad \lambda = -1$$

Both of these values of λ are less than $\frac{1}{24}$, so both correspond to real pairs of lines inclined at $\frac{\pi}{4}$ to each other.

3) A curve has equation $x^2 + kxy + 9y^2 + 3x - 4y + 2 = 0$.
(a) Determine the value(s) of k for which it has real asymptotes.
(b) If $k = 10$ find the equations of lines parallel to its asymptotes.
(c) Comment on the case when $k = 6$.

(a) When $x \to \infty$ and $y \to \infty$, the equation of the given curve approximates to

$$x^2 + kxy + 9y^2 = 0$$

This equation represents a line pair through the origin if

$$\left(\frac{k}{2}\right)^2 - (1)(9) \geqslant 0$$

i.e. if $k^2 \geqslant 36$.

Therefore if k has any value *except* $-6 < k < 6$, the given curve has real asymptotes.

(b) If $k = 10$ (a value for which there are real asymptotes), the asymptotes are parallel to the line pair $x^2 + 10xy + 9y^2 = 0$

$$\Rightarrow \qquad (x + 9y)(x + y) = 0$$

So the asymptotes are parallel to the lines

$$x + 9y = 0 \quad \text{and} \quad x + y = 0$$

(c) If $k = 6$ the asymptotes are real and their combined equation is

$$x^2 + 6xy + 9y^2 = 0$$

$$\Rightarrow \qquad (x + 3y)^2 = 0$$

So for this value of k we have a pair of coincident lines, and there is only one asymptote, a line parallel to $x + 3y = 0$.

EXERCISE 13h

1) Which of the following equations represent a pair of straight lines, distinct or identical, through the origin?

(a) $x^2 + 2xy + y^2 = 0$ (b) $2x^2 - xy + 3y^2 = 0$

(c) $x^2 - y^2 = 0$ (d) $x^2 + y^2 = 0$

(e) $2x^2 + 3xy + y^2 = 0$ (f) $x^2 + 4xy + 3y^2 = 0$.

2) In those parts of Question 1 where a pair of straight lines is represented, write down the equations of these lines.

3) Find the equation of each of the following line pairs through O:

(a) lines with gradients 2 and 3,

(b) lines through (2, 3) and (1, −4) respectively,

(c) lines inclined at $\pm\dfrac{\pi}{3}$ to the x axis.

4) Determine which of the following pairs of lines are

(a) perpendicular (b) coincident

(c) equally inclined to Ox (but not coincident):

(i) $x^2 - y^2 = 0$ (ii) $y^2 - 4x^2 = 0$

(iii) $x^2 + 4xy + 4y^2 = 0$ (iv) $x^2 + 9y^2 = 0$

(v) $3x^2 + 10xy + 3y^2 = 0$.

5) Find the angle between the following pairs of lines:
(a) $2x^2 + 5xy + 2y^2 = 0$ (b) $x^2 - 9y^2 = 0$
(c) $x^2 + 2xy - 3y^2 = 0$ (d) $x^2 + pxy + qy^2 = 0$.

6) Find the value of n if $nx^2 + 2xy + y^2 = 0$ represents
(a) two identical lines (b) two perpendicular lines
(c) two lines inclined at $60°$.

7) Find the condition(s) that the equation

$$px^2 + qxy + ry^2 = 0$$

shall represent (a) two distinct lines,
(b) a pair of perpendicular lines.
If $2p = 2r = q$, what does the equation represent?

8) By considering the set of points $P(x, y)$ that are equidistant from the two lines represented by the equation $ax^2 + 2hxy + by^2 = 0$, show that the equation of the pair of angle bisectors of these lines is

$$hx^2 + (b-a)xy - hy^2 = 0.$$

9) Show that, if α is the acute angle between the lines that form the line pair

$$ax^2 + 2hxy + by^2 = 0$$

then $\tan\alpha = \left|\dfrac{2\sqrt{(h^2-ab)}}{a+b}\right|$. Deduce the condition that the lines shall be

(a) perpendicular (b) coincident.

10) Find the equations of the asymptotes of the hyperbola $\dfrac{x^2}{a^2} - \dfrac{y^2}{b^2} = 1$, and find an expression for the tangent of the angle between them. Hence find the condition for the hyperbola to be rectangular.

11) Without assuming any knowledge of its shape, prove that the ellipse $\dfrac{x^2}{a^2} + \dfrac{y^2}{b^2} = 1$ has no real asymptotes.

12) Determine whether the following curves have real asymptotes. If they have, find the equations of the lines parallel to the asymptotes.
(a) $3x^2 - xy + y^2 - x + 2y = 3$ (b) $x^2 + 2xy - 3y^2 - 2 = 0$
(c) $4x^2 + xy - 5y^2 + 2y = x$ (d) $(x-2)^2 + (y-3)^2 = 1$
(e) $y^2 = 4(x^2 + 1)$ (f) $(2x-1)^2 + (3y+2)^2 = 7$.

THE LINE PAIR THROUGH THE POINTS COMMON TO A GIVEN LINE AND A GIVEN CURVE

Consider a line with equation $L = 0$ [1]

and a curve with equation $C = 0$ [2]

where L is a linear function of x and y, and C is of second degree in x and y.
The coordinates of any point on the line satisfy [1]
and the coordinates of any point on the curve satisfy [2].
So the coordinates of a point common to the line and the curve will give a zero value to any combination of L and C.
The particular combination we are going to consider is the one that uses equation [1] to make equation [2] into a homogeneous equation.
Suppose, for example, that the equations of a given line and curve are

$$2x + y - 4 = 0 \quad [1]$$

and

$$x^2 + 3y^2 - 2x - y - 1 = 0 \quad [2]$$

From [1] we use $1 = \dfrac{2x + y}{4}$ so that any term in [2] which is not already of degree two, can be multiplied by $\left(\dfrac{2x + y}{4}\right)$ or $\left(\dfrac{2x + y}{4}\right)^2$ to give a homogeneous equation which is satisfied by the coordinates of A and B, the points where the given line and curve meet.

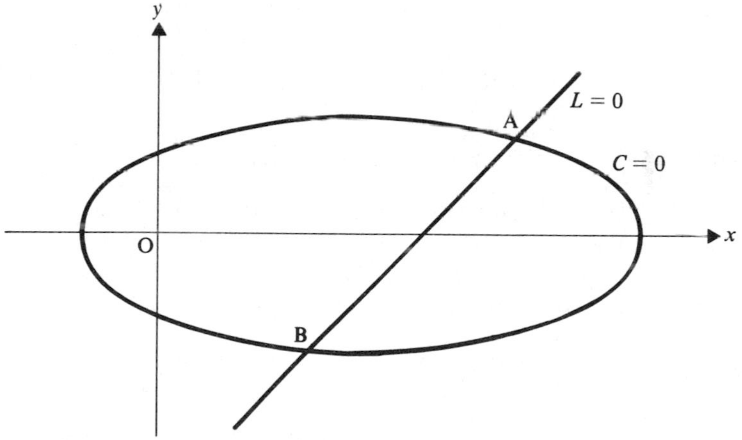

So A and B are points on the conic

$$x^2 + 3y^2 - 2x\left(\frac{2x + y}{4}\right) - y\left(\frac{2x + y}{4}\right) - 1\left(\frac{2x + y}{4}\right)^2 = 0$$

i.e.

$$4x^2 + 20xy - 43y^2 = 0 \quad [3]$$

But this equation is homogeneous of degree two and satisfies the condition $h^2 - ab \geqslant 0$, so it represents a pair of lines through O.
We already know that A and B lie on the conic section with equation [3].
Therefore equation [3] represents the line pair OA and OB.

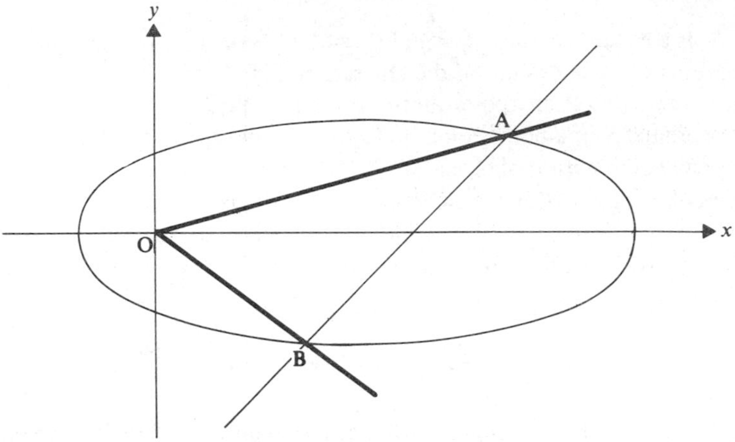

In general, if the line $lx + my + n = 0$ meets the curve $ax^2 + 2hxy + by^2 + 2gx + 2fy + c = 0$ at real points A and B, then the equation of the line pair OA and OB is given by

$$ax^2 + 2hxy + by^2 + (2gx + 2fy)\left(\frac{lx + my}{-n}\right) + c\left(\frac{lx + my}{-n}\right)^2 = 0$$

Note that

(a) if this equation does not satisfy the condition for real factors, we deduce that the given line and curve *do not intersect*.

(b) if the *given line touches the given curve*, then A and B coincide. In this case the homogeneous equation [3], above, represents *two identical lines* and so satisfies the condition for *equal factors*.

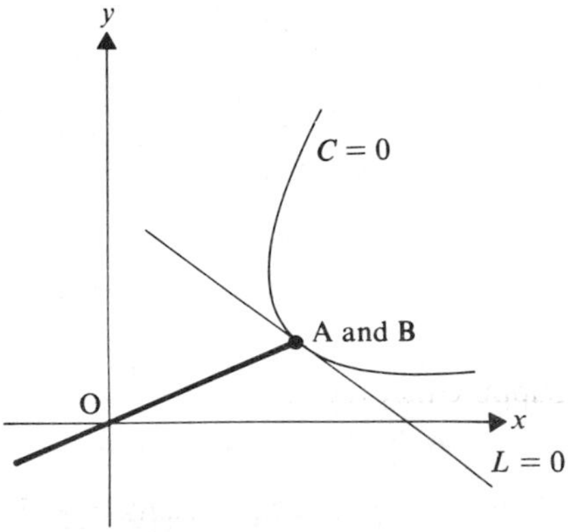

This property provides an alternative method for handling problems on tangency.

POLE AND POLAR

Suppose that, from a point $P(p, q)$ two tangents are drawn to a conic, touching it at $A(x_1, y_1)$ and $B(x_2, y_2)$.

Then AB is the *chord of contact* of the tangents from P.
AB is also called the *polar of P* with respect to the given conic and the point P is called the *pole*.

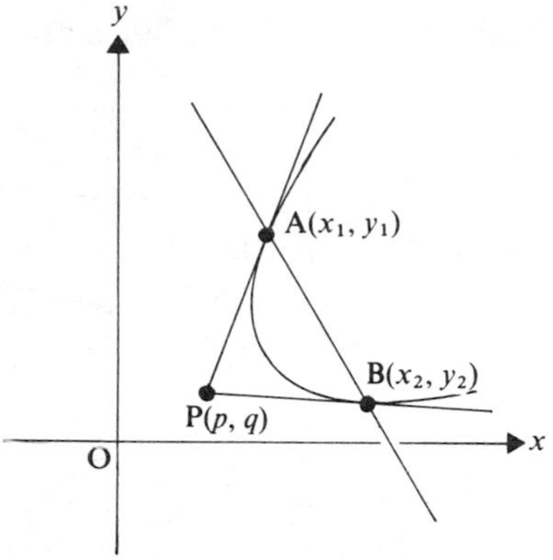

Attempting to find the equation of AB by determining the actual coordinates of A and B in each individual problem is tedious. The equation of the chord of contact is better found by the deductive method illustrated in the following example.

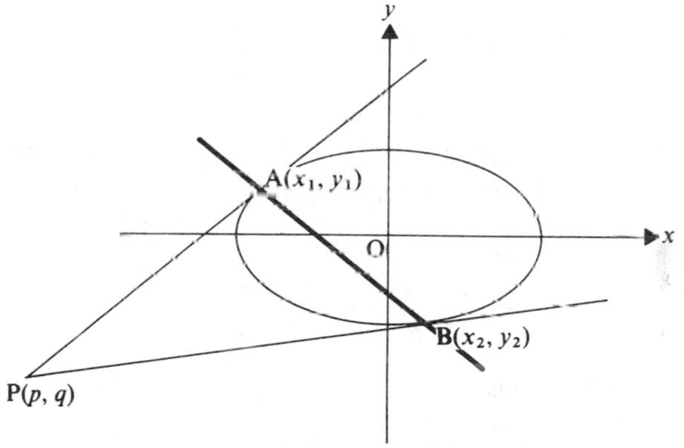

Suppose that the given conic is the ellipse $\frac{x^2}{a^2}+\frac{y^2}{b^2}=1$.

Then at $A(x_1, y_1)$ the equation of the tangent PA is

$$\frac{xx_1}{a^2}+\frac{yy_1}{b^2}=1$$

As $P(p, q)$ is on this tangent,

$$\frac{px_1}{a^2}+\frac{qy_1}{b^2} = 1 \qquad [1]$$

Similarly, considering the tangent PB, we have

$$\frac{px_2}{a^2}+\frac{qy_2}{b^2} = 1 \qquad [2]$$

Now observe the equation

$$\frac{px}{a^2}+\frac{qy}{b^2} = 1 \qquad [3]$$

and note that

(a) it is the equation of a line

(b) it is satisfied by (x_1, y_1) (see equation [1])

(c) it is satisfied by (x_2, y_2) (see equation [2])

(d) it is *not* satisfied by (p, q) $\frac{p^2}{a^2}+\frac{q^2}{b^2} \neq 1$ as (p, q) is not on the ellipse.

So equation [3] represents a line through A and B and is therefore the equation of AB,
i.e. the equation of the *polar* of $P(p, q)$ with respect to the ellipse
$\frac{x^2}{a^2}+\frac{y^2}{b^2} = 1$ is

$$\frac{px}{a^2}+\frac{qy}{b^2} = 1$$

Extending this method to a general conic (i.e. a general second order curve) we find that the polar of $P(p, q)$ to the conic

$$ax^2 + 2hxy + by^2 + 2gx + 2fy + c = 0$$

is
$$apx + h(py + qx) + bqy + g(p + x) + f(q + y) + c = 0$$

Note that although this equation has the *same form* as that of a tangent at a given point on the conic, *the pole, $P(p, q)$ is not on the conic.*
The equation of the polar of a given point with respect to a given conic can be quoted unless derivation is required;
e.g. the polar of the point $(2, 3)$ with respect to the hyperbola $4x^2 - y^2 = 4$ is given by

$$4xp - yq - 4 = 0 \quad \Rightarrow \quad 8x - 3y = 4$$

EXAMPLES 13i

1) Find the equation of the line pair through the origin and the points of intersection of the line $y = mx + 2$ and the parabola $y^2 = 4(x-1)$. Hence find the value(s) of m for which the given line touches the parabola.

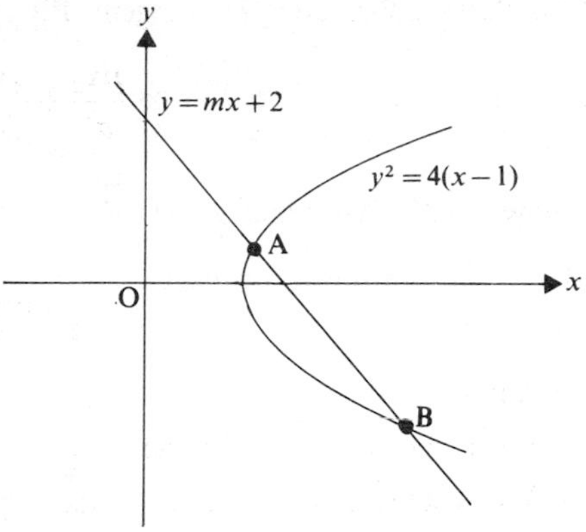

If the line and the parabola meet at A and B, then at A and B

$$\frac{y - mx}{2} = 1$$

and $y^2 - 4x + 4 = 0$.

So the coordinates of A and B satisfy

$$y^2 - 4x\left(\frac{y-mx}{2}\right) + 4\left(\frac{y-mx}{2}\right)^2 = 0$$

i.e. $$(2m + m^2)x^2 - 2(m+1)xy + 2y^2 = 0$$

As this is a homogeneous equation of order two, it can represent a line pair through O.

Thus it represents the pair of lines OA and OB, provided that A and B are real.

If the line $y = mx + 2$ touches the parabola $y^2 = 4(x-1)$ so that the points A and B coincide, then the lines OA and OB also coincide. In this case

$$(2m + m^2)x^2 - 2(m+1)xy + 2y^2 = 0$$

has equal factors (i.e. $h^2 = ab$).

Thus $$(m+1)^2 = (2m + m^2)(2)$$

$\Rightarrow$ $$m^2 + 2m - 1 = 0$$

$\Rightarrow$ $$m = -1 \pm \sqrt{2}$$

2) Write down the equation of the polar of the point (p, q) with respect to the circle $x^2 + y^2 - 6x - 4y + 12 = 0$. Use your result to derive the equation of the pair of tangents from the origin to this circle.

The polar of (p, q) with respect to the given circle is

$$px + qy - 3(x + p) - 2(y + q) + 12 = 0$$

This line is the chord of contact of the tangents drawn to the circle from the point (p, q), so, for tangents from the origin we use $p = 0$, $q = 0$ giving the equation of the chord of contact, AB, as $3x + 2y - 12 = 0$.

Now the equation of the line pair OA, OB is given by using $\dfrac{3x+2y}{12} = 1$ to make the equation of the circle into a homogeneous equation,

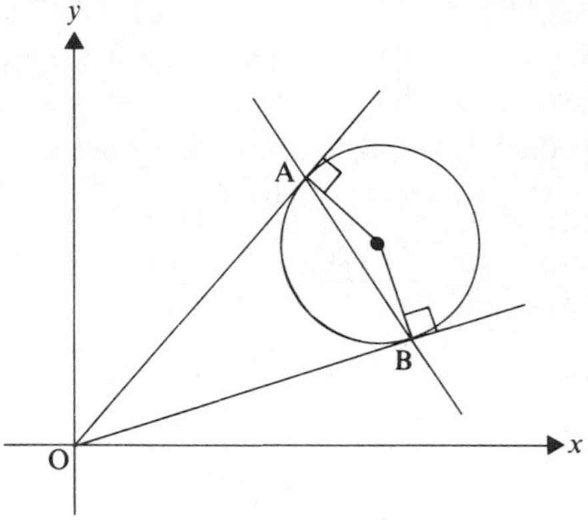

i.e. $$x^2 + y^2 - 6x\left(\frac{3x+2y}{12}\right) - 4y\left(\frac{3x+2y}{12}\right) + 12\left(\frac{3x+2y}{12}\right)^2 = 0$$

$$\Rightarrow \qquad 3x^2 - 12xy + 8y^2 = 0$$

This equation is therefore the equation of the tangents from O to the circle.

EXERCISE 13i

Find the equation of the line pair from the origin through the points of intersection of the following lines and curves:

1) $x + y = 3, \quad 16x^2 + 25y^2 = 400.$

2) $2x - 1 = 4y, \quad x^2 + y^2 = 9.$

3) $4x + 3y = 5, \quad x^2 + y^2 - 4x - 6y = 0.$

4) $y = 2x - 5, \quad y^2 = 9x.$

5) $y = mx + c, \quad x^2 + y^2 = a^2.$

Write down the polar with respect to the given curve of the given point in each of the following cases:

6) $\dfrac{x^2}{a^2} + \dfrac{y^2}{b^2} = 1, \quad (2a, 2b).$

7) $x^2 + y^2 - 2x - 4y + 4 = 0, \quad (3, 5).$

8) $y^2 = 4x, \quad (-2, 1).$

9) $xy = 16$, $(1, 1)$.

10) $x^2 + 2y^2 + 3xy - 4x = 5$, (p, q).

By considering the equation of the pair of lines from the origin through the possible points of intersection of the given line and the given curve, determine in the following questions whether the given line crosses, touches or misses the given curve:

11) $x^2 + y^2 = 4$, $y = 2x + 3$.

12) $x^2 + 2y^2 = 4$, $2y = 5x + 9$.

13) $\dfrac{x^2}{4} + \dfrac{y^2}{9} = 1$, $x + y = 7$.

14) $xy = 16$, $2x + 3y = 4$.

15) $x^2 - y^2 = 9$, $y = 3x - 6\sqrt{2}$.

MISCELLANEOUS EXERCISE 13

1) Prove that the equation of the chord joining the points $P(ap^2, 2ap)$ and $Q(aq^2, 2aq)$ of the parabola $y^2 = 4ax$ is

$$2x - (p + q)y + 2apq = 0$$

S is the focus of the parabola and M is the midpoint of PQ.
The line through S perpendicular to PQ meets the directrix at R.
Prove that

$$2RM = SP + SQ.$$

(U of L)

2) Find the equation of the tangent at $P(at^2, 2at)$ to the parabola $y^2 = 4ax$. The tangent and the normal to the parabola at P meet the x axis at T and G respectively and M is the foot of the perpendicular from P to the line $x = -a$; S is the point $(a, 0)$.
Show that $ST = PM = SP$ and deduce that PT bisects the angle SPM. Show also that PG bisects the angle between PS and MP produced. (U of L)

3) Find the equation of the chord joining the points $P(at_1^2, 2at_1)$ and $Q(at_2^2, 2at_2)$ on the parabola $y^2 = 4ax$.
If the chord PQ passes through the focus of the parabola and the normals at P and Q meet at R, show that the locus of R is given by $y^2 + 3a^2 = ax$.
(U of L)

4) P and Q are the points $(ap^2, 2ap)$, $(aq^2, 2aq)$ respectively on the parabola $y^2 = 4ax$, and M is the midpoint of the chord PQ.
(a) Show that the area, A, enclosed by the curve and the chord PQ is given by $9A^2 = a^4(p - q)^6$.
(b) If $q = p - 4$, give the coordinates of M in terms of p only, and find the equation of the locus of M as the value of p varies continuously. Sketch the two curves on the same diagram.

5) A point P on the parabola $(x-a)^2 = 4ay$ has coordinates $x = a + 2at$, $y = at^2$. Find the equations of the tangent and the normal to the parabola at P.
If the tangent and normal cut the x axis at the points T and N respectively, prove that

$$PT^2/TN = at$$

Find the coordinates of the point Q in which the normal at P intersects the parabola again. (U of L)

6) Find the equation of the tangent to the parabola $y^2 = \frac{1}{2}x$ at the point $(2t^2, t)$.
Hence, or otherwise, find the equations of the common tangents of the parabola $y^2 = \frac{1}{2}x$ and the ellipse $5x^2 + 20y^2 = 4$.
Draw a sketch to illustrate the results. (O)

7) The tangent at $P(a\cos\theta, b\sin\theta)$ to the ellipse

$$b^2x^2 + a^2y^2 = a^2b^2$$

cuts the y axis at Q. The normal at P is parallel to the line joining Q to one focus S'. If S is the other focus, show that PS is parallel to the y axis.
(U of L)

8) Show that for all values of m the straight lines with equation $y = mx \pm \sqrt{(b^2 + a^2m^2)}$ are tangents to the ellipse

$$\frac{x^2}{a^2} + \frac{y^2}{b^2} = 1$$

Hence show that, if the tangents from an external point P to the ellipse meet at right angles, the locus of P is a circle and find its equation. (U of L)

9) It is given that the line $y = mx + c$ is a tangent to the ellipse

$$\frac{x^2}{a^2} + \frac{y^2}{b^2} = 1 \quad \text{if} \quad a^2m^2 = c^2 - b^2$$

Show that if the line $y = mx + c$ passes through the point $(\frac{5}{4}, 5)$ and is a tangent to the ellipse $8x^2 + 3y^2 = 35$, then $c = \frac{35}{3}$ or $\frac{35}{9}$.
Find the coordinates of the points of contact of the tangents from the point $(\frac{5}{4}, 5)$ to the curve $8x^2 + 3y^2 = 35$. (U of L)

10) Show that the equation of the normal at $P(a\cos\theta, b\sin\theta)$ to the ellipse $b^2x^2 + a^2y^2 = a^2b^2$ is

$$ax\sin\theta - by\cos\theta = (a^2 - b^2)\sin\theta\cos\theta$$

The normal at P meets the x axis at A and the y axis at B.
Show that the area of the triangle OAB, where O is the origin, cannot exceed $(a^2 - b^2)^2/(4ab)$. Find the equation of the locus of the centroid of the triangle OAB. (U of L)

11) Prove that the equation of the chord of the ellipse $\dfrac{x^2}{a^2}+\dfrac{y^2}{b^2}=1$ joining the points $(a\cos\theta, b\sin\theta)$ and $(a\cos\phi, b\sin\phi)$ is

$$\frac{x}{a}\cos\tfrac{1}{2}(\theta+\phi)+\frac{y}{b}\sin\tfrac{1}{2}(\theta+\phi) = \cos\tfrac{1}{2}(\theta-\phi)$$

Prove that, if this chord touches the ellipse

$$\frac{x^2}{a^2}+\frac{y^2}{b^2}=\frac{1}{2}$$

θ and ϕ differ by an odd multiple of $\dfrac{\pi}{2}$. (AEB, '71)

12) The line $y = mx + c$ cuts the ellipse $x^2 + 4y^2 = 16$ in the points P and Q. Show that the coordinates of M, the midpoint of PQ, are

$$x = -4mc/(4m^2+1),\quad y = c/(4m^2+1)$$

If the chord PQ passes through the point $(2, 0)$, show that M lies on the ellipse $x^2 + 4y^2 = 2x$. Sketch the two ellipses in the same diagram.

(U of L)

13) The tangent and the normal at a point $P(3\sqrt{2}\cos\theta, 3\sin\theta)$ on the ellipse $x^2/18 + y^2/9 = 1$ meet the y axis at T and N respectively. If O is the origin, prove that OT . ON is independent of the position of P. Find the coordinates of X, the centre of the circle through P, T and N. Find also the equation of the locus of the point Q on PX produced such that X is the midpoint of PQ. (AEB, '72)

14) Show that the equation of the chord joining the points $P(a\cos\phi, b\sin\phi)$ and $Q(a\cos\theta, b\sin\theta)$ on the ellipse $b^2x^2 + a^2y^2 = a^2b^2$ is

$$bx\cos\tfrac{1}{2}(\theta+\phi)+ay\sin\tfrac{1}{2}(\theta+\phi) = ab\cos\tfrac{1}{2}(\theta-\phi)$$

Prove that, if the chord PQ subtends a right angle at the point $(a, 0)$, then PQ passes through a fixed point on the x axis. (U of L)

15) A point P moves so that its distances from $A(a, 0)$, $A'(-a, 0)$, $B(b, 0)$, $B'(-b, 0)$ are related by the equation $AP.PA' = BP.PB'$. Show that the locus of P is a hyperbola and find the equations of its asymptotes. (U of L)

16) Two diameters of the ellipse $b^2x^2 + a^2y^2 - a^2b^2 = 0$ are *conjugate* if each bisects all chords parallel to the other. If the gradients of two conjugate diameters are m and m', prove that $a^2mm' + b^2 = 0$.
If the conjugate diameters are AB, CD prove that

$$AB^2 + CD^2 = 4(a^2+b^2)$$

Prove that the area of the parallelogram formed by tangents to the ellipse at A, B, C, D is $4ab$. (U of L)

17) The tangents to the hyperbola $b^2x^2 - a^2y^2 = a^2b^2$ at points A and B on the curve meet at T. If M is the midpoint of AB, prove that TM passes through the centre of the hyperbola. Prove that the product of the gradients of AB and TM is constant. (U of L)

18) Show that the equation of the tangent to the ellipse $x^2/a^2 + y^2/b^2 = 1$ at the point $(a \cos \theta, b \sin \theta)$ is

$$\frac{x \cos \theta}{a} + \frac{y \sin \theta}{b} = 1$$

P is any point on the ellipse and the tangent at P meets the coordinate axes at Q, R. If P is the midpoint of QR, show that P lies on a diagonal of the rectangle which circumscribes the ellipse and has its sides parallel to the axes of coordinates.
Find the equation of the locus of the midpoint of QR. (U of L)

19) Write down the equations of the two asymptotes of the hyperbola $x^2/9 - y^2/16 = 1$.
The tangent to the hyperbola at the point $P(3 \sec \theta, 4 \tan \theta)$ meets the asymptotes at X and Y. Show that
(a) P is the midpoint of XY,
(b) If O is the origin, the area of the $\triangle$XOY is independent of θ. (U of L)

20) Obtain the equations of the tangent and the normal to the ellipse $b^2x^2 + a^2y^2 = a^2b^2$ at the point $P(a \cos \phi, b \sin \phi)$.
The normal at P meets the axes at Q and R, and the midpoint of QR is M. Show that the locus of M is an ellipse.
Sketch the two ellipses and show that they have the same eccentricity. (U of L)

21) Show that the straight line

$$x \cos \alpha + y \sin \alpha = p$$

is a tangent to the hyperbola

$$\frac{x^2}{a^2} - \frac{y^2}{b^2} = 1$$

if

$$a^2 \cos^2\alpha - b^2 \sin^2\alpha = p^2$$

Find the coordinates of the point of contact.
Obtain the equations of the tangents to the hyperbola $9x^2 - 16y^2 = 144$ which touch the circle $x^2 + y^2 = 9$. (U of L)

22) Prove that the ellipse $4x^2 + 9y^2 = 36$ and the hyperbola $4x^2 - y^2 = 4$ have the same foci, and that they intersect at right angles.
Find the equation of the circle through the points of intersection of the two conics. (U of L)

23) Prove that the gradient of the tangent at an extremity of a latus rectum of the hyperbola

$$\frac{x^2}{9}-\frac{y^2}{7} = 1$$

is equal to the eccentricity of the hyperbola. (AEB, '67)p

24) Find the coordinates of the points of intersection of the ellipse $\frac{x^2}{a^2}+\frac{y^2}{b^2}= 1\ (a > b)$ and the hyperbola $x^2 - y^2 = c^2$.
If the curves cut at right angles at each of these points, find the relation between a, b and c. (O)

25) Find the equation of the tangent to the hyperbola $b^2x^2 - a^2y^2 = a^2b^2$ at the point $(a \sec t, b \tan t)$ and show that the equation of the normal to the curve at this point is

$$ax \sin t + by = (a^2 + b^2) \tan t$$

Show that the product of the areas of the two triangles formed by each of these lines and the coordinate axes is independent of t.
Find the locus of the circumcentre of the triangle formed by the tangent and the coordinate axes. (U of L)

26) A hyperbola of the form $x^2/\alpha^2 - y^2/\beta^2 = 1$ has asymptotes $y^2 = m^2x^2$ and passes through the point $(a, 0)$. Find the equation of the hyperbola in terms of x, y, a and m.
A point P on this hyperbola is equidistant from one of its asymptotes and the x axis. Prove that, for all values of m, P lies on the curve

$$(x^2 - y^2)^2 = 4x^2(x^2 - a^2)$$ (U of L)

27) Find the equation of the normal to the hyperbola $x^2/a^2 - y^2/b^2 = 1$ at the point $P(a \sec \theta, b \tan \theta)$.
If there is a value of θ such that the normal at P passes through the point $(2a, 0)$, show that the eccentricity of the hyperbola cannot be greater than $\sqrt{2}$.
Show that in this case, the parameter ϕ of the point on the hyperbola where the normal passes through the point $(0, -2b)$, is such that $-1 \leqslant \tan \phi < 0$.
If the normal at any point P meets the y axis at L, find the locus, as θ varies, of the midpoint of PL. (AEB '73)

28) If a given line l meets the hyperbola

$$\frac{x^2}{a^2}-\frac{y^2}{b^2} = \lambda$$

at P_1 and P_2, show that the midpoint M of P_1P_2 is independent of λ.
If the coordinates of M are (x_0, y_0), find the equation of l. (O)

29) Find the gradient of the tangent to the hyperbola $x^2/a^2 - y^2/b^2 = 1$ at the point $(a \sec\theta, b \tan\theta)$ and deduce that the equation of this tangent is $bx \sec\theta - ay \tan\theta = ab$.
If this tangent passes through a focus of the ellipse $x^2/a^2 + y^2/b^2 = 1$, show that it is parallel to one of the lines $y = x$, $y = -x$ and that its point of contact with the hyperbola lies on a directrix of the ellipse. (U of L)

30) Show that the equation

$$\frac{x^2}{29-c} + \frac{y^2}{4-c} = 1$$

represents:
(a) an ellipse if c is any constant less than 4,
(b) a hyperbola if c is any constant between 4 and 29.
Show that the foci of each ellipse in (a) and each hyperbola in (b) are independent of the value of c.
If $c = 13$ find the coordinates of the vertices A and B of the hyperbola.
If P and Q are points on this hyperbola such that PQ is a double ordinate, prove that the locus of the intersection of AP and BQ is an ellipse. (AEB '72)

31) Find the equation of the normal to the hyperbola $xy = c^2$ at the point P, whose coordinates are $(ct, c/t)$. Find the coordinates of the point P′ in which the normal at P cuts the hyperbola again, and write down the coordinates of the point P″ in which the normal at P′ cuts the hyperbola again.
Find the equation of the locus of the midpoint of PP′. (U of L)

32) Prove that the chord joining the points $P(cp, c/p)$ and $Q(cq, c/q)$ on the rectangular hyperbola $xy = c^2$ has the equation

$$x + pqy = c(p+q)$$

Three points P, Q, R are given on the rectangular hyperbola $xy = c^2$. Prove that
(a) if PQ and PR are equally inclined to the axes of coordinates, then QR passes through the origin O,
(b) if angle QPR is a right angle, then QR is perpendicular to the tangent at P. (C)

33) Two points $P(4p, 4/p)$ and $Q(4q, 4/q)$ lie on one branch of the rectangular hyperbola $xy = 16$. If the line LPQM meets the axes at L and M, show that LP = QM.
The tangent at a point T on the other branch of the rectangular hyperbola meets the axes at R and S. Prove that TR = TS.
The tangents to the hyperbola at P and Q meet at U. Show that if PQ and RS are parallel, the points T and U are collinear with the origin. (U of L)

34) The tangent and normal at the point $P(ct, c/t)$ of the hyperbola $xy = c^2$ meet the x axis at Q and R. Find the equation of the circle on QR as diameter (which, of course, passes through P).
If this circle passes through the point of the hyperbola with parameter $-t$, prove that $t^4 = \frac{1}{3}$, and that the circle then passes also through the points with parameters $1/t$ and $-1/t$. (O)

35) The line $y = mx$ meets the hyperbola $xy = c^2$, where $c > 0$, at the points R and S. Prove that the tangents to the hyperbola at R and S are parallel. Find the distance between the parallel tangents and show that, as m varies, the maximum distance between them is $2c\sqrt{2}$.
The tangents and normals at R and S together form a rectangle.
Find the area of the rectangle and show that, when $m = 3$, the area is $6.4c^2$.
(U of L)

36) Prove that the equation of the chord joining the points $P(cp, c/p)$ and $Q(cq, c/q)$ on the rectangular hyperbola $xy = c^2$ is

$$x + pqy = c(p+q)$$

If this chord is also the normal at P, prove that $p^3q + 1 = 0$.
If, in this case, the normal at Q cuts the hyperbola again at R, prove that PR has the equation

$$x + p^{10}y = cp(1+p^8).$$ (C)

37) Write down the coordinates of the centre and the equations of the asymptotes of the rectangular hyperbola

$$(x-h)(y-k) = c^2$$

Sketch the hyperbolae $2x(y-2) = 3$ and $2y(x-1) = 3$ and find the coordinates of the points P and Q in which they intersect.
Show that the tangents to the hyperbolae at P and Q form a parallelogram.
(U of L)

38) Find the equation of the tangent to the curve $xy = c^2$ at the point $P(cp, c/p)$.

Show that the tangents at P and $Q\left(cq, \dfrac{c}{q}\right)$ can never be perpendicular and

that when they are parallel the line PQ passes through the origin, O.
Find the coordinates of R, the point of intersection of the tangents at P and Q, and show that the line RO, produced if necessary, passes through the midpoint of PQ. (AEB, '78)

39) Find the equation of the normal to the rectangular hyperbola $xy = c^2$

at the point $P\left(ct, \dfrac{c}{t}\right)$.

The normal at P meets the curve again at Q. Determine the coordinates of Q. QR is the diameter through Q of the hyperbola. Show that the locus of the midpoint of PR as P varies is

$$4x^3y^3 = c^2(x^2+y^2)^2.$$ (U of L)

40) The straight line $y = mx + b$ meets the coordinate axes at P_1 and Q_1; it also meets the rectangular hyperbola $xy = c^2$ at P and Q. Prove that P_1Q_1 and PQ have the same midpoint.
If a set of parallel lines is drawn to cut the hyperbola, prove that the midpoints of the chords so formed lie on a straight line through the origin. (C)

41) The foot of the perpendicular from a point P to the straight line

$$x+y = \sqrt{2}$$

is the point R, and Q is the point with coordinates $(\sqrt{2}, \sqrt{2})$. If P varies in such a way that $PQ^2 = 2PR^2$, show that its locus is the rectangular hyperbola

$$xy = 1$$

Find the equation of the tangent to this hyperbola at the point $(t, 1/t)$. This tangent cuts the x axis at A and the y axis at B, and C is the point on AB such that $AC : CB = a : b$. Show that the locus of C as t varies is the rectangular hyperbola

$$xy = \frac{4ab}{(a+b)^2}$$

Determine the two possible values of the ratio $a : b$ such that the straight line

$$x+y = \sqrt{2}$$

is a tangent to the locus of C. (JMB)

42) The chord through two variable points P and Q on the rectangular hyperbola $xy = c^2$ cuts the x axis at R. If S is the mid-point of PQ and O is the origin, prove that the triangle OSR is isosceles. Show that, if OP, OQ and OS make angles θ_1, θ_2 and θ_3 respectively with OR, then

$$\tan^2\theta_3 = \tan\theta_1 \tan\theta_2.$$ (AEB, '71)

43) The line of gradient m $(\neq 0)$ through the point $A(a, 0)$ is a tangent to the rectangular hyperbola $xy = c^2$ at the point P. Find m in terms of a and c, and show that the coordinates of P are $(\frac{1}{2}a, 2c^2/a)$.
The line through A parallel to the y axis meets the hyperbola at Q, and the line joining Q to the origin O intersects AP at R. Given that OQ and AP are perpendicular to each other, find the numerical value of c^2/a^2 and the numerical value of the ratio AR : RP. (JMB)

44) Find the equation of the pair of lines joining the origin to the points where the line $x + 2y - 5 = 0$ meets the pair of lines

$$4x^2 - 15xy - 4y^2 + 39x + 65y - 169 = 0$$

Find also the angle between the pair of lines through the origin.

45) Write down the equation of the chord of contact of the tangents from the origin to the circle

$$x^2 + y^2 + 6x + 4y + 4 = 0$$

Prove that the area of the triangle formed by these tangents and the chord of contact is $\frac{24}{13}$ square units.

46) Prove that the line $lx + my + n = 0$ touches the parabola $y^2 = 4ax$ if $am^2 = nl$. If the polar of a point P with respect to the ellipse $4(x+1)^2 + y^2 = 4$ touches the parabola $y^2 = 4x$, show that the equation of the locus of P is $16x(x+1) = y^2$.

47) Find the equation of the pair of lines that pass through the origin and through the points of intersection of the line $y = \lambda(x+1)$ and the circle $(x-2)^2 + (y-2)^2 = 9$. If these lines are perpendicular, find λ.
For what value of λ do these lines coincide? Deduce the equations of the tangents to the circle from the point $(1, 0)$.

48) Show that the equation of the pair of straight lines joining the origin to the points A and B where the line $lx + my = 1$ meets the conic $ax^2 + by^2 = 1$, is

$$(a - l^2)x^2 - 2lmxy + (b - m^2)y^2 = 0$$

If angle AOB is a right angle, show that AB touches the circle

$$(a + b)(x^2 + y^2) = 1.$$

49) Find the equation of the line pair obtained by rotating the pair of lines with equation $x^2 + 3xy + 2y^2 = 0$ through an angle of $\frac{\pi}{3}$ about the origin.

50) One of the medians of the triangle formed by the line pair $ax^2 + 2hxy + by^2 = 0$ and the line $px + qy = r$ lies along the y axis. If neither a nor r is zero, prove that $bp - hq = 0$. (U of L)

51) Form the equation of the straight lines joining the origin to the points of intersection of $ax^2 + 2hxy + by^2 + 2gx + 2fy + c = 0$ and $px + qy + r = 0$, and write down the condition that these lines should be at right angles.
If this condition is satisfied, show that the equation of the locus of the foot of the perpendicular from the origin to the line $px + qy + r = 0$ is

$$(a + b)(x^2 + y^2) + 2gx + 2fy + c = 0.$$ (U of L)

CHAPTER 14

GROUPS

SYMMETRY

Groups arise in many contexts in mathematics and perhaps the simplest interesting examples are symmetry groups of plane figures.
Consider an equilateral triangle ABC in the plane

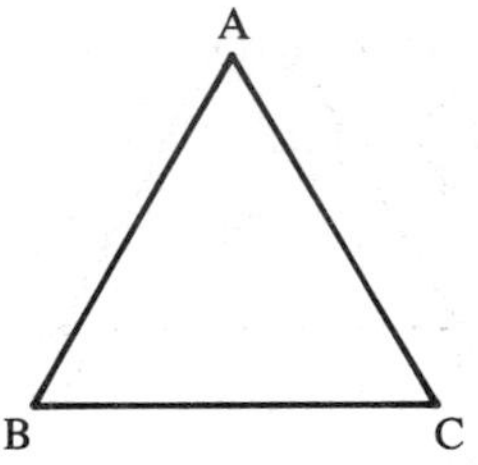

(where the letters A, B, C are labels for points of the plane).
It has several 'symmetries'; for example it is symmetrical about the median through A (perpendicular to BC),

i.e.

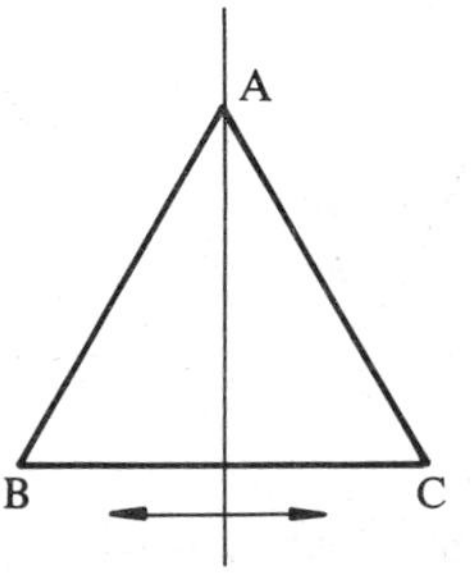

and it has rotational symmetry about its centre

i.e.

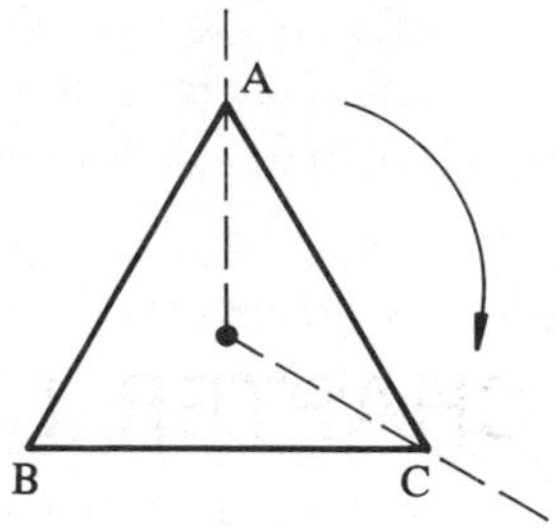

By a *symmetry* of a figure we mean any rigid transformation which leaves the figure 'looking the same' i.e. occupying the same region of the plane. Rigid transformations are rotations, reflections, translations or combinations of these. The symmetries mentioned above are the transformations 'reflect in the median through A' and 'rotate through 120° clockwise about the centre'.

There are in fact *six* symmetries of the triangle ABC, as follows.
Three reflective symmetries,

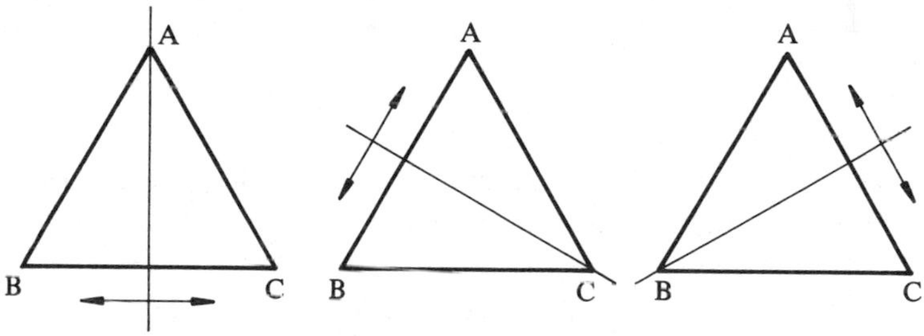

two rotational ones,

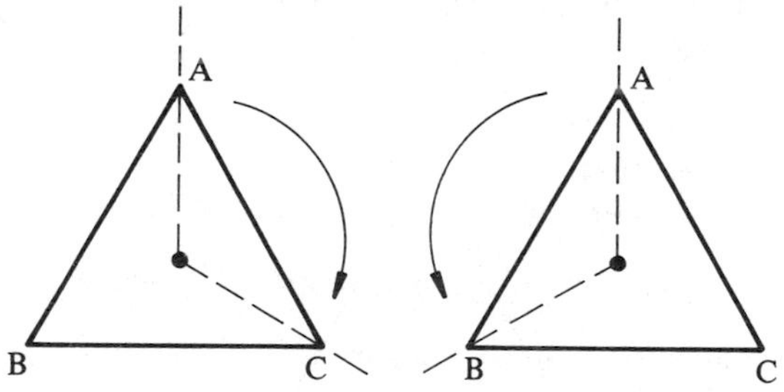

and one which is the transformation which leaves the triangle alone (i.e. the instruction 'do nothing'). This symmetry is called the *identity* symmetry.
It may seem odd that we include the identity symmetry 'do nothing' in our list

of symmetries of the triangle. After all *every* figure has this symmetry, even unsymmetrical ones! But, as we shall see later, this symmetry is very important; leaving it out would be like trying to do addition without the number 0.

Since symmetries are transformations, they can be *combined*. Given two symmetries we can perform one of them first and then follow it with the other one; the result is another symmetry. For example, suppose we perform the symmetry reflect in the median through A' and then follow this by 'rotate through $120°$ clockwise'. The result is illustrated in the following diagrams (where we have labelled the triangle with a cross and a dot in order to follow the movements).

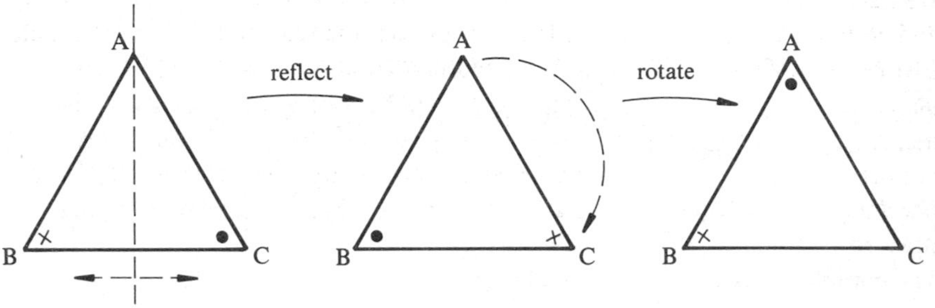

The fact that the letters A, B, C are labels for points of the *plane* is now important because it implies that *the letters do not move with the triangle.* Looking at the end result, we see that the effect is exactly the same as if we had reflected in the median through B. Expressing this in symbols we write

$$r \circ a = b$$

where a means 'reflect in the median through A'
r means 'rotate through $120°$ clockwise'
b means 'reflect in the median through B'
and the symbol $\circ$ means 'following'.

The collection of symbols $r \circ a$ means 'r following a',
i.e. *first* reflect (a) *then* rotate (r).
In other words, $r \circ a$ is performed *from right to left.*

$r \circ a$ is called the *composition* of r with a.

Clearly we can combine (or compose) our six symmetries in a total of 36 ways and in order to display the results it is convenient to use a table. Let us use the symbols e, s, c for the other symmetries
i.e. s means 'rotate through $120°$ anticlockwise'
c means 'reflect in the median through C'
and e represents the identity symmetry 'do nothing'.

$\circ$	e	r	s	a	b	c
e						
r				b		
s						
a						
b						
c						

We have found that $r \circ a = b$ so we write b in the square formed by row r and column a. As an exercise it is suggested that the reader completes the table before continuing. It will probably be found helpful to use a model triangle made out of card and marked as in the diagrams above. *Both* sides should be marked because a reflection is performed by turning the triangle over. It is important to bear in mind the two conventions: firstly, the letters A, B, C are *fixed* and *do not move with the triangle* and secondly, $x \circ y$ means y *first, then* x.
The completed table should look like this:

$\circ$	e	r	s	a	b	c
e	e	r	s	a	b	c
r	r	s	e	b	c	a
s	s	e	r	c	a	b
a	a	c	b	e	s	r
b	b	a	c	r	e	s
c	c	b	a	s	r	e

Here is the working for some of the cases:

$s \circ a$:

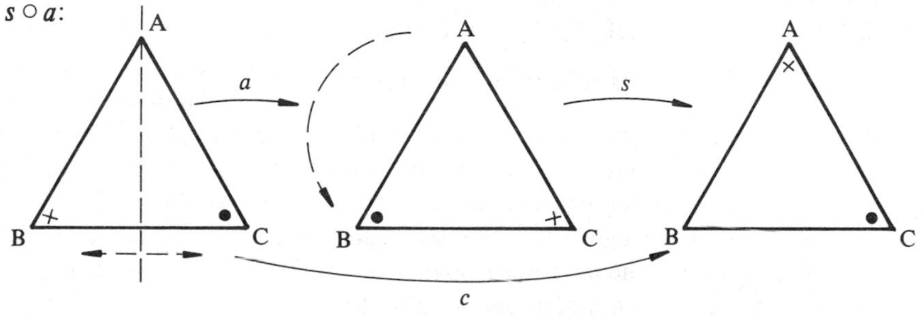

$a \circ e$:

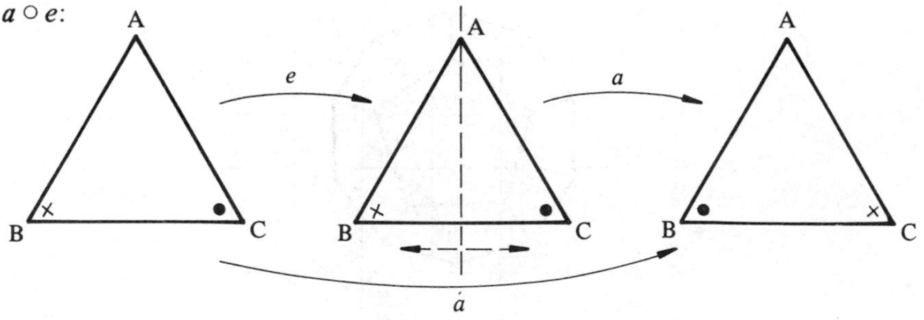

$a \circ c$:

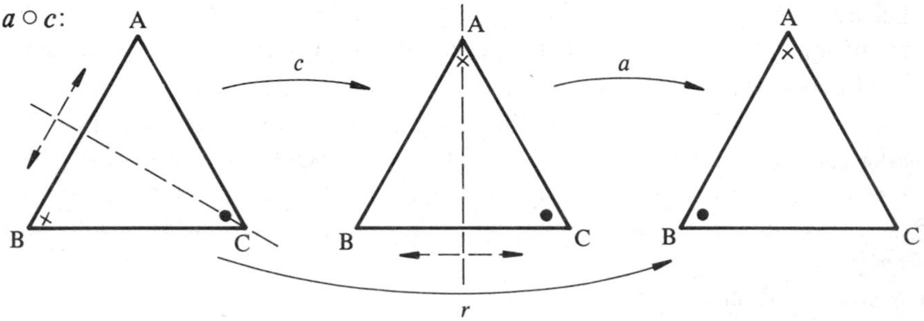

Note that it is by no means universally accepted that $x \circ y$ means '*y first, then x*' or '*x* following *y*': a considerable number of authors use it to mean '*x* then *y*'. Although this latter convention may seem more natural, it is at variance with the usual notation for composing transformations; we have already used the convention that $x \circ y$ means '*y first, then x*' in our work on transformations and matrices in Chapters 1 and 2 and in composition of functions. Notice that the order of composition *is* important: *$x \circ y$ is usually different from $y \circ x$.* For example (referring to the table) $a \circ b = s \neq b \circ a = r$.

The set of six symmetries of the triangle together with the operation of composition is called the *symmetry group* of the triangle ABC.

The completed table of compositions is called the *group table* for this symmetry group; other names often used are 'multiplication' table, 'combination' table and 'Cayley' table.

Symmetry groups for solid figures can be constructed in exactly the same way as for plane figures. Although we shall not deal with them here, the reader may like to investigate some of these. The resulting tables can be quite large. The dodecahedron for instance has 120 symmetries.

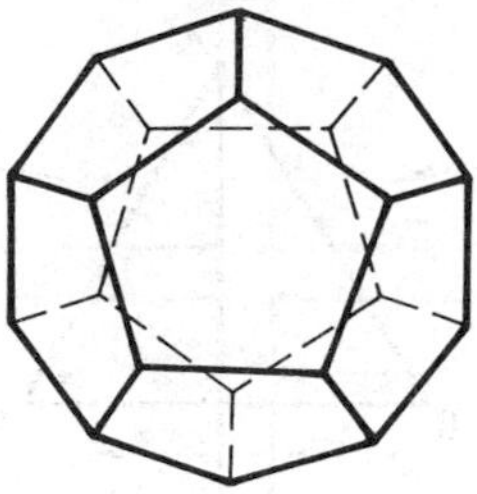

EXERCISE 14a

Construct group tables for the symmetry groups of the following plane figures. In each case it is necessary to first write down all the symmetries (using suitable notation) and then to complete the table (using a cut-out model if this is found helpful).

1) An isosceles triangle:

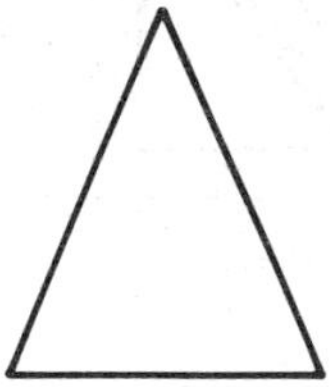

2) A rectangle:

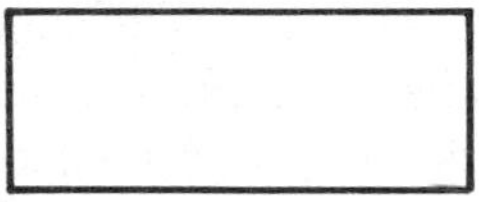

3) A 'dog-leg':

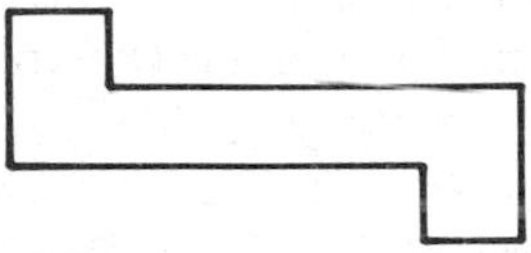

4) A three-bladed 'windmill':

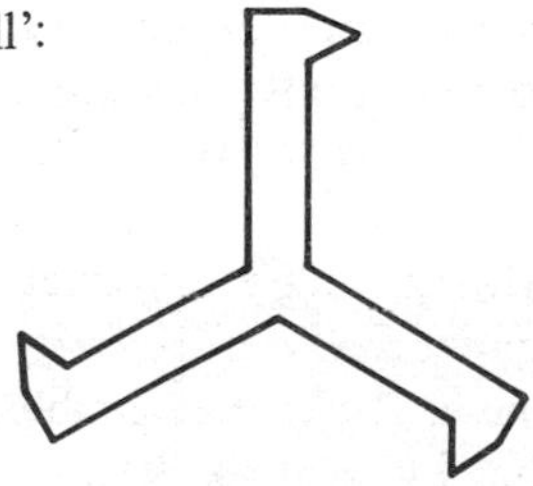

5) A four-bladed 'windmill':

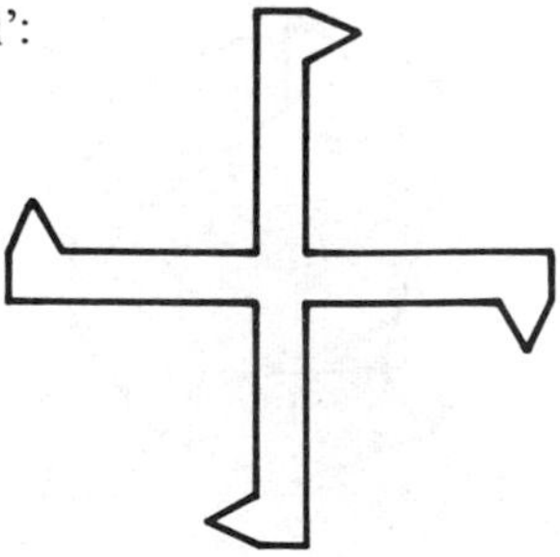

6) A five-bladed 'windmill':

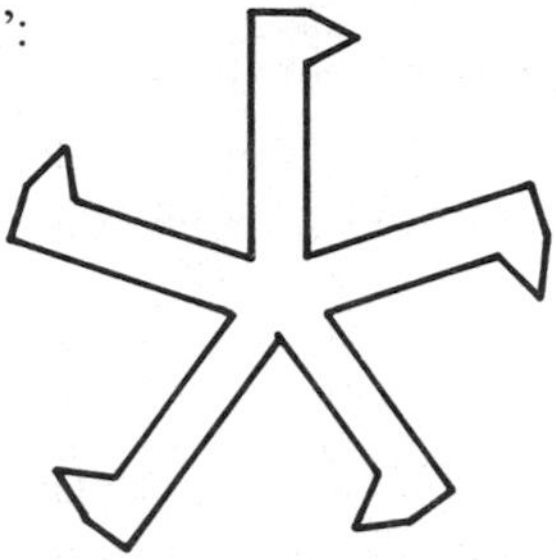

7) A square:

8) An irregular triangle:

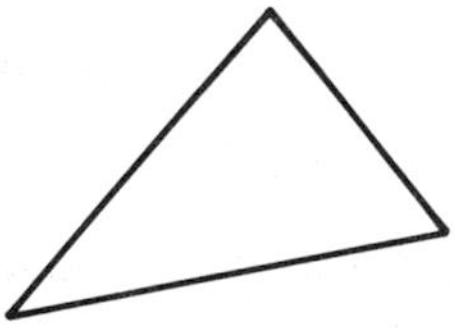

GROUPS

A *group* comprises a *set* G (whose members are the *elements* of the group) and an *operation* $*$ on this set which *together* satisfy four properties:

(1) *Closure*

Given any two elements $g, h \in G$ then the operation gives us an element $g * h$ which is required to be *an element of* G.

(2) *Identity*

There is an element $e \in G$ which is an *identity* for $*$. This means that for each element $g \in G$ we have

$$g * e = e * g = g$$

(3) *Inverse*

Given any element $g \in G$, there is an element g^{-1} (called the *inverse* of g) which satisfies

$$g * g^{-1} = g^{-1} * g = e$$

(4) *Associativity*

Given any three elements g, h, k of G we have

$$g * (h * k) = (g * h) * k$$

If the set G and operation $*$ forms a group then we say

$$(G, *) \text{ is a group.}$$

In order to show that a set of elements together with an operation on the set forms a group, it is necessary to verify all four of the properties listed above.

Symmetry Groups

We shall verify that the set of symmetries of a plane figure forms a group under the operation $\circ$ (composition). In other words we shall prove that symmetry groups are examples of groups. We shall give the proof for a general plane figure F and at the same time we shall refer to the symmetry group of the equilateral triangle as an example.
The *set* G is the set of symmetries of F. Thus in the case of the triangle we have

$$G = \{e, r, s, a, b, c\}$$

The *operation* $*$ is the operation $\circ$, composition of symmetries.

Closure

The closure property says that given any two symmetries g, h of F then $g \circ h$ is also a symmetry of F. In the case of the triangle, we have checked this explicitly by calculating all the compositions (see the table on page 569). In general, since a symmetry of F is just some combination of rigid motions which leaves F occupying the same space, any combination of symmetries must be another symmetry.

Identity

The identity symmetry e is clearly an identity for $\circ$: the equations

$$e \circ g = g \circ e = g$$

simply say that the transformation g followed or preceded by 'do nothing' results in g. In the case of the triangle, we can see that e is an identity by inspecting the group table: the e-row and e-column are exactly the same as the borders.

Inverses

Given any symmetry g of F we can perform the rigid motions which gave g *in reverse*, and this gives us the inverse symmetry g^{-1}. For example, if g is the transformation 'reflect in the line l and then rotate through $x°$ anticlockwise about O' then g^{-1} is the transformation 'rotate through $x°$ *clockwise* about O and then reflect in l'. Thus the effect of g^{-1} is to 'undo' what g does and hence $g \circ g^{-1} = g^{-1} \circ g = e$. In the case of the triangle we can actually read off the inverses from the table. Thus $r^{-1} = s$ since $r \circ s = s \circ r = e$ (from the table) and $a^{-1} = a$, $b^{-1} = b$, $c^{-1} = c$, $s^{-1} = r$ and $e^{-1} = e$. In fact, inspecting the table carefully, we see that element e occurs exactly once in each row x of the table and it occurs in column x^{-1}.

Associativity

Both the compositions

$$(g \circ h) \circ k \quad \text{and} \quad g \circ (h \circ k)$$

mean 'do k first, then h and then g', so they are equal. We can verify some of the cases for the triangle by reference to the table:

$$\begin{cases} (r \circ s) \circ a = e \circ a = a \\ r \circ (s \circ a) = r \circ c = a \end{cases}$$

$$\begin{cases} (a \circ b) \circ r = s \circ r = e \\ a \circ (b \circ r) = a \circ a = e \end{cases}$$

EXAMPLES 14b

1) *The integers* $(\mathbb{Z}, +)$
Prove that the set of integers together with the operation of addition forms a group; i.e. show that $(\mathbb{Z}, +)$ is a group.

This is a very familiar set and operation. The *set* is the set $\mathbb{Z}$ of integers

$$\mathbb{Z} = \{\ldots, -3, -2, -1, 0, 1, 2, 3, \ldots\}$$

and the *operation* $*$ is the familiar addition operation $+$. We need to verify the four properties of a group.

Closure

Given $n, m \in \mathbb{Z}$ (n and m are integers) then $n + m$ is an integer (well-known property of integers).

Identity

The identity in this group is the integer 0 since

$$n+0 = 0+n = n \quad \text{for each integer } n.$$

Inverses

The inverse of n is $-n$ since

$$n+(-n) = (-n)+n = 0 \quad \text{for each integer } n.$$

Associativity

The associativity property of addition

$$n+(m+p) = (n+m)+p$$

for any three integers n, m and p, is another well-known property of integers.

2) *Integers modulo* 3
Show that $(\{0, 1, 2\}, +_3)$ is a group (where $+_3$ is the operation defined below).

In this example the *set* is the set comprising three integers

$$\{0, 1, 2\}$$

and the *operation* $+_3$ is *addition modulo* 3. This is defined by adding and then taking the remainder on division by 3 (the remainder *modulo* 3) so for instance

$$2 +_3 2 = 1$$

since $2+2=4$ which gives remainder 1 on division by 3. We can now construct a table for $+_3$:

$+_3$	0	1	2
0	0	1	2
1	1	2	0
2	2	0	1

and the table can be used to verify the four properties of a group.

Closure

Follows at once from the table, since only members of the set appear in it.

Identity

0 is an identity since

$$0 +_3 l = l +_3 0 = l$$

for each $l = 0, 1$ or 2.

Inverses

$$0 +_3 0 = 0$$

and
$$1 +_3 2 = 2 +_3 1 = 0$$

hence 0 is the inverse of 0, 2 is the inverse of 1, and 1 is the inverse of 2.

Associativity

Checking all the cases of associtivity using the table is very tedious, so we give a 'general' argument instead. Both $l +_3 (m +_3 n)$ and $(l +_3 m) +_3 n$ are the remainder of $l + m + n$ modulo 3 and hence they are equal.

Non-examples of Groups

In order for a set and operation to *fail* to form a group, it is only necessary that *one* of the four properties should fail. As an example we shall show that $(\mathbb{Z}, \times)$ is not a group.
The set is the set of integers and the operation is multiplication. The *closure* and *associative* properties are well-known properties of multiplication. There is an *identity*, namely 1, since

$$n \times 1 = 1 \times n = n$$

for each integer n. But the *inverses* property fails because, for example, 2 has no inverse which is an integer, i.e. there is no integer n such that $n \times 2 = 2 \times n = 1$ (since $2n$ is an even integer for every integer n). Hence $(\mathbb{Z}, \times)$ is *not* a group.

EXAMPLES 14b (continued)

3) Show that $(\mathbb{Z}, *)$ is not a group where $*$ is the operation defined by

$$n * m = nm - n - m + 1$$

The *identity* property fails. Suppose that q is an identity, then we have

$$q * n = n \quad \text{for } each \text{ integer } n.$$

But taking $n = 0$ we have

$$q \times 0 - q - 0 + 1 = 0$$

$$\Rightarrow \qquad q = 1$$

But then

$$q * 1 = 1 - 1 - 1 + 1 = 0 \neq 1$$

so q is not an identity after all! Thus there is *no* identity and hence $(\mathbb{Z}, *)$ is *not* a group.

EXERCISE 14b

1) Using *either* the group tables constructed in Exercise 14a *or* geometrical considerations, write down the inverse of each element in the symmetry groups of (a) the four-bladed windmill and (b) the square. Use the group table to verify three cases of associativity for the symmetry group of the square.

2) Prove that the rationals $\mathbb{Q}$ form a group under addition, i.e. that $(\mathbb{Q}, +)$ is a group.

3) Show that $(\mathbb{R}, +)$ is a group.

4) Does $(\mathbb{Q}, \times)$ form a group?

5) Show that $(\mathbb{Q}^*, \times)$ is a group where $\mathbb{Q}^*$ means the set of rationals *excluding* zero.

6) Show that $(\mathbb{R}^+, \times)$ is a group where $\mathbb{R}^+$ means the set of *strictly positive* real numbers $(r > 0)$.

7) Prove that $(\{0, 1, 2, 3\}, +_4)$ is a group, where $+_4$ (addition modulo 4) is defined by adding and then taking the remainder on division by 4.

8) The operation $\times_3$ on the set $\{0, 1, 2\}$ is defined by multiplying and then taking the remainder modulo 3. Construct a multiplication table for $(\{0, 1, 2\}, \times_3)$ and decide whether it is a group.

9) Is $(\{1, 2\}, \times_3)$ a group?

10) Is $(\{1, 2, 3\}, \times_4)$ a group?
($\times_4$ means multiplication modulo 4).

11) The table defines an operation $\square$ on the set $\{e, a, b\}$. Decide whether $(\{e, a, b\}, \square)$ is a group.

$\square$	e	a	b
e	e	a	b
a	a	e	b
b	b	b	e

12) The table defines an operation $*$ on the set $\{x, y, z\}$. Decide whether $(\{x, y, z\}, *)$ is a group.

$*$	x	y	z
x	z	x	y
y	x	y	z
z	y	z	x

13) Show that $(\mathbb{Q}^*, \div)$ is not a group.

14) Decide whether $(\mathbb{Z}, *)$ is a group where $*$ is defined by

$$n * m = n + m - 1.$$

INTEGERS MODULO *n*

In Example 14b, No. 2 we proved that the set $\{0, 1, 2\}$ with the operation $+_3$ (addition modulo 3) forms a group and in Exercise 14b, Question 7 we saw that the set $\{0, 1, 2, 3\}$ with the operation $+_4$ (addition modulo 4) also forms a group. We shall now show that there is a group of this type *for each integer* $n \geqslant 1$ called the *group of integers modulo* n.

The set is the set

$$\mathbb{Z}_n = \{0, 1, 2, \ldots, n-1\}$$

and the operation $+_n$ (addition modulo n) is defined by adding and then taking the remainder after division by n.

Closure

The remainder is a non-negative integer less than n; i.e. it lies in the set $\mathbb{Z}_n$.

Identity

0 is an identity for $+_n$ since $0 + p = p + 0 = p$ for each $p \in \mathbb{Z}_n$ and hence $0 +_n p = p +_n 0 = p$.

Inverses

0 is clearly the inverse of 0. If $p > 0$ and $p \in \mathbb{Z}_n$, then $n - p \in \mathbb{Z}_n$ and

$$(n-p) + p = p + (n-p) = n$$

which gives remainder 0 on division by n;

i.e. $$(n-p) +_n p = p +_n (n-p) = 0$$

Associativity

Both $p +_n (q +_n r)$ and $(p +_n q) +_n r$ can be found by taking the remainder of $p + q + r$ after division by n. Hence they are equal.

COMMUTATIVITY

We have now seen several examples of groups. In some of these examples the order in which the elements are combined by the operation matters, and in other examples the order does not matter. In the latter case we have a fifth property in addition to the four basic properties of a group given on page 572. as follows.

5) *Commutativity*

Given any two elements g, h of G we have

$$g * h = h * g$$

A group which satisfies this property is called a *commutative* group (or an *Abelian* group). So in a commutative group, the order in which elements are combined by the group operation does not matter. For example the group $(\mathbb{Z}, +)$ is commutative because for each pair of integers $n, m \in \mathbb{Z}$, we have $n + m = m + n$ (another well-known property of numbers).

To show that a group is not commutative it is necessary only to find *one* pair of elements $g, h \in G$ which do not commute $g * h \neq h * g$. Such a group is said to be *non-commutative*. For example the symmetry group of the equilateral triangle is non-commutative because we have seen that $a \circ b = s \neq b \circ a = r$ (page 570).

Commutativity can be interpreted in terms of the group table. The entries $x * y$ and $y * x$ are symmetrically placed with respect to the diagonal:

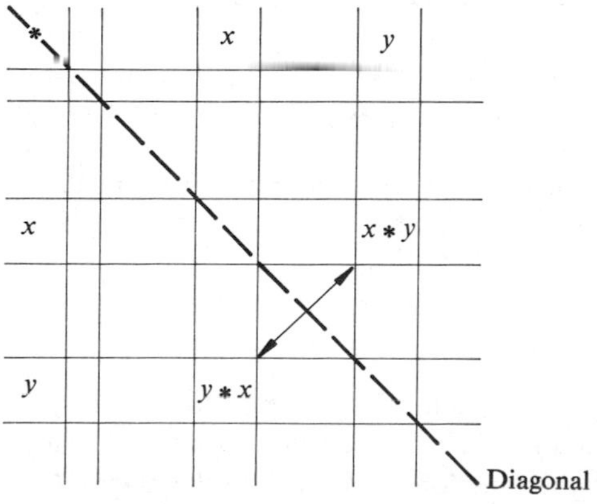

Thus a group is commutative if and only if its group table is symmetric about the diagonal. For example the group table for the group $(\mathbb{Z}_3, +_3)$ is symmetric about the diagonal:

$+_3$	0	1	2
0	0	1	2
1	1	2	0
2	2	0	1

Therefore $(\mathbb{Z}_3, +_3)$ is commutative.

EXERCISE 14c

1) Construct group tables for the groups $(\mathbb{Z}_2, +_2)$ and $(\mathbb{Z}_7, +_7)$ and verify that they are commutative.

2) Prove that the group $(\mathbb{R}, +)$ is commutative.

3) Prove that the symmetry group of the square is non-commutative.

4) Inspect the tables constructed in Exercise 14a and classify the groups as commutative or non-commutative.

5) Prove that the following groups are all commutative:
(a) $(\mathbb{Q}^*, \times)$, (b) $(\mathbb{R}^+, \times)$, (c) $(\mathbb{Z}_4, +_4)$, (d) $(\{1, 2\}, \times_3)$,
(e) the group given by the table in Exercise 14b, Question 12,
(f) the group $(\mathbb{Z}, *)$ of Exercise 14b, Question 14.

6) Construct a table for $(\{0, 2, 4\}, +_6)$ and prove that it is a commutative group (it is necessary to verify all *five* properties for a commutative group).

7) Prove that the group $(\mathbb{Z}_n, +_n)$ is commutative for each $n \geqslant 1$.

PERMUTATIONS

A *permutation* of a set is a function which reorders (or 'shuffles') the members of the set, i.e. a function which 'moves them around'. For instance, consider the set $\{1, 2, 3\}$. Then the function f defined by the diagram

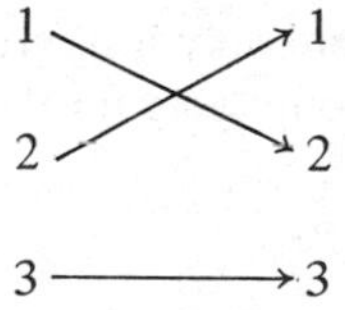

f

is a permutation (f interchanges 1 and 2 and leaves 3 alone).
In order to give a more precise definition of a permutation, we introduce a convenient symbolic notation for permutation.
Write the elements 1, 2, 3 in a row and write the image of each element underneath. Thus f is written

$$f = \begin{pmatrix} 1 & 2 & 3 \\ 2 & 1 & 3 \end{pmatrix}$$

since

$$f\colon \begin{matrix} 1 & 2 & 3 \\ \downarrow & \downarrow & \downarrow \\ 2 & 1 & 3 \end{matrix}.$$

> $\begin{pmatrix} 1 & 2 & 3 \\ 2 & 1 & 3 \end{pmatrix}$ is called the *permutation symbol* for f.

Note that we can recover f from the symbol since each element in the top row is mapped to the element underneath it in the bottom row.
In general, a permutation symbol is any symbol of this type such that the elements 1, 2, 3 *each occur exactly once* in the second row. Thus $\begin{pmatrix} 1 & 2 & 3 \\ 3 & 1 & 2 \end{pmatrix}$ *is* a permutation symbol whilst $\begin{pmatrix} 1 & 2 & 3 \\ 1 & 2 & 2 \end{pmatrix}$ *is not.*

We can now define a permutation precisely as any function given by a permutation symbol. Thus *all* possible reorderings of the elements 1, 2, 3 are regarded as permutations.
Given two permutations f and g then we can *compose* them to obtain a third permutation. For instance, suppose that f is as above and that g is given by the diagram

g: 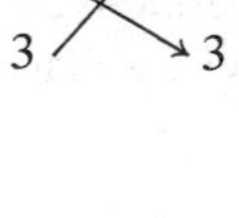 i.e. $g = \begin{pmatrix} 1 & 2 & 3 \\ 2 & 3 & 1 \end{pmatrix}$

then $f \circ g$ is given by the diagram

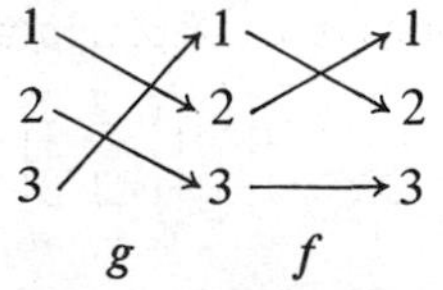

(remember g comes first) which, on straightening the arrows, gives:

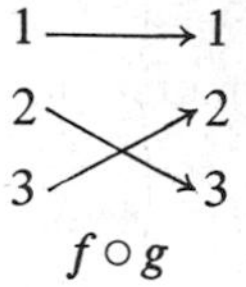

i.e. $f \circ g = \begin{pmatrix} 1 & 2 & 3 \\ 1 & 3 & 2 \end{pmatrix}$

Compositions can be calculated using permutation symbols.

EXAMPLES 14d

1) Calculate the composition $g \circ f$, i.e.

$$\begin{pmatrix} 1 & 2 & 3 \\ 2 & 3 & 1 \end{pmatrix} \circ \begin{pmatrix} 1 & 2 & 3 \\ 2 & 1 & 3 \end{pmatrix}$$

Working from right to left we see that f maps 1 to 2 and then g maps 2 to 3 so $g \circ f$ maps 1 to 3 (see dotted arrows in diagram) and we write 3 under 1:

$$\begin{pmatrix} 1 & 2 & 3 \\ 3 & . & . \end{pmatrix}$$

Similarly $2 \overset{f}{\longmapsto} 1 \overset{g}{\longmapsto} 2$ and $3 \overset{f}{\longmapsto} 3 \overset{g}{\longmapsto} 1$, so we have

$$g \circ f = \begin{pmatrix} 1 & 2 & 3 \\ 3 & 2 & 1 \end{pmatrix}$$

2) Calculate the composition

$$\begin{pmatrix} a & b & c & d \\ d & c & b & a \end{pmatrix} \circ \begin{pmatrix} a & b & c & d \\ a & c & b & d \end{pmatrix}$$

of permutations of the set $\{a, b, c, d\}$.

Again working from right to left we have
$a \longmapsto a \longmapsto d$ (dotted arrows) and similarly $b \longmapsto c \longmapsto b$,
$c \longmapsto b \longmapsto c$ and $d \longmapsto d \longmapsto a$, so the result is

$$\begin{pmatrix} a & b & c & d \\ d & b & c & a \end{pmatrix}$$

EXERCISE 14d

1) Calculate the following compositions of permutation of the set $\{1, 2, 3\}$:

(a) $\begin{pmatrix}1 & 2 & 3\\2 & 3 & 1\end{pmatrix} \circ \begin{pmatrix}1 & 2 & 3\\1 & 3 & 2\end{pmatrix}$ (b) $\begin{pmatrix}1 & 2 & 3\\1 & 2 & 3\end{pmatrix} \circ \begin{pmatrix}1 & 2 & 3\\3 & 1 & 2\end{pmatrix}$

(c) $\begin{pmatrix}1 & 2 & 3\\2 & 1 & 3\end{pmatrix} \circ \begin{pmatrix}1 & 2 & 3\\3 & 1 & 2\end{pmatrix}$ (d) $\begin{pmatrix}1 & 2 & 3\\2 & 1 & 3\end{pmatrix} \circ \begin{pmatrix}1 & 2 & 3\\3 & 2 & 1\end{pmatrix}$

(e) $\begin{pmatrix}1 & 2 & 3\\3 & 1 & 2\end{pmatrix} \circ \begin{pmatrix}1 & 2 & 3\\2 & 3 & 1\end{pmatrix}$ (f) $\begin{pmatrix}1 & 2 & 3\\3 & 2 & 1\end{pmatrix} \circ \begin{pmatrix}1 & 2 & 3\\1 & 2 & 3\end{pmatrix}$

2) Calculate the compositions of permutations of $\{a, b, c, d\}$:

(a) $\begin{pmatrix}a & b & c & d\\a & c & d & b\end{pmatrix} \circ \begin{pmatrix}a & b & c & d\\b & a & d & c\end{pmatrix}$ (b) $\begin{pmatrix}a & b & c & d\\b & a & d & c\end{pmatrix} \circ \begin{pmatrix}a & b & c & d\\a & c & d & b\end{pmatrix}$

3) Calculate the compositions of permutation of $\{i, j\}$ and display your results in a table:

(a) $\begin{pmatrix}i & j\\i & j\end{pmatrix} \circ \begin{pmatrix}i & j\\i & j\end{pmatrix}$ (b) $\begin{pmatrix}i & j\\i & j\end{pmatrix} \circ \begin{pmatrix}i & j\\j & i\end{pmatrix}$ (c) $\begin{pmatrix}i & j\\j & i\end{pmatrix} \circ \begin{pmatrix}i & j\\i & j\end{pmatrix}$ (d) $\begin{pmatrix}i & j\\j & i\end{pmatrix} \circ \begin{pmatrix}i & j\\j & i\end{pmatrix}$

(all possible compositions are listed).

4) How many permutations are there of the set $\{1, 2, 3\}$? List them.

5) Construct a composition table for the permutations on the set $\{1, 2, 3\}$.

PERMUTATION GROUPS

The set of *all* permutations of the set $\{1, 2, 3\}$ with the operation of composition forms a group called the *permutation group* on $\{1, 2, 3\}$. This can be checked using the composition table constructed in the last exercise. The argument that we shall use works just as well for permutations on any set.

Closure

$f \circ g$ means 'first permute by g and then permute by f' and the result is again a permutation. One way to see that this is true in general, is to notice that the symbol for $f \circ g$ can be obtained by combining those of f and g: reorder the entries of f to match the bottom line of g and then the bottom line of f becomes the bottom line of $f \circ g$.

E.g. $f{:}\downarrow\begin{pmatrix}1 & 2 & 3\\3 & 2 & 1\end{pmatrix}$ $g{:}\downarrow\begin{pmatrix}1 & 2 & 3\\3 & 1 & 2\end{pmatrix}$

reorder $f: \downarrow \begin{pmatrix} 3 & 1 & 2 \\ 1 & 3 & 2 \end{pmatrix}$

then $\begin{matrix} & 1 & 2 & 3 \\ g\downarrow & & & \\ & 3 & 1 & 2 \\ f\downarrow & & & \\ & 1 & 3 & 2 \end{matrix} \Bigg\downarrow f \circ g$ i.e. $f \circ g = \begin{pmatrix} 1 & 2 & 3 \\ 1 & 3 & 2 \end{pmatrix}$

Since the bottom line of $f \circ g$ is a reordered version of that for f, each element 1, 2, 3 occurs exactly once, i.e. $f \circ g$ is a permutation.

Identity

The identity permutation e is the one which does nothing i.e. $e = \begin{pmatrix} 1 & 2 & 3 \\ 1 & 2 & 3 \end{pmatrix}$

and clearly $f \circ e = e \circ f = f$ for any f.

Inverses

The permutation f^{-1} is the reverse of f (think of reversing the arrows in the diagram for f). Thus if $f = \begin{pmatrix} 1 & 2 & 3 \\ 2 & 3 & 1 \end{pmatrix}$

or

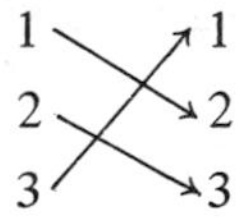

f

then

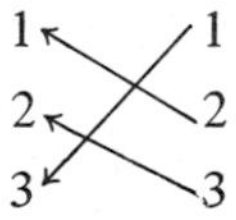

f^{-1}

i.e. $f^{-1} = \begin{pmatrix} 1 & 2 & 3 \\ 3 & 1 & 2 \end{pmatrix}$.

Then f^{-1} is another permutation which 'undoes' what f does. Hence $f \circ f^{-1} = f^{-1} \circ f = e$.

Associativity

Permutations are associative for the same reason as symmetries. Each of

$$(f \circ g) \circ h \quad \text{and} \quad f \circ (g \circ h)$$

means 'do h first *then* g then f'; hence they are equal.

We use the notation S_3 for the set of all permutations of $\{1, 2, 3\}$.

$(S_3, \circ)$ is a group, called the *permutation group* on 3 symbols.

There is a similar group $(S_n, \circ)$ for each $n = 1, 2, 3, \ldots$, namely the group of permutations on $\{1, 2, 3, \ldots, n\}$.

EXAMPLE 14e

Find the inverse of $\begin{pmatrix}1 & 2 & 3\\ 2 & 1 & 3\end{pmatrix}$.

We can read the inverse directly from the symbol. Since $2 \longmapsto 1$ (dotted arrow), we must have $1 \longmapsto 2$ under the inverse. So we write 2 under 1:

$$\begin{pmatrix}1 & 2 & 3\\ 2 & . & .\end{pmatrix}$$

Similarly $2 \longmapsto 1$ and $3 \longmapsto 3$ under the inverse:

$$\begin{pmatrix}1 & 2 & 3\\ 2 & 1 & 3\end{pmatrix}$$

EXERCISE 14e

1) Write down the inverses of

(a) $\begin{pmatrix}1 & 2 & 3\\ 3 & 1 & 2\end{pmatrix}$ (b) $\begin{pmatrix}1 & 2 & 3\\ 1 & 3 & 2\end{pmatrix}$ (c) $\begin{pmatrix}1 & 2 & 3\\ 3 & 2 & 1\end{pmatrix}$.

2) Check associativity for

$$\begin{pmatrix}1 & 2 & 3\\ 3 & 1 & 2\end{pmatrix} \circ \begin{pmatrix}1 & 2 & 3\\ 2 & 1 & 3\end{pmatrix} \circ \begin{pmatrix}1 & 2 & 3\\ 3 & 2 & 1\end{pmatrix}.$$

3) Is $(S_3, \circ)$ a commutative group?

4) Is $(S_2, \circ)$ a commutative group?
(The group table is essentially the same as the one constructed in Exercise 14d, Question 3.)

5) (a) How many elements are there in S_4?
(b) Calculate the compositions:

$$\begin{pmatrix}1 & 2 & 3 & 4\\ 2 & 4 & 3 & 1\end{pmatrix} \circ \begin{pmatrix}1 & 2 & 3 & 4\\ 2 & 1 & 3 & 4\end{pmatrix}$$

$$\begin{pmatrix}1 & 2 & 3 & 4\\ 2 & 1 & 3 & 4\end{pmatrix} \circ \begin{pmatrix}1 & 2 & 3 & 4\\ 2 & 4 & 3 & 1\end{pmatrix}.$$

(c) Is $(S_4, \circ)$ a commutative group?
(d) Find the inverses of

$$\begin{pmatrix}1 & 2 & 3 & 4\\ 3 & 4 & 2 & 1\end{pmatrix} \text{ and } \begin{pmatrix}1 & 2 & 3 & 4\\ 1 & 3 & 4 & 2\end{pmatrix}.$$

ISOMETRY AND MATRIX GROUPS

Rigid transformations of the plane (i.e. rotations, reflections, translations and combinations of these) are called *isometries*. Sets of isometries often give further examples of groups and when these are expressed using matrices, they give matrix groups.
Consider the following isometries of the plane:

$$e = \text{identity}$$

$$r = \text{rotation through } 90^\circ$$

$$s = \text{rotation through } 180^\circ$$

$$t = \text{rotation through } 270^\circ$$

(all rotations being anticlockwise about the origin).

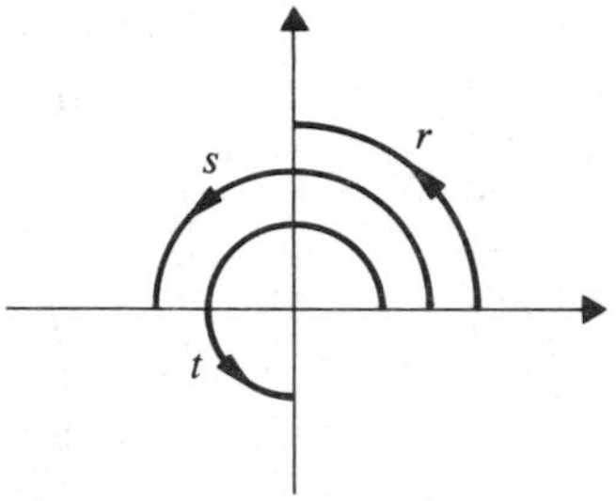

We shall prove that $\{e, r, s, t\}$ is a group under composition. If we perform r following s then we obtain t ('rotate through 180°' followed by 'rotate through 90°' results in 'rotate through 270°') and in a similar way we can compute all the compositions, giving this table:

$\circ$	e	r	s	t
e	e	r	s	t
r	r	s	t	e
s	s	t	e	r
t	t	e	r	s

The table proves *closure*. e is clearly an *identity* and we can read off the *inverses*: $e^{-1} = e$, $r^{-1} = t$, $t^{-1} = r$, $s^{-1} = s$.
Associativity holds for the usual reason: each of $(f \circ g) \circ h$ and $f \circ (g \circ h)$ means 'h first, then g then f'.
We have proved that $(\ e, r, s, t\ , \circ)$ is a group.

A set of isometries which forms a group under composition is called an *isometry group*.

If we express each of the transformations in the example above as a matrix, then, since composition of transformations corresponds to matrix multiplication, we obtain a set of matrices which forms a group under matrix multiplication.

A set of matrices which forms a group under matrix multiplication is called a *matrix group.*

EXAMPLES 14f

1) Write down the matrix group corresponding to the group in the above example and verify that $r \circ s = t$ and $t^{-1} = r$ using the matrices.

The matrices which give the four transformations are:

$$e = \begin{pmatrix} 1 & 0 \\ 0 & 1 \end{pmatrix} \qquad r = \begin{pmatrix} 0 & -1 \\ 1 & 0 \end{pmatrix}$$

$$s = \begin{pmatrix} -1 & 0 \\ 0 & -1 \end{pmatrix} \qquad t = \begin{pmatrix} 0 & 1 \\ -1 & 0 \end{pmatrix}$$

so the corresponding matrix group is

$$\left(\left\{\begin{pmatrix} 1 & 0 \\ 0 & 1 \end{pmatrix}, \begin{pmatrix} 0 & -1 \\ 1 & 0 \end{pmatrix}, \begin{pmatrix} -1 & 0 \\ 0 & -1 \end{pmatrix}, \begin{pmatrix} 0 & 1 \\ -1 & 0 \end{pmatrix}\right\}, \times\right)$$

where $\times$ denotes the operation of matrix multiplication.

$r \circ s$ is represented by the matrix

$$\begin{pmatrix} 0 & -1 \\ 1 & 0 \end{pmatrix}\begin{pmatrix} -1 & 0 \\ 0 & -1 \end{pmatrix} = \begin{pmatrix} 0 & 1 \\ -1 & 0 \end{pmatrix} = t$$

$$t^{-1} = \begin{pmatrix} 0 & 1 \\ -1 & 0 \end{pmatrix}^{-1} = \begin{pmatrix} 0 & -1 \\ 1 & 0 \end{pmatrix} = r$$

Note that not all matrix groups arise from isometry groups, as we see in the following example.

2) Show that the set of matrices of the form $\begin{pmatrix} 1 & n \\ 0 & 1 \end{pmatrix}$, where n is an integer, forms a group under matrix multiplication. Interpret the group geometrically.

Calculating a general product we have:

$$\begin{pmatrix} 1 & n \\ 0 & 1 \end{pmatrix}\begin{pmatrix} 1 & m \\ 0 & 1 \end{pmatrix} = \begin{pmatrix} 1 & n+m \\ 0 & 1 \end{pmatrix}$$

So the set is *closed* (if n and m are integers, so is $n+m$). The unit matrix $\begin{pmatrix}1 & 0\\0 & 1\end{pmatrix}$ is in the set, so we have an *identity*. Examining the product above we can see that $\begin{pmatrix}1 & -n\\0 & 1\end{pmatrix}$ is the inverse of $\begin{pmatrix}1 & n\\0 & 1\end{pmatrix}$ and since $-n$ is an integer, we have inverses as well.
Finally matrix multiplication is associative. Thus we have the four properties of a group.

The matrix $\begin{pmatrix}1 & n\\0 & 1\end{pmatrix}$ represents a shear parallel to the x axis which moves the line $y=1$ n units to the right. Thus the group consists of all such shear transformations where n is an integer.

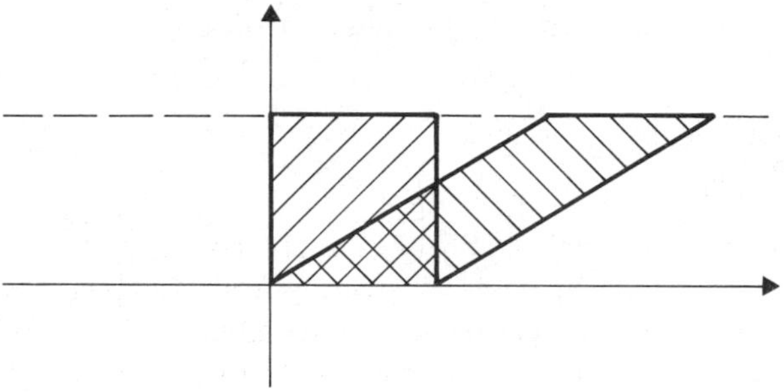

EXERCISE 14f

1) Verify that the transformations

$$\begin{aligned} e &= \text{identity} \\ r &= \text{rotate through } 120^\circ \\ s &= \text{rotate through } 240^\circ \end{aligned} \Bigg\} \text{ anticlockwise about the origin}$$

form a group. Express the transformations as matrices and verify that $r \circ s = e$ using the matrices.

2) Prove that

$$\left(\left\{\begin{pmatrix}1 & 0\\0 & 1\end{pmatrix}, \begin{pmatrix}0 & -1\\-1 & 0\end{pmatrix}, \begin{pmatrix}0 & 1\\1 & 0\end{pmatrix}, \begin{pmatrix}-1 & 0\\0 & -1\end{pmatrix}\right\}, \times\right)$$

is a matrix group. Interpret the matrices as isometries and hence write down the corresponding isometry group.

3) Prove that the transformations

$$e = \text{identity}$$

$$a = \text{reflect in the } x \text{ axis}$$

$$b = \text{reflect in the } y \text{ axis}$$

$$r = \text{rotate through } 180^\circ \text{ about the origin}$$

form an isometry group and write down the corresponding matrix group.

4) Prove that the set of matrices of the form $\begin{pmatrix} r & 0 \\ 0 & r \end{pmatrix}$, where r is a positive real number, forms a group under matrix multiplication. Interpret the group geometrically.

5) Prove that the set of matrices of the form $\begin{pmatrix} r & 0 \\ 0 & \frac{1}{r} \end{pmatrix}$, when r is a positive real number, forms a matrix group. Interpret it geometrically.

SUBGROUPS

We have seen a considerable number of examples of groups: symmetry groups, arithmetic groups (e.g. integers, rationals, reals, integers modulo n), permutation groups, isometry groups and matrix groups. Now we will start to consider some of the general concepts associated with groups and which introduce the *theory* of groups.
The first concept is that of a subgroup.

When a subset of a group forms a group in its own right (using the *same* operation), then we say that the subset is a *subgroup*.

Consider again the symmetry group of the equilateral triangle

$$(\{e, r, s, a, b, c\}, \circ)$$

(notation as on page 568). Then, if we look at the table of compositions (page 569), the compositions involving the elements e, r, s are as follows:

$\circ$	e	r	s
e	e	r	s
r	r	s	e
s	s	e	r

We can see that these elements form a group in their own right as follows.

Closure

The table shows closure.

Identity

e is in the subset and since e is an identity in the whole symmetry group, then it must be an identity in the subset: $e \circ f = f \circ e = f$ for all symmetries f and hence the equations certainly hold for all symmetries in the subset.

Inverses

$e^{-1} = e$, $r^{-1} = s$, $s^{-1} = r$. So the inverses of all elements in the subset lie in the subset.

Associativity

Since associativity holds in the whole group, it must hold in the subset.

Thus we have proved that $(\{e, r, s\}, \circ)$ is a group with the same operation, i.e. that it is a subgroup of $(\{e, r, s, a, b, c\}, \circ)$ and we write $(\{e, r, s\}, \circ) \subseteq (\{e, r, s, a, b, c\}, \circ)$.

Since this subgroup consists of the symmetries which are rotations, we call it the *rotation subgroup* of the symmetry group.

In general, if $(H, *)$ is a subgroup of $(G, *)$ then we write $(H, *) \subseteq (G, *)$.

The example shows that in order to check that $(H, *)$ is a subgroup of $(G, *)$ (where H is a subset of G) we need to check three things:

Closure

H is closed under $*$.

Identity

The identity of G lies in H.

Inverses

The inverses of elements of H also lie in H.

(There is no need to check associativity since it follows from associativity in G.)

EXAMPLE 14g

Show that $(2\mathbb{Z}, +)$ is a subgroup of $(\mathbb{Z}, +)$ when $2\mathbb{Z}$ means the set of even integers.

Closure

A typical element of $2\mathbb{Z}$ is $2n$ when n is any integer. But

$$2n + 2m = 2(n + m)$$

(the sum of even integers is even). Thus the subset is closed.

Identity

0 is the identity in $(\mathbb{Z}, +)$ and since $0 = 2 \times 0$ it lies in the subset.

Inverses

The inverse of $2n$ is $-2n = 2(-n)$ which is also even.

EXERCISE 14g

1) Verify that $(\{e, a\}, \circ)$ is a subgroup of the symmetry group of the equilateral triangle.

2) Prove that $(3\mathbb{Z}, +) \subseteq (\mathbb{Z}, +)$ where $3\mathbb{Z}$ is the set of all integers of the form $3n$ (all multiples of 3).

3) Verify that $\left(\left\{\begin{pmatrix}1 & 0\\0 & 1\end{pmatrix}, \begin{pmatrix}-1 & 0\\0 & -1\end{pmatrix}\right\}, \times\right)$ is a subgroup of each of the matrix groups in Example 14f, No 1 and Exercise 14f, Question 2.

4) Find all the subgroups of the symmetry group of the equilateral triangle.

5) Find all the subgroups of

(a) $(\mathbb{Z}_4, +_4)$ (b) $(\mathbb{Z}_6, +_6)$.

6) Suppose that g is a member of an isometry group or a symmetry group and that g is a reflection. Prove that $(\{e, g\}, \circ)$ is a subgroup of the isometry or symmetry group.

7) Find all the subgroups of the symmetry groups of

(a) the rectangle (b) the square.

8) Verify that, in each case of a subgroup in Questions 4, 5 and 7, the number of elements in the subgroups is a *factor* of the number of elements in the whole group. (This is a *general fact* about groups with a finite number of elements in them; it is *Lagrange's Theorem* and it is investigated at the end of the chapter.)

ISOMORPHISM

When constructing group tables earlier in the chapter, the reader may have been struck by the similarities between the tables for different groups. For example, the symmetry group of the rectangle has this table:

$\circ$	e	a	b	r
e	e	a	b	r
a	a	e	r	b
b	b	r	e	a
r	r	b	a	e

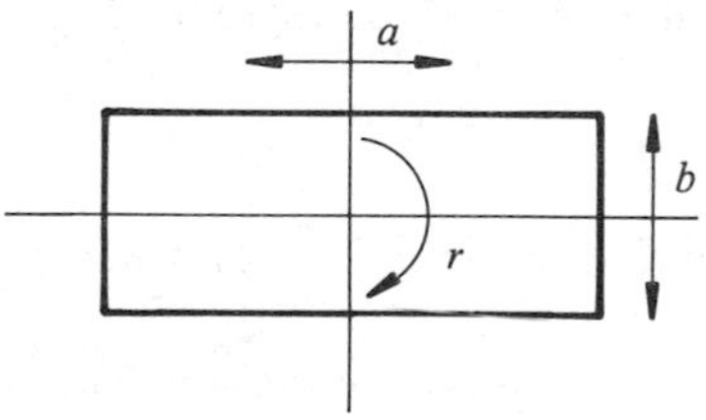

and the matrix group

$$\left(\left\{\begin{pmatrix}1&0\\0&1\end{pmatrix},\begin{pmatrix}0&1\\1&0\end{pmatrix},\begin{pmatrix}0&-1\\-1&0\end{pmatrix},\begin{pmatrix}-1&0\\0&-1\end{pmatrix}\right\},\times\right)$$

has this table:

$\times$	$\begin{pmatrix}1&0\\0&1\end{pmatrix}$	$\begin{pmatrix}0&1\\1&0\end{pmatrix}$	$\begin{pmatrix}0&-1\\-1&0\end{pmatrix}$	$\begin{pmatrix}-1&0\\0&-1\end{pmatrix}$
$\begin{pmatrix}1&0\\0&1\end{pmatrix}$	$\begin{pmatrix}1&0\\0&1\end{pmatrix}$	$\begin{pmatrix}0&1\\1&0\end{pmatrix}$	$\begin{pmatrix}0&-1\\-1&0\end{pmatrix}$	$\begin{pmatrix}-1&0\\0&-1\end{pmatrix}$
$\begin{pmatrix}0&1\\1&0\end{pmatrix}$	$\begin{pmatrix}0&1\\1&0\end{pmatrix}$	$\begin{pmatrix}1&0\\0&1\end{pmatrix}$	$\begin{pmatrix}-1&0\\0&-1\end{pmatrix}$	$\begin{pmatrix}0&-1\\-1&0\end{pmatrix}$
$\begin{pmatrix}0&-1\\-1&0\end{pmatrix}$	$\begin{pmatrix}0&-1\\-1&0\end{pmatrix}$	$\begin{pmatrix}-1&0\\0&-1\end{pmatrix}$	$\begin{pmatrix}1&0\\0&1\end{pmatrix}$	$\begin{pmatrix}0&1\\1&0\end{pmatrix}$
$\begin{pmatrix}-1&0\\0&-1\end{pmatrix}$	$\begin{pmatrix}-1&0\\0&-1\end{pmatrix}$	$\begin{pmatrix}0&-1\\-1&0\end{pmatrix}$	$\begin{pmatrix}0&1\\1&0\end{pmatrix}$	$\begin{pmatrix}1&0\\0&1\end{pmatrix}$

To make the similarity obvious, consider the following correspondence between elements of the two groups:

$$e \leftrightarrow \begin{pmatrix}1&0\\0&1\end{pmatrix}$$

$$a \leftrightarrow \begin{pmatrix}0&1\\1&0\end{pmatrix}$$

$$b \leftrightarrow \begin{pmatrix}0&-1\\-1&0\end{pmatrix}$$

$$c \leftrightarrow \begin{pmatrix}-1&0\\0&-1\end{pmatrix}$$

Clearly the second table is the same as the first table but with the elements relabelled according to the correspondence. We say that the two groups are *isomorphic* and we write

$$(\{e, a, b, r\}, \circ) \cong \left(\left\{\begin{pmatrix}1&0\\0&1\end{pmatrix},\begin{pmatrix}0&1\\1&0\end{pmatrix},\begin{pmatrix}0&-1\\-1&0\end{pmatrix},\begin{pmatrix}-1&0\\0&-1\end{pmatrix}\right\},\times\right)$$

Since we have an exact correspondence between the group tables, we can use *either* table to work out combinations of elements in either group. In other words combinations of elements (compositions) in the first group correspond to combinations (matrix products) of the corresponding elements in the second group.

In general, groups $(G, *)$ and $(H, \Box)$ are *isomorphic* if it is possible to set up a one-to-one correspondence between the elements of G and those of H in such a way that combinations in G correspond to combinations in H, i.e.

$$\left.\begin{array}{l} g \longleftrightarrow g' \\ h \longleftrightarrow h' \end{array}\right\} \Rightarrow \quad g * h \longleftrightarrow g' \Box h'$$

We write $(G, *) \cong (H, \Box)$.

Isomorphic groups are interchangeable as far as group theory is concerned, they share all the same group-theory properties. We shall be more precise about this later.

When showing an isomorphism by using group tables, it may be necessary to *reorder* the elements of one of the groups in order to match the table for the other one.

EXAMPLES 14h

1) Prove that the symmetry group of the equilateral triangle is isomorphic to the permutation group $(S_3, \circ)$ on the set $\{1, 2, 3\}$.

The group table for $(S_3, \circ)$ is this:

$\circ$	$\begin{pmatrix}1&2&3\\1&2&3\end{pmatrix}$	$\begin{pmatrix}1&2&3\\1&3&2\end{pmatrix}$	$\begin{pmatrix}1&2&3\\3&2&1\end{pmatrix}$	$\begin{pmatrix}1&2&3\\2&1&3\end{pmatrix}$	$\begin{pmatrix}1&2&3\\2&3&1\end{pmatrix}$	$\begin{pmatrix}1&2&3\\3&1&2\end{pmatrix}$
$\begin{pmatrix}1&2&3\\1&2&3\end{pmatrix}$	$\begin{pmatrix}1&2&3\\1&2&3\end{pmatrix}$	$\begin{pmatrix}1&2&3\\1&3&2\end{pmatrix}$	$\begin{pmatrix}1&2&3\\3&2&1\end{pmatrix}$	$\begin{pmatrix}1&2&3\\2&1&3\end{pmatrix}$	$\begin{pmatrix}1&2&3\\2&3&1\end{pmatrix}$	$\begin{pmatrix}1&2&3\\3&1&2\end{pmatrix}$
$\begin{pmatrix}1&2&3\\1&3&2\end{pmatrix}$	$\begin{pmatrix}1&2&3\\1&3&2\end{pmatrix}$	$\begin{pmatrix}1&2&3\\1&2&3\end{pmatrix}$	$\begin{pmatrix}1&2&3\\2&3&1\end{pmatrix}$	$\begin{pmatrix}1&2&3\\3&1&2\end{pmatrix}$	$\begin{pmatrix}1&2&3\\3&2&1\end{pmatrix}$	$\begin{pmatrix}1&2&3\\2&1&3\end{pmatrix}$
$\begin{pmatrix}1&2&3\\3&2&1\end{pmatrix}$	$\begin{pmatrix}1&2&3\\3&2&1\end{pmatrix}$	$\begin{pmatrix}1&2&3\\3&1&2\end{pmatrix}$	$\begin{pmatrix}1&2&3\\1&2&3\end{pmatrix}$	$\begin{pmatrix}1&2&3\\2&3&1\end{pmatrix}$	$\begin{pmatrix}1&2&3\\2&1&3\end{pmatrix}$	$\begin{pmatrix}1&2&3\\1&3&2\end{pmatrix}$
$\begin{pmatrix}1&2&3\\2&1&3\end{pmatrix}$	$\begin{pmatrix}1&2&3\\2&1&3\end{pmatrix}$	$\begin{pmatrix}1&2&3\\2&3&1\end{pmatrix}$	$\begin{pmatrix}1&2&3\\3&1&2\end{pmatrix}$	$\begin{pmatrix}1&2&3\\1&2&3\end{pmatrix}$	$\begin{pmatrix}1&2&3\\1&3&2\end{pmatrix}$	$\begin{pmatrix}1&2&3\\3&2&1\end{pmatrix}$
$\begin{pmatrix}1&2&3\\2&3&1\end{pmatrix}$	$\begin{pmatrix}1&2&3\\2&3&1\end{pmatrix}$	$\begin{pmatrix}1&2&3\\2&1&3\end{pmatrix}$	$\begin{pmatrix}1&2&3\\1&3&2\end{pmatrix}$	$\begin{pmatrix}1&2&3\\3&2&1\end{pmatrix}$	$\begin{pmatrix}1&2&3\\3&1&2\end{pmatrix}$	$\begin{pmatrix}1&2&3\\1&2&3\end{pmatrix}$
$\begin{pmatrix}1&2&3\\3&1&2\end{pmatrix}$	$\begin{pmatrix}1&2&3\\3&1&2\end{pmatrix}$	$\begin{pmatrix}1&2&3\\3&2&1\end{pmatrix}$	$\begin{pmatrix}1&2&3\\2&1&3\end{pmatrix}$	$\begin{pmatrix}1&2&3\\1&3&2\end{pmatrix}$	$\begin{pmatrix}1&2&3\\1&2&3\end{pmatrix}$	$\begin{pmatrix}1&2&3\\2&3&1\end{pmatrix}$

To see the equivalence with the table for the symmetry group, we reorder the symmetries to give:

$\circ$	e	a	b	c	s	r
e	e	a	b	c	s	r
a	a	e	s	r	b	c
b	b	r	e	s	c	a
c	c	s	r	e	a	b
s	s	c	a	b	r	e
r	r	b	c	a	e	s

(Check with the table on page 569.) Now the correspondence:

$$e \leftrightarrow \begin{pmatrix} 1 & 2 & 3 \\ 1 & 2 & 3 \end{pmatrix} \quad a \leftrightarrow \begin{pmatrix} 1 & 2 & 3 \\ 1 & 3 & 2 \end{pmatrix} \quad b \leftrightarrow \begin{pmatrix} 1 & 2 & 3 \\ 3 & 2 & 1 \end{pmatrix}$$

$$c \leftrightarrow \begin{pmatrix} 1 & 2 & 3 \\ 2 & 1 & 3 \end{pmatrix} \quad s \leftrightarrow \begin{pmatrix} 1 & 2 & 3 \\ 2 & 3 & 1 \end{pmatrix} \quad r \leftrightarrow \begin{pmatrix} 1 & 2 & 3 \\ 3 & 1 & 2 \end{pmatrix}$$

makes the tables correspond exactly (as the reader should check). Therefore the groups are isomorphic.

Note that it is not always practicable to use tables to prove isomorphism. The group $(\mathbb{Z}, +)$ for instance has an *infinite* table! In this case we must use an argument which shows *in general* how to make combinations correspond, as illustrated in the next example.

2) Prove that $(\mathbb{Z}, +)$ is isomorphic to the matrix group consisting of all matrices of the form $\begin{pmatrix} 1 & n \\ 0 & 1 \end{pmatrix}$ where n is an integer (see Example 14f, No. 2).

The correspondence is obvious:

$$n \leftrightarrow \begin{pmatrix} 1 & n \\ 0 & 1 \end{pmatrix}$$

where n is a typical integer. Now if $n \leftrightarrow \begin{pmatrix} 1 & n \\ 0 & 1 \end{pmatrix}$ and $m \leftrightarrow \begin{pmatrix} 1 & m \\ 0 & 1 \end{pmatrix}$ then the

the combinations are $n+m$ in $(\mathbb{Z}, +)$ and $\begin{pmatrix} 1 & n \\ 0 & 1 \end{pmatrix}\begin{pmatrix} 1 & m \\ 0 & 1 \end{pmatrix} = \begin{pmatrix} 1 & n+m \\ 0 & 1 \end{pmatrix}$ in the matrix group, and indeed we do have $n+m \longleftrightarrow \begin{pmatrix} 1 & n \\ 0 & 1 \end{pmatrix}\begin{pmatrix} 1 & m \\ 0 & 1 \end{pmatrix}$

i.e. the groups are isomorphic.

Note that often it is possible to see isomorphisms between groups *without* reference to the tables, by using either a geometrical argument or an argument using 'labels', as illustrated in the next two examples.

3) Prove that the isometry group in Example 14f, No. 1 $(\{e, r, s, t\}, \circ)$ is isomorphic to the symmetry group of the 'four-bladed windmill' (Exercise 14a, Question 5):

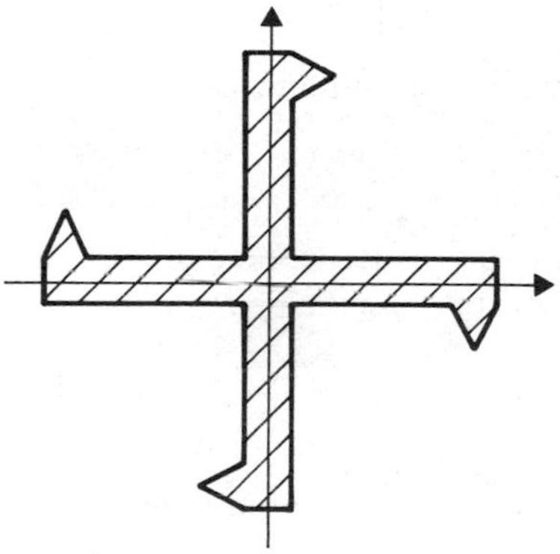

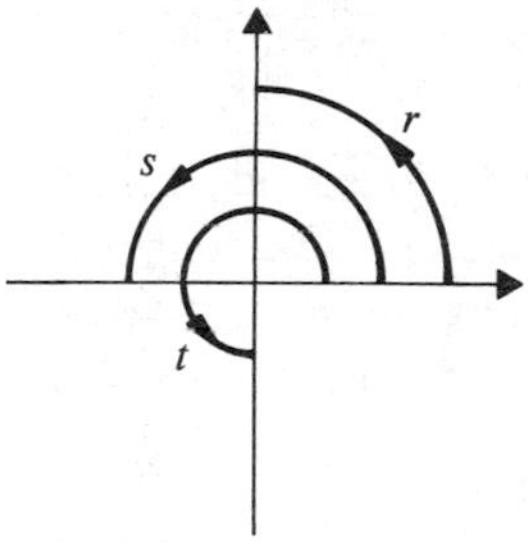

Position the windmill (as above left) with its centre at the origin. The four isometries $\{e, r, s, t\}$ (above right) then actually realize the four symmetries of the windmill. Clearly composition of isometries corresponds to composition of symmetries, i.e. the groups are isomorphic.

4) Use a 'labelling' argument to establish the isomorphism between the symmetry group of the equilateral triangle and $(S_3, \circ)$.

Use the labels 1, 2, 3 for the positions of the corners (rather than A, B, C as earlier):

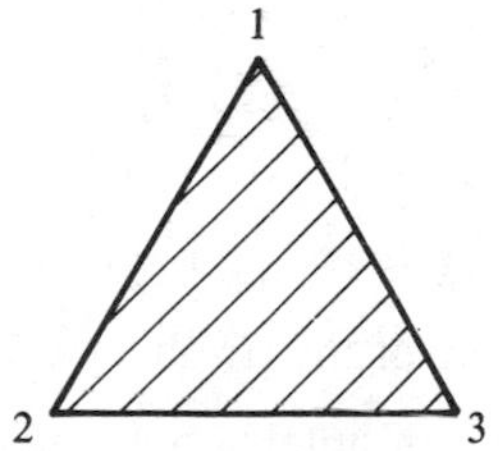

Then each symmetry determines a permutation of $\{1, 2, 3\}$ by sending i to j where the corner at i is mapped to the corner at j under the symmetry.

Thus

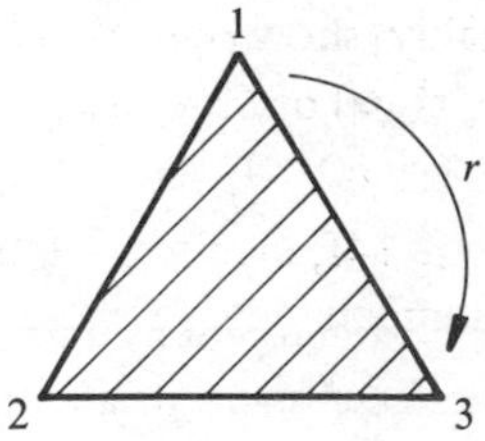

gives $\begin{pmatrix} 1 & 2 & 3 \\ 3 & 1 & 2 \end{pmatrix}$

and

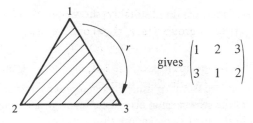

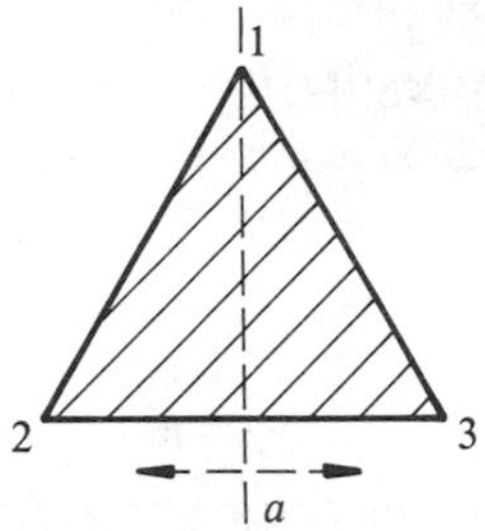

gives $\begin{pmatrix} 1 & 2 & 3 \\ 1 & 3 & 2 \end{pmatrix}$

(in fact the correspondence is precisely the one in Example 1 above).
Now the effect of the composition of two symmetries is to perform the two permutations one after the other, i.e. to compose the permutations. Thus the correspondence is an isomorphism.

EXERCISE 14h

1) Use group tables to establish the following group isomorphisms:

(a) $(\mathbb{Z}_3, +_3) \cong (\{0, 2, 4\}, +_6)$.

(b) $(\mathbb{Z}_3, +_3)$ is isomorphic to the symmetry group of the three-bladed windmill.

(c) $(\{e, r, s\}, \circ)$ (the isometry group in Exercise 14f, Question 1) is isomorphic to the group in Exercise 14b, Question 12. (In fact, as we shall see later, *any* two groups with three elements in them must be isomorphic.)

(d) $(\mathbb{Z}_4, +_4)$ is isomorphic to $(\{e, r, s, t\}, \circ)$ (the group discussed in Example 14h, No. 3).

(e) The matrix groups:

$$\left(\left\{\begin{pmatrix} 1 & 0 \\ 0 & 1 \end{pmatrix}, \begin{pmatrix} 0 & 1 \\ 1 & 0 \end{pmatrix}, \begin{pmatrix} 0 & -1 \\ -1 & 0 \end{pmatrix}, \begin{pmatrix} -1 & 0 \\ 0 & -1 \end{pmatrix}\right\}, \times\right)$$

$$\cong \left(\left\{\begin{pmatrix} 1 & 0 \\ 0 & 1 \end{pmatrix}, \begin{pmatrix} -1 & 0 \\ 0 & 1 \end{pmatrix}, \begin{pmatrix} 1 & 0 \\ 0 & -1 \end{pmatrix}, \begin{pmatrix} -1 & 0 \\ 0 & -1 \end{pmatrix}\right\}, \times\right)$$

(f) $(\mathbb{Z}_5, +_5) \cong$ symmetry group of the five-bladed 'windmill'.

2) By positioning the rectangle in suitable positions in the plane, establish that the symmetry group of the rectangle is isomorphic to each of the matrix groups in Exercise 1(e) above.

3) By positioning the three-bladed windmill suitably, show that its symmetry group is isomorphic to the isometry group $(\{e, r, s\}, \circ)$ of Exercise 14f, Question 1.

4) Prove that each of the matrix groups in Exercise 14f, Questions 4 and 5 is isomorphic to the group $(\mathbb{R}^+, \times)$ (positive real numbers under multiplication).

5) Use the function e^x and the exponential law $e^x e^y = e^{x+y}$ to prove that $(\mathbb{R}, +) \cong (\mathbb{R}^+, \times)$.

6) By labelling the corners of the square as shown, write down a set of eight elements of S_4 which form a subgroup of $(S_4, \circ)$:

7) Check, by inspecting tables, that all groups with two elements in them that we have seen are isomorphic. Can you find a *proof* that *all* groups with two elements in them are isomorphic?

8) Check by using tables that the subgroup $(\{e, r, s, t\}, \circ)$ of the symmetry group of the square is isomorphic to $(\mathbb{Z}_4, +_4)$ and by using Question 6 above, write down a subgroup of $(S_4, \circ)$ which is also isomorphic to $(\mathbb{Z}_4, +_4)$ and check using tables.

ALGEBRA IN GROUPS

We shall return to the idea of isomorphism in the next section, when we discuss features and properties of groups 'preserved' by an isomorphism. But first we shall develop some of the general notation for groups and investigate the *algebra of groups*. When dealing with general groups it is usual to omit the symbol for the group operation. Thus we shall say 'G is a group' rather than '$(G, *)$ is a group' and we shall write gh for the combination in G of the elements g, h in G (rather than $g * h$ or $g \times h$ or $g \circ h$, etc.), and we call gh the *product* of g and h rather than 'combination' or 'composition', etc.

There is an exception to this; in arithmetic groups where the operation is addition (such as $(\mathbb{Z}, +)$ or $(\mathbb{R}, +)$) we *retain* the symbol for the operation and write $g + h$ for the combination of g and h in the group and call it the *sum* of g and h. This is because the notation gh is easily confused with $g \times h$ (the result of *multiplying* g and h rather than *adding*).

We also drop the operation symbol from the notation for particular groups *wherever there is no possibility of confusion*. Thus the groups $\mathbb{Z}, \mathbb{Q}, \mathbb{R}$ means 'the groups $(\mathbb{Z}, +)$, $(\mathbb{Q}, +)$, $(\mathbb{R}, +)$' since $+$ is the only sensible operation for these groups. Similarly we shall write $\mathbb{Z}_n$ instead of $(\mathbb{Z}_n, +_n)$ for the group of integers modulo n and S_n instead of $(S_n, \circ)$ for the permutation group on n symbols.

We shall also use the notation $S(F)$ for the symmetry group of the plane figure F. For example, $S(\Delta), S(\mathrm{R}), S(\square), S(\mathrm{WIND}_3)$ denote the symmetry groups of the equilateral triangle, the rectangle, the square and the three-bladed windmill, respectively. Using our condensed notation we can write for instance $S_3 \cong S(\Delta)$ (the permutation group on three symbols is isomorphic to the symmetry group of the equilateral triangle).

We shall use lower-case (small) letters, $a, b, c, \ldots, g, h, k, l, \ldots, r, s, t, \ldots$ for elements in a general group G (as we have been doing for many of our particular groups); so, for example, we may write

$$(a(bc))b \qquad [1]$$

for an element in G obtained by combining the elements a, b, c in a certain order. The associativity property implies that the position of the brackets in an expression such as [1] is immaterial as, for example,

$$(ab)(cb) = ((ab)c)b = (a(bc))b$$

by two applications of the associativity law. Thus we may safely omit the brackets brackets and write [1] simply as:

$$abcb$$

We shall use the symbol e for the identity of a general group G; if we need to be more precise we shall write e_G. We use index notation for the products of an element with itself. Thus:

$$a^2 = aa, \quad a^3 = aaa, \text{ etc.}$$

and

$$a^{-2} = a^{-1}a^{-1}$$

$$a^{-3} = a^{-1}a^{-1}a^{-1}, \qquad \text{etc.}$$

and by convention:

$$a^0 = e$$

Then the usual laws for indices apply, for example:

$$\begin{aligned} a^2a^{-3} &= aaa^{-1}a^{-1}a^{-1} = aea^{-1}a^{-1} && (\text{since } aa^{-1} = e) \\ &= aa^{-1}a^{-1} && (\text{since } ae = a) \\ &= ea^{-1} && (\text{since } aa^{-1} = e) \\ &= a^{-1} && (= a^{2+(-3)}) \end{aligned}$$

Note that in an arithmetic group in which the operation is addition (for example in the group $\mathbb{Z}$) this notation needs to be interpreted *with great care*:

$$a^2 \text{ is short for } a * a$$

and in $\mathbb{Z}$ this *means* $a + a = 2a$.
Similarly a^3 *means* $a + a + a = 3a$.

Thus a^2, a^3, etc., *do not have their usual meanings in* $\mathbb{Z}$.

Similarly a^{-1} means the *inverse* of a, i.e. $-a$ *not* $\dfrac{1}{a}$.

(To avoid confusion we usually write $2a, 3a, -a, -2a$, etc., for the 'powers' of a in arithmetic groups such as $\mathbb{Z}$.)

Solution of Equations

We shall now use this general notation to explain how to *solve equations* in groups.

EXAMPLES 14i

1) Solve for x the equation $ax = b$.

Multiply both sides by a^{-1} on the left:

$$a^{-1}ax = a^{-1}b$$

$$\Rightarrow \quad ex = a^{-1}b \qquad \text{(since } a^{-1}a = e\text{)}$$

$$\Rightarrow \quad x = a^{-1}b \qquad \text{(since } ex = x\text{)}$$

and we have 'solved' for x.
We do indeed have a *solution* because:

$$ax = aa^{-1}b = eb = b$$

2) Show that the solution $x = a^{-1}b$ for $ax = b$ is *unique*.

Suppose that $ay = b$, then since $ax = b$ we have

$$ax = ay$$

multiply by a^{-1} on the left:

$$a^{-1}ax = a^{-1}ay$$

$$\Rightarrow \quad ex = ey \qquad \text{(since } a^{-1}a = e\text{)}$$

$$\Rightarrow \quad x = y \qquad \text{(since } ex = x,\ ey = y\text{)}$$

i.e. any other solution of the given equation is equal to x, i.e. $x = a^{-1}b$ is the *unique* solution.

The bracketed argument above, showing that $ax = ay \Rightarrow x = y$, is summarized by saying 'cancel a on the left'. This type of cancellation argument occurs very frequently in group theory. There is a similar argument which allows cancellation on the right (i.e. an argument which shows that $xa = ya \Rightarrow x = y$) and the two are often referred to as the *group cancellation laws.*

3) Given that $ab = e$, prove that

$$a = b^{-1} \quad \text{and} \quad b = a^{-1}$$

Multiply $ab = e$ on the right by b^{-1}:

$$abb^{-1} = eb^{-1}$$

$$\Rightarrow \qquad ae = b^{-1}$$

$$\Rightarrow \qquad a = b^{-1}$$

Similarly multiplying on the left by a^{-1}:

$$a^{-1}ab = a^{-1}e$$

$$\Rightarrow \qquad eb = a^{-1}$$

$$\Rightarrow \qquad b = a^{-1}$$

This example explains why the element e occurs *exactly once* in any row g of a group table and that it occurs in column g^{-1} (a feature we observed on page 574): if e occurs in column h then $gh = e$ and therefore $h = g^{-1}$, i.e. e occurs only in column g^{-1}. In fact *each* element of a group occurs *exactly once* in any row of a group table (a feature that the reader may well have observed).

4) Prove that each element of a group occurs *exactly once* in any given row of the group table.

The statement that element b occurs in row a and column x is exactly the same as the equation

$$ax = b$$

But we have seen above that there is a unique solution for x, i.e. given a row a and given an element b, there is precisely one column x in which it occurs, i.e. each element occurs exactly once in each row.

5) Given that the equation

$$(ab)^{-1} = a^{-1}b^{-1}$$

holds for each pair of elements a, b in a group G, prove that G is commutative.

Multiply the given equation on the left by (ab):

$$(ab)(ab)^{-1} = aba^{-1}b^{-1}$$

$$\Rightarrow \qquad e = aba^{-1}b^{-1}$$

Now multiply on the right by b:

$$eb = aba^{-1}b^{-1}b$$

$\Rightarrow$ $$b = aba^{-1}e$$

$\Rightarrow$ $$b = aba^{-1}$$

Now multiply by a on the right:

$$ba = aba^{-1}a$$

$\Rightarrow$ $$ba = abe$$

$\Rightarrow$ $$ba = ab$$

Since this holds *for each* pair of elements a, b in G, the group is commutative.

EXERCISE 14i

1) Check the following cases of the law of indices:
(a) $a^{-2}a^3 = a$ (b) $a^4a^{-2} = a^2$.

2) (a) Simplify: $a^2ea^{-1}a^2a^4$.
(b) Given that $ab = e$, simplify:

$$a^2ba^{-1}bea^2.$$

3) Prove that the equation $ya = b$ has a unique solution for y in a group G.

4) Use the result proved in Example 14i, No. 3 to prove that
(a) $(a^{-1})^{-1} = a$ (b) $(ab)^{-1} = b^{-1}a^{-1}$.

5) Given *either* that $fa = a$ or $af = a$, prove $f = e$. Deduce that the identity in a group is unique.

6) Prove that each element in a group occurs exactly once in each column of the group table.

7) Use the fact that each element in a group occurs exactly once in each row and column to complete the group table:

	e	a	b
e	e	a	b
a	a		
b	b		

Deduce that any two groups with three elements in them are isomorphic.

8) Complete the group tables:

(a)

	e	a	b	c
e	e	a	b	c
a	a	e		
b	b		e	
c	c			e

(b)

	e	a	b	c
e	e	a	b	c
a	a			e
b	b		e	
c	c	e		

Deduce that any group with four elements in it must be isomorphic to one of these two groups. To which of the two is (i) $\mathbb{Z}_4$, (ii) $S(\mathrm{R})$ (where R is a rectangle) isomorphic?

9) Given that $a^2b^2 = (ab)^2$ holds for each pair of elements a, b in a group G, prove that G is commutative.

10) Given that $a^3b^3 = (ab)^3$ for each pair of elements a, b in a group G, prove that $a^2b^2 = (ba)^2$. Deduce (or prove otherwise) that $a^4b^4 = (ab)^4$ and $(ab)^3 = (ba)^3$. (C)

PROPERTIES PRESERVED BY ISOMORPHISM

We now return to our discussion of isomorphisms. There are some useful techniques for showing that two groups *are* isomorphic of which the most elementary is to write out the group tables in an appropriate pattern in order to display the correspondence. However, this is not a good technique for proving that two groups are *not* isomorphic because it would be necessary to check all possible reorderings for the tables and this would be tedious. What is needed is some property of one of the groups which is preserved by isomorphism and which is not shared by the other group. Then the groups cannot be isomorphic.

Order of a Group

One such property is the number of elements in the group, called the *order* of the group.

A group G with a finite number of elements in it is said to be a *finite* group and the number of elements in the group is called the *order* of G and is denoted by $|G|$. A group with an infinite number of elements is said to be an *infinite* group.

If groups H and G are isomorphic then clearly they are either both finite groups or both infinite groups, and if they are both finite then $|H| = |G|$.

Isomorphism preserves order.

EXAMPLES 14j

1) Write down:

(a) $|S(\Delta)|$ (b) $|S(\mathrm{R})|$ (c) $|S(\square)|$

(Δ denotes an equilateral triangle, R a rectangle and $\square$ a square.) Deduce that no two of these groups are isomorphic.

$|S(\Delta)| = 6$ $|S(\mathrm{R})| = 4$ and $|S(\square)| = 8$ (see solution to Exercise 14a, Question 7).
Since the three groups are all of different orders, no two can be isomorphic.

Commutativity

The second property of a group which is preserved by isomorphism is commutativity. In terms of the group tables, we have seen that a commutative group has a symmetric table while a non-commutative group does not. Since isomorphic groups have similar tables, if one is commutative then so must be the other one.
We can give a general proof of this fact as follows. Suppose $G \cong G'$ and under the correspondence

$$g \longleftrightarrow g' \quad \text{and} \quad h \longleftrightarrow h'$$

then $$gh \longleftrightarrow g'h' \quad \text{and} \quad hg \longleftrightarrow h'g'$$
Therefore $gh = hg \Rightarrow g'h' = h'g'$.
Since this is true for each pair of elements $g, h \in G$, we have proved

$$G \text{ commutative } \Rightarrow G' \text{ commutative.}$$

Isomorphism preserves commutativity.

EXAMPLES 14j (continued)

2) Prove that $\mathbb{Z}_6$ is not isomorphic to $S(\Delta)$.

Both groups have order 6 so we cannot use order to prove that they are non-isomorphic. But we *can* use commutativity: $\mathbb{Z}_6$ is commutative but $S(\Delta)$ is not commutative. Therefore they cannot be isomorphic.

Subgroups

We next show that isomorphisms preserve *subgroups*. We do this in three stages. Suppose that we have two groups G, G' and an isomorphism $G \cong G'$.

Under the isomorphism the identity in G corresponds to the identity in G'.

Proof: Let e be the identity in G and let a be any element of G. Suppose $e \longleftrightarrow x'$ and $a \longleftrightarrow a'$. We have to show that $x' = e'$ (the identity in G').
Since $e \longleftrightarrow x'$ and $a \longleftrightarrow a'$ we have $ea \longleftrightarrow x'a'$.

But $ea = a$, therefore $x'a' = a' = e'a'$.
But this implies $x' = e'$ by cancelling a' on the right).

Under the isomorphism, inverses in G correspond to inverses in G'.

Proof: Suppose $a \longleftrightarrow a'$ and $a^{-1} \longleftrightarrow b'$, then $e = aa^{-1} \longleftrightarrow a'b'$. But we have just proved that $e \longleftrightarrow e'$. Therefore $a'b' = e'$ which implies $b' = (a')^{-1}$ (from Example 14i, No. 3).

Now a subgroup in G is a subset which is closed under products, contains the identity and contains the inverse of each element in the subset. Since the isomorphism preserves *all* of these, we have:

Under the isomorphism subgroups of G correspond to subgroups of G'.

Thus if G has a certain number of different subgroups of a particular size, then G' must have the *same* number of subgroups of that size.

EXAMPLES 14j (continued)

3) Prove that $\mathbb{Z}_4$ is not isomorphic to $S(\text{R})$.

Both groups have order 4 and both are commutative so we cannot use either of these properties to distinguish the groups. Instead we use *subgroups.* Referring to Exercise 14h, Questions 5(a) and 7(a), we have the following subgroups of order 2:

$$(\{0, 2\}, +_4) \subseteq \mathbb{Z}_4$$

$$(\{e, a\}, \circ), \quad (\{e, b\}, \circ), \quad (\{e, r\}, \circ) \subseteq S(\text{R})$$

Since $\mathbb{Z}_4$ has only *one* subgroup of order 2 while $S(\text{R})$ has *three*, the groups cannot be isomorphic.

EXERCISE 14j

1) Write down:

$$|S(\square)|, \ |\mathbb{Z}_6|, \ |\mathbb{Z}_9|, \ |S(\text{DOG-LEG})|, \ |\mathbb{Z}_7|, \ |S(\text{R})|$$

Deduce that no two of the groups are isomorphic.

2) Prove that $\mathbb{Z}$ is not isomorphic to $\mathbb{Z}_{10}$.

3) Prove that $S(\square)$ is not isomorphic to either of $\mathbb{Z}$ or $\mathbb{Z}_8$.

4) Count the number of subgroups of order 2 in (a) $S(\Delta)$, (b) $\mathbb{Z}_6$, and hence find an alternative proof that these groups are not isomorphic (we gave one proof in Example 14j, No. 2).

5) The group Q of 'unit quaternions' is a group of order 8 with the following table:

	1	-1	i	$-i$	j	$-j$	k	$-k$
1	1	-1	i	$-i$	j	$-j$	k	$-k$
-1	-1	1	$-i$	i	$-j$	j	$-k$	k
i	i	$-i$	-1	1	k	$-k$	$-j$	j
$-i$	$-i$	i	1	-1	$-k$	k	j	$-j$
j	j	$-j$	$-k$	k	-1	1	i	$-i$
$-j$	$-j$	j	k	$-k$	1	-1	$-i$	i
k	k	$-k$	j	$-j$	$-i$	i	-1	1
$-k$	$-k$	k	$-j$	j	i	$-i$	1	-1

Prove that Q is not isomorphic to *either* (a) $\mathbb{Z}_8$ *or* (b) $S(\square)$.

THE ORDER OF AN ELEMENT

Suppose we have an element a of a group G and suppose that we list the powers of a:

$$a^0 = e, \quad a^1 = a, \quad a^2, \quad a^3, \quad a^4, \ \ldots$$

(remember that a^2 means $a * a$, a^3 means $a * a * a$, etc.)
then *either* all these elements are different elements of G (in which case G must be infinite) and we say 'a has infinite order' *or* two at least of these powers must be the same (and if G is finite then this *must* happen) and we say 'a has finite order'.
Suppose that a does have finite order, then we have

$$a^p = a^q \quad \text{where} \quad 0 \leqslant p < q$$

and multiplying on the right by a^{-p} we have

$$a^p a^{-p} = a^q a^{-p}$$

$$\Rightarrow \qquad e = a^{q-p}$$

i.e. $$a^n = e \quad \text{where} \quad n = q - p > 0$$

The smallest number n $(n > 0)$ such that $a^n = e$ is called the *order* of a.

For example, consider the element r in $S(\Delta)$ then

$$r^0 = e, \quad r^1 = r, \quad r^2 = s, \quad r^3 = sr = e$$

(see the table on page 569). Therefore r has order 3.

Since an isomorphism preserves products, it preserves powers as well:

$$a \longleftrightarrow a' \Rightarrow a^2 \longleftrightarrow (a')^2, \quad a^3 \longleftrightarrow (a')^3, \ \ldots$$

and therefore it must preserve the order of an element. Thus we have another property preserved by an isomorphism.

An isomorphism preserves the order of an element.

EXAMPLE 14k

Find the orders of the elements of $S(\square)$. Find the order of the element 1 in $\mathbb{Z}_8$. Deduce (again) that these groups are non-isomorphic.

Listing the powers of the elements of $S(\square)$ we have (using the notation in the solution to Exercise 14a, No. 7)

$$e^0 = e, \quad e^1 = e \quad \Rightarrow \quad e \text{ has order } 1$$

$$a^0 = e, \quad a^1 = a, \quad a^2 = e \quad \Rightarrow \quad a \text{ has order } 2$$

Similarly b, c, d have order 2.

$r^0 = e, \quad r^1 = r, \quad r^2 = t, \quad r^3 = rt = s, \quad r^4 = rs = e \quad \Rightarrow \quad r$ has order 4.

Similarly s has order 4.

$$t^0 = e, \quad t^1 = t, \quad t^2 = e \quad \Rightarrow \quad t \text{ has order } 2$$

Thus $S(\square)$ has *one* element of order 1
five elements of order 2
and *two* elements of order 4.

When listing the powers of 1 in $\mathbb{Z}_8$ it is important to remember that the group operation in $\mathbb{Z}_8$ is *addition* and therefore $(1)^n$ *means* $1 + 1 + \ldots + 1$ (n times), i.e. $n \times 1$. Therefore the powers of 1 are:

$$0 \times 1 = 0, \ 1 \times 1 = 1, \ 2 \times 1 = 2, \ 3, \ 4, \ 5, \ 6, \ 7, \ 8 \times 1 = 0$$

i.e. 1 has order 8 in $\mathbb{Z}_8$.
Since $S(\square)$ has no element of order 8, there cannot be an isomorphism between the groups.

EXERCISE 14k

1) Find the orders of all the elements of $S(\Delta)$ and find the order of 1 in $\mathbb{Z}_6$. Deduce (again) that the groups are non-isomorphic.

2) Find the orders of all the elements of $S(\mathrm{R})$ and $S(\mathrm{W}_4)$ (W_4 denotes the four-bladed windmill) and deduce that the groups are non-isomorphic.

3) Find the order of the following elements (or state if they have infinite order) in the given group:

(a) 1 in $\mathbb{Z}$

(b) 3 in $\mathbb{Z}_8$

(c) 5 in $\mathbb{Z}_{10}$

(d) $\frac{1}{2}$ in $\mathbb{Q}$

(e) $\frac{1}{2}$ in $(\mathbb{Q}^*, \times)$
(f) 1 in $(\mathbb{R}^+, \times)$ } (see Exercise 14b for notation)

(g) $\begin{pmatrix} 1 & 2 & 3 & 4 \\ 2 & 3 & 4 & 1 \end{pmatrix}$ in S_4

(h) $\begin{pmatrix} 1 & 2 & 3 & 4 & 5 \\ 3 & 1 & 2 & 4 & 5 \end{pmatrix}$ in S_5.

4) Prove that an element $g \neq e$ in a group G has order 2 if and only if e appears on the diagonal in row g of the group table.

GROUPS DEFINED BY EQUATIONS

It is often convenient to specify a group using 'equations'. We shall explain the idea by examples.

EXAMPLES 14I

1) Construct a table for the group

$$G = \{a \mid a^5 = e\}$$

To which familiar group is it isomorphic?

The notation means that the group consists of all the powers of a, subject to the restriction that $a^5 = e$. Thus for instance $a^6 = a^5a = ea = a$ and $a^{-1} = a^5a^{-1} = a^4$. Therefore the elements of the group are

$$e,\ a,\ a^2,\ a^3,\ a^4$$

since any other power can be changed to one of these. In a similar way we can calculate products, e.g.:

$$a^4a^3 = a^7 = a^5a^2 = ea^2 = a^2$$

$$a^2a^4 = a^6 = a^5a = ea = a$$

And we can readily fill in the table:

	e	a	a^2	a^3	a^4
e	e	a	a^2	a^3	a^4
a	a	a^2	a^3	a^4	e
a^2	a^2	a^3	a^4	e	a
a^3	a^3	a^4	e	a	a^2
a^4	a^4	e	a	a^2	a^3

Clearly the group is isomorphic to $\mathbb{Z}_5$.

2) How many distinct elements are there in the group

$$\{a, b \mid a^3 = b^2 = e, \quad ab = ba^{-1}\}?$$

Simplify: a^2baba^{-1}, $ba^{-1}ba^{-1}ba^2$.

The notation means that the group consists of all the powers of a and b and all the combinations of those powers, subject to the relations given by the equations.

We shall do the simplifications first:

$a^2b\underbrace{ab}a^{-1} = a^2b^2a^{-1}a^{-1}$ (we have used the equation $ab = ba^{-1}$ to change the middle ab (bracketed) to ba^{-1})

$= a^2ea^{-2}$ (using the equation $b^2 = e$)

$= e$

$ba^{-1}ba^{-1}ba^2 = ba^2b\underbrace{a^2b}a^2$ (using $a^3 = e$ which implies $a^{-1} = a^2$, twice)

$= ba^2b^2a^{-2}a^2$ (moving the b left across the a^2 (bracketed) using $ab = ba^{-1}$ twice)

$= ba^2ea^{-2}a^2$ (since $b^2 = e$)

$= ba^2$

For a general element of the group we can use the equation $ab = ba^{-1}$ to move all the powers of b to the left (as we did in the simplifications) and express any given element as b^pa^q for some p and q. But then we can use the equations $a^3 = e$, $b^2 = e$ to reduce p to 1 or 0 and to reduce q to 0, 1 or 2. Thus the distinct elements are:

$$e, a, a^2, b, ba, ba^2$$

EXERCISE 14l

1) Complete the group table for the group $\{a, b \mid a^3 = b^2 = e,\ ab = ba^{-1}\}$ and verify that it is isomorphic to $S(\Delta)$.

2) Write down the distinct elements of the group

$$\{a, b \mid a^2 = b^2 = e,\ ab = ba\}$$

Construct a group table. To which familiar group is it isomorphic?

3) Write down the distinct elements of the group

$$\{a, b \mid a^3 = b^2 = e, \quad ab = ba\}$$

Complete a group table and prove that the group is isomorphic to $\mathbb{Z}_6$.

4) Write down the distinct elements of

$$\{a, b \mid a^4 = b^2 = e, \quad ab = ba\}$$

Complete a group table and prove that the group is not isomorphic to any of the groups

(a) $\mathbb{Z}_8$, (b) $S(\square)$, (c) Q

(Q denotes the unit quarternions, see Exercise 14j, Question 5).

DIRECT PRODUCTS

Suppose we have two groups G, H, then we can construct a new group which is called the *direct product* of G and H and is written $G \times H$. The elements of $G \times H$ are pairs (g, h) where g is an element of G and h is an element of H. The group operation is defined by

$$(g, h)(g', h') = (gg', hh')$$

where gg', hh' denote products in G and H respectively. We shall check that $G \times H$ is a group.

Closure

Since gg', hh' are elements of G, H respectively, $(g, h)(g', h')$ is an element of $G \times H$.

Identity

The identity of $G \times H$ is (e_G, e_H), where e_G, e_H are identities in G and H respectively, since

$$(g, h)(e_G, e_H) = (ge_G, he_H) = (g, h)$$

and

$$(e_G, e_H)(g, h) = (e_G g, e_H h) = (g, h)$$

Inverses:

The inverse of (g, h) is (g^{-1}, h^{-1}) since

$$(g, h)(g^{-1}, h^{-1}) = (gg^{-1}, hh^{-1}) = (e_G, e_H)$$

and

$$(g^{-1}, h^{-1})(g, h) = (g^{-1}g, h^{-1}h) = (e_G, e_H)$$

Associativity

$$((g, h)(g', h'))(g'', h'') = (gg', hh')(g'', h'') = (gg'g'', hh'h'')$$

and similarly

$$(g, h)((g', h')(g'', h'')) = (gg'g'', hh'h'')$$

EXAMPLES 14m

1) Write out a group table for $\mathbb{Z}_2 \times \mathbb{Z}_2$. To which familiar group is it isomorphic?

The elements of $\mathbb{Z}_2 \times \mathbb{Z}_2$ are pairs (i, j) where $i, j \in \mathbb{Z}_2$, i.e. they are:

$$(0, 0),\ (1, 0),\ (0, 1),\ (1, 1)$$

and the group operation is given by

$$(i, j)(i', j') = (i +_2 i', j +_2 j')$$

for example

$$(0, 0)(1, 0) = (1, 0) \quad \text{and} \quad (1, 0)(1, 1) = (0, 1)$$

and completing the table we have:

	(0, 0)	(1, 0)	(0, 1)	(1, 1)
(0, 0)	(0, 0)	(1, 0)	(0, 1)	(1, 1)
(1, 0)	(1, 0)	(0, 0)	(1, 1)	(0, 1)
(0, 1)	(0, 1)	(1, 1)	(0, 0)	(1, 0)
(1, 1)	(1, 1)	(0, 1)	(1, 0)	(0, 0)

Clearly this is yet another group isomorphic to $S(\mathrm{R})$.

2) Describe how to combine elements in $\mathbb{R} \times \mathbb{R}$. This group has a very familiar interpretation, what is it?

A typical element of $\mathbb{R} \times \mathbb{R}$ is a pair of real numbers (r, s). The group operation is defined by

$$(r_1, s_1)(r_2, s_2) = (r_1 + r_2, s_1 + s_2)$$

(remember the operation in $\mathbb{R}$ is addition).

This is just the familiar *vector law of addition* for pairs of real numbers (usually written $(r_1, s_1) + (r_2, s_2)$). Thus this group is the set of two dimensional based vectors with vector addition as the group operation. The identity is $(0, 0)$ i.e. the zero vector and the inverse of (r, s) is $(-r, -s) = -(r, s)$ the opposite vector.

EXERCISE 14m

1) Construct a group table for $\mathbb{Z}_4 \times \mathbb{Z}_2$ and verify that this group is isomorphic to the group constructed in Exercise 14l, Question 4.

2) Construct a group table for $\mathbb{Z}_3 \times \mathbb{Z}_2$ and by observing that this group is essentially the same as the group of Exercise 14l, Question 3, show that $\mathbb{Z}_3 \times \mathbb{Z}_2 \cong \mathbb{Z}_6$.

3) Construct a table for $\mathbb{Z}_2 \times \mathbb{Z}_2 \times \mathbb{Z}_2$ (i.e. the set of ordered *triples* of elements of $\mathbb{Z}_2$ with operation defined in the same way as for $G \times H$) and prove that this group (of order 8) is not isomorphic to *any* of the groups:

(a) $\mathbb{Z}_8$, (b) $\mathbb{Z}_4 \times \mathbb{Z}_2$, (c) $S(\square)$, (d) Q.

4) Prove that every element other than the identity in $\mathbb{Z}_3 \times \mathbb{Z}_3$ has order 3. Deduce that $\mathbb{Z}_3 \times \mathbb{Z}_3$ is not isomorphic to $\mathbb{Z}_9$.

5) If G and H are finite groups, what is the order of $G \times H$?

6) Prove that if G and H are commutative then $G \times H$ is commutative.

TWO THEOREMS OF GROUP THEORY

To end this chapter we shall prove two theorems of group theory, Cayley's Theorem and Lagrange's Theorem. These theorems are included in order to give a taste of the 'abstract' development of group theory. This section may be omitted at the discretion of the reader.

Cayley's Theorem

Any finite group is isomorphic to a subgroup of a permutation group.

The theorem gives a good reason for studying permutation groups: once these groups are understood, then all finite groups are understood up to isomorphism.

PROOF OF CAYLEY'S THEOREM

In essence the proof is nothing more than the observation that any row of the group table for a group contains each element of the group exactly once (see Exercise 14i, No. 4). What this observation implies is that we can associate to any element $g \in G$ a permutation: namely the permutation of the elements of G whose permutation symbol has the g-row of the table as its bottom line. For example, consider the group $\{e, a, b\}$ with table:

	e	a	b
e	e	a	b
a	a	b	e
b	b	e	a

Then the a-row is:

a	b	e

which gives the permutation

$$\begin{pmatrix} e & a & b \\ a & b & e \end{pmatrix}$$

and similarly e and b give the permutations:

$$e \longleftrightarrow \begin{pmatrix} e & a & b \\ e & a & b \end{pmatrix} \qquad b \longleftrightarrow \begin{pmatrix} e & a & b \\ b & e & a \end{pmatrix}$$

In a general group g determines the permutation $x \longmapsto gx$ since gx is the entry in the g-row below x:

	e		x
e	e	$\dots$	x
	$\downarrow$	$\vdots$	$\downarrow$
g	g	$\dots$	gx

Let us use the notation $\widehat{g}$ for this permutation, then $\widehat{g}(x) = gx$. Notice that *different* elements g, h of G give rise to *different* permutations, since $\widehat{g}\colon e \longmapsto g$ and $\widehat{h}\colon e \longmapsto h$; so the set of all the permutations arising in this way is in one-to-one correspondence with G.

Notice also that $(\widehat{h} \circ \widehat{g})$ maps x to hgx since $x \overset{\widehat{g}}{\longmapsto} gx \overset{\widehat{h}}{\longmapsto} hgx$ in other words the composition of the permutations corresponds to the product of the corresponding elements of G. Thus the permutations form a group isomorphic to G and we have found the required subgroup of a permutation group isomorphic to G.

Lagrange's Theorem

The order of any subgroup of a finite group G is a factor of the order of G.

PROOF OF LAGRANGE'S THEOREM

The proof hinges on the idea of a *coset*. Suppose H is a subgroup of the finite group G and suppose that the elements of H are listed first in the list of elements of G in the group table:

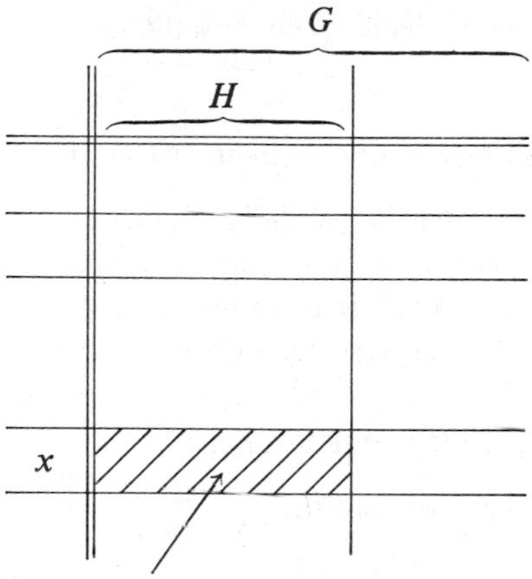

coset corresponding to x

If we consider one row of the table (say row x) then all the entries in the row which correspond to H form a *coset* of H in G. For example, if G is $S(\Delta)$ and H is the subgroup $\{e, r, s\}$ then we can read off the cosets from the table for $S(\Delta)$:

	e (H)	r (H)	s (H)	a	b	c
e	e	r	s	a	b	c
r	r	s	e	b	c	a
s	s	e	r	c	a	b
a	a	c	b	e	s	r
b	b	a	c	r	e	s
c	c	b	a	s	r	e

The coset in row a is $\{a, b, c\}$ and the same coset occurs in rows b and c (the different orders in which the elements occur is ignored as far as the cosets are concerned). $\{e, r, s\}$ is also a coset and it also appears three times (in rows e, r and s).
In general, the coset in row x consists of all the elements xh as h runs through the various elements of H.
The first fact about cosets that we need comes from the observation that we used in Cayley's Theorem: since a row of the table contains each element of G exactly once, the elements in the coset must *all be different.* Thus each coset contains the *same* number of elements namely the number of elements in H, i.e. the order of H, $|H|$.

Fact 1 Each coset has exactly $|H|$ elements of G in it.

Next notice that in the example above there are just two different cosets and they do not overlap at all. We can prove this 'non-overlapping' property of cosets in general, and this is the second fact that we need. Suppose that the cosets in rows x and y overlap. Since typical elements of these cosets are xh_1 and yh_2 respectively, where $h_1, h_2 \in H$, the overlapping of the cosets implies that

$$xh_1 = yh_2$$

where h_1, h_2 are certain elements of H.
Therefore

$$x = yh_2h_1^{-1}$$

and if h is *any* element of H then

$$xh = yh_2h_1^{-1}h = yh' \qquad [1]$$

where $h' = h_2h_1^{-1}h$..
Since H is a subgroup, h' is also an element of H (H contains inverses and is closed under products).
Therefore equation [1] shows that any element of the coset in row x lies in the coset in row y. Similarly any element of the coset in row y lies in the coset in row x. Thus the two cosets *coincide exactly*. Thus if two cosets overlap at all then they are in fact the *same* coset.

Fact 2 Different cosets of H in G do not overlap.

Finally notice that all the elements of G occur in some coset since they all occur in each column (in particular they all occur in the first column of the table).

Fact 3 Every element of G occurs in some coset of H in G.

Now combine these three facts: the elements of G all occur in a coset, the different cosets do not overlap, and each coset has exactly $|H|$ elements in it. In other words the elements of G are 'parcelled up' into parcels of size $|H|$.

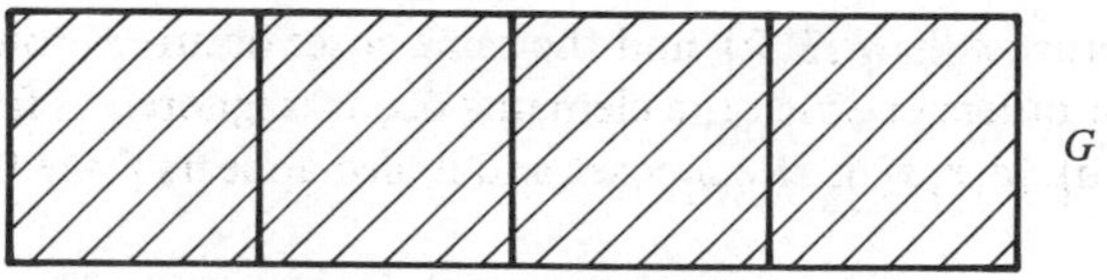

cosets each of size $|H|$

Therefore $|G| = |H| \times$ (no. of cosets)
i.e. $|H|$ is a factor of $|G|$.

MISCELLANEOUS EXERCISE 14

1) Write out the tables of the following groups:
(i) $G_1 = \{S_1, +\}$ where $S_1 = \{0, 1, 2, 3\}$ and $+$ is addition modulo 4;
(ii) G_2, the group of symmetries of the rectangle, explaining clearly the geometrical meaning of the elements of the group.
Are these groups isomorphic? If your answer is yes, you should give a correspondence between the clements of the two groups that establishes the isomorphism. If your answer is no, you should explain why the groups are not isomorphic. (O)

2) Prove that the integers $\{2, 4, 6, 8\}$ form a group under the operation of multiplication modulo 10.
Prove that the numbers $\{1, -1, \mathrm{i}, -\mathrm{i}\}$ form a group under the operation of complex multiplication.
Prove that these groups are isomorphic. (C)

3) Interpret the transformations $z \longmapsto z$, $z \longmapsto -z$, $z \longmapsto \mathrm{i}z$, $z \longmapsto -\mathrm{i}z$ of the complex plane geometrically.
Hence give a geometric argument that the group $(\{1, -1, \mathrm{i}, -\mathrm{i}\}, \times)$ is isomorphic to the symmetry group of the four-bladed windmill.

4) Prove that the complex numbers $\{1, \omega, \omega^2\}$ form a group under complex multiplication where

$$\omega = \cos\left(\frac{2\pi}{3}\right) + \mathrm{i}\sin\left(\frac{2\pi}{3}\right).$$

Interpret the group geometrically and hence prove that it is isomorphic to a symmetry group $S(F)$. (You should sketch a suitable figure F.)

5) Prove that the complex numbers $\{1, \alpha, \alpha^2, \alpha^3, \alpha^4\}$ form a group under complex multiplication where

$$\alpha = \cos\left(\frac{2\pi}{5}\right) + \mathrm{i}\sin\left(\frac{2\pi}{5}\right)$$

and prove that the group is isomorphic to $\mathbb{Z}_5$.
Write down a group of complex numbers which is isomorphic to $\mathbb{Z}_n$.

6) (i) Compile the group tables for
(a) the symmetries of the rhombus,
(b) the mappings f_1, f_2, f_3, f_4 where

$$f_1\colon x \to x,\ f_2\colon x \to \frac{1}{x},\ f_3\colon x \to -x,\ f_4\colon x \to -\frac{1}{x}$$

Show that (a) and (b) are isomorphic.

(ii) Show that the set of complex numbers with modulus 1 forms an Abelian (commutative) group under multiplication. Find the identity element and the inverse element of $\cos\theta + i\sin\theta$. (U of L)

7) (a) Prove that the set $\mathbb{C}$ of complex numbers forms a group under *addition* of complex numbers. Show that

$$(\mathbb{C}, +) \cong (\mathbb{R} \times \mathbb{R}, +)$$

(see Example 14m, No. 2).

(b) Prove that the set $\mathbb{C}^*$ of non-zero complex numbers forms a group under multiplication of complex numbers. Write down subgroups of $(\mathbb{C}^*, \times)$ of (i) order 3, (ii) order 4, (iii) order 5.

8) The elements e, a, b, u, v, w of a group represent the isometries of an equilateral triangle, where e is the identity, a and b represent rotations in the plane of the triangle, and u, v and w represent reflections in its axes of symmetry. The operation is the composition of isometries

	e	a	b	u	v	w
e						
a		I			II	
b						
u						
v		III			IV	
w						

From geometrical considerations, complete section I of the group table shown and enter e in the appropriate places in section IV.
Given that $au = w$ and $uv = a$, complete sections II and IV.
By considering $(au)^{-1}$, or otherwise, find ub and hence complete section III. (U of L)

9) Define a permutation group.
Prove that the set of all permutation of the three digits 1, 2, 3, under a law of combination to be specified, does indeed form a group.
Find all the subgroups. (C)

10) A set S of elements $\{a, b, c, d\}$ is closed under an associative binary operation $*$: that is, if $x, y, z \in S$, then $(x*y)*z = x*(y*z)$. One of the elements of S is the identity element under the operation $*$. Given that

$$a*a = a, \quad b*b = d, \quad c*c = d, \quad d*d = a$$

determine whether $(S, *)$ is a group. (C)

11) The matrices (with complex elements)

$$\mathbf{E} = \begin{pmatrix} 1 & 0 \\ 0 & 0 \end{pmatrix}, \quad \mathbf{A} = \begin{pmatrix} -1 & 0 \\ 0 & 0 \end{pmatrix}, \quad \mathbf{B} = \begin{pmatrix} \mathrm{i} & 0 \\ 0 & 0 \end{pmatrix}, \quad \mathbf{C} = \begin{pmatrix} x & y \\ z & t \end{pmatrix}$$

are given to form a group under matrix multiplication. Find x, y, z and t; confirm that $\mathbf{A(BC)}$ is then equal to $\mathbf{(AB)C}$ and exhibit the group operation table. (C)

12) Prove that the set of matrices $\begin{pmatrix} \lambda & 0 \\ 0 & 0 \end{pmatrix}$ where $\lambda \in \mathbb{R}^*$ (the set of non-zero real numbers) forms a group under matrix multiplication.
Prove that the group is isomorphic to $(\mathbb{R}^*, \times)$.

13) In the set G of symmetry transformations of the square ABCD,
let **R** denote an anticlockwise rotation of $\pi/2$ radians about the centre O of the square,
I denote the identity transformation,
L denote a reflection in the axis BC and
M denote a reflection in the axis BD.
Form a combination table for the elements $\mathbf{I}, \mathbf{R}^2, \mathbf{L}, \mathbf{M}$.
Deduce that these elements form a group of order 4. (Associativity may be assumed.)
State another set of elements of G which form a group of order 4. (U of L)

14) The Cartesian product of a set with itself, $G \times G$, is the set of all ordered pairs (x, y) with $x \in G$ and $y \in G$. Prove that if G is a multiplicative group, then $(G \times G, \circ)$ is a group, where the binary operation $\circ$ on $G \times G$ is defined by

$$(p, q) \circ (r, s) = (pr, qs)$$

In the case when G is the group with exactly two elements, determine whether or not $(G \times G, \circ)$ is isomorphic to the group with the four elements $\{1, -1, \mathrm{i}, -\mathrm{i}\}$, the operation being multiplication of complex numbers. (C)

15) Prove that the set of matrices of the form $\begin{pmatrix} \lambda & 0 \\ 0 & \mu \end{pmatrix}$ where $\lambda, \mu \in \mathbb{R}^*$ (non-zero reals) forms a group under matrix multiplication. Show that the group is isomorphic to $\mathbb{R}^* \times \mathbb{R}^*$ and find two different subgroups isomorphic to $\mathbb{R}^*$.

16) Write down the distinct elements of the group

$$G = \{a, b \mid a^5 = e = b^2,\ ab = ba^{-1}\}$$

and by considering the following geometrical transformation:

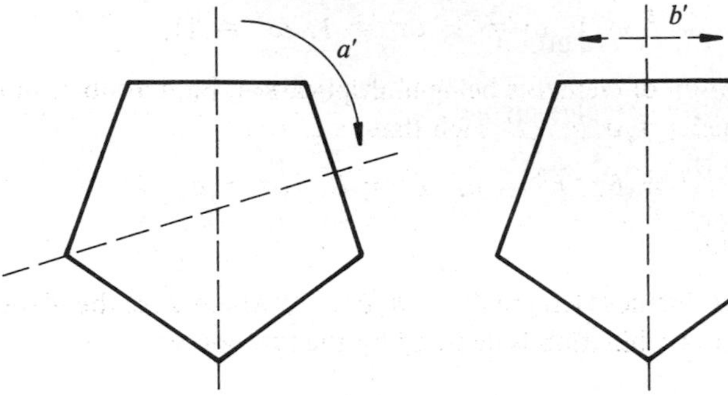

or otherwise, prove that $G \cong S(\text{PENT})$ (where PENT denotes a regular pentagon).

17) Use a similar proof to Question 16 to show

$$\{a, b \mid a^6 = e = b^2,\ ab = ba^{-1}\} \cong S(\text{HEX})$$

(HEX denotes a regular hexagon).

18) By placing the pentagon in a suitable position, or otherwise, write down a matrix group isomorphic to the group G in Question 16.

19) Describe a similar group to those in Questions 16 and 17 of order $2n$ for each n and interpret it geometrically.

20) The figure F is the infinite saw-blade illustrated:

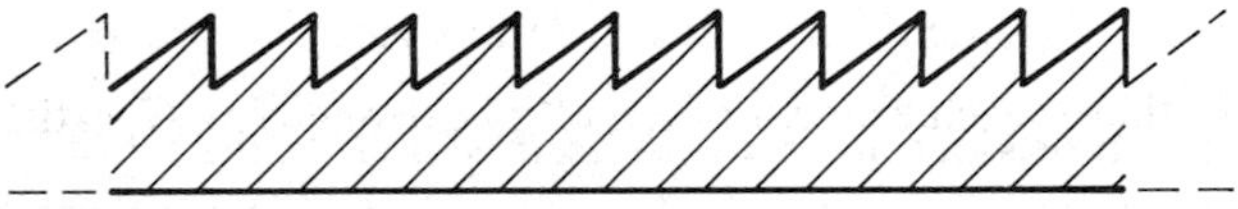

F

Prove that $S(F) \cong \mathbb{Z}$.

21) By considering the labellings (or otherwise):

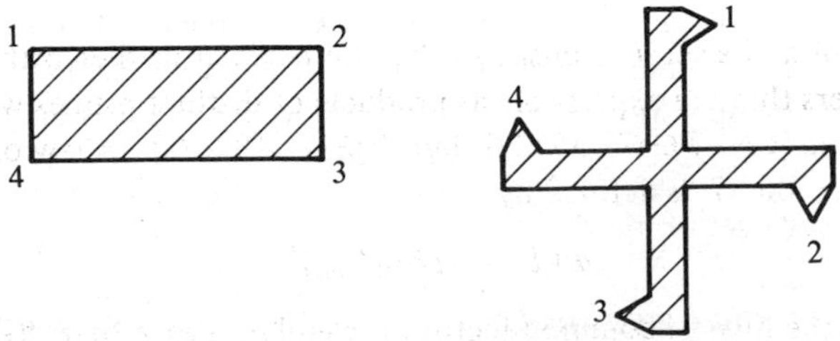

Write down subgroups of S_4 isomorphic to $\mathbb{Z}_2 \times \mathbb{Z}_2$ and $\mathbb{Z}_4$.

22) (a) A group H has fifteen elements,

$$1, \omega, \omega^2, \ldots, \omega^{14},$$

where ω is a complex 15th root of unity

$$(\omega^{15} = 1,\ \omega \neq 1,\ \omega^3 \neq 1,\ \omega^5 \neq 1),$$

the law of combination of elements being multiplication. Find a sub-group of H having five elements $1, a, b, c, d$ such that

$$a^2 = b,\quad b^2 = c,\quad c^2 = d,\quad d^2 = a,$$

and prove your result.

(b) A group G has distinct elements $e, a, b, c, \ldots,$ where e is the identity element. The law of combination is denoted by the symbol $*$.

(i) Prove that, if

$$a * a = b,\quad b * b = a,$$

then the elements e, a, b form a sub-group of G.

(ii) Prove that, if

$$a * a = b,\quad b * b = c,\quad c * c = a,$$

then the elements e, a, b, c do not form a sub-group of G. (C)

23) The identity element of a multiplicative group G is e, and a $(a \neq e)$ is a given element of the group. A is the subset of G consisting of all elements of the form xax^{-1}, where x is any element of G. Prove (i) $a \in A$, (ii) $e \notin A$.

If $y = xax^{-1}$ is an element of A, prove that aya^{-1} is also an element of A. Show that, if $a^{-1} \in A$ then an element $x \in G$ exists such that $x = axa$.

(C)

24) (a) Prove that if each element of a multiplicative group G satisfies the relation

$$a^2 = e,$$

where e is the identity element of the group, then the group is commutative (Abelian).

(b) The elements of a set G consist of the number 1 together with all the positive integers that are expressible as products of distinct primes with non-repeated factors (e.g. $7 \in G$, $35 \in G$ but $9 \notin G$, $18 \notin G$). A law of combination $*$ on G is defined by

$$a * b = (ab)/(D_{ab})^2,$$

where D_{ab} is the highest common factor of a and b. Prove that $\{G, *\}$ is a group. (C)

25) Let S be the set of non-zero rational numbers (in their lowest terms), and let $\times$ be ordinary multiplication followed by reduction to lowest terms. Prove that $\{S, \times\}$ is a group G.
Let S_0 be the subset of S consisting of those members of S whose numerator and denominator are both odd numbers. Prove that $\{S_0, \times\}$ is a subgroup G_0 of G.
If x is any member of S, show that $x = 2^r x_0$, where r is an integer and x_0 is a suitably chosen member of S_0.
If S_r is defined to be the set of all members of S that correspond to the same r and if $S_r \times S_s$ is defined to be the set of all members of S obtained by multiplying any member of S_r by any member of S_s, show that
$S_r \times S_s = S_{r+s}$. (O)

26) Evaluate the residues modulo 13 of the powers of 2 up to and including 2^{12}. Verify that 2^{13} is congruent to 2, and that any non-zero residue is congruent to a power of 2.
Find all the subgroups of G, where G is the group formed by the non-zero residues mod 13 with the operation of multiplication mod 13. (O)
Note that 'residues mod 13' means remainder on division by 13; 'congruent' means equal modulo 13.

27) Let G be a group and $g \in G$ be any element. Prove that the set of elements

$$\ldots, g^{-2}, g^{-2}, g^{-1}, e, g, g^2, g^3, \ldots$$

forms a subgroup of G.
Identify the subgroup in the following cases:
(a) $G = S(\Delta)$, $g = r$ (rotation)
(b) $G = S(\Delta)$, $g = a$ (reflection)
(c) $G = \mathbb{Z}$, $g = 1$
(d) $G = \mathbb{Z}_6$, $g = 2$.
(The subgroup is called the subgroup *generated* by g and is denoted $\langle g \rangle$).

28) Prove that $\langle g \rangle$ (see Question 27) is
(a) $\cong \mathbb{Z}$ if g has infinite order
(b) $\cong \mathbb{Z}_n$ if g has finite order n.
Deduce from Lagrange's Theorem that the order of any element in a finite group G is a factor of $|G|$.

29) By using Question 28 and considering the order of an element other than the identity (or otherwise) prove the following theorem:

> Any group of order p where p is a prime number is isomorphic to $\mathbb{Z}_p$.

TABLE OF ISOMORPHISM CLASSES

As a summary of the chapter on groups we give here a list of most of the groups introduced in the chapter listed by *isomorphism class* (groups in the same class are all isomorphic, those in different classes are not isomorphic). The table is 'complete' up to order 11: that is to say every possible isomorphism class of group is represented.

Order	Class	Commutative or non-commutative	(*Name where appropriate*) Examples
1	1	C	(*Trivial group*) S(irregular Δ), $\mathbb{Z}_1$
2	1	C	(*Cyclic group of order 2*) S(DOG-LEG), $\mathbb{Z}_2$
3	1	C	(*Cyclic group of order 3*) $S(\text{WIND}_3)$, $\mathbb{Z}_3$, Rotation subgroup of $S(\Delta)$, $(\{1, \omega, \omega^2\}, \times)$ where $\omega = \cos \frac{2\pi}{3} + i \sin \frac{2\pi}{3}$.
4	1	C	(*Cyclic group of order 4*) $S(\text{WIND}_4)$, $\mathbb{Z}_4$, $(\{1, -1, i, -i\}, \times)$ $\left(\left\{\begin{pmatrix}1 & 0\\0 & 1\end{pmatrix}, \begin{pmatrix}0 & -1\\1 & 0\end{pmatrix}, \begin{pmatrix}-1 & 0\\0 & -1\end{pmatrix}, \begin{pmatrix}0 & 1\\-1 & 0\end{pmatrix}\right\}, \times\right)$ Rotation subgroup of $S(\square)$.
	2	C	(*Klein four group*) S(R), $\mathbb{Z}_2 \times \mathbb{Z}_2$, $\left(\left\{\begin{pmatrix}1 & 0\\0 & 1\end{pmatrix}, \begin{pmatrix}0 & -1\\-1 & 0\end{pmatrix}, \begin{pmatrix}0 & 1\\1 & 0\end{pmatrix}, \begin{pmatrix}-1 & 0\\0 & -1\end{pmatrix}\right\}, \times\right)$ $\left(\left\{\begin{pmatrix}1 & 0\\0 & 1\end{pmatrix}, \begin{pmatrix}-1 & 0\\0 & 1\end{pmatrix}, \begin{pmatrix}1 & 0\\0 & -1\end{pmatrix}, \begin{pmatrix}-1 & 0\\0 & -1\end{pmatrix}\right\}, \times\right)$
5	1	C	(*Cyclic group of order 5*) $S(\text{WIND}_5)$, $\mathbb{Z}_5$, $\{a \mid a^5 = e\}$, $(\{1, \alpha, \alpha^2, \alpha^3, \alpha^4\}, \times)$, where $\alpha = \cos \frac{2\pi}{5} + i \sin \frac{2\pi}{5}$.

Order	Class	Commutative or non-commutative	(*Name where appropriate*) Examples
6	1	C	(*Cyclic group of order 6*) $\mathbb{Z}_6$, $\mathbb{Z}_2 \times \mathbb{Z}_3$, $\{a, b \mid a^3 = b^2 = e,\ ab = ba\}$
	2	NC	(*Dihedral group of order 6*) $S(\Delta)$, S_3
7	1	C	(*Cyclic group of order 7*) $\mathbb{Z}_7$
8	1	C	(*Cyclic group of order 8*) $\mathbb{Z}_8$
	2	C	$\mathbb{Z}_4 \times \mathbb{Z}_2$
	3	C	$\mathbb{Z}_2 \times \mathbb{Z}_2 \times \mathbb{Z}_2$
	4	NC	(*Dihedral group of order 8*) $S(\square)$, $\{a, b \mid a^4 = b^2 = e,\ ab = ba^{-1}\}$
	5	NC	(*Quaternionic group*) Q (see Exercise 14j, Q5)
9	1	C	(*Cyclic group of order 9*) $\mathbb{Z}_9$
	2	C	$\mathbb{Z}_3 \times \mathbb{Z}_3$
10	1	C	(*Cyclic group of order 10*) $\mathbb{Z}_{10}$
	2	NC	(*Dihedral group of order 10*) S(PENT), $\{a, b \mid a^5 = b^2 = e,\ ab = ba^{-1}\}$
11	1	C	(*Cyclic group of order 11*) $\mathbb{Z}_{11}$

General Classes of Finite Groups

n	C	$\mathbb{Z}_n$ (*Cyclic group of order n*)
$2n$	NC	(*Dihedral group of order 2n*) S(Regular n-gon), $\{a, b \mid a^n = b^2 = e,\ ab = ba^{-1}\}$
$n!$	NC	(*Permutation group on n symbols*) S_n

Infinite Groups

∞	C	(*Integers*) $(\mathbb{Z},+)$, S(Infinite saw blade) $\left(\left\{\begin{pmatrix}1 & n\\0 & 1\end{pmatrix}, n\in\mathbb{Z}\right\},\times\right)$
∞	C	(*Rationals*) $(\mathbb{Q},+)$
∞	C	$(\mathbb{Q}^*,\times)$
∞	C	(*Reals*) $(\mathbb{R},+)$, $(\mathbb{R}^+,\times)$ / $(\mathbb{C},+)$, $\mathbb{R}\times\mathbb{R}$ — The isomorphism between $\mathbb{R}$ and $\mathbb{R}\times\mathbb{R}$ is *very* difficult to describe and that is why it has not been covered in the chapter
∞	C	$(\mathbb{R}^*,\times)$, $\left(\left\{\begin{pmatrix}r & 0\\0 & r\end{pmatrix}, r\in\mathbb{R}^*\right\},\times\right)$ $\left(\left\{\begin{pmatrix}r & 0\\0 & \frac{1}{r}\end{pmatrix}, r\in\mathbb{R}^*\right\},\times\right)$ $\left(\left\{\begin{pmatrix}r & 0\\0 & 0\end{pmatrix}, r\in\mathbb{R}^*\right\},\times\right)$
∞	C	(*Circle group*) ({complex numbers of modulus 1}, $\times$)
∞	C	$(\mathbb{C}^*,\times)$

CHAPTER 15

MATRIX ALGEBRA

Matrices were introduced in Chapters 1 and 2 in the contexts of transformations and the solution of linear equations, and some of the algebra of matrices was also introduced in these contexts. The purpose of this chapter is to study the algebra of matrices in a more general and systematic way and we therefore start the chapter with a brief reminder of some of the earlier work.

Matrices and Notation for Matrices

A *matrix* is a rectangular array of numbers. Usually the numbers are *real numbers*, but occasionally more general numbers, for example complex numbers, are used; in this chapter we shall study *matrices of real numbers.*

An $n \times m$ *matrix* $\mathbf{A}$ is a matrix with n rows and m columns.
An $n \times n$ matrix is a *square* matrix.
A $1 \times n$ matrix is a *row vector* and an $n \times 1$ matrix is a column vector.
A 1×1 matrix is a single real number (a), or simply a, and is often called a *scalar.*

The $n \times n$ square matrix $\mathbf{I} = \begin{pmatrix} 1 & 0 & \dots & 0 \\ 0 & 1 & \dots & 0 \\ \vdots & \vdots & & \vdots \\ 0 & 0 & \dots & 1 \end{pmatrix}$

is called the *unit matrix* of size n

and the $n \times m$ matrix $\mathbf{O} = \begin{pmatrix} 0 & 0 & \dots & 0 \\ 0 & 0 & \dots & 0 \\ \vdots & \vdots & & \vdots \\ 0 & 0 & \dots & 0 \end{pmatrix}$

is the *zero* $n \times m$ matrix.

There are various convenient notations in use for general matrices. For small matrices we can use a separate letter for each entry, for instance the general 2×3 matrix can be denoted by $\begin{pmatrix} a & b & c \\ d & e & f \end{pmatrix}$ where a, b, c, d, e, f are real numbers. For larger matrices it is often convenient to use *subscript* notation. In this notation the (i,j)th entry of the matrix $\mathbf{A}$ is denoted by a_{ij}; i.e. the entry in row i and column j:

$$\mathbf{A} = i \begin{pmatrix} & \overset{j}{\vdots} & \\ \cdots & a_{ij} & \cdots \\ & \vdots & \end{pmatrix}$$

So the whole matrix is written

$$\mathbf{A} = \begin{pmatrix} a_{11} & a_{12} & \cdots & a_{1m} \\ a_{21} & a_{22} & \cdots & a_{2m} \\ \vdots & & & \vdots \\ a_{n1} & a_{n2} & \cdots & a_{nm} \end{pmatrix}$$

There is a condensed version of this notation which is also used

$$\mathbf{A} = (a_{ij})$$

and the subscripts i, j are then understood to run between 1 and n, and 1 and m respectively.
In a similar way a column vector can be denoted by

$$\mathbf{x} = \begin{pmatrix} x_1 \\ x_2 \\ \vdots \\ x_n \end{pmatrix} \quad \text{or} \quad (x_i)$$

and a similar notation may be used for row vectors.

Addition

Addition is defined for matrices of the same size. Suppose $\mathbf{A} = (a_{ij})$ and $\mathbf{B} = (b_{ij})$ are both $n \times m$ matrices, then their *sum* $\mathbf{A} + \mathbf{B}$ is the $n \times m$ matrix whose (i,j)th entry is $a_{ij} + b_{ij}$ (the sum of the corresponding entries of $\mathbf{A}$ and $\mathbf{B}$) i.e.

$$(a_{ij}) + (b_{ij}) = (a_{ij} + b_{ij})$$

Scalar Multiplication

If $\mathbf{A} = (a_{ij})$ is an $n \times m$ matrix and λ is a real number (a 'scalar') then $\lambda\mathbf{A}$ is the $n \times m$ matrix with (i,j)th entry λa_{ij},
i.e.

$$\lambda(a_{ij}) = (\lambda a_{ij})$$

Matrix Multiplication

The dot (or scalar) product of a $1 \times n$ matrix with an $n \times 1$ matrix is defined by

$$\begin{pmatrix} a_1 a_2 & \ldots & a_n \end{pmatrix} \begin{pmatrix} b_1 \\ b_2 \\ \vdots \\ b_n \end{pmatrix} = a_1 b_1 + a_2 b_2 + \ldots + a_n b_n$$

We express the product using sigma notation as

$$\sum_{i=1}^{n} a_i b_i.$$

In general the matrix product of an $n \times m$ matrix $\mathbf{A}$ with an $m \times p$ matrix $\mathbf{B}$ is the $n \times p$ matrix $\mathbf{AB}$ whose (i,j)th entry is the (dot) product of the ith *row* in $\mathbf{A}$ with the jth *column* in $\mathbf{B}$:

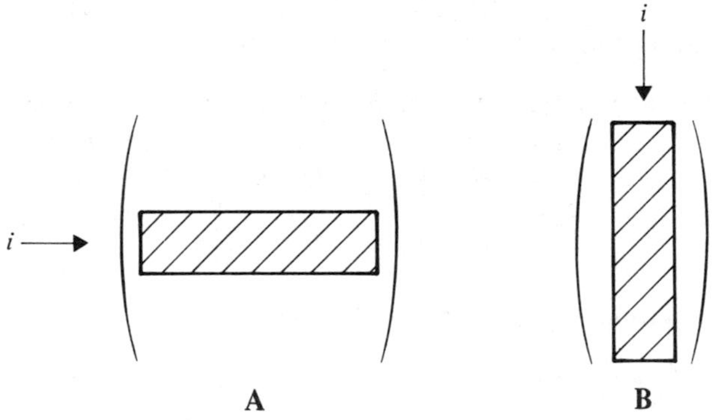

In other words, if $\mathbf{AB} = \mathbf{C} = (c_{ij})$ then

$$\begin{aligned} c_{ij} &= a_{i1} b_{1j} + a_{i2} b_{2j} + \ldots + a_{im} b_{mj} \\ &= \sum_{q=1}^{m} a_{iq} b_{qj} \end{aligned}$$

We can express the matrix product in terms of vectors as follows. Let **A** have rows $\mathbf{r}_1, \mathbf{r}_2, \ldots, \mathbf{r}_n$ (each being an m-vector) and let **B** have columns $\mathbf{c}_1, \mathbf{c}_2, \mathbf{c}_p$ (also m-vectors). Then the (i,j)th entry of **AB** is the dot product $\mathbf{r}_i \cdot \mathbf{c}_j$, i.e.:

$$\mathbf{AB} = \begin{pmatrix} \mathbf{r}_1 \cdot \mathbf{c}_1 & \mathbf{r}_1 \cdot \mathbf{c}_2 & \ldots & \mathbf{r}_1 \cdot \mathbf{c}_p \\ \mathbf{r}_2 \cdot \mathbf{c}_1 & \mathbf{r}_2 \cdot \mathbf{c}_2 & \ldots & \mathbf{r}_2 \cdot \mathbf{c}_p \\ \vdots & \vdots & & \vdots \\ \mathbf{r}_n \cdot \mathbf{c}_1 & \mathbf{r}_n \cdot \mathbf{c}_2 & \ldots & \mathbf{r}_n \cdot \mathbf{c}_p \end{pmatrix} \qquad [1]$$

Note that the jth column of **AB** is

$$\begin{pmatrix} \mathbf{r}_1 \cdot \mathbf{c}_j \\ \mathbf{r}_2 \cdot \mathbf{c}_j \\ \vdots \\ \mathbf{r}_n \cdot \mathbf{c}_j \end{pmatrix} = \mathbf{A}\mathbf{c}_j$$

The jth column of **AB** is **A** times the jth column of **B**.

Algebraic Laws

We now state three laws which the operations of matrix addition and multiplication satisfy. The proofs of these laws, although not difficult, involve messy summations and we have therefore left them to the end of the chapter (see pages 679–81).

Firstly matrix multiplication is *associative.* Suppose we are given three matrices $\mathbf{A}\,(n \times m)$, $\mathbf{B}\,(m \times p)$ and $\mathbf{C}\,(p \times q)$, then we can form the products $\mathbf{AB}\,(n \times p)$ and $\mathbf{BC}\,(m \times q)$ and then we can form $\mathbf{(AB)C}$ and $\mathbf{A(BC)}$ (both $n \times q$). The resulting matrices are equal.

ASSOCIATIVE LAW OF MATRIX MULTIPLICATION

Suppose $\mathbf{A}, \mathbf{B}, \mathbf{C}$ are respectively $n \times m$, $m \times p$ and $p \times q$ matrices then

$$\mathbf{(AB)C} = \mathbf{A(BC)}$$

Secondly, matrix multiplication is *distributive over addition.* Suppose **A** is an $n \times m$ matrix and **B** and **C** are both $m \times p$ matrices, then we can form the sum $\mathbf{B} + \mathbf{C}$ $(m \times p)$ and then the product $\mathbf{A(B + C)}$ $(n \times p)$; the result is the same as multiplying first and *then* adding. There is a similar distributive law if the addition takes place on the left.

DISTRIBUTIVE LAWS

(1) Suppose $\mathbf{A}$ is an $n \times m$ matrix and $\mathbf{B}, \mathbf{C}$ are both $m \times p$ matrices then

$$\mathbf{A}(\mathbf{B}+\mathbf{C}) = \mathbf{AB}+\mathbf{BC}$$

(2) Suppose $\mathbf{A}, \mathbf{B}$ are both $n \times m$ matrices and $\mathbf{C}$ is an $m \times p$ matrix then

$$(\mathbf{A}+\mathbf{B})\mathbf{C} = \mathbf{AC}+\mathbf{BC}$$

Thirdly, the unit matrices of the appropriate sizes act as 'identities' for multiplication.

UNIT LAW

Suppose $\mathbf{A}$ is an $n \times m$ matrix and let $\mathbf{I}_n, \mathbf{I}_m$ denote the unit matrices of size n, m respectively, then

$$\mathbf{I}_n\mathbf{A} = \mathbf{A}\mathbf{I}_m = \mathbf{A}$$

There are other laws satisfied by the operations of addition and matrix multiplication and also scalar multiplication, which have straightforward proofs. One of these is illustrated in the following example and others are left for the reader to prove in Exercise 15a, Questions 8 to 12.

EXAMPLES 15a

1) Prove the *associative law of matrix addition.* That is, suppose $\mathbf{A}, \mathbf{B}, \mathbf{C}$ are all $n \times m$ matrices, then

$$\mathbf{A}+(\mathbf{B}+\mathbf{C}) = (\mathbf{A}+\mathbf{B})+\mathbf{C}$$

The (i,j)th entry in $\mathbf{B}+\mathbf{C}$ is $b_{ij}+c_{ij}$ and therefore the (i,j)th entry in $\mathbf{A}+(\mathbf{B}+\mathbf{C})$ is $a_{ij}+(b_{ij}+c_{ij}) = a_{ij}+b_{ij}+c_{ij}$.
But in a similar way, this is the (i,j)th entry in $(\mathbf{A}+\mathbf{B})+\mathbf{C}$.
Hence the matrices are equal.

MATRICES AS TRANSFORMATIONS

A $n \times m$ matrix $\mathbf{A}$ determines a transformation from m dimensional space to n dimensional space by the rule

$$\begin{pmatrix} x_1 \\ x_2 \\ \vdots \\ x_m \end{pmatrix} \longmapsto \begin{pmatrix} X_1 \\ X_2 \\ \vdots \\ X_n \end{pmatrix}$$

where

$$\begin{pmatrix} X_1 \\ X_2 \\ \vdots \\ X_n \end{pmatrix} = \mathbf{A} \begin{pmatrix} x_1 \\ x_2 \\ \vdots \\ x_m \end{pmatrix}$$

or in other words $\mathbf{x} \longmapsto \mathbf{Ax}$ where $\mathbf{x} = \begin{pmatrix} x_1 \\ x_2 \\ \vdots \\ x_m \end{pmatrix}$. A transformation given in this way by a matrix is called a *linear transformation.*

For example the matrix $\mathbf{A} = \begin{pmatrix} 2 & 3 \\ -1 & 7 \\ 9 & 4 \end{pmatrix}$ determines the linear transformation $\begin{pmatrix} x \\ y \end{pmatrix} \longmapsto \begin{pmatrix} X \\ Y \\ Z \end{pmatrix}$ from two dimensional space to three dimensional space, where

$$\begin{pmatrix} X \\ Y \\ Z \end{pmatrix} = \begin{pmatrix} 2 & 3 \\ -1 & 7 \\ 9 & 4 \end{pmatrix} \begin{pmatrix} x \\ y \end{pmatrix} = \begin{pmatrix} 2x+3y \\ -x+7y \\ 9x+4y \end{pmatrix}$$

i.e.

$$\begin{pmatrix} x \\ y \end{pmatrix} \longmapsto \begin{pmatrix} 2x+3y \\ -x+7y \\ 9x+4y \end{pmatrix}$$

Notice that in this example $\begin{pmatrix} 1 \\ 0 \end{pmatrix}$ is mapped to $\begin{pmatrix} 2 \\ -1 \\ 9 \end{pmatrix}$ and $\begin{pmatrix} 0 \\ 1 \end{pmatrix}$ is mapped to $\begin{pmatrix} 3 \\ 7 \\ 4 \end{pmatrix}$, i.e. the columns of $\mathbf{A}$ are the images under the transformation of the Cartesian unit vectors $\begin{pmatrix} 1 \\ 0 \end{pmatrix}$ and $\begin{pmatrix} 0 \\ 1 \end{pmatrix}$.

This is true in general.

The columns of **A** are the images under the transformation $\mathbf{x} \longmapsto \mathbf{Ax}$ of the *Cartesian unit vectors*

$$\begin{pmatrix}1\\0\\ \vdots\\0\end{pmatrix}, \begin{pmatrix}0\\1\\ \vdots\\0\end{pmatrix}, \ldots, \begin{pmatrix}0\\0\\ \vdots\\1\end{pmatrix}$$

Now suppose that **A** is an $n \times m$ matrix and **B** is an $m \times q$ matrix then **B** gives the transformation $\mathbf{x} \longmapsto \mathbf{Bx}$ from q-space to m-space and **A** gives the transformation $\mathbf{y} \longmapsto \mathbf{Ay}$ from m-space to n-space.
But we can *compose* the transformation to obtain one from q-space to n-space,
i.e. $\mathbf{x} \longmapsto \mathbf{Bx}$ *then* $\mathbf{y} \longmapsto \mathbf{Ay}$:

$$\mathbf{x} \longmapsto \mathbf{Bx} = \mathbf{y} \longmapsto \mathbf{Ay} = \mathbf{A}(\mathbf{Bx})$$

So the result is $\mathbf{x} \longmapsto \mathbf{A}(\mathbf{Bx})$.
But since matrix multiplication is associative this is the same as $(\mathbf{AB})\mathbf{x}$ i.e. the transformation determined by the product of **A** and **B**.
Note carefully the order in which **A** and **B** occur. The transformation **AB** is the transformation **B** *first then* **A**.

Matrix multiplication corresponds to composition of transformations.

We can use this fact to verify the unit law of matrix multiplication. If $\mathbf{B} = \mathbf{I}_m$, the $m \times m$ unit matrix, then **AB** represents the transformation 'identity on m-space followed by **A**' i.e. **A** itself.
Therefore $\mathbf{AI}_m = \mathbf{A}$ and in a similar way $\mathbf{I}_n\mathbf{A} = \mathbf{A}$.

Determinants

The determinant $|\mathbf{A}|$ of a 2×2 matrix is defined by

$$|\mathbf{A}| = \begin{vmatrix} a & b \\ c & d \end{vmatrix} = ad - bc$$

and of a 3×3 matrix by

$$|\mathbf{A}| = \begin{vmatrix} a & b & c \\ d & e & f \\ g & h & i \end{vmatrix} = a\begin{vmatrix} e & f \\ h & i \end{vmatrix} - b\begin{vmatrix} d & f \\ g & i \end{vmatrix} + c\begin{vmatrix} d & e \\ g & h \end{vmatrix}$$

This expression uses the top row and its cofactors but $|\mathbf{A}|$ can be expanded by using any row or column and the corresponding cofactors.

In general there is a similar definition for the determinant of an $n \times n$ matrix (using cofactors) i.e.

$$|\mathbf{A}| = a_{11}|\mathbf{A}_{11}| - a_{12}|\mathbf{A}_{12}| + \ldots \pm a_{1n}|\mathbf{A}_{1n}|$$

(expansion by top row), where $\mathbf{A}_{ij}$ is the $(n-1) \times (n-1)$ matrix obtained by deleting the ith row and jth column of $\mathbf{A}$.

For example:

$$\begin{vmatrix} 1 & 2 & 0 & 4 \\ 5 & 2 & 1 & -3 \\ 2 & 6 & 5 & 0 \\ 1 & 1 & -1 & 4 \end{vmatrix} = 1 \begin{vmatrix} 2 & 1 & -3 \\ 6 & 5 & 0 \\ 1 & -1 & 4 \end{vmatrix} - 2 \begin{vmatrix} 5 & 1 & -3 \\ 2 & 5 & 0 \\ 1 & -1 & 4 \end{vmatrix}$$

$$+ 0 \begin{vmatrix} 5 & 2 & -3 \\ 2 & 6 & 0 \\ 1 & -1 & 4 \end{vmatrix} - 4 \begin{vmatrix} 5 & 2 & 1 \\ 2 & 6 & 5 \\ 1 & 1 & -1 \end{vmatrix}$$

Determinants have a number of properties which were given in Chapters 1 and 2. Here are some of those properties and also some new ones. All the properties are stated for 3×3 determinants. There are analogous properties for $n \times n$ determinants.

(1) $|\mathbf{A}| = 0$ if and only if $\mathbf{A}$ is singular, that is to say $\mathbf{A}$ collapses three dimensional space on to a plane (two dimensional space) or a space of lower dimension.

(2) $|\mathbf{A}|$ can be interpreted as a *scale factor* for volume: that is to say $\mathbf{A}$ scales all volumes by the factor $|\mathbf{A}|$.

(3) $|\mathbf{AB}| = |\mathbf{A}| \times |\mathbf{B}|$. This property follows from property (2) since $\mathbf{AB}$ is the transformation $\mathbf{B}$ followed by $\mathbf{A}$, which scales volumes by the factor $|\mathbf{B}| \times |\mathbf{A}| = |\mathbf{A}| \times |\mathbf{B}|$.

(4) $|\mathbf{I}| = 1$. This also follows from (2) since $\mathbf{I}$ represents the identity transformation which leaves volumes unaltered.

Inverting Matrices

A 3×3 matrix $\mathbf{A}$ is said to be *invertible* (or to possess an inverse) if there is a 3×3 matrix $\mathbf{A}^{-1}$ such that

$$\mathbf{AA}^{-1} = \mathbf{A}^{-1}\mathbf{A} = \mathbf{I}$$

We now continue with our list of properties of determinant.

(5) $\mathbf{A}$ is invertible if and only if $\mathbf{A}$ is non-singular (or equivalently $|\mathbf{A}| \neq 0$).

Proof of Property (5)

If $\mathbf{A}$ is invertible then

$$|\mathbf{A}| \times |\mathbf{A}^{-1}| = |\mathbf{I}| = 1 \qquad [1]$$

by properties (3) and (4), hence $|\mathbf{A}| \neq 0$.
Conversely if $|\mathbf{A}| \neq 0$ then we can define

$$\mathbf{A}^{-1} = \frac{1}{|\mathbf{A}|}\operatorname{adj}\mathbf{A}$$

where $\operatorname{adj}\mathbf{A}$ is the adjoint of $\mathbf{A}$ (see page 91).
The proof that $\mathbf{A}^{-1}$ is an inverse for $\mathbf{A}$ is to be found at the end of Chapter 2.

Formula [1] in the above proof gives a formula for $|\mathbf{A}^{-1}|$:

(6) If $\mathbf{A}$ is non-singular then

$$|\mathbf{A}^{-1}| = \frac{1}{|\mathbf{A}|}$$

(7) $|\mathbf{A}| = 0$ if and only if there is a vector $\mathbf{x} \neq \mathbf{0}$ such that $\mathbf{Ax} = \mathbf{0}$.

Proof of Property (7)

If $|\mathbf{A}| = 0$ then, by property (1), $\mathbf{A}$ collapses three-dimensional space on to a plane or space of lower dimension;
therefore there is at least a line of vectors which is collapsed to $\mathbf{0}$ by $\mathbf{A}$;
therefore there is an $\mathbf{x} \neq \mathbf{0}$ such that $\mathbf{Ax} = \mathbf{0}$.
Conversely, suppose that $\mathbf{Ax} = \mathbf{0}$ where $\mathbf{x} \neq \mathbf{0}$. We shall prove $|\mathbf{A}| = 0$ *by contradiction.*
Suppose $|\mathbf{A}| \neq 0$, then $\mathbf{A}$ is invertible so we can multiply $\mathbf{Ax} = \mathbf{0}$ on the left by $\mathbf{A}^{-1}$

i.e. $$\mathbf{A}^{-1}(\mathbf{Ax}) = \mathbf{A}^{-1}\mathbf{0} = \mathbf{0}$$

$\Rightarrow$ $$\mathbf{Ix} = \mathbf{0} \Rightarrow \mathbf{x} = \mathbf{0}$$

which is a contradiction. Therefore $|\mathbf{A}| = 0$.

EXAMPLES 15a (continued)

2) Prove that if $\mathbf{A}$ and $\mathbf{B}$ are square matrices such that $\mathbf{AB}$ is non-singular then each of $\mathbf{A}$ and $\mathbf{B}$ is non-singular.

$$\mathbf{AB} \text{ non-singular} \Rightarrow |\mathbf{AB}| \neq 0$$

$\Rightarrow$ $$|\mathbf{A}| \times |\mathbf{B}| \neq 0 \quad \text{by property (3)}$$

$\Rightarrow$ $$|\mathbf{A}| \neq 0 \quad \text{and} \quad |\mathbf{B}| \neq 0$$

$\Rightarrow$ $\mathbf{A}$ and $\mathbf{B}$ are non-singular.

3) Prove that if **A** and **B** are square matrices such that $\mathbf{AB} = \mathbf{I}$, then $\mathbf{A} = \mathbf{B}^{-1}$ *and* $\mathbf{B} = \mathbf{A}^{-1}$.

$$\mathbf{AB} = \mathbf{I} \Rightarrow \mathbf{AB} \text{ non-singular}$$

$\Rightarrow$ each of **A, B** is non-singular (by the last example)

$\Rightarrow$ each of **A, B** is invertible.

Now multiply the equation $\mathbf{AB} = \mathbf{I}$ on the left by $\mathbf{A}^{-1}$:

$$\mathbf{A}^{-1}(\mathbf{AB}) = \mathbf{A}^{-1}\mathbf{I}$$

$$\Rightarrow \quad (\mathbf{A}^{-1}\mathbf{A})\mathbf{B} = \mathbf{A}^{-1}$$

$$\Rightarrow \quad \mathbf{IB} = \mathbf{A}^{-1}$$

$$\Rightarrow \quad \mathbf{B} = \mathbf{A}^{-1}$$

Similarly, multiplying on the *right* by $\mathbf{B}^{-1}$ we find $\mathbf{A} = \mathbf{B}^{-1}$.

EXERCISE 15a

1) Express the following as a single matrix or vector, as appropriate.

(a) $\begin{pmatrix} 2 & 1 \\ 0 & -7 \\ 3 & 0 \end{pmatrix} + 5\begin{pmatrix} -\frac{1}{5} & -\frac{1}{10} \\ 0 & 2 \\ -1 & 7 \end{pmatrix}$

(b) $\begin{pmatrix} 0 & 1 & 0 \\ 1 & 0 & 1 \end{pmatrix}\begin{pmatrix} 2 & 1 \\ -1 & -1 \\ -1 & 2 \end{pmatrix}\begin{pmatrix} 0 \\ -1 \end{pmatrix}$

(c) $\begin{pmatrix} 3 & 7 & 4 \end{pmatrix}\begin{pmatrix} 1 & 0 \\ 0 & 2 \\ 3 & 0 \end{pmatrix}\begin{pmatrix} 8 & -1 & 0 \\ 0 & 7 & 1 \end{pmatrix}$.

2) Evaluate the determinants

(a) $\begin{vmatrix} 1 & 3 \\ -2 & -1 \end{vmatrix}$ (b) $\begin{vmatrix} 0 & 1 & 0 \\ 1 & 0 & 0 \\ 0 & 0 & 1 \end{vmatrix}$ (c) $\begin{vmatrix} 7 & 0 & 0 & 1 \\ 0 & 6 & 3 & 4 \\ 0 & 0 & 1 & 0 \\ 0 & 1 & 0 & 0 \end{vmatrix}$.

3) Invert the matrices

(a) $\begin{pmatrix} 0 & 1 \\ -1 & 0 \end{pmatrix}$ (b) $\begin{pmatrix} 2 & 0 & 1 \\ 0 & 1 & 0 \\ 0 & -1 & 1 \end{pmatrix}$.

4) Interpret the transformation $\begin{pmatrix} 0 & 1 \\ -1 & 0 \end{pmatrix}$ geometrically and verify that the inverse found in Question 3(a) is the inverse transformation.

5) Prove that $|\mathbf{AB}| = |\mathbf{BA}|$, where $\mathbf{A}$ and $\mathbf{B}$ are square matrices of the same size.

6) Prove that $|\mathbf{A}^n| = |\mathbf{A}|^n$, where $\mathbf{A}$ is a square matrix.

7) Prove that $|\lambda\mathbf{A}| = \lambda^n|\mathbf{A}|$ where $\mathbf{A}$ is an $n \times n$ matrix and λ is a scalar.

8) Prove that the set of 2×3 matrices forms a group under the operation of matrix addition (the *identity* is the zero 2×3 matrix $\mathbf{0}$).

9) Prove that the set of 3×3 non-singular matrices forms a group under matrix multiplication (the identity is the 3×3 unit matrix $\mathbf{I}$).

10) Prove that if $\mathbf{A}$ and $\mathbf{B}$ are 3×3 non-singular matrices then
(a) $(\mathbf{A}^{-1})^{-1} = \mathbf{A}$ (b) $(\mathbf{AB})^{-1} = \mathbf{B}^{-1}\mathbf{A}^{-1}$.

11) (a) Prove that if $\mathbf{A}$ is any $n \times m$ matrix and if $\mathbf{0}_{mp}$ is the zero $m \times p$ matrix then $\mathbf{A0}_{mp} = \mathbf{0}_{np}$.
(b) Prove that if $\mathbf{A}$ is any $n \times m$ matrix then $\mathbf{0}_{qn}\mathbf{A} = \mathbf{0}_{qm}$.

12) (a) Prove that if $\mathbf{A}$ and $\mathbf{B}$ are $n \times m$ and $m \times p$ matrices and λ is a scalar then

$$\mathbf{A}(\lambda\mathbf{B}) = (\lambda\mathbf{A})\mathbf{B} = \lambda(\mathbf{AB})$$

(b) Prove that $$\lambda(\mu\mathbf{A}) = (\lambda\mu)\mathbf{A}$$

and $$\lambda(\mathbf{A} + \mathbf{B}) = \lambda\mathbf{A} + \lambda\mathbf{B}$$

where λ, μ are scalars and $\mathbf{A}, \mathbf{B}$ are $n \times m$ matrices.

THE TRANSPOSE OF A MATRIX

The *transpose* $\mathbf{A}^{\mathrm{T}}$ of a matrix $\mathbf{A}$ is obtained by interchanging the rows and columns of $\mathbf{A}$.

For example if $\mathbf{A} = \begin{pmatrix} 1 & 3 \\ 2 & 4 \\ 7 & -9 \end{pmatrix}$ then $\mathbf{A}^{\mathrm{T}} = \begin{pmatrix} 1 & 2 & 7 \\ 3 & 4 & -9 \end{pmatrix}$.

Thus if $\mathbf{A}$ is an $n \times m$ matrix then $\mathbf{A}^{\mathrm{T}}$ is an $m \times n$ matrix and if $\mathbf{A}$ is square then $\mathbf{A}^{\mathrm{T}}$ is also square.
More formally we define the (i,j)th entry in $\mathbf{A}^{\mathrm{T}}$ to be the (j,i)th entry in $\mathbf{A}$:

$$\mathbf{A} = \begin{pmatrix} & & & & & + & & & \\ & & & & & + & & & \\ \times & \times & \times & \times & \times & a_{ji} & \times & \times & \times \\ & & & & & + & & & \\ & & & & & + & & & \\ & & & & & + & & & \\ & & & & & + & & & \end{pmatrix} j\text{th row} \qquad \mathbf{A}^{\mathrm{T}} = \begin{pmatrix} & & \times & & & & \\ & & \times & & & & \\ & & \times & & & & \\ & & \times & & & & \\ & & \times & & & & \\ + & + & a_{ij} & + & + & + & + \\ & & \times & & & & \\ & & \times & & & & \\ & & \times & & & & \end{pmatrix} i\text{th row}$$

ith column $\qquad\qquad$ jth column

i.e.

$$a_{ij}^{\mathrm{T}} = a_{ji}$$

Properties of the Transpose

(1) Suppose $\mathbf{A}$ and $\mathbf{B}$ are both $n \times m$ matrices then

$$(\mathbf{A}+\mathbf{B})^{\mathrm{T}} = \mathbf{A}^{\mathrm{T}} + \mathbf{B}^{\mathrm{T}}$$

Proof of Property (1)

$$\begin{aligned} (i,j)\text{th entry in } (\mathbf{A}+\mathbf{B})^{\mathrm{T}} &= (j,i)\text{th entry in } (\mathbf{A}+\mathbf{B}) \\ &= a_{ji} + b_{ji} \\ &= a_{ij}^{\mathrm{T}} + b_{ij}^{\mathrm{T}} \\ &= (i,j)\text{th entry in } \mathbf{A}^{\mathrm{T}} \\ &\quad + (i,j)\text{th entry in } \mathbf{B}^{\mathrm{T}} \\ &= (i,j)\text{th entry in } \mathbf{A}^{\mathrm{T}} + \mathbf{B}^{\mathrm{T}} \end{aligned}$$

Thus $(\mathbf{A}+\mathbf{B})^{\mathrm{T}}$ and $\mathbf{A}^{\mathrm{T}} + \mathbf{B}^{\mathrm{T}}$ have the same entries, i.e. they are equal.

(2) $(\mathbf{A}^{\mathrm{T}})^{\mathrm{T}} = \mathbf{A}$
(Property (2) is obvious.)

(3) Suppose $\mathbf{A}$ is $n \times m$ and $\mathbf{B}$ is $m \times q$ then

$$(\mathbf{AB})^{\mathrm{T}} = \mathbf{B}^{\mathrm{T}}\mathbf{A}^{\mathrm{T}}$$

i.e. transposing reverses the order of products.

Proof of Property (3)

The (i, j)th entry in $\mathbf{AB}$ is

$$\sum_{q=1}^{m} a_{iq}b_{qj} \qquad [1]$$

Now $a_{iq} = a_{qi}^{\mathrm{T}}$ and $b_{qj} = b_{jq}^{\mathrm{T}}$.
So [1] is

$$\sum_{q=1}^{m} a_{qi}^{\mathrm{T}}\, b_{jq}^{\mathrm{T}} = \sum_{q=1}^{m} b_{jq}^{\mathrm{T}} a_{qi}^{\mathrm{T}}$$

and this is a row-column product which is the (j, i)th entry of $\mathbf{B}^{\mathrm{T}}\mathbf{A}^{\mathrm{T}}$, i.e. the (i, j)th entry of $\mathbf{AB}$ is the (j, i)th entry of $\mathbf{B}^{\mathrm{T}}\mathbf{A}^{\mathrm{T}}$.
But the (i, j)th entry of $\mathbf{AB}$ is the (j, i)th entry of $(\mathbf{AB})^{\mathrm{T}}$,
i.e. $(\mathbf{AB})^{\mathrm{T}} = \mathbf{B}^{\mathrm{T}}\mathbf{A}^{\mathrm{T}}$.

(4) If $\mathbf{A}$ is square then $|\mathbf{A}^{\mathrm{T}}| = |\mathbf{A}|$ and therefore $\mathbf{A}$ is non-singular if and only if $\mathbf{A}^{\mathrm{T}}$ is non-singular.

For 2×2 matrices, property (4) follows immediately from the definition and for 3×3 matrices we gave a proof on page 44. In general, the result can be proved by induction on n where n is the size of the matrix. We shall not give details.

(5) $\mathbf{I}^{\mathrm{T}} = \mathbf{I}$

(Property (5) is obvious.)

(6) If $\mathbf{A}$ is square and non-singular then

$$(\mathbf{A}^{\mathrm{T}})^{-1} = (\mathbf{A}^{-1})^{\mathrm{T}}$$

Proof of Property (6)

$$\begin{aligned}(\mathbf{A}^{-1})^{\mathrm{T}}\mathbf{A}^{\mathrm{T}} &= (\mathbf{A}\mathbf{A}^{-1})^{\mathrm{T}} && \text{(property (3))}\\ &= \mathbf{I}^{\mathrm{T}} = \mathbf{I} && \text{(property (5))}\end{aligned}$$

Therefore $(\mathbf{A}^{-1})^{\mathrm{T}} = (\mathbf{A}^{\mathrm{T}})^{-1}$ by Example 1a, No. 3.

EXERCISE 15b

1) Write down the transposes of the following matrices:

(a) $\begin{pmatrix} 1 & 3 \\ 0 & 4 \end{pmatrix}$ (b) $\begin{pmatrix} 1 & 2 & 7 & 9 \end{pmatrix}$ (c) $\begin{pmatrix} -1 \\ 0 \\ 1 \end{pmatrix}$ (d) $\begin{pmatrix} 0 & 1 & 2 \\ 1 & 0 & 2 \\ 2 & 2 & 0 \end{pmatrix}$

(e) $\begin{pmatrix} 0 & 1 & 2 \\ -1 & 0 & -2 \\ -2 & 2 & 0 \end{pmatrix}$

2) Let **A** be a non-singular square matrix. $\mathbf{A}^!$ is defined as follows:

$$\mathbf{A}^! = (\mathbf{A}^T)^{-1} \quad (= (\mathbf{A}^{-1})^T)$$

Prove that $(\mathbf{AB})^! = \mathbf{A}^!\mathbf{B}^!$.

3) Prove that $(\lambda\mathbf{A})^T = \lambda(\mathbf{A}^T)$ where **A** is an $n \times m$ matrix and λ is a scalar.

SYMMETRIC AND SKEW-SYMMETRIC MATRICES

A square matrix **A** is said to be *symmetric* if $\mathbf{A}^T = \mathbf{A}$ and is said to be *skew-symmetric* if $\mathbf{A}^T = -\mathbf{A}$.
A matrix is symmetric if it is unchanged under reflection in the main diagonal, i.e. the off-diagonal entries are equal in pairs:

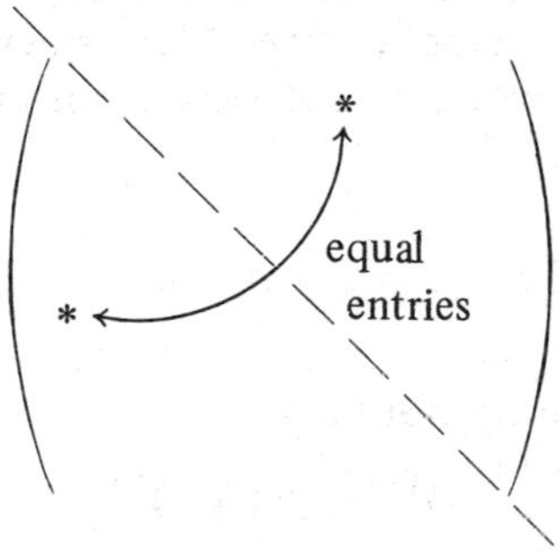

It is skew-symmetric if the diagonal entries are zero and the off-diagonal entries are equal but opposite in sign:

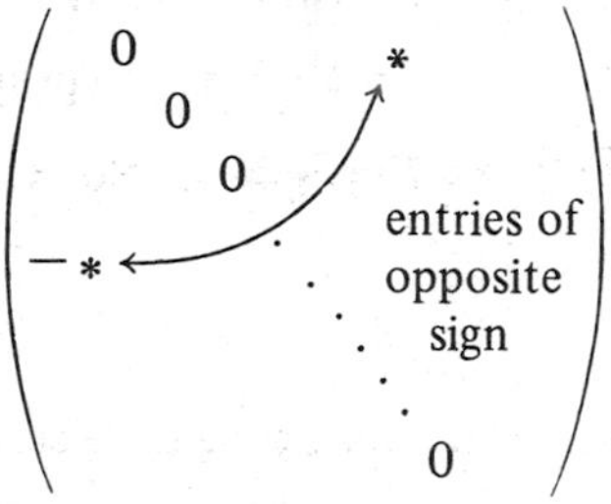

To see that the diagonal entries of a skew-symmetric matrix are zero we note that

$$a_{ii} = a_{ii}^T = -a_{ii}$$

$$\Rightarrow \qquad 2a_{ii} = 0 \Rightarrow a_{ii} = 0$$

EXAMPLES 15c

1) Which of the following matrices are symmetric, skew-symmetric, neither or both?

(a) $\begin{pmatrix} 1 & 3 \\ 0 & 4 \end{pmatrix}$ (b) $\begin{pmatrix} 1 & -2 & 7 & 0 \\ -2 & 0 & 0 & 4 \\ 7 & 0 & -3 & 0 \\ 0 & 4 & 0 & 0 \end{pmatrix}$ (c) $\begin{pmatrix} 0 & -1 & 2 \\ 1 & 0 & -2 \\ -2 & 2 & 0 \end{pmatrix}$

(a) neither; (b) symmetric; (c) skew-symmetric.

2) Suppose that $\mathbf{A}$ is a symmetric $n \times n$ matrix and that $\mathbf{P}$ is any $n \times n$ matrix. Prove that $\mathbf{P}^T\mathbf{AP}$ is symmetric.
Deduce that $\mathbf{P}^T\mathbf{P}$ is symmetric for any $n \times n$ matrix $\mathbf{P}$.

$$(\mathbf{P}^T\mathbf{AP})^T = \mathbf{P}^T\mathbf{A}^T(\mathbf{P}^T)^T \quad \text{(by property (3), page 635)}$$

$$= \mathbf{P}^T\mathbf{AP} \quad \text{since } \mathbf{A} \text{ is symmetric and } (\mathbf{P}^T)^T = \mathbf{P}$$

i.e. $\mathbf{P}^T\mathbf{AP}$ is symmetric.
Now $\mathbf{I}$ is obviously symmetric and therefore $\mathbf{P}^T\mathbf{IP} = \mathbf{P}^T\mathbf{P}$ is symmetric by the first part.

EXERCISE 15c

1) In each case state whether the matrix is symmetric, skew-symmetric, neither or both:

(a) $\begin{pmatrix} 0 & 1 \\ 1 & 0 \end{pmatrix}$ (b) $\begin{pmatrix} 0 & 1 \\ -1 & 0 \end{pmatrix}$ (c) $\begin{pmatrix} 0 & 0 \\ 0 & 0 \end{pmatrix}$

(d) $\begin{pmatrix} 1 & 0 \\ 0 & 1 \end{pmatrix}$ (e) $\begin{pmatrix} 1 & 0 & 0 \\ 0 & 1 & 0 \end{pmatrix}$ (f) $\begin{pmatrix} -1 & -1 \\ 1 & 1 \end{pmatrix}$

(g) $\begin{pmatrix} 2 & 3 & 7 & 4 \\ 3 & 0 & 1 & 6 \\ 7 & 1 & 0 & 0 \\ 4 & 6 & 0 & 2 \end{pmatrix}$ (h) $\begin{pmatrix} 0 & -1 & 2 \\ 1 & 0 & 7 \\ -2 & -7 & 0 \end{pmatrix}$ (i) $\begin{pmatrix} 2 & 8 & 9 & 7 \\ 9 & 6 & 4 & 8 \\ 8 & 4 & 0 & 3 \\ 7 & 8 & 3 & 9 \end{pmatrix}$.

2) Prove that a square matrix which is both symmetric and skew-symmetric must be the zero matrix.

3) Prove that if $\mathbf{A}$ is a skew-symmetric $n \times n$ matrix and $\mathbf{P}$ is any $n \times n$ matrix then $\mathbf{P}^T\mathbf{AP}$ is skew-symmetric.

4) Suppose that $\mathbf{A}, \mathbf{B}$ are symmetric $n \times n$ matrices such that $\mathbf{AB} = \mathbf{BA}$. Prove that $\mathbf{AB}$ is symmetric.

5) Suppose that $\mathbf{A}$ is a skew-symmetric matrix. Prove that $\mathbf{A}^2$ is symmetric.

SYMMETRIC MATRICES AND QUADRATIC FORMS

We have seen one use of matrices: to describe linear transformations of space. Another quite different use is to describe *quadratic forms* and, for this purpose, *symmetric* matrices are especially useful.

A *quadratic form* is a homogeneous second degree polynomial in a set of variables.

For example:

(1) $x^2 + y^2$ is a quadratic form in the two variables x and y,

(2) $x_1^2 + 2x_1x_3 + 3x_2x_3 - x_2^2$ is a quadratic form in the three variables x_1, x_2 and x_3,

(3) $3x^2 - 4xy + 2z^2$ is a quadratic form in x, y and z,

(4) The *general* quadratic form in the variables x and y is

$$ax^2 + bxy + cy^2$$

where a, b and c are constants (real numbers).

Given an $n \times n$ matrix $\mathbf{A}$, then we can use $\mathbf{A}$ to define a quadratic form in n variables $x_1, x_2, \ldots, x_n$. We shall illustrate the 2×2 case. Let

$\mathbf{A} = \begin{pmatrix} a & b \\ c & d \end{pmatrix}$ and let $\mathbf{x} = \begin{pmatrix} x \\ y \end{pmatrix}$, then the expression

$$\mathbf{x}^{\mathrm{T}}\mathbf{A}\mathbf{x}$$

is a quadratic form namely:

$$\begin{pmatrix} x, y \end{pmatrix} \begin{pmatrix} a & b \\ c & d \end{pmatrix} \begin{pmatrix} x \\ y \end{pmatrix}$$

$$= \begin{pmatrix} x, y \end{pmatrix} \begin{pmatrix} ax + by \\ cx + dy \end{pmatrix} = ax^2 + bxy + cxy + dy^2$$

$$= ax^2 + (b + c)xy + dy^2$$

Note that different matrices can give the same quadratic form, e.g.

$$(x, y)\begin{pmatrix}1 & -2\\ 0 & 1\end{pmatrix}\begin{pmatrix}x\\ y\end{pmatrix} = (x, y)\begin{pmatrix}1 & -1\\ -1 & 1\end{pmatrix}\begin{pmatrix}x\\ y\end{pmatrix}$$
$$= x^2 - 2xy + y^2$$

But if we insist that the matrix is *symmetric*, then we have a *unique* representation. For example

$$(x, y)\begin{pmatrix}a & b\\ b & c\end{pmatrix}\begin{pmatrix}x\\ y\end{pmatrix} = ax^2 + 2bxy + cy^2$$

and we can clearly read the entries of $\mathbf{A} = \begin{pmatrix}a & b\\ b & c\end{pmatrix}$ from the quadratic form.

EXAMPLES 15d

1) Expand $(x, y, z)\begin{pmatrix}3 & 0 & 2\\ 0 & -1 & 4\\ 1 & 1 & -2\end{pmatrix}\begin{pmatrix}x\\ y\\ z\end{pmatrix}$.

$$(x, y, z)\begin{pmatrix}3 & 0 & 2\\ 0 & -1 & 4\\ 1 & 1 & -2\end{pmatrix}\begin{pmatrix}x\\ y\\ z\end{pmatrix} = (x, y, z)\begin{pmatrix}3x \quad +2z\\ -y +4z\\ x +y -2z\end{pmatrix}$$
$$= x(3x + 2z) + y(-y + 4z) + z(x + y - 2z)$$
$$= 3x^2 - y^2 - 2z^2 + 3xz + 5yz$$

2) Find the symmetric 2×2 matrix $\mathbf{A}$ which gives the quadratic form $3x^2 - 2xy + 7y^2$.

Since $(x, y)\begin{pmatrix}a & b\\ b & c\end{pmatrix}\begin{pmatrix}x\\ y\end{pmatrix} = ax^2 + 2bxy + cy^2$ we need $a = 3,\ b = -1,\ c = 7$,

i.e. $\mathbf{A} = \begin{pmatrix}3 & -1\\ -1 & 7\end{pmatrix}$.

Conics and Quadrics

Quadratic forms in two or three variables can be pictured in terms of conics (two variables) and quadrics (three variables). As an example, consider the quadratic form xy (in two variables). The equation $xy = 1$ is a rectangular hyperbola with the x and y axes as asymptotes. More generally, the equation $xy = k$, as k varies, describes a *family* of rectangular hyperbolae with the same asymptotes.

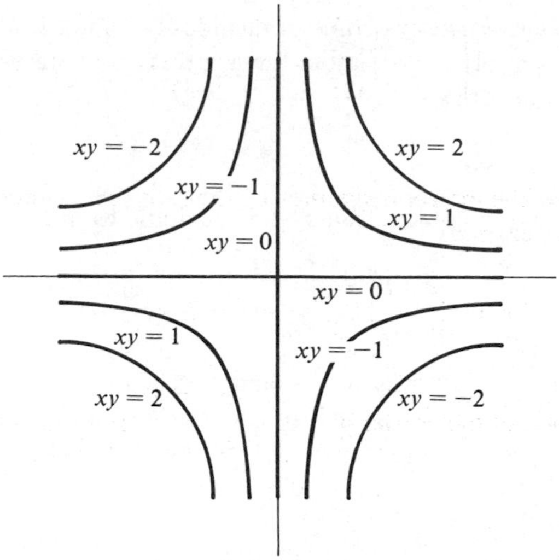

As an example in three variables, consider the quadratic form $x^2 + y^2 - z^2$.
The equation $x^2 + y^2 - z^2 = k$ is a *quadric surface* in three dimensions (called a hyperboloid).
If $k > 0$ then the surface has one sheet while if $k < 0$ the surface has two sheets.
The case $k = 0$ is 'singular': the surface is a circular cone in this case.

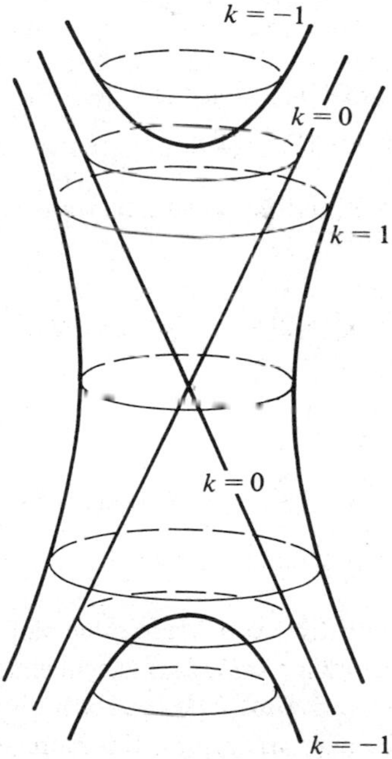

(The reader is *not* expected to be able to produce sketches of quadric surfaces!)

Notice that in each example the conic or quadric is symmetric about the origin (the origin is a *centre*). For this reason, these curves and surfaces are called *central* conics and quadrics.

EXAMPLES 15d (continued)

3) Sketch the family of conics given by the quadratic form

$$\begin{pmatrix} x, y \end{pmatrix} \begin{pmatrix} 2 & \frac{1}{2} \\ \frac{1}{2} & 1 \end{pmatrix} \begin{pmatrix} x \\ y \end{pmatrix}.$$

The quadratic form is $2x^2 + xy + y^2$ and the equation $2x^2 + xy + y^2 = k$ gives a family of *ellipses* as k varies through positive values.
(For $k < 0$ the locus is empty since $2x^2 + xy + y^2 = (\frac{1}{2}x + y)^2 + \frac{7}{4}x^2$ which is always $\geqslant 0$).

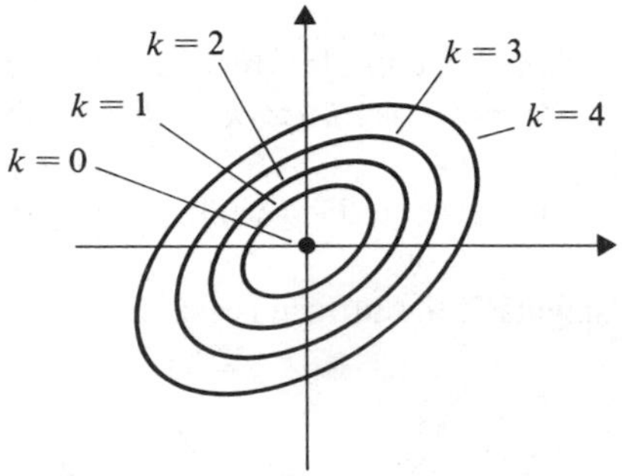

EXERCISE 15d

1) Expand $\begin{pmatrix} x, y, z \end{pmatrix} \begin{pmatrix} a & b & c \\ d & e & f \\ g & h & i \end{pmatrix} \begin{pmatrix} x \\ y \\ z \end{pmatrix}.$

2) Find the symmetric matrix $\mathbf{A}$ such that

$$\begin{pmatrix} x, y, z \end{pmatrix} \mathbf{A} \begin{pmatrix} x \\ y \\ z \end{pmatrix} = 3x^2 + 2xy + 5yz + 7y^2 - 2z^2.$$

3) Find the symmetric matrices $\mathbf{A}$ which give the following quadratic forms
(a) $x^2 + y^2 - z^2$ (b) $x^2 + 3xy + 7yz - 10zx + z^2$
(c) $2x^2 - 4y^2 + 10xy - 7yz - 11xz + 3z^2$.

4) Sketch the family of conics

$$x^2 - 2xy = k.$$

5) Sketch the family of conics corresponding to the quadratic form

$$(x, y)\begin{pmatrix}1 & 1\\1 & 0\end{pmatrix}\begin{pmatrix}x\\y\end{pmatrix}.$$

6) Write down the symmetric matrix **A** which gives the general quadratic form in three variables:

$$ax^2+by^2+cz^2+2fyz+2gzx+2hxy.$$

DIAGONAL MATRICES

A square matrix whose entries are zero *except* on the main diagonal is called a *diagonal* matrix.

For example $\begin{pmatrix}2 & 0\\0 & 3\end{pmatrix}$ is diagonal but $\begin{pmatrix}0 & 1 & 0\\0 & 0 & 0\\0 & 0 & 0\end{pmatrix}$ is *not*.

The identity and zero 3×3 matrices $\begin{pmatrix}1 & 0 & 0\\0 & 1 & 0\\0 & 0 & 1\end{pmatrix}$ and $\begin{pmatrix}0 & 0 & 0\\0 & 0 & 0\\0 & 0 & 0\end{pmatrix}$ are both diagonal.

A *general* diagonal 3×3 matrix has the form $\begin{pmatrix}\lambda & 0 & 0\\0 & \mu & 0\\0 & 0 & \nu\end{pmatrix}$

where λ, μ, ν are real numbers.
Diagonal matrices are important because they multiply in a very simple way and in particular they *commute* (the order of multiplication is immaterial).

EXERCISE 15e

1) Suppose $\mathbf{A}=\begin{pmatrix}2 & 0\\0 & -1\end{pmatrix}$, $\mathbf{B}=\begin{pmatrix}1 & 0\\0 & -7\end{pmatrix}$.

Verify that $\mathbf{AB}=\mathbf{BA}$.

In Questions 2 to 4, $\mathbf{A}, \Lambda$ are general 3×3 diagonal matrices defined as follows:

$$\mathbf{A}=\begin{pmatrix}a & 0 & 0\\0 & b & 0\\0 & 0 & c\end{pmatrix},\quad \Lambda=\begin{pmatrix}\lambda & 0 & 0\\0 & \mu & 0\\0 & 0 & \nu\end{pmatrix}$$

2) Verify that $\mathbf{A\Lambda} = \mathbf{\Lambda A} = \begin{pmatrix} a\lambda & 0 & 0 \\ 0 & b\mu & 0 \\ 0 & 0 & c\nu \end{pmatrix}$.

3) Prove by induction that $\Lambda^n = \begin{pmatrix} \lambda^n & 0 & 0 \\ 0 & \mu^n & 0 \\ 0 & 0 & \nu^n \end{pmatrix}$.

4) Prove that $|\mathbf{A}| = abc$ and deduce that $\mathbf{A}$ is non-singular if and only if none of the diagonal entries is zero.

5) Write down a *general* 2×2 diagonal matrix.

6) What is the condition for a 2×2 diagonal matrix to be non-singular?

7) Prove that the set of non-singular diagonal 2×2 matrices forms a group under matrix multiplication.

ORTHOGONAL MATRICES

A non-singular square matrix $\mathbf{A}$ is said to be *orthogonal* if

$$\mathbf{A}^{\mathrm{T}}\mathbf{A} = \mathbf{I} \qquad [1]$$

Note that from Example 15a, No. 3, equation [1] is equivalent to

$$\mathbf{A}^{\mathrm{T}} = \mathbf{A}^{-1} \qquad [2]$$

A non-singular square matrix is orthogonal if and only if its transpose is equal to its inverse.

EXAMPLES 15f

1) Decide which of the following matrices are orthogonal:

(a) $\begin{pmatrix} 0 & 1 \\ -1 & 0 \end{pmatrix}$ (b) $\begin{pmatrix} 1 & 1 \\ 0 & 1 \end{pmatrix}$ (c) $\begin{pmatrix} 0 & 1 \\ 0 & 1 \end{pmatrix}$ (d) $\begin{pmatrix} 1 & 0 & 0 \\ 0 & -\frac{1}{2} & \frac{\sqrt{3}}{2} \\ 0 & -\frac{\sqrt{3}}{2} & -\frac{1}{2} \end{pmatrix}$.

(a) $\mathbf{A} = \begin{pmatrix} 0 & 1 \\ -1 & 0 \end{pmatrix}$ $\mathbf{A}^{\mathrm{T}} = \begin{pmatrix} 0 & -1 \\ 1 & 0 \end{pmatrix}$

$$\Rightarrow \qquad \mathbf{A}^{\mathrm{T}}\mathbf{A} = \begin{pmatrix} 0 & -1 \\ 1 & 0 \end{pmatrix}\begin{pmatrix} 0 & 1 \\ -1 & 0 \end{pmatrix} = \begin{pmatrix} 1 & 0 \\ 0 & 1 \end{pmatrix} = \mathbf{I}$$

So $\mathbf{A}$ *is* orthogonal.

(b) $\mathbf{A} = \begin{pmatrix} 1 & 1 \\ 0 & 1 \end{pmatrix}$ $\mathbf{A}^T = \begin{pmatrix} 1 & 0 \\ 1 & 1 \end{pmatrix}$

$$\Rightarrow \qquad \mathbf{A}^T\mathbf{A} = \begin{pmatrix} 1 & 0 \\ 1 & 1 \end{pmatrix}\begin{pmatrix} 1 & 1 \\ 0 & 1 \end{pmatrix} = \begin{pmatrix} 1 & 1 \\ 1 & 2 \end{pmatrix} \neq \mathbf{I}$$

So **A** is *not* orthogonal.

(c) $\mathbf{A} = \begin{pmatrix} 0 & 1 \\ 0 & 1 \end{pmatrix}$ $\mathbf{A}^T = \begin{pmatrix} 0 & 0 \\ 1 & 1 \end{pmatrix}$

$$\Rightarrow \qquad \mathbf{A}^T\mathbf{A} = \begin{pmatrix} 0 & 1 \\ 0 & 1 \end{pmatrix}\begin{pmatrix} 0 & 0 \\ 1 & 1 \end{pmatrix} = \begin{pmatrix} 1 & 1 \\ 1 & 1 \end{pmatrix} \neq \mathbf{I}$$

So **A** is *not* orthogonal.
Alternatively: **A** is singular, therefore **A** is *not* orthogonal.

(d) $\mathbf{A} = \begin{pmatrix} 1 & 0 & 0 \\ 0 & -\frac{1}{2} & \frac{\sqrt{3}}{2} \\ 0 & -\frac{\sqrt{3}}{2} & -\frac{1}{2} \end{pmatrix}$ $\mathbf{A}^T = \begin{pmatrix} 1 & 0 & 0 \\ 0 & -\frac{1}{2} & -\frac{\sqrt{3}}{2} \\ 0 & \frac{\sqrt{3}}{2} & -\frac{1}{2} \end{pmatrix}$

$$\Rightarrow \qquad \mathbf{A}^T\mathbf{A} = \begin{pmatrix} 1 & 0 & 0 \\ 0 & \frac{1}{4}+\frac{3}{4} & \frac{\sqrt{3}}{4}-\frac{\sqrt{3}}{4} \\ 0 & \frac{\sqrt{3}}{4}-\frac{\sqrt{3}}{4} & \frac{3}{4}+\frac{1}{4} \end{pmatrix} = \begin{pmatrix} 1 & 0 & 0 \\ 0 & 1 & 0 \\ 0 & 0 & 1 \end{pmatrix}$$

So **A** *is* orthogonal.

Properties of Orthogonal Matrices

(1) Suppose **A** and **B** are orthogonal matrices, then **AB** is an orthogonal matrix.

Proof

$$(\mathbf{AB})^T = \mathbf{B}^T\mathbf{A}^T$$

$$\Rightarrow \qquad (\mathbf{AB})^T\mathbf{AB} = \mathbf{B}^T\mathbf{A}^T\mathbf{AB}$$

$$= \mathbf{B}^T\mathbf{IB} \quad \text{since } \mathbf{A}^T\mathbf{A} = \mathbf{I}$$

$$= \mathbf{B}^T\mathbf{B} = \mathbf{I}$$

So **AB** is orthogonal.

(2) If $\mathbf{A}$ is an orthogonal matrix then $|\mathbf{A}| = \pm 1$.

Proof

We know that
$$|\mathbf{A}^{-1}| = \frac{1}{|\mathbf{A}|} \qquad [1]$$
(property (6) of determinants, page 632)

and
$$|\mathbf{A}^{\mathrm{T}}| = |\mathbf{A}| \qquad [2]$$
(property (4) of transpose, page 636).

But $\mathbf{A}^{\mathrm{T}} = \mathbf{A}^{-1}$ since $\mathbf{A}$ is orthogonal

$$\Rightarrow \qquad |\mathbf{A}^{\mathrm{T}}| = |\mathbf{A}^{-1}| \qquad [3]$$

Combining [1], [2] and [3] we have

$$|\mathbf{A}| = |\mathbf{A}^{\mathrm{T}}| = |\mathbf{A}^{-1}| = \frac{1}{|\mathbf{A}|}$$

$$\Rightarrow \qquad |\mathbf{A}|^2 = 1 \quad \Rightarrow \quad |\mathbf{A}| = \pm 1.$$

(3) If $\mathbf{A}$ is orthogonal then $\mathbf{A}^{-1}$ and $\mathbf{A}^{\mathrm{T}}$ are each orthogonal.

Proof

Since $\mathbf{A}^{-1} = \mathbf{A}^{\mathrm{T}}$, both parts of the property are the same!

Now
$$(\mathbf{A}^{\mathrm{T}})^{\mathrm{T}} = \mathbf{A}$$

therefore
$$(\mathbf{A}^{\mathrm{T}})^{\mathrm{T}}\mathbf{A}^{\mathrm{T}} = \mathbf{A}\mathbf{A}^{\mathrm{T}} = \mathbf{A}\mathbf{A}^{-1} = \mathbf{I}$$

i.e. $\mathbf{A}^{\mathrm{T}}$ is orthogonal.

Interpretation of Orthogonality

Orthogonal 2×2 and 3×3 matrices have a very simple interpretation in terms of transformations: they correspond to *rigid transformations of the plane or three dimensional space which fix the origin.* A *rigid* transformation is one which preserves distances and angles: for example, a rotation or a reflection or a combination of these. A transformation which shears or stretches is *not* rigid. We shall discuss the 3×3 case, the 2×2 case is similar.

Suppose that $\mathbf{A}$ is a 3×3 matrix and let the columns of $\mathbf{A}$ be $\mathbf{c}_1$, $\mathbf{c}_2$ and $\mathbf{c}_3$. Then $\mathbf{c}_1, \mathbf{c}_2, \mathbf{c}_3$ are *rows* of $\mathbf{A}^{\mathrm{T}}$,

therefore
$$\mathbf{A}^{\mathrm{T}}\mathbf{A} = \begin{pmatrix} \mathbf{c}_1.\mathbf{c}_1 & \mathbf{c}_1.\mathbf{c}_2 & \mathbf{c}_1.\mathbf{c}_3 \\ \mathbf{c}_2.\mathbf{c}_1 & \mathbf{c}_2.\mathbf{c}_2 & \mathbf{c}_2.\mathbf{c}_3 \\ \mathbf{c}_3.\mathbf{c}_1 & \mathbf{c}_3.\mathbf{c}_2 & \mathbf{c}_3.\mathbf{c}_3 \end{pmatrix}$$

(see equation [1] on page 627).

Therefore $\mathbf{A}$ is orthogonal, i.e. $\mathbf{A}^{\mathrm{T}}\mathbf{A} = \mathbf{I}$, if and only if we have the following values for the dot products:

$$\mathbf{c}_1.\mathbf{c}_1 = \mathbf{c}_2.\mathbf{c}_2 = \mathbf{c}_3.\mathbf{c}_3 = 1$$

$$\mathbf{c}_1.\mathbf{c}_2 = \mathbf{c}_1.\mathbf{c}_3 = \mathbf{c}_2.\mathbf{c}_3 = 0$$

but these equations simply say that $\mathbf{c}_1$, $\mathbf{c}_2$ *and* $\mathbf{c}_3$ *are mutually perpendicular vectors of unit length.*
Thus we have a further property.

(4) A matrix is orthogonal if and only if its columns are mutually perpendicular vectors of unit length.

Note that property (4) can be used to test for orthogonality.

The interpretation of orthogonal matrices as rigid transformations follows very easily from this property. For remember that the columns are the images under the transformation of the Cartesian unit vectors $\begin{pmatrix}1\\0\\0\end{pmatrix}, \begin{pmatrix}0\\1\\0\end{pmatrix}, \begin{pmatrix}0\\0\\1\end{pmatrix}$. Therefore the transformation given by an orthogonal matrix $\mathbf{A}$ carries the 'standard' unit cube (with edges the Cartesian unit vectors) to the unit cube with edges $\mathbf{c}_1, \mathbf{c}_2, \mathbf{c}_3$ i.e. it is a rigid transformation.

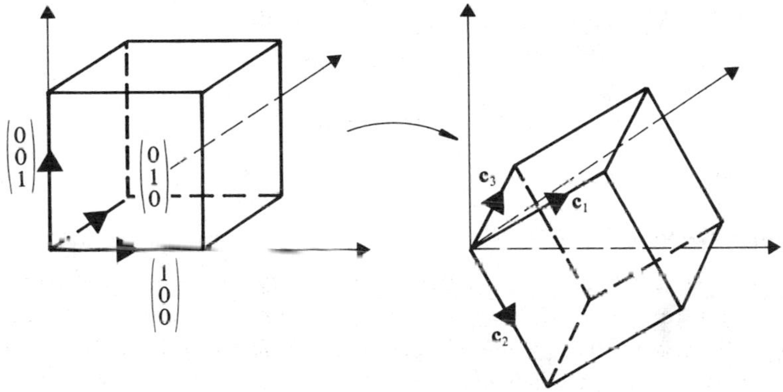

Conversely a rigid transformation which preserves the origin carries the standard unit cube to another unit cube with corner at O and hence is represented by a matrix whose columns are mutually perpendicular vectors of unit length, i.e. an orthogonal matrix.

EXAMPLES 15f (continued)

2) Find an orthogonal matrix with first two columns $\begin{pmatrix}\frac{\sqrt{2}}{2}\\ \frac{\sqrt{2}}{2}\\ 0\end{pmatrix}, \begin{pmatrix}0\\0\\1\end{pmatrix}$.

The third column $\begin{pmatrix} a \\ b \\ c \end{pmatrix}$ must be perpendicular to the two given vectors and must have unit length.
Therefore

$$a\frac{\sqrt{2}}{2} + b\frac{\sqrt{2}}{2} = 0 \quad \text{(dot product with first given vector)} \qquad [1]$$

$$c = 0 \quad \text{(dot product with second given vector)} \qquad [2]$$

$$a^2 + b^2 + c^2 = 1 \quad \text{(unit length)} \qquad [3]$$

From [1] $a = -b$ and hence from [3]

$$2a^2 = 1$$

$$\Rightarrow \qquad a = \pm\frac{\sqrt{2}}{2} \quad \text{and} \quad b = \mp\frac{\sqrt{2}}{2}$$

Therefore there are two solutions

$$\begin{pmatrix} \frac{\sqrt{2}}{2} & 0 & \frac{\sqrt{2}}{2} \\ \frac{\sqrt{2}}{2} & 0 & -\frac{\sqrt{2}}{2} \\ 0 & 1 & 0 \end{pmatrix} \quad \text{and} \quad \begin{pmatrix} \frac{\sqrt{2}}{2} & 0 & -\frac{\sqrt{2}}{2} \\ \frac{\sqrt{2}}{2} & 0 & \frac{\sqrt{2}}{2} \\ 0 & 1 & 0 \end{pmatrix}$$

EXERCISE 15f

1) Decide whether the given matrix is orthogonal:

(a) $\begin{pmatrix} 0 & 0 \\ 0 & 0 \end{pmatrix}$ (b) $\begin{pmatrix} \frac{\sqrt{2}}{2} & \frac{\sqrt{2}}{2} \\ \frac{\sqrt{2}}{2} & \frac{\sqrt{2}}{2} \end{pmatrix}$ (c) $\begin{pmatrix} 1 & 0 \\ 0 & -1 \end{pmatrix}$

(d) $\begin{pmatrix} \frac{\sqrt{2}}{2} & -\frac{\sqrt{2}}{2} & 0 \\ 0 & 0 & 1 \\ \frac{\sqrt{2}}{2} & \frac{\sqrt{2}}{2} & 0 \end{pmatrix}$ (e) $\begin{pmatrix} -1 & 0 & 0 \\ 0 & \frac{1}{2} & -\frac{1}{2} \\ 0 & \frac{\sqrt{3}}{2} & -\frac{\sqrt{3}}{2} \end{pmatrix}$ (f) $\begin{pmatrix} 0 & 1 & 0 \\ \frac{3}{5} & 0 & \frac{4}{5} \\ \frac{4}{5} & 0 & \frac{3}{5} \end{pmatrix}$

(g) $\begin{pmatrix} \frac{3}{5} & 0 & -\frac{3}{5} \\ 0 & -1 & 0 \\ \frac{4}{5} & 0 & \frac{4}{5} \end{pmatrix}$ (h) $\begin{pmatrix} \frac{3}{5} & 0 & -\frac{4}{5} \\ \frac{4}{5} & 0 & \frac{3}{5} \\ 0 & -1 & 0 \end{pmatrix}$.

2) Prove that if $\mathbf{A}$ is an orthogonal 3×3 matrix then the rows of $\mathbf{A}$ are mutually perpendicular vectors of unit length.

3) Given that the matrix is orthogonal, find values for a, b, c, d:

(a) $\begin{pmatrix} \frac{\sqrt{2}}{2} & 0 & a \\ 0 & 1 & b \\ \frac{\sqrt{2}}{2} & 0 & c \end{pmatrix}$ (b) $\begin{pmatrix} a & b & c \\ \frac{3}{5} & 0 & \frac{4}{5} \\ 0 & -1 & 0 \end{pmatrix}$ (c) $\begin{pmatrix} -1 & 0 & 0 \\ 0 & a & b \\ 0 & c & d \end{pmatrix}$

(d) $\begin{pmatrix} \cos\theta & a \\ \sin\theta & b \end{pmatrix}$.

4) Find all the 3×3 orthogonal *diagonal* matrices.

5) Prove that the set of orthogonal 2×2 matrices forms a group under matrix multiplication. Interpret this group geometrically.

EIGENVECTORS AND EIGENVALUES

An *eigenvector* of a square matrix $\mathbf{A}$ is a non-zero column vector $\mathbf{x}$ such that $\mathbf{A}\mathbf{x} = \lambda\mathbf{x}$ where λ is a scalar, called the corresponding *eigenvalue* of $\mathbf{A}$. For example

$$\begin{pmatrix} 2 & 0 & 1 \\ -1 & 2 & 3 \\ 1 & 0 & 2 \end{pmatrix}\begin{pmatrix} -2 \\ -4 \\ -2 \end{pmatrix} = \begin{pmatrix} -6 \\ -12 \\ -6 \end{pmatrix} = 3\begin{pmatrix} -2 \\ -4 \\ -2 \end{pmatrix}$$

Therefore $\begin{pmatrix} -2 \\ -4 \\ -2 \end{pmatrix}$ is an *eigenvector* of $\begin{pmatrix} 2 & 0 & 1 \\ -1 & 2 & 3 \\ 1 & 0 & 2 \end{pmatrix}$ with corresponding *eigenvalue* $\lambda = 3$.

If we multiply an eigenvector by any scalar then it is still an eigenvector and the *corresponding eigenvalue is unaltered.*

For suppose $\mathbf{x}$ is an eigenvector of $\mathbf{A}$ with eigenvalue λ, then if α is any scalar,

$$\mathbf{A}(\alpha\mathbf{x}) = \alpha\mathbf{A}\mathbf{x} = \alpha\lambda\mathbf{x} = \lambda(\alpha\mathbf{x})$$

i.e. $\alpha\mathbf{x}$ is also an eigenvector with eigenvalue λ.

For example $\begin{pmatrix}1\\2\\1\end{pmatrix} = -\frac{1}{2}\begin{pmatrix}-2\\-4\\-2\end{pmatrix}$ is also an eigenvector of $\begin{pmatrix}2 & 0 & 1\\-1 & 2 & 3\\1 & 0 & 2\end{pmatrix}$ with eigenvalue 3 as we can check:

$$\begin{pmatrix}2 & 0 & 1\\-1 & 2 & 3\\1 & 0 & 2\end{pmatrix}\begin{pmatrix}1\\2\\1\end{pmatrix} = \begin{pmatrix}3\\6\\3\end{pmatrix} = 3\begin{pmatrix}1\\2\\1\end{pmatrix}$$

Thinking in terms of transformations, eigenvectors have a simple interpretation. Suppose $\mathbf{x}$ is an eigenvector of $\mathbf{A}$ then $\mathbf{x}$ is mapped to a scalar multiple of itself. In fact, as we saw above, any scalar multiple of $\mathbf{x}$ is mapped to a scalar multiple of $\mathbf{x}$. In other words the line through the origin passing through $\mathbf{x}$ is mapped into itself. So an eigenvector corresponds to a *preserved line* through the origin.

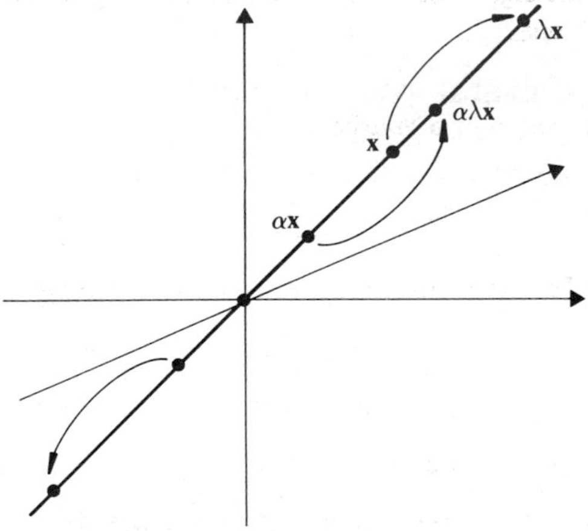

By choosing an ***appropriate*** scalar multiple of $\mathbf{x}$ we can find an eigenvector *of unit length* with the same eigenvalue (in other words we choose one of the points at unit distance on the preserved line).

For example $$\begin{pmatrix}\frac{1}{\sqrt{6}}\\ \frac{\sqrt{2}}{\sqrt{3}}\\ \frac{1}{\sqrt{6}}\end{pmatrix} = \frac{1}{\sqrt{6}}\begin{pmatrix}1\\2\\1\end{pmatrix}$$

is again an eigenvector of $\begin{pmatrix} 2 & 0 & 1 \\ -1 & 2 & 3 \\ 1 & 0 & 2 \end{pmatrix}$ with eigenvalue 3 as we can check

$$\begin{pmatrix} 2 & 0 & 1 \\ -1 & 2 & 3 \\ 1 & 0 & 2 \end{pmatrix} \begin{pmatrix} \frac{1}{\sqrt{6}} \\ \frac{\sqrt{2}}{\sqrt{3}} \\ \frac{1}{\sqrt{6}} \end{pmatrix} = \begin{pmatrix} \frac{\sqrt{3}}{\sqrt{2}} \\ \sqrt{6} \\ \frac{\sqrt{3}}{\sqrt{2}} \end{pmatrix} = 3 \begin{pmatrix} \frac{1}{\sqrt{6}} \\ \frac{\sqrt{2}}{\sqrt{3}} \\ \frac{1}{\sqrt{6}} \end{pmatrix},$$

and $\begin{pmatrix} \frac{1}{\sqrt{6}} \\ \frac{\sqrt{2}}{\sqrt{3}} \\ \frac{1}{\sqrt{6}} \end{pmatrix}$ has unit length.

In order to find the eigenvectors and eigenvalues of a given matrix $\mathbf{A}$ we start by finding the *eigenvalues*.

Finding the Eigenvalues of a Matrix

Rearranging the equation

$$\mathbf{Ax} = \lambda\mathbf{x}$$

by writing $\mathbf{x} = \mathbf{Ix}$ and collecting terms on the left hand side gives

$$\mathbf{Ax} - \lambda\mathbf{Ix} = \mathbf{0}$$

$$\Rightarrow \qquad (\mathbf{A} - \lambda\mathbf{I})\mathbf{x} = \mathbf{0} \qquad [1]$$

(where $\mathbf{x}$ is an eigenvector and λ is the corresponding eigenvalue).

Now we saw (property (7) on page 632) that a matrix $\mathbf{B}$ is *singular* if and only if there is a non-zero vector $\mathbf{x}$ such that $\mathbf{Bx} = \mathbf{0}$. Therefore from [1] we see that $\mathbf{A} - \lambda\mathbf{I}$ is singular

$$\Rightarrow \qquad |\mathbf{A} - \lambda\mathbf{I}| = 0$$

Conversely, if $|\mathbf{A} - \lambda\mathbf{I}| = 0$ then there is a non-zero vector $\mathbf{x}$ such that $(\mathbf{A} - \lambda\mathbf{I})\mathbf{x} = \mathbf{0}$ which implies $\mathbf{Ax} = \lambda\mathbf{x}$, i.e. λ is an eigenvalue of $\mathbf{A}$. Thus

λ is an eigenvalue of $\mathbf{A}$ if and only if $|\mathbf{A} - \lambda\mathbf{I}| = 0$.

The equation $|\mathbf{A} - \lambda\mathbf{I}| = 0$ is an equation in the variable λ called the *characteristic equation* of $\mathbf{A}$. Its roots are the eigenvalues of $\mathbf{A}$.

EXAMPLES 15g

1) Find the eigenvalues of the matrix

$$\mathbf{A} = \begin{pmatrix} 2 & 1 \\ 3 & 0 \end{pmatrix}.$$

$$\mathbf{A} - \lambda\mathbf{I} = \begin{pmatrix} 2 & 1 \\ 3 & 0 \end{pmatrix} - \lambda\begin{pmatrix} 1 & 0 \\ 0 & 1 \end{pmatrix} = \begin{pmatrix} 2-\lambda & 1 \\ 3 & -\lambda \end{pmatrix}$$

so the characteristic equation is

$$\begin{vmatrix} 2-\lambda & 1 \\ 3 & -\lambda \end{vmatrix} = 0$$

$\Rightarrow$ $$(2-\lambda)(-\lambda)-3 = 0$$

$\Rightarrow$ $$\lambda^2-2\lambda-3 = 0$$

$\Rightarrow$ $$(\lambda-3)(\lambda+1) = 0$$

$\Rightarrow$ $$\lambda = 3 \quad \text{or} \quad -1$$

The eigenvalues are $\lambda = 3$ *and* $\lambda = -1$.

2) We already know (see page 649) that $\lambda = 3$ is an eigenvalue of $\begin{pmatrix} 2 & 0 & 1 \\ -1 & 2 & 3 \\ 1 & 0 & 2 \end{pmatrix}$. Find the other eigenvalues.

Here $\mathbf{A} = \begin{pmatrix} 2 & 0 & 1 \\ -1 & 2 & 3 \\ 1 & 0 & 2 \end{pmatrix}$, therefore

$$\mathbf{A} - \lambda\mathbf{I} = \begin{pmatrix} 2 & 0 & 1 \\ -1 & 2 & 3 \\ 1 & 0 & 2 \end{pmatrix} - \lambda\begin{pmatrix} 1 & 0 & 0 \\ 0 & 1 & 0 \\ 0 & 0 & 1 \end{pmatrix} = \begin{pmatrix} 2-\lambda & 0 & 1 \\ -1 & 2-\lambda & 3 \\ 1 & 0 & 2-\lambda \end{pmatrix}$$

So the characteristic equation is

$$\begin{vmatrix} 2-\lambda & 0 & 1 \\ -1 & 2-\lambda & 3 \\ 1 & 0 & 2-\lambda \end{vmatrix} = 0$$

$$\Rightarrow \quad (2-\lambda)\begin{vmatrix} 2-\lambda & 3 \\ 0 & 2-\lambda \end{vmatrix} - 0\begin{vmatrix} -1 & 3 \\ 1 & 2-\lambda \end{vmatrix} + 1\begin{vmatrix} -1 & 2-\lambda \\ 1 & 0 \end{vmatrix}$$

$$\Rightarrow \quad (2-\lambda)^3 - 0 + (-(2-\lambda)) = 0$$

$$\Rightarrow \quad (2-\lambda)^3 - (2-\lambda) = 0$$

$$\Rightarrow \quad (2-\lambda)(\lambda^2 - 4\lambda + 4 - 1) = 0$$

$$\Rightarrow \quad (2-\lambda)(\lambda^2 - 4\lambda + 3) = 0$$

$$\Rightarrow \quad (2-\lambda)(\lambda - 3)(\lambda - 1) = 0$$

$$\Rightarrow \quad \lambda = 1,\ 2 \text{ or } 3$$

The other eigenvalues are $\lambda = 1$ *and* $\lambda = 2$.

Finding Eigenvectors of a Matrix

Having found an eigenvalue λ for the matrix $\mathbf{A}$, then to find a corresponding eigenvector, we use equation [1] (page 651) again:

$$(\mathbf{A} - \lambda\mathbf{I})\mathbf{x} = 0$$

where $\mathbf{x}$ is an eigenvector and λ is the corresponding eigenvalue. In other words we *solve this for* $\mathbf{x}$, and notice that this is in fact a *set* of equations in the components of $\mathbf{x}$.

Note that we do not expect a unique solution because we know that any scalar multiple of a solution is another solution.

EXAMPLES 15g (continued)

3) Find eigenvectors corresponding to the eigenvalues $\lambda = 1$ and $\lambda = 2$ of the matrix $\mathbf{A} = \begin{pmatrix} 2 & 0 & 1 \\ -1 & 2 & 3 \\ 1 & 0 & 2 \end{pmatrix}$.

When $\lambda = 1$, $\mathbf{A} - \lambda\mathbf{I} = \begin{pmatrix} 2-1 & 0 & 1 \\ -1 & 2-1 & 3 \\ 1 & 0 & 2-1 \end{pmatrix} = \begin{pmatrix} 1 & 0 & 1 \\ -1 & 1 & 3 \\ 1 & 0 & 1 \end{pmatrix}$

so we solve

$$\begin{pmatrix} 1 & 0 & 1 \\ -1 & 1 & 3 \\ 1 & 0 & 1 \end{pmatrix}\begin{pmatrix} x \\ y \\ z \end{pmatrix} = \begin{pmatrix} 0 \\ 0 \\ 0 \end{pmatrix}$$

for x, y and z.

$$\Rightarrow \qquad \begin{aligned} x \quad + z &= 0 \qquad [1] \\ -x + y + 3z &= 0 \qquad [2] \\ x \quad + z &= 0 \qquad [3] \end{aligned}$$

Equation [3] is redundant (it is the same as equation [1]) and this was to be expected since only *two* equations are needed to solve a set of *homogeneous* equations in three variables.

Now suppose that $x = k$, then from equation [1], $z = -k$ and from equation [2] $y = x - 3z = k + 3k = 4k$. Thus for any value of k we have the solution

$$\begin{aligned} x &= k \\ y &= 4k \\ z &= -k \end{aligned}$$

i.e. $\begin{pmatrix} k \\ 4k \\ -k \end{pmatrix}$ is an eigenvector of $\mathbf{A}$ with eigenvalue $\lambda = 1$. For example $\begin{pmatrix} 1 \\ 4 \\ -1 \end{pmatrix}$ is one such eigenvector.

When $\lambda = 2$, $\mathbf{A} - \lambda\mathbf{I} = \begin{pmatrix} 2-2 & 0 & 1 \\ -1 & 2-2 & 3 \\ 1 & 0 & 2-2 \end{pmatrix} = \begin{pmatrix} 0 & 0 & 1 \\ -1 & 0 & 3 \\ 1 & 0 & 0 \end{pmatrix}$

So we solve

$$\begin{pmatrix} 0 & 0 & 1 \\ -1 & 0 & 3 \\ 1 & 0 & 0 \end{pmatrix} \begin{pmatrix} x \\ y \\ z \end{pmatrix} = \begin{pmatrix} 0 \\ 0 \\ 0 \end{pmatrix}$$

$$\Rightarrow \qquad \begin{aligned} z &= 0 \\ -x + 3z &= 0 \\ x &= 0 \end{aligned}$$

$$\Rightarrow \qquad x = z = 0$$

So y can take any value and any vector of the form $\begin{pmatrix} 0 \\ k \\ 0 \end{pmatrix}$, e.g. $\begin{pmatrix} 0 \\ 1 \\ 0 \end{pmatrix}$, is an eigenvector with eigenvalue $\lambda = 2$.

4) Find eigenvectors of unit length for the matrix $\mathbf{A} = \begin{pmatrix} 2 & 1 \\ 3 & 0 \end{pmatrix}$.

We found the eigenvalues in Example No. 1, $\lambda = 3$ and $\lambda = -1$.

$\lambda = 3$:

$$\mathbf{A} - \lambda\mathbf{I} = \begin{pmatrix} 2 & 1 \\ 3 & 0 \end{pmatrix} - 3\begin{pmatrix} 1 & 0 \\ 0 & 1 \end{pmatrix} = \begin{pmatrix} -1 & 1 \\ 3 & -3 \end{pmatrix}$$

So we solve $\begin{pmatrix} -1 & 1 \\ 3 & -3 \end{pmatrix}\begin{pmatrix} x \\ y \end{pmatrix} = \begin{pmatrix} 0 \\ 0 \end{pmatrix}$

or

$$-x + y = 0 \qquad [1]$$

$$3x - 3y = 0 \qquad [2]$$

Equation [2] is redundant (it is the same as equation [1] multiplied by -3) and equation [1] has general solution $x = k$, $y = k$, i.e. any vector of the form $\begin{pmatrix} k \\ k \end{pmatrix}$ is an eigenvector with eigenvalue 3.

For unit length we need $k^2 + k^2 = 1$ i.e. $k = \pm\sqrt{2}$. Thus there are *two* possible solutions:

$$\begin{pmatrix} \sqrt{2} \\ \sqrt{2} \end{pmatrix} \quad \text{and} \quad \begin{pmatrix} -\sqrt{2} \\ -\sqrt{2} \end{pmatrix}$$

$\lambda = -1$:

$$\mathbf{A} - \lambda\mathbf{I} = \begin{pmatrix} 2 & 1 \\ 3 & 0 \end{pmatrix} + 1\begin{pmatrix} 1 & 0 \\ 0 & 1 \end{pmatrix} = \begin{pmatrix} 3 & 1 \\ 3 & 1 \end{pmatrix}$$

So the equations are

$$3x + y = 0$$

$$3x + y = 0$$

and the general solution is $\begin{pmatrix} x \\ y \end{pmatrix} = \begin{pmatrix} k \\ -3k \end{pmatrix}$.

For unit length $k^2 + (-3k)^2 = 1$

$$\Rightarrow \qquad 10k^2 = 1$$

$$k = \pm\frac{1}{\sqrt{10}}$$

So again there are two solutions:

$$\begin{pmatrix} \frac{1}{\sqrt{10}} \\ \frac{-3}{\sqrt{10}} \end{pmatrix} \quad \text{and} \quad \begin{pmatrix} -\frac{1}{\sqrt{10}} \\ \frac{3}{\sqrt{10}} \end{pmatrix}$$

Note that the two eigenvectors in this example correspond to two preserved lines through the origin, namely $y = x$ and $y = -3x$:

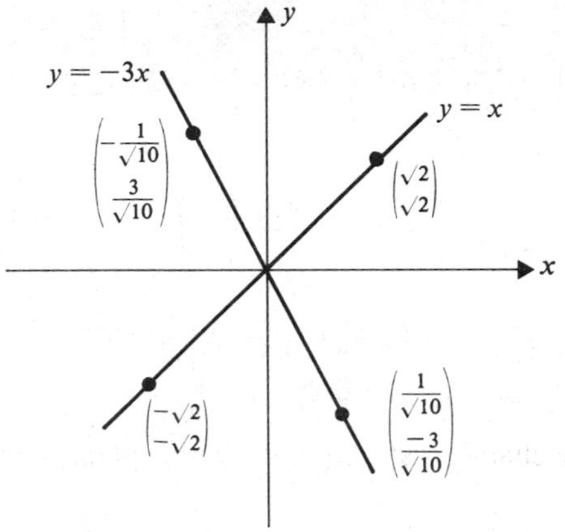

This confirms the result found in Example 1c, No. 4 in Chapter 1 (page 15).

EXERCISE 15g

1) Find eigenvalues of the matrices:

(a) $\begin{pmatrix} -5 & 3 \\ 6 & -2 \end{pmatrix}$ (b) $\begin{pmatrix} 4 & 0 \\ 2 & -1 \end{pmatrix}$ (c) $\begin{pmatrix} 2 & 1 \\ 0 & 2 \end{pmatrix}$

(d) $\begin{pmatrix} 3 & 2 & 2 \\ -2 & -2 & -2 \\ 1 & 2 & 2 \end{pmatrix}$.

2) Given that $\lambda = 2$ is an eigenvalue of $\begin{pmatrix} 0 & 0 & 2 \\ 1 & -1 & 2 \\ 1 & 0 & 1 \end{pmatrix}$, find the other eigenvalues.

3) Find an eigenvectoı corresponding to each eigenvalue for the matrices of Questions 1 and 2.

4) Find all the eigenvectors of unit length for the matrix $\begin{pmatrix} 1 & 1 & 0 \\ 0 & 1 & 0 \\ 0 & 0 & 2 \end{pmatrix}$.

5) Given that $\lambda = -2$ is an eigenvalue of $\begin{pmatrix} 5 & -1 & 1 \\ -1 & 1 & -3 \\ 1 & -3 & 1 \end{pmatrix}$, find the eigenvectors of unit length corresponding to each eigenvalue.

6) Verify that $\lambda = 2$ is an eigenvalue of $\begin{pmatrix} 4 & 2 & 2 \\ -1 & 1 & -1 \\ 0 & 0 & 2 \end{pmatrix}$ and find two *independent* eigenvectors (*not* scalar multiples of each other) corresponding to this eigenvalue.

7) Write down the characteristic equation and explain geometrically why $\begin{pmatrix} 0 & 1 \\ -1 & 0 \end{pmatrix}$ has no real eigenvalues.

8) Find all the eigenvalues and eigenvectors of $\begin{pmatrix} 1 & 1 & 0 \\ 0 & 1 & 1 \\ 0 & 0 & 1 \end{pmatrix}$.

DIAGONALIZATION

Consider the matrix $\mathbf{A} = \begin{pmatrix} 2 & 0 & 1 \\ -1 & 2 & 3 \\ 1 & 0 & 2 \end{pmatrix}$.

We have found the eigenvalues of $\mathbf{A}$ and for each eigenvalue we have found corresponding eigenvectors, namely:

$$\lambda = 3, \begin{pmatrix} 1 \\ 2 \\ 1 \end{pmatrix}; \quad \lambda = 1, \begin{pmatrix} 1 \\ 4 \\ -1 \end{pmatrix}; \quad \lambda = 2, \begin{pmatrix} 0 \\ 1 \\ 0 \end{pmatrix}$$

(See pages 652–4.)

Now consider the matrix $\mathbf{P} = \begin{pmatrix} 1 & 1 & 0 \\ 2 & 4 & 1 \\ 1 & -1 & 0 \end{pmatrix}$

whose columns are these eigenvectors of **A**.

Calculating **AP** we find

$$\mathbf{AP} = \begin{pmatrix} 2 & 0 & 1 \\ -1 & 2 & 3 \\ 1 & 0 & 2 \end{pmatrix}\begin{pmatrix} 1 & 1 & 0 \\ 2 & 4 & 1 \\ 1 & -1 & 0 \end{pmatrix} = \begin{pmatrix} 3 & 1 & 0 \\ 6 & 4 & 2 \\ 3 & -1 & 0 \end{pmatrix}$$

Inspecting the columns of **AP** we see that each is in fact the corresponding column of **P** *multiplied by the corresponding eigenvalue.*

This is true in general. Suppose that **A** is any 3×3 matrix and that **P** is any 3×3 matrix whose columns are eigenvectors **c, d, e** of **A** with eigenvalues λ, μ, ν respectively. Then **AP** has columns **Ac, Ad, Ae** (see the observation in the middle of page 627).
But $\mathbf{Ac} = \lambda\mathbf{c}$, $\mathbf{Ad} = \mu\mathbf{d}$, $\mathbf{Ae} = \nu\mathbf{e}$. That is, *each column of* **AP** *is the corresponding column of* **P** *multiplied by the corresponding eigenvalue.*

Note that throughout this section we shall state and prove all general results for 3×3 matrices. But there are analogous results for 2×2 matrices and indeed for square matrices of any size.

Continuing now with our numerical example, consider the diagonal matrix

$$\Lambda = \begin{pmatrix} 3 & 0 & 0 \\ 0 & 1 & 0 \\ 0 & 0 & 2 \end{pmatrix}$$

whose diagonal entries are the eigenvalues of **A** (in the order which corresponds to the columns of **P**).

Calculating $\mathbf{P}\Lambda$ we find

$$\mathbf{P} \quad \mathbf{P}\Lambda = \begin{pmatrix} 1 & 1 & 0 \\ 2 & 4 & 1 \\ 1 & -1 & 0 \end{pmatrix}\begin{pmatrix} 3 & 0 & 0 \\ 0 & 1 & 0 \\ 0 & 0 & 2 \end{pmatrix} = \begin{pmatrix} 3 & 1 & 0 \\ 6 & 4 & 2 \\ 3 & -1 & 0 \end{pmatrix} = \mathbf{AP}$$

i.e. $\mathbf{AP} = \mathbf{P}\Lambda$.

This result is again true in general. Let **P**, as before, have columns **c, d, e** which are eigenvectors of **A** with eigenvalues λ, μ, ν respectively and let

$$\Lambda = \begin{pmatrix} \lambda & 0 & 0 \\ 0 & \mu & 0 \\ 0 & 0 & \nu \end{pmatrix}$$

Consider the first column of $\mathbf{P\Lambda}$, this is $\mathbf{P}\begin{pmatrix}\lambda\\0\\0\end{pmatrix} = \mathbf{P}\lambda\begin{pmatrix}1\\0\\0\end{pmatrix} = \lambda\mathbf{P}\begin{pmatrix}1\\0\\0\end{pmatrix}$.

But $\lambda\mathbf{P}\begin{pmatrix}1\\0\\0\end{pmatrix} = \lambda(\text{first column of } \mathbf{P})$, (see top of page 630)

$= \lambda\mathbf{c}$

and $\lambda\mathbf{c}$ is the first column of $\mathbf{AP}$.
So the first columns of $\mathbf{P\Lambda}$ and $\mathbf{AP}$ are identical. In a similar way we can show that the other columns of $\mathbf{AP}$ and $\mathbf{P\Lambda}$ are the same, i.e. $\mathbf{AP} = \mathbf{P\Lambda}$.

Now in our numerical example $\mathbf{P}$ is in fact a *non-singular matrix* because

$|\mathbf{P}| = \begin{vmatrix}1 & 1 & 0\\2 & 4 & 1\\1 & -1 & 0\end{vmatrix} = 2$ therefore we can multiply the equation $\mathbf{AP} = \mathbf{P\Lambda}$

on the left by $\mathbf{P}^{-1}$:

$$\mathbf{P}^{-1}\mathbf{AP} = \mathbf{P}^{-1}\mathbf{P\Lambda}$$

$$\Rightarrow \quad \mathbf{P}^{-1}\mathbf{AP} = \mathbf{I\Lambda} \quad (\text{since } \mathbf{P}^{-1}\mathbf{P} = \mathbf{I})$$

$$\Rightarrow \quad \mathbf{P}^{-1}\mathbf{AP} = \mathbf{\Lambda}$$

Where $\mathbf{\Lambda} = \begin{pmatrix}3 & 0 & 0\\0 & 1 & 0\\0 & 0 & 2\end{pmatrix}$, the diagonal matrix whose diagonal entries are the eigenvalues of $\mathbf{A}$.

Clearly we can rearrange the equation $\mathbf{AP} = \mathbf{P\Lambda}$ to give $\mathbf{P}^{-1}\mathbf{AP} = \mathbf{\Lambda}$ in general, provided that $\mathbf{P}$ is non-singular. In fact we have now proved the following theorem.

Diagonalization Theorem

Suppose that $\mathbf{P}$ is a non-singular 3×3 matrix whose columns are eigenvectors of the 3×3 matrix $\mathbf{A}$ with eigenvalues λ, μ, ν (in order), then $\mathbf{P}^{-1}\mathbf{AP} = \mathbf{\Lambda}$ where $\mathbf{\Lambda}$ is the diagonal matrix

$$\begin{pmatrix}\lambda & 0 & 0\\0 & \mu & 0\\0 & 0 & \nu\end{pmatrix}$$

We say that $\mathbf{A}$ is *diagonalizable* and that $\mathbf{P}$ *diagonalizes* $\mathbf{A}$. The diagonal matrix $\mathbf{\Lambda}$ is the *diagonal form* of $\mathbf{A}$ and we say that $\mathbf{A}$ is *similar to* $\mathbf{\Lambda}$.

Technique for Diagonalization

In order to diagonalize a given square matrix $\mathbf{A}$ we need to find a matrix $\mathbf{P}$ which is non-singular and has eigenvectors of $\mathbf{A}$ for columns. So we proceed as follows:

(1) Find the eigenvalues of $\mathbf{A}$.
(2) Find corresponding eigenvectors.
(3) Construct, if possible, the required non-singular matrix $\mathbf{P}$ with eigenvectors of $\mathbf{A}$ as columns.

Note that the diagonal form of $\mathbf{A}$ is then the diagonal matrix whose entries are the eigenvalues of $\mathbf{A}$ in the order which corresponds to the columns of $\mathbf{P}$.

If it is not possible to construct $\mathbf{P}$, then $\mathbf{A}$ is *not* diagonalizable.

Note that there are some general results given in the next section which simplify this technique in certain cases (see page 667).

EXAMPLES 15h

In each case decide whether the given matrix $\mathbf{A}$ is diagonalizable and, if so, find a matrix $\mathbf{P}$ which diagonalizes $\mathbf{A}$ and write down the corresponding diagonal form.

1) $\mathbf{A} = \begin{pmatrix} 1 & 2 \\ 2 & 1 \end{pmatrix}$.

We follow the technique given above. To find the eigenvalues of $\mathbf{A}$ we write down the characteristic equation:

$$|\mathbf{A} - \lambda\mathbf{I}| = 0 \quad \text{where} \quad \mathbf{A} - \lambda\mathbf{I} = \begin{pmatrix} 1-\lambda & 2 \\ 2 & 1-\lambda \end{pmatrix}$$

$$\Rightarrow \quad \begin{vmatrix} 1-\lambda & 2 \\ 2 & 1-\lambda \end{vmatrix} = 0$$

$$\Rightarrow \quad (1-\lambda)^2 - 4 = 0$$

$$\Rightarrow \quad \lambda^2 - 2\lambda - 3 = 0$$

$$\Rightarrow \quad (\lambda - 3)(\lambda + 1) = 0, \quad \lambda = 3 \quad \text{or} \quad \lambda = -1.$$

Next we find corresponding eigenvectors.

When $\lambda = 3$, $\mathbf{A} - \lambda\mathbf{I} = \begin{pmatrix} -2 & 2 \\ 2 & -2 \end{pmatrix}$

and we need to solve $(\mathbf{A}-\lambda\mathbf{I})\begin{pmatrix}x\\y\end{pmatrix}=\begin{pmatrix}0\\0\end{pmatrix}$ for x, y.

$$\Rightarrow \qquad -2x+2y = 0$$

and $$2x-2y = 0$$

$$\Rightarrow \qquad x = y$$

i.e. any vector of the form $\begin{pmatrix}k\\k\end{pmatrix}$ is an eigenvector, e.g. $\begin{pmatrix}1\\1\end{pmatrix}$.

When $\lambda=-1$, $\mathbf{A}-\lambda\mathbf{I}=\begin{pmatrix}2 & 2\\2 & 2\end{pmatrix}$ so the equations are

$$2x+2y = 0$$

and $$2x+2y = 0$$

$$\Rightarrow \qquad x = -y$$

i.e. $\begin{pmatrix}k\\-k\end{pmatrix}$ is an eigenvector for any value of k, e.g. $\begin{pmatrix}1\\-1\end{pmatrix}$.

Now we consider the matrix $\mathbf{P}$ whose columns are the eigenvectors we have found, i.e. $\mathbf{P}=\begin{pmatrix}1 & 1\\1 & -1\end{pmatrix}$.

Then $|\mathbf{P}|=-2$, so $\mathbf{P}$ is non-singular and hence $\mathbf{P}$ does diagonalize $\mathbf{A}$.

The diagonal form is $\begin{pmatrix}3 & 0\\0 & -1\end{pmatrix}$ (where $3, -1$ are the eigenvalues of $\mathbf{A}$ in the order corresponding to the columns of $\mathbf{P}$).

2) $\mathbf{A}=\begin{pmatrix}1 & 1\\0 & 1\end{pmatrix}$.

The characteristic equation is

$$|\mathbf{A}-\lambda\mathbf{I}| = \begin{vmatrix}1-\lambda & 1\\0 & 1-\lambda\end{vmatrix} = 0$$

$$\Rightarrow \qquad (1-\lambda)^2 = 0 \quad \Rightarrow \quad \lambda = 1 \qquad \text{(repeated root)}$$

i.e. $\lambda=1$ is the only eigenvalue.

An eigenvector $\begin{pmatrix}x\\y\end{pmatrix}$ satisfies $(\mathbf{A}-\lambda\mathbf{I})\begin{pmatrix}x\\y\end{pmatrix}=0$

$$\Rightarrow \qquad \begin{pmatrix} 0 & 1 \\ 0 & 0 \end{pmatrix}\begin{pmatrix} x \\ y \end{pmatrix} = 0$$

$$\Rightarrow \qquad \begin{aligned} y &= 0 \\ 0 &= 0 \end{aligned}$$

i.e. $y = 0$, $x = k$, say.

So $\begin{pmatrix} k \\ 0 \end{pmatrix}$ is a general eigenvector, e.g. $\begin{pmatrix} 1 \\ 0 \end{pmatrix}$ or $\begin{pmatrix} -1 \\ 0 \end{pmatrix}$.

Hence it is *not possible* to find a non-singular matrix **P** whose columns are eigenvectors of **A** since any such matrix must have its second row zero. So in this case **A** is *not* diagonalizable.

3) $\mathbf{A} = \begin{pmatrix} 2 & 0 & 0 \\ 0 & 1 & 1 \\ 0 & 1 & 1 \end{pmatrix}$

The characteristic equation is

$$\begin{vmatrix} 2-\lambda & 0 & 0 \\ 0 & 1-\lambda & 1 \\ 0 & 1 & 1-\lambda \end{vmatrix} = 0$$

$$\Rightarrow \qquad (2-\lambda)((1-\lambda)^2 - 1) = 0$$

$$\Rightarrow \qquad (2-\lambda)(\lambda^2 - 2\lambda) = 0$$

$$\Rightarrow \qquad \lambda = 2 \quad \text{(repeated root)} \quad \text{or} \quad \lambda = 0.$$

When $\lambda = 2$, the equations for an eigenvector are

$$\begin{aligned} 0 &= 0 \\ -y + z &= 0 \\ y - z &= 0 \end{aligned}$$

i.e. $x = k$, $y = l$ and $z = -l$ when k and l are any real numbers. Thus there is a very large choice of solutions. Two particular solutions are

$\begin{pmatrix} 1 \\ 0 \\ 0 \end{pmatrix}$ and $\begin{pmatrix} 0 \\ 1 \\ 1 \end{pmatrix}$.

When $\lambda = 0$, the equations are

$$\begin{aligned} x \qquad &= 0 \\ y + z &= 0 \\ y + z &= 0 \end{aligned}$$

i.e. $\begin{pmatrix} 0 \\ k \\ -k \end{pmatrix}$ is a general solution, e.g. $\begin{pmatrix} 0 \\ 1 \\ -1 \end{pmatrix}$ is a particular solution.

If we now consider the matrix $\mathbf{P} = \begin{pmatrix} 1 & 0 & 0 \\ 0 & 1 & 1 \\ 0 & 1 & -1 \end{pmatrix}$ with columns the eigenvectors that we have found, then $|\mathbf{P}| = -2$, i.e. $\mathbf{P}$ is non-singular.
Thus $\mathbf{P}$ diagonalizes $\mathbf{A}$ and the diagonal form is

$$\mathbf{P}^{-1}\mathbf{A}\mathbf{P} = \begin{pmatrix} 2 & 0 & 0 \\ 0 & 2 & 0 \\ 0 & 0 & 0 \end{pmatrix}$$

where 2, 2, 0 are the eigenvalues which correspond to the columns of $\mathbf{P}$.

4) $\mathbf{A} = \begin{pmatrix} 0 & 1 \\ -1 & 0 \end{pmatrix}$.

The characteristic equation is

$$\begin{vmatrix} \lambda & 1 \\ -1 & \lambda \end{vmatrix} = 0 \quad \Rightarrow \quad \lambda^2 + 1 = 0$$

for which there are *no real roots.* So $\mathbf{A}$ is not diagonalizable.

Note that $\mathbf{A}$ can in fact be diagonalized using a complex matrix $\mathbf{P}$, but in this chapter we are dealing only with *real* matrices.

Orthogonal Diagonalization

In Example 1 above, the chosen eigenvectors were $\begin{pmatrix} 1 \\ 1 \end{pmatrix}$ and $\begin{pmatrix} 1 \\ -1 \end{pmatrix}$, which are *perpendicular*. By scaling them suitably we can find *perpendicular eigenvectors of unit length.* Indeed $\begin{pmatrix} \frac{\sqrt{2}}{2} \\ \frac{\sqrt{2}}{2} \end{pmatrix}$ and $\begin{pmatrix} \frac{\sqrt{2}}{2} \\ -\frac{\sqrt{2}}{2} \end{pmatrix}$

are such eigenvectors. If we choose these eigenvectors for the columns of $\mathbf{P}$ then

$\mathbf{P} = \begin{pmatrix} \frac{\sqrt{2}}{2} & \frac{\sqrt{2}}{2} \\ \frac{\sqrt{2}}{2} & -\frac{\sqrt{2}}{2} \end{pmatrix}$ is an *orthogonal matrix* which diagonalizes $\mathbf{A} = \begin{pmatrix} 1 & 2 \\ 2 & 1 \end{pmatrix}$.

Notice that since $\mathbf{P}$ is orthogonal, $\mathbf{P}^{\mathrm{T}} = \mathbf{P}^{-1}$ and therefore the equation

$$\mathbf{P}^{-1}\mathbf{AP} = \Lambda$$

can be written

$$\mathbf{P}^{\mathrm{T}}\mathbf{AP} = \Lambda$$

where $\Lambda = \begin{pmatrix} 3 & 0 \\ 0 & -1 \end{pmatrix}$.

In general we have the following result.

Orthogonal Diagonalization Theorem

Suppose that $\mathbf{A}$ is a 3×3 matrix and that $\mathbf{P}$ is a 3×3 matrix whose columns $\mathbf{c}, \mathbf{d}, \mathbf{e}$ are mutually perpendicular eigenvectors of $\mathbf{A}$ of unit length, then $\mathbf{P}$ is an orthogonal matrix which diagonalizes $\mathbf{A}$. Since $\mathbf{P}^{\mathrm{T}} = \mathbf{P}^{-1}$ we can write the equation $\mathbf{P}^{-1}\mathbf{AP} = \Lambda$ in the form $\mathbf{P}^{\mathrm{T}}\mathbf{AP} = \Lambda$ where

$\Lambda = \begin{pmatrix} \lambda & 0 & 0 \\ 0 & \mu & 0 \\ 0 & 0 & \nu \end{pmatrix}$ and λ, μ, ν are the eigenvalues corresponding to $\mathbf{c}, \mathbf{d}, \mathbf{e}$

(in order).

We say that $\mathbf{P}$ *orthogonally diagonalizes* $\mathbf{A}$ and that $\mathbf{A}$ is *orthogonally equivalent* to the diagonal form Λ.

EXAMPLES 15h (continued)

5) Find an orthogonal matrix $\mathbf{P}$ which diagonalizes $\mathbf{A} = \begin{pmatrix} 2 & 0 & 0 \\ 0 & 1 & 1 \\ 0 & 1 & 1 \end{pmatrix}$

(the matrix in Example 3 above).

The chosen eigenvectors were

$$\begin{pmatrix} 1 \\ 0 \\ 0 \end{pmatrix}, \begin{pmatrix} 0 \\ 1 \\ 1 \end{pmatrix} \text{ and } \begin{pmatrix} 0 \\ 1 \\ -1 \end{pmatrix}$$

which are in fact mutually perpendicular and the first one also has unit length.

So we have to scale the other two to have unit length:

$$\begin{pmatrix} 1 \\ 0 \\ 0 \end{pmatrix}, \begin{pmatrix} 0 \\ \frac{\sqrt{2}}{2} \\ \frac{\sqrt{2}}{2} \end{pmatrix}, \begin{pmatrix} 0 \\ \frac{\sqrt{2}}{2} \\ -\frac{\sqrt{2}}{2} \end{pmatrix}$$

Thus $\mathbf{P} = \begin{pmatrix} 1 & 0 & 0 \\ 0 & \frac{\sqrt{2}}{2} & \frac{\sqrt{2}}{2} \\ 0 & \frac{\sqrt{2}}{2} & -\frac{\sqrt{2}}{2} \end{pmatrix}$ is an orthogonal matrix which diagonalizes **A**.

EXERCISE 15h

1) Verify by matrix multiplication that $\mathbf{P}^T\mathbf{A}\mathbf{P}$ is diagonal where **P** and **A** are given in the example above.

2) In each case write down a matrix **P** which diagonalizes the given matrix and also write down the diagonal form Λ. (Eigenvalues and eigenvectors for each matrix were calculated in Examples 15g and Exercise 15g, incomplete in the case of part (e).)

(a) $\begin{pmatrix} 2 & 1 \\ 3 & 0 \end{pmatrix}$ (b) $\begin{pmatrix} -5 & 3 \\ 6 & -2 \end{pmatrix}$ (c) $\begin{pmatrix} 4 & 0 \\ 2 & -1 \end{pmatrix}$

(d) $\begin{pmatrix} 3 & 2 & 2 \\ -2 & -2 & -2 \\ 1 & 2 & 2 \end{pmatrix}$ (e) $\begin{pmatrix} 4 & 2 & 2 \\ -1 & 1 & 1 \\ 0 & 0 & 2 \end{pmatrix}$.

3) Decide whether or not the given matrix is diagonalizable (eigenvalues and eigenvectors were calculated for parts (a) and (e) in Exercise 15g).

(a) $\begin{pmatrix} 2 & 1 \\ 0 & 2 \end{pmatrix}$ (b) $\begin{pmatrix} 2 & 1 & 0 \\ 0 & 1 & 0 \\ 0 & 0 & 1 \end{pmatrix}$ (c) $\begin{pmatrix} 1 & 1 & 0 \\ 0 & 1 & 0 \\ 0 & 0 & 0 \end{pmatrix}$

(d) $\begin{pmatrix} 3 & 4 & 7 \\ 0 & 2 & 2 \\ 0 & 0 & 2 \end{pmatrix}$ (e) $\begin{pmatrix} 0 & 0 & 2 \\ 1 & -1 & 2 \\ 1 & 0 & 1 \end{pmatrix}$ (f) $\begin{pmatrix} \frac{\sqrt{2}}{2} & \frac{\sqrt{2}}{2} \\ -\frac{\sqrt{2}}{2} & \frac{\sqrt{2}}{2} \end{pmatrix}$.

4) Diagonalize the matrix by a suitable orthogonal matrix **P** and verify that $\mathbf{P}^T\mathbf{A}\mathbf{P}$ is diagonal by matrix multiplication

(a) $\begin{pmatrix} 1 & 1 \\ 1 & 1 \end{pmatrix}$ (b) $\begin{pmatrix} 0 & 1 \\ 1 & 0 \end{pmatrix}$.

5) Verify that the matrix in Exercise 15g, Question 5 is orthogonally diagonalizable and check by matrix multiplication.

TWO GENERAL RESULTS

In this section we shall state two general results which tell us that diagonalization is always possible under certain circumstances. The proofs of these results are beyond the scope of this book – the interested reader should consult a text on linear algebra for these proofs. As in the last section, the results are stated for 3×3 matrices, but similar results are true for 2×2 matrices and indeed for square matrices of any size.
We have seen several examples of matrices, some of which are diagonalizable and some of which are not. In particular we have seen the following examples:

(a) $\begin{pmatrix} 2 & 0 & 1 \\ -1 & 2 & 3 \\ 1 & 0 & 2 \end{pmatrix}$ Eigenvalues $\lambda = 1, 2, 3$. Diagonalizable.

(b) $\begin{pmatrix} 2 & 1 \\ 3 & 0 \end{pmatrix}$ Eigenvalues $\lambda = 3, -1$. Diagonalizable.

(c) $\begin{pmatrix} 0 & 1 \\ -1 & 0 \end{pmatrix}$ No real eigenvalues. *Not* diagonalizable.

(d) $\begin{pmatrix} 1 & 1 \\ 0 & 1 \end{pmatrix}$ Eigenvalue $\lambda = 1$ (repeated root). *Not* diagonalizable.

(e) $\begin{pmatrix} 2 & 0 & 0 \\ 0 & 1 & 1 \\ 0 & 1 & 1 \end{pmatrix}$ Eigenvalues $\lambda = 0$ and $\lambda = 2$ (repeated root). Orthogonally diagonalizable.

(f) $\begin{pmatrix} 2 & 1 & 0 \\ 0 & 1 & 0 \\ 0 & 0 & 1 \end{pmatrix}$ Eigenvalues $\lambda = 1$ (repeated) and $\lambda = 2$. Diagonalizable.

(g) $\begin{pmatrix} 1 & 2 \\ 2 & 1 \end{pmatrix}$ Eigenvalues $\lambda = 3, -1$. Orthogonally diagonalizable.

In examples (a), (b) and (g), the eigenvalues are real and distinct. In each case the matrix is diagonalizable. This is true in general and it is our first general result:

Result 1

A square matrix with real distinct eigenvalues is diagonalizable. More precisely, suppose $\mathbf{A}$ is a 3×3 matrix with distinct eigenvalues λ, μ, ν, then if we take eigenvectors corresponding to λ, μ, ν (in order) as columns of $\mathbf{P}$ then $\mathbf{P}$ is always non-singular and therefore $\mathbf{P}$ diagonalizes $\mathbf{A}$:

$$\mathbf{P}^{-1}\mathbf{AP} = \begin{pmatrix} \lambda & 0 & 0 \\ 0 & \mu & 0 \\ 0 & 0 & \nu \end{pmatrix}$$

In examples (e) and (g), the matrix is *symmetric* and in each case it is diagonalizable, indeed it is *orthogonally* diagonalizable. This is again true in general and it is our second general result.

Result 2

Any symmetric matrix $\mathbf{A}$ is diagonalizable. Moreover we can always find an *orthogonal* matrix whose columns are eigenvectors of $\mathbf{A}$, i.e. $\mathbf{A}$ is orthogonally diagonalizable.

Both the results give *sufficient* conditions for diagonalizability; that is to say: *if* the matrix $\mathbf{A}$ is *either* symmetric *or* has real distinct eigenvalues *then* $\mathbf{A}$ is diagonalizable. However, it is important to realize that neither of these conditions is *necessary*; in example (f) the matrix is *neither* symmetric *nor* does it have distinct eigenvalues but nevertheless it *is* diagonalizable.

Note carefully examples (c) and (d). In example (c) the eigenvalues are not real and in example (d) there is a repeated real eigenvalue. Neither matrix is symmetric. In neither example is the matrix diagonalizable.

Result 2 does in fact give a necessary *and* sufficient condition for *orthogonal* diagonalizability. We shall show in Example 15i, No. 3 that if a matrix is orthogonally diagonalizable then it must be symmetric; hence a matrix is orthogonally diagonalizable *if and only if* it is symmetric.

Technique for Diagonalization When Eigenvalues are Real and Distinct

The results simplify our technique for diagonalization (page 660) in the case where the eigenvalues of $\mathbf{A}$ are real and distinct. We simply choose an eigenvector for each eigenvalue and take these eigenvectors as columns of $\mathbf{P}$

and then **P** is *automatically* non-singular. Moreover if, in addition, **P** is symmetric, then the eigenvectors will always be mutually perpendicular and hence, by scaling them to have unit length, **P** will be an orthogonal matrix.

EXAMPLES 15i

1) The matrix $\mathbf{A} = \begin{pmatrix} -1 & 7 & 9 \\ 0 & 1 & 4 \\ 0 & 0 & 3 \end{pmatrix}$ has eigenvalues $\lambda = -1, 1, 3$. Verify that **A** is diagonalizable.

We find an eigenvector corresponding to each eigenvalue.

When $\lambda = -1$, $\mathbf{A} - \lambda\mathbf{I} = \begin{pmatrix} 0 & 7 & 9 \\ 0 & 2 & 4 \\ 0 & 0 & 4 \end{pmatrix}$

So we solve
$$\begin{aligned} 7y + 9z &= 0 \\ 2y + 4z &= 0 \\ 4z &= 0 \end{aligned}$$

$\Rightarrow$ $y = z = 0$

So $\begin{pmatrix} 1 \\ 0 \\ 0 \end{pmatrix}$ is a suitable eigenvector.

When $\lambda = 1$, $\mathbf{A} - \lambda\mathbf{I} = \begin{pmatrix} -2 & 7 & 9 \\ 0 & 0 & 4 \\ 0 & 0 & 2 \end{pmatrix}$

So we solve
$$\begin{aligned} -2x + 7y + 9z &= 0 \\ 4z &= 0 \\ 2z &= 0 \end{aligned}$$

$\Rightarrow$ $z = 0$ and $2x = 7y$.

So $\begin{pmatrix} 7 \\ 2 \\ 0 \end{pmatrix}$ is a suitable eigenvector.

When $\lambda = 3$, $\mathbf{A} - \lambda\mathbf{I} = \begin{pmatrix} -4 & 7 & 9 \\ 0 & -2 & 4 \\ 0 & 0 & 0 \end{pmatrix}$

So we solve

$$\begin{aligned} -4x + 7y + 9z &= 0 \\ -2y + 4z &= 0 \\ 0 &= 0 \end{aligned}$$

$\Rightarrow$ $\quad y = 2z \quad \text{and} \quad 4x = 7y + 9z$

$$\begin{aligned} &= 14z + 9z \\ &= 23z \end{aligned}$$

So $\begin{pmatrix} 23 \\ 8 \\ 4 \end{pmatrix}$ is a suitable eigenvector.

Taking these three vectors as columns we have $\mathbf{P} = \begin{pmatrix} 1 & 7 & 23 \\ 0 & 2 & 8 \\ 0 & 0 & 4 \end{pmatrix}$

and $|\mathbf{P}| = 8 \neq 0$ i.e. $\mathbf{P}$ *is* non-singular and hence diagonalizes $\mathbf{A}$.

2) The symmetric matrix $\mathbf{A} = \begin{pmatrix} 1 & 0 & 2 \\ 0 & 2 & 0 \\ 2 & 0 & 1 \end{pmatrix}$

has eigenvalues $\lambda = -1, 2$ and 3. Verify that $\mathbf{A}$ is orthogonally diagonalizable.

Eigenvectors for $\mathbf{A}$ are:

$$\lambda = -1, \begin{pmatrix} 1 \\ 0 \\ -1 \end{pmatrix}; \quad \lambda = 2, \begin{pmatrix} 0 \\ 1 \\ 0 \end{pmatrix}; \quad \lambda = 3, \begin{pmatrix} 1 \\ 0 \\ 1 \end{pmatrix}$$

which are mutually perpendicular.
Scaling them to have unit length we find the orthogonal matrix

$$\mathbf{P} = \begin{pmatrix} \frac{\sqrt{2}}{2} & 0 & \frac{\sqrt{2}}{2} \\ 0 & 1 & 0 \\ -\frac{\sqrt{2}}{2} & 0 & \frac{\sqrt{2}}{2} \end{pmatrix}$$

which diagonalizes $\mathbf{A}$.

3) The matrix $\mathbf{A}$ is orthogonally diagonalizable (i.e. $\mathbf{P}^T\mathbf{A}\mathbf{P} = \Lambda$, where Λ is diagonal and $\mathbf{P}$ is orthogonal). Prove that $\mathbf{A} = \mathbf{Q}^T\Lambda\mathbf{Q}$ where $\mathbf{Q}$ is also orthogonal and deduce that $\mathbf{A}$ is symmetric.

Multiplying the equation

$$\mathbf{P}^T\mathbf{A}\mathbf{P} = \Lambda$$

on the left by $\mathbf{P}$ and on the right by $\mathbf{P}^T$ we have

$$\mathbf{P}\mathbf{P}^T\mathbf{A}\mathbf{P}\mathbf{P}^T = \mathbf{P}\Lambda\mathbf{P}^T \qquad [1]$$

But $\mathbf{P}\mathbf{P}^T = \mathbf{I}$, since $\mathbf{P}$ is orthogonal, so [1] reduces to

$$\mathbf{A} = \mathbf{P}\Lambda\mathbf{P}^T \qquad [2]$$

and writing $\mathbf{Q} = \mathbf{P}^T$ we have

$$\mathbf{A} = \mathbf{Q}^T\Lambda\mathbf{Q} \qquad [3]$$

Now Λ is obviously symmetric, so equation [3] implies that $\mathbf{A}$ is symmetric using Example 15c, No. 2.

EXERCISE 15i

1) The eigenvalues of $\mathbf{A} = \begin{pmatrix} -1 & 2 & 2 \\ 2 & 2 & 2 \\ -3 & -6 & -6 \end{pmatrix}$

are $\lambda = 0$, $\lambda = -2$ and $\lambda = -3$. Verify that $\mathbf{A}$ is diagonalizable.

2) The eigenvalues of $\mathbf{A} = \begin{pmatrix} 3 & -4 & 2 \\ -4 & -1 & 6 \\ 2 & 6 & -2 \end{pmatrix}$

are $\lambda = -9$, $\lambda = 3$ and $\lambda = 6$. Verify that eigenvectors corresponding to different eigenvalues are mutually perpendicular and write down an orthogonal matrix which diagonalizes $\mathbf{A}$.

3) The eigenvalues of $\mathbf{A} = \begin{pmatrix} 1 & 2 & 0 \\ 2 & 1 & 0 \\ 0 & 0 & -1 \end{pmatrix}$

are $\lambda = 3$ and $\lambda = -1$ (repeated root). Find the general form of an eigenvector corresponding to $\lambda = -1$ and verify that it is always perpendicular to one corresponding to $\lambda = 3$. Find an orthogonal matrix which diagonalizes $\mathbf{A}$.

4) Given that $\lambda = 4$ is an eigenvalue of $\begin{pmatrix} -50 & 12 & 138 \\ -36 & 10 & 96 \\ -17 & 4 & 47 \end{pmatrix}$

find the other eigenvalues and, without finding any eigenvectors, write down the diagonal form.

5) Given that $\lambda = 9$ is an eigenvalue of $\begin{pmatrix} 7 & 4 & -4 \\ 4 & -8 & -1 \\ -4 & -1 & -8 \end{pmatrix}$

find the other eigenvalues and write down the diagonal form.

6) Check that each matrix in Exercise 15h, which either has distinct eigenvalues, or is symmetric (or both), is diagonalizable.

INTERPRETATION OF DIAGONALIZATION

We have given two contexts for matrices:
(1) Transformations of space.
(2) Quadratic forms (and in this context *symmetric* matrices were especially useful).
We pictured quadratic forms in terms of conics and quadrics.
The diagonalization process that we have been studying can be interpreted in *both* contexts. As usual we shall take the case of 3×3 matrices as typical.

Interpretation in Terms of Transformations

Suppose that $\mathbf{A}$ is a 3×3 matrix. Then $\mathbf{A}$ can be interpreted as a transformation of three dimensional space *to itself*. Namely, the transformation $\mathbf{x}$ maps to $\mathbf{X}$ where $\mathbf{X} = \mathbf{A}\mathbf{x}$.
Now suppose that we *change coordinates* by means of a non-singular 3×3 matrix $\mathbf{P}$. Precisely, we choose new coordinates $\mathbf{x}'$ which are related to the old ones $\mathbf{x}$ by the equations

$$\mathbf{x} = \mathbf{P}\mathbf{x}' \quad \text{or} \quad \mathbf{x}' = \mathbf{P}^{-1}\mathbf{x}. \qquad [1]$$

How does the transformation appear in the new coordinates?
Let the new coordinates corresponding to $\mathbf{X}$ be $\mathbf{X}'$ then

$$\mathbf{X}' = \mathbf{P}^{-1}\mathbf{X} = \mathbf{P}^{-1}\mathbf{A}\mathbf{x} = \mathbf{P}^{-1}\mathbf{A}\mathbf{P}\mathbf{x}'$$

In the new coordinates, the transformation is given by the matrix $\mathbf{P}^{-1}\mathbf{A}\mathbf{P}$.

Now remember that diagonalization means finding a matrix **P** such that $\mathbf{P}^{-1}\mathbf{AP}$ is a diagonal matrix; so in terms of transformations it means *choosing a new coordinate ststem in which the transformation is given by a diagonal matrix.* We can interpret this new coordinate system in terms of the eigenvectors of the original matrix **A**. The new coordinate axes lie along the directions of the vectors $\begin{pmatrix}1\\0\\0\end{pmatrix}'$, $\begin{pmatrix}0\\1\\0\end{pmatrix}'$ and $\begin{pmatrix}0\\0\\1\end{pmatrix}'$ (where the primes remind us that we are using *new* coordinates). But, referring back to equations [1], we can express these in *old* coordinates as

$$\mathbf{P}\begin{pmatrix}1\\0\\0\end{pmatrix}', \quad \mathbf{P}\begin{pmatrix}0\\1\\0\end{pmatrix}' \quad \text{and} \quad \mathbf{P}\begin{pmatrix}0\\0\\1\end{pmatrix}'$$

and these vectors are the *columns* of **P**. But when we diagonalize, we choose *eigenvectors of* **A** to be *columns of* **P**. Thus *the new coordinate system has its axes lying along eigenvectors of* **A**. That is to say *the new coordinate axes are lines through the origin preserved by* **A**.

As an example consider the matrix

$$\mathbf{A} = \begin{pmatrix}2 & 1\\3 & 0\end{pmatrix}$$

We saw in Example 15g, No. 4 that $\begin{pmatrix}1\\1\end{pmatrix}$ and $\begin{pmatrix}-1\\3\end{pmatrix}$ are eigenvectors of this matrix with eigenvalues 3 and -1 respectively.

Therefore the matrix $\mathbf{P} = \begin{pmatrix}1 & -1\\1 & 3\end{pmatrix}$ diagonalizes **A** and

$$\mathbf{P}^{-1}\mathbf{AP} = \begin{pmatrix}3 & 0\\0 & -1\end{pmatrix}$$

If we use **P** to change coordinates, then in the new coordinates (x', y') the transformation has the matrix $\begin{pmatrix}3 & 0\\0 & -1\end{pmatrix}$, i.e. $(x', y') \longmapsto (3x', -y')$.

This is very easy to picture: it is a stretch with factor 3 in the x' direction combined with a reflection in the x' axis:

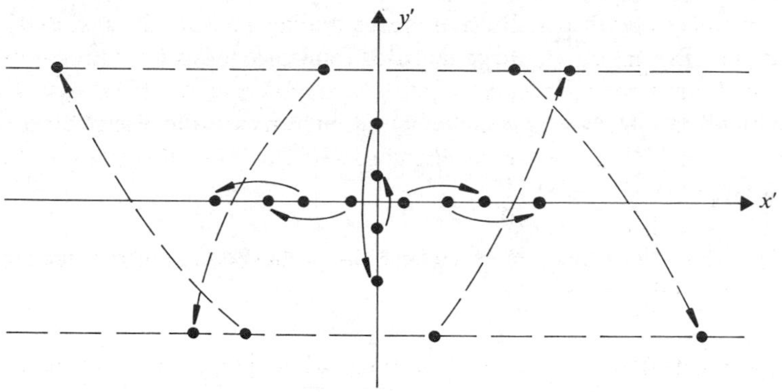

We can use this picture to see what that transformation looks like on the *old* coordinates. The x' and y' axes lie along the directions of the eigenvectors $\begin{pmatrix}1\\1\end{pmatrix}$ and $\begin{pmatrix}-1\\3\end{pmatrix}$ (they are lines through the origin *preserved* by the transformation) and the transformation is a 'twisted' and 'skewed' version of that shown above:

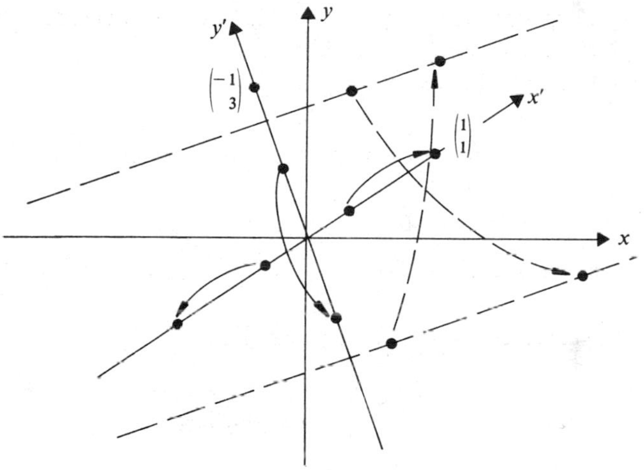

EXAMPLES 15j

1) Use the diagonal form of the matrix $\mathbf{A} = \begin{pmatrix}\frac{5}{4} & -\frac{3}{4}\\ -\frac{3}{4} & \frac{5}{4}\end{pmatrix}$ to sketch the transformation.

Using the standard procedure we find that the eigenvalues of $\mathbf{A}$ are $\lambda = \frac{1}{2}$ and $\lambda = 2$ and that corresponding eigenvectors are $\begin{pmatrix}1\\1\end{pmatrix}$ and $\begin{pmatrix}-1\\1\end{pmatrix}$.

Therefore $\mathbf{P}=\begin{pmatrix}1 & -1\\ 1 & 1\end{pmatrix}$ diagonalizes $\mathbf{A}$ and $\mathbf{P}^{-1}\mathbf{AP}=\begin{pmatrix}\frac{1}{2} & 0\\ 0 & 2\end{pmatrix}$.

If we use $\mathbf{P}$ to change coordinates, then in the new coordinates the transformation is given by $\begin{pmatrix}x'\\ y'\end{pmatrix}\longmapsto\begin{pmatrix}\frac{1}{2} & 0\\ 0 & 2\end{pmatrix}\begin{pmatrix}x'\\ y'\end{pmatrix}=\begin{pmatrix}\frac{x'}{2}\\ 2y'\end{pmatrix}$

which is a *stretch* of the y' axis combined with a *shrink* of the x' axis:

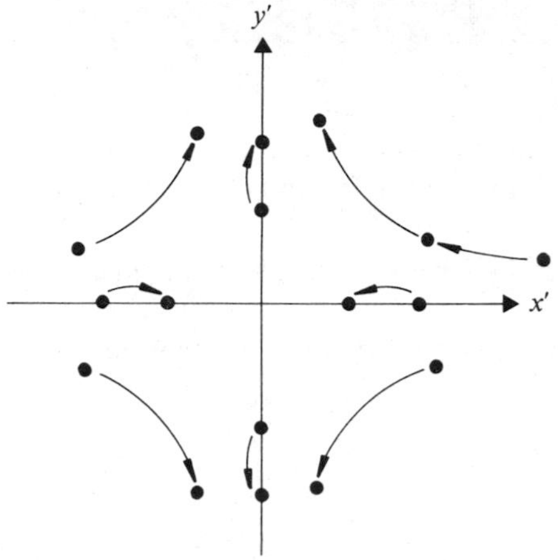

In the *old* coordinates, the new axes point in the directions of the eigenvectors $\begin{pmatrix}1\\ 1\end{pmatrix}$ and $\begin{pmatrix}-1\\ 1\end{pmatrix}$, so the transformation has this picture:

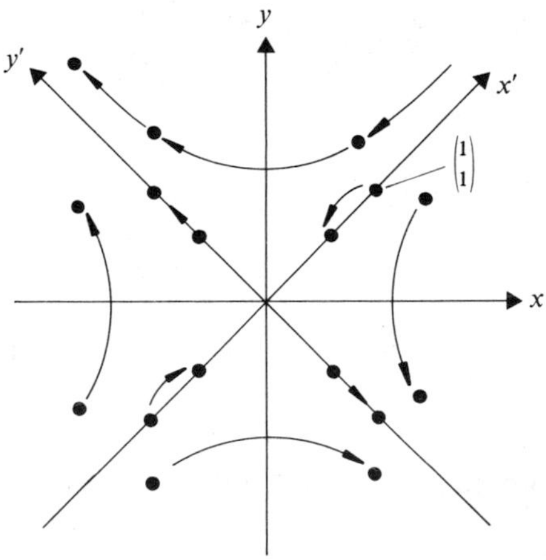

i.e. it is a 'rotated' version of the picture in the new coordinates.

Note that the eigenvectors in this example are perpendicular and hence by *scaling* them suitably we can choose **P** to be the *orthogonal* matrix

$$\begin{pmatrix} \frac{\sqrt{2}}{2} & -\frac{\sqrt{2}}{2} \\ \frac{\sqrt{2}}{2} & \frac{\sqrt{2}}{2} \end{pmatrix}$$

Then the new coordinates are obtained ***precisely*** by rotating the old ones through 45° (as the illustration at the bottom of the previous page suggests).

2) Sketch the transformations of the plane give by the matrices

(a) $\begin{pmatrix} 1 & 1 \\ 0 & 1 \end{pmatrix}$ (b) $\begin{pmatrix} 0 & 1 \\ -1 & 0 \end{pmatrix}$

and explain why neither is diagonalizable.

(a) The transformation is a shear

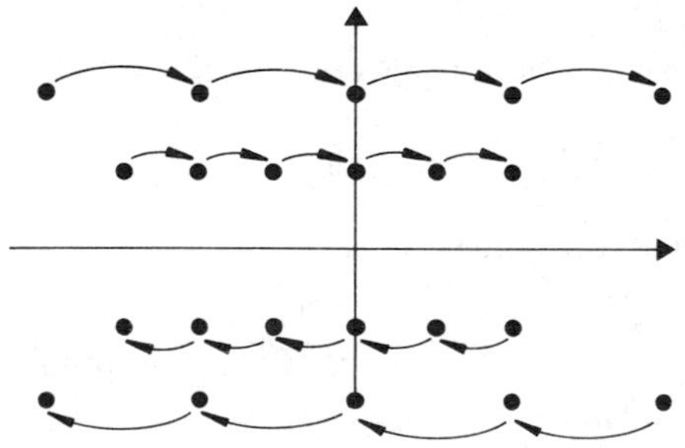

and there is *only one* preserved line through the origin (the x axis) therefore it is not possible to choose new coordinates with axes being lines through the origin preserved by the transformation.

(b) The transformation is a rotation through $\pi/2$ anticlockwise:

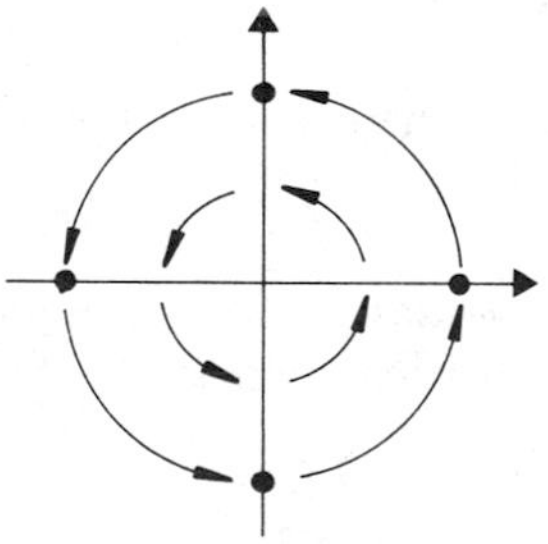

There are *no* lines through the origin preserved by the transformation.

Interpretation in Terms of Quadratic Forms

Suppose that $\mathbf{A}$ is a symmetric 3×3 matrix, then $\mathbf{A}$ determines the quadratic form $\mathbf{x}^T\mathbf{A}\mathbf{x}$, where $\mathbf{x} = \begin{pmatrix} x \\ y \\ z \end{pmatrix}$ i.e. the quadratic form

$$(x, y, z) \begin{pmatrix} a & h & g \\ h & b & f \\ g & f & c \end{pmatrix} \begin{pmatrix} x \\ y \\ z \end{pmatrix} = ax^2 + by^2 + cz^2 + 2fyz + 2gzx + 2hxy$$

where $\mathbf{A} = \begin{pmatrix} a & h & g \\ h & b & f \\ g & f & c \end{pmatrix}$.

Now suppose again that we choose new coordinates $\mathbf{x}'$ related to the old ones $\mathbf{x}$ by

$$\mathbf{x} = \mathbf{P}\mathbf{x}' \quad \text{or} \quad \mathbf{x}' = \mathbf{P}^{-1}\mathbf{x} \qquad [1]$$

How does the quadratic form appear in the new coordinates?
Since $\mathbf{x} = \mathbf{P}\mathbf{x}'$ we have

$$\mathbf{x}^T = \mathbf{x}'^T\mathbf{P}^T \quad \text{(transposing both sides)}$$

Therefore $\mathbf{x}^T\mathbf{A}\mathbf{x} = \mathbf{x}'^T\mathbf{P}^T\mathbf{A}\mathbf{P}\mathbf{x}'$

$$= \mathbf{x}'^T\mathbf{B}\mathbf{x}'$$

where $\mathbf{B} = \mathbf{P}^T\mathbf{A}\mathbf{P}$

In the new coordinates, the quadratic form is given by the matrix $\mathbf{P}^T\mathbf{A}\mathbf{P}$.

Now by the second general result (see page 667) we can always find an orthogonal matrix $\mathbf{P}$ which diagonalizes $\mathbf{A}$; in fact such that $\mathbf{P}^T\mathbf{A}\mathbf{P}$ is the diagonal matrix $\begin{pmatrix} \lambda & 0 & 0 \\ 0 & \mu & 0 \\ 0 & 0 & \nu \end{pmatrix}$ where λ, μ, ν are the eigenvalues of $\mathbf{A}$.

Thus *we can always change coordinates so that in the new coordinates the quadratic form is given by a diagonal matrix.*

Now the diagonal matrix $\begin{pmatrix} \lambda & 0 & 0 \\ 0 & \mu & 0 \\ 0 & 0 & \nu \end{pmatrix}$ gives the quadratic form

$\lambda x^2 + \mu y^2 + \nu z^2$, so, in the new coordinates (x',y',z'), the quadratic form is $\lambda x'^2 + \mu y'^2 + \nu z'^2$. The equation for the corresponding family of conics or quadrics also has a simple form in the new coordinates:

$$\lambda x'^2 + \mu y'^2 + \nu z'^2 = k$$

and this is easy to sketch; and then the original conic is also easy to sketch. In fact, since we are dealing with an *orthogonal* change of coordinates, the conic or quadric is a *rotated* copy of the diagonal version.
The new coordinate directions (the x', y' and z' axes) are called the *principal axes* for the conic or quadric. *Notice that the principal axes lie in the directions of the eigenvectors of the matrix* **A**.

EXAMPLES 15j (continued)

3) Find the principal axes and sketch the conics:
(a) $\frac{5}{4}x^2 - \frac{3}{2}xy + \frac{5}{4}z^2 = 1$
(b) $6xy - x^2 - y^2 = 2$.

(a) The symmetric matrix which gives the quadratic form $\frac{5}{4}x^2 - \frac{3}{2}xy + \frac{5}{4}z^2$ is $\begin{pmatrix} \frac{5}{4} & -\frac{3}{4} \\ -\frac{3}{4} & \frac{5}{4} \end{pmatrix}$ which we studied in Example 1 above. The eigenvectors are $\begin{pmatrix} 1 \\ 1 \end{pmatrix}$ and $\begin{pmatrix} -1 \\ 1 \end{pmatrix}$ so the principal axes lie in these directions, i.e. along the lines $x = y$ and $x = -y$. In new coordinates the conic has the equation $\frac{x'^2}{2} + 2y'^2 = 1$ which is easy to sketch (it is an ellipse), so the original conic is a rotated version:

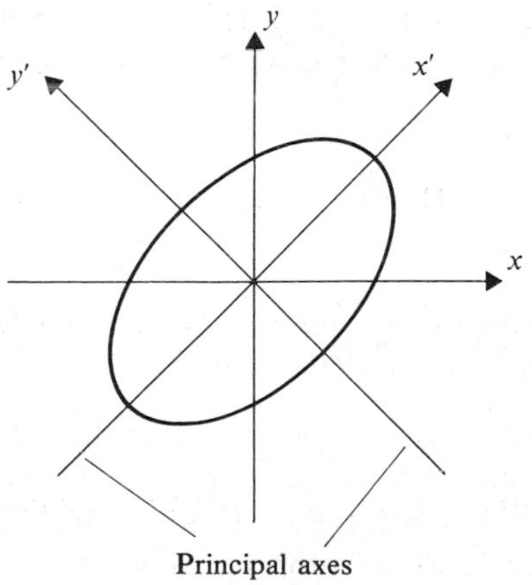

(b) The relevant matrix $\mathbf{A}$ is $\begin{pmatrix} -1 & 3 \\ 3 & -1 \end{pmatrix}$

which has eigenvalues $\lambda = -2$ and $\lambda = 4$. Corresponding eigenvectors are $\begin{pmatrix} 1 \\ 1 \end{pmatrix}$, $\begin{pmatrix} 1 \\ -1 \end{pmatrix}$ so the principal axes are the same as in the previous example. In new coordinates the equation becomes $-2x'^2 + 4y'^2 = 2$ which is a hyperbola. So the original conic is this hyperbola rotated through $\pi/4$:

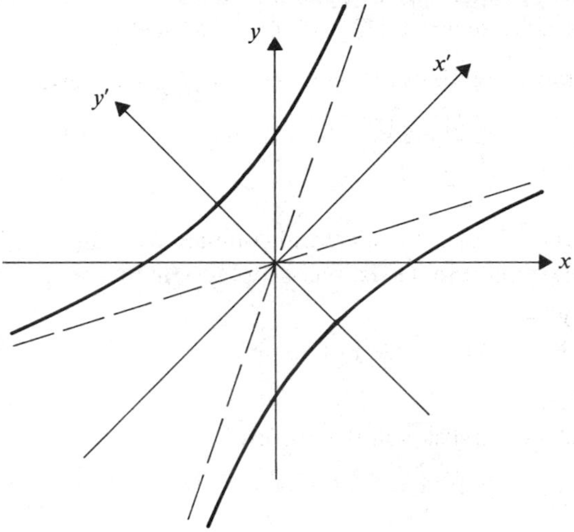

Note that any *scalar multiple* of the matrix $\mathbf{A}$ in either example gives a conic in the *same* family and hence has the *same* principal axes.

4) Find the direction cosines of the principal axes of the quadric

$$\frac{5}{2}x^2 + \frac{y^2}{2} + \frac{z^2}{2} - xy + xz - 3yz = 7$$

The relevant matrix is

$$\begin{pmatrix} \frac{5}{2} & -\frac{1}{2} & \frac{1}{2} \\ -\frac{1}{2} & \frac{1}{2} & -\frac{3}{2} \\ \frac{1}{2} & -\frac{3}{2} & \frac{1}{2} \end{pmatrix} = \frac{1}{2}\begin{pmatrix} 5 & -1 & 1 \\ -1 & 1 & -3 \\ 1 & -3 & 1 \end{pmatrix}$$

But we found unit eigenvectors for the matrix on the right in Exercise 15g, Question 5 and since this is a scalar multiple, these eigenvectors give the directions of the required principal axes. From the solution to Exercise 15g, Question 5, the principal axes lie in the directions of the unit vectors.

$$\begin{pmatrix} 0 \\ \frac{\sqrt{2}}{2} \\ \frac{\sqrt{2}}{2} \end{pmatrix}, \quad \begin{pmatrix} \frac{\sqrt{3}}{3} \\ \frac{\sqrt{3}}{3} \\ -\frac{\sqrt{3}}{3} \end{pmatrix} \quad \text{and} \quad \begin{pmatrix} \frac{\sqrt{6}}{3} \\ -\frac{\sqrt{6}}{6} \\ \frac{\sqrt{6}}{6} \end{pmatrix}$$

The coordinates of these vectors are the required direction cosines.

EXERCISE 15j

Relevant matrices can be found in Examples and Exercises 15h and 15i for all questions in this exercise, except Questions 2 and 4(a).

1) Sketch the transformations of the plane given by the matrices:

(a) $\begin{pmatrix} -5 & 3 \\ 6 & -2 \end{pmatrix}$ (b) $\begin{pmatrix} 4 & 0 \\ 2 & -1 \end{pmatrix}$.

2) Find principal axes for the conics, and compare with the sketches in Example 15d, No. 3 and Exercise 15d, Questions 4 and 5:

(a) $2x^2 + xy + y^2 = 3$

(b) $x^2 - 2xy = 1$

(c) $x^2 + 2xy - \quad 2$.

3) Find principal axes and sketch the conics:

(a) $\dfrac{x^2}{2} + \dfrac{y^2}{2} + 2xy = 3$

(b) $2x^2 + 2y^2 + 4xy = 7$.

4) Find the direction cosines of the principal axes of the quadrics:

(a) $\dfrac{x^2}{4} + \dfrac{y^2}{4} + \dfrac{z^2}{4} + xy = 10$

(b) $6x^2 - 2y^2 - 4z^2 - 16xy + 8xz + 24yz = 93$.

5) Write down the equation of the quadric

$$7x^2 - 8y^2 - 8z^2 + 8xy - 8xz - 2yz = 105$$

referred to principal axes as new coordinate axes.

PROOFS OF THE ALGEBRAIC LAWS

We now give the proofs of the three algebraic laws which were stated without proof on pages 627 and 628.

Proof of the Associative Law of Matrix Multiplication

The (i, j)th entry in **AB** is

$$\sum_{q=1}^{m} a_{iq}b_{qj}$$

and therefore the (i, l)th entry in $(\mathbf{AB})\mathbf{C}$ is

$$\sum_{j=1}^{p}\left(\sum_{q=1}^{m} a_{iq}b_{qj}\right)c_{jl} \qquad [1]$$

Similarly the (i, l)th entry in $\mathbf{A}(\mathbf{BC})$ is

$$\sum_{q=1}^{m} a_{iq}\left(\sum_{j=1}^{p} b_{qj}c_{jl}\right) \qquad [2]$$

But each of [1] and [2] is equal to the double sum

$$\sum_{q=1, j=1}^{m, n} a_{iq}b_{qj}c_{jl}$$

and hence they are equal.
Thus a typical entry in $(\mathbf{AB})\mathbf{C}$ is the same as the corresponding entry in $\mathbf{A}(\mathbf{BC})$. Hence $(\mathbf{AB})\mathbf{C} = \mathbf{A}(\mathbf{BC})$.

Proof of the Distributive Laws

We shall prove the first distributive law. The proof of the second law is similar and will be omitted.
The (q, j)th entry in $\mathbf{B} + \mathbf{C}$ is $b_{qj} + c_{qj}$, therefore the (i, j)th entry in $\mathbf{A}(\mathbf{B} + \mathbf{C})$ is

$$\sum_{q=1}^{m} a_{iq}(b_{qj} + c_{qj}) \qquad [1]$$

But the (q, j)th entry in $\mathbf{AB} + \mathbf{BC}$ is

$$\sum_{q=1}^{m} a_{iq}b_{qj} + \sum_{q=1}^{m} a_{iq}c_{qj} \qquad [2]$$

and [1] and [2] are each equal to

$$\sum_{q=1}^{m} (a_{iq}b_{qj} + a_{iq}c_{qj})$$

Therefore $\mathbf{A}(\mathbf{B} + \mathbf{C}) = \mathbf{AB} + \mathbf{AC}$.

Proof of the Unit Law

Write δ_{ij} for the (i,j)th entry of the unit $n \times n$ matrix $\mathbf{I}_n$. Then $\delta_{ij} = 0$ unless $i=j$ when $\delta_{ij} = 1$.
The (i,j)th entry of $\mathbf{I}_n\mathbf{A}$ is

$$\sum_{q=1}^{n} \delta_{iq}a_{qj} \qquad [1]$$

But there is only one non-zero term in the sum [1], namely when $q = i$.
Therefore [1] is equal to

$$\delta_{ii}a_{ij} = a_{ij}$$

which is the (i,j)th entry in $\mathbf{A}$.
Thus $\mathbf{I}_n\mathbf{A} = \mathbf{A}$.
The proof that $\mathbf{AI}_m = \mathbf{A}$ is similar.

MISCELLANEOUS EXERCISE 15

1) (a) Given that the matrix

$$\mathbf{A} = \begin{pmatrix} 1 & 0 & 0 \\ -1 & -2 & -1 \\ 2 & 3 & 2 \end{pmatrix},$$

show that

$$\mathbf{A}^3 - \mathbf{A} = \mathbf{A}^2 - \mathbf{I}.$$

Use the method of induction to prove that

$$\mathbf{A}^n - \mathbf{A}^{n-2} = \mathbf{A}^2 - \mathbf{I}$$

for every integer $n \geqslant 3$. Hence find the matrix $\mathbf{A}^{100}$.

(b) Obtain a matrix $\mathbf{M}$ such that $\mathbf{D}$ is a diagonal matrix and

$$\mathbf{B} + \mathbf{B}^T = \mathbf{MDM}^{-1},$$

where

$$\mathbf{B} = \begin{pmatrix} 0 & 1 \\ 0 & 0 \end{pmatrix}.$$ (U of L)

2) Prove that, if $\mathbf{A}$ is a non-singular square matrix, then $(\mathbf{A}^T)^{-1} = (\mathbf{A}^{-1})^T$.
A square matrix $\mathbf{A}$ with real elements is said to be orthogonal if $\mathbf{A}^T = \mathbf{A}^{-1}$. Show that the transpose and the inverse of an orthogonal matrix also are orthogonal matrices.

The square matrix $\mathbf{B}$ is skew-symmetric (i.e. $\mathbf{B} = -\mathbf{B}^T$) and the matrix $\mathbf{I}+\mathbf{B}$, where $\mathbf{I}$ is the unit matrix, is non-singular. Using the identity $(\mathbf{I}-\mathbf{B})(\mathbf{I}+\mathbf{B}) \equiv (\mathbf{I}+\mathbf{B})(\mathbf{I}-\mathbf{B})$, or otherwise, show that the matrix $(\mathbf{I}-\mathbf{B})(\mathbf{I}+\mathbf{B})^{-1}$ is orthogonal. (U of L)

3) The matrices $\mathbf{A}$ and $\mathbf{B}$ are symmetric. Prove that

(a) if $\mathbf{AB} = \mathbf{BA}$, then $\mathbf{AB}$ is symmetric,
(b) if $\mathbf{AB}$ is symmetric, then $\mathbf{AB} = \mathbf{BA}$.

Find the set of all matrices which commute, under the operation of matrix multiplication, with the matrix

$$\begin{pmatrix} 1 & 0 & 1 \\ 0 & 1 & 0 \\ 1 & 0 & 1 \end{pmatrix}.$$

Investigate which of the group axions are satisfied by this set under the operation of matrix multiplication. (U of L)

4) Find the eigenvalues of the matrix $\mathbf{A}$, where

$$\mathbf{A} = \begin{pmatrix} 2 & 0 & -2 \\ 0 & 4 & 0 \\ -2 & 0 & 5 \end{pmatrix}$$

and obtain direction cosines for each corresponding eigenvector. Verify that these directions are mutually perpendicular.
Hence write down a matrix $\mathbf{P}$ such that $\mathbf{P}^{-1}\mathbf{AP}$ is a diagonal matrix $\mathbf{D}$. Obtain $\mathbf{P}^{-1}$ and calculate the elements of the leading diagonal of $\mathbf{D}$. (U of L)

5) Prove that the eigenvalues, $\lambda_1, \lambda_2, \lambda_3$, of the matrix

$$\mathbf{A} = \begin{pmatrix} 3 & -6 & -4 \\ -6 & 4 & 2 \\ -4 & 2 & -1 \end{pmatrix}$$

are $-1, -4$ and 11. Find the corresponding unit eigenvectors.
If $\mathbf{B}$ is the matrix whose columns from left to right are the unit eigenvectors of $\mathbf{A}$ corresponding to $\lambda_1, \lambda_2, \lambda_3$ respectively, show that $\mathbf{AB} = \mathbf{BL}$, where $\mathbf{L}$ is the 3×3 matrix that has $\lambda_1, \lambda_2, \lambda_3$ down the leading diagonal and zero elsewhere. (O)

6) Find the eigenvalues λ_1, λ_2 of the matrix

$$\mathbf{A} = \begin{pmatrix} 1 & 2 & 2 \\ 0 & 2 & 1 \\ -1 & 2 & 2 \end{pmatrix}.$$

Find the corresponding eigenvectors, that is, vectors $\mathbf{v}_1, \mathbf{v}_2$ such that $\mathbf{A}\mathbf{v}_1 = \lambda_1\mathbf{v}_1$ and $\mathbf{A}\mathbf{v}_2 = \lambda_2\mathbf{v}_2$.

Express the vector $\begin{pmatrix} 6 \\ 4 \\ -2 \end{pmatrix}$ in terms of $\mathbf{v}_1$ and $\mathbf{v}_2$ and hence find a vector $\mathbf{v}$

such that $\mathbf{A}\mathbf{v} = \begin{pmatrix} 6 \\ 4 \\ -2 \end{pmatrix}$. (O)

7) Show that the eigenvalues of the matrix

$$\mathbf{A} = \begin{pmatrix} -2 & -3 & -1 \\ 1 & 2 & 1 \\ 3 & 3 & 2 \end{pmatrix}.$$

are $1, -1, 2$ and find the associated eigenvectors. Verify that there is a matrix $\mathbf{P}$, formed from the eigenvectors, such that $\mathbf{P}^{-1}\mathbf{A}\mathbf{P}$ is a diagonal matrix in which the elements in the leading diagonal are the eigenvalues of $\mathbf{A}$. (U of L)

8) Find the eigenvalues of the symmetric matrix $\mathbf{A}$ given by

$$\begin{pmatrix} x & y & z \end{pmatrix} \mathbf{A} \begin{pmatrix} x \\ y \\ z \end{pmatrix} = x^2 + 4y^2 + z^2 + 4xz.$$

Using the eigenvectors of $\mathbf{A}$, or otherwise, obtain an orthogonal matrix $\mathbf{P}$ such that $\mathbf{P}^{-1}\mathbf{A}\mathbf{P}$ is a diagonal matrix. Verify that the diagonal elements of the matrix $\mathbf{P}^{-1}\mathbf{A}\mathbf{P}$ are the eigenvalues of $\mathbf{A}$.
Show that the transformation

$$\begin{pmatrix} X \\ Y \\ Z \end{pmatrix} = \mathbf{P}^{-1} \begin{pmatrix} x \\ y \\ z \end{pmatrix}$$

reduces

$$x^2 + 4y^2 + z^2 + 4xz$$

to the form

$$aX^2 + bY^2 + cZ^2.$$

(U of L)

9) Find the eigenvalues of the matrix $\begin{pmatrix} 3 & 2 \\ 2 & 6 \end{pmatrix}$.

Find the corresponding unit eigenvectors.
Find the equation of the conic $3x^2 + 4xy + 6y^2 = 14$ referred to coordinate axes in the direction of these eigenvectors. (O)

10) Find the direction cosines for a set of principal axes for the quadric

$$14x^2 + 17y^2 + 14z^2 + 4yz - 8xz + 4xy = 9.$$

How much choice is there in such a set of principal axes?
Write down the equation of the quadric referred to a set of principal axes as new coordinate axes.

In Questions 11 to 17 of this miscellaneous exercise we introduce *complex* matrices. These are very much like the real matrices that we have been studying, but there is one new operation, *complex conjugation* defined in Question 11.

11) Define the operation of complex conjugation on matrices with complex entries as follows:
the (i,j)th entry of $\overline{\mathbf{A}}$ is $\bar{a}_{ij}$ (where a_{ij} is the (i,j)th entry of $\mathbf{A}$).
Prove the following properties of this operation:

(a) $\overline{\mathbf{AB}} = \overline{\mathbf{A}}\,\overline{\mathbf{B}}$, where $\mathbf{A}$ is $n \times m$ and $\mathbf{B}$ is $m \times p$,

(b) $\overline{\mathbf{A} + \mathbf{B}} = \overline{\mathbf{A}} + \overline{\mathbf{B}}$, where $\mathbf{A}$ and $\mathbf{B}$ are both $m \times n$,

(c) $\overline{\mathbf{A}} = \mathbf{A}$ if and only if each entry of $\mathbf{A}$ is real,

(d) $\overline{\lambda \mathbf{A}} = \bar{\lambda}\,\overline{\mathbf{A}}$ where $\mathbf{A}$ is $n \times m$ and λ is a scalar (complex number),

(e) $\overline{\mathbf{A}^{\mathrm{T}}} = \overline{\mathbf{A}}^{\mathrm{T}}$, where $\mathbf{A}$ is $n \times m$,

(f) $|\overline{\mathbf{A}}| = \overline{|\mathbf{A}|}$, where $\mathbf{A}$ is $n \times n$.

12) Define the operation $*$ on complex matrices by $\mathbf{A}^* = \overline{\mathbf{A}}^{\mathrm{T}}$. Prove:

(a) $(\mathbf{AB})^* = \mathbf{B}^*\mathbf{A}^*$, where $\mathbf{A}$ is $n \times m$ and $\mathbf{B}$ is $m \times p$,

(b) $(\mathbf{A} + \mathbf{B})^* = \mathbf{A}^* + \mathbf{B}^*$, where $\mathbf{A}$ and $\mathbf{B}$ are both $m \times n$,

(c) $(\mathbf{A}^*)^{-1} = (\mathbf{A}^{-1})^*$, where $\mathbf{A}$ is $n \times n$ and non-singular,

(d) $|\mathbf{A}^*| = \overline{|\mathbf{A}|}$, where $\mathbf{A}$ is $n \times n$.

13) An $n \times n$ complex matrix $\mathbf{H}$ is said to be *Hermitian* if $\mathbf{H}^* = \mathbf{H}$. Prove:

(a) the diagonal entries of a Hermitian matrix are all real,

(b) a matrix all of whose entries are real is Hermitian if and only if it is symmetric,

(c) if $\mathbf{P}$ is any $n \times m$ matrix and $\mathbf{H}$ is a Hermitian $n \times n$ matrix, then $\mathbf{P}^*\mathbf{HP}$ is Hermitian,

(d) if $\mathbf{A}$ is any complex matrix then $\mathbf{A}^*\mathbf{A}$ and $\mathbf{AA}^*$ are both Hermitian matrices,

(e) if **H** and **G** are Hermitian matrices such that **HG** = **HG** then **HG** is Hermitian.

14) An $n \times n$ non-singular matrix **U** is said to be unitary if $\mathbf{U}^*\mathbf{U} = \mathbf{I}$ (or equivalently if $\mathbf{U}^* = \mathbf{U}^{-1}$).
Prove that the set of $n \times n$ unitary matrices forms a group under matrix multiplication.

15) Find (complex) eigenvalues and eigenvectors for the matrix $\mathbf{A} = \begin{pmatrix} 0 & 1 \\ -1 & 0 \end{pmatrix}$

and hence find a (complex) matrix **P** which diagonalizes **A**.

16) Find a unitary matrix **U** which diagonalizes the Hermitian matrix

$\mathbf{H} = \begin{pmatrix} 0 & i \\ -i & 0 \end{pmatrix}$ and verify that $\mathbf{U}^*\mathbf{H}\mathbf{U}$ is diagonal by direct multiplication.

17) A 3×3 matrix **H**, with complex elements h_{ij}, is said to be Hermitian when $\mathbf{H} = \mathbf{H}^*$. Here $\mathbf{H}^*$ denotes the transposed conjugate of **H**, so that $h_{ji}^* = \bar{h}_{ij}$, where $\bar{h}_{ij}$ denotes the conjugate of h_{ij}. Show that the set of all 3×3 Hermitian matrices forms a group under the operation of matrix addition. Show also that the set does not form a group under the operation of matrix multiplication.

Given that **z** denotes the column vector $\begin{pmatrix} z_1 \\ z_2 \\ z_3 \end{pmatrix}$, $\mathbf{z}^*$ denotes the transposed conjugate of **z**, and **H** is a 3×3 Hermitian matrix, show that $\mathbf{z}^*\mathbf{z}$ is real and that the Hermitian form $\mathbf{z}^*\mathbf{H}\mathbf{z}$ is also real.
Deduce that the eigenvalues of a 3×3 Hermitian matrix are real, and hence that the eigenvalues of a 3×3 symmetric matrix are real. (U of L)

18) Let **A** be an $n \times n$ matrix (real or complex) and μ a non-zero scalar. Prove that the eigenvectors of **A** are the same as those of $\mu\mathbf{A}$. How are the *eigenvalues* related?

19) The $n \times n$ matrix **B** is *similar* to the $n \times n$ matrix **A** if there is an $n \times n$ non-singular matrix **P** such that $\mathbf{B} = \mathbf{P}^{-1}\mathbf{A}\mathbf{P}$. Prove that

(a) if **A** is similar to **B** then **B** is similar to **A**,

(b) if **A** is similar to **B** and **B** is similar to **C**, then **A** is similar to **C**.

20) Prove that similar matrices have the same characteristic equation and hence the same eigenvalues.
How are their eigen*vectors* related?

ANSWERS

Exercise 1a – p. 2

1) $\begin{pmatrix}2\\9\end{pmatrix}$ 2) $\begin{pmatrix}0\\8\end{pmatrix}$ 3) $\begin{pmatrix}1\\7\end{pmatrix}$

4) $\begin{pmatrix}1\\6\end{pmatrix}$ 5) $\begin{pmatrix}\frac{1}{2}\\1\end{pmatrix}$ 6) $\begin{pmatrix}0\\2\end{pmatrix}$

7) $\begin{pmatrix}\frac{1}{3}\\2\end{pmatrix}$

Exercise 1b – p. 5

1) a) 11 b) – 22 c) – 3

2) a) $\begin{pmatrix}8\\2\end{pmatrix}$ b) $\begin{pmatrix}-5\\-9\end{pmatrix}$ c) $\begin{pmatrix}-4\\6\end{pmatrix}$

3) $\begin{pmatrix}1\\0\end{pmatrix}, \begin{pmatrix}2\\0\end{pmatrix}, \begin{pmatrix}2\\-3\end{pmatrix}$; reflection in Ox

4) $\begin{pmatrix}-1\\0\end{pmatrix}, \begin{pmatrix}-2\\0\end{pmatrix}, \begin{pmatrix}-2\\3\end{pmatrix}$;
reflection in Oy

Exercise 1c – p. 15

1) a) $\begin{pmatrix}-1 & 0\\0 & 1\end{pmatrix}$ b) $\begin{pmatrix}\frac{1}{2}\sqrt{2} & -\frac{1}{2}\sqrt{2}\\\frac{1}{2}\sqrt{2} & \frac{1}{2}\sqrt{2}\end{pmatrix}$

c) $\begin{pmatrix}1 & 0\\0 & 2\end{pmatrix}$ d) $\begin{pmatrix}-\frac{3}{5} & \frac{4}{5}\\\frac{4}{5} & \frac{3}{5}\end{pmatrix}$

e) $\begin{pmatrix}1 & 0\\\frac{1}{3}\sqrt{3} & 1\end{pmatrix}$ f) $\begin{pmatrix}3 & 0\\0 & 1\end{pmatrix}$

g) $\begin{pmatrix}3 & 0\\0 & 3\end{pmatrix}$ h) $\begin{pmatrix}2 & 0\\0 & -1\end{pmatrix}$

2) a) reflection in y axis
b) rotation of 180°
c) plane maps to the line $y = x$
d) no change
e) enlargement and shear ∥ to Ox
f) plane maps to x axis
g) plane maps to the origin
h) rotation of $90° - \theta$
i) shear ∥ to negative y axis

3) a) A′(2, 2) B′(4, 2) C′(4, 8)
b) A′(– 3, – 3) B′(– 6, – 3) C′(– 6, – 12)
c) A′(– 1, 3) B′(– 2, 3) C′(– 2, 12)
d) A′(1, 1) B′(2, 2) C′(2, 2)
e) A′(3, 7) B′(4, 10) C′(10, 22)
f) A′(2, – 3) B′(2, – 3) C′(8, – 12)

4) a) $\begin{pmatrix}-\frac{7}{25} & \frac{24}{25}\\\frac{24}{25} & \frac{7}{25}\end{pmatrix}$ b) $\begin{pmatrix}-\frac{3}{5} & -\frac{4}{5}\\-\frac{4}{5} & \frac{3}{5}\end{pmatrix}$

5) a) $\begin{pmatrix}\frac{1}{2} & -\sqrt{3}/2\\\sqrt{3}/2 & \frac{1}{2}\end{pmatrix}$

b) $\begin{pmatrix}\sqrt{2}/2 & \sqrt{2}/2\\-\sqrt{2}/2 & \sqrt{2}/2\end{pmatrix}$

6) a) $y = 2x$ b) $y = -2$
c) $x + y - 2 = 0$

7) a) $y = x + c,\ y + 5x = c$
b) no real lines
c) $x + y = c,\ 3y = x + c$

Exercise 1d – p. 19

2) **A** and **E**, **b** and **h**, **C** and **D**, **F** and **G**

3) $\begin{pmatrix}-2 & 6 & 10\\14 & -8 & 4\end{pmatrix}$, $\begin{pmatrix}-7 & 1 & 0\end{pmatrix}$,

$\begin{pmatrix} -4 & 12 \\ 2 & 2 \end{pmatrix}$, does not exist,

does not exist, $\begin{pmatrix} -2 & 3 \\ -2 & 12 \\ 7 & -6 \end{pmatrix}$

4) a) $\begin{pmatrix} 4 & 24 & 0 \\ 8 & 4 & 20 \end{pmatrix}$

b) $\begin{pmatrix} 6 & 0 & -20 \\ -24 & 18 & 2 \end{pmatrix}$

c) $\begin{pmatrix} -25 & 45 \\ 0 & -15 \end{pmatrix}$ d) $\begin{pmatrix} 0 & \lambda & 3\lambda \end{pmatrix}$

5) 1, 3, 7, 0, −1, 2, −5, 0

6) a) $\begin{pmatrix} -7 & 12 & 30 \\ 40 & -25 & 7 \end{pmatrix}$

b) $\begin{pmatrix} -11 & 9 \\ 14 & 1 \\ 21 & 32 \end{pmatrix}$ c) $\begin{pmatrix} 6 & -6 \\ 2 & 8 \end{pmatrix}$

Exercise 1e – p. 21

1) (−1) 2) (−11) 3) (13)

4) $(2p^2 - q + 3r^2)$ 5) $\begin{pmatrix} -9 \\ 3 \\ -12 \end{pmatrix}$

6) $\begin{pmatrix} 27 \\ -1 \end{pmatrix}$ 7) $\begin{pmatrix} 15 \\ -18 \\ 6 \end{pmatrix}$ 8) $\begin{pmatrix} -10 \\ 15 \\ 6 \\ -32 \end{pmatrix}$

9) $\begin{pmatrix} -8 \\ -2 \\ 1 \\ 0 \end{pmatrix}$ 10) $\begin{pmatrix} 1 \\ 0 \end{pmatrix}$

11) $\begin{pmatrix} 3x^2+6xy \\ 6xy+2y^2 \end{pmatrix}$ 12) $\begin{pmatrix} 0 \\ \sin 2\theta - 2 \end{pmatrix}$

13) $\begin{pmatrix} 0 \\ -3t^2 \\ 3t^2 \end{pmatrix}$ 14) $\begin{pmatrix} i^2+j^2 \\ i+j \end{pmatrix}$

Exercise 1f – p. 25

1) $\mathbf{AB} = \begin{pmatrix} -4 & 3 \\ 0 & -6 \end{pmatrix}$, $\mathbf{BA} = \begin{pmatrix} -2 & 1 \\ -8 & -8 \end{pmatrix}$

2) $\mathbf{AB} = \begin{pmatrix} 20 & 0 \\ 22 & -6 \end{pmatrix}$, $\mathbf{BA} = \begin{pmatrix} -6 & -14 \\ 0 & 20 \end{pmatrix}$

3) $\mathbf{AB} = \begin{pmatrix} -18 & 8 & -25 \\ 24 & -7 & 26 \end{pmatrix}$

4) $\mathbf{AB} = \begin{pmatrix} 13 & -22 \\ -8 & -2 \\ 4 & -13 \end{pmatrix}$

5) $\mathbf{AB} = \begin{pmatrix} -10 & 5 \\ 26 & -13 \\ 56 & -28 \end{pmatrix}$

6) $\mathbf{AB} = \begin{pmatrix} -3 & 1 & 0 \\ 2 & 0 & 4 \\ 6 & -5 & 3 \end{pmatrix}$,

$\mathbf{BA} = \begin{pmatrix} 1 & -2 & 0 \\ 1 & 0 & 8 \\ 3 & -1 & -1 \end{pmatrix}$

7) $\mathbf{AB} = \begin{pmatrix} 10 & -12 & 5 & 2 \\ -3 & 4 & -2 & 0 \end{pmatrix}$

8) $\mathbf{AB} = \begin{pmatrix} -2 & 3 & 13 \\ 1 & 4 & 10 \\ -2 & 0 & 4 \end{pmatrix}$, $\mathbf{BA} = \begin{pmatrix} 4 & 5 \\ -3 & 2 \end{pmatrix}$

9) $\mathbf{AB} = \begin{pmatrix} 3 & 1 & 3 \\ -2 & -1 & 2 \\ 5 & 6 & 1 \end{pmatrix}$,

$\mathbf{BA} = \begin{pmatrix} 6 & -1 & 11 \\ 0 & -8 & 16 \\ 0 & -2 & 5 \end{pmatrix}$

10) $\mathbf{AB} = \mathbf{BA} = \begin{pmatrix} 1 & 2 & 3 \\ 4 & 5 & 6 \\ 7 & 8 & 9 \end{pmatrix}$

11) $\mathbf{AB} = \begin{pmatrix} 0 & 0 \\ 0 & 0 \end{pmatrix}$,

$\mathbf{BA} = \begin{pmatrix} 50 & 20 \\ -125 & -50 \end{pmatrix}$

12) $\mathbf{AB} = \begin{pmatrix} 6 & 4 \\ 21 & 14 \end{pmatrix}$

$\mathbf{BA} = (20)$

Exercise 1g – p. 30

1) $\begin{pmatrix} 0 & 1 \\ 1 & 0 \end{pmatrix}$ 2) $\begin{pmatrix} 0 & -1 \\ -1 & 0 \end{pmatrix}$

3) $\begin{pmatrix} 2 & 0 \\ 0 & 6 \end{pmatrix}$ 4) $\begin{pmatrix} -1 & 0 \\ 0 & 1 \end{pmatrix}$

5) $\begin{pmatrix} 2 & 0 \\ 0 & -2 \end{pmatrix}$ 6) $\begin{pmatrix} 0 & -2 \\ 2 & 0 \end{pmatrix}$

7) $\begin{pmatrix} 0 & 2 \\ 2 & 0 \end{pmatrix}$

Exercise 1h – p. 34

1) $\mathbf{A}^2 = \begin{pmatrix} 1 & 0 \\ 0 & 1 \end{pmatrix}$, $\mathbf{A}^3 = \begin{pmatrix} 1 & 0 \\ 2 & -1 \end{pmatrix}$

2) $\mathbf{B}^2 = \begin{pmatrix} 2 & 0 & -1 \\ -1 & 1 & 1 \\ -1 & 0 & 1 \end{pmatrix}$,

$\mathbf{B}^3 = \begin{pmatrix} 3 & 0 & -2 \\ -2 & 1 & 2 \\ -2 & 0 & 1 \end{pmatrix}$

4) $\mathbf{C} = \begin{pmatrix} 0 & -1 \\ 1 & 0 \end{pmatrix}$

5) $\mathbf{A}^2 = \begin{pmatrix} -1 & 0 \\ 0 & -1 \end{pmatrix}$, $\mathbf{A}^4 = \begin{pmatrix} 1 & 0 \\ 0 & 1 \end{pmatrix}$

6) $\begin{pmatrix} \cos^2\theta & 0 & 0 \\ 0 & \sin^2\theta & 0 \\ 0 & 0 & \cos^2\theta \end{pmatrix}$

7) If $\mathbf{A}$ is $m \times n$ then $\mathbf{B}$ is $n \times m$. They are square.

8) $\mathbf{M}$ maps all points to the line $X = Y$.
$\mathbf{M}^2$ maps all points to the line $X = Y$.

9) $\mathbf{AB}$ maps all points to the origin.
$\mathbf{BA}$ maps all points to the line $X + Y = 0$.

10) $\begin{pmatrix} 6 & -2 \\ -10 & 6 \end{pmatrix}$

11) $\begin{pmatrix} \pm 3 & 0 \\ 0 & \pm 3 \end{pmatrix}$ or $\begin{pmatrix} a & b \\ \dfrac{9-a^2}{b} & -a \end{pmatrix}$

for all real values of a and b.

Exercise 1i – p. 39

1) 13 2) -5 3) 0 4) -6

5) A parallelogram of area 13 sq. units, a parallelogram of area 5 sq. units (turned over), a line (area destroyed), a parallelogram of area 6 sq. units (turned over).

6) a) $\begin{pmatrix} -1 & 19 \\ -5 & -3 \end{pmatrix}$ b) 7 c) 14
d) 98; yes

7) 72, 72

8) a) true b) true c) true
d) false.

11) a) 1 b) $xy(y-x)$
c) $a^2 - b^2$ d) $a^2 - ab - b^2$
e) $\ln 2 \ln \frac{6}{25}$ f) $\cos\theta$

13) a) $\frac{7}{11}, \frac{5}{11}$ b) $\frac{17}{29}, \frac{9}{29}$ c) $\frac{5}{26}, \frac{29}{26}$

Exercise 1j – p. 43

1) a) $\begin{vmatrix} 6 & -12 & -8 \\ -5 & 11 & 7 \\ -8 & 18 & 12 \end{vmatrix}$

b) $\begin{vmatrix} -2 & -29 & -7 \\ 0 & -12 & 0 \\ -2 & -11 & -1 \end{vmatrix}$

2) a) 25 b) 14 c) -51
3) a) 16 b) 12
4) a) yes b) no c) yes
5) a) $(\cos\theta - \sin\theta)$
$\times(\cos\theta + \sin\theta - \cos\theta\sin\theta - 1)$
b) $(a-b)(b-c)(c-a)abc$
$= abc[bc(c-b) + ac(c-a)$
$+ ab(b-a)]$
c) $1 - \sin 2\theta$ d) 0

Exercise 1k – p. 51

1) 384 2) -320 3) 505×10^3
4) 1190 5) $x^2(1-x)^2(1+2x)$
6) $-4x$
7) $\cos\theta\sin\theta(\cos\theta - \sin\theta)(1-\sin\theta)$
$\times(1-\cos\theta)$
8) $2(a+1)(a^2-a+1)$ 9) -2
10) -3

Multiple Choice Exercise 1 – p. 53

1) c 2) c 3) a 4) d
5) a 6) e 7) c 8) c
9) c 10) b 11) a, b 12) c
13) a 14) a, b, c 15) b
16) *B* 17) *C* 18) *A* 19) *D*
20) *D* 21) *E* 22) *A* 23) *D*
24) a 25) b, c 26) *I* 27) *F*
28) *F* 29) *T* 30) *T* 31) *F*
32) *T* 33) *F*

Miscellaneous Exercise 1 – p. 58

1) 54 2) $p=2,\ q=-1$
4) $y=x,\ y=6x$
6) $\begin{pmatrix}0 & 1\\ 1 & 0\end{pmatrix}$, $\theta = 0,\ 180°,\ 360°$
7) a) $\begin{pmatrix}-\frac{1}{2} & \sqrt{3}/2\\ \sqrt{3}/2 & \frac{1}{2}\end{pmatrix}$ b) $\begin{pmatrix}0 & -1\\ 1 & 0\end{pmatrix}$
c) $\begin{pmatrix}-\frac{1}{2} & -\sqrt{3}/2\\ -\sqrt{3}/2 & \frac{1}{2}\end{pmatrix}$
rotation through 30° about O.

9) $-(a-b)(b-c)(c-a)$
10) $2\begin{vmatrix}1 & 0 & z\\ z & 1 & 0\\ 0 & z & 1\end{vmatrix}$
11) $(\sin 4\theta - \sin\theta)(\sin\theta - \frac{1}{2})(\sin 4\theta - \frac{1}{2})$,
$\theta = \frac{2n\pi}{3},\ (2n+1)\frac{\pi}{5},\ [6n + (-1)^n]\frac{\pi}{6},$
$[6n + (-1)^n]\frac{\pi}{24}$
14) (1, 3), (2, 1), No
15) $\mathbf{R}_\alpha\mathbf{S}\mathbf{R}_{-\alpha} = \begin{pmatrix}\cos 2\alpha & \sin 2\alpha\\ \sin 2\alpha & -\cos 2\alpha\end{pmatrix}$
16) $\lambda = 1$ or 5, $x+y=0,\ 3y-x=0$
17) $\mathbf{u} = \begin{pmatrix}a\\ c\end{pmatrix},\ \mathbf{v} = \begin{pmatrix}b\\ d\end{pmatrix}$
18) False 20) $-9xy^2$ 21) $4n(n^2-1)$
23) $p = y/b,\ q = x - ay/b$ 24) -9
25) $\begin{pmatrix}a & b\\ 4b & 3b+a\end{pmatrix}$, $\mathbf{P} = \begin{pmatrix}0 & 1\\ 4 & 3\end{pmatrix}$
27) $3y' = 8x' - 5$

Exercise 2a – p. 67

1) a) $\mathbf{a}\times\mathbf{b}$ b) 0 c) $2\mathbf{a}\times\mathbf{b}$
d) $\mathbf{a}\times\mathbf{c}.\mathbf{b}$ e) 0 f) 0
2) a) $-3\mathbf{j} - 3\mathbf{k}$ b) $-3\mathbf{j} - 3\mathbf{k}$
3) $(3\mathbf{i} - \mathbf{j} + \mathbf{k})/\sqrt{11};\ \sqrt{(\frac{11}{12})}$
4) $-\mathbf{i} - 3\mathbf{j} + 2\mathbf{k};\ \mathbf{i} - \mathbf{j}$
5) $\sqrt{(\frac{3}{5})};\ \pm(\mathbf{i} - \mathbf{j} - 2\mathbf{k})/\sqrt{6}$
6) $\pm(4\mathbf{i} - 5\mathbf{j} + 7\mathbf{k})/3\sqrt{10}$
8) $k\mathbf{a} = 3\mathbf{b} - 2\mathbf{c}$
10) $2\mathbf{k} + t(\mathbf{i} - 3\mathbf{k})$

Exercise 2b – p. 71

1) $\frac{1}{2}\sqrt{35}$ 2) $\frac{1}{2}\sqrt{5}$ 4) 1
5) 7 6) 3 7) $\frac{1}{3}$
8) $\frac{11}{6}$
9) $\frac{1}{6}|\mathbf{b}\times\mathbf{c}.\mathbf{a} + \mathbf{b}\times\mathbf{a}.\mathbf{d} + \mathbf{c}\times\mathbf{b}.\mathbf{d}$
$+ \mathbf{a}\times\mathbf{c}.\mathbf{d}|$

Exercise 2c – p. 80

1) a) $\begin{pmatrix} \sqrt{3}/2 & 0 & \frac{1}{2} \\ 0 & 1 & 0 \\ -\frac{1}{2} & 0 & \sqrt{3}/2 \end{pmatrix}$

b) $\begin{pmatrix} 1 & 0 & 0 \\ 0 & \sqrt{2}/2 & -\sqrt{2}/2 \\ 0 & \sqrt{2}/2 & \sqrt{2}/2 \end{pmatrix}$

c) $\begin{pmatrix} 1 & 0 & 0 \\ 0 & 1 & 0 \\ 0 & 0 & -1 \end{pmatrix}$ d) $\begin{pmatrix} 1 & 0 & 0 \\ 0 & -1 & 0 \\ 0 & 0 & 1 \end{pmatrix}$

e) $\begin{pmatrix} 0 & 1 & 0 \\ 1 & 0 & 0 \\ 0 & 0 & -1 \end{pmatrix}$ f) $\begin{pmatrix} \frac{1}{3} & \frac{2}{3} & \frac{2}{3} \\ \frac{2}{3} & \frac{1}{3} & -\frac{2}{3} \\ \frac{2}{3} & -\frac{2}{3} & \frac{1}{3} \end{pmatrix}$

g) $\begin{pmatrix} 2 & 0 & 0 \\ 0 & 2 & 0 \\ 0 & 0 & 2 \end{pmatrix}$ h) $\begin{pmatrix} \frac{1}{3} & 0 & 0 \\ 0 & \frac{1}{3} & 0 \\ 0 & 0 & \frac{1}{3} \end{pmatrix}$

i) $\begin{pmatrix} 2 & 0 & 0 \\ 0 & 1 & 0 \\ 0 & 0 & 2 \end{pmatrix}$

2) a) reflection in the plane $x - z = 0$
b) a rotation of π about Oz together with a stretch by a factor of 2 parallel to Ox and Oy
c). all points are mapped to the xy plane

3) a) $\begin{pmatrix} 1 & 0 & 0 \\ 0 & \sqrt{2}/2 & -\sqrt{2}/2 \\ 0 & -\sqrt{2}/2 & -\sqrt{2}/2 \end{pmatrix}$

b) $\begin{pmatrix} 1 & 0 & 0 \\ 0 & \sqrt{2}/2 & \sqrt{2}/2 \\ 0 & \sqrt{2}/2 & -\sqrt{2}/2 \end{pmatrix}$

c) $\begin{pmatrix} \sqrt{3} & 0 & 1 \\ 0 & 2 & 0 \\ -1 & 0 & \sqrt{3} \end{pmatrix}$ d) $\begin{pmatrix} 0 & 2 & 0 \\ 1 & 0 & 0 \\ 0 & 0 & -2 \end{pmatrix}$

4) **AB** is a stretch by a factor of 3 parallel to the y axis followed by a rotation of 180° about Oz and a reflection in the xy plane
AC is an enlargement with scale factor 2 and centre O, followed by a rotation of 180° about Oz and a reflection in the xy plane

Exercise 2d – p. 83

1) a) enlarges volume by a factor 16
b) enlarges volume by a factor 8
c) does not alter volume
d) enlarges volume by a factor 12

2) a) $x + y - z = 0$ b) $x + 4y - 2z = 0$
c) $z = 0$ d) $5x + 7y - 11z = 0$

3) a) $x = \dfrac{y}{2} = \dfrac{z}{-1}$ b) $\dfrac{x}{5} = \dfrac{y}{-2} = z$
c) $x = z = 0$ d) $x = \dfrac{y}{3} = \dfrac{z}{-1}$

4) a) maps all points to the plane $x + y - z = 0$
b) maps all points to the plane $x - 17y - 7z = 0$
c) changes a unit cube to a rhomboid of volume 2 cu. units
d) maps all points to the line $x = \dfrac{y}{-3} = \dfrac{z}{2}$
e) maps all points to the origin
f) maps all points to the x axis.

5) $1 \pm \sqrt{2}$ 6) (c)

Exercise 2e – p. 89

1) a) $\begin{pmatrix} \frac{1}{2} & 0 \\ 0 & 1 \end{pmatrix}$ b) $\begin{pmatrix} \frac{1}{5} & \frac{1}{10} \\ -\frac{2}{5} & \frac{3}{10} \end{pmatrix}$

c) – d) – e) $\begin{pmatrix} \frac{1}{3} & -\frac{1}{3} \\ -\frac{2}{3} & \frac{5}{3} \end{pmatrix}$

f) $\begin{pmatrix} -\frac{3}{10} & -\frac{2}{5} \\ \frac{1}{10} & -\frac{1}{5} \end{pmatrix}$

g) $\begin{pmatrix} \sin\theta & -\cos\theta \\ \cos\theta & \sin\theta \end{pmatrix}$

h) $\dfrac{1}{ps - rq}\begin{pmatrix} s & -q \\ -r & p \end{pmatrix}$

2) a) $\begin{pmatrix} \cos\theta & \sin\theta \\ -\sin\theta & \cos\theta \end{pmatrix}$ b) $\begin{pmatrix} \frac{1}{2} & 0 \\ 0 & \frac{1}{2} \end{pmatrix}$

c) $\begin{pmatrix} \frac{1}{3} & 0 \\ 0 & 1 \end{pmatrix}$ d) $\begin{pmatrix} 1 & 0 \\ -1 & 1 \end{pmatrix}$

3) a) $14X + Y + 22 = 0$
b) $13X - 3Y + 44 = 0$

5) $2Y + X = 0,\ 2y = x - 3$

Exercise 2f – p. 93

1) $\mathbf{A}^T = \begin{pmatrix} 1 & -1 & 2 \\ 2 & 1 & 5 \\ 0 & 0 & 1 \end{pmatrix}$,

$\text{adj}\,\mathbf{A} = \begin{pmatrix} 1 & -2 & 0 \\ 1 & 1 & 0 \\ -7 & -1 & 3 \end{pmatrix}$, $\mathbf{A}^{-1} = \frac{1}{3}\,\text{adj}\,\mathbf{A}$

2) $\mathbf{B}^T = \begin{pmatrix} 2 & 5 & -1 \\ 3 & 3 & 2 \\ 1 & 4 & 5 \end{pmatrix}$,

$\text{adj}\,\mathbf{B} = \begin{pmatrix} 7 & -13 & 9 \\ -29 & 11 & -3 \\ 13 & -7 & -9 \end{pmatrix}$,

$\mathbf{B}^{-1} = -\frac{1}{60}\,\text{adj}\,\mathbf{B}$

3) $\mathbf{C}^T = \begin{pmatrix} 1 & 2 & 6 \\ -1 & 0 & -2 \\ 3 & 4 & 22 \end{pmatrix}$,

$\text{adj}\,\mathbf{C} = \begin{pmatrix} 8 & 16 & -4 \\ -20 & 4 & 2 \\ -4 & -4 & 2 \end{pmatrix}$,

$\mathbf{C}^{-1} = \frac{1}{16}\,\text{adj}\,\mathbf{C}$

4) $\mathbf{D}^T = \begin{pmatrix} -1 & 0 & 1 \\ 0 & 5 & 4 \\ 4 & -2 & -1 \end{pmatrix}$,

$\text{adj}\,\mathbf{D} = \begin{pmatrix} 3 & 16 & -20 \\ -2 & -3 & -2 \\ -5 & 4 & -5 \end{pmatrix}$,

$\mathbf{D}^{-1} = -\frac{1}{23}\,\text{adj}\,\mathbf{D}$

5) $\mathbf{E}^T = \begin{pmatrix} 2 & 3 & 5 \\ -1 & 2 & -2 \\ 4 & -1 & 9 \end{pmatrix}$,

$\text{adj}\,\mathbf{E} = \begin{pmatrix} 16 & 1 & -7 \\ -32 & -2 & 14 \\ -16 & -1 & 7 \end{pmatrix}$,

$\mathbf{E}^{-1}$ does not exist

9) $\mathbf{A}^T = \begin{pmatrix} 2 & 3 \\ 1 & -1 \\ 4 & 0 \end{pmatrix}$,

$\mathbf{B}^T = \begin{pmatrix} 4 & 0 & 1 \\ 1 & 2 & 0 \end{pmatrix}$, $\mathbf{AB} = \begin{pmatrix} 12 & 4 \\ 12 & 1 \end{pmatrix}$

12) $\mathbf{A}^{-1} = \frac{1}{2}\begin{pmatrix} -4 & 2 \\ 3 & -1 \end{pmatrix}$,

$a = -2,\ b = 5$

Exercise 2g – p. 99

1) $(3, 4, -6)$

2) $\dfrac{1}{16}\begin{pmatrix} -7 & 6 & -5 \\ -5 & 2 & 1 \\ 13 & -2 & 7 \end{pmatrix}$,

$x = \frac{7}{16},\ y = \frac{5}{16},\ z = \frac{19}{16}$

3) $\dfrac{1}{12}\begin{pmatrix} 2 & -3 & 1 \\ 8 & -12 & -8 \\ -2 & 9 & 5 \end{pmatrix}$,

$x = \frac{1}{3},\ y = -\frac{5}{3},\ z = \frac{5}{3}$

4) $\begin{pmatrix} -2 & 8 & -19 \\ 1 & -4 & 10 \\ 1 & -3 & 7 \end{pmatrix}$

a) $X - 3Y + 6Z = -4$ b) $Y = 2$
c) $2X - 11Y + 27Z = 1$

5) a) $\mathbf{r} = \mathbf{i} - 2\mathbf{j} - \mathbf{k} + \lambda(7\mathbf{i} + 9\mathbf{j} + 3\mathbf{k})$
   b) $\mathbf{r} = 5\mathbf{j} + 2\mathbf{k} + \mu(9\mathbf{i} + 5\mathbf{j} + \mathbf{k})$

6) $x - z = D$ for all values of D

7) The y axis and any line through O in the xz plane

8) $\mathbf{A}^{-1} = \begin{pmatrix} 0 & 0 & 1 \\ 1 & 0 & 0 \\ 0 & -1 & 0 \end{pmatrix}$

and represents a reflection in the plane $x - z = 0$ followed by a rotation of $-90°$ about Ox;
$\lambda = 1,\ x = -y = -z$

Exercise 2h – p. 106

1) $x = -\frac{1}{5}, y = -\frac{8}{5}, z = -2$
2) $x = \frac{3}{2}, y = \frac{3}{4}, z = \frac{7}{4}$
3) $x = 2, y = 5, z = 1$
4) $x = 1, y = -4, z = -4$
5) $x = 3, y = 2, z = -1$
6) $x = -2, y = 1, z = -1$
7) $x = 3, y = 5, z = -2$
8) $x = 3, y = 12, z = 9$

Exercise 2i – p. 113

1) an infinite set of solutions dependent on one parameter
2) unique solution
3) no solution
4) an infinite set of solutions dependent on one parameter
5) an infinite set of solutions dependent on one parameter
6) an infinite set of solutions dependent on two parameters
7) unique solution
8) no solution
9) $x = \lambda, y = -1 - \frac{1}{2}\lambda, z = 1 - \frac{5}{2}\lambda$;
   $x = \frac{6}{5}, y = -3, z = -\frac{7}{5}$;
   $x = -1, y = \lambda, z = \lambda - 3$;
   $x = \lambda,\ y = 3\lambda,\ z = 2\lambda$;
   $x = \lambda,\ y = \mu,\ z = 1 - (\lambda + \mu)$;
   $x = 1, y = 0, z = 0$
10) The set of points on the line $x = \frac{5}{3}$, $3y = 3z + 1$
11) The set of points on the line $\dfrac{x + \frac{5}{3}}{-14} = \dfrac{y}{3} = \dfrac{z - \frac{2}{3}}{5}$
13) The set of points in the plane $x - y + 2z = 1$

14) $X = \dfrac{Y}{2} = \dfrac{Z}{3}$; all points on the plane $x + 2y - z = a$ map to a point on the line for all values of a.

15) $\frac{1}{4}(-3 \pm \sqrt{17})$

16) $\begin{vmatrix} a_1 & b_1 & c_1 \\ a_2 & b_2 & c_2 \\ a_3 & b_3 & c_3 \end{vmatrix} = 0$

17) $\begin{vmatrix} a_1 & b_1 & c_1 \\ a_2 & b_2 & c_2 \\ a_3 & b_3 & c_3 \end{vmatrix} = 0$;

the planes represented by [1] have a line in common.

18) $\begin{vmatrix} a_1 & b_1 & c_1 \\ a_2 & b_2 & c_2 \\ a_3 & b_3 & c_3 \end{vmatrix} = 0$;

[1] represent a set of concurrent lines.

19) $a = -1$

20) Any matrix of the form

$$\begin{pmatrix} a_1 & b_1 & c_1 \\ a_2 & b_2 & c_2 \\ -(a_1+a_2) & -(b_1+b_2) & -(c_1+c_2) \end{pmatrix}$$

21) $\begin{vmatrix} a_1 & b_1 \\ a_2 & b_2 \end{vmatrix} \neq 0$

Exercise 2j – p. 118

1) $\dfrac{1}{19}\begin{pmatrix} -1 & -14 & 6 \\ -4 & 1 & 5 \\ 7 & 3 & -4 \end{pmatrix}$

2) $\dfrac{1}{22}\begin{pmatrix} 11 & -5 & 1 \\ -11 & -3 & 5 \\ -11 & 7 & 3 \end{pmatrix}$

3) $\dfrac{1}{75}\begin{pmatrix} 0 & -15 & 6 \\ 75 & 75 & -75 \\ -50 & -50 & 55 \end{pmatrix}$

4) $\dfrac{1}{5}\begin{pmatrix} 13 & -11 & 1 \\ 1 & -2 & 2 \\ -5 & 5 & 0 \end{pmatrix}$

5) No inverse

6) $\dfrac{1}{32}\begin{pmatrix} 55 & 33 & -27 \\ 19 & 5 & -7 \\ -64 & -32 & 32 \end{pmatrix}$

Exercise 2k – p. 121

1) Vector space 2) Vector space
3) Vector space 4) Vector space
5) Vector space: identity is the A.P. whose first term is zero and whose common difference is zero
6) Not a vector space: addition is not closed
7) Not a vector space: addition is not closed
8) Not a vector space: addition is not closed
9) Vector space

Exercise 2l – p. 131

1) Linearly independent
2) $\mathbf{b} = 2\mathbf{a}$ 3) $2\mathbf{a} + \mathbf{b} = 0$
4) Linearly independent
5) $-2\mathbf{a} + \mathbf{b} + 3\mathbf{c} = 0$
6) $\mathbf{c} = \mathbf{b} - \mathbf{a}$
7) Linearly independent
8) $3\mathbf{a} + \mathbf{b} - 2\mathbf{c} + \mathbf{d} = 0$
9) $5x + 3 = \frac{13}{3}(2x - 1) + \frac{11}{3}(-x + 2)$, $\{x, 1\}$
10) 4, $\{x^3, x^2, x, 1\}$
11) i), ii), v) 12) i), iii), iv)
13) V is of dimension one. $\{(1, 1)\}$ is a basis for V. If $(a, a+1)$ and $(b, b+1)$ are two points on the line, adding gives $(a+b, a+b+2)$ which is not on the line: addition is not closed
14) A is a vector space, B is not (addition of coordinates of points on $x + y + z = 2$ is not closed). A is of dimension two and a basis is $\{(1, 0, -1), (0, 1, -1)\}$.
15) a) V' dimension 3, V'' dimension 2
b) V': $\{(1, 0, 0), (0, 1, 0), (0, 0, 1)\}$
V'': $\{(1, 0, 1), (0, 1, 1)\}$
c) $f_1 f_2$: $(x_1, x_2, x_3) \longmapsto (2x_1, 2x_2, 2(x_1 + x_2))$
$f_2 f_1$: $(x_1, x_2, x_3) \longmapsto (2x_1, 2x_2, 2(x_1 + x_2))$
d) dimension 2
16) There are no inverses as the components of all vectors (except **O**) are positive
17) a) one b) three
c) three d) not a vector space
18) $\{(1, 0, 0, 0), (0, 1, 0, 0), (0, 0, 1, 0), (0, 0, 0, 1)\}$
$\mathbf{x} \longmapsto \{x_1 + x_4, x_2 + x_4, x_3 + x_4, x_1 + x_4\}$

Multiple Choice Exercise 2 – p. 135

1) b	2) d	3) b
4) a	5) d	6) d
7) a	8) a, c	9) c
10) a, c	11) b, c	12) *B*
13) *C*	14) *C*	15) *C*
16) *C*	17) *T*	18) *T*
19) *F*	20) *F*	21) *T*
22) *F*	23) *T*	

Miscellaneous Exercise 2 – p. 139

1) $\left.\begin{aligned} 2X + Y + Z &= 4 \\ 7X - Y + 2Z &= 6 \\ 7X + 2Y - Z &= 18 \end{aligned}\right\}$ meet at $(1\frac{2}{3}, 2\frac{1}{3}, -1\frac{2}{3})$
2) $\lambda = 3$ or 6; $(1, 2)$ or $(-\frac{1}{5}, \frac{1}{5})$
3) $\mathbf{M}^2 = \begin{pmatrix} -1 & 0 & 0 \\ 0 & 1 & 0 \\ 0 & 0 & 1 \end{pmatrix}$

M represents a rotation of $-90°$ about Oz, $\mathbf{M}^2$ a rotation of 180° about Oz
4) a) $x = 2, y = 1, z = -10$
5) a) $3X - 2Y = 0$ b) $X = \dfrac{Y}{3} = \dfrac{Z}{5}$
c) $\begin{pmatrix} 0 & 1 & 0 \\ -1 & 0 & 0 \\ 0 & 0 & 1 \end{pmatrix}$
6) $a = 2$, $b = -\frac{4}{5}$, $c = \frac{3}{5}$; $\mathbf{x} = \begin{pmatrix} \frac{1}{5} \\ 1 \\ -\frac{2}{5} \end{pmatrix}$

7) $\mathbf{A}^{-1} = \begin{pmatrix} 1 & 0 & 0 \\ 1 & 1 & 0 \\ -5 & -2 & 1 \end{pmatrix}$.

$\mathbf{B}^{-1} = \begin{pmatrix} 1 & -4 & 14 \\ 0 & 1 & -3 \\ 0 & 0 & 1 \end{pmatrix}$,

$\begin{pmatrix} x_1 \\ x_2 \\ x_3 \end{pmatrix} = \begin{pmatrix} 5 \\ -1 \\ 0 \end{pmatrix}$

8) $-\dfrac{1}{6}\begin{pmatrix} 0 & -3 & 0 \\ 0 & -3 & 6 \\ -2 & 2 & -4 \end{pmatrix}$,

$\begin{pmatrix} 7 & -3 & 0 \\ 0 & 4 & 6 \\ -2 & 2 & 3 \end{pmatrix}$, $\begin{pmatrix} -6 & 14 & 21 \\ 14 & -6 & 0 \\ 7 & -7 & -6 \end{pmatrix}$

9) $\mathbf{A}^{-1} = \begin{pmatrix} \frac{1}{3} & -\frac{1}{3} & \frac{1}{3} \\ -\frac{1}{2} & \frac{1}{2} & 0 \\ \frac{4}{3} & -\frac{1}{3} & -\frac{2}{3} \end{pmatrix}$; 1, −1, 6

10) $\begin{pmatrix} \frac{5}{2} & -1 \\ -4 & 2 \\ 3 & -\frac{5}{4} \end{pmatrix}$

11) $55(k-1)$,
a) $x = -3, y = 2, z = 3$,
b) $\dfrac{x}{10} = -y = -\dfrac{z}{11}$;
three planes through O with a line in common.

12) 5; $x = \lambda, y = 3 - 2\lambda, z = \lambda$;
three planes intersecting in a line.

13) a) $\begin{pmatrix} -13 \\ 8 \end{pmatrix}$, b) $\begin{pmatrix} 0 \\ 0 \end{pmatrix}$,

c) $\begin{pmatrix} 1-2\lambda \\ \lambda \end{pmatrix}$, d) no solution

$\mathbf{M}^{-1}$ exists in a) and b) but not in c) and d).

14) $\begin{pmatrix} -2 & -2 \\ 6 & 5 \end{pmatrix}$

15) $x = -5, y = 3, z = -4$

16) $-\dfrac{1}{160}\begin{pmatrix} 6 & -22 & -18 \\ 23 & -31 & 11 \\ -31 & 7 & 13 \end{pmatrix}$;

$x = 0.6, y = -0.2, z = 0.4$

17) $\frac{2}{3}\pi$

18) $\lambda = 1;\ x = \alpha, y = \beta, z = -\alpha - \beta$
$\lambda = 4;\ x = y = z = \gamma$

19) $-(\mathbf{A} + p\mathbf{I})/q$ 20) $(-z, y, x)$

21) 1, −1, 0

23) (a) $\mathbf{d} = 2\mathbf{a} - 3\mathbf{b} + \mathbf{c}$

24) $\alpha = x_1 - x_2,\ \beta = x_2 - x_3,\ \gamma = x_3$;
b, **c** and **d** are linearly dependent
$(\mathbf{c} = \mathbf{b} + \mathbf{d})$; $\begin{pmatrix} 2 \\ 2 \\ 1 \end{pmatrix} = 2\mathbf{b} + \mathbf{d}$; any linear combination of **b**, **c** and **d** is of the form $\begin{pmatrix} x_1 \\ x_1 \\ x_2 \end{pmatrix}$.

25) Three

26) $\mathbf{X} = 2\mathbf{A} - 3\mathbf{B} + \mathbf{C}$, three

27) $(0, 1, 2, 3) = \frac{1}{2}(1, 2, 5, 6) - \frac{1}{2}(1, 0, 1, 0)$ and $(2, -1, 0, -3) = \frac{5}{2}(1, 0, 1, 0) - \frac{1}{2}(1, 2, 5, 6)$; i.e. each of the second pair of vectors is included in S.

28) $\{(1, -6, 2, 0),\ (-1, -2, 0, 2)\}$

29) (i) No (ii) Yes (iii) No

31) $\{(2, 1, 0),\ (2, 0, 1)\}$, two

32) (i) yes: $\{x^2, x\}$
(ii) No, addition is not closed, e.g. $(x^2 + x + 1) + (-x^2 - x + 1) = 2$ and $2 \notin P_2$
(iii) Yes, $\{x^2 - 1,\ x - 1\}$
(iv) Yes, $\{x^2, 1\}$

33) $\begin{pmatrix} 1 & 2 \\ -2 & 1 \end{pmatrix}$

(i) $\begin{pmatrix} -3a+4b \\ -4a-3b \end{pmatrix}$ (ii) $\begin{pmatrix} \frac{1}{5}(a-2b) \\ \frac{1}{5}(2a+b) \end{pmatrix}$

34) $\{(-3, 1, 0, 1),\ (1, 0, -1, 0)\}$

Exercise 3a – p. 150

1) b 2) a 3) c
4) b 5) a 6) b
7) b 8) b 9) false
10) false 11) true 12) true

Exercise 3b – p. 152

1) c 2) b 3) c 4) b
5) c 6) d 7) d 8) e

Exercise 3c – p. 154

1) true 2) false 3) true
4) false 5) true 6) false
7) false 8) true 9) false
10) true

Exercise 3d – p. 159

1) $p \Leftrightarrow q$ 2) $p \Leftarrow q$ 3) $p \Rightarrow q$
4) $p \Leftrightarrow q$ 5) $p \Leftarrow q$ 6) $p \Leftrightarrow q$
7) $p \Leftarrow q$ 8) $p \Rightarrow q$ 9) $p \Rightarrow q$
10) $p \Leftarrow q$ 11) $p \Rightarrow q$ 12) $p \Leftarrow q$
13) $p \Leftrightarrow q$ 14) $p \Leftrightarrow q$ 15) $p \Leftarrow q$

16) a) If this car is a good car it is made in Britain
b) If this car is not a good car it is not made in Britain
c) If this car is not made in Britain it is not a good car
d) This car is made in Britain and it is not a good car

17) a) If $b^2 - 4ac \geqslant 0$ then $ax^2 + bx + c = 0$ has two real roots
b) If $b^2 - 4ac < 0$ then $ax^2 + bx + c = 0$ does not have two real roots
c) If $ax^2 + bx + c = 0$ does not have two real roots then $b^2 - 4ac < 0$
d) $ax^2 + bx + c = 0$ has two real roots and $b^2 - 4ac < 0$

18) a) If Henry is doing A-level maths he is mad
b) Henry is doing A-level maths because he is mad
c) If Henry is not doing A-level maths he is mad
d) Henry is doing A-level maths because he is not mad
e) It is not true that if Henry is doing A-level maths he is mad
f) Henry is doing A-level maths if and only if he is mad
g) If Henry is not doing A-level maths he is not mad
h) Henry is not doing A-level maths because he is not mad

19) a) $p \Rightarrow q$ b) $\sim p \Leftarrow q$
c) $\sim p \Rightarrow \sim q$ d) $\sim (p \Leftarrow q)$
e) $\sim p \Leftarrow q$ f) $p \Leftrightarrow q$
g) $\sim p \Leftarrow \sim q$
True statements are a), d), g).

20) a) If x has no real value then $x^2 = -1$
b) It if not true that if $x^2 = -1$ then x has no real value
c) If $x^2 \neq -1$ then it is not true that x has no real value
d) If x can have at least one real value then $x^2 \neq -1$

21) (c)

Exercise 3e – p. 166

1) converse true 2) converse true
3) converse false 4) converse false
11) $u_2 = \frac{1}{3}$, $u_3 = -\frac{1}{9}$, $u_4 = -\frac{11}{27}$
15) yes

Exercise 3f – p. 171

9) $2 \div 1 \neq 1 \div 2$
10) $\mathbf{i} \times \mathbf{j} = \mathbf{k}$ but $\mathbf{j} \times \mathbf{i} = -\mathbf{k}$
11) $(0) - (-1) > 0$ but $(0)^2 - (-1)^2 < 0$
12) $x^2 + x + 1 = 0$ does not have real roots
13) $\mathbf{i} \times 2\mathbf{i} = \mathbf{0}$ and $\mathbf{i} \neq \mathbf{0}$, $2\mathbf{i} \neq \mathbf{0}$
14) $f(x) = x^4$: $f''(0) = 0$, $f(0)$ is a minimum value
15) $\mathbf{A} = \begin{pmatrix} 1 & 1 \\ 1 & 1 \end{pmatrix}$, $\mathbf{B} = \begin{pmatrix} 1 & 0 \\ 0 & 1 \end{pmatrix}$, $\mathbf{C} = \begin{pmatrix} 0 & 1 \\ 1 & 0 \end{pmatrix}$
16) (i) $(\sqrt{2}) + (2 - \sqrt{2})$ is rational
(ii) $(\sqrt{8}) \times (\sqrt{2})$ is rational

Exercise 4a – p. 188

1) a) $\{x: x \in \mathbb{R}\}$ b) $\{x: x \in \mathbb{R}\}$
c) $\{x: x \in \mathbb{R}\}$ d) $\{x: x \in \mathbb{R}\}$
e) $\{x: x \in \mathbb{R}\}$ f) $\{x: x \in \mathbb{R}\}$
g) $\{x: x \in \mathbb{R},\ x \neq 0\}$
h) $\{x: x \in \mathbb{R}\}$ i) $\{x: x \in \mathbb{R}\}$
j) $\{x: x \in \mathbb{R}\}$ k) $\{x: x \in \mathbb{R}\}$

2) (b), (e), (g), (i)
3) (a), (c), (j)
4) a) 2π, b) 2π, k) 2
5) all except (h)
6) (e) at $x = 0$,
(h) at all integral values of x,
(k) at all integral values of x
7) (g)
8) (e) at $x = 0$,
(h) at all integral values of x,
(k) at all integral values of x
10) a) odd b) odd
c) even and periodic
d) even but not periodic
11) a) 3 b) 3 c) 24
12) a) $f(x) = 2 - x$ for $0 < x \leqslant 4$; $f(x) = 1$ for $4 < x \leqslant 5$; $f(x+5) = f(x)$ for all values of x
b) $f(x) = 2x$ for $0 < x \leqslant 1$; $f(x) = 5 - 2x$ for $1 < x \leqslant 2$; $f(x) = -x$ for $2 < x \leqslant 3$; $f(x+3) = f(x)$ for all values of x
13) d, e (in two parts) 14) 1
15) 1 16) 4 17) 1, 2
18) 0 19) 0, 1 20) 0
21) 0
22) a) $f(x)$ has a singularity at $x = 0$; $f(x)$ and $f'(x)$ continuous
b) $f(x)$ continuous, $f'(x)$ has singularities at $x = n\pi + \pi/2$
c) $f(x)$ continuous, $f'(x)$ undefined at $x = 0$

Exercise 4b – p. 195

5) 0.6933 6) 0.095 31 7) 0.182 32

Exercise 4c – p. 199

7) $\sinh(x+y) \equiv \sinh x \cosh y + \cosh x \sinh y$
8) $\cosh(x-y) \equiv \cosh x \cosh y - \sinh x \sinh y$
9) $\sinh 3x \equiv 3 \sinh x + 4 \sinh^3 x$
10) $\cosh x + \cosh y \equiv 2 \cosh \dfrac{x+y}{2} \cosh \dfrac{x-y}{2}$
11) $\sinh 2x \equiv \dfrac{2 \tanh x}{1 - \tanh^2 x}$
12) $\cosh 2x \equiv \dfrac{1 + \tanh^2 x}{1 - \tanh^2 x}$
13) $\frac{4}{5}, \frac{3}{5}, \frac{17}{8}, \frac{15}{17}$
14) $\frac{12}{5}, \frac{13}{12}, \frac{65}{72}, \frac{65}{97}$
15) a) $\operatorname{sech} x$, $\sinh^2 x$
b) $\coth x$, $\operatorname{cosech} x$, $\tanh(-x)$, $\sinh(-x)$
c) $\coth x$, $\operatorname{sech} x$, $\cosh x + \sinh x$, $\sinh^2 x$, $\tanh(-x)$, $\sinh(-x)$

Exercise 4d – p. 202

1) $0, \ln \frac{5}{3}$ 2) $\ln 4$
3) $\ln 2, -\ln 4$ 4) $\ln 2, -\ln 3$
5) $\operatorname{arcosh} \frac{3}{2}$, $\operatorname{arcosh} 2$
6) $\operatorname{arcosh} \frac{4}{3}$ 7) $0, -\ln 2$
8) 0, $\operatorname{arcosh} 4$ 9) $\ln \frac{2}{5}$
10) $\ln 5, -\ln 4$ 11) $r = 3$, $\tanh y = \frac{4}{5}$
12) 12

Exercise 4e – p. 204

1) $-\operatorname{sech} x \tanh x$ 2) $2 \operatorname{sech}^2 2x$
3) $-\operatorname{cosech}^2 x$ 4) $-2 \operatorname{sech}^2 x \tanh x$
5) $4 \cosh 4x$ 6) $6 \cosh^2 2x \sinh 2x$
7) $\sinh x + x \cosh x$
8) $\sinh x\,(1 + \operatorname{sech}^2 x)$
9) $e^x(\cosh x + \sinh x) = e^{2x}$
10) $\dfrac{5 \sinh 5x}{2\sqrt{\cosh 5x}}$
11) $2x \tanh 3x(3x \operatorname{sech}^2 3x + \tanh 3x)$
12) $e^x \operatorname{cosech} 2x(1 - 2 \coth 2x)$
13) $\sinh x\, e^{\cosh x}$ 14) $\coth x$
15) $\frac{1}{2} \operatorname{sech}^2 x \sqrt{\coth x}$
16) $2 \tanh x \operatorname{sech}^2 x\, e^{\tanh^2 x}$
17) $\frac{1}{5} \cosh 5x + K$ 18) $\frac{1}{4} \sinh 4x + K$
19) $\frac{1}{2} \ln(K \cosh 2x)$ 20) $\ln(K \sinh x)$
21) $\frac{1}{4}(\sinh 2x - 2x) + K$
22) $\frac{1}{12}(\sinh 3x + 9 \sinh x) + K$
23) $\frac{1}{4}(2x \cosh 2x - \sinh 2x) + K$
24) $K - \operatorname{sech} x$
25) $\frac{1}{4}(e^{2x} + 2x) + K$
26) $\frac{1}{8}(\sinh 4x + 2 \sinh 2x) + K$
27) $\frac{1}{16}(\cosh 8x + 4 \cosh 2x) + K$
28) $\frac{1}{3} \sinh 12$
29) $\frac{1}{4}(e^2 - 2)(e^2 + 1)$ 30) $\frac{1}{4} \ln(\cosh 4)$
31) $\frac{1}{5}(\tanh 20 - \tanh 5)$
32) $bx \cosh u - ay \sinh u = ab$,
$ax \sinh u + by \cosh u = (a^2 + b^2) \sinh u \cosh u$
33) 24, minimum
34) $y = 2 \cosh 2x + \sinh 2x$

Exercise 4f – p. 208

1) $\ln(3 + \sqrt{8})$, $\ln(\sqrt{2} - 1)$, $\ln \sqrt{3}$
2) $x \geqslant \frac{1}{3}$ 3) $-2 < x < 2$

4) $y = \frac{1}{2} \ln \dfrac{1+x}{1-x}$ 5) $\ln (2 + \sqrt{3})$

7) a) and b) $\ln \left(\dfrac{1 + \sqrt{1-x^2}}{x} \right)$

c) $\ln (x^2 - 1 + \sqrt{x^4 - 2x^2 + 2})$

8) $x = \frac{1}{2}$ 9) $x = \frac{1}{3}$

10) $x = \pm 2$

Exercise 4g – p. 211

2) a) $\operatorname{arsinh} \dfrac{x}{2} + K$ b) $\operatorname{arcosh} \dfrac{x}{3} + K$

c) $\frac{1}{2} \operatorname{arcosh} \dfrac{x}{2} + K$ d) $\frac{1}{2} \operatorname{arsinh} \dfrac{2x}{3} + K$

3) a) $\dfrac{1}{\sqrt{(x^2 + 2x)}}$ b) $\dfrac{2}{\sqrt{(1 + 4x^2)}}$

c) $\dfrac{1}{1 - x^2}$

d) $\operatorname{arcosh} x + \dfrac{x}{\sqrt{(x^2 - 1)}}$

e) $\dfrac{-1}{x\sqrt{(x^2 + 1)}}$ f) $\dfrac{1}{2\sqrt{(x^2 - x)}}$

g) $\dfrac{1}{x\sqrt{(1 - x^2)}}$

h) $\dfrac{1}{(1-x)\sqrt{(x^2 - 2x + 2)}}$

i) $\dfrac{2 \cosh 2x}{\sqrt{(\sinh^2 2x - 1)}}$

j) $\dfrac{1}{\sqrt{(x^2 - 1)}} e^{\operatorname{arcosh} x}$

Add K to each answer in questions 4–15.

4) $\frac{2}{3} \operatorname{arsinh} \dfrac{3x}{2} + \dfrac{x}{2}\sqrt{9x^2 + 4}$

5) $\dfrac{x}{8}(8x^2 - 5)\sqrt{4x^2 - 1} + \frac{3}{16} \operatorname{arcosh} 2x$

6) $\frac{1}{2} \{\operatorname{arsinh} (x + 2)$
$+ (x + 2)\sqrt{x^2 + 4x + 5}\}$

7) $\frac{1}{2} \{(x + 2)\sqrt{x^2 + 4x + 3}$
$- \operatorname{arcosh} (x + 2)\}$

8) $\operatorname{arsinh} \left(\dfrac{x\sqrt{2}}{2} \right)$

9) $\sqrt{2}/2 \operatorname{arcosh} x\sqrt{2}$

10) $\operatorname{arsinh} (x + 1)$

11) $\operatorname{arcosh} (x + 1)$

12) $\frac{1}{2}(x + 2)\sqrt{x^2 + 4x} - 2 \operatorname{arcosh} \left(\dfrac{x+2}{2} \right)$

13) $\frac{1}{2} \operatorname{arcosh} (8x - 1)$

14) $\sqrt{x^2 + 1} + \operatorname{arsinh} x$

15) $\sqrt{x^2 - 1} - \operatorname{arcosh} x$

16) $\frac{1}{3}(\operatorname{arcosh} 6 - \operatorname{arcosh} 3)$
$= \frac{1}{3} \ln \left(\dfrac{6 + \sqrt{35}}{3 + \sqrt{8}} \right)$

17) $\operatorname{arsinh} \frac{1}{2}$

18) $\operatorname{arcosh} 2 - \operatorname{arcosh} \frac{3}{2}$
$= \ln \left(\dfrac{4 + 2\sqrt{3}}{3 + \sqrt{5}} \right)$

19) $\sqrt{5} + \operatorname{arcosh} \frac{3}{2}$

20) $2\sqrt{2} + 2 \operatorname{arsinh} 1$

21) $2\sqrt{5} - \frac{8}{3} \operatorname{arcosh} \frac{3}{2}$

Add K to each answer in questions 22–33

22) $\frac{1}{4} \arcsin 2x + \dfrac{x}{2}\sqrt{1 - 4x^2}$

23) $\arcsin \dfrac{x}{2}$ 24) $\operatorname{arcosh} \dfrac{x}{3}$

25) $\frac{1}{6} \ln \dfrac{3x - 1}{3x + 1}$ 26) $\frac{1}{3} \arctan 3x$

27) $\frac{1}{3} \operatorname{arsinh} 3x$

28) $\sqrt{x^2 - 4} + 2 \operatorname{arcosh} \dfrac{x}{2}$

29) $\ln (x - 2)$ 30) $\frac{1}{2} \ln (x^2 - 4)$

31) $\sqrt{x^2 - 4}$ 32) $2\sqrt{x - 1}$

33) $\sqrt{x^2 + 1} - \operatorname{arcosech} x$

34) 3 35) 1 36) ∞ 37) 0

Multiple Choice Exercise 4 – p. 213

1) d	2) d	3) b
4) c	5) c	6) d
7) d	8) d	9) c
10) a	11) a	12) b, c
13) b, d	14) a	15) a, c
16) b	17) a, c	18) b, d
19) b, c	20) E	21) A
22) B	23) B	24) A
25) F	26) T	27) T
28) F	29) T	30) F
31) F	32) T	33) F

Miscellaneous Exercise 4 – p. 217

1) (i) 6 (ii) $-\frac{1}{3}$ 2) 16

3) (i) $3 < f(x) \leqslant 12$ (ii) $2 \leqslant g(x) \leqslant 6$
(iii) $-\sqrt{5} \leqslant h(x) \leqslant \sqrt{5}$
f^{-1} exists as f is a one-one mapping
g^{-1} does not exist as g is not a one-one mapping
h^{-1} does not exist as h is not a one-one mapping

5) $\frac{5}{6}$ 6) both continuous
9) $n\pi$; (i) $\frac{1}{2}(\pi^2-4)$ (ii) $\frac{1}{8}(5\pi^2-24)$
14) 4 15) 0, ln 4
16) a) $\frac{1}{3}\text{arsinh } 3x + K$
b) $\frac{1}{12}\sinh 6x - \frac{x}{2} + K$
18) $\pm\sqrt{\frac{1}{2}}$ 19) a) $\frac{1}{2}\ln\frac{1}{5}$
20) $\coth\theta$ 21) $(-\frac{3}{4}, \ln\frac{1}{2})$
22) a) $\frac{e^x - e^{-x}}{e^x + e^{-x}}$;
$-\frac{1}{2}$ (Note. $\tanh y \neq -2$)
b) $\text{arcosh}\,\frac{5}{3}$; $1-\sqrt{2} + \text{arsinh } 1$
23) $\cosh x \cosh y + \sinh x \sinh y$;
$\sinh x \cosh y + \cosh x \sinh y$;
$2\cosh^2 x - 1$; $4\cosh^3 x - 3\cosh x$
24) $\{\ln(1+\sqrt{2}), -1\}$, minimum
25) $\frac{1}{10}(e^5 - e^{-5}) + \frac{1}{6}(e^3 - e^{-3}) - (e - e^{-1})$
26) $\arctan(\tanh x) + K$
28) $\ln\left(\frac{1+\sqrt{x^2+1}}{x}\right)$; $\sqrt{3}$
29) $x = \ln\frac{1}{2}$, $y = \ln\frac{3}{2}$
30) a) $\text{arsinh}\left(\frac{x-1}{3}\right) + K$
b) $\frac{1}{3}\arctan\left(\frac{x-1}{3}\right) + K$
31) $\ln(7+5\sqrt{2})$
33) $\frac{3}{4}$; $\frac{\pi}{32}(16\ln 2 + 15)$
34) $x = \ln(1+\sqrt{2})$, $y = \ln(5+2\sqrt{6})$;
$x = \ln\left(\frac{1+\sqrt{37}}{6}\right)$
$y = \ln\left(\frac{25+\sqrt{481}}{12}\right)$

Exercise 5a – p. 225

2) a) $13 - 6\sqrt{2}$ b) 13
c) 20 d) 4 e) 36
3) right-angled a), d); isosceles b), g);
neither c), e), f)
4) $x^2 + y^2 = a^2$
5) $(x^2+y^2)^{3/2} = a(x^2 - y^2)$
6) $(x^2+y^2)^2 = 2a^2xy$
7) $(x^2+y^2)^3 = 4a^2x^4$
8) $x\cos\alpha + y\sin\alpha = d$
9) $r^2 + a^2 = 2ar(\cos\theta + \sin\theta)$
10) $r^2(a^2\sin^2\theta + b^2\cos^2\theta) = a^2b^2$
11) $\theta = \tan^{-1} 2$ 12) $r = \tan\theta\sec\theta$
13) $r^2\sin 2\theta = 8$

Exercise 5b – p. 232

3) $\theta = 0, \pi$
4) $\theta = 0, \frac{1}{3}\pi, \frac{2}{3}\pi, \pi, \frac{4}{3}\pi, \frac{5}{3}\pi$
5) $\theta = \frac{1}{6}\pi, \frac{1}{2}\pi, \frac{5}{6}\pi, \frac{7}{6}\pi, \frac{3}{2}\pi, \frac{11}{6}\pi$
6) $\theta = \frac{1}{4}\pi, \frac{3}{4}\pi, \frac{5}{4}\pi, \frac{7}{4}\pi$
7) $\theta = \frac{1}{2}\pi, \frac{3}{2}\pi$ 8) $\theta = \frac{1}{2}\pi, \frac{3}{2}\pi$
9) a) $\theta = 0$, b) $\theta = \pi$
10) $\theta = \pm\cos^{-1}\frac{3}{2}$
12) $\theta = \frac{1}{4}\pi, \frac{3}{4}\pi, \frac{5}{4}\pi, \frac{7}{4}\pi$

Exercise 5c – p. 236

1) $\frac{1}{2}\pi a^2$ 2) $\frac{1}{2}\pi a^2$ 3) $\frac{3}{2}\pi a^2$
4) $\frac{25}{48}\pi^3$ 5) $\frac{1}{3}a^2$ 6) $\frac{1}{2}\pi a^2$
7) $\frac{43}{2}\arccos\frac{3}{5} - 18$ 8) $\frac{1}{2}\pi a^2$
9) $\frac{32\sqrt{3}}{3}$ 10) $\frac{41}{2}\pi a^2$
11) Each loop is traced out twice, once when r is +ve and once when r is −ve

Miscellaneous Exercise 5 – p. 239

1) $a^2\tan\frac{\alpha}{2}\left[1 + \frac{1}{3}\tan^2\frac{\alpha}{2}\right]$
2) $(3\pi+8)/6\pi$ 3) $\frac{4}{3}$
5) a) $\frac{1}{12}\pi a^2$ b) a^2 7) $1, \sqrt{6}/6$
8) $(6, \pi/3)$, $(6, -\pi/3)$; $9\pi + \frac{3\sqrt{3}}{2}$
9) a) $\pi/2$ b) $\frac{1}{2}a^2$
11) $\left(\frac{3}{2}, \frac{2\pi}{3}\right), \left(\frac{3}{2}, -\frac{2\pi}{3}\right)$
12) $\pi/2, 3\pi/2$; $\left(\frac{3\sqrt{3}}{2}, \frac{11\pi}{6}\right), \left(-\frac{3\sqrt{3}}{2}, \frac{7\pi}{6}\right)$
$(0, \pi/2)$
13) $\frac{1}{16}\pi a^2$

Exercise 6a – p. 250

1) convergent 2) convergent
3) divergent 4) divergent
5) divergent 6) divergent
7) convergent 8) convergent
9) divergent 10) divergent
11) divergent 12) divergent
13) divergent 14) convergent
15) convergent

Exercise 6b – p. 254

1) $1 - x + \dfrac{x^2}{2!} - \dfrac{x^3}{3!} + \dfrac{x^4}{4!}$

2) $x - \dfrac{x^3}{3!} + \dfrac{x^5}{5!} - \dfrac{x^7}{7!}$

3) $-x - \dfrac{x^2}{2} - \dfrac{x^3}{3} - \dfrac{x^4}{4}$

4) $1 + nx + \dfrac{n(n-1)}{2!}x^2$

$+ \dfrac{n(n-1)(n-2)}{3!}x^3$

5) $x + \frac{1}{3}x^3$
6) $-x - \frac{3}{2}x^2$
7) $1 + x + x^2 + x^3 + x^4$
8) $1 + x - \dfrac{x^3}{3}$
9) $\tan 1^c + x \sec^2 1^c$

$+ \dfrac{x^2}{2!}\sec^2 1^c(1 + 2\tan 1^c)$

Exercise 6c – p. 259

1) $1 + 3x + \frac{9}{2}x^2 + \frac{9}{2}x^3 + \frac{27}{8}x^4, \dfrac{3^r x^r}{r!}$, all x

2) $1 - \dfrac{x^2}{8} + \dfrac{x^4}{384}, \dfrac{(-1)^r x^{2r}}{(2r)!4^r}$, all x

3) $2x - \frac{4}{3}x^3, \dfrac{(-1)^r 2^{2r+1} x^{2r+1}}{(2r+1)!}$, all x

4) $-x - \dfrac{x^3}{3}, \dfrac{-(x)^{2r+1}}{2r+1}, -1 < x < 1$

5) $2 - 2x + 2x^2 - \frac{4}{3}x^3 + \frac{2}{3}x^4$

$\dfrac{(-1)^r 2^r x^r}{r!}$, all x

6) $-3x - \frac{9}{2}x^2 - 5x^3 - \frac{33}{4}x^4$,

$\dfrac{x^r}{r}\left[(-1)^{r+1} - 2^{r+1}\right], -\frac{1}{2} \leqslant x < \frac{1}{2}$

7) $-1 + x - x^2 + \frac{1}{2}x^3 - \frac{1}{6}x^4$,

$\dfrac{(-1)^{r-1}}{(r-1)!}x^r$, all x

8) $1 - 4x^2 + \frac{16}{3}x^4, \dfrac{(-1)^r(4x)^{2r}}{2(2r)!}$, all x

9) $1 + 2x + \frac{7}{2}x^2 + \frac{19}{6}x^3 + \frac{11}{8}x^4, \left(\dfrac{3^r}{r!} - \dfrac{2^{r-1}}{r}\right)x^r$,

$-\frac{1}{2} \leqslant x < \frac{1}{2}$

10) $x^3, \dfrac{(-1)^r}{4(2r+1)!}(3 - 3^{2r+1})x^{2r+1}$, all x

11) $x + \dfrac{x^3}{3!}, \dfrac{x^{2r+1}}{(2r+1)!}$, all x

12) $1 + \dfrac{x^2}{2!} + \dfrac{x^4}{4!}, \dfrac{x^{2r}}{(2r)!}$, all x

13) $y + \frac{1}{2}y^2$
14) y^2
15) $1 + y$
16) $\sin a + y\cos a - \dfrac{y^2}{2}\sin a$

17) $\dfrac{1}{x} - \dfrac{1}{2x^2} + \dfrac{1}{3x^3}, x \geqslant 1, x < -1$

18) $\ln 2 - \dfrac{1}{2x} + \dfrac{3}{8x^2} - \dfrac{7}{24x^3}, \quad -1 < x \leqslant 1$

19) 2.710 282
21) 0.693
22) 0.0953, −0.1054
23) $1 + x$
24) $x + \dfrac{x^3}{3!}$, 3.1
26) $x = 0$
27) $x - \frac{1}{6}x^3$

Exercise 6d – p. 262

1) $\cos a - (x - a)\sin a - \frac{1}{2}(x - a)^2 \cos a$
2) $e^a + e^a(x - a) + \frac{1}{2}e^a(x - a)^2$
3) $\cosh a + (x - a)\sinh a + \frac{1}{2}(x - a)^2 \cosh a$
4) $\dfrac{1}{a} - \dfrac{1}{a^2}(x - a) + \dfrac{1}{a^3}(x - a)^2$
5) $\ln(1 + a) + \dfrac{x - a}{1 + a} - \dfrac{(x - a)^2}{2(1 + a)^2}$
6) $\frac{1}{2}\sin 2a + (x - a)\cos 2a - (x - a)^2 \sin 2a$
7) $(x - 1)\cos 1^c - \frac{1}{2}(x - 1)^2(2\sin 1^c + \cos 1^c)$
8) $\frac{1}{2} + \frac{1}{2}\sqrt{3}(x - \pi/6) - \frac{1}{4}(x - \pi/6)^2$, 0.515

Miscellaneous Exercise 6 – p. 265

1) Divergent

3) $\Sigma\frac{1}{r}$ is divergent, $x > 1$

4) a) convergent, b) convergent
5) (i), (ii) and (iv) converge; (iii) diverges
6) (a) and (c) converge; (b) diverges
7) (i) $k > 1$, (ii) (a) diverges, (b) and (c) converge, (iii) $\frac{11}{96}$
8) (i) converges (ii) a) all values of p, b) $p < -1$ c) no values of p
9) $a = 1,\ b = 0$, or $a = -1,\ b = 2$

10) a) $2x + \frac{2x^3}{3} + \frac{2x^5}{5}$

11) $-\frac{3x}{2} - \frac{3x^2}{4} - \frac{3x^3}{2},\ -\frac{1}{2} \leqslant x < \frac{1}{2}$

12) $1 - 2x + 2x^2 - \frac{4}{3}x^3$,
$(-2x)^{r-1}/(r-1)!$,
$-\frac{2x}{r},\ k = 2,\ a = -1$

13) $|y| > 1$ 14) $\frac{2}{3}x^3$

15) $-\frac{x^2}{2} - \frac{x^4}{12}$

16) 0.955 17) $\frac{1}{6}\pi - \frac{1}{6}\sqrt{3}x^2$
18) $x + \frac{1}{2}x^2 + \frac{1}{2}x^3 + \frac{13}{24}x^4$
19) $-\frac{1}{2}x^2 - \frac{1}{12}x^4$, $\ln 2 \simeq 0.68$, all terms negative
20) $2x + \frac{4}{3}x^3$
21) $(x-1) + \frac{3}{2}(x-1)^2 + \frac{1}{3}(x-1)^3 - \frac{1}{12}(x-1)^4$
22) $(x-1) + \frac{1}{2}(x-1)^2 - \frac{1}{6}(x-1)^3 + \frac{1}{12}(x-1)^4 - \frac{1}{20}(x-1)^5$; 0.105 17

23) $\frac{\sqrt{3}\pi}{6}$

Exercise 7a – p. 273

1) $xy = e^x + A$
2) $3y \cos x = x^3 + A$
3) $2x \ln y = x^2 + 2x + A$
4) $y = Ax - x \cos x$
5) $ye^x = 2x + A$
6) $xe^y = e^x + A$
7) $4y \ln x = 2x^2 \ln x - x^2 + A$
8) $4y(1 + x) = x^4 + A$
9) $x \tan y = \ln \sec x + A$
10) $2e^{(x+y)} = e^{2x} + A$
11) $ye^{3x} = x + A$
12) $y \sin x = x + A$
13) $xy = x + \ln Ax$
14) $2y = (x+1)^3(x^2 + 2x + A)$
15) $2y \sin x = e^x(\sin x - \cos x) + A$
16) $(2v - 1)e^{t^2} = A$
17) $x^2y = A - \cos x$
18) $y = x(e^{-x} + A)$
19) $2r \sin\theta = \theta - \sin\theta \cos\theta + A$
20) $2y(x-1) + xe^{-x^2} = Ax$
21) $y = x(\sin x - \cos x - 1)$
22) $(1 - y)e^{\frac{1}{2}x^2} = 1$

Exercise 7b – p. 277

1) $y = Ae^x + Be^{2x}$
2) $y = Ae^x + Be^{\frac{4x}{3}}$
3) $y = e^x(A + Bx)$

4) $y = Ae^{-\frac{1}{2}x} \cos\left(\frac{\sqrt{3}x}{2} + \epsilon\right)$

5) $y = Ae^{4x} + Be^x$
6) $y = Ae^{2x} + Be^{-2x}$
7) $y = A \cos(2x + \epsilon)$

8) $y = Ae^{-\frac{x}{4}} \cos\left(\frac{\sqrt{15}}{4}x + \epsilon\right)$

9) $y = A + Be^{2x}$
10) $y = e^{\frac{1}{3}x}(A + Bx)$

Exercise 7c – p. 284

1) $y = Ae^x + Be^{5x} + \frac{3}{5}$
2) $y = e^x(A + Bx) + e^{2x}$
3) $y = e^x(A + Bx + \frac{1}{2}x^2)$
4) $y = Ae^{-x} + Be^{-2x} + \frac{1}{10}(\sin x - 3\cos x)$

5) $y = Ae^{-\frac{x}{2}} \cos\left(\frac{\sqrt{3}x}{2} + \epsilon\right) + x$

6) $y = Ae^{2x} + (B - \frac{3}{4}x)e^{-2x}$
7) $y = Ae^{\frac{x}{4}} + Be^x - \frac{1}{17}(4\cos x + \sin x)$
8) $y = e^{-\frac{x}{3}}(A + Bx) + x^2 - 10x + 45$
9) $y = A\cos(5x + \epsilon) + \frac{1}{629}(50 - 4i)e^{(1+i)x}$
10) $y = Ae^x + Be^{3x} + (8\cos 2x - \sin 2x)$
11) $s = e^{2t}(A + Bt) + \frac{5}{4}$

12) $x = Ae^{-\frac{1}{2}t} \cos\left(\frac{\sqrt{7}t}{2} + \epsilon\right) - \frac{1}{226}(1 - 15i)e^{(2-3i)t}$

13) $y = Ae^{-x} + Be^{-2x} + e^{x}$
$+ \frac{1}{10}(\sin x - 3\cos x)$
14) $\theta = A\cos(t + \epsilon) + e^{t}(\sin t - 2\cos t)$
15) $y = 2e^{5x} - e^{-2x}$
16) $y = \sqrt{325}\, e^{-4x}\cos(3x + \epsilon) + 2\cos x$
where $\tan\epsilon = \frac{-17}{6}$
17) $f(x) = xe^{-x}$; $y = e^{2x} - e^{3x} + x\,e^{-x}$
18) C.F. is $e^{-2x}(A + Bx)$; P.I. is $-\cos 2x$;
$y = 2e^{-2x}(1 + 2x) - \cos 2x$

Miscellaneous Exercise 7 – p. 285

1) $8y\sin x = 2\cos 2x - \cos 4x + K$
2) $x = Ae^{2t} + B\,e^{-t} - 3\sin t + \cos t$;
$x = 3e^{-t} - 3\sin t + \cos t$
3) a) $y = \tan\{\ln(1 + \cos^2 x)\}$
b) $xy + x + 1 = Ae^{x}$
4) $x^2 y = 2(x^2 - 1)^2$
5) a) $\tan y = 2\arcsin\left(\dfrac{2x}{3} - 1\right) + A$
b) $4y = 2(1 + x)$
$+ (1 - x^2)\ln\dfrac{A(1 - x)}{(1 + x)}$
6) $x = \cos t - 3\sin t$ in each case
7) $y = x\cos x - \sin x\cos^2 x + \cos x$
8) a) $\sqrt{(1 + y^2)} = \frac{1}{2}\ln\dfrac{A(1 + x^2)}{x^2}$
b) $y = 4e^{x} - x - 1$
9) a) $y = \sin x(1 + \sin x)$
b) $2ye^{x} = 1 - \cos 2x - \sin 2x$
10) $3y\sin^2 x = \sin^3 x + A$
11) a) $y = 2x^4 - 1$
b) $2y = e^{2x}(1 - \cos 2x - \sin 2x)$
12) a) $2\sqrt{2}\cos x = 1 + e^{-y}$
b) $ye^{x} = A(1 + x) - 1$
13) (i) $a = 4$, $b = 3$;
$y = Ae^{-x}\cos(2x + \epsilon) + 4 + 3x$
(ii) $y = x\tan(\ln x)$
14) $a = 2$, $y = -2e^{-2x} + 3e^{-x} + 2xe^{-x}$
15) (i) $P = 1, Q = 1, y = Ae^{5x/3} + Be^{-x}$
$+ \cos 2x + \sin 2x$
(ii) $\dfrac{d^2y}{dx^2} - 9y = -9x^2 - 16$
16) $y = 4\cos(3x + \pi/2) + 2; -2$;
$\frac{1}{6}(4n - 3)\pi \pm \pi/9$
17) $\dfrac{dx}{dt} = (\alpha - \beta)x$
19) $x = Ae^{2t} + Be^{-t} + \cos t - 3\sin t$;
$x = 3e^{-t} + \cos t - 3\sin t$

Exercise 8a – p. 298

1) a) $\cos 7\theta + i\sin 7\theta$
b) $\cos(-3\theta) + i\sin(-3\theta)$
$\equiv \cos 3\theta - i\sin 3\theta$
c) $\cos(\frac{1}{2}\theta) + i\sin(\frac{1}{2}\theta)$
d) $\cos\pi + i\sin\pi \equiv -1$
e) $\cos\left(-\dfrac{\pi}{2}\right) + i\sin\left(-\dfrac{\pi}{2}\right) \equiv -i$
f) $\cos\dfrac{\pi}{3} + i\sin\dfrac{\pi}{3} \equiv \frac{1}{2} + i\dfrac{\sqrt{3}}{2}$
2) a) $(\cos\theta + i\sin\theta)^5$
b) $(\cos\theta + i\sin\theta)^{-2}$
c) $(\cos\theta + i\sin\theta)^{\frac{1}{3}}$
d) $(\cos\theta + i\sin\theta)^{-\frac{1}{2}}$
3) $\cos 7\theta + i\sin 7\theta$
4) $\cos 9\theta + i\sin 9\theta$
5) $\cos 2\theta + i\sin 2\theta$
6) $\cos 3\theta - i\sin 3\theta$
7) $\cos 7\theta + i\sin 7\theta$
8) $\cos 3\theta + i\sin 3\theta$
9) $\cos\left(\dfrac{10\pi}{3}\right) + i\sin\left(\dfrac{10\pi}{3}\right) = -\frac{1}{2} - \dfrac{\sqrt{3}}{2}i$
10) 1   11) $\cos\theta + i\sin\theta$
12) $\dfrac{\sqrt{3}}{2} + \dfrac{1}{2}i$
13) $\cos\dfrac{\pi}{5} + i\sin\dfrac{\pi}{5}$
14) $\cos\dfrac{\theta}{n} + i\sin\dfrac{\theta}{n}$
21) $-1.00, 0.268, 3.73$
22) a) $\frac{1}{4}\sin 4\theta + 2\sin 2\theta + 3\theta + K$
b) $\frac{3}{2}\sin 4\theta + \frac{15}{2}\sin 2\theta + 10\theta + K$
c) $2\sin 2\theta - 3\theta + K$

Exercise 8b – p. 308

1) a) $\pm 2^{1/4}\left\{\cos\dfrac{\pi}{8} - i\sin\dfrac{\pi}{8}\right\}$
b) $\pm\sqrt{5}\,\{\cos(0.46^c) + i\sin(0.46^c)\}$
c) $\pm\sqrt{13}\{\cos(0.98^c) + i\sin(0.98^c)\}$
d) $\pm\sqrt{2}\left\{\cos\dfrac{\pi}{12} + i\sin\dfrac{\pi}{12}\right\}$
e) $\pm 8^{1/4}\left\{\cos\left(\dfrac{5\pi}{8}\right) + i\sin\left(\dfrac{5\pi}{8}\right)\right\}$

f) $\pm\left(\cos\frac{\pi}{4} - i\sin\frac{\pi}{4}\right) = \pm\frac{1}{\sqrt{2}}(1+i)$

g) $\pm\left(\cos\frac{\pi}{2} + i\sin\frac{\pi}{2}\right) = \pm i$

2) a) $2^{1/6}\left\{\cos\left(\frac{2}{3}n\pi - \frac{\pi}{12}\right) + i\sin\left(\frac{2}{3}n\pi - \frac{\pi}{12}\right)\right\},\ n = -1, 0, 1$

b) $5^{1/3}\{\cos(\frac{2}{3}n\pi + 0.31^c) + i\sin(\frac{2}{3}n\pi + 0.31^c)\},\ n = -1, 0, 1$

c) $(13)^{1/3}\{\cos(\frac{2}{3}n\pi + 0.65^c) + i\sin(\frac{2}{3}n\pi + 0.65^c)\},\ n = -2, -1, 0$

d) $2^{1/3}\left\{\cos\left(\frac{2}{3}n\pi + \frac{\pi}{18}\right) + i\sin\left(\frac{2}{3}n\pi + \frac{\pi}{18}\right)\right\},\ n = -1, 0, 1$

e) $\sqrt{2}\left\{\cos\left(\frac{2}{3}n\pi - \frac{\pi}{4}\right) + i\sin\left(\frac{2}{3}n\pi - \frac{\pi}{4}\right)\right\},\ n = -1, 0, 1$

f) $\cos\left(\frac{2}{3}n\pi + \frac{\pi}{6}\right) + i\sin\left(\frac{2}{3}n\pi + \frac{\pi}{6}\right)$, $n = -1, 0, 1$

g) $\cos\left(\frac{2}{3}n\pi + \frac{\pi}{3}\right) + i\sin\left(\frac{2}{3}n\pi + \frac{\pi}{3}\right)$, $n = -1, 0, 1$

5) 31

6) $-\alpha^3(1+\alpha)$

7) $1 - \beta^2$

8) $-1,\ \cos\frac{\pi}{3} + i\sin\frac{\pi}{3},\ \cos\left(-\frac{\pi}{3}\right) + i\sin\left(-\frac{\pi}{3}\right)$

9) $3abc - (a^3 + b^3 + c^3)$

12) 2^{13}

Exercise 8c – p. 312

1) a) $\sqrt{2}e^{i\pi/4}$ b) $e^{i\pi/2}$
c) $4e^{-i\pi/3}$ d) $\sqrt{2}e^{i3\pi/4}$
e) $4e^{0i}$ f) $5e^{0.93i}$
g) $e^{i2\pi/3}$

2) a) $\frac{1}{2} - \frac{\sqrt{3}}{2}i$ b) $-\sqrt{3} + i$
c) -5 d) $-i$ e) -4

3) $e^{i\pi/3}, e^{-i\pi/3}, e^{i\pi}$

4) $2e^{-i3\pi/4}, 2e^{-i7\pi/20}, 2e^{i\pi/20}, 2e^{i9\pi/20}, 2e^{i17\pi/20}$

5) $e^{0i}, e^{\pm i2\pi/7}, e^{\pm i4\pi/7}, e^{\pm i6\pi/7}$

6) $\sqrt{2}e^{-0.18i}, \sqrt{2}e^{1.39i}, \sqrt{2}e^{2.96i}, \sqrt{2}e^{4.53i}$

7) $-\frac{1}{2}\{1 - e^{-\pi/2}\}$

8) $\frac{1}{5}\{2e^{\pi/4} - 1\}$

Exercise 8d – p. 323

15) a) $|z - 3 - 4i| = 5$ b) $|z - 5| = 2$
c) $|z - 4i| = 4$

16) a) $r = 1$ b) $r = 4$ c) $\theta = \frac{1}{4}\pi$
d) $\theta = -\frac{2}{3}\pi$

17) a) $x^2 + y^2 = 1$ b) $x^2 + y^2 = 16$
c) $y = x\ (x > 0)$ d) $y = \sqrt{3}x\ (x < 0)$

18) a) (i) $|z - 4| = |z + 8|$ (ii) $x = -2$
b) (i) $|z - 1 - 2i| = |z - 7 + 4i|$
(iii) $y = x - 5$
c) (i) $|z - 6i| = |z - 6|$ (iii) $y = x$

19) a) $\arg(z - 2) = \frac{2}{3}\pi$
b) $\arg(z + 1) = -\frac{1}{2}\pi$
c) $\arg(z + 1 - 2i) = -\frac{3}{4}\pi$

20) $3x^2 + 3y^2 - 16x - 8y = 0$

21) $x = 0, y = 0$

22) $\theta = 0$ and $r = 2$

23) $r = 2\cos\theta$

24) $\theta = \pi/2$ and $r = 1$

Exercise 8e – p. 327

3) a) $\sqrt{2}$ b) $\sqrt{17} - 2$

4) a) $2\sqrt{2}(1 + i)$ b) $1 + 3i, 1 - 5i$

5) a) $(2, 1), (0, -1)$ b) $(\sqrt{2}, -\sqrt{2})$

Exercise 8f – p. 335

1) a) a circle, centre (4, 0), radius 3
b) the part-line from (4, 0) at $\frac{\pi}{3}$ to Ox
c) a circle, centre 0, radius 5
d) a circle, centre (4, 0), radius 1
e) the line through (4, 0) at arctan 2 to Ox
f) a parabola $v^2 = 4(u - 4)$

2) a) $u^2 - v^2 = 4$

b) $\arg(w) = -\dfrac{\pi}{2}$, i.e. the negative v axis

c) $|w| = 8$ d) $v = 2u - 6$

3) a) $|w| = \frac{1}{6}$ b) $\arg(w) = \dfrac{\pi}{4}$

c) $u^2 + 4v^2 = 4(u^2 + v^2)^2$

d) $4(u^2 + v^2)^2 + uv = 0$

e) $3u + v + 4(u^2 + v^2) = 0$

4) a) $(u^2 + v^2)^2 = 4uv$

b) the u axis ($v = 0$)

c) the line $u + v = 0$

5) a) $u^2 + v^2 - 2u + v = 0$

b) $u^2 + v^2 - u = 0$

c) the v axis ($u = 0$)

Miscellaneous Exercise 8 – p. 336

1) -1 2) $9, 6\theta$ 5) -2^{11}

6) $\cos\left\{\dfrac{p(\theta + 2n\pi)}{q}\right\} + i\sin\left\{\dfrac{p(\theta + 2n\pi)}{q}\right\}$,

$(n = -4, -3, \ldots, 3, 4)$;

$\pm i, \frac{1}{2}(\pm\sqrt{3} \pm i)$

10) a) $|u| = 1$ or $\arg u = 0$ or π

b) $(-\frac{1}{2} + i\sqrt{3}/2), (-\frac{1}{2} - i\sqrt{3}/2)$

11) a) 3 or 0

12) $1, -\frac{1}{2} \pm \dfrac{\sqrt{3}}{2}i$

13) $-7 - 24i$; $\frac{1}{25}(3 + 4i)$; $\pm(2 - i)$

14) $32\cos^6\theta - 48\cos^4\theta + 18\cos^2\theta - 1$; $\frac{1}{60}$

15) $-1, \frac{1}{2} \pm i\sqrt{3}/2$

16) a) (i) 0 (ii) 3 b) $(-\frac{5}{4}, 0), \frac{3}{4}$

17) $\tan\dfrac{r\pi}{15}, r = 4, 7, 13$

18) $\sqrt{3}; -1$ 19) $2e^{-i\pi/6}$

21) $0, 3; 4\omega^2, 4\omega$

22) $2e^{i\pi/6}, 2e^{i5\pi/6}, e^{-i\pi/2}$

23) $8e^{i\pi/6}$, b) $2e^{i\pi/18}, 2e^{i13\pi/18}, 2e^{-i11\pi/18}$

24) a) $z = 1, e^{\pm i2\pi/5}, e^{\pm i4\pi/5}$

25) $u = \dfrac{2x}{x^2 + (y+1)^2}$, $v = \dfrac{x^2 + y^2 - 1}{x^2 + (y+1)^2}$

26) $2 - i, 3 - 4i, (2, 1)$

27) $\frac{1}{5}(-2 + 6i)$

28) $x^2 + y^2 + 2x - 4y = 15$

29) $(\sqrt{2} + 1)/(\sqrt{2} - 1)$

30) $4x^2 + 4y^2 + x + 9y + 4 = 0$; $(\frac{1}{4}, -\frac{3}{4})$, $(-\frac{1}{2}, -\frac{3}{2})$

31) $(x + 1)^2 + y^2 = 4$

33) (i) $3 + 2i, -3 - 2i$

34) $\frac{1}{2}\sqrt{3} + \frac{1}{2}i, -\frac{1}{2}\sqrt{3} + \frac{1}{2}i$

35) $a = 4, b = 9, c = -4$; (i) $(4, 0), 5$

Exercise 9a – p. 348

1) $-x - 4$ 2) 13

3) $p = 3, q = 5$

4) $a = 9, b = -2, c = -11$

5) $-7x + 10$ 6) $m = \frac{9}{2}, n = -\frac{7}{2}$

7) a) no b) $(x - 3), (x + 3)$

c) $(x - 1)$ d) no

8) $-5, -4$ 9) 3

10) 2 13) 3 14) $p = \frac{3}{2}, q = 2$

16) $(q^2 - p)(p^2 - q) = (pq - 1)^2$

17) -2

Exercise 9b – p. 351

1) 1, homogeneous and cyclic
2) 2, homogeneous
3) 2, homogeneous and cyclic
4) 2, homogeneous
5) 3, homogeneous
6) 3, cyclic
7) 3, homogeneous and cyclic
8) 2, homogeneous
9) 2, homogeneous and cyclic
10) 2, homogeneous and cyclic

11) a) $\alpha(\alpha^2 - \beta^2) + \beta(\beta^2 - \gamma^2) + \gamma(\gamma^2 - \alpha^2)$

b) $xy^2 + yz^2 + zx^2$

c) $\alpha(\beta^2 - \gamma^2) + \beta(\gamma^2 - \alpha^2) + \gamma(\alpha^2 - \beta^2)$

d) $ab^2 + bc^2 + cd^2 + da^2$

12) a) $\sum x^2(y^2 + z^2)$ b) $\sum \alpha^2(\beta + \gamma)$

c) $\sum \alpha$ d) $\sum a^2 b$

Exercise 9c – p. 354

1) $2a(a^2 + 3b^2)$ 2) $(x + y + z)^2$

3) $-(x - y)(y - z)(z - x)$

4) $(a - b)(b - c)(c - a)$

5) $3(a - b)(b - c)(c - a)$

6) $-(x - y)(y - z)(z - x) \times (x^2 + y^2 + z^2 + xy + yz + zx)$

7) $(a - b)(a + b)(a^2 + ab + b^2) \times (a^2 - ab + b^2)$

8) $(x-2)(x+2)(x^2+2x+4)$
$\times (x^2-2x+4)$

9) $-(p-q)(q-r)(r-p)$

10) $(a+b+c)(a^2+b^2+c^2+2bc$
$-ab-ac)$

11) $\dfrac{a^n-x^n}{a-x}$;
$(x-2)(x^4+2x^3+4x^2+8x+16)$;
$(a-b)(a^4+a^3b+a^2b^2+ab^3+b^4)$

13) $(p-q)^2(p+q)$

14) $(a+2b+3c)$
$\times (a^2+4b^2+9c^2-2ab-6bc-3ac)$

9) $(x^2-(1+\sqrt{3})x+2)$
$(x^2-(1-\sqrt{3})x+2)$
(x^2+2x+2)
(x^2-4x+5)
(x^2+4x+5)
$(x-1)(x+1)(x^2+x+1)(x^2-x+1)$
$(x-3)(x^2-3x+3)$
$(x+1)(x^2-x+1)(x^2+x+1)$
$(x+1)(x^2+1)$
$(x-1)(x^2+1)$
$(x^2-\frac{1}{2}(1+\sqrt{5})x+1)$
$(x^2-\frac{1}{2}(1-\sqrt{5})x+1)$

Exercise 9d – p. 362

1) a) $\frac{1}{4}; \frac{1}{2}; \frac{7}{4}$ b) $0; -3; -1$
c) $0; 0; \frac{1}{8}$ d) $0; -1; 0$
e) $-4; 0; 5$

2) a) $7; 31; 18; -\frac{1}{5}$ b) $\frac{25}{9}; \frac{14}{9}; \frac{13}{9}; 4$
c) $-\frac{3}{2}; \frac{9}{16}; \frac{21}{4}; -\frac{3}{7}$
d) $1; 10; 6; -\frac{2}{3}$ e) $0; 0; 3; 0$
f) $0; 0; 2; 1$
g) 1; 0; 0; undefined

3) a) $x^3+8x^2+15x+7=0$
b) $x^3-5x^2+2x+1=0$
c) $x^3+4x^2-20x+8=0$
d) $x^3-14x^2+21x-1=0$
e) $x^3+5x^2+2x-1=0$
f) $x^3+4x^2-x-11=0$
g) $x^3-12x^2+29x-7=0$

4) $\frac{33}{7}; \frac{32}{49}; 7\,\Sigma\alpha^5 = 11\,\Sigma\alpha^2 + 4\,\Sigma\alpha^3$

5) a) $b^3d = c^3a$
b) $4abc = b^3 + 8a^2d$
c) $9ad = bc$ and $b^2 = 3ac$

6) $x^3-2x^2-5x+6=0; -2, 1, 3$

7) $-\frac{4}{3}, -\frac{7}{3}, -\frac{5}{3}, -1;\ \frac{58}{9}, \frac{5}{3}$;
$48x^4+32x^3-28x^2+10x=3$

8) $2, 2, \frac{1}{3}(-2 \pm \sqrt{10})$

10) $X^4-5X^3-7X^2+5X+6=0$;
$x=-3, -3, -1, 4$

Exercise 9e – p. 366

1) $\sqrt{2}e^{\pm i\pi/12}, \sqrt{2}e^{\pm i3\pi/4}, \sqrt{2}e^{\pm i7\pi/12}$

2) $\pm\sqrt{5}e^{\pm i(0.46^c)}$

3) $\pm 1, \pm e^{\pm i\pi/3}$

4) $3, 2 + e^{\pm i2\pi/3}$

5) $-1, e^{\pm i\pi/3}, e^{\pm i2\pi/3}$

6) $-1, \pm e^{i\pi/2}$

7) $1, \pm e^{i\pi/2}$

8) $e^{\pm i\pi/5}, e^{\pm i3\pi/5}$

Multiple Choice Exercise 9 – p. 367

1) b	2) b	3) a
4) b	5) d	6) b
7) b, c	8) a, b, c	9) a
10) b, c	11) *B*	12) *B*
13) *E*	14) *B*	15) *A*
16) *C*	17) *A*	18) *B*
19) *E*	20) *F*	21) *T*
22) *F*	23) *T*	24) *T*
25) *F*		

Miscellaneous Exercise 9 – p. 369

1) $a=1, b=-37$

2) $m=-1, n=-2$

3) $2x^2-5x-3;\ (x+1)^2(2x^2-5x-3)$

4) $[(R_1-R_2)x+aR_2-bR_1]/(a-b)$

5) $3, -1;\ 0, -3$

6) $a=\frac{1}{2}f(1), b=-f(2), c=\frac{1}{2}f(3);\ -90$

8) a) 0 b) $a^{n-1}x-a^n$

9) $\sqrt{7}, \frac{1}{2}(-\sqrt{7}\pm\sqrt{5})$

10) $(x-y)(y-z)(z-x)(x+y+z)$

11) $(a+b+c)(a-b+c)(a-b-c)$
$\times (a+b-c)$

13) $(a+b+c)(a-b+c)(a+b-c)$
$\times (b+c-a)$; positive

14) $-(a-b)(b-c)(c-a)$;
$-(a-b)(b-c)(c-a)(a^2+b^2+c^2)$

15) $-\frac{3}{4}, \frac{1}{4}, \frac{1}{2}$ 16) $2x^3+3x^2+8=0$

17) $-\frac{4}{3}, -\frac{1}{2}, \frac{1}{3}$

18) a) $x^3-2x^2-16x+40=0$
b) $5x^3-4x^2-x+1=0$
c) $x^3-2x^2-3x-1=0;\ 10$

19) -216

20) $p=10;\ q=31$

21) $3, \frac{3}{2}, -\frac{1}{2}, -2$

23) $-2pq$; 0; $x = -2, 3, 6$

24) $(0, 0)$ min; $(-\frac{1}{4}, \frac{5}{256})$ max;
$(-2, -4)$ min
a) $\frac{1}{2} \pm i\sqrt{3}/2$
b) -3; 1 and 2 c) 7

25) $a = 25$, $b = -54$; $1 - 3i$,
$2 \pm \sqrt{3}i$

26) $e^{\pm i\pi/9}$, $e^{\pm i7\pi/9}$, $e^{\pm i5\pi/9}$
$(x^2 - 2x\cos\pi/9 + 1)$
$(x^2 - 2x\cos 5\pi/9 + 1)$
$(x^2 - 2x\cos 7\pi/9 + 1)$

27) $\phi = 2\pi/9, 4\pi/9, 8\pi/9, 10\pi/9, 14\pi/9, 16\pi/9$; $\theta = 2\pi/9, 4\pi/9, 8\pi/9, 10\pi/9, 14\pi/9, 16\pi/9$

28) $|z| = 2$, $\arg z = \pm\pi/5, \pm 3\pi/5, \pi$;
$(z^2 - 4z\cos\pi/5 + 1)$
$(z^2 - 4z\cos 3\pi/5 + 1)$

29) $\cos\pi/4 \pm i\sin\pi/4$;
$|z| = \sqrt{2}$, $\arg z = \pm\pi/4, \pm 3\pi/4$;
$(z^2 - 2z + 2)(z^2 + 2z + 2)$

30) $e^{\pm i2\pi/5}$, $e^{\pm i4\pi/5}$, $e^{i2\pi}$, $e^{\pm i\pi/4}$, $e^{\pm i3\pi/4}$,
$e^{\pm i2\pi/5}$, $e^{\pm i4\pi/5}$, $e^{i2\pi}$

Exercise 10a – p. 377

1) $nI_n = (n-1)I_{n-2} - \cos x \sin^{n-1}x$
2) $2(n-1)(I_n + I_{n-2}) = \tan^{n-1}2x$
3) $2I_n = (x+1)^n e^{2x} - nI_{n-1}$
4) $2_nI_n = 2(n-1)I_{n-2} + \sin 2\theta \cos^{n-1}2\theta$
5) $aI_n = x^n e^{ax} - nI_{n-1}$
6) $2I_n = x^2(\ln x)^n - nI_{n-1}$
7) $nI_n = \sinh x \cosh^{n-1}x + (n-1)I_{n-2}$
8) $(n-1)I_n = \tan x \sec^{n-2}x + (n-2)I_{n-2}$
9) $\frac{1}{6}\sin x\cos^5 x + \frac{5}{24}\sin x\cos^3 x + \frac{1}{16}\sin x\cos x + \frac{5}{16}x + K$
10) $K - \frac{1}{7}\sin^6 x\cos x - \frac{6}{35}\sin^4 x\cos x - \frac{8}{35}\sin^2 x\cos x - \frac{16}{35}\cos x$
11) $e^{4x}\{\frac{1}{4}(1-x)^3 + \frac{3}{16}(1-x)^2 + \frac{3}{32}(1-x) + \frac{3}{128}\} + K$
12) $\frac{1}{4a}\sin(ax+b)\cos^3(ax+b) + \frac{3}{8a}\sin(ax+b)\cos(ax+b) + \frac{3x}{8} + K$
13) $\frac{1}{12}\cos\left(\frac{\pi}{4} - 3\theta\right)\sin^3\left(\frac{\pi}{4} - 3\theta\right) + \frac{1}{8}\cos\left(\frac{\pi}{4} - 3\theta\right)\sin\left(\frac{\pi}{4} - 3\theta\right) + \frac{3}{8}\theta + K$
14) $\frac{1}{4}\tan^4 x - \frac{1}{2}\tan^2 x + \ln(\sec x) + K$
15) $K - x^3\cos x + 3x^2\sin x + 6x\cos x - 6\sin x$
16) $\frac{1}{5}\cosh x\sinh^4 x - \frac{4}{15}\cosh x\sinh^2 x + \frac{8}{15}\cosh x + K$
17) $\frac{1}{3}\tan x\sec^2 x + \frac{2}{3}\tan x + K$
18) $K - e^{-x}(x^6 + 6x^5 + 30x^4 + 120x^3 + 360x^2 + 720x + 720)$
19) $x^2\{\frac{1}{2}(\ln x)^3 - \frac{3}{4}(\ln x)^2 + \frac{3}{4}(\ln x) - \frac{3}{8}\} + K$
20) $\frac{1}{12}\sinh 3x\cosh^3 3x + \frac{1}{8}\sinh 3x\cosh 3x + \frac{3}{8}x + K$
21) $\frac{1}{216}(8x^3 - 1)(1 + x^3)^8 + K$
22) $\ln|\cot x - \operatorname{cosec} x| + \cos x + \frac{1}{3}\cos^3 x + \frac{1}{5}\cos^5 x + K$

Exercise 10b – p. 385

1) $\frac{256}{693}$ 2) $\frac{63\pi}{512}$ 3) 0
4) $\frac{5\pi}{16}$ 5) $\frac{105\pi}{256}$ 6) $\frac{8}{15}$
7) $\frac{3\pi}{4}$ 8) $\frac{231\pi}{512}$ 9) 8!
10) $\frac{13}{15} - \frac{\pi}{4}$ 11) $\frac{7\pi}{512}$ 12) $\frac{8}{35}$
13) 0 14) $\frac{3\pi}{64}$
15) $I_n = e - nI_{n-1}$; $120 - 44e$
16) $\frac{3^n n!}{(3n+2)(3n-1)\dots(8)(5)(2)}$
17) a) $37\sinh 1 - 28\cosh 1$
b) $7\sinh 1 - 9\cosh 1 + 6$
18) $(\frac{18}{21})(\frac{16}{19})\dots(\frac{4}{7})(\frac{2}{5})(\frac{2}{3})$
19) $\left(\frac{\pi}{2}\right)^6 - 30\left(\frac{\pi}{2}\right)^4 + 360\left(\frac{\pi}{2}\right)^2 - 720$
20) $\frac{1}{5}\sinh^5 1 - \frac{1}{3}\sinh^3 1 + \sinh 1 - 2\arctan e + \frac{\pi}{2}$
21) $\frac{5!\,6!\,2^{12}}{12!}$

Exercise 10c – p. 392

3) $\frac{1}{2}$ 5) $\frac{3}{2}\sqrt[3]{4}$ 8) $\frac{1}{2}$
11) $\pi/2 - 1$ 13) $-\frac{1}{4}$ 15) $\frac{1}{2a}\ln 3$

Nos 1, 2, 4, 6, 7, 9, 10, 12, 14 do not exist

Exercise 10d – p. 396

1) $-\frac{1}{6}$ 2) $2\ln 2$ 3) $\frac{1}{4}(e^{-1} - e^{-5})$
4) $(\ln 2)2/\pi$ 5) $\frac{1}{2}$ 6) $\ln\frac{4}{3}$

7) $4/3\pi$ 8) e^2
9) $\pi/4$ 10) $\pi/3$ 11) $\frac{1}{2}$
12) $\sqrt{(4/\pi - 1)}$ 13) $\sqrt{(\frac{3}{2} + 4/\pi)}$
14) $\frac{e\sqrt{2}}{2}\sqrt{(e^2 - 1)}$ 15) $\frac{2\sqrt{6}}{3}$
16) $\frac{1}{2}\sqrt{(\sinh 4 - \sinh 2 - 2)}$
17) 0.813 18) 0.434 19) $\sqrt{(943/15)}$

Exercise 10e – p. 403

1) π 2) $c \sinh \frac{x}{c}$
3) $\frac{1}{27}(13\sqrt{13} - 8)$
4) $\frac{1}{2}(\sqrt{2} + \operatorname{arsinh} 1)$ 5) $\frac{1}{2}\ln 3$
6) $2a\sqrt{2}$
7) $a\{\operatorname{arsinh} t + t\sqrt{(t^2 + 1)}\}$
8) $2\pi a$ 9) $\arcsin(\tanh 1)$
10) $\frac{k}{2}\{\operatorname{arsinh}\frac{\pi}{2} + \frac{\pi}{4}\sqrt{(\pi^2 + 4)}\}$
11) $8a$ 12) $\frac{1}{2}\pi a$ 13) $a\sqrt{2}(e^{\pi} - 1)$

Exercise 10f – p. 409

1) $2\pi a^2$ 2) $\frac{12}{5}\pi a^2$
3) $4\pi\sqrt{5}$ 4) $4\pi a^2$
5) $4\pi a^2$ 6) $\frac{8\pi}{3}(17^{\frac{3}{2}} - 1)$
7) $\frac{2\pi\sqrt{2}}{5}(2e^{\pi} + 1)$
8) $\frac{1}{2}\pi a^2\{4\sqrt{7} - 2 + \sqrt{2}(\operatorname{arcosh}\sqrt{2} - \operatorname{arcosh} 2\sqrt{2})\}$
9) $\frac{1}{2}\pi(2 + \sinh 2)$
10) $\pi\{\operatorname{arsinh} e - \operatorname{arsinh} 1 + e\sqrt{(1 + e^2)} - \sqrt{2}\}$
11) $\frac{6\pi}{5}\{247\sqrt{13} + 64\}$
12) a) $2^{10}\pi$ b) $\frac{2^{10}\pi}{5}$
13) $\frac{16\pi}{105}$ 14) $5\pi^2 a^3$
15) $\frac{128\pi\sqrt{2}}{15}$
16) a) $\pi(31 - 32\ln 2)$
b) $\pi(129 - 32\ln 2)$
c) 127π d) 29π

Exercise 10g – p. 416

1) $\sqrt{2} + \operatorname{arsinh} 1$ 2) $\frac{1}{12}(17\sqrt{17} - 5\sqrt{5})$
3) $\frac{17}{32}\operatorname{arsinh} 2 + \frac{7\sqrt{5}}{16}$
4) a) 16 b) $\frac{128}{5}$
5) a) $\frac{1}{4}\pi$ b) $\frac{1}{4}(e^4 - 1)$
c) $\frac{8a^3}{105}$ d) $\frac{1}{4}$
6) a) 2 b) $e^2 - 1$
c) $\frac{3\pi a^2}{32}$ d) $1 - \ln 2$
7) a) $\frac{1}{8}\pi$ b) $\frac{1}{4}(e^2 + 1)$
c) $\frac{256a}{315\pi}$ d) $\frac{1}{4(1 - \ln 2)}$
8) a) $\frac{1}{2}\pi^2$ b) $\frac{1}{2}\pi(e^4 - 1)$
c) $\frac{16\pi a^3}{105}$ d) $\frac{\pi}{2}$
9) $2; \pi; \frac{2}{\pi}; 4\pi$
10) $\frac{8}{3}\pi$; $\left(0, \frac{4}{3\pi}\right)$ 11) $(2, \frac{3}{2})$; 15π

Exercise 10h – p. 424

1) $\frac{17\sqrt{17}}{4}$ 2) $\frac{7\sqrt{7}}{4}$ 3) $2\sqrt{2}$
4) $\frac{5\sqrt{5}}{2}$ 5) $4a\sqrt{2}$ 6) $\sqrt{2}$
7) 1 8) $\cosh^2 1$ 9) $4\sqrt{2}$
10) $\frac{(a^2 + 3b^2)^{\frac{3}{2}}}{8ab}$ 11) $\frac{17c\sqrt{17}}{16}$
12) $\frac{3a}{2}$ 13) ∞ 14) a
15) $\frac{6\sqrt{13}}{169}$ 16) $\frac{\sqrt{5}}{100}$ 17) $\frac{2}{a}$
18) $\frac{14\sqrt{5}}{25a}$ 19) $\frac{3}{4a}$
20) $\frac{3\sqrt{(\cos 2\theta)}}{a}$
21) $\arctan(-\frac{1}{2}\cot 2\theta)$
22) $\arctan\left(\frac{\sin\theta - 1}{\cos\theta}\right)$ 23) $\frac{\pi}{4}$
24) $\arctan(\frac{1}{3}\tan 3\theta)$

Miscellaneous Exercise 10 – p. 424

1) $\frac{16}{105}$ 2) $\frac{22}{65}e^{\pi} + \frac{6}{65}$ 4) $\frac{1}{4}\ln 3 + \frac{1}{3}$
5) $I_m = \frac{x^4}{4}(\ln x)^m - \frac{m}{4}I_{m-1}$;
$\frac{x^4}{128}\{32(\ln x)^3 - 24(\ln x)^2 + 12\ln x - 3\}$
6) $I_n = \frac{1}{n-1}\tan^{n-1} x - I_{n-2}$

7) a) $37 \sinh 1 - 28 \cosh 1$
$= \frac{1}{2e}(9e^2 - 65)$

b) $7 \sinh 1 - 9 \cosh 1 + 6$
$= 6 - \frac{1}{e}(e^2 + 8)$

8) $I_n = \frac{1}{2n+1}(2^n + 2nI_{n-1})$

9) $\frac{1}{2}\pi$

10) $2y = \sin x(\sin 4x + 2 \sin 2x + 2x + A)$;
$24y \sin x = 3 \sin 4x - 2 \sin 6x + A$

11) (i) π (ii) $\ln(1 + \sqrt{2})$ (iii) 1

12) a) $\pi/3$ b) $\pi\sqrt{3}/9$
c) $\ln(2 + \sqrt{3})$

13) $2(e - 1)$

14) $\frac{49\pi}{3}$

16) $\frac{3}{4}$; a) $3 + \ln 2$ b) $\frac{\pi}{2}(\ln 2 + \frac{15}{16})$

19) $8a$; $\frac{64}{5}\pi a^2$

20) a) $\frac{12\pi}{5}$ b) $2 \cosh^2 1$; $4 \sinh 1$

21) $\frac{6\pi a^2}{5}$ 22) $\frac{58a}{35}$

23) $\frac{335a}{27}$ 24) $\pi + 2\sqrt{2}$; $\frac{1}{3}\pi$

25) $\frac{\sqrt{2}}{2}$ 26) $\frac{2r}{\pi}$

27) $\left(0, \frac{4b}{3\pi}\right)$; $\frac{2}{3}ab^3$

28) a) 21π b) 84π

29) $(0, 0)$, (π, π), $(2\pi, 2\pi)$, $(3\pi, 3\pi)$,
$(4\pi, 4\pi)$; $\frac{1}{6}(32\pi^2 + 15)$

Exercise 11a – p. 434

Answers in this chapter will vary slightly depending on the number of decimal places used in the working.

1) a) $1 - x^2$; 0.75 b) $2 + \frac{1}{4}x^2$; 2.0625
c) $-1 + 2x - 3x^2 + \frac{16}{3}x^3$; -0.0833

2) a) accurate to 1 s.f.: true $y = 0.8$
b) accurate to 0 s.f.: true $y = 4.25$
c) accurate to 0 s.f.: true $y = -0.3993$

3) $3 + x - \frac{1}{2}x^2 + \frac{2}{3}x^3$; 3.4583; 4.1667

4) $1 + x - \frac{1}{2}x^2 - \frac{1}{6}x^3 + \frac{1}{6}x^4$; 1.09485;
1.3646

5) $1 - (x-1) + (x-1)^2 - (x-1)^3$;
$y \simeq -x^3 + 4x^2 - 6x + 4$; 0.909;
0.6250; 0.181

6) $xy = 1$;
when $x = 1.1$, y accurate to 3 s.f.;
when $x = 1.5$, y accurate to 1 s.f.;
when $x = 1.9$, error is approx. 66%

7) $y \simeq -x^2 + x$

8) 2.32 9) 1.9 10) -0.875

11) 2.1327 12) 0.68 13) 1.1876

14) $(1 - \pi/2)x + \frac{\pi}{12}x^3 - \frac{\pi}{240}x^5 + \frac{\pi}{10\,080}x^7$;
$-0.325\,322\,08$;
$y = \frac{\pi}{2}\cos\left(x + \frac{\pi}{2}\right) + x$;
y accurate to 6 s.f.

Exercise 11b – p. 443

1) -0.8253 2) 1.9766
3) 0.8290 4) 0.25
5) -0.8087 (with $y_{-1} = -1$)
6) 30.5 ($y_{-1} = -1$)
7) 2.6833 ($y_{-1} = 0.7$)
8) 1.244, 1.6639 ($y_{-1} = 0.9$)
9) $\frac{1}{y} + x^2 + 1 = 0$, -0.8
10) 1.11, 1.24205
11) 0.0824, 0.6391

Exercise 11c – p. 447

1) 1.6737 2) 1.2804, 1.5064
3) 1.7857 4) 1.1414
5) 1.0810 6) 0.9262
7) 1.6801 8) 0.4817
9) 0.9252, 0.9255
10) 0.2502, 0.2684

Exercise 12a – p. 455

2) $f(x) \leqslant \frac{2}{5}$, $f(x) > 2$
4) $\{x : x \in \mathbb{R}\}$, $\frac{1}{2} - \sqrt{\frac{5}{2}} \leqslant f(x) \leqslant \frac{1}{2} + \sqrt{\frac{5}{2}}$

Exercise 12b – p. 464

1) $x = 0$ 2) $y + 3x = 0$
3) $y = x$ 4) $2y = 3x$
5) $x = 0$ 6) $y = 0$
7) $y = \pm x$ 8) $2y = \pm x$
9) $(1, 0)$ 10) none
11) none 12) $(0, -1)$
13) $(0, 0)$ 14) $(a, 0)$; $(a, -8a)$
15) a) $y^3 = x$ b) $y^3 = x^2$
c) $y^3 = x$ at $(0, 0)$; $x = 0$
16) $y = x^3 - 6x^2 - 3x$
17) $4y = 4x^2 + 5x$ 18) $3y = 8x$

Exercise 12c – p. 476

23) $\begin{cases} x^2+y^2 \leqslant 4 \\ y < 1 \end{cases}$

24) $\begin{cases} x^2+y^2 \leqslant 9 \\ 4x^2+9y^2 \geqslant 36 \end{cases}$

25) $y \geqslant |x|$
26) $1 \leqslant (x-2)^2+(y-1)^2 \leqslant 4$

27) $\begin{cases} x^2+y^2 \geqslant 9 \\ x^2+4y^2 \leqslant 16 \end{cases}$

28) $3y \geqslant 8-2x,\ y \leqslant 2x,\ y \leqslant 8-2x$

Multiple Choice Exercise 12 – p. 478

1) a 2) d 3) b
4) a 5) e 6) d
7) a, b 8) a 9) *E*
10) *A* 11) *A*

Miscellaneous Exercise 12 – p. 480

1) $(\pi, \frac{5}{6})$ 2) $y = 4x$
3) $(0, 0), (\sqrt{3}, \frac{1}{4}\sqrt{3}), (-\sqrt{3}, -\frac{1}{4}\sqrt{3})$
4) $(0, 1), (-1, 1); (-\frac{1}{2}, \frac{3}{2})$
5) $(n\pi, n\pi), n = 0, 1, 2, 3, 4$
6) $(0, 1), (-3, 0), (-6, -1)$
7) a) -1 b) $\frac{1}{2}$ c) $-\frac{5}{2}$ or 0
8) $\pm\dfrac{3x-4}{2\sqrt{(x-2)}}$ 9) $\frac{13}{6}a^2$
10) $\frac{5}{6}+2\ln 2$ 14) a) 1, b) $-\sqrt{2}$
15) a) $y = \pm x$, b) $y = \pm 1$
17) $x = 1, x = 3, y = 1; (0,0), (2,0)$;
a) $x < 0, 1 < x < 2, x > 3$;
b) $1 < x < \frac{3}{2}, x > 3$
18) $y = 5x$; 90°
20) $\frac{1}{3} \leqslant \lambda \leqslant 3$

Exercise 13a – p. 491

1) $y^2 = a(x-3a); (3a, 0)$
2) $(at^2, 2at)$
5) $\dfrac{a}{3}\left(p^2+\dfrac{1}{p^2}\right), \dfrac{2a}{3}\left(p-\dfrac{1}{p}\right)$;
$9y^2 = 12ax - 8a^2$
9) b) $x = -4a$ 10) $3y^2 = 16x$
11) $y^2 = 2a(x-a)$

Exercise 13b – p. 499

1) 3, 2 2) 4, 3 3) 2, 1
4) 3, 2 5) 1, $\frac{1}{5}$
6) $4x+3\sqrt{5}y = 18; 9\sqrt{5}x-12y = 10\sqrt{5}$
7) $\sqrt{7}x+4y = 16; 16x-4\sqrt{7}y = 7\sqrt{7}$
8) $\sqrt{3}x+2y = 4; 4x-2\sqrt{3}y = 3\sqrt{3}$
9) a) $4x^2+9y^2 = 36$
b) $4x^2+(y-1)^2 = 16$
c) $3(x-2)^2+4y^2 = 12a^2$
d) $16(x-3)^2+9(y-4)^2 = 144$
10) a) $(0, 0); 6, 8; \frac{1}{4}\sqrt{7}$
b) $(0, 1); 4, 2; \frac{1}{2}\sqrt{3}$
c) $(0, 0); 10, 8; \frac{3}{5}$
d) $(0, 0); 2, \frac{8}{5}; \frac{3}{5}$
e) $(2, 0); 8, 4; \frac{1}{2}\sqrt{3}$
f) $(a, b); 2a, 2b; \frac{1}{a}\sqrt{a^2-b^2}$
g) $(-3, 4); 2\sqrt{3}, 2\sqrt{2}; \frac{1}{3}\sqrt{3}$
11) a) $(\pm\sqrt{5}, 0); x = \pm\dfrac{9\sqrt{5}}{5}$
b) $(0, 1\pm 2\sqrt{3}); y = 1\pm\dfrac{8\sqrt{3}}{3}$
c) $(2\pm a, 0); x = 2\pm 4a$
d) $(3, 4\pm\sqrt{7}); y = 4\pm\dfrac{16\sqrt{7}}{7}$
12) $\frac{1}{2}$; 6; $3x^2+4y^2 = 27$
13) a) $7x^2+7y^2-46x-62y-2xy + 199 = 0$
b) $9x^2+8y^2-36y+36 = 0$

Exercise 13c – p. 508

1) $\pm\sqrt{37}; \pm\left(\dfrac{12\sqrt{37}}{37}, -\dfrac{\sqrt{37}}{37}\right)$
3) a) misses
b) meets in two distinct points
c) touches
4) $9y+8x = 0; 2y = x$;
$45y+40x = \pm 30$
6) $5\pm\frac{4}{5}y_1$ 7) 7 units
9) $5x^2+9y^2 = 180$
11) $(\pm\sqrt{a^2-b^2}, 0)$ 12) $x^2+y^2 = 25$

Exercise 13d – p. 516

2) a) $\frac{1}{4}\sqrt{7}; (0, 0); (\pm\sqrt{7}, 0)$
b) $\frac{1}{2}\sqrt{3}; (0, 0); (0, \pm\sqrt{3})$
c) $\frac{1}{4}\sqrt{7}; (5, 2); (5\pm\sqrt{7}, 2)$
d) $\frac{1}{2}\sqrt{3}; (-1, 0); (-1, \pm\sqrt{3})$
3) a) $9x^2+16y^2 = 144$
b) $4x^2+y^2 = 4$
c) $9(x-5)^2+16(y-2)^2 = 144$
d) $4(x+1)^2+y^2 = 4$

4) a) $x = 1 + 3\cos\theta$ $y = 3 + 2\sin\theta$
b) $x = \sqrt{2}\cos\theta - 2$
$y = \sqrt{3}\sin\theta - 1$
c) $x = 3\cos\theta$ $y = 4\sin\theta$
d) $x = 2\cos\theta - 1$ $y = \sin\theta + 1$
5) a) $3x + 4y\sqrt{3} = 24$
b) $2x\sqrt{3} - y = 4$
c) $x - 4y + 4\sqrt{2} = 0$
d) $3x + 5\sqrt{3}y + 30 = 0$
6) $x\cos\theta + 2y\sin\theta = 2$;
$y\cos\theta + 2x\sin\theta + 3\sin\theta\cos\theta = 0$
7) $2x(\cos\theta - \sin\theta)$
$+ 3y(\cos\theta + \sin\theta) = 6$
8) a) (i) $(a\sec\theta, 0)$ (ii) $(0, b\operatorname{cosec}\theta)$
b) (i) $\left\{\frac{(a^2-b^2)}{a}\cos\theta, 0\right\}$
(ii) $\left\{0, \frac{(b^2-a^2)}{b}\sin\theta\right\}$

9) b) $x = \cos\theta - 2$, $y = 2\sin\theta + 5$
10) $4(x + 2\sqrt{5})^2 + 9y^2 = 324$
11) $\tan^2\theta$ 12) a^2/b^2
13) a) $\cos\theta\,(1 - \cos\theta)$: $\sin\theta\,(1 - \sin\theta)$
b) $\begin{cases} x = 2(1 - \tan\theta + \sec\theta) \\ y = \frac{3}{2}(1 - \cot\theta + \operatorname{cosec}\theta) \end{cases}$
14) $\pm\left(\frac{4}{\sqrt{5}}, \frac{4}{\sqrt{5}}\right)$,
$\pm\left(\frac{8}{\sqrt{5}}, -\frac{2}{\sqrt{5}}\right)$
15) a) $-\frac{5\pi}{6}$ b) $3y + \sqrt{3}x = \pm 10$
c) $\sqrt{3}y + x = 0$

Exercise 13e – p. 521

1) a) $3x^2 - y^2 = 27$ b) $8x^2 - y^2 = 288$
c) $72(x-2)^2 - 9y^2 = 32$
d) $5(y-1)^2 - 4x^2 = 1$
2) a) (0, 0); 4 b) (0, 0); 8
c) (1, 0); 2 d) (– 1, – 2); 4
e) (0, 0); 2 f) (0, 0); 4
3) a) $(\pm\sqrt{5}, 0)$; $x = \pm 4/\sqrt{5}$
b) $(\pm\sqrt{41}, 0)$; $x = \pm 16/\sqrt{41}$
c) $(1 \pm \sqrt{10}, 0)$; $x = 1 \pm 1/\sqrt{10}$
d) $(-2, -1 \pm \sqrt{13})$; $y = -1 \pm 4/\sqrt{13}$
e) $(\pm\sqrt{2}, 0)$; $x = \pm 1/\sqrt{2}$
f) $(0, \pm 2\sqrt{2})$; $y = \pm\sqrt{2}$

4) $x^2 + y^2 - 4xy + 4x + 2y = 5$
5) $8x^2 - y^2 + 8x - 16 = 0$

Exercise 13f – p. 532

1) a) $y + 3 = \sqrt{3}x$ b) $4x - 5y = 9$
c) $2x - y = 1$
2) a) $2y = \pm\sqrt{3}x$ b) $y = \pm x$
c) $y = \pm\sqrt{3}x$
3) $\frac{1}{3}$ 4) $b^2 + c^2 = a^2m^2$; $ay = \pm bx$
5) $(-4/\sqrt{3}, -1/\sqrt{3})$ 6) $\pm\frac{1}{2}\sqrt{10}$
8) $8y = x$ 12) 2
13) $4(x-a)^2 - 4y^2 = a^2$; $(a, 0)$; $\left(\frac{a}{2}, 0\right)$;
$\left(\frac{3a}{2}, 0\right)$; ± 1
14) $y = x \pm 1$; (3, 2), (– 3, – 2); $\sqrt{6}$

Exercise 13g – p. 542

1) The answers given are suitable parametric equations but there are other possibilities.
a) $x = 4t, y = 4/t$
b) $x = 5t, y = -5/t$
c) $x = t + 2, y = 1/t$
d) $x = 3t, y = 3/t$
e) $x = 2t + 1, y = 2/t$
f) $x = t/2, y = 1/2t$
2) a) $xy = 4$ b) $xy + 9 = 0$
c) $xy = 1$ d) $y(x-1) = 1$
e) $x(1-y) = 16$ f) $xy + 4 = 0$
3) (i) a) (0, 0); $(4\sqrt{2}, 4\sqrt{2})$,
$(-4\sqrt{2}, -4\sqrt{2})$; (4, 4),
(– 4, – 4)
b) (0, 0); $(5\sqrt{2}, -5\sqrt{2})$,
$(-5\sqrt{2}, 5\sqrt{2})$; (5, – 5),
(– 5, 5)
c) (2, 0); $(2 + \sqrt{2}, \sqrt{2})$,
$(2 - \sqrt{2}, -\sqrt{2})$; (3, 1), (1, – 1)
d) (0, 0); $(3\sqrt{2}, 3\sqrt{2})$,
$(-3\sqrt{2}, -3\sqrt{2})$; (3, 3),
(– 3, – 3)
e) (1, 0); $(1 + 2\sqrt{2}, 2\sqrt{2})$,
$(1 - 2\sqrt{2}, -2\sqrt{2})$; (3, 2),
(– 1, – 2)
f) (0, 0); $(\sqrt{2}/2, \sqrt{2}/2)$,
$(-\sqrt{2}/2, -\sqrt{2}/2)$; $(\frac{1}{2}, \frac{1}{2})$,
$(-\frac{1}{2}, -\frac{1}{2})$

(ii) a) $(0, 0); (2\sqrt{2}, 2\sqrt{2}), (-2\sqrt{2}, -2\sqrt{2}); (2, 2), (-2, -2)$
b) $(0, 0); (3\sqrt{2}, -3\sqrt{2}), (-3\sqrt{2}, 3\sqrt{2}); (3, -3), (-3, 3)$
c) $(0, 0); (\sqrt{2}, \sqrt{2}), (-\sqrt{2}, -\sqrt{2}); (1, 1), (-1, -1)$
d) $(1, 0); (1+\sqrt{2}, \sqrt{2}), (1-\sqrt{2}, -\sqrt{2}); (0, -1), (2, 1)$
e) $(0, 1); (4\sqrt{2}, 1-4\sqrt{2}), (-4\sqrt{2}, 1+4\sqrt{2}); (4, -3), (-4, 5)$
f) $(0, 0); (2\sqrt{2}, -2\sqrt{2}), (-2\sqrt{2}, 2\sqrt{2}); (2, -2), (-2, 2)$

4) a) $t^2y + x = 8t; t^3x - ty = 4(t^4 - 1)$
b) $4y + x = 4; 8x - 2y = 15$
c) $t^2y - x = 6t; t^3x + ty = 3(1 - t^4)$
d) $x + y = \pm 4; y = x$ is the normal in both cases

5) a) $2y + x = \pm 4\sqrt{2}$
b) $y + 4x = 0; y = 0$

6) $(-\frac{3}{8}, -24); \dfrac{51\sqrt{17}}{8}$ 7) $2\sqrt{29}$

8) $\frac{21}{10}\sqrt{29}$ 9) $\left(\dfrac{c}{m}, cm\right)$ 10) $\dfrac{17^{3/2}}{4}$

13) $x + 3y = 9$ 14) $n^2 = 4mlc^2$

17) $(x^2 + y^2)^2 = 4c^2xy$

Exercise 13h – p. 549

1) a, c, e, f

2) a) $x + y = 0, x + y = 0$
c) $x + y = 0, x - y = 0$
e) $2x + y = 0, x + y = 0$
f) $x + 3y = 0, x + y = 0$

3) a) $6x^2 - 5xy + y^2 = 0$
b) $2y^2 + 5xy - 12x^2 = 0$
c) $y^2 = 3x^2$

4) a) (i) b) (iii) c) (ii), (v)
(*Note.* (iv) does not represent a line pair through O)

5) a) $\arctan \frac{3}{4}$ b) $\arctan \frac{3}{4}$
c) $\arctan 2$ d) $\arctan \dfrac{\sqrt{p^2 - 4q}}{1 + q}$

6) a) 1 b) -1 c) $\frac{1}{3}(-5 \pm 2\sqrt{7})$

7) a) $q^2 > 4pr$ b) $p + r = 0$;
two coincident lines, $x + y = 0$

9) a) $a + b = 0$ b) $h^2 = ab$

10) $ay = \pm bx$; $\arctan \left|\dfrac{2ab}{b^2 - a^2}\right|$; $a^2 = b^2$

11) $a^2y^2 + b^2x^2 = 0$ does not represent a line pair through O.

12) a) no real asymptote through O
b) $x + 3y = 0, x - y = 0$
c) $x - y = 0, 4x + 5y = 0$
d) no real asymptotes through O
e) $2x + y = 0, 2x - y = 0$
f) no real asymptotes through O

Exercise 13i – p. 556

1) $256x^2 + 800xy + 175y^2 = 0$
2) $35x^2 - 144xy + 143y^2 = 0$
3) $11x^2 + 36xy + 13y^2 = 0$
4) $5y^2 = 18x^2 - 9xy$
5) $(c^2 - m^2a^2)x^2 + 2a^2mxy + (c^2 - a^2)y^2 = 0$
6) $2bx + 2ay = ab$
7) $2x + 3y = 9$
8) $y = 2(x - 2)$
9) $x + y = 32$
10) $(2p + 3q - 4)x + (4q + 3p)y = 10 + 4p$
11) crosses 12) crosses
13) misses 14) misses
15) touches

Miscellaneous Exercise 13 – p. 557

2) $ty = x + at^2$
3) $(t_1 + t_2)y = 2x + 2at_1t_2$
4) b) $\{a(p^2 - 4p + 8), a(2p - 4)\}$; $y^2 = 4a(x - 4a)$
5) $y = tx - at - at^2$;
$x + ty = a + 2at + at^3$;
$\left\{a - 2at - \dfrac{4a}{t}, at^2 + 4a + \dfrac{4a}{t^2}\right\}$
6) $x - 4ty + 2t^2 = 0$; $x \pm 4y + 2 = 0$
8) $x^2 + y^2 = a^2 + b^2$
9) $(2, 1), (-1, 3)$
10) $9a^2x^2 + 9b^2y^2 = (a^2 - b^2)^2$
13) $\left\{0, \dfrac{3}{2}\left(\dfrac{1}{\sin\theta} - \sin\theta\right)\right\}$;
$\dfrac{2x^2}{9} + \dfrac{9}{16y^2} = 1$

15) $y = \pm x$

18) $\dfrac{a^2}{x^2} + \dfrac{b^2}{y^2} = 4$

19) $3y = \pm 4x$

21) $\left(\dfrac{a^2}{p}\cos\alpha, -\dfrac{b^2}{p}\sin\alpha\right)$;

$\pm 3\sqrt{2}x \pm \sqrt{7}y = 15$

22) $x^2 + y^2 = 5$

24) $bx \sec t - ay \tan t = ab$;
$a^2y^2 = 4x^2y^2 + b^2x^2$

25) $m^2x^2 - y^2 = m^2a^2$

26) $\dfrac{x}{b}\tan\theta + \dfrac{y}{a}\sec\theta = \dfrac{a^2 + b^2}{ab}\sec\theta\tan\theta$;

$\left(\dfrac{2x}{a}\right)^2 - \left(\dfrac{2yb}{a^2 + 2b^2}\right)^2 = 1$

27) $a^2y_0(y - y_0) = b^2x_0(x - x_0)$

29) $(\pm 4, 0)$; $9x^2 + 16y^2 = 144$

30) $\left(-\dfrac{c}{t^3}, -ct^3\right)$; $\left(ct^9, \dfrac{c}{t^9}\right)$;

$4x^3y^3 + c^2(x^2 - y^2)^2 = 0$

33) $4c\sqrt{\dfrac{m}{1 + m^2}}$; $8c^2\left(\dfrac{1 - m^2}{1 + m^2}\right)$

35) (h, k); $x = h, y = k$; $(\frac{3}{2}, 3)$, $(-\frac{1}{2}, -1)$

36) $p^2y + x = 2cp$; $\left(\dfrac{2cpq}{p + q}, \dfrac{2c}{p + q}\right)$

37) $ty - t^3x = c(1 - t^4)$; $\left(-\dfrac{c}{t^3}, -ct^3\right)$

39) $x + t^2y = 2t$; $3 \pm 2\sqrt{2}$

41) $-\dfrac{4c^2}{a^2}$; $\frac{1}{2}$; $2:3$

42) $3x^2 - 8xy - 3y^2 = 0$; $90°$

43) $3x + 2y + 4 = 0$

45) $y^2(\lambda^2 - 4\lambda - 1) + 2\lambda xy(2\lambda - 1)$
$+ 4\lambda^2x^2 = 0$; 1 or $-\frac{1}{5}$; ∞ or $-\frac{5}{12}$;
$12y + 5x + 5 = 0, x = -1$

47) $(5 + 3\sqrt{3})y^2 - (6 + 2\sqrt{3})xy$
$+ (7 - 3\sqrt{3})x^2 = 0$

49) $x^2(ar^2 - 2gpr + p^2c)$
$+ 2xy(hr^2 - fpr - gqr + pqc)$
$+ y^2(br^2 - 2fqr + q^2c) = 0$;
$r^2(a + b) - 2r(gp + fq)$
$+ c(p^2 + q^2) = 0$

Exercise 14a – p. 571

In each case the symmetries are listed first and the table follows:

1) e (identity) and a (reflect in vertical median).

∘	e	a
e	e	a
a	a	e

2) e (identity), a (reflect in vertical axis of symmetry 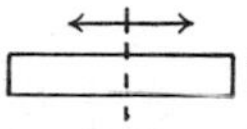), b (reflect on horizontal axis of symmetry) and r (rotate through 180°).

∘	e	a	b	r
e	e	a	b	r
a	a	e	r	b
b	b	r	e	a
r	r	b	a	e

3) e (identity) and r (rotate through 180°).

∘	e	r
e	e	r
r	r	e

4) e (identity), r (rotate through 120° clockwise) and s (rotate through 120° anticlockwise).

∘	e	r	s
e	e	r	s
r	r	s	e
s	s	e	r

5) e (identity), r_1 (rotate through 90° clockwise), r_2 (rotate through 180°) and r_3 (rotate through 270°).

$\circ$	e	r_1	r_2	r_3
e	e	r_1	r_2	r_3
r_1	r_1	r_2	r_3	e
r_2	r_2	r_3	e	r_1
r_3	r_3	e	r_1	r_2

6) e (identity), r_1 (rotate through 72°), r_2 (through 144°), r_3 (through 216°), r_4 (through 288°).

$\circ$	e	r_1	r_2	r_3	r_4
e	e	r_1	r_2	r_3	r_4
r_1	r_1	r_2	r_3	r_4	e
r_2	r_2	r_3	r_4	e	r_1
r_3	r_3	r_4	e	r_1	r_2
r_4	r_4	e	r_1	r_2	r_3

7) e (identity), four reflections:

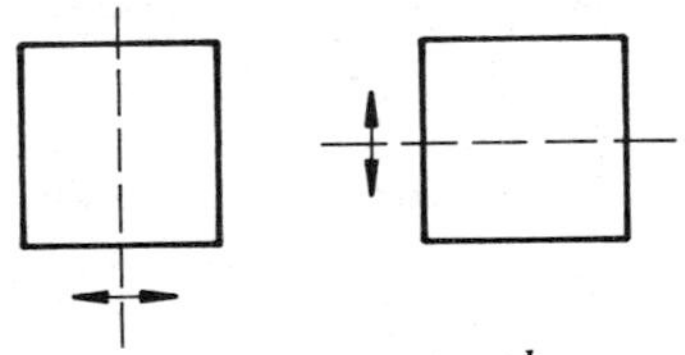

a *b*

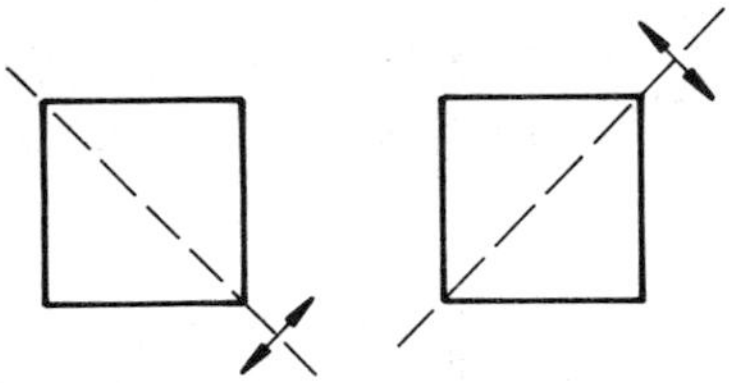

c *d*

and three rotations:

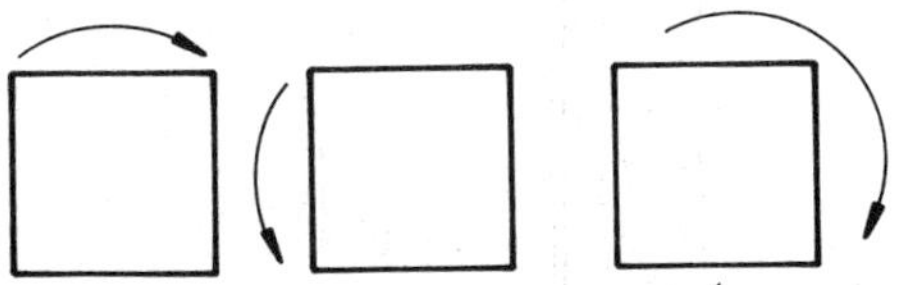

r *s* *t*

$\circ$	e	r	s	t	a	b	c	d
e	e	r	s	t	a	b	c	d
r	r	t	e	s	d	c	a	b
s	s	e	t	r	c	d	b	a
t	t	s	r	e	b	a	d	c
a	a	c	d	b	e	t	r	s
b	b	d	c	a	t	e	s	r
c	c	b	a	d	s	r	e	t
d	d	a	b	c	r	s	t	e

8) The only symmetry is the identity e:

$\circ$	e
e	e

Exercise 14b – p. 577

(Notation as in Exercise 14a)

1) a) $e^{-1}=e, r_1^{-1}=r_3, r_2^{-1}=r_2, r_3^{-1}=r_1$
   b) $e^{-1}=e, r^{-1}=s, s^{-1}=r, a^{-1}=a, b^{-1}=b, c^{-1}=c, d^{-1}=d, t^{-1}=t$
2) Closure and associativity are well-known properties of rational numbers; 0 is the identity and $-q$ is the inverse of q
3) Similar to Q2
4) No. *Inverses* fails because $0\times q=0$ for each q and hence 0 has no inverse (the identity is 1)
5) *Closure* if $q, r \in \mathbb{Q}^*$, then $q\times r\neq 0$, i.e. $q\times r\in\mathbb{Q}^*$. 1 is the *identity*. If $q\in\mathbb{Q}^*$ then $1/q\in\mathbb{Q}^*$ and $q\times 1/q=1/q\times q=1$. Associativity is well-known
6) Similar to Q5
7) Proof is very similar to Example 14b(2)
8)

$\times_3$	0	1	2
0	0	0	0
1	0	1	2
2	0	2	1

Not a group. 1 is the identity. There is no inverse for zero

9) Yes: closure can be seen from the table in Q8. 1 is the identity. $1^{-1} = 1$ and $2^{-1} = 2$. Associativity is proved as for $+_3$ (see Example 14b(2))
10) No: *closure* fails because $2 \times_4 2 = 0$ ($2 \times 2 = 4$ which gives remainder 0 on division by 4) and 0 is not in the set
11) Not a group. *Associativity* fails; e.g. $(a \square b) \square b = b \square b = e$ whereas $a \square (b \square b) = a \square e = a$
12) This is a group. Closure follows from the table. y is the identity. $y^{-1} = y$, $x^{-1} = z$ and $z^{-1} = x$. Associativity can be checked
13) *Identity* fails: $x \div q = x \Rightarrow q = 1$ but then $q \div x = x$ is false, for example $1 \div 2 = \frac{1}{2} \neq 2$. (*Associativity* also fails)
14) This *is* a group. Closure is obvious; 1 is the identity; $-n + 2$ is the inverse of n and associativity is readily checked

4) All commutative except the square
5) (a), (b): well-known properties of numbers; (c), (d), (e): tables are symmetric; (f): $n * m = n + m - 1 = m + n - 1 = m * n$
6)

$+_6$	0	2	4
0	0	2	4
2	2	4	0
4	4	0	2

Closure follows from table. 0 is the identity. 0 is the inverse of 0, 2 is the inverse of 4 and vice versa. Associativity is proved in the usual way (Example 14b(2)). Commutativity follows from symmetry of table

7) Each of $p +_n q$ and $q +_n p$ is the remainder of $p + q$ $(= q + p)$ on division by n. Hence they are equal

Exercise 14c – p. 580

1)

$+_2$	0	1
0	0	1
1	1	0

$+_7$	0	1	2	3	4	5	6
0	0	1	2	3	4	5	6
1	1	2	3	4	5	6	0
2	2	3	4	5	6	0	1
3	3	4	5	6	0	1	2
4	4	5	6	0	1	2	3
5	5	6	0	1	2	3	4
6	6	0	1	2	3	4	5

Tables are symmetric

2) The property $r + s = s + r$ for each pair $r, s \in \mathbb{R}$ is a well-known property of numbers
3) (Notation of solution to Exercise 14a, Q7). There are several possible choices for pairs which do not commute, example $a \circ c = r \neq c \circ a = s$

Exercise 14d – p. 583

1) a) $\begin{pmatrix}1 & 2 & 3\\2 & 1 & 3\end{pmatrix}$ b) $\begin{pmatrix}1 & 2 & 3\\3 & 1 & 2\end{pmatrix}$

c) $\begin{pmatrix}1 & 2 & 3\\3 & 2 & 1\end{pmatrix}$ d) $\begin{pmatrix}1 & 2 & 3\\3 & 1 & 2\end{pmatrix}$

e) $\begin{pmatrix}1 & 2 & 3\\1 & 2 & 3\end{pmatrix}$ f) $\begin{pmatrix}1 & 2 & 3\\3 & 2 & 1\end{pmatrix}$

2) a) $\begin{pmatrix}a & b & c & d\\c & a & b & d\end{pmatrix}$ b) $\begin{pmatrix}a & b & c & d\\b & d & c & a\end{pmatrix}$

3)

$\circ$	$\begin{pmatrix}i & j\\i & j\end{pmatrix}$	$\begin{pmatrix}i & j\\j & i\end{pmatrix}$
$\begin{pmatrix}i & j\\i & j\end{pmatrix}$	$\begin{pmatrix}i & j\\i & j\end{pmatrix}$	$\begin{pmatrix}i & j\\j & i\end{pmatrix}$
$\begin{pmatrix}i & j\\j & i\end{pmatrix}$	$\begin{pmatrix}i & j\\j & i\end{pmatrix}$	$\begin{pmatrix}i & j\\i & j\end{pmatrix}$

4) Six: $\begin{pmatrix}1&2&3\\1&2&3\end{pmatrix}, \begin{pmatrix}1&2&3\\1&3&2\end{pmatrix},$

$\begin{pmatrix}1&2&3\\3&2&1\end{pmatrix}, \begin{pmatrix}1&2&3\\2&1&3\end{pmatrix},$

$\begin{pmatrix}1&2&3\\2&3&1\end{pmatrix}, \begin{pmatrix}1&2&3\\3&1&2\end{pmatrix}$

5) The table appears on page 593

Exercise 14e – p. 585

1) a) $\begin{pmatrix}1&2&3\\2&3&1\end{pmatrix}$ b) $\begin{pmatrix}1&2&3\\1&3&2\end{pmatrix}$

c) $\begin{pmatrix}1&2&3\\3&2&1\end{pmatrix}$

3) No 4) Yes

5) a) 24;

b) $\begin{pmatrix}1&2&3&4\\4&2&3&1\end{pmatrix}, \begin{pmatrix}1&2&3&4\\1&4&3&2\end{pmatrix}$

c) No (for example the elements in b) do not commute;

d) $\begin{pmatrix}1&2&3&4\\4&3&1&2\end{pmatrix}, \begin{pmatrix}1&2&3&4\\1&4&2&3\end{pmatrix}$

Exercise 14f – p. 588

1) Table is

$\circ$	e	r	s
e	e	r	s
r	r	s	e
s	s	e	r

Proof is similar to Example 14f(2)

Matrices are $\begin{pmatrix}1&0\\0&1\end{pmatrix},$

$\begin{pmatrix}-\frac{1}{2}&-\frac{\sqrt{3}}{2}\\\frac{\sqrt{3}}{2}&-\frac{1}{2}\end{pmatrix}, \begin{pmatrix}-\frac{1}{2}&\frac{\sqrt{3}}{2}\\-\frac{\sqrt{3}}{2}&-\frac{1}{2}\end{pmatrix}$

2) Table is

$\times$	$\begin{pmatrix}1&0\\0&1\end{pmatrix}$	$\begin{pmatrix}0&-1\\-1&0\end{pmatrix}$	$\begin{pmatrix}0&1\\1&0\end{pmatrix}$	$\begin{pmatrix}-1&0\\0&-1\end{pmatrix}$
$\begin{pmatrix}1&0\\0&1\end{pmatrix}$	$\begin{pmatrix}1&0\\0&1\end{pmatrix}$	$\begin{pmatrix}0&-1\\-1&0\end{pmatrix}$	$\begin{pmatrix}0&1\\1&0\end{pmatrix}$	$\begin{pmatrix}-1&0\\0&-1\end{pmatrix}$
$\begin{pmatrix}0&-1\\-1&0\end{pmatrix}$	$\begin{pmatrix}0&-1\\-1&0\end{pmatrix}$	$\begin{pmatrix}1&0\\0&1\end{pmatrix}$	$\begin{pmatrix}-1&0\\0&-1\end{pmatrix}$	$\begin{pmatrix}0&1\\1&0\end{pmatrix}$
$\begin{pmatrix}0&1\\1&0\end{pmatrix}$	$\begin{pmatrix}0&1\\1&0\end{pmatrix}$	$\begin{pmatrix}-1&0\\0&-1\end{pmatrix}$	$\begin{pmatrix}1&0\\0&1\end{pmatrix}$	$\begin{pmatrix}0&-1\\-1&0\end{pmatrix}$
$\begin{pmatrix}-1&0\\0&-1\end{pmatrix}$	$\begin{pmatrix}-1&0\\0&-1\end{pmatrix}$	$\begin{pmatrix}0&1\\1&0\end{pmatrix}$	$\begin{pmatrix}0&-1\\-1&0\end{pmatrix}$	$\begin{pmatrix}1&0\\0&1\end{pmatrix}$

Group properties follow from the table in the usual way. Transformations are: identity, reflect in $y = -x$, reflect in $y = x$, rotate through 180°

3) Table is

$\circ$	e	a	b	r
e	e	a	b	r
a	a	e	r	b
b	b	r	e	a
r	r	b	a	e

Group properties follow. Matrix group is

$\left(\left\{\begin{pmatrix}1&0\\0&1\end{pmatrix}, \begin{pmatrix}1&0\\0&-1\end{pmatrix}, \begin{pmatrix}-1&0\\0&1\end{pmatrix}, \begin{pmatrix}-1&0\\0&-1\end{pmatrix}\right\}, \times\right)$

4) $\begin{pmatrix}r&0\\0&r\end{pmatrix}\begin{pmatrix}s&0\\0&s\end{pmatrix} = \begin{pmatrix}rs&0\\0&rs\end{pmatrix}$. Proof is similar to Example 14f(2) using the fact that positive real numbers from a group under multiplication. $\begin{pmatrix}r&0\\0&r\end{pmatrix}$ represents a dilation with scale factor r, centre the origin: so the group is the group of all such dilations.

5) $\begin{pmatrix}r&0\\0&1/r\end{pmatrix}\begin{pmatrix}s&0\\0&1/s\end{pmatrix} = \begin{pmatrix}rs&0\\0&1/rs\end{pmatrix}$. Proof

as in last question. $\begin{pmatrix} r & 0 \\ 0 & 1/r \end{pmatrix}$ stretches the x axis and shrinks the y axis by the same factor: group consists of all such transformations

Exercise 14g – p. 591

1) From the table for the whole group we have:

$\circ$	e	a
e	e	a
a	a	e

Closure: see table. *Identity*: e is in the subset. *Inverses*: $e^{-1} = e$, $a^{-1} = a$
2) Similar to Example 14g
3) Similar to Q1
4) $(\{e\}, \circ)$, $(\{e, a\}, \circ)$, $(\{e, b\}, \circ)$, $(\{e, c\}, \circ)$, $(\{e, r, s\}, \circ)$, $(\{e, r, s, a, b, c\}, \circ)$
Note that the whole group *is* regarded as subgroup
5) a) $(\{0\}, +_4)$, $(\{0, 2\}, +_4)$, $(\mathbb{Z}_4, +_4)$
b) $(\{0\}, +_6)$, $(\{0, 3\}, +_6)$, $(\{0, 2, 4\}, +_6)$, $(\mathbb{Z}_6, +_6)$
6) Since $g \circ g = e$ we have the table

$\circ$	e	g
e	e	g
g	g	e

and the proof is similar to Q1
7) (Notation as in Exercise 14a)
a) $(\{e\}, \circ)$, $(\{e, a\}, \circ)$, $(\{e, b\}, \circ)$, $(\{e, r\}, \circ)$, $(\{e, a, b, r\}, \circ)$
b) $(\{e\}, \circ)$, $(\{e, t\}, \circ)$, $(\{e, a\}, \circ)$, $(\{e, b\}, \circ)$, $(\{e, c\}, \circ)$, $(\{e, d\}, \circ)$, $(\{e, r, s, t\}, \circ)$, $(\{e, t, a, b\}, \circ)$, $(\{e, t, c, d\}, \circ)$, $(\{e, r, s, t, a, b, c, d\}, \circ)$

Exercise 14h – p. 596

2)

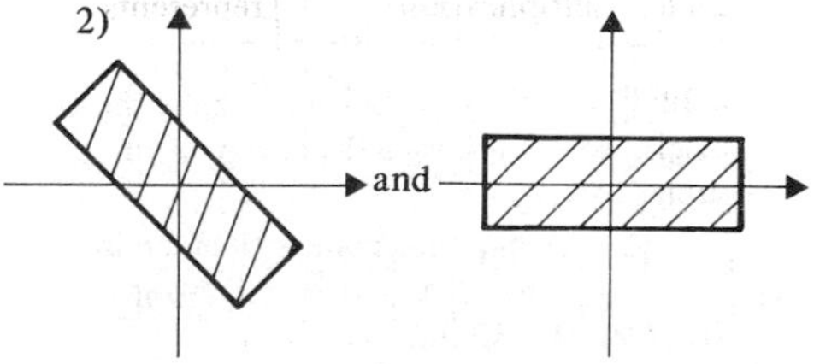

3) Place centre at origin
4) Similar to Example 14h(2)
5) The correspondence is $x \longleftrightarrow e^x$ and then $x + y \longleftrightarrow e^{x+y} = e^x e^y$, which shows the correspondence is an isomorphism
6) The symmetries give rise to the following eight permutations (the order is as in the table in Exercise 14a, Q7):

$$\begin{pmatrix} 1 & 2 & 3 & 4 \\ 1 & 2 & 3 & 4 \end{pmatrix} \begin{pmatrix} 1 & 2 & 3 & 4 \\ 4 & 1 & 2 & 3 \end{pmatrix}$$
$$\begin{pmatrix} 1 & 2 & 3 & 4 \\ 2 & 3 & 4 & 1 \end{pmatrix} \begin{pmatrix} 1 & 2 & 3 & 4 \\ 3 & 4 & 1 & 2 \end{pmatrix}$$
$$\begin{pmatrix} 1 & 2 & 3 & 4 \\ 4 & 3 & 2 & 1 \end{pmatrix} \begin{pmatrix} 1 & 2 & 3 & 4 \\ 2 & 1 & 4 & 3 \end{pmatrix}$$
$$\begin{pmatrix} 1 & 2 & 3 & 4 \\ 1 & 4 & 3 & 2 \end{pmatrix} \begin{pmatrix} 1 & 2 & 3 & 4 \\ 3 & 2 & 1 & 4 \end{pmatrix}$$

which therefore form a group isomorphic to the symmetry group
7) All the tables have the form:

$*$	e	g
e	e	g
g	g	e

Since e is the identity the entries are forced except possibly for the bottom right hand corner. But g must have an inverse, therefore e must appear in row g i.e. the table must be like the above
8) It is necessary to reorder the elements thus $\{e, r, t, s\}$ to make the table correspond. The four elements in the top two rows of the solutions to Q6 form the required subgroup

Exercise 14i – p. 601

1) a) $a^{-2}a^3 = a^{-1}a^{-1}aaa = a^{-1}eaa = a^{-1}aa = ea = a$
b) $a^4a^{-2} = aaaaa^{-1}a^{-1} = aaaea^{-1} = aaaa^{-1} = aae = aa = a^2$
2) a) a^7, b) a
3) Similar to Examples 14i(1) and (2). Solution $y = ba^{-1}$
4) a) $a^{-1}a = e \Rightarrow (a^{-1})^{-1} = a$
b) $(ab)(b^{-1}a^{-1}) = abb^{-1}a^{-1} = aea^{-1} = aa^{-1} = e \Rightarrow (b^{-1}a^{-1}) = (ab)^{-1}$

5) $fa = a \Rightarrow f = aa^{-1} = e$ (by 3)
$af = a \Rightarrow f = a^{-1}a = e$ (by Examples 14i (1) and (2)
6) Similar to Example 14i(4) using Q3)
7)

	e	a	b
e	e	a	b
a	a	b	e
b	b	e	a

Since this is the only way to complete the table any group with three elements must have essentially the same table

8)

a)

	e	a	b	c
e	e	a	b	c
a	a	e	c	b
b	b	c	e	a
c	c	b	a	e

b)

	e	a	b	c
e	e	a	b	c
a	a	b	c	e
b	b	c	e	a
c	c	e	a	b

The given tables are the only ways to place e once in each row and column (any other way is equivalent to b) after reordering elements). Hence any group with four elements must have essentially one of these two tables. (i) table b), (ii) table a)

9) $a^2b^2 = (ab)^2 = abab \Rightarrow a^{-1}a^2b^2b^{-1} = a^{-1}abab^{-1} \Rightarrow ab = ba$
10) $a^3b^3 = (ab)^3 = ababab \Rightarrow a^{-1}a^3b^3b^{-1} = a^{-1}abababb^{-1} \Rightarrow a^2b^2 = baba = (ba)^2$
$a^4b^4 = (a^2)^2(b^2)^2 = (b^2a^2)^2 = ((ab)^2)^2 = (ab)^4$ (applying the first result firstly to a^2, b^2, secondly to b, a); $a^4b^4 = (ab)^4 = abababab = a(ba)^3b \Rightarrow a^3b^3 = (ba)^3 = (ab)^3$

Exercise 14j – p. 604

1) 8, 6, 9, 2, 7, 4
2) $\mathbb{Z}$ is infinite, $\mathbb{Z}_{10}$ is finite
3) $S(\square)$ is non-commutative; each of $\mathbb{Z}$ and $\mathbb{Z}_8$ is commutative
4) a) 3, b) 1
5) a) Q is non-commutative, $\mathbb{Z}_8$ is commutative
b) Q has three subgroups of order 4, $S(\square)$ has only one.
Alternatively a) Q has three subgroups of order 4 and $\mathbb{Z}_8$ has only one. b) Q has one subgroup of order 2 and $S(\square)$ has five

Exercise 14k – p. 606

1) One of order 1 (e), three of order 2 (a, b, c), two of order 3 (r, s). 1 has order 6 in $\mathbb{Z}_6$
2) $S(R)$: all order 2 except e (order 1); $S(W_4)$: one order 1, one order 2, two order 4
3) a) infinite, b) 8, c) 2, d) infinite, e) infinite, f) 1, g) 4, h) 3
4) e appears on diagonal $\Longleftrightarrow g^2 = e$. Since $g = g^1 \neq e$ (given) this is equivalent to g having order 2

Exercise 14l – p. 609

1)

	e	a	a^2	b	ba	ba^2
e	e	a	a^2	b	ba	ba^2
a	a	a^2	e	ba^2	b	ba
a^2	a^2	e	a	ba	ba^2	b
b	b	ba	ba^2	e	a	a^2
ba	ba	ba^2	b	a^2	e	a
ba^2	ba^2	b	ba	a	a^2	e

Interchanging the elements b and c in the table for $S(\Delta)$ (page 569) makes the tables correspond

2) $e, a, b, ab\ (= ba)$

	e	a	b	ab
e	e	a	b	ab
a	a	e	ab	b
b	b	ab	e	a
ab	ab	b	a	e

$S(R)$

3) e, a, a^2, b, ba, ba^2.
The table has been constructed using an ordering which makes the isomorphism clear

	e	ba	a^2	b	a	ba^2
e	e	ba	a^2	b	a	ba^2
ba	ba	a^2	b	a	ba^2	e
a^2	a^2	b	a	ba^2	e	ba
b	b	a	ba^2	e	ba	a^2
a	a	ba^2	e	ba	a^2	b
ba^2	ba^2	e	ba	a^2	b	a

4) $e, a, a^2, a^3, b, ba, ba^2, ba^3$

	e	a	a^2	a^3	b	ba	ba^2	ba^3
e	e	a	a^2	a^3	b	ba	ba^2	ba^3
a	a	a^2	a^3	e	ba	ba^2	ba^3	b
a^2	a^2	a^3	e	a	ba^2	ba^3	b	ba
a^3	a^3	e	a	a^2	ba^3	b	ba	ba^2
b	b	ba	ba^2	ba^3	e	a	a^2	a^3
ba	ba	ba^2	ba^3	b	a	a^2	a^3	e
ba^2	ba^2	ba^3	b	ba	a^2	a^3	e	a
ba^3	ba^3	b	ba	ba^2	a^3	e	a	a^2

a) The group has three elements of order 2 (a^2, b, ba^2) while $\mathbb{Z}_8$ only has one (4). Alternatively, the group has no element of order 8.

b) and c) The group is commutative and neither $S(\square)$ nor Q is

Exercise 14m – p. 611

1)

	(0, 0)	(1, 0)	(2, 0)	(3, 0)	(0, 1)	(1, 1)	(2, 1)	(3, 1)
(0, 0)	(0, 0)	(1, 0)	(2, 0)	(3, 0)	(0, 1)	(1, 1)	(2, 1)	(3, 1)
(1, 0)	(1, 0)	(2, 0)	(3, 0)	(0, 0)	(1, 1)	(2, 1)	(3, 1)	(0, 1)
(2, 0)	(2, 0)	(3, 0)	(0, 0)	(1, 0)	(2, 1)	(3, 1)	(0, 1)	(1, 1)
(3, 0)	(3, 0)	(0, 0)	(1, 0)	(2, 0)	(3, 1)	(0, 1)	(1, 1)	(2, 1)
(0, 1)	(0, 1)	(1, 1)	(2, 1)	(3, 1)	(0, 0)	(1, 0)	(2, 0)	(3, 0)
(1, 1)	(1, 1)	(2, 1)	(3, 1)	(0, 1)	(1, 0)	(2, 0)	(3, 0)	(0, 0)
(2, 1)	(2, 1)	(3, 1)	(0, 1)	(1, 1)	(2, 0)	(3, 0)	(0, 0)	(1, 0)
(3, 1)	(3, 1)	(0, 1)	(1, 1)	(2, 1)	(3, 0)	(0, 0)	(1, 0)	(1, 1)

2)

	(0, 0)	(1, 0)	(2, 0)	(0, 1)	(1, 1)	(2, 1)
(0, 0)	(0, 0)	(1, 0)	(2, 0)	(0, 1)	(1, 1)	(2, 1)
(1, 0)	(1, 0)	(2, 0)	(0, 0)	(1, 1)	(2, 1)	(0, 1)
(2, 0)	(2, 0)	(0, 0)	(1, 0)	(2, 1)	(0, 1)	(1, 1)
(0, 1)	(0, 1)	(1, 1)	(2, 1)	(0, 0)	(1, 0)	(2, 0)
(1, 1)	(1, 1)	(2, 1)	(0, 1)	(1, 0)	(2, 0)	(0, 0)
(2, 1)	(2, 1)	(0, 1)	(1, 1)	(2, 0)	(0, 0)	(1, 0)

The correspondence $a^i b^j \longleftrightarrow (i, j)$ is an isomorphism between this group and the one in Exercise 14l, Q3

3) The elements of $\mathbb{Z}_2 \times \mathbb{Z}_2 \times \mathbb{Z}_2$ are triples (i, j, k) where i, j, k are 0 or 1.

	(0,0,0)	(1,0,0)	(0,1,0)	(1,1,0)	(0,0,1)	(1,0,1)	(0,1,1)	(1,1,1)
(0,0,0)	(0,0,0)	(1,0,0)	(0,1,0)	(1,1,0)	(0,0,1)	(1,0,1)	(0,1,1)	(1,1,1)
(1,0,0)	(1,0,0)	(0,0,0)	(1,1,0)	(0,1,0)	(1,0,1)	(0,0,1)	(1,1,1)	(0,1,1)
(0,1,0)	(0,1,0)	(1,1,0)	(0,0,0)	(1,0,0)	(0,1,1)	(1,1,1)	(0,0,1)	(1,0,1)
(1,1,0)	(1,1,0)	(0,1,1)	(1,0,0)	(0,0,0)	(1,1,1)	(0,1,1)	(1,0,1)	(0,0,1)
(0,0,1)	(0,0,1)	(1,0,1)	(0,1,1)	(1,1,1)	(0,0,0)	(1,0,0)	(0,1,0)	(1,1,0)
(1,0,1)	(1,0,1)	(0,0,1)	(1,1,1)	(0,1,1)	(1,0,0)	(0,0,0)	(1,1,0)	(0,1,0)
(0,1,1)	(0,1,1)	(1,1,1)	(0,0,1)	(1,0,1)	(0,1,0)	(1,1,0)	(0,0,0)	(1,0,0)
(1,1,1)	(1,1,1)	(0,1,1)	(1,0,1)	(0,0,1)	(1,1,0)	(0,1,0)	(1,0,0)	(0,0,0)

a) and b): $\mathbb{Z}_2 \times \mathbb{Z}_2 \times \mathbb{Z}_2$ has seven elements of order 2 while $\mathbb{Z}_8$ has one and $\mathbb{Z}_4 \times \mathbb{Z}_2$ has three

c) and d) $\mathbb{Z}_2 \times \mathbb{Z}_2 \times \mathbb{Z}_2$ is commutative

4) $(i, j)^3$ means $(3i, 3j)$ which is $(0, 0)$. Therefore (i, j) has order 3 or 1. $\mathbb{Z}_9$ has an element of order 9

5) $|G| \times |H|$

6) $(g, h)(g', h') = (gg', hh') = (g'g, h'h)$
$= (g', h')(g, h)$

Miscellaneous Exercise 14 – p. 615

1) See Exercises 14b, Q7; 14a, Q2, and Example 14j (3)

2) 6 is the identity!

$\times_{10}$	6	2	4	8
6	6	2	4	8
2	2	4	8	6
4	4	8	6	2
8	8	6	2	4

$\times$	1	i	-1	$-i$
1	1	i	-1	$-i$
i	i	-1	$-i$	1
-1	-1	$-i$	1	i
$-i$	$-i$	1	i	-1

3) Identity, rotate through 180° about origin, rotate through 90° anticlockwise, ditto clockwise. Now see Example 14h(3)

4)

$\times$	1	ω	ω^2
1	1	ω	ω^2
ω	ω	ω^2	1
ω^2	ω^2	1	ω

(Note $\omega^3 = 1$.) $z \longmapsto \omega z$ is 'rotate through 120° anticlockwise' and $z \longmapsto \omega^2 z$ is the same clockwise. Now use the argument in Exercise 14h, Q3

5)

	1	α	α^2	α^3	α^4
1	1	α	α^2	α^3	α^4
α	α	α^2	α^3	α^4	1
α^2	α^2	α^3	α^4	1	α
α^3	α^3	α^4	1	α	α^2
α^4	α^4	1	α	α^2	α^3

(Note $\alpha^5 = 1$.) Isomorphism clear from table.

$(\{1, \beta, \beta^2, \dots, \beta^{n-1}\}, \times)$ where

$\beta = \cos\dfrac{2\pi}{n} + i \sin\dfrac{2\pi}{n}$

6) (i) a)

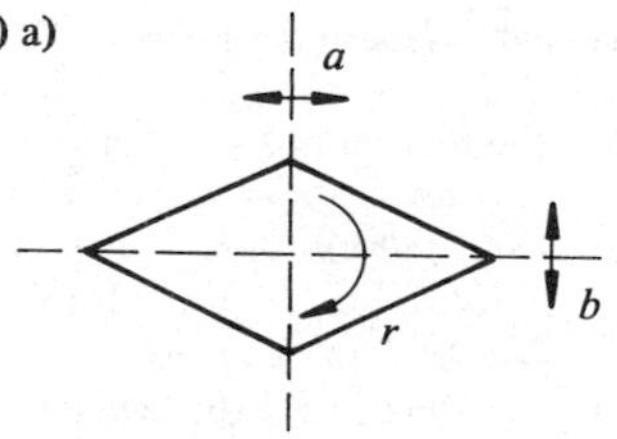

Table is same as $S(R)$

b) The operation is not given! From context we assume it is composition of mappings:

$\circ$	f_1	f_2	f_3	f_4
f_1	f_1	f_2	f_3	f_4
f_2	f_2	f_1	f_4	f_3
f_3	f_3	f_4	f_1	f_2
f_4	f_4	f_3	f_2	f_1

Isomorphism clear from tables

(ii) A typical complex number of modulus 1 is $\cos\theta + i\sin\theta$ and by de Moivre's Theorem $(\cos\theta + i\sin\theta)(\cos\psi + i\sin\psi) = \cos(\theta+\psi) + i\sin(\theta+\psi) \Rightarrow$ *closure.* $1 = \cos 0 + i\sin 0$ is the *identity* and $\cos(-\theta) + i\sin(-\theta) = \cos\theta - i\sin\theta$ is the inverse of $\cos\theta + i\sin\theta$. Associativity is a well-known property of complex multiplication

7) a) Exactly like the proof for $(\mathbb{R}, +)$ (Exercise 14b, Q3). Isomorphism is given by $(x + iy) \longleftrightarrow (x, y)$.

b) Exactly like the proof for $(\mathbb{Q}^*, \times)$ (Exercise 14b, Q5. Subgroups are the groups in Questions 4, 2 and 5

8)

	e	a	b	u	v	w
e	ⓔ	ⓐ	ⓑ	ⓤ	○	ⓦ
a	ⓐ	b	e	ⓦ	u	v
b	○	e	a	v	w	u
u	ⓤ	v	w	e	ⓐ	b
v	ⓥ	w	u	b	e	a
w	ⓦ	u	v	a	b	e

The circled entries are either given or follow from the fact that e is the identity. Geometric considerations imply that section I consists of rotations, II and III of reflections and IV of rotations and also that $u^2 = v^2 = w^2 = e$. It is now easy to complete sections I, II and IV using the fact that each element must appear *once* in each row and column.
Now $w = w^{-1} = (au)^{-1} = u^{-1}a^{-1} = ub$ and it is now easy to complete section II

9) See pages 580 to 584. Subgroups are:

$$S_3: \left\{\begin{pmatrix}1&2&3\\1&2&3\end{pmatrix}, \begin{pmatrix}1&2&3\\2&3&1\end{pmatrix}, \begin{pmatrix}1&2&3\\3&1&2\end{pmatrix}\right\},$$

$$\left\{\begin{pmatrix}1&2&3\\1&2&3\end{pmatrix}, \begin{pmatrix}1&2&3\\1&3&2\end{pmatrix}\right\},$$

$$\left\{\begin{pmatrix}1&2&3\\1&2&3\end{pmatrix}, \begin{pmatrix}1&2&3\\3&2&1\end{pmatrix}\right\},$$

$$\left\{\begin{pmatrix}1&2&3\\1&2&3\end{pmatrix}, \begin{pmatrix}1&2&3\\2&1&3\end{pmatrix}\right\}, \left\{\begin{pmatrix}1&2&3\\1&2&3\end{pmatrix}\right\}$$

10) We are given three of the group properties; we only have to find *inverses.* But a is clearly the identity, and $a^{-1} = a, d^{-1} = d, b^{-1} = b^3$, $c^{-1} = c^3$ are inverses (e.g. $b^3b = bb^3 = b^2b^2 = dd = a \Rightarrow$ $(S, *)$ *is a group*)

11) $EA = A \Rightarrow E$ must be identity.
$\Rightarrow EC = C$

i.e. $\begin{pmatrix}x&y\\0&0\end{pmatrix} = \begin{pmatrix}x&y\\z&t\end{pmatrix} \Rightarrow z = t = 0.$

Similarly $y = t = 0 \Rightarrow C = \begin{pmatrix}x&0\\0&0\end{pmatrix}$ i.e. of same 'form' as the others.

Note $\begin{pmatrix}x&0\\0&0\end{pmatrix}\begin{pmatrix}y&0\\0&0\end{pmatrix} = \begin{pmatrix}xy&0\\0&0\end{pmatrix}$

$\therefore \{1, -1, i, x\}$ form a group under complex multiplication. Clearly $x = -i$. The remainder is simple

12) Use calculation for products in last question, which $\Rightarrow \begin{pmatrix}\lambda&0\\0&0\end{pmatrix} \longleftrightarrow \lambda$ is an isomorphism

13) See Exercises 14a, Q7 and 14g, Q7

14) See pages 609 to 611 and Example 14m(1). Not isomorphic, see Example 14j(3)

15) Subgroups are $\left\{\begin{pmatrix}\lambda & 0\\ 0 & 1\end{pmatrix}\right\}$ and $\left\{\begin{pmatrix}1 & 0\\ 0 & \lambda\end{pmatrix}\right\}$, rest similar to Exercise 14f, Q4 and Example 14m(2)

16) $e, a, a^2, a^2, a^4, b, ba, ba^2, ba^3, ba^4$

18) $\left\{\begin{pmatrix}\cos\theta & -\sin\theta\\ \sin\theta & \cos\theta\end{pmatrix}, \begin{pmatrix}\cos\theta & \sin\theta\\ -\sin\theta & \cos\theta\end{pmatrix}: \theta = 0, \frac{2\pi}{5}, \frac{4\pi}{5}, \frac{6\pi}{5}, \frac{8\pi}{5}\right\}$

19) $\{a, b \mid a^n = b^2 = e, ab = ba^{-1}\}$. S (regular n-gon)

20) Symmetries are translations through n notches to the right where $n \in \mathbb{Z}$ (negative n means movement to left). Composition adds movements

21) $\left\{\begin{pmatrix}1&2&3&4\\1&2&3&4\end{pmatrix}, \begin{pmatrix}1&2&3&4\\2&1&4&3\end{pmatrix}, \begin{pmatrix}1&2&3&4\\4&3&2&1\end{pmatrix}, \begin{pmatrix}1&2&3&4\\3&4&1&2\end{pmatrix}\right\}$

and $\left\{\begin{pmatrix}1&2&3&4\\1&2&3&4\end{pmatrix}, \begin{pmatrix}1&2&3&4\\2&3&4&1\end{pmatrix}, \begin{pmatrix}1&2&3&4\\3&4&1&2\end{pmatrix}, \begin{pmatrix}1&2&3&4\\4&1&2&3\end{pmatrix}\right\}$

22) a) Subgroup is $\{1, \omega^3, \omega^6, \omega^9, \omega^{12}\}$ with the labels $a = \omega^3$, $b = \omega^6$, $c = \omega^{12}$, $d = \omega^9$ since $a^2 = \omega^6 = b$, $b^2 = \omega^{12} = c$, $c^2 = \omega^{24} = \omega^9 = d$, $d^2 = \omega^{18} = \omega^3 = a$

b) (i) Observe $a * a * b = b * b = a$ $\Rightarrow a * b = e$, similarly $b * a = e$ and hence table can be constructed and proof is then easy

(ii) $a * b = a \Rightarrow b = e$; $a * b = b$ $\Rightarrow a = e$; $a * b = c \Rightarrow a * a * b$ $= a * c \Rightarrow b * b = a * c \Rightarrow c = a * c$ $\Rightarrow a = e$; $a * b = e \Rightarrow a * a * b = a$ $\Rightarrow b * b = a \Rightarrow c = a$. Therefore $a * b$ is not in the set, so it is not closed

23) (i) $eae^{-1} = eae = ae = a \Rightarrow a \in A$

(ii) $xax^{-1} = e \Rightarrow ax^{-1} = x^{-1} \Rightarrow a = e$ (contradiction) $\Rightarrow e \notin A$.
$y = xax^{-1} \Rightarrow aya^{-1} = axax^{-1}a^{-1}$
$= (ax)a(ax)^{-1} \in A$
$a^{-1} \in A \Rightarrow a^{-1} = xax^{-1}$ for some x
$\Rightarrow a^{-1}x = xa \Rightarrow x = axa$

24) a) $a^2 = e \Rightarrow a = a^{-1}$ for each $a \in G$ $\Rightarrow ab = (ab)^{-1} = b^{-1}a^{-1} = ba$ for each $a, b \in G$

b) *Closure:* $a * b$ has no repeated factors: indeed if a and b have p as a common factor then p is *not* a factor of $a * b$.
Identity: 1 is the identity.
Inverses: each element is its own inverse since

$$a * a = \frac{a^2}{(D_{aa})^2} = \frac{a^2}{a^2} = 1$$

Associativity: both $a * (b * c)$ and $(a * b) * c$ have a prime factor p *either* if p is a factor of only one of a, b, c or if p is a factor of all three. Hence they are equal

25) See Exercise 14b, Q5 ('lowest terms' makes no difference) for first part. S_0 is closed under products and inverses and contains $1 = \frac{1}{1}$. Let $x = \frac{p}{q}$ in lowest terms then p and q are not both even. If p is even then $p = 2^r p_0$ and $x = 2^r \frac{p_0}{q}$. If q is even then $q = 2^n q_0$ and $x = 2^{-n}\frac{p_0}{q_0} = 2^r \frac{p}{q_0}$ where $r = -n$. If neither is even then $x = 2^0 \frac{p}{q}$. Let $x \in S_r$, $y \in S_s$ then

$$xy = 2^r\frac{p}{q}\,2^s\frac{p'}{q'} = 2^r 2^s \frac{pp'}{qq'} = 2^{r+s}\frac{p''}{q''},$$

say, where p, p', p'', etc., are all odd

26) Remainder of 2^n on division by 13 is given:

r	1	2	3	4	5	6	7
rem.	2	4	8	3	6	12	11
r	8	9	10	11	12	13	
rem.	9	5	10	7	1	2	

Subgroups are $\{1\}$, G, $\{1, 4, 3, 12, 9, 10\}$, $\{1, 8, 12, 5\}$, $\{1, 3, 9\}$, $\{1, 12\}$. (Obtained by observing that the correspondence above gives an isomorphism with $\mathbb{Z}_{12}$)

27) *Closure*: $g^n g^m = g^{n+m}$
Identity: $g^0 = e \in \langle g \rangle$
Inverses: $(g^n)^{-1} = g^{-n}$ since $g^{-n}g^n = g^n = e$.
a) Rotation subgroup;
b) $\{e, a\}$;
c) $\mathbb{Z}$ (whole group);
d) $\{(0, 2, 4), +_6\}$

28) a) If $g^n = g^m$, $n < m$, then $g^{m-n} = e$ i.e. g has finite order. Thus g infinite order $\Rightarrow$ all g^ns are different. Clearly $g^n \longleftrightarrow n$ is an isomorphism
b) $e, g, \ldots, g^{n-1}$ are distinct (as in part a)). Clearly $g^n \longleftrightarrow n$ is again an isomorphism. Order $(g) = |\langle g \rangle|$ which is a factor of $|G|$ by Lagrange's Theorem

29) Choose $g \neq e$. Then $|\langle g \rangle| = p$ (since p is prime) $\Rightarrow \langle g \rangle = G \Rightarrow G \cong \mathbb{Z}_p$ by Question 23

Exercise 15a – p. 633

1) a) $\begin{pmatrix} 1 & \frac{1}{2} \\ 0 & 3 \\ -2 & 35 \end{pmatrix}$; b) $\begin{pmatrix} 1 \\ -3 \end{pmatrix}$

c) $(120 \quad 83 \quad 14)$

2) a) 5; b) -1; c) -28

3) a) $\begin{pmatrix} 0 & -1 \\ 1 & 0 \end{pmatrix}$; b) $\begin{pmatrix} \frac{1}{2} & -\frac{1}{2} & -\frac{1}{2} \\ 0 & 1 & 0 \\ 0 & 1 & 1 \end{pmatrix}$

4) Rotation through $\pi/2$ clockwise about the origin. The inverse gives the *anti*clockwise rotation

5) $|\mathbf{AB}| = |\mathbf{A}||\mathbf{B}|$ (property 3)
$= |\mathbf{B}||\mathbf{A}| = |\mathbf{BA}|$

6) By induction on n. Case $n = 1$ is obvious. Assume true for $n-1$. Then
$|\mathbf{A}^n| = |\mathbf{A}||\mathbf{A}^{n-1}|$ (property 3)
$= |\mathbf{A}||\mathbf{A}|^{n-1}$ (by assumption)
$= |\mathbf{A}|^n$
which proves the induction step

7) By induction on n.

Case $n = 2$: $\left|\lambda\begin{pmatrix} a & b \\ c & d \end{pmatrix}\right| = \begin{vmatrix} \lambda a & \lambda b \\ \lambda c & \lambda d \end{vmatrix}$
$= \lambda a \lambda d - \lambda b \lambda c$
$= \lambda^2(ad - bc)$
$= \lambda^2 \begin{vmatrix} a & b \\ c & d \end{vmatrix}$.

Now assume true for $n-1$. Then
$|\lambda\mathbf{A}| = \lambda a_{11}|\lambda\mathbf{A}_{11}| - \lambda a_{12}|\lambda\mathbf{A}_{12}| + \ldots$
$= \lambda a_{11}\lambda^{n-1}|\mathbf{A}_{11}| - \lambda a_{12}\lambda^{n-1}|\mathbf{A}_{12}| + \ldots$ (by assumption)
$= \lambda^n(a_{11}|\mathbf{A}_{11}| - a_{12}|\mathbf{A}_{12}| + \ldots)$
$= \lambda^n|\mathbf{A}|$,
which completes the induction step

8) *Closure*: follows from the definition of matrix addition.

Identity: $\begin{pmatrix} a & b & c \\ d & e & f \end{pmatrix} + \begin{pmatrix} 0 & 0 & 0 \\ 0 & 0 & 0 \end{pmatrix}$
$= \begin{pmatrix} 0 & 0 & 0 \\ 0 & 0 & 0 \end{pmatrix} + \begin{pmatrix} a & b & c \\ d & e & f \end{pmatrix} = \begin{pmatrix} a & b & c \\ d & e & f \end{pmatrix}$.

Inverses: the inverse of $\mathbf{A}$ is $-\mathbf{A} = (-1)\mathbf{A}$ since

$\mathbf{A} + (-\mathbf{A}) = \begin{pmatrix} a & b & c \\ d & e & f \end{pmatrix}$
$+ (-1)\begin{pmatrix} a & b & c \\ d & e & f \end{pmatrix} = \begin{pmatrix} a & b & c \\ d & e & f \end{pmatrix}$
$+ \begin{pmatrix} -a & -b & -c \\ -d & -e & -f \end{pmatrix} = \begin{pmatrix} 0 & 0 & 0 \\ 0 & 0 & 0 \end{pmatrix}$
$= \mathbf{0}$
and similarly $(-1)\mathbf{A} + \mathbf{A} = \mathbf{0}$.
Associativity: see Example 15a(1)

9) *Closure*: $\mathbf{A}, \mathbf{B}$ non-singular $\Rightarrow |\mathbf{A}|, |\mathbf{B}| \neq 0 \Rightarrow |\mathbf{AB}| \neq 0$ (by property 3), $\Rightarrow \mathbf{AB}$ non-singular
Identity: unit law
Inverses: $\mathbf{A}^{-1}$ is the inverse of $\mathbf{A}$ since $\mathbf{AA}^{-1} = \mathbf{A}^{-1}\mathbf{A} = \mathbf{I}$
Associativity: associativity law of matrix multiplication

10) a) $\mathbf{AA}^{-1} = \mathbf{I} \Rightarrow (\mathbf{A}^{-1})^{-1} = \mathbf{A}$ by Example 15a(3)
b) $(\mathbf{AB})(\mathbf{B}^{-1}\mathbf{A}^{-1}) = \mathbf{ABB}^{-1}\mathbf{A}^{-1} = \mathbf{AIA}^{-1}$ $= \mathbf{AA}^{-1} = \mathbf{I} \Rightarrow \mathbf{B}^{-1}\mathbf{A}^{-1} = (\mathbf{AB})^{-1}$ by Example 15a(3)

11) a) Each entry is of the form $\sum_{j=1}^{m} a_{ij}0_{jk}$ where $0_{jk} = 0$ for each j, k; i.e. each entry is zero
b) Similar

12) a) (i, k)th entry of $\mathbf{A}(\lambda\mathbf{B})$ is $\sum_{j=1}^{m} a_{ij}(\lambda b_{jk}) = \lambda \sum_{j=1}^{m} a_{ij}b_{jk} = \lambda$ $\times (i, k)$th entry of $\mathbf{AB}$; i.e. $\mathbf{A}(\lambda\mathbf{B}) = \lambda(\mathbf{AB})$ and the other equality is similar
b) (i, j)th entry of $\lambda(\mu\mathbf{A})$ is $\lambda(\mu a_{ij}) = (\lambda\mu)a_{ij} = (i, j)$th entry of $(\lambda\mu)\mathbf{A}$. (i, j)th entry of $\lambda(\mathbf{A} + \mathbf{B})$ is $\lambda(a_{ij} + b_{ij}) = \lambda a_{ij} + \lambda b_{ij} = (i, j)$th entry of $\lambda\mathbf{A} + \lambda\mathbf{B}$

Exercise 15b – p. 636

1) a) $\begin{pmatrix} 1 & 0 \\ 3 & 4 \end{pmatrix}$ b) $\begin{pmatrix} 1 \\ 2 \\ 7 \\ 9 \end{pmatrix}$

c) $\begin{pmatrix} -1 & 0 & 1 \end{pmatrix}$ $\begin{pmatrix} 0 & 1 & 2 \\ 1 & 0 & 2 \\ 2 & 2 & 0 \end{pmatrix}$

e) $\begin{pmatrix} 0 & -1 & -2 \\ 1 & 0 & 2 \\ 2 & -2 & 0 \end{pmatrix}$

2) $(\mathbf{AB})^! = ((\mathbf{AB})^{\mathrm{T}})^{-1} = (\mathbf{B}^{\mathrm{T}}\mathbf{A}^{\mathrm{T}})^{-1}$ (property 3)
$= (\mathbf{A}^{\mathrm{T}})^{-1}(\mathbf{B}^{\mathrm{T}})^{-1}$ (Exercise 15a, Q10 b))
$= \mathbf{A}^!\mathbf{B}^!$

3) The (i, j)th entry of $(\lambda\mathbf{A})^{\mathrm{T}}$ is the (j, i)th entry of $\lambda\mathbf{A}$ i.e. λa_{ji}. But the (i, j)th entry of $\lambda(\mathbf{A}^{\mathrm{T}})$ is λ times the (i, j)th entry of $\mathbf{A}^{\mathrm{T}}$ i.e. λa_{ji}. Therefore $(\lambda\mathbf{A})^{\mathrm{T}} = \lambda(\mathbf{A}^{\mathrm{T}})$

Exercise 15c – p. 638

1) a) symmetric b) skew-symmetric
c) both d) symmetric
e) neither (not square!)
f) neither g) symmetric
h) skew-symmetric
i) neither

2) $a_{ij} = a_{ji}$ (symmetric) and $a_{ij} = -a_{ji}$ (skew-symmetric) $\Rightarrow a_{ij} = -a_{ij}$ $\Rightarrow 2a_{ij} = 0 \Rightarrow a_{ij} = 0$

3) $(\mathbf{P}^{\mathrm{T}}\mathbf{AP})^{\mathrm{T}} = \mathbf{P}^{\mathrm{T}}\mathbf{A}^{\mathrm{T}}(\mathbf{P}^{\mathrm{T}})^{\mathrm{T}} = \mathbf{P}^{\mathrm{T}}(-\mathbf{A})\mathbf{P}$ $= -\mathbf{P}^{\mathrm{T}}\mathbf{AP}$

4) $(\mathbf{AB})^{\mathrm{T}} = (\mathbf{BA})^{\mathrm{T}}$ since $\mathbf{AB} = \mathbf{BA}$ $= \mathbf{A}^{\mathrm{T}}\mathbf{B}^{\mathrm{T}} = \mathbf{AB}$ since $\mathbf{A}, \mathbf{B}$ are symmetric

5) $(\mathbf{A}^2)^{\mathrm{T}} = (\mathbf{AA})^{\mathrm{T}} = \mathbf{A}^{\mathrm{T}}\mathbf{A}^{\mathrm{T}} = (-\mathbf{A})(-\mathbf{A})$ $= \mathbf{A}^2$

Exercise 15d – p. 642

1) $ax^2 + ey^2 + iz^2 + (f + h)yz + (c + g)xz + (b + d)xy$

2) $\begin{pmatrix} 3 & 1 & 0 \\ 1 & 7 & \frac{5}{2} \\ 0 & \frac{5}{2} & -2 \end{pmatrix}$

3) a) $\begin{pmatrix} 1 & 0 & 0 \\ 0 & 1 & 0 \\ 0 & 0 & -1 \end{pmatrix}$ b) $\begin{pmatrix} 1 & \frac{3}{2} & -5 \\ \frac{3}{2} & 0 & \frac{7}{2} \\ -5 & \frac{7}{2} & 1 \end{pmatrix}$

c) $\begin{pmatrix} 2 & 5 & -\frac{11}{2} \\ 5 & -4 & -\frac{7}{2} \\ -\frac{11}{2} & -\frac{7}{2} & 3 \end{pmatrix}$

4)

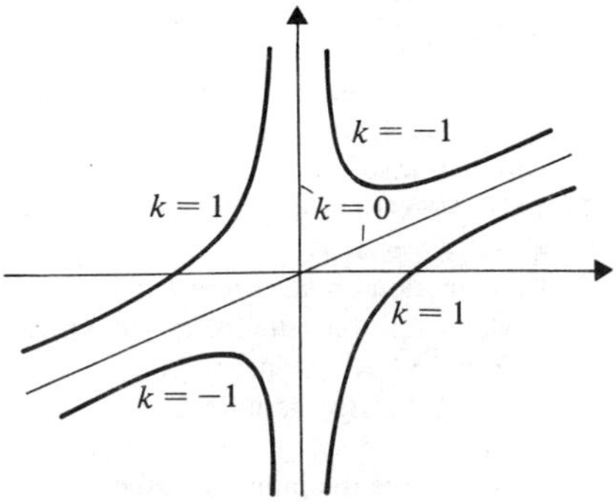

5) $x^2 + 2xy = k$ which is a reflected
   version of the last example:

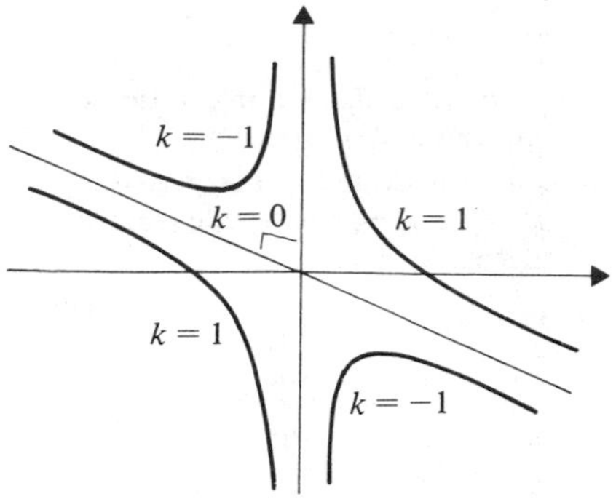

6) $\begin{vmatrix} a & h & g \\ h & b & f \\ g & f & c \end{vmatrix}$

Exercise 15e — p. 643

3) Case $n = 1$ is obvious. Assume true for
   $n - 1$. Then

$$\Lambda^n = \Lambda(\Lambda^{n-1})$$

$$= \begin{vmatrix} \lambda & 0 & 0 \\ 0 & \mu & 0 \\ 0 & 0 & \nu \end{vmatrix} \begin{vmatrix} \lambda^{n-1} & 0 & 0 \\ 0 & \mu^{n-1} & 0 \\ 0 & 0 & \nu^{n-1} \end{vmatrix}$$

$$= \begin{vmatrix} \lambda^n & 0 & 0 \\ 0 & \mu^n & 0 \\ 0 & 0 & \nu^n \end{vmatrix}, \text{ which completes}$$

the induction step

4) $|A| = a \begin{vmatrix} b & 0 \\ 0 & c \end{vmatrix} \dots 0 \begin{vmatrix} 0 & 0 \\ 0 & c \end{vmatrix} + 0 \begin{vmatrix} 0 & b \\ 0 & 0 \end{vmatrix}$

   $= abc.$

   A non-singular $\Longleftrightarrow |A| \neq 0 \Longleftrightarrow abc \neq 0$
   $\Longleftrightarrow a, b, c$ each non-zero

5) $\begin{vmatrix} a & 0 \\ 0 & b \end{vmatrix}$

6) a, b each non-zero

7) *Closure:* $\begin{vmatrix} a & 0 \\ 0 & b \end{vmatrix} \begin{vmatrix} \lambda & 0 \\ 0 & \mu \end{vmatrix} = \begin{vmatrix} a\lambda & 0 \\ 0 & b\mu \end{vmatrix}$

   and a, b, λ, μ all non-zero $\Rightarrow a\lambda, b\mu$
   non-zero.

   Identity: $\begin{vmatrix} 1 & 0 \\ 0 & 1 \end{vmatrix}$ is a 2 $\times$ 2 non-singular

   diagonal matrix and is an identity by
   the unit law.

   Inverse: $\begin{vmatrix} \frac{1}{a} & 0 \\ 0 & \frac{1}{b} \end{vmatrix}$ is the inverse of

   $\begin{vmatrix} a & 0 \\ 0 & b \end{vmatrix}.$

   Associativity: associative law of matrix
   multiplication

Exercise 15f — p. 648

1) c), d) and h) are orthogonal
2) A orthogonal $\Rightarrow A^T$ orthogonal
   (property 3). But the *rows* of A are
   the *columns* of A^T so the result follows
   from property 4.

3) a) $a = \dfrac{\sqrt{2}}{2}, b = 0, c = -\dfrac{\sqrt{2}}{2}$ or

   $a = -\dfrac{\sqrt{2}}{2}, b = 0, c = \dfrac{\sqrt{2}}{2}$

   b) $a = \frac{4}{5}, b = 0, c = -\frac{3}{5}$ or

   $a = -\frac{4}{5}, b = 0, c = \frac{3}{5}$

   c) $\begin{vmatrix} a & b \\ c & d \end{vmatrix}$ can be *any* 2 $\times$ 2 orthogonal

   matrix e.g. $a = \pm 1, d = \pm 1,$
   $b = c = 0$

   d) $a = \sin\theta, b = -\cos\theta$ or
   $a = -\sin\theta, b = \cos\theta$

4) $\begin{vmatrix} 1 & 0 & 0 \\ 0 & 1 & 0 \\ 0 & 0 & 1 \end{vmatrix}, \begin{vmatrix} -1 & 0 & 0 \\ 0 & 1 & 0 \\ 0 & 0 & 1 \end{vmatrix}, \begin{vmatrix} 1 & 0 & 0 \\ 0 & -1 & 0 \\ 0 & 0 & 1 \end{vmatrix},$

   $\begin{vmatrix} 1 & 0 & 0 \\ 0 & 1 & 0 \\ 0 & 0 & -1 \end{vmatrix}, \begin{vmatrix} -1 & 0 & 0 \\ 0 & -1 & 0 \\ 0 & 0 & 1 \end{vmatrix},$

$$\begin{pmatrix} -1 & 0 & 0 \\ 0 & 1 & 0 \\ 0 & 0 & -1 \end{pmatrix}, \begin{pmatrix} 1 & 0 & 0 \\ 0 & -1 & 0 \\ 0 & 0 & -1 \end{pmatrix} \text{ and } \begin{pmatrix} -1 & 0 & 0 \\ 0 & -1 & 0 \\ 0 & 0 & -1 \end{pmatrix}$$

5) *Closure*: **A, B** orthogonal ⇒ **AB** orthogonal by property 1.
Identity: **I** is orthogonal since $\mathbf{II}^T = \mathbf{II} = \mathbf{I}$.
Inverses: **A** orthogonal ⇒ $\mathbf{A}^{-1}$ orthogonal by property 3.
Associativity: associative law of matrix multiplication.
The group is the group of all rigid transformations of the plane which fix the origin, i.e. all rotations about the origin and reflections in lines through the origin.

Exercise 15g – p. 656

1) a) $\lambda = 1$ or -8;
b) $\lambda = 4$ or -1;
c) $\lambda = 2$ (repeated root);
d) $\lambda = 0$, 1 or 2

2) $\lambda = -1$ is the only other eigenvalue (it is a repeated root)

3) *Question 1*: The eigenvectors are given in the order corresponding to the eigenvalues in solution to Q1:

a) $\begin{pmatrix} 1 \\ 2 \end{pmatrix}, \begin{pmatrix} 1 \\ -1 \end{pmatrix}$; b) $\begin{pmatrix} 5 \\ 2 \end{pmatrix}, \begin{pmatrix} 0 \\ 1 \end{pmatrix}$;

c) $\begin{pmatrix} 1 \\ 0 \end{pmatrix}$;

d) $\begin{pmatrix} 0 \\ 1 \\ -1 \end{pmatrix}, \begin{pmatrix} 1 \\ 0 \\ -1 \end{pmatrix}, \begin{pmatrix} -2 \\ 1 \\ 0 \end{pmatrix}$

Question 2: $\lambda = 2 \begin{pmatrix} 1 \\ 1 \\ 1 \end{pmatrix}$, $\lambda = -1 \begin{pmatrix} 0 \\ 1 \\ 0 \end{pmatrix}$

or $\begin{pmatrix} 2 \\ 0 \\ -1 \end{pmatrix}$ or any combination.

Note any scalar multiple of the above answers is also correct

4) Eigenvalues are $\lambda = 1$ (repeated) and $\lambda = 2$. Corresponding unit length eigenvectors are:

$\lambda = 1$: $\begin{pmatrix} 1 \\ 0 \\ 0 \end{pmatrix}$ or $\begin{pmatrix} -1 \\ 0 \\ 0 \end{pmatrix}$,

$\lambda = 2$: $\begin{pmatrix} 0 \\ 0 \\ 1 \end{pmatrix}$ or $\begin{pmatrix} 0 \\ 0 \\ -1 \end{pmatrix}$

5) Other eigenvalues are $\lambda = 3$ and $\lambda = 6$. Unit length eigenvectors are:

$\lambda = -2$: $\begin{pmatrix} 0 \\ \frac{\sqrt{2}}{2} \\ \frac{\sqrt{2}}{2} \end{pmatrix}$ or $\begin{pmatrix} 0 \\ -\frac{\sqrt{2}}{2} \\ -\frac{\sqrt{2}}{2} \end{pmatrix}$;

$\lambda = 3$: $\begin{pmatrix} \frac{\sqrt{3}}{3} \\ \frac{\sqrt{3}}{3} \\ -\frac{\sqrt{3}}{3} \end{pmatrix}$ or $\begin{pmatrix} -\frac{\sqrt{3}}{3} \\ -\frac{\sqrt{3}}{3} \\ \frac{\sqrt{3}}{3} \end{pmatrix}$;

$\lambda = 6$: $\begin{pmatrix} \frac{\sqrt{6}}{3} \\ -\frac{\sqrt{6}}{6} \\ \frac{\sqrt{6}}{6} \end{pmatrix}$ or $\begin{pmatrix} -\frac{\sqrt{6}}{3} \\ \frac{\sqrt{6}}{6} \\ -\frac{\sqrt{6}}{6} \end{pmatrix}$

6) Any vector $\begin{pmatrix} x \\ y \\ z \end{pmatrix}$ satisfying $x + y + z = 0$ is an eigenvector. Suitable independent choices are $\begin{pmatrix} 1 \\ -1 \\ 0 \end{pmatrix}$ and $\begin{pmatrix} 1 \\ 0 \\ -1 \end{pmatrix}$, but there are many others!

7) $\lambda^2 + 1 = 0$ (no real roots). The matrix represents rotation through $\pi/2$ (clockwise) about the origin and there are no preserved lines through the origin

8) $\lambda = 1$ (triple root), corresponding eigenvectors are of the form $\begin{pmatrix} k \\ 0 \\ 0 \end{pmatrix}$ for *any* value of k

Exercise 15h – p. 665

2) a) $\mathbf{P} = \begin{pmatrix} 1 & 1 \\ 1 & -3 \end{pmatrix}$, $\Lambda = \begin{pmatrix} 3 & 0 \\ 0 & -1 \end{pmatrix}$

(Note that we have taken scalar multiples of the eigenvectors found in Example 15g(4) in order to simplify **P**.)

b) $\mathbf{P} = \begin{pmatrix} 1 & 1 \\ 2 & -1 \end{pmatrix}$, $\Lambda = \begin{pmatrix} 1 & 0 \\ 0 & -8 \end{pmatrix}$;

c) $\mathbf{P} = \begin{pmatrix} 5 & 0 \\ 2 & 1 \end{pmatrix}$, $\Lambda = \begin{pmatrix} 4 & 0 \\ 0 & -1 \end{pmatrix}$;

d) $\mathbf{P} = \begin{pmatrix} 0 & 1 & -2 \\ 1 & 0 & 1 \\ -1 & -1 & 0 \end{pmatrix}$, $\Lambda = \begin{pmatrix} 0 & 0 & 0 \\ 0 & 1 & 0 \\ 0 & 0 & 2 \end{pmatrix}$

Note that **P** is non-singular in each case.

e) The eigenvalue $\lambda = 2$ was given in Exercise 15g, Q6. The remaining eigenvalue is $\lambda = 3$ with eigenvector $\begin{pmatrix} 2 \\ -1 \\ 0 \end{pmatrix}$ and using the two eigenvectors found in Exercise 15g we have the non-singular matrix $\mathbf{P} = \begin{pmatrix} 1 & 1 & 2 \\ -1 & 0 & -1 \\ 0 & -1 & 0 \end{pmatrix}$ with corresponding diagonal form $\Lambda = \begin{pmatrix} 2 & 0 & 0 \\ 0 & 2 & 0 \\ 0 & 0 & 3 \end{pmatrix}$

3) a) Not diagonalizable for similar reasons to Example 15h(2)

b) Diagonalizable. Eigenvalues are $\lambda = 1$ (repeated) and $\lambda = 2$. Two independent eigenvectors for $\lambda = 1$ are $\begin{pmatrix} -1 \\ 1 \\ 0 \end{pmatrix}$ and $\begin{pmatrix} 0 \\ 0 \\ 1 \end{pmatrix}$ and an eigenvector for $\lambda = 2$ is $\begin{pmatrix} 1 \\ 0 \\ 0 \end{pmatrix}$. $\mathbf{P} = \begin{pmatrix} -1 & 0 & 1 \\ 1 & 0 & 0 \\ 0 & 1 & 0 \end{pmatrix}$ is non-singular and hence

$$\mathbf{P}^{-1}\mathbf{AP} = \begin{pmatrix} 1 & 0 & 0 \\ 0 & 1 & 0 \\ 0 & 0 & 2 \end{pmatrix}$$

c) Not diagonalizable. Eigenvalues are $\lambda = 1$ (repeated) and $\lambda = 0$. Corresponding eigenvectors are of the form $\begin{pmatrix} k \\ 0 \\ 0 \end{pmatrix}$ and $\begin{pmatrix} 0 \\ 0 \\ k \end{pmatrix}$ and it is not possible to find a non-singular matrix whose columns are eigenvectors.

d) Not diagonalizable: similar to c)

e) Diagonalizable: similar to b)

f) Not diagonalizable. Characteristic equation is $\lambda^2 - \sqrt{2}\lambda + 1 = 0$ which has no real roots.

4) a) Eigenvalues and corresponding unit eigenvectors are:

$$\lambda = 0, \begin{pmatrix} \frac{\sqrt{2}}{2} \\ -\frac{\sqrt{2}}{2} \end{pmatrix}; \quad \lambda = 2, \begin{pmatrix} \frac{\sqrt{2}}{2} \\ \frac{\sqrt{2}}{2} \end{pmatrix}$$

$\mathbf{P} = \begin{pmatrix} \frac{\sqrt{2}}{2} & \frac{\sqrt{2}}{2} \\ -\frac{\sqrt{2}}{2} & \frac{\sqrt{2}}{2} \end{pmatrix}$ is orthogonal and

$$\mathbf{P}^{\mathrm{T}}\mathbf{AP} = \begin{pmatrix} 0 & 0 \\ 0 & 2 \end{pmatrix}$$

b) $\lambda = 1, \begin{pmatrix} \frac{\sqrt{2}}{2} \\ \frac{\sqrt{2}}{2} \end{pmatrix}; \lambda = -1, \begin{pmatrix} \frac{\sqrt{2}}{2} \\ -\frac{\sqrt{2}}{2} \end{pmatrix}$

$$\mathbf{P} = \begin{pmatrix} \frac{\sqrt{2}}{2} & \frac{\sqrt{2}}{2} \\ \frac{\sqrt{2}}{2} & -\frac{\sqrt{2}}{2} \end{pmatrix}, \mathbf{P}^{\mathrm{T}}\mathbf{AP} = \begin{pmatrix} 1 & 0 \\ 0 & -1 \end{pmatrix}$$

5) We found eigenvectors of unit length in Exercise 15g, Q5 and taking these as columns we obtain the orthogonal matrix

$$\mathbf{P} = \begin{pmatrix} 0 & \frac{\sqrt{3}}{3} & \frac{\sqrt{3}}{6} \\ \frac{\sqrt{2}}{2} & \frac{\sqrt{3}}{2} & -\frac{\sqrt{6}}{6} \\ \frac{\sqrt{2}}{2} & -\frac{\sqrt{3}}{3} & \frac{\sqrt{6}}{6} \end{pmatrix}$$

which diagonalizes **A**.

$$\mathbf{P}^{\mathrm{T}}\mathbf{AP} = \begin{pmatrix} -2 & 0 & 0 \\ 0 & 3 & 0 \\ 0 & 0 & 6 \end{pmatrix}$$

Exercise 15i – p. 670

1) Corresponding eigenvectors are $\begin{pmatrix} 0 \\ 1 \\ -1 \end{pmatrix}$, $\begin{pmatrix} 2 \\ -1 \\ 0 \end{pmatrix}$ and $\begin{pmatrix} 1 \\ 0 \\ -1 \end{pmatrix}$.

If $\mathbf{P} = \begin{pmatrix} 0 & 2 & 1 \\ 1 & -1 & 0 \\ -1 & 0 & -1 \end{pmatrix}$ then $|\mathbf{P}| = 1$.

Therefore **P** is non-singular and **A** is diagonalizable.

2) Corresponding eigenvectors are $\begin{pmatrix} 1 \\ 2 \\ -2 \end{pmatrix}$, $\begin{pmatrix} 2 \\ 1 \\ 2 \end{pmatrix}$ and $\begin{pmatrix} -2 \\ 2 \\ 1 \end{pmatrix}$ which are mutually perpendicular. The required orthogonal matrix is $\begin{pmatrix} \frac{1}{3} & \frac{2}{3} & -\frac{2}{3} \\ \frac{2}{3} & \frac{1}{3} & \frac{2}{3} \\ -\frac{2}{3} & \frac{2}{3} & \frac{1}{3} \end{pmatrix}$

3) General form of eigenvector for $\lambda = -1$ is $\begin{pmatrix} k \\ -k \\ l \end{pmatrix}$ where k, l are real numbers.

Eigenvector for $\lambda = 3$ is $\begin{pmatrix} 1 \\ 1 \\ 0 \end{pmatrix}$ and these are perpendicular for *any* choices of k and l. Choosing two perpendicular eigenvectors for $\lambda = -1$ and scaling all three to unit length we find the orthogonal matrix $\begin{pmatrix} \frac{\sqrt{2}}{2} & 0 & \frac{\sqrt{2}}{2} \\ -\frac{\sqrt{2}}{2} & 0 & \frac{\sqrt{2}}{2} \\ 0 & 1 & 0 \end{pmatrix}$

which diagonalizes **A**

4) Others are $\lambda = 1$ and $\lambda = 2$. Diagonal form is $\begin{pmatrix} 1 & 0 & 0 \\ 0 & 2 & 0 \\ 0 & 0 & 4 \end{pmatrix}$ (although the eigenvalues may appear in *any* order on the diagonal)

5) Other eigenvalue is $\lambda = -9$ (repeated).

Diagonal form is $\begin{pmatrix} 9 & 0 & 0 \\ 0 & -9 & 0 \\ 0 & 0 & -9 \end{pmatrix}$

Exercise 15j – p. 679

1) a)

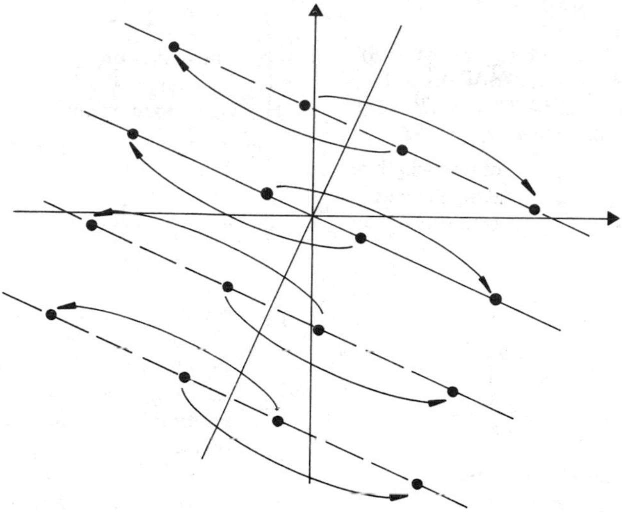

(The line $y = 2x$ is fixed (point wise) and lines parallel to $x = -y$ are reflected and stretched outwards (factor -8))

b)

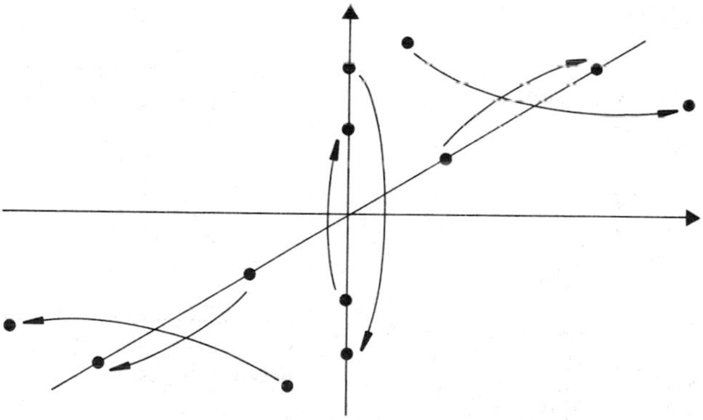

(A reflection of the y axis combined with an outwards stretch (factor 4) of the line $y = \frac{2}{5}x$)

2) a) $(1+\sqrt{2})x+y=0$ and $(1-\sqrt{2})x+y=0$;

b) $2x+(1+\sqrt{5})y=0$ and $2x+(1-\sqrt{5})y=0$;

c) $2x-(1+\sqrt{5})y=0$ and $2x-(1-\sqrt{5})y=0$

3) a) $x=\pm y$ are the principal axes and referred to these axes we have the hyperbola $3x'^2-y'^2=3$. So the conic is this hyperbola rotated through $\pi/4$

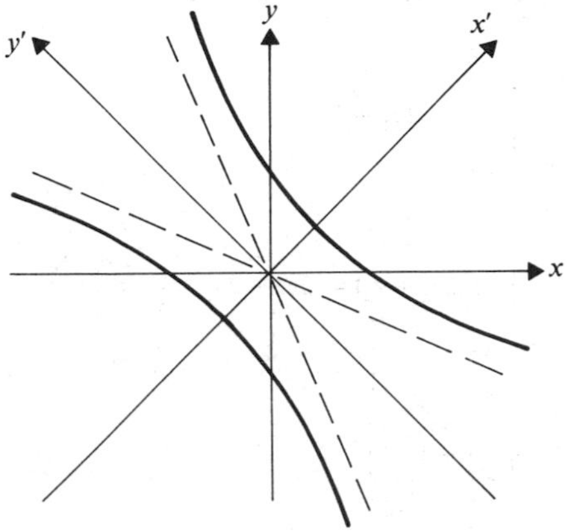

b) The principal axes are *again* $x=\pm y$ and the diagonalized equation is $2x'^2=7$ which is a pair of parallel lines (parallel to the y' axis). So the picture is this:

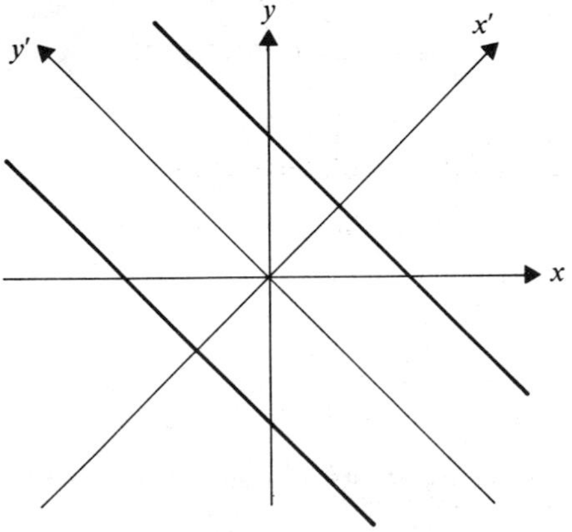

4) a) The relevant matrix is

$$\begin{pmatrix} \frac{1}{4} & \frac{1}{2} & 0 \\ \frac{1}{2} & \frac{1}{4} & 0 \\ 0 & 0 & \frac{1}{4} \end{pmatrix} = \frac{1}{4}\begin{pmatrix} 1 & 2 & 0 \\ 2 & 1 & 0 \\ 0 & 0 & 1 \end{pmatrix}$$

The matrix on the right has eigenvalues $\lambda=-1$, 1 and 3 and corresponding unit eigenvectors

$$\begin{pmatrix} \frac{\sqrt{2}}{2} \\ -\frac{\sqrt{2}}{2} \\ 0 \end{pmatrix}, \begin{pmatrix} 0 \\ 0 \\ 1 \end{pmatrix} \text{ and } \begin{pmatrix} \frac{\sqrt{2}}{2} \\ \frac{\sqrt{2}}{2} \\ 0 \end{pmatrix}$$ and the coordinates of these are the required direction cosines

b) The relevant matrix is $2\mathbf{A}$ where $\mathbf{A}$ is as in Exercise 15i, Q2. So from solution 15i the required direction cosines are $\frac{1}{3}, \frac{2}{3}, -\frac{2}{3}$; $\frac{2}{3}, \frac{1}{3}, \frac{2}{3}$; $-\frac{2}{3}, \frac{2}{3}, \frac{1}{3}$

5) $9x'^2-9y'^2-9z'^2=105$

Miscellaneous Exercise 15 – p 681

1) (i) $\mathbf{A}(\mathbf{A}^2-\mathbf{I})=\mathbf{A}^2-\mathbf{I}=\begin{pmatrix} 0 & 0 & 0 \\ -1 & 0 & 0 \\ 3 & 0 & 0 \end{pmatrix}$.

Induction step: $\mathbf{A}^n(\mathbf{A}^2-\mathbf{I})=\mathbf{A}^2-\mathbf{I}$
$\Rightarrow \mathbf{A}^{n+1}(\mathbf{A}^2-\mathbf{I})=\mathbf{A}(\mathbf{A}^2-\mathbf{I})=\mathbf{A}^2-\mathbf{I}$

$$\mathbf{A}^{100}=\mathbf{A}^{98}+(\mathbf{A}^2-\mathbf{I})$$
$$=\mathbf{A}^{96}+2(\mathbf{A}^2-\mathbf{I})=\dots$$
$$=\mathbf{I}+50(\mathbf{A}^2-\mathbf{I})$$
$$=\begin{pmatrix} 1 & 0 & 0 \\ -50 & 1 & 0 \\ 150 & 0 & 1 \end{pmatrix}$$

(ii) $\mathbf{A}=\mathbf{B}+\mathbf{B}^{\mathrm{T}}=\begin{pmatrix} 0 & 1 \\ 1 & 0 \end{pmatrix}$.

Diagonalizing in the usual way we find $\mathbf{M}=\begin{pmatrix} 1 & 1 \\ 1 & -1 \end{pmatrix}$ diagonalizes $\mathbf{A}$,

i.e. $\mathbf{M}^{-1}\mathbf{AM} = \mathbf{D}$ (diagonal)
$\Rightarrow \mathbf{A} = \mathbf{MDM}^{-1}$

2) For the first two parts see property 6, page 636, and property 3, page 646.

$((\mathbf{I}-\mathbf{B})(\mathbf{I}+\mathbf{B})^{-1})^{\mathrm{T}}$
$= ((\mathbf{I}+\mathbf{B})^{-1})^{\mathrm{T}}(\mathbf{I}-\mathbf{B})^{\mathrm{T}}$
$= ((\mathbf{I}+\mathbf{B})^{\mathrm{T}})^{-1}(\mathbf{I}-\mathbf{B})^{\mathrm{T}}$
$= (\mathbf{I}-\mathbf{B})^{-1}(\mathbf{I}+\mathbf{B})$ since $\mathbf{B}^{\mathrm{T}} = -\mathbf{B}$
$= (\mathbf{I}-\mathbf{B})^{-1}(\mathbf{I}+\mathbf{B})(\mathbf{I}-\mathbf{B})(\mathbf{I}-\mathbf{B})^{-1}$
$= (\mathbf{I}-\mathbf{B})^{-1}(\mathbf{I}-\mathbf{B})(\mathbf{I}+\mathbf{B})(\mathbf{I}-\mathbf{B})^{-1}$
by the given identity
$= (\mathbf{I}+\mathbf{B})(\mathbf{I}-\mathbf{B})^{-1}$
$= ((\mathbf{I}-\mathbf{B})(\mathbf{I}+\mathbf{B})^{-1})^{-1}$

3) a) See Exercise 15c, Q4.
b) $(\mathbf{AB})^{\mathrm{T}} = \mathbf{AB} \Rightarrow \mathbf{B}^{\mathrm{T}}\mathbf{A}^{\mathrm{T}} = \mathbf{AB}$
$\Rightarrow \mathbf{BA} = \mathbf{AB}$ since $\mathbf{A}^{\mathrm{T}} = \mathbf{A}$ and $\mathbf{B}^{\mathrm{T}} = \mathbf{B}$.
The set consists of all matrices of the form $\begin{pmatrix} a & 0 & b \\ 0 & c & 0 \\ b & 0 & a \end{pmatrix}$ where a, b, c are real numbers. The *closure*, *identity* and *associativity* axioms are true but *inverses* fails because e.g.

$\begin{pmatrix} 1 & 0 & 1 \\ 0 & 1 & 0 \\ 1 & 0 & 1 \end{pmatrix}$ is singular

4) $\lambda = 1, 4$ or 6. Direction cosines are $\frac{2\sqrt{5}}{5}, 0, \frac{\sqrt{5}}{5}$; $0, 1, 0$; $\frac{\sqrt{5}}{5}, 0, -\frac{2\sqrt{5}}{5}$.

$$\mathbf{P} = \begin{pmatrix} \frac{2\sqrt{5}}{2} & 0 & \frac{\sqrt{5}}{5} \\ 0 & 1 & 0 \\ \frac{\sqrt{5}}{5} & 0 & -\frac{2\sqrt{5}}{5} \end{pmatrix}, \ \mathbf{P} \text{ is orthogonal}$$

hence $\mathbf{P}^{-1} = \mathbf{P}^{\mathrm{T}}$ and $\mathbf{P}^{\mathrm{T}}\mathbf{AP} = \begin{pmatrix} 1 & 0 & 0 \\ 0 & 4 & 0 \\ 0 & 0 & 6 \end{pmatrix}$

5) Eigenvectors are

$$\begin{pmatrix} \frac{1}{3} \\ \frac{2}{3} \\ -\frac{2}{3} \end{pmatrix}, \begin{pmatrix} \frac{2}{3} \\ \frac{1}{3} \\ \frac{2}{3} \end{pmatrix}, \begin{pmatrix} -\frac{2}{3} \\ \frac{2}{3} \\ \frac{1}{3} \end{pmatrix}$$

6) $\lambda = 1$ or $\lambda = 2$ (repeated root)

$$\mathbf{v}_1 = \begin{pmatrix} 1 \\ 1 \\ -1 \end{pmatrix}, \ \mathbf{v}_2 = \begin{pmatrix} 2 \\ 1 \\ 0 \end{pmatrix}$$

$$\begin{pmatrix} 6 \\ 4 \\ -2 \end{pmatrix} = 2\mathbf{v}_1 + 2\mathbf{v}_2$$

$$\Rightarrow \mathbf{A}(2\mathbf{v}_1 + \mathbf{v}_2) = 2\mathbf{v}_1 + 2\mathbf{v}_2 = \begin{pmatrix} 6 \\ 4 \\ -2 \end{pmatrix}$$

i.e. $\mathbf{v} = \begin{pmatrix} 4 \\ 3 \\ -2 \end{pmatrix}$

7) Eigenvectors are $\begin{pmatrix} 1 \\ -1 \\ 0 \end{pmatrix}, \begin{pmatrix} 1 \\ 0 \\ -1 \end{pmatrix}, \begin{pmatrix} 1 \\ -1 \\ -1 \end{pmatrix}$

respectively. $\mathbf{P} = \begin{pmatrix} 1 & 1 & 1 \\ -1 & 0 & -1 \\ 0 & -1 & -1 \end{pmatrix}$ is non-singular and hence diagonalizes $\mathbf{A}$

8) $\mathbf{A} = \begin{pmatrix} 1 & 0 & 2 \\ 0 & 4 & 0 \\ 2 & 0 & 1 \end{pmatrix}$. Eigenvalues $-1, 3$ and 4.

$$\mathbf{P} = \begin{pmatrix} \frac{\sqrt{2}}{2} & \frac{\sqrt{2}}{2} & 0 \\ 0 & 0 & 1 \\ \frac{\sqrt{2}}{2} & -\frac{\sqrt{2}}{2} & 0 \end{pmatrix}.$$

$$\mathbf{P}^{-1}\mathbf{AP} = \mathbf{P}^T\mathbf{AP} = \begin{pmatrix} -1 & 0 & 0 \\ 0 & 3 & 0 \\ 0 & 0 & 4 \end{pmatrix}.$$

For the remainder, see page 676

9) $\lambda = 2$ or 7
Corresponding unit eigenvectors are

$$\begin{pmatrix} \frac{2\sqrt{5}}{5} \\ -\frac{\sqrt{5}}{2} \end{pmatrix} \text{ and } \begin{pmatrix} \frac{\sqrt{5}}{5} \\ \frac{2\sqrt{5}}{5} \end{pmatrix}.$$

Equation is $2x'^2 + 7y'^2 = 14$

10) Relevant matrix is

$$\mathbf{A} = \begin{pmatrix} 14 & 2 & -4 \\ 2 & 17 & 2 \\ -4 & 2 & 14 \end{pmatrix}$$

which has eigenvalues $\lambda = 9$ and $\lambda = 18$ (repeated root). For $\lambda = 9$ a unit length eigenvector is $\pm \begin{pmatrix} \frac{2}{3} \\ -\frac{1}{3} \\ \frac{2}{3} \end{pmatrix}$. For $\lambda = 18$, the eigenvectors fill out the plane $2x - y + 2z = 0$, so we may choose *any* two perpendicular unit eigenvectors in this plane, e.g. $\begin{pmatrix} -\frac{1}{3} \\ \frac{2}{3} \\ \frac{2}{3} \end{pmatrix}$ and $\begin{pmatrix} \frac{2}{3} \\ \frac{2}{3} \\ -\frac{1}{3} \end{pmatrix}$. The coordinates of these vectors are the required direction cosines. The equation becomes $x'^2 + 2y'^2 + 2z'^2 = 1$

12) c) $(\mathbf{A}^{-1})^*\mathbf{A}^* = (\mathbf{A}^{-1}\mathbf{A})^* = \mathbf{I}^* = \mathbf{I}$
$\Rightarrow (\mathbf{A}^{-1})^* = (\mathbf{A}^*)^{-1}$

13) a) $\bar{h}_{ij} = h_{ji} \Rightarrow \bar{h}_{ii} = h_{ii} \Rightarrow h_{ii}$ real
b) $h_{ij} = h_{ji} \Longleftrightarrow \bar{h}_{ij} = h_{ji}$ since h_{ij} is real
c), d) and e) similar to Example 15c(2) and Exercise 15c, Q3

14) Similar to the proof for orthogonal matrixes (see Exercise 15f, Q5)

15) $\lambda = \pm$ i. $\mathbf{P} = \begin{pmatrix} 1 & \mathrm{i} \\ \mathrm{i} & 1 \end{pmatrix}$

16) $\mathbf{P} = \begin{pmatrix} \mathrm{i}\frac{\sqrt{2}}{2} & \frac{\sqrt{2}}{2} \\ \frac{\sqrt{2}}{2} & \mathrm{i}\frac{\sqrt{2}}{2} \end{pmatrix}$

17) Group under + is straightforward
Group under $\times$ fails because *closure* and *inverses* both fail; e.g.

$\begin{pmatrix} 1 & 0 & 0 \\ 0 & 0 & 0 \\ 0 & 0 & 0 \end{pmatrix}$ is Hermitian but singular.

$\mathbf{z}^*\mathbf{z}$ is a Hermitian 1×1 matrix by Q13(c). But the diagonal entries are real, and since the only entry is a diagonal entry, $\mathbf{z}^*\mathbf{z}$ is real. $\mathbf{z}^*\mathbf{Hz}$is Hermitian (by Q13(c)). Hence $\mathbf{z}^*\mathbf{Hz}$ is real for the same reasons.
$\mathbf{Hz} = \lambda\mathbf{z} \Rightarrow \mathbf{z}^*\mathbf{Hz} = \lambda\mathbf{z}^*\mathbf{z}$. But $\mathbf{z}^*\mathbf{Hz}$ and $\mathbf{z}^*\mathbf{z}$ are both real, therefore λ is real. A symmetric matrix *is* a Hermitian matrix (Q13(b))

18) $\mathbf{Ax} = \lambda\mathbf{x} \Rightarrow (\mu\mathbf{A})\mathbf{x} = \mu\lambda\mathbf{x} \Rightarrow$ eigenvectors are the same but eigenvalues are all multiplied by μ

19) a) $\mathbf{A} = \mathbf{P}^{-1}\mathbf{BP} \Rightarrow \mathbf{B} = \mathbf{PAP}^{-1}$
$\Rightarrow \mathbf{B} = \mathbf{Q}^{-1}\mathbf{AQ}$ (where $\mathbf{Q} = \mathbf{P}^{-1}$)
b) $\mathbf{A} = \mathbf{P}^{-1}\mathbf{BP}$ and $\mathbf{B} = \mathbf{Q}^{-1}\mathbf{CQ}$
$\Rightarrow \mathbf{A} = \mathbf{P}^{-1}\mathbf{Q}^{-1}\mathbf{CQP} = \mathbf{R}^{-1}\mathbf{CR}$
where $\mathbf{R} = \mathbf{QP}$

20) $|\mathbf{P}^{-1}\mathbf{AP} - \lambda\mathbf{I}|$

$$= |\mathbf{P}^{-1}\mathbf{AP} - \lambda\mathbf{P}^{-1}\mathbf{P}|$$

$$= |\mathbf{P}^{-1}\mathbf{I}|\,|\mathbf{A} - \lambda\mathbf{I}|\,|\mathbf{P}|$$

$$= \frac{1}{|\mathbf{P}|}\,|\mathbf{A} - \lambda\mathbf{I}|\,|\mathbf{P}|$$

$$= |\mathbf{A} - \lambda\mathbf{I}|$$

Eigenvector $\mathbf{x}$ becomes $\mathbf{P}^{-1}\mathbf{x}$ since $(\mathbf{P}^{-1}\mathbf{AP})(\mathbf{P}^{-1}\mathbf{x}) = \mathbf{P}^{-1}\mathbf{Ax} = \mathbf{P}^{-1}\lambda\mathbf{x} = \lambda(\mathbf{P}^{-1}\mathbf{x})$

INDEX